SEDIMENTATION ENGINEERING

SEDIMENTATION ENGINEERING

Vito A. Vanoni, Editor

PREPARED BY
THE ASCE TASK COMMITTEE
FOR THE PREPARATION OF
THE MANUAL ON SEDIMENTATION
OF THE SEDIMENTATION COMMITTEE
OF THE HYDRAULICS DIVISION

HEADQUARTERS OF THE SOCIETY
345 EAST 47TH ST. REPRINTED 1977 NEW YORK, N.Y. 10017

Library of Congress Catalog Card Number: 75-7751

ISBN: 0-87262-001-8

Contents

Foreword

This book is concerned mainly with sediment problems involved in the development, use, control and conservation of water and land resources. Its aim is to give an understanding of the nature and scope of sedimentation problems, of methods for their investigation, and of practical approaches to their solution. In essence, it is a textbook on sedimentation engineering. As such, it must necessarily treat sedimentation in broad perspective, considering the interrelated processes of erosion, sediment transportation by water and air, and sediment deposition where it creates problems of practical importance. On the other hand, limitations of space and of objective preclude consideration of certain important aspects of sedimentation that are adequately treated in other publications. This treatment is concerned, for example, only with the dynamic phases of sedimentation excluding mass movement and not with the historical aspects that are described in many geological reference works. It deals with sediment control methods for watersheds, streams, canals, and reservoirs, but not with all of the details of sediment removal in the process of water purification—a subject extensively treated in books on waterworks practice and sanitary engineering. Nor does it consider the important subject of littoral and ocean sedimentation. Some economic and legal aspects of sedimentation and a section on transportation of particulate matter in pipes are also included.

In preparing the text it was assumed that the reader has a knowledge of elementary fluid mechanics, calculus, and a general technical background. The material is presented in sufficient detail so the reader can follow developments from basic principles. For those readers faced with practical problems the text may not give enough information and it may be necessary to seek additional material in some of the references listed at the end of each chapter.

The text was prepared from reports published in the ASCE *Journal of the Hydraulics Division* and the discussions of these reports. Each of the published reports was prepared by a specialist in the field and reviewed by at least three experts. By this procedure the material in the book has been reviewed and checked by many specialists.

The book was prepared under the supervision of the Task Committee for the Preparation of the Manual on Sedimentation of the Sedimentation Committee of the Hydraulics Division of ASCE. The idea for the work was conceived by the late Carl B. Brown in 1949 when he was the first Chairman of the Sedimentation Committee and the work was carried on by the Committee until 1954. At that time the Task Committee on Preparation of the Manual on Sedimentation was formed with V. A. Vanoni as Chairman. Others who served on the Task Committee were D. C. Bondurant, C. S. Howard, T. Means, J. Smallshaw, J. E. McKee, L. M. Glymph, Jr., P. C. Benedict, G. R. Hall, Jr., C. R. Miller, and R. F. Piest. Several people assisted the Task Committee in reviewing material, attending meetings, and carrying out other work normally done by Committeemen. Among these, special mention is due T. Maddock, Jr., J. W. Roehl, H. P. Guy, and J. N. Holeman. During the period 1964-1975 when most of the book was completed the Committee consisted of P. C. Benedict, D. C. Bondurant, J. E. McKee, R. F. Piest and V. A. Vanoni.

The principal contributors to the book are listed as follows after the chapter and section number to which they contributed.

Chapter I—L. C. Gottschalk.

Chapter II—(A, B, D, E, G, and H) V. A. Vanoni; (C) A. G. Anderson*; (F) J. F. Kennedy; (I) N. P. Woodruff, W. S. Chepil, and A. W. Zingg; (J) H. W. Shen, S. Karaki, A. R. Chamberlain, and M. L. Albertson; (K) D. R. F. Harleman; and (L) S. C. Happ.

Chapter III—(A) P. C. Benedict; (B) J. M. Lara; (C) J. W. Roehl, J. N. Holeman, and V. H. Jones; (D) N. P. Woodruff, W. S. Chepil, and A. W. Zingg; and (E) H. P. Guy.

Chapter IV—R. F. Piest and C. R. Miller.

Chapter V—(A,B) J. W. Roehl and others in the United States Department of Agriculture; (C and E) D. C. Bondurant; and (D) J. A. Hufferd, R. L. Vance, and J. J. Watkins.

Chapter VI—T. Maddock, Jr.

Chapter VII—C. E. Busby.

Many others contributed to this work by reviewing and criticizing manuscripts, discussing ideas with authors, and contributing written discussions. One must also acknowledge the financial support contributed by the organizations who paid the salaries and other expenses of the contributors. And finally, the encouragement and support of the many people in ASCE must be acknowledged for without their patient support, this long-overdue work would not have been completed.

---

*Deceased July 1, 1975. Mr. Anderson received the 1965 J. C. Stevens Award for his discussion of Chapter II, Section C.

Chapter I.—Nature of Sedimentation Problems

A. Introduction

1. General.—Sedimentation embodies the processes of erosion, entrainment, transportation, deposition, and the compaction of sediment. These are natural processes that have been active throughout geological times and have shaped the present landscape of our world. The principal external dynamic agents of sedimentation are water, wind, gravity, and ice. Although each may be important locally, only the hydrospheric forces of rainfall, runoff, streamflow, and wind forces are considered herein. Thus, erosion will be defined as the detachment and removal of rock particles by the action of water and wind. Fluvial sediment is a collective term meaning an accumulation of rock and mineral particles transported or deposited by flowing water. Aeolian sediment is that moved or deposited by wind.

The processes of erosion, entrainment, transportation, and deposition of fluvial sediment are complex. The detachment of particles in the erosion process occurs through the kinetic energy of raindrop impact, or by the forces generated by flowing water. Once a particle has been detached, it must be entrained before it can be transported away. Both entrainment and transport depend on the shape, size, and weight of the particle and the forces exerted on the particle by the flow. When these forces are diminished to the extent that the transport rate is reduced or transport is no longer possible, deposition occurs.

Water erosion may be quite obvious and even spectacular as seen in the Grand Canyon or Bryce Canyon in the western United States. It may also be relatively imperceptible, e.g., the sheet erosion that occurs on vast areas of cleared or poorly vegetated land in the United States and elsewhere in the world. The muddy waters of streams provide evidence, even to the uninformed, that erosion is occurring upstream and that sediment is being transported. Mud on sidewalks and streets, sand splays on fertile bottom lands, and the great deltas of the Nile and Mississippi Rivers are all examples of deposition of eroded and transported materials.

The deposition of sediment may produce new land or enhance existing land used for various purposes; it also may create serious problems. In fact, the processes of sedimentation can create severe problems. Erosion, besides producing harmful sediment, may cause serious on-site damage to agricultural land by reducing the fertility and productivity of soils. In more advanced stages, it may even modify farm fields to the extent that cultivation is no longer profitable or even physically possible. Sediment in transport effects the quality of water and its suitability for human consumption or use in various enterprises. Some industries cannot tolerate even the smallest amounts of sediment in water used for certain manufacturing processes, and the public pays a large price for removing sediment from water used in everyday life.

Problems created by the deposition of sediment are many and varied. Harmful materials deposited on farm lands at the foot of slopes or on fertile flood plains may reduce the fertility of soils, impair surface drainage, and even completely bury valuable crops. Sediment deposited in stream channels reduces the flood-carrying capacity, resulting in more frequent overflows and greater floodwater damage to adjacent properties. The deposition of sediment in irrigation and drainage canals, in navigation channels and floodways, in reservoirs and harbors, on streets and highways, and in buildings not only creates a nuisance but also inflicts a high public cost in maintenance removal or in reduced services (Bennett, 1939) (Brune, 1958).[1]

Sedimentation is of vital concern in the conservation, development, and utilization of our soil and water resources. Without these resources man cannot exist, and when they are of poor quality or of insufficient quantity, man's lot is a sad one, indeed. With a rapidly expanding population and an ever-increasing demand for food and products derived from soil and water, exploitation and apathy must rapidly be replaced by wise planning and circumspection if future generations are to maintain standards of living prevalent today (in 1972). History records many instances of civilizations that declined because people could not cope with problems of sedimentation (Lowdermilk, 1935). In the United States, early settlements along the shores of the Chesapeake Bay fell into decay because they were unable to surmount the problems of sediment deposition in harbors (Gottschalk, 1945).

Considerable knowledge relative to the nature and influences of the forces involved in sedimentation has been accumulated during the past few decades. Some of the more important elements are summarized in the subsequent chapters. Out of this knowledge has emerged the realization that man, through his activities and endeavors, can locally influence the so-called "natural sedimentation processes" and that the results may be beneficial or harmful to his own well being. Out of this knowledge has also come a better understanding of the fundamentals involved in the creation or alleviation of sedimentation problems. This information is basic to the proper planning, design, installation, and maintenance of works of improvement for the development, use, and conservation of our soil and water resources to assure that they will function as designed and will not create new problems or accentuate existing ones.

B. Problems of Erosion

2. Geologic Erosion.—Geologic erosion is defined as erosion of the surface of the earth under natural or undisturbed conditions. The vast extent and great thicknesses of sedimentary rocks found on the earth's surface provide evidence of the tremendous amount of geologic erosion and deposition that has occurred throughout geologic times. Whereas the most recent deposits of the Mississippi

---

[1] References are given in Chapter I, Section G.

River delta are measured in terms of feet, during Paleocene and Eocene epochs alone, sediment measuring nearly 4 miles in thickness was deposited in the ancient estuary of the Mississippi River (Termier, 1963). Erosion is not a process of recent origin nor is it restricted to the ancient past. Geologic erosion exists today (in 1972) as it did in the past (Leopold, et al., 1964).

Geologic erosion varies in different places on the earth's surface because of differences in character of rocks and climatic and vegetative conditions. In some areas, e.g., in parts of the western United States where there are extensive deposits of geologically young and poorly indurated rocks, and where vegetation is sparse and rainfall torrential, the rate of geologic erosion is high. Thus, the Missouri and Colorado Rivers and many of their principal tributaries are heavily laden with sediment and have been for a long time before man occupied the area. Conversely, streams such as those in New England are relatively clear because they drain areas where the surface formations consist of ancient well-indurated, and erosion-resistant rocks that are covered with a more luxuriant canopy of vegetation.

The control of geologic erosion, by and large, is often difficult to achieve because the natural conditions that have prevailed over centuries of time cannot be changed significantly to effect any great reductions in erosion. Under certain local conditions, however, improvements can be made to reduce erosion. For example, long, gently-sloping flats that have sparse vegetation can often be chiseled and scarified or contour furrowed to increase water intake. The resulting improvement in soil moisture conditions will increase the density of existing vegetation or permit the introduction of better species for protection against erosion. On high and steep mountain slopes, e.g., those along the Wasatch front in Utah, the San Gabriel Mountains in California, and elsewhere, the Forest Service of the United States Department of Agriculture has installed contour trenches to reduce rapid runoff and improve vegetative conditions that have proved most effective in reducing erosion and sediment movement (Noble, 1965).

Where geologic erosion is confined primarily to channel erosion, stabilizing structures can often be built to halt such erosion. These include mostly grade-control structures, e.g., check dams and sills and water-level control structures that drown out the head cuts of natural arroyos or barancas. The difficulty in controlling geologic erosion comes mostly from the standpoint of economic feasibility. This is because the land on which serious geologic erosion often occurs is usually of low value for agricultural purposes. The justification for structural measures, therefore, depends mainly on the reduction of severe sediment damages immediately downstream from the source.

3. Accelerated Erosion.—Accelerated erosion is defined as the increased rate of erosion over the normal or geologic erosion brought about by man's activity. That man can exercise any great influence on the natural process of erosion seems inconceivable, yet locally it is possible. The erodibility of natural materials can be altered by disturbing the soil structure through plowing or other tillage activities. The protective vegetative canopy can be changed by grubbing, cutting, or burning existing vegetation and by introducing new species. Recent developments indicate that man may be able to change rainfall conditions by cloud seeding. The runoff conditions from land surfaces and the hydraulic characteristics of flow in channels can be changed by structural works of improvement or by altering the natural characteristics of stream channels, e.g., channel cross sections, alinements, or gradients. All of these activities can have a profound influence on erosion. Under some circumstances, it is not uncommon to increase erosion rates by more than 100 fold over the normal or geological rate.

4. Agricultural Activities.—In the short life-span of man's occupancy of the earth's surface, many changes have been brought about that have influenced erosion. Of these, the widespread use of land and associated practices for agricultural purposes leads all other activities resulting in accelerated erosion and the production of sediment so far as the world as a whole is concerned. Historically, accelerated erosion from the widespread use of land for the production of food began some 7,000 yr ago (Lowdermilk, 1935) (Toynbee, 1948). Since that time, the decline of numerous civilizations has been recorded because erosion so badly deteriorated the land that it became impossible to produce enough food to support the population. Dale and Carter (1955) report that civilizations in the past were able to survive, on the average, only 40–60 generations before they declined, perished, or were forced to move to new lands. Of all the original lands opened for cultivation during the dawn of civilization, only the alluvial plains of the Indus Valley, Mesopotamia, and the Nile Valley remain in sustained crop production today. These are noneroding areas that are enriched from time to time by periodic deposits of fertile sediment.

The opening of new lands for agricultural purposes necessarily disturbs the natural conditions. The indigenous forests are removed by ax and fire, while native grasslands are burned, overgrazed, or the sod turned and broken by plow. Except under rare conditions, e.g., the drainage of absolutely flat areas, the removal of protective vegetation and loosening of the soil in cultivation results in speeding up the erosion processes. In areas with only limited depths of fertile topsoil, or where erosion is progressing at an unusually high rate, the productivity of the soil is soon exhausted and cultivation is no longer practiced. The land may then be abandoned, but the erosion continues. Today (in 1972) only a small patch of the original verdant forest of the Cedars of Lebanon can be found. In the wake of the removal of this forest for cultivation is a desolate, bare, and rocky eroded land.

In the United States, it is not at all unusual to find cultivated farm fields that are losing soil by water erosion at rates greater than 200 tons/acre/yr (Gottschalk and Jones, 1955). Gottschalk and Brune (1950) report rates of erosion ranging as high as 127,000 tons/sq mile annually from small, intensely cultivated watersheds in western Iowa.

The rate of loss of productivity of such soils depends on the depth of erosion of plant-producing topsoil. Where topsoil depth is limited (which is true in most parts of the United States), damage by erosion of agricultural land may be severe. Stallings (1950) has reported the effects of erosion on reductions of corn yields at various locations in the midwest. A number of measurements herein indicate that corn yields are reduced from 5.3%/in.–8.8%/in. of topsoil lost. Gottschalk (1962) has shown that the loss of gross income from reduced corn yields from 6,246 acres of corn land in the Spring Creek watershed in Illinois will cost the farmers $1,873,800 for a 50-yr evaluation period. This cornland is located in the 20.2-sq mile watershed above Spring Lake, water-supply reservoir for Macomb, Ill., which is losing its capacity by sediment deposited at the rate of 2.3% annually.

The results of plot studies in the United States indicate that the removal of forest cover and conversion of land to intertilled crops can accelerate erosion from

100 to 1,000 fold (Musgrave, 1957) (Gottschalk, 1958). The conversion of grassland to cultivated crops also may increase erosion from 10 to 50 fold.

The total land area of conterminous United States is approx 1.9 billion acres. Of this, about 1.4 billion acres, or 74%, is land in farms. An inventory of soil and water conservation needs developed by the United States Department of Agriculture (1971) indicates that in 1967 approx 30% of the land in the United States was used for cropland, 34% for pasture and range, 32% for forest and woodland, and 4% for urban and other nonfarm uses.

From these statistics, it may be seen that demands for agricultural land to support an ever-increasing population have already disturbed the natural conditions of significant acreages of land in the United States. This has greatly accelerated erosion and contributed a vast amount of sediment to the streams and waterways of the nation. Although no detailed inventory of the total soil loss in the United States has been compiled, it is estimated that erosion from croplands alone amounts to about 4 billion tons/yr (Stallings, 1957). It is estimated that the streams and rivers of the United States carry approx 1 billion tons of sediment to the oceans each year (Gottschalk and Jones, 1955).

Since the beginning of the 1930's, much progress has been made in accumulating knowledge of the erosion processes and developing farm-field measures for reducing accelerated erosion and associated damages. These measures include land-use conversions based on use of land according to its capabilities and the development of conservation measures, e.g., contour plowing, strip cropping, terracing, and other measures to conserve soil and reduce loss of income by erosion. In the United States, the Soil Conservation Service provides technical assistance to local landowners in the planning and application of conservation measures (see Chapter V). This is a voluntary program on the part of landowners, however, and with some 3,200,000 farms and ranches of the nation involved, various degrees of participation and success might be expected in specific areas.

According to the United States Department of Agriculture (1971), 64% of the cropland, 69% of the pasture and rangeland, and more than half of the forest and woodland in the United States still need conservation treatment. Continued efforts in education in respect to the desirability of proper land use and conservation and provisions for technical assistance in the planning and application of measures are needed to hold accelerated erosion of farmland to an acceptable maximum.

5. Urbanization.—There are other types of activities that create accelerated erosion conditions which, if not as important to the total overall sediment-producing activities of agriculture, nevertheless can create serious localized problems. One of these is the effect of urbanization on accelerating erosion. Urbanized areas, when fully developed, are actually low sediment-producing areas because a large percentage of the land is protected against erosion by roofs, streets, parking lots or well-cared-for lawns and parks, and curbs, gutters, and storm sewers. During actual construction, however, erosion rates are high. This is a result of the removal of trees and other vegetation, and excavation and grading activities. According to Holeman and Geiger (1959), urbanization involving large blocks of housing developments and shopping centers appears to be of primary concern because large tracts of land are exposed to serious erosion for periods of 2 yr–3 yr before final completion and stabilization. Individual homes with sodding and landscaping, drives, etc. usually are completed in a matter of 3 months–4 months and thus do not produce as much erosion and sediment per unit of area as do the larger developments.

Holeman and Geiger (1959) report that urbanization in the 14.3-sq mile watershed of Holmes Run in Fairfax County, Va., a suburban area of metropolitan Washington, D.C., resulted in more than doubling the rate of sediment deposition in Lake Barcroft. Similar measurements elsewhere in the United States show even greater effects of urbanization (Guy, 1963). In the United States, according to the United States Department of Agriculture (1971), there are approx 61,000,000 acres of urban and built-up land. Between 1958 and 1975, it is estimated that about 20,000,000 acres of land will be taken out of agriculture and shifted into urban and other nonfarm uses. Therefore, urbanization will become increasingly important in the future with respect to erosion and sediment production.

6. Road and Highway Construction.—Serious erosion can occur during the construction of roads and highways when protective vegetation is removed and steeply sloping cuts and fills are left unprotected (Diseker and Richardson, 1963). Such erosion can create local problems of serious downstream sediment damages. In the construction of major thoroughfares, e.g., freeways, interstate roads, etc., large areas may be exposed to erosive conditions for long periods of time. However, much greater attention is now being given to stabilizing exposed surfaces along the major thoroughfares so that they do not become serious long-term sediment-producing areas. This is not true for many local roads constructed and maintained by township and county governments. Many old roads built in the hillier sections of the United States 100 or more years ago are deeply incised by erosion and still represent important, unprotected, sediment-source areas.

7. Mining Operations.—Mining operations may introduce large volumes of sediment directly into streams. Mine dumps and spoil banks, which are left ungraded and unvegetated, often continue to erode by natural rainfall for many years after mining operations have ceased. A study (Collier, et al., 1964) on the effects of strip mining on floods and water quality in McCreary County, Ky., shows the significance of this type of mining on the sediment load of streams. Cane Branch, which drains an area where strip mining is active, discharged sediment at the rate of 2,800 tons/sq mile in the 1957 and 1958 water years. Nearby Helton Branch discharged sediment at the rate of only 49 tons/sq mile during the same period. There is no mining in the Helton Branch Watershed. There are, in the United States, approx 1,500 large bituminous coal strip mines representing some 800,000 acres of stripped coal lands. Much of this stripped land is eroding rapidly and many states have found it necessary to enact legislation requiring the reclamation of strip-mined areas to reduce damages resulting therefrom.

Other types of mining, e.g., placer-mining, can produce tremendous amounts of sediment unless special precautions are taken to keep mine wastes out of streams. Extensive hydraulic placer-mining operations on the west slopes of the Sierra Nevada Mountains in northern California, between 1849 and 1914, dumped more than 1.5 billion cu yd of sediment into the Sacramento River and tributaries (Gilbert, 1917). This caused serious sediment damages on downstream flood plains and in navigation channels of the Sacramento and in San Francisco Bay approx 100 miles downstream from the location of mining operations. These damages had become so severe by 1884 that valley interests obtained injunctions against the mining interests to prevent them from further sluicing mining debris into the channels. Many of today's (1972) drainage and flood problems of the Sacramento Valley and problems of construction and maintenance of navigation channels can be traced back directly to hydraulic mining activities a century ago (in the 1850's).

Gravel and sand pit operations may also represent the source of large amounts of sediment unless adequate control measures are applied. The primary problem involves the disposal of waste from gravel washing plants. Where these wastes are sluiced directly into streams, they may create severe downstream sediment damages.

8. Altering Runoff Conditions.—Erosive energy is a function of the volume of water and its velocity. When the volume of runoff water is increased, or is concentrated in natural or artificial channels, its erosive energy is increased. Many human activities as applied to the land generally either increase the amount of runoff water or concentrate the flow. Thus the clearing of land increases the percentage of rainfall that runs off while water-collecting and disposal systems concentrate the volume of flow. Many of the gullies of the Piedmont area of southeastern United States, which follow no previously existing drainage ways, were formed as a direct result of concentrated flow from early terracing systems that came into wide use there in the 1890's (Ireland, et al., 1939). The concentration of flow by collecting and disposal systems in highway construction, e.g., diversions, road ditches, culverts, and other means often creates serious erosion conditions that not only constitute important sediment sources but also cause high maintenance costs.

An example of serious erosion initiated by increasing the volume of waterflow in a natural channel is that of Fivemile Creek near Riverton, Wyo. According to Miller and Borland (1963) and Colby, et al. (1956), the natural annual discharge from the 397-sq mile watershed of the stream was approx 5,000 acre-ft. Irrigation wastewater was first discharged into this stream channel in 1925, with the initiation of the irrigation project upstream. The discharge through the channel in 1952 was 92,000 acre-ft. Erosion of the channel began in about 1935, and between 1935 and 1950 an estimated 27,000 acre-ft of sediment was eroded from the channel.

According to a 1960 report of the Bureau of Reclamation, United States Department of Interior (United States Bureau of Reclamation, 1960), a total of 726,060 acre-ft of water was conveyed through the channel of Fivemile Creek during the 7 yr between October, 1951 and September, 1958. This is an average of somewhat over 100,000 acre-ft/yr, or, 20 times the natural discharge. As a result, the channel of Fivemile Creek has been widened and deeply incised, in some places to bedrock. The resulting sediment has been transported to the 970,000-acre-ft Boysen Reservoir on the Wind River into which Fivemile Creek empties. A total of 7,497,200 tons of sediment was carried into Boysen Reservoir.

9. Stream and River Control Works.—Stream and river control works may have a serious local influence on accelerating channel erosion. Any structural work of improvement that changes the direction of flow or increases the depth, duration, and velocity of flow may result in erosion. Channel straightening, which increases the channel gradient and flow velocity, may initiate channel erosion. Likewise, constricting the cross section of channels usually increases velocity and depth of flow and the eroding and transport power of streams. In sediment-bearing streams where movable beds are in poised equilibrium with flow conditions, e.g., in sand-bed streams, erosion is initiated promptly and proceeds rapidly. The lowering of the bed of a main stream results in lowering the bed of tributary streams. In many instances, such bed degradation is beneficial because it restores the flood-carrying capacity of channels to original channel conditions. However, where parent materials are eroded by a new set of hydraulic conditions, degradation may proceed far beyond the original bed levels and actually initiate an entirely new cycle by erosion in a watershed. Besides destroying productive land by bed and bank erosion of main stream and tributary channels, the degradation of channels may seriously depreciate adjacent land areas by lowering the ground-water tables needed to maintain productive capacity.

The construction of a dam influences downstream channel stability in two ways. It traps the sediment load and it changes the downstream natural flow characteristics. Both the sediment load and flow conditions were responsible for establishing the natural regime of the channel prior to construction of the dam. Clear water released from a reservoir immediately picks up a new load downstream if the discharge is sufficient to erode the bed and transport the sediment. When this condition prevails the net result is the erosion of the channel and lowering of the bed that will continue until the stream is again in balance with the new flow characteristics.

Degradation of channels below dams has been noted on numerous occasions (Hathaway, 1948). The seriousness of the problem depends on the erodibility of downstream channel materials in respect to the hydraulic characteristics of outflow from the dam. Where resistant channel materials prevail, erosion is limited solely to "plunge pools" immediately below the outlets. But if the channel materials are such that they are readily eroded by the regulated outflow, erosion of the channel may occur for many miles downstream. For example, studies of channel changes below major dams on the Colorado River, reported by Borland and Miller (1960), indicate that 151,800,000 cu yd were eroded from the channel below Hoover Dam, for a distance of 91.8 miles to the headwaters of Lake Havasu, between 1935 and 1951. Some concern is being expressed for the possible effects of degradation below the major dams on the Missouri River. In respect to water treatment, downstream benefits have been derived by reduction in costs of water treatment as a result of lower turbidity and increased water hardness (Neel, 1958). But, operational problems appear to have arisen as a result of the release of the clear water. This clear water permits deeper light penetration and promotes more rapid growth of plankton algae and other aquatic plants. Because of this, purification methods for Missouri River water must now be changed to remove objectionable tastes and odors from the water (Neel, 1963).

Where navigation is important, the regulated clear flows below a dam may not maintain adequate depth in channels formed by the river in its unregulated state. Shifting bars that are a hazard to navigation and expensive to regulate may develop.

10. Quality of Water.—Sediment is not only the major water pollutant by weight and volume (United States Senate Select Committee on National Water Resources, 1960) but it also serves as a catalyst, carrier, and storage agent of other forms of pollution. Desirable qualities of water vary according to use and there are a few uses in which sediment in the water is desirable. Usually, however, the greater the sediment concentration, the poorer the quality. Sediment alone degrades water specifically for municipal supply, recreation, industrial consumption and cooling, hydroelectric facilities, and aquatic life (Cordone and Kelley, 1961). In addition, chemicals and wastes are assimilated onto and into sediment particles. Ion exchange occurs between solutes and sediments (Kennedy, 1963). Thus, sediment has become a source of increased concern as a carrier and storage agent of pesticide residue, adsorbed phosphorus, nitrogen and other organic compounds, and pathogenic bacteria and viruses. Additional information is needed on the chemical and biological relationships of sediment. Studies are underway to determine more precisely the behavior of pesticides and other chemicals in soil, water, and other segments of the environment (Glymph and Storey, 1966).

C. Problems of Sediment Transport

11. Movement of Sediment.—Sediment is transported in suspension, as bed load rolling or sliding along the bed and interchangeably by suspension and bed load. The nature of movement depends on the particle size, shape, and specific gravity in respect to the associated velocity and turbulence. Under some conditions of high velocity and turbulence e.g., high flows in steep-gradient mountain streams, cobbles actually are carried intermittently in suspension. Conversely, silt size particles may move as bed load in low-gradient, low-velocity channels, e.g., drainage ditches. In waters of certain electrochemical composition, some clay minerals, e.g., montmorillonite and illite may remain in suspension for days or even months in relatively quiet water. Even in transport, whether as bed load or in suspension, sediment may cause problems. The aggregate damages in no way approach those of either accelerated erosion or deposition; however, sediment in transport may create severe problems.

12. Impingement of Sediment Particles.—Damage caused by the striking of objects by sediment particles in transport is relatively rare, but it does occur in isolated cases. Bridge abutments and trestle piers on torrential streams transporting cobbles may be severely damaged by boulders chipping away at wood or concrete unless protection is afforded. Paved flood channels that carry coarse sand and gravel are subject to wear and need periodic repairs. Torrential flood flows from mountain streams, e.g., those along the Wasatch Mountains in Utah and the San Gabriel Mountains in California have been known to set huge boulders in motion, some weighing several tons, that roll and bounce down alluvial fans causing severe damage to trees, fences, telephone lines, buildings, and other objects in their path.

Damage by impingement is not limited to the large rocks but may also result from transport of smaller particles. Chief damage of this nature is to turbines and pumps where sediment-laden water causes excessive wear on runners, vanes, propellers, and other appurtenant parts (Murthy and Madhaven, 1959). Damage in such instances is most often caused by sand-size particles that are transported at high velocity yet have sufficent mass to dislodge metal on impact. Considerable damage of this nature can occur to hydroelectric generating equipment using sand-laden run-of-river flow. In these plants, frequent replacement of worn parts may be necessary to maintain efficient operation.

13. Sediment in Suspension.—The more important problems related to sediment in transport involves fine-grained sediment carried in suspension. Most people have a natural antipathy of "muddy streams." This is particularly evident in fishermen. Aside from the fact that few people care to fish in a muddy stream, there is a definite effect of suspended sediment on the size, population, and species of fish in a stream (Ellis, 1936). Suspended sediment effects the light penetration in water and thus reduces the growth of microscopic organisms on which insects and fish feed. Suspended sediment also injures the gills and breathing structure of certain types of fish. Thus continually muddy streams harbor only what is known as "rough fish," because they can tolerate these conditions.

A study of the effects of the number of muddy flows of the Merimac River, in Missouri, on the income from recreation, made in 1940 by the Missouri State Planning Board, showed that this income is reduced as a result of muddy water flows. The study (Brown, 1945), indicated a loss of 49,090 visitor-days of recreation because of muddy streamflow. This resulted in an average annual loss of $60,775, which would have been spent locally if the stream had not been too muddy for fishing.

The removal of suspended matter from water to render it suitable for industrial uses and for human consumption is a costly process in the United States. The cost of water treatment is closely related to water turbidity, particularly in the use of alum for flocculating sediment. According to the Senate Select Committee on Natural Water Resources (1961), a total of 98.6 billion gal of water was withdrawn from streamflow in 1954 for municipal and industrial purposes. Of this amount, 16.7 billion gal are for municipal use alone. The total annual cost of this water is over $1 billion, of which more than $2,000,000 represents treatment costs; much of this is due to removal of suspended sediment. These costs will rise in the future in accordance with progressively increasing population and per capita demands for water.

D. Problems of Sediment Deposition

14. General.—Deposition is the counterpart of erosion. The products of erosion may be deposited immediately below their sources, or may be transported considerable distances to be deposited in channels, on flood plains, or in lakes, reservoirs, estuaries, and oceans. Sediment may cause severe damages depending on the amount, character, and place of deposition.

All sediment deposition is not injurious. Some deposits may form fertile flood-plain or delta soils and some may be reclaimed and put back to beneficial uses, e.g., land fills or even as topsoil for lawns and gardens. Those deposits that are carried far out into the oceans have little effect on mankind, but less than 1 ton in 4 tons of upland eroded materials ever reaches the ocean (Gottschalk and Jones, 1955).

15. Deposits at Base of Eroding Slopes.—These are usually heterogeneous deposits of rock and soil that have not been transported through any well-defined channel system. They are found mostly at the breaks of eroding slopes and gravity often plays an important role in the transport of particles. These deposits, when occurring in farm fields, may completely bury crops and fence lines. Such deposits may seriously damage high-cost property when, e.g., they cover highways or in urban areas, streets, lawns, and gardens. Sometimes cars, buildings, and other property are partially buried.

16. Flood-Plain Deposits.—Deposits that occur on flood plains create numerous types of damages to crops and developments. They may completely bury crops. Infertile materials reduce the fertility and long-term productivity of flood-plain soils. In the processes of flood-plain deposition, natural levees of coarser materials form adjacent to the channels while the fine-grained materials are transported and deposited in areas behind the levees. These levees impair the surface drainage of floodwaters back to the channel, and the fine-grained deposits reduce the permeability and restrict vertical drainage behind the levees. The net result is the swamping of flood-plain lands. In the early stages of detrimental flood-plain deposition of this kind, owners find it necessary to convert periodically wet croplands to less intensive pasture use. Increased swamping makes such lands permanently wet and unfit even for pasture purposes. The land may then be converted to woodland or abandoned entirely. Even woodland, in time, will be drowned and in the final stages the land becomes covered with standing water and desolate dead snags. Many thousands of acres of flood-plain lands in the Peidmont valleys of the southeastern United States have been completely drowned out in this fashion in the past, and thousands of acres elsewhere in the United States are being damaged annually through these processes.

The deposition of sediment on flood plains in urban areas can create serious local deposition on streets, highways, railroads, and in buildings and other developments. High maintenance and cleanup costs from overbank deposition, year after year, often result in high damages.

17. Channel Deposits.—The deposition of sediment in drainage ditches, irrigation canals, and in navigation and natural stream channels creates serious problems in loss of services and cleanout costs. Sediment deposits in ditches result in the raising of water tables and damage to crops. Sediment normally deposited under the generally low velocity of flow in drainage ditches is fine-grained. As such, it is inherently fertile and helps to promote rapid vegetative growth. This further serves to slow down the flow and results in higher rates of deposition and damage.

Sediment in irrigation ditches reduces the rate and volume of water delivered to irrigated areas. Serious crop damage can result when water cannot be delivered in sufficient amounts to meet irrigation needs. The clogging of diversions and turnouts with sediment also reduces the amount of water available in time of need. The flushing of coarse-grained sediment onto irrigated areas reduces soil fertility, while colloidal sediment reduces permeability, both of which result in reduced productivity of irrigated lands. The cleanout and maintenance costs of irrigation ditches and reduced productivity of irrigable soils represent a sizable damage in the United States each year.

The sediment deposited in our navigable channels, waterways, and harbors must be removed periodically to maintain required depths. Each year the Federal government and private interests remove from navigation channels of the United States a volume of sediment equivalent to the total amount excavated in the construction of the Panama Canal. According to the annual reports of the Chief of Engineers, United States Army, the Federal government, during the years 1960, 1961, and 1962, removed approx 194,000,000 cu yd of sediment to maintain existing navigable channels. This work cost in excess of $62,000,000 annually. New channels are being dredged each year at a cost of about $74,000,000. The maintenance of these additional as well as existing channels will progressively increase the present bill for maintenance dredging costs unless some other means can be developed to appreciably reduce erosion and the transport of sediment into them.

The deposition of sediment in our natural stream channels has greatly aggravated floodwater damages. The deposition of sediment in channels decreases the channel capacity and the flood-carrying capacity. This results in higher and more frequent overflows.

Floods are quite obvious to most people, but the influence of sediment is often masked and overlooked. Thus, many channels are cleaned out annually in the United States to improve their floodwater-carrying capacity. Little attention is given, however, to the erosion and source of the sediment that created the problem in the first place and that will continue to encroach on the improved channel.

Planners often conclude that the solution of all floodwater problems is channel cleanout or the construction of upstream retarding reservoirs to regulate the flow and stage at damage areas. Although this provides an immediate solution to the flood problems, it is not always the most economical or lasting solution. The reservoirs like the improved channels trap sediment, and in time they too will become filled and useless. Furthermore, as indicated previously, the release of clear water from such reservoirs can initiate serious downstream erosion conditions that in time could exceed in cost the floodwater damage reduction brought about by the reservoir. In many instances, the proper solution to the problem involves the control of accelerated erosion and reduction of sediment movement that has caused channel deterioration.

18. Deposits in Lakes and Reservoirs.—When streamflow enters a natural lake or reservoir, its velocity and transport capacity is reduced and its sediment load is deposited. In natural lakes that have no outlets the total incoming sediment load is deposited. In artificial lakes with outlets, e.g., reservoirs, the amount deposited depends on the detention storage time, the shape of the reservoir, operating procedures, and other factors. In most storage reservoirs of modern design, more than 90% of the incoming load is usually trapped (Brune, 1953).

The deposition of sediment in natural lakes and its influence on recreation and other uses has been observed in numerous instances, particularly where such lakes are surrounded by intensive farming activities. In most natural lakes, the total rate of filling with sediment is generally of less concern than the location of deposits. Certain species of aquatic plants thrive only in water depths of less than 7 ft or 8 ft. Incoming fine-grained sediment, derived particularly from prolonged sheet erosion, is inherently fertile so that accumulations not only shoal specific areas but also provide a fertile bottom encouraging prolific plant growth. As a result, spawning areas for certain species of fish are destoyed (Langlois, 1941) as well as fishing areas, and the operation of boats in such areas becomes difficult.

Waterfront properties situated along weedy shores are greatly depreciated so far as land values are concerned. Such conditions have developed along the southern shore lines of the once-popular resort lake, Lake Chatauqua, at Jamestown, N.Y. Here serious problems of small boat operation exist in some parts of the lake where large sight-seeing vessels formerly sailed. Many natural lakes in the humid agricultural areas of the midwest have also had similar problems develop, and spawning and fishing grounds in parts of the Great Lakes have been damaged by sediment.

In artificial reservoirs, both the location of deposits and the loss of storage capacity are of concern. As in natural lakes, the location of deposits is important, particularly in reservoirs built for recreation or esthetic purposes, e.g., real-estate developments (Holeman and Geiger, 1959). In this respect, the shoaling of areas, particularly by deltaic deposits may seriously affect adjacent property values, convert attractive beach areas into undesirable mud flats, and damage fishing and boating activities. Recreation is an important element in all reservoir developments, but the primary purpose of most reservoirs is to store water for flood prevention, water supply, irrigation, power, or other uses. In such reservoirs, loss of storage capacity is usually a more important factor than the location of deposits, because the loss of carryover storage effects the proper functioning of these reservoirs. However, reservoir evaporation losses may be increased by changes in the area-capacity relation in a reservoir caused by sediment accumulation. Density or turbidity currents that carry fine sediments to the deepest parts of the reservoir can be generated, and depending on the size, purpose and operation of the reservoir, can create excessively turbid water for irrigation, municipal, or industrial use. Such turbid currents can also affect the temperature and quality of the water released to the extent that fish life may be damaged and recreational use endangered. The progressive deposition of sediment in reservoir delta areas can affect the backwater profiles. The resultant changes of the incoming channel geometry alters the hydraulic properties of the cross sections used in computing backwater profiles.

In the early history of reservoir development in the United States, little attention was given to sedimentation as a factor in design. Reservoirs built below watersheds with low rates of erosion, e.g., the well-forested lands of New England, did not suffer appreciable damage because of sediment accumulation. But in the humid agricultural areas of the midwest, in the Great Plains states, in the southeastern United States, and other problem areas where erosion rates are relatively higher, serious deposition in reservoirs has occurred. A review of the annual rates of deposition in reservoirs built prior to 1935 in these problem areas, as reported by the Sedimentation Committee, Water Resources Council (United States Department of Agriculture, 1973), indicates that 33% have lost from one-fourth to one-half of their original capacity; about 14% have lost from one-half to three-quarters of their original capacity; and approx 10% have had their capacity reduced to run-of-river conditions with all usable storage completely depleted.

No reliable inventory of the total stored water resources in reservoirs and the annual rate of depletion by sediment deposits has yet been made in the United States. From existing information however, it appears that the total original capacity of the larger reservoirs built to date is approx 400,000,000 acre-ft. The rate of deposition in these reservoirs is estimated to be about 1.5 billion cu yd or 1,000,000 acre-ft of sediment annually. This sediment each year replaces water storage worth close to $100,000,000, based on original development cost.

E. Solution of Sediment Problems

19. General.—Many problems arising from sediment transport and deposition would be solved by stopping upstream erosion completely. But stopping erosion completely is not always physically possible, or economically feasible in the solution of specific downstream sediment problems. Furthermore, the solution of one problem in this manner can often lead to the creation of others, and, in such instances, alternatives must be considered.

Often various methods of reducing specific sediment damages must be considered in project formulation. These include, in addition to erosion control of primary upstream sediment-source areas, such measures as trapping sediment in sedimentation basins, allowance for storage of sediment in structure design, dredging, sediment bypassing, and other methods. The determination of the measure, or measures, to use in a specific case and the evaluation of the effectiveness of the measure, or combinations thereof, requires a thorough understanding of the fundamental sedimentation processes and principles involved.

20. Basic Considerations in Solution of Sediment Problems.—It is necessary to fully recognize and understand the nature of a sediment problem and the part that fundamental sedimentation processes play in creating the problem before an adequate project for alleviation of the problem can be formulated. For example, it must be determined whether the problem is caused by the total amount of sediment delivered to the problem area or only some part of it, e.g., that contained within certain limitations of grain sizes. Thus, a channel stability problem might be solely caused by gravel-size particles, while a problem of drainage impairment might be caused by only the clay-size particles. The total sediment load may be of concern in other problems, e.g., the loss of storage capacity in reservoirs.

Once the nature of damage is determined and the characteristics (usually the particle size) of critical sediment are established, the source of this sediment must be determined. It is in this realm of sedimentology that the specialist must draw heavily on his knowledge of the processes and principles of sedimentation. This is because the amount and character of sediment delivered to any point in a stream system is dependent on the nature of materials being eroded, the type of erosion, and the hydraulic characteristics of flow transporting the sediment and influencing its deposition.

The determination of the type, or types, of measures needed for the alleviation of damages follows the delineation of the source of damaging sediment. As indicated previously, it is usually necessary to consider and compare alternative methods to arrive at measures that are both economically possible and physically feasible. Thus, the analysis of a specific problem might indicate that the damaging sediment is derived from upstream erosion that could be readily and cheaply controlled at the source. However, those sustaining injury may have no control over the land on which the erosion is occurring; and, if they cannot convince the landowners to install erosion prevention measures, the control of erosion at its source cannot be affected even though this is highly desirable. Thus, some other method must be selected though it proves to be more costly and perhaps less effective in the long run.

F. Conclusions

The damages created by sediment are varied and extensive. The kinds of damages created by sediment depend on the amount and character of sediment that, in turn, is influenced by the processes of erosion, transport, and deposition. A knowledge of these processes is necessary to understand the underlying causes of sediment damages, to determine the sources of critical sediment causing such damages, and to formulate feasible control measures.

Considerable knowledge is still needed relative to the various aspects of erosion, transport, and deposition of sediment before accurate predictions of causes and effects can be made. However, great strides have been made in this field, particularly during the last four decades (since the 1930's), and much useful information has been developed that would be helpful in the planning, design, and maintenance of land and water resource development projects. A great deal of this information is summarized in the subsequent chapters.

G. References

Bennett, H. H., *Soil Conservation*, McGraw-Hill Book Co., Inc., New York, N.Y., 1939.

Borland, W. M., and Miller, C. R., "Sediment Problems of the Lower Colorado River," *Journal of the Hydraulics Division*, ASCE, Vol. 86, No. HY4, Proc. Paper 2452, Apr. 1960, pp. 61–87.

Brown, C. B., "Floods and Fishing," *The Land*, Vol. 4, No. 1, 1945, pp. 78–79.

Brune, G. M., "Trap Efficiency of Reservoirs," *Transactions of the American Geophysical Union*, Vol. 34, No. 3, June, 1953, pp. 407–448.

Brune, G. M., "Sediment is Your Problem," *Agricultural Information Bulletin No. 174*, Soil Conservation Service, Mar., 1958.

Colby, B. R., Hembree, C. H., and Rainwater, F. H., "Sedimentation and Chemical Quality of Surface Waters in the Wind River Basin, Wyoming," *Water–Supply Paper 1373*, United States Geological Survey, 1956.

Collier, C. R., et al., "Influences of Strip Mining on the Hydrologic Environment of Parts of Beaver Creek Basin Kentucky, 1955-1959," *Professional Paper 427-B*, United States Geological Survey, 1964.

Cordone, A. J., and Kelley, D. W., "The Influences of Inorganic Sediment on the Aquatic Life of Streams" *California Fish and Game*, Vol. 47, No. 2, pp. 189–228.

Dale, T., and Carter, V. G., *Topsoil and Civilization*, University of Oklahoma Press, Norman, Okla., 1955.

Diseker, E. G., and Richardson, E. C., "Roadbank Erosion and Its Control in the Piedmont Upland of Georgia," *ARS 41-73*, Agricultural Research Service, Aug., 1963.

Ellis, M. M., "Erosion Silt as a Factor in Aquatic Environments," *Ecology*, Vol. 17, No. 1, Jan., 1936, pp. 29–42.

Gilbert, G. K., "Hydraulic–Mining Debris in the Sierra Nevada," *Professional Paper 105*, United States Geological Survey, 1917.

Glymph, L. M., and Storey, H. C., "Sediment—Its Consequences and Control," AAAS Symposium on Agriculture and the Quality of Our Environment, Washington, D.C., Dec., 1966.

Gottschalk, L. C., "Effects of Soil Erosion on Navigation in Upper Chesapeake Bay," *Geographical Review*, Vol. 35, No. 2, 1945, pp. 219–238.

Gottschalk, L. C., "Predicting Erosion and Sediment Yields," International Union of Geodesy and Geophysics, Association of Scientific Hydrology, XIth General Assembly, Toronto, Canada, Tome I, Vol. 1, 1958, pp. 146–153.

Gottschalk, L. C., "Effects of Watershed Protection Measures on Reduction of Erosion and Sediment Damages in the United States," International Union of Geodesy and Geophysics, Association of Scientific Hydrology, Symposium on Continental Erosion, Bari, Italy, Publication No. 59, 1962, pp 426–450.

Gottschalk, L. C., and Brune, G. M., "Sediment Design Criteria for the Missouri Basin Loess Hills," *SCS-TP 97*, Soil Conservation Service, Oct., 1950.

Gottschalk, L. C., and Jones, V. H., "Valleys and Hills, Erosion and Sedimentation," United States Department of Agriculture, 1955 Yearbook of Agriculture, 1955, pp. 135–143.

Guy, H. P., et al., "A Program for Sediment Control in the Washington Metropolitan Region," *Technical Bulletin 1963-1*, Interstate Commission on Potomac River, May, 1963.

Hathaway, G. A., "Observations on Channel Changes, Degradation, and Scour Below

Dams," International Association on Hydraulic Structures, Research Project No. 2, Appendix 16, Dec., 1948, pp. 287–307.

Holeman, J. N., and Geiger, A. F., "Sedimentation of Lake Barcroft, Fairfax County, Va," *SCS-TP-136,* Soil Conservation Service, Mar., 1959.

Ireland, H. A., Sharpe C. F. S., and Eargle, D. H., "Principles of Gully Erosion in the Piedmont of South Carolina," *Technical Bulletin 633,* United States Department of Agriculture, Jan., 1939.

Kennedy, V. C., "Mineralogy and Cation Exchange Capacity of Sediment from Selected Streams," *United States Geological Survey Professional Paper 433D,* 1963.

Langlois, T. H., "Two Processes Operating for the Reduction of Fish Species from Certain Types of Water Areas," *Transactions,* North American Wildlife Conference, No. 6, 1941, pp. 189–201.

Leopold, L. B., Wolman, M. G., and Miller, W. H., *Fluvial Processes in Geomorphology,* W. H. Freeman and Co., San Francisco, Calif., 1964.

Lowdermilk, W. C., "Conquest of the Land Through 7,000 Years," *Agricultural Information Bulletin No. 99,* Soil Conservation Service, Aug., 1935.

Miller, C. R., and Borland, W. M., "Stabilization of Fivemile and Muddy Creeks," *Journal of the Hydraulics Division,* ASCE, Vol. 89, No. HY1, Proc. Paper 3392, Jan., 1963, pp. 67–98.

Murthy, Y. K., and Madhaven, K., "Silt Erosion in Turbines and Its Prevention," *Water Power,* 1959, pp. 70–75.

Musgrave, G. W., "The Quantitative Evaluation of Factors in Water Erosion— A First Approximation," *Journal of Soil and Water Conservation,* Vol. 2, No. 3, July, 1957, pp. 133–138.

Neel, J. K., "Reservoir Influences on Central Missouri River," *Journal of the American Water Works Association,* Vol. 50, No. 9, Sept., 1958, pp. 1188–1200.

Neel, J. K., "Impact of Reservoirs," *Limnology in North America,* D. G. Frey, ed., University of Wisconsin Press, Madison, Wisc., 1963, pp. 575–593.

Noble, E. L., "Sediment Reduction Through Watershed Rehabilitation," *Miscellaneous Publication 970, Proceedings of Federal Interagency Sedimentation Conference,* United States Department of Agriculture, Jackson, Miss., 1965, pp. 114–123.

Stallings, J. H., "Erosion of Topsoil Reduces Productivity," *SCS-TP-98,* Soil Conservation Service, 1950.

Stallings, J. H., *Soil Conservation,* Prentice-Hall, Inc., Englewood Cliffs, N.J., 1957.

Termier, H., and Termier, G., *Erosion and Sedimentation,* translated by D. W. and E. E. Humphries, D. Van Nostrand Co., Inc., New York, N.Y., 1963.

Toynbee, A. J., *A Study of History,* 2nd ed., Oxford University Press, London, England, 1948.

United States Bureau of Reclamation, "Sedimentation and Delta Formation in Boysen Reservoir, Wyoming," Hydrology Branch, Oct., 1960.

United States Department of Agriculture, "Basic Statistics of the National Inventory of Soil and Water Conservation Needs, 1967," *Statistical Bulletin 461,* Washington, D.C., 1971.

United States Department of Agriculture, "Summary of Reservoir Sediment Deposition Surveys made in the United States through 1970," *Miscellaneous Publication* 1266, Agricultural Research Service, Water Resources Council, July, 1973.

United States Senate Select Committee on National Water Resources, "Pollution Abatement," *Committee Print No. 9,* 86th Congress, 2nd Session, Jan., 1960.

United States Senate Select Committee on National Water Resources, "Report of Select Committee," No. 29, 87th Congress, 1st Session, Jan., 1961.

Chapter II.—Sediment Transportation Mechanics

A. Introduction

1. Description of Sediment Motion.—To define the terms to be used in considering sediment movement, it is convenient to describe the motion of grains caused by water flowing over a bed of sediment that was first flattened artificially. At very low velocities no sediment will move, but at some higher velocity individual grains will roll and slide intermittently along the bed. The material so moved is defined as the contact load of the stream. At a still higher velocity, some grains will make short jumps, leaving the bed for short instants of time and returning either to come to rest or to continue in motion on the bed or by executing further jumps. The material moved in this manner is said to saltate and is called the saltation load of the stream. If now the flow velocity is raised gradually, the jumps executed by the grains will occur more frequently and some of the grains will be swept into the main body of the flow by the upward components of the turbulence and kept in suspension for appreciable lengths of time. Sediment that is carried in suspension in this manner is known as the suspended load of a stream. Sediment movement by wind and by water are qualitatively similar. However, because the density of air is so much less than that of water and the sediment density is the same, the detailed behavior of the two systems differs.

2. Bed Configuration.—Under certain flow conditions with relatively low sediment-transport rates, the sediment bed will deform into wavelike forms with small slopes on their upstream faces, a sharp crest, and steep downstream faces. At higher velocities and depths the length of these forms may increase many fold, still keeping the sharp crest. These forms move downstream at velocities that are small compared to the flow velocity. Sharp-crested forms of this kind with wave lengths less than approx 1 ft are called ripples; those with wave lengths longer than approx 1 ft are called dunes. At some higher velocity the ripples or dunes, or both, disappear and the bed becomes flat. At some still higher velocity a wave of sinusoidal shape develops, which usually moves upstream and is accompanied by waves on the water surface. Sand waves of the latter type are known as antidunes. Antidunes are always accompanied by waves on the water surface that are called surface waves or sand waves. Surface waves are often unsteady; they usually move upstream slowly, increase in amplitude until they become unstable, then break by curling over in the upstream direction, and disappear only to form again and repeat the cycle. At even higher velocities the form known as chutes and pools occurs. This subject is examined in more detail in Sections F and G of Chapter II. Diagrams of bed forms are shown in Fig. 2.73 and photographs of ripples, dunes, and antidunes are shown in Figs. 2.74, 2.75, and 2.76, respectively.

Ripples and dunes roughly similar to those forming in water also form in wind transportation. Antidunes and chutes and pools do not form in air. Other forms, not observed in water, are formed in air.

3. Classification of Sediment Load.—While material is transported in suspension, saltation and rolling and sliding on the bed is also occurring, so that all three modes of transportation occur simultaneously. Apparently then, the different modes of transportation are closely related and it is difficult, if not impossible, to separate them completely. The borderline between contact load and saltation load is certainly not well defined, because it is indeed hard to imagine a particle rolling on the bed without at some time losing contact with the bed and executing short jumps, and, according to definition, becoming saltation load. In a similar manner, the distinction between saltation and suspension is also not definite.

These difficulties are avoided in a practical way by introducing the term "bed load," which is defined as material moving on or near the bed so that the total load is now made up of the bed load and suspended load. In addition, the total load is divided into "bed sediment load" and "wash load," which are defined as being, respectively, of particle sizes found in appreciable quantities and in very small quantities in the shifting portions of the bed. Obviously, both the bed sediment load and the wash load may move partially as bed load and partially as suspended load, although by definition, practically all the wash load is carried in suspension.

Finally, it is convenient to introduce the term "sediment discharge," which is defined as the quantity of sediment per unit time carried past any cross section of a stream. The term should be qualified. For example, one may refer to the bed load discharge, the bed sediment discharge, or the total sediment discharge. It will be noted that, as used herein, the term "load" denotes the material that is being transported, whereas the term "sediment discharge" denotes the rate of transport of the material.

4. Status of Knowledge of Sediment Transportation.—The early studies of sediment transportation dealt with coarse material carried as bed load, and were not concerned with suspended sediment. In approximately 1925, work began on the problem of suspension and has developed to the point at which this phase of the sedimentation problem is actually understood better than the bed load transport phase. However, it is well to point out that it is not yet (1972) possible to predict the suspended-load discharge with any greater degree of certainty than the bed load discharge. This is because the quantity of material that goes into suspension depends on the fluid forces at the bed that also transport the bed load.

Despite the substantial volume of study devoted to sedimentation mechanics, it is still not possible to predict the sediment discharge of an alluvial stream with a degree of certainty that is satisfactory for most engineering purposes. It is also not possible, e.g., to predict accurately such common but important occurrences as the amount of aggradation or degradation that will occur in a stream when a dam is installed. Before important problems of this kind can be dealt with satisfactorily, the investigations and sedimentation must be extended to include further studies of such factors as the range in grain size of the sediment, the alinement of the stream, and the scale of the flow system. These limitations are outlined herein in an attempt to caution the reader lest he rely too heavily on the results given by transport formulas. Despite the limitations previously outlined, sedimentation mechanics is a useful tool in solving practical problems.

Transportation of material by wind is similar to transportation by water, in that the sediment is subjected to the same kinds of forces in the two cases. However, the great difference between the densities and viscosities of air and water seems to introduce important differences in the relative importance of the forces and, thus, in the transport relations. Because of this, it has not been possible to apply the results of studies with water to the wind problem and knowledge of wind transport has developed more or less independently. Because work is this field has not been as intensive as in water transport, there is less information on the subject of wind transport.

B. Properties of Sediment

5. Pertinent Properties.—The entrainment, transportation, and subsequent deposition of a sediment depend not only on the characteristics of the flow involved, but also on the properties of the sediment itself. Those properties of most importance in the sedimentation processes can be divided into properties of the particles and of the sediment as a whole. The most important property of the sediment particle or grain is its size. In much of the river sediment studies of the past, average size alone has been used even to describe the sediment as a whole. This procedure can give reasonable results only if the shape, density, and size distribution of natural sediments do not have major variations between river systems. To obtain more accurate results, a more precise description of the sediment is necessary.

The settling velocity of a particle directly characterizes its reaction to flow, and ranks next to size in importance. Frequency distributions of properties, e.g., size and settling velocity are necessary to the description of sediments. Flocculation is of importance in the behavior of fine sediments and, in many cases, may be the major factor in determining the settling velocity and the specific weight of a deposit. Specific weight of a deposit is determined by the sediment properties and by the environment and manner in which it was laid down; thus specific weight is not strictly a true sediment property. However, it is included in this section because it appears to depend more on sediment properties than on other factors.

6. Size of Sediment Particles.—Because the size and shape of grains making up a sediment vary over wide ranges, it is meaningless to consider in detail the properties of an individual particle, and it is necessary to determine averages or statistical values. For this reason, it has been convenient to group sediments into different size classes or grades. Because such classifications are essentially arbitrary, many grading systems are to be found in the engineering and geologic literature (Krumbein and Pettijohn, 1938).[1] Table 2.1 shows a grade scale proposed by the subcommittee on Sediment Terminology of the American Geophysical Union (Lane, 1947). This scale, which is an extension of the Wentworth scale, has proven advantageous in sediment work because the sizes are arranged in a geometric series with a ratio of two, and because the sizes correspond closely to the mesh opening in sieves in common use as shown in the

[1]References are given in Chapter II, Section M.

TABLE 2.1.—Sediment Grade Scale

Class name (1)	Size Range				Approximate Sieve Mesh Openings per inch	
	Millimeters		Microns (4)	Inches (5)	Tyler (6)	United States standard (7)
	(2)	(3)				
Very large boulders		4,096–2,048		160–80		
Large boulders		2,048–1,024		80–40		
Medium boulders		1,024–512		40–20		
Small boulders		512–256		20–10		
Large cobbles		256–128		10–5		
Small cobbles		128–64		5–2.5		
Very coarse gravel		64–32		2.5–1.3		
Coarse gravel		32–16		1.3–0.6		
Medium gravel		16–8		0.6–0.3	2-1/2	
Fine gravel		8–4		0.3–0.16	5	5
Very fine gravel		4–2		0.16–0.08	9	10
Very coarse sand	2–1	2.000–1.000	2,000–1,000		16	18
Coarse sand	1–1/2	1.000–0.500	1,000–500		32	35
Medium sand	1/2–1/4	0.500–0.250	500–250		60	60
Fine sand	1/4–1/8	0.250–0.125	250–125		115	120
Very fine sand	1/8–1/16	0.125–0.062	125–62		250	230
Coarse silt	1/16–1/32	0.062–0.031	62–31			
Medium silt	1/32–1/64	0.031–0.016	31–16			
Fine silt	1/64–1/128	0.016–0.008	16–8			
Very fine silt	1/128–1/256	0.008–0.004	8–4			
Coarse clay	1/256–1/512	0.004–0.0020	4–2			
Medium clay	1/512–1/1,024	0.0020–0.0010	2–1			
Fine clay	1/1,024–1/2,048	0.0010–0.0005	1–0.5			
Very fine clay	1/2,048–1/4,096	0.0005–0.00024	0.5–0.24			

table. Note that the smallest sieve has a mesh size of 1/16 of a millimeter, which, by definition, is the size dividing the sands and silts. This also corresponds roughly to the finest sediment found in appreciable quantities in the beds of most streams.

Natural sediment particles are of irregular shape and, therefore, any single length or diameter that is to characterize the size of a group of grains must be chosen either arbitrarily or according to some convenient method of measurement. Three such diameters recommended for use by the subcommittee on Sediment Terminology of the American Geophysical Union (Lane, 1947) are defined as follows:

1. *Sieve diameter* is the length of the side of a square sieve opening through which the given particle will just pass.
2. *Sedimentation diameter* is the diameter of a sphere of the same specific weight and the same terminal settling velocity as the given particle in the same sedimentation fluid.
3. *Nominal diameter* is the diameter of a sphere of the same volume as the given particle.

The sieve and sedimentation diameters obviously have come into common use because of the convenience in measuring them. The size of sands is commonly measured by sieving, although sometimes the sedimentation diameter is also obtained by determining the settling velocity. The size of silts and clays is generally expressed as a sedimentation diameter and determined by sedimentation methods because of convenience. The sedimentation diameter is actually a fictitious size that enables the calculation of the settling velocity. For this reason, it has greater physical significance than the other two diameters. The nominal diameter has little significance in sediment transportation, but it is useful in discussing the nature of sedimentary deposits.

7. Shape of Sediment Particles.—The size of a sediment particle alone is usually not sufficient to describe it. A number of systems for describing the additional characteristics of sediment grains (Krumbein and Pettijohn, 1938; Interagency Committee, 1957) have been devised by geologists and engineers. Those characteristics that seem most important to engineers concerned with sediment transportation are shape and roundness, as conceived by H. Wadell (Wadell, 1932, 1933, 1935; Krumbein and Pettijohn, 1938). Shape describes the form of the particle without reference to the sharpness of its edges, while roundness depends on the sharpness or radius of curvature of the edges. The shape of a grain has been expressed in terms of true sphericity, which has been defined as the ratio of the surface area of the sphere with the same volume as the grain to the surface area of the particle.

The difficulty of determining the surface area and volume of small grains led Wadell to adopt an approximate but more convenient expression for sphericity. His expression for sphericity is the ratio of the diameter of the circle with an area equal to that of the projection of the grain when it rests on its largest face to the diameter of the smallest circle circumscribing this projection. The roundness of a grain is defined as the ratio of the average radius of curvature of individual edges to the radius of the largest circle that can be inscribed within either the projected area or a cross section of the grain. According to these concepts, a particle, e.g., a hexahedron, can have high sphericity and very low roundness. Conversely, a shape such as a circular cylinder with hemispherical ends can have low sphericity and high roundness.

In studying the fall velocity of geometric shapes (McNown and Malaika, 1950) and sand grains (Albertson, 1953), the shapes of the particles have been expressed by a shape factor, SF, given by

$$SF = \frac{c}{\sqrt{ab}} \qquad (2.1)$$

in which, a, b, and c are, respectively, the lengths of the longest, intermediate, and shortest mutually perpendicular axes of the particle. It was found that the fall velocity of particles could be expressed in terms of the nominal diameter, SF, and Reynolds number.

8. Specific Weight of Sediment Particles.—Practically all sediments, whether borne by wind or water, have their origin in rock material, and all constituents of the parent material can usually be found in the sediment. However, as the materials become finer due to weathering and abrasion, the less stable minerals tend to weather faster and be carried away as fine particles or in solution, leaving behind the more stable components. The highest degree of sorting of minerals is to be expected in the fine fractions of sediment. Coarse material, e.g., boulders, may be a part of the parent rock and contain all the constituents of the original material.

Although quartz, because of its great stability, is by far the commonest mineral found in sediments moved by water and wind, numerous other minerals also are present. Table 2.2 shows the mineral composition of a sand taken from the bed of the Missouri River at Omaha, Neb. (This information was obtained by the Corps of Engineers, United States Department of the Army, and is published with their permission.) In this case, it is seen that only approximately half of the sand is quartz. Although other material besides quartz may be present in appreciable quantities, the average specific gravity of sands is very close to that of quartz, i.e., 2.65, and this value is used often in calculations and analysis. The specific gravity will subsequently be shown to be an important quantity. It should, therefore, be measured at each site of sedimentation investigation.

9. Fall Velocity of Spheres.—The simplest shape for which information on fall velocity exists is the sphere. Because natural sediment particles are not spherical, their fall velocity cannot be calculated directly from sphere data. Use is nevertheless made of these data, which are presented herein.

TABLE 2.2.–Mineral Composition of Missouri River Sand at Omaha, Neb.

Mineral (1)	Specific gravity (2)	Percentage by weight (3)
Miscellaneous igneous and metamorphic rocks	–	8
Shale	–	2
Carbonate particles	2.85	3
Chert	2.65	7
Quartz	2.65	49
Feldspar	2.55-2.76	18
Heavy mineral (except magnetite)	–	8
Magnetite	5.17	2
Other constituents	–	3

For a sphere of diameter d, the fall velocity, w, for values of Reynolds number $\mathsf{R} = wd/\nu$ less than approx 0.1 is given by Stokes law

$$w = \frac{gd^2}{18\nu}\left(\frac{\gamma_s - \gamma}{\gamma}\right) \qquad (2.2)$$

in which ν and γ are, respectively, the kinematic viscosity and specific weight of the fluid; g denotes the acceleration of gravity; and γ_s = the specific weight of the sphere. The fall velocity over the entire range of Reynolds numbers, in terms of the drag coefficient, C_D, is given by

$$w^2 = \frac{4}{3}\frac{gd}{C_D}\left(\frac{\gamma_s - \gamma}{\gamma}\right) \qquad (2.3)$$

The drag coefficient in the Stokes range, i.e., $\mathsf{R} < 0.1$ is given by $C_D = 24/\mathsf{R}$. For larger Reynolds numbers C_D is still a function of R but it cannot be expressed analytically and has, therefore, been determined experimentally by observing fall velocities in still fluids or by measuring the drag of spheres in wind and water tunnels. Fig. 2.1 shows the drag coefficient, C_D, for spheres as a function of **R** over a wide range of R-values as given by Rouse (1937b).

From Fig. 2.1 one has a relation between C_D and R that will make possible the solution of Eq. 2.3 for w, once d, γ_s, ν, and γ are given. This solution involves tedious trial and error and is inconvenient. The tedious calculation can be avoided by using the auxiliary scale of $F/\rho\nu^2$ in Fig. 2.1, in which F = the submerged weight of the sphere given by

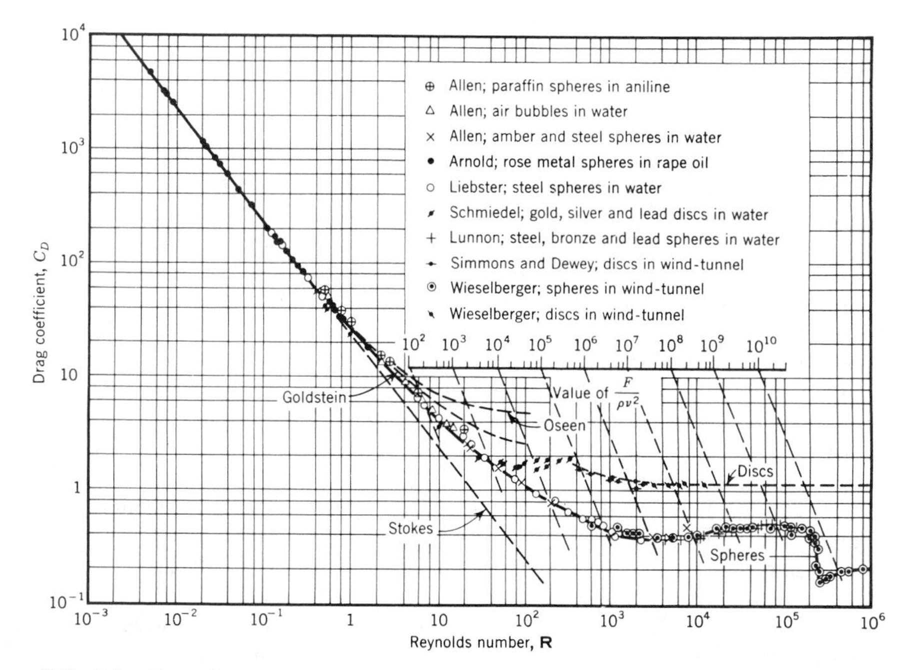

FIG. 2.1.—Drag Coefficient of Spheres as Function of Reynolds Number (Rouse, 1937a)

$$F = \frac{\pi d^3}{6}(\gamma_s - \gamma) \qquad (2.4)$$

and ρ = the density of the fluid in mass per unit volume. To calculate w, once d, γ_s, ν, and γ are known, one calculates ratio $F/\rho\nu^2$, locates the value of the ratio on the auxiliary scale, moves parallel to the sloping lines to the C_D-R curve and reads R, and then calculates w from the expression for R. If the fluid is either air or water and the sphere has a specific gravity of 2.65 relative to water, w can be read directly from Fig. 2.2, which is a chart directly giving the fall velocity of quartz spheres in fresh water and in air under a pressure of one atmosphere (14.696 psi) for a range of temperatures. For other fluids or for spheres with specific gravities different from that of quartz, i.e., 2.65, it is necessary to use Fig. 2.1 to solve for the fall velocity.

10. Effect of Shape of Particle on Fall Velocity.—Sediment grains are never truly spherical and their shape varies over a wide range from rodlike to spherelike to disklike. Therefore, in connection with determining fall velocity of sediments, it is of interest to study the effect of shape. McNown, et al. (1951) studied this problem using machined geometric shapes instead of natural particles. The results of these studies for Reynolds numbers less than 0.1 are shown in Fig. 2.3 in terms of a resistance factor, K; the shape factor, SF = $a/\sqrt{bc}$; and a length ratio b/c in which a, b, and c are the lengths of the mutually perpendicular axes of the particle. The motion is parallel to the axis, a, and K is defined by

$$F = K\,(3\,\pi\,\mu\,w\,d_n) \qquad (2.5)$$

in which F = the submerged weight of the particle; w = its terminal fall velocity; and d_n = its nominal diameter. McNown defines the Reynolds number of the particle as $\mathsf{R} = wd_n/\nu$. The curves in Fig. 2.3 represent theoretical results (McNown and Malaika, 1950) for ellipsoids. The numbers beside the data points in the graph give the values of b/c and the shapes used are indicated in the figure. It is seen that, despite the differences in shape of particles, there is little difference between the theoretical values for ellipsoids and the observed values for the various other shapes. The values of K based on the two ratios, $a/\sqrt{bc}$ and b/c are all within 10% of the theoretical values for ellipsoids, thus indicating that the ratios represent the principal hydrodynamic features of the particle shapes. It is pertinent to observe that for low Reynolds numbers for which Eq. 2.2 is valid, the coefficient, K, is equal to the ratio of the fall velocity of a sphere with the same volume and weight as the particle to the fall velocity of the particle.

Fig. 2.4 shows (McNown, et al., 1951) graphs of K against R for large values of R for various shapes. The shapes, as well as the axis ratios a to b to c, are given for each curve. In these experiments, the particles were oriented so that they fell in a direction parallel to their shortest axis. This was the only orientation that gave stable motion; however, the motion became unstable even for this orientation at values of R between 100 and 1,000, the value at which instability developed being different for each shape. The particles settling at low Reynolds numbers (Fig. 2.3) were quite different in this respect because they fell with whatever orientation they had when released. The values of K for spheres for high Reynolds numbers are given by the relation, $K = C_D\,\mathsf{R}/24$, in which C_D is taken from the experimental C_D-R curve for spheres (Fig. 2.1). The value of K is seen to increase as R increases. The data for particles with the same axis ratios are seen to follow close

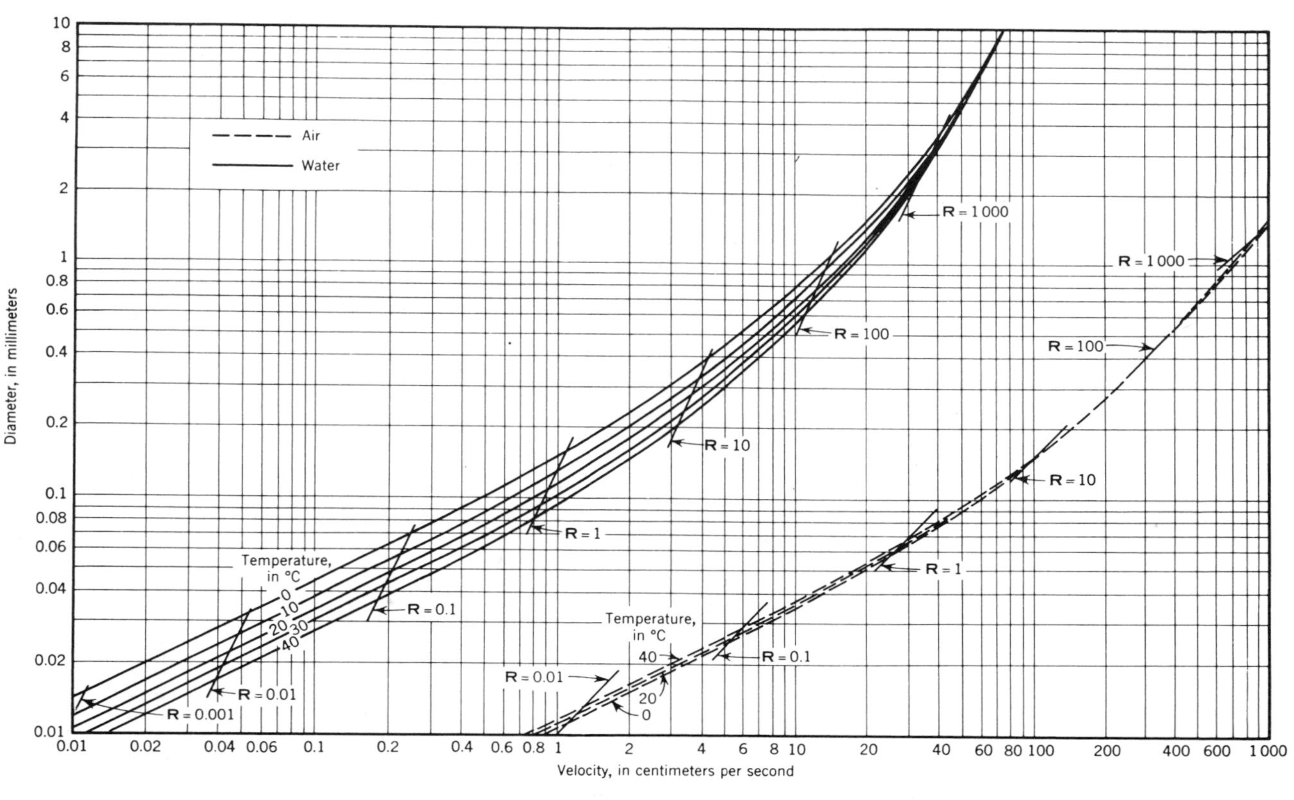

FIG. 2.2.—Fall Velocity of Quartz Spheres in Air and Water (Rouse, 1937b)

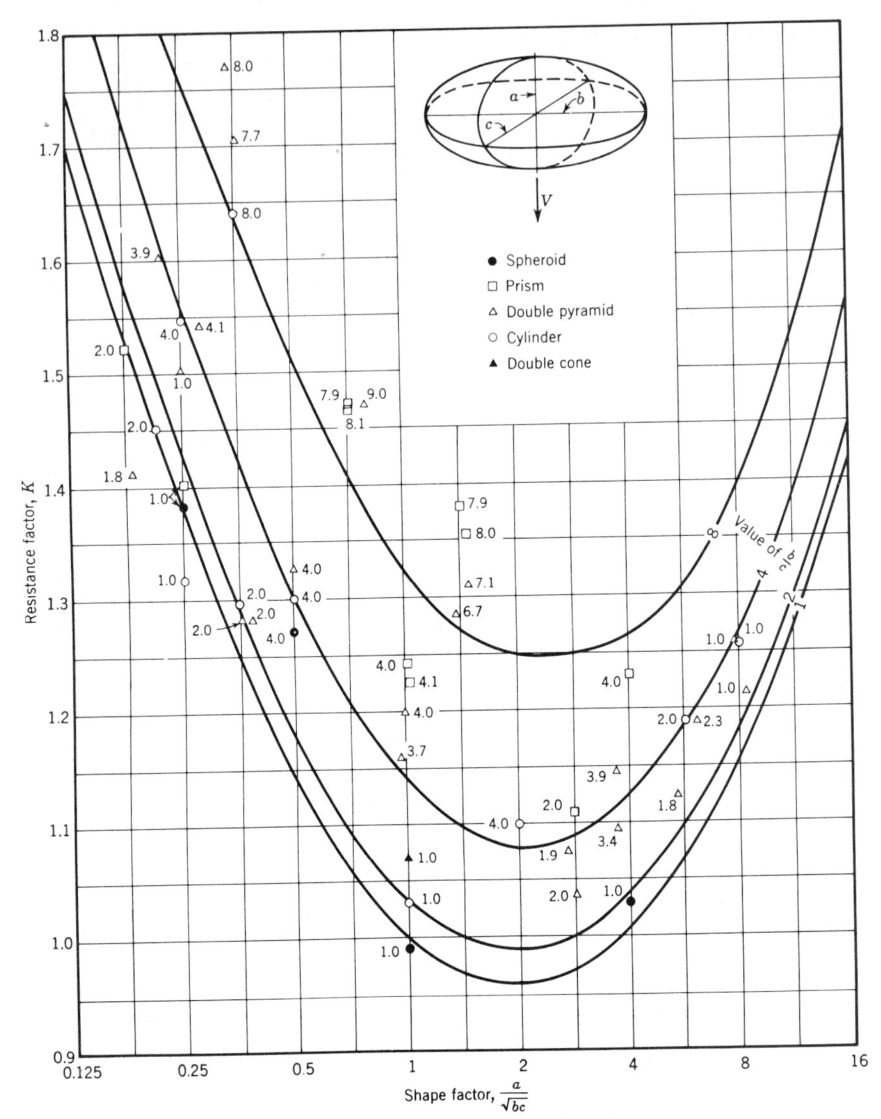

FIG. 2.3.—Comparison of Theoretical Values of K for Ellipsoids and Observed Values for Ellipsoids and Several Other Shapes for Reynolds Numbers Less than 0.1 (McNown, et al., 1951)

to one line despite differences in shape (e.g., see points for prisms and double pyramids with axis ratios 8 to 2 to 1). This indicates that, as for low Reynolds numbers, the important hydrodynamic feature of the particle is expressed by its axis ratio. In an attempt to simplify the use of the data (McNown, et al., 1951), K was plotted against R with SF $= a/\sqrt{bc}$ as a parameter. This method introduced errors in K as large as 40% that result from neglecting the effect of the ratio, b/c.

11. Effect of Sediment Concentration on Fall Velocity.—The fall velocities

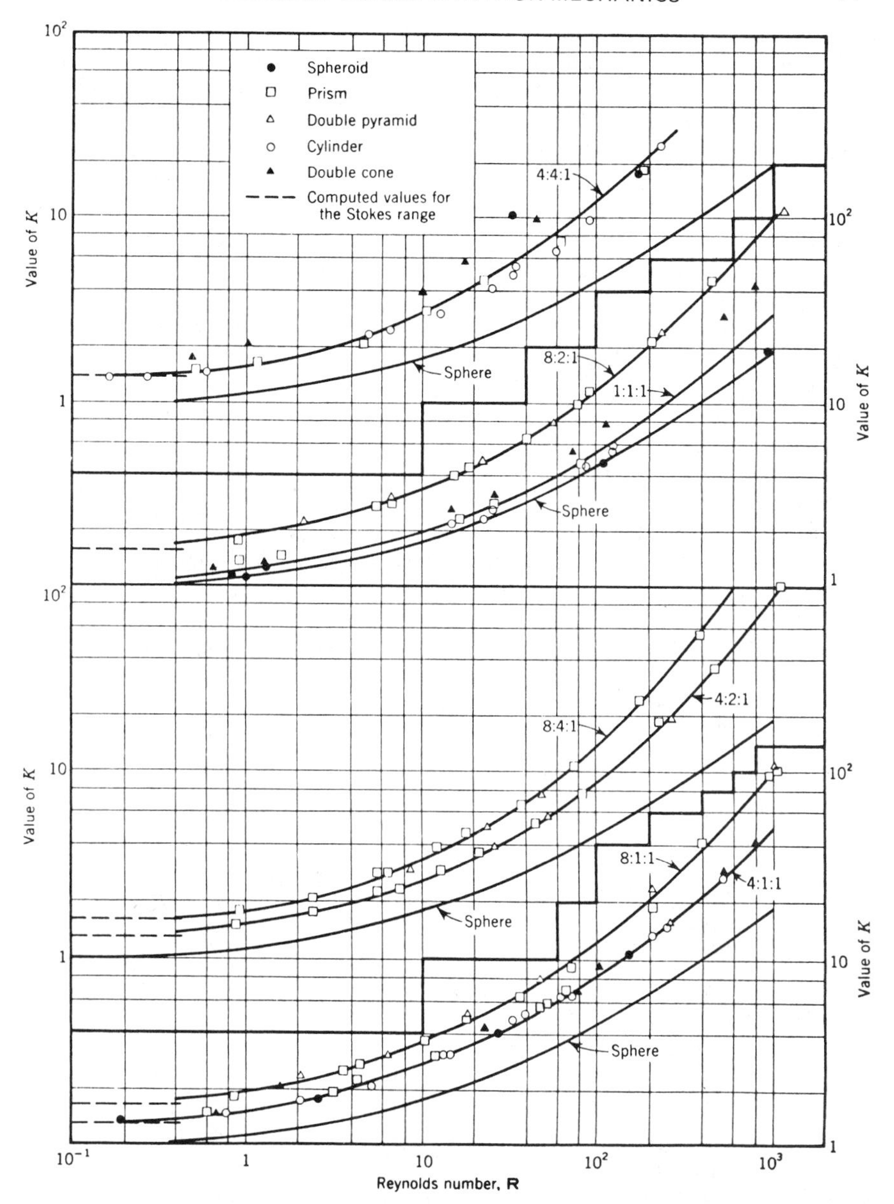

FIG. 2.4.—Variation of K with Reynolds Number for Several Particle Shapes Falling with Their Short Axes Vertical (McNown, et al., 1951)

determined from Figs. 2.1 and 2.2 are for a single spherical particle in an infinite fluid. When there are a number of particles dispersed in a fluid, the fall velocity will differ from that of a single particle, due to the mutual interference of the particles. If only a few closely spaced particles are in a fluid, they will fall in a group with a velocity that is higher than that of a particle falling alone. On the other hand, if particles are dispersed throughout the fluid, the interference between neighboring particles will tend to reduce their fall velocity. This interference is said to hinder the settling, and the process is often referred to as hindered settling. This problem has been studied theoretically and experimentally by McNown and Lin (1952), whose theoretical results are shown in Fig. 2.5, in which w_o is the settling velocity in clear fluid and w_c is the velocity in a suspension of concentration c. The curves shown are for an approximate theory based on the Oseen modification of Stokes' theory for the motion of a sphere in a viscous liquid at low velocity. The curves are not expected to apply for Reynolds numbers in excess of two. The theory agreed well with experiments with uniform quartz spheres and sand settling in water with a range of concentration for 0.1%–6% by weight. Fig. 2.5 shows that even for moderate concentrations, the correction in the fall velocity becomes significant.

The particles used by McNown and Lin (1952) were coarse enough so that no flocculation occurred. In some suspensions of clay and silt, electrochemical forces tend to hold particles together once they come into contact. If turbulence is present, the particles can be brought into contact by the mixing that always occurs in such an environment. However, the agitation due to turbulence may also tear apart agglomerations of particles that were brought together initially by turbulence mixing. In this manner, the size of agglomerations, or flocs, as they are often called, tends to reach a limit. If a suspension of fine material is settling in a quiescent fluid, the flocculation occurs by particles with higher fall velocities

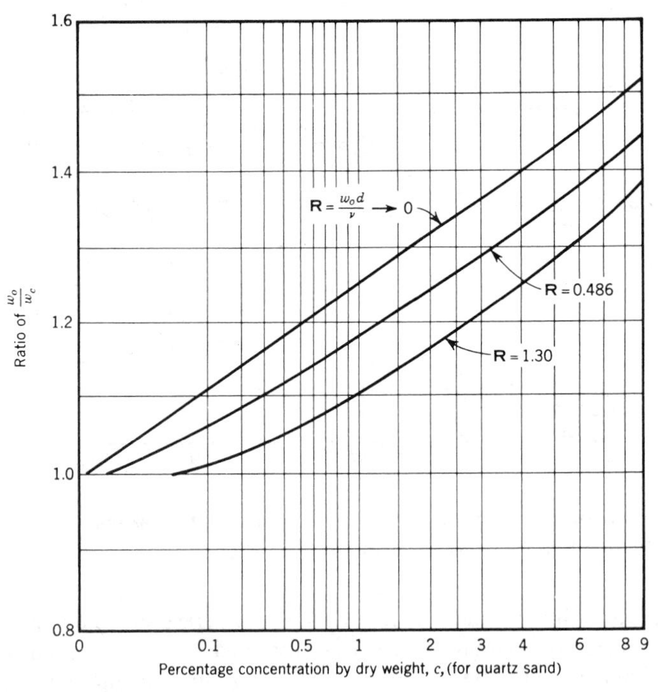

FIG. 2.5.—Effect of Concentration on Fall Velocity of Uniform Quartz Spheres (McNown and Lin, 1952)

overtaking and capturing slower ones. Once two or more particles combine, they will settle as a group with higher velocity than any of the individual particles of the group falling alone. Therefore, in a flocculent suspension, the mean fall velocity of the material will increase with time, or in other words, the fall velocity is a function of time.

McLaughlin (1959) devised a method for measuring the mean fall velocity, w, of a suspension of fine particles that he then used to measure the properties of such suspensions. This method involves taking sets of simultaneous samples at several levels of a suspension settling quiescently in a column. The procedure in sampling is the same as in a standard pipette analysis for determining size distribution of fine sediments (Chapter III, Section F), except that the samples are taken simultaneously at several levels instead of at only one level. To illustrate the procedure, consider the hypothetical settling tube in Fig. 2.6 in which three sampling levels are indicated, located at distances $z = D_1$, D_2, and D_3, respectively, below the liquid surface. At the beginning of the test ($t = 0$) the concentration is uniform and the concentration profile is represented on the diagram by the $t = 0$ line. At time $t = T_1$, a set of samples is withdrawn, giving the concentration profile marked $t = T_1$. The profiles marked $t = T_2$ and $t = T_3$ represent similar measurements at time T_2 and T_3, respectively. The concentration is seen to vary with position and time, and is denoted mathematically as $C\ (z, t)$. The products, $\bar{w}C$, shown in the diagram of the settling tube represent the rate at which sediment settles through unit horizontal areas at various levels and actually defines w.

To proceed with the calculation of $\bar{w}$, use is made of the equation

$$\frac{\partial C}{\partial t} + \frac{\partial(\bar{w}C)}{\partial z} = 0 \qquad (2.6)$$

which derives from the condition that, in a short time interval, the quantity of sediment settling into an elementary prism, less that settling out of it, is equal to the increase in concentration in the prism in the same short time interval. Integration of Eq. 2.6 with respect to z gives

$$(\bar{w}C)_{z=D} = -\frac{\partial}{\partial t}\int_0^D C dz \qquad (2.7)$$

which is used to calculate $\bar{w}$. To calculate $\bar{w}$ at $z = D_1$, e.g., one measures the areas 025, 024, and 023 on the concentration profiles of Fig. 2.6 and the values are plotted against the corresponding times T_1, T_2, and T_3. The slope of this curve is the value of the right side of Eq. 2.7 and can be used to calculate $\bar{w}$ at various times because C is known. The procedure can be repeated for any depth at which samples are taken. Eq. 2.7 can be interpreted in simple physical terms by noting that the integral gives the amount of sediment above the level, D, in a column of unit cross section. The rate of decrease of this quantity must be equal to the rate of the settling of material as expressed by Eq. 2.7.

A concentration profile diagram obtained by McLaughlin (1959) for Bentonite clay and alum in water, for which the initial concentration was 655 mg/l, is shown in Fig. 2.8, and the mean fall velocity $\bar{w}(z, t)$ calculated as previously described for three values of z as functions of time are shown in Fig. 2.8. The fall velocity at each of the three levels is seen to increase with time to a maximum value and then to decrease. For the smaller times during which the fall velocity was increasing,

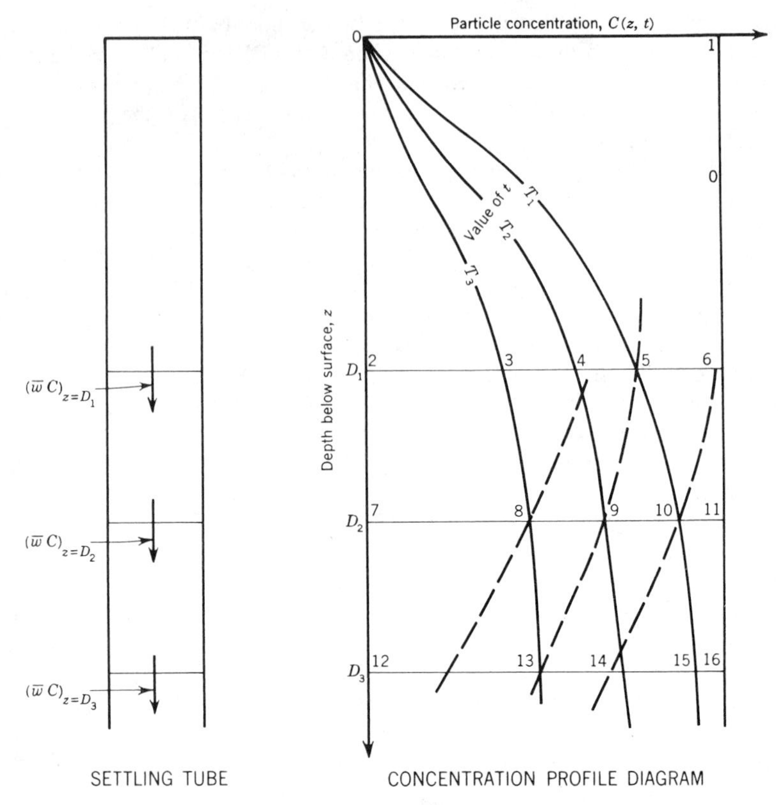

FIG. 2.6.—Diagram of Settling Tube and Concentration Profile for Measuring Settling Properties of Suspensions (McLaughlin, 1959)

the concentration changed only a few percent so that the effect of hindered settling did not change appreciably. It may be concluded that most of the increase in $\overline{w}$ was due to flocculation.

The decrease in $\overline{w}$ after it has reached a maximum occurs because the faster particles are continually settling out and their loss offsets the effect of flocculation. It is also possible to determine that flocculation is increasing w by looking at the dashed lines, z/t = constant, in Fig. 2.7. These lines give the value of the local concentration encountered by an observer beginning from the water surface at time $t = 0$ and descending with a uniform velocity, z/t.

If there are no effects of hindered settling and no flocculation occurs, the concentration at the level of the descending observer will be that of the particles with fall velocity equal to or less than that of the observer. However, this concentration is constant so the lines z/t = constant are parallel to the z-axis. If hindered settling slows down the particles more than flocculation speeds them up, the concentration seen by the observer will increase with z so that the lines will slope away from the z-axis. Finally, if the flocculation effect is larger than that of hindered settling, the concentration will decrease with depth and the lines will slope towards the z-axis. Therefore, it is seen that, for the case represented in Fig. 2.7, flocculation was the predominant effect.

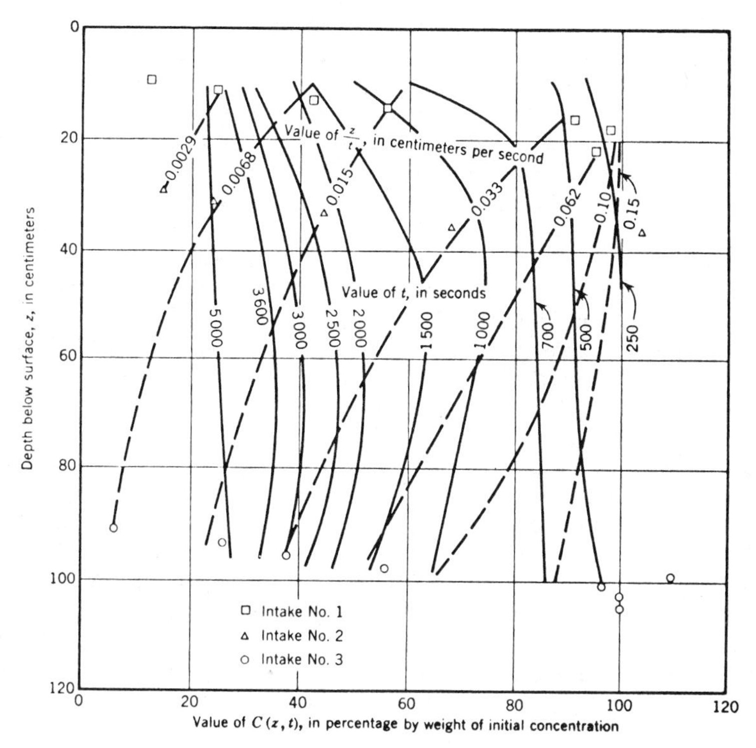

FIG. 2.7.—Observed Concentration Profile for Bentonite Clay and Alum in Water (McLaughlin, 1959)

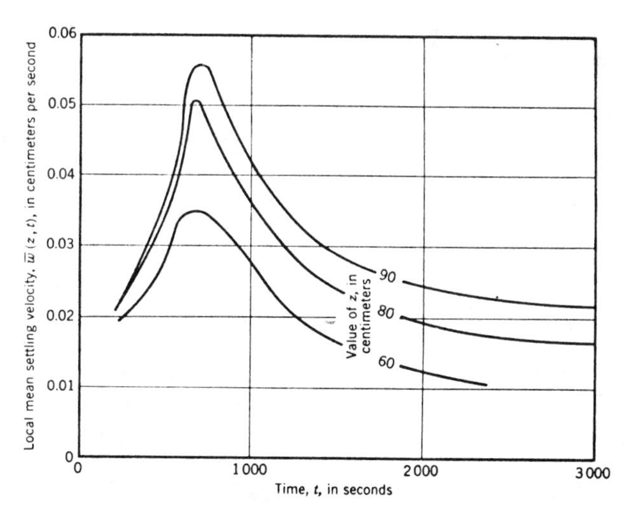

FIG. 2.8.—Local Mean Fall Velocity as Function of Time at Three Depths for Bentonite Clay and Alum in Water (McLaughlin, 1959)

The shear stress in fluids, e.g., water under conditions of laminar flow is proportional to the velocity gradient; the constant of proportionality being the dynamic viscosity, μ. Fluids that behave in this manner are known as Newtonian fluids (van Olphen, 1963). The ratio of shear stress to velocity gradient (apparent dynamic viscosity) on highly concentrated suspensions of clay in water is not constant but varies with the velocity gradient. Fluids or suspensions, or both, displaying this kind of behavior are said to be non-Newtonian.

As pointed out by Howard (1962a) the prediction of the fall velocity of a particle in a non-Newtonian clay suspension cannot be based on the apparent viscosity. The viscosity and shear stress in the vicinity of the particle would vary with velocity gradient and thus also with the size and density of the particle.

Because the concentrations of clay in McLaughlin's experiments (Figs. 2.7 and 2.8) were so low the suspension would behave as a Newtonian fluid.

12. Fall Velocity of Sand Grains.—The sedimentation diameter of a particle is determined from a measured fall velocity with the aid of Fig. 2.1 and Eq. 2.3, or if the particle is quartz and it is falling in water or air, Fig. 2.2 may be used. Once the sedimentation diameter is known, one can then use Fig. 2.1 and Eq. 2.3 or Fig. 2.2 to calculate the fall velocity. However, it has been found that the sedimentation diameter of any particle that is not spherical is not constant, but varied with the Reynolds number it attains while falling. This means that the sedimentation diameter of a particle will vary with the specific weight, density, and viscosity of the fluid.

For instance, it means that in water the sedimentation diameter will change with temperature, and that the fall velocity calculated from a particular value of sedimentation diameter is valid only for the temperature at which the original determination was made. The relation between sedimentation diameter and water temperature is illustrated by the data of Table 2.3 taken from Report No. 12 (Interagency Committee, 1957) for the case of a natural quartz grain with a nominal diameter of 1 mm. It is seen that the sedimentation diameter decreases as

TABLE 2.3.—Variation of Fall Velocity and Sedimentation Diameter of Quartz Particle[a] with Temperature of Distilled Water in Which it Is Falling

Water temperature, in degrees Celsius (1)	Fall velocity, in centimeters per second (2)	Sedimentation diameter, in millimeters (3)	Reynolds number, R (4)
0	10.4	0.820	58
24	12.3	0.761	135
40	13.0	0.725	200

[a]Having nominal diameter of 1 mm and shape factor (SF) of 0.7.

the temperature and Reynolds number increase. The sedimentation diameter will also vary from fluid to fluid, and is not a property of the grain in the same sense that the nominal diameter and shape factor are particle properties. This has led to the introduction of the standard fall velocity and the standard fall diameter (Interagency Committee, 1957) to characterize the fall velocity of a particle. These two quantities are defined as follows:

1. The *standard fall velocity* of a particle is the average rate of fall that it would attain if falling alone in quiescent distilled water of infinite extent at a temperature of 24°C.
2. The *standard fall diameter* of a particle is the diameter of a sphere that has the same specific weight and has the same standard fall velocity as the given particle.

Although sedimentation diameter varies with water temperature, its rate of variation is not large and, for purposes of calculating fall velocity, it can be considered constant over small ranges of temperature without introducing excessive error. For instance, it has been found (Interagency Committee, 1957) that, if one knows the fall velocity of a natural sand in water for a particular temperature in the range 20°C–30°C, one can then determine the fall velocity for another temperature as much as 5° different from the first, assuming constant sedimentation diameter, without exceeding errors of 2%.

Such a calculation can be made by using Fig. 2.2, if the sediment has the same specific gravity as quartz (i.e., 2.65), or by Fig. 2.1 if the specific gravity is different from 2.65. The problem is to determine sedimentation diameter when w, γ, γ_s, and ν are given. If Fig. 2.1 is used, a trial and error solution of Eq. 2.3 is required. The procedure is to assume a value of sedimentation diameter, calculate R, read C_D from Fig. 2.1, calculate w from Eq. 2.2, and repeat until the given and calculated w-values are the same. If Fig. 2.2 is used, sedimentation diameter is read directly from the graph for the given w and temperature, and the value of w for the new temperature is read from the figure for the same value of sedimentation diameter.

Sizes of sand are usually determined by sieving and are given in terms of sieve diameter because of the convenience of this method. Therefore, it is also convenient to have a relation between sieve diameter and fall velocity. Such a relation is given in Fig. 2.9 for naturally worn quartz particles falling in distilled water, and for a range of water temperature and particle shape factor, SF, as given by Eq. 2.1. These curves give average values that should be considered as estimates. When fall velocity is of major importance it should be measured for the sediment of the stream under study. Fig. 2.10 gives the relation between standard fall diameter and sieve diameter (Interagency Committee, 1957) for a range of grain shape factors for naturally worn quartz particles. These figures are useful in obtaining settling velocity of quartz particles once one has the sieve diameter and shape factor. The sieve diameter as defined herein is difficult to obtain. For practical purposes the sieve diameter of the sediment retained in one of a nest of sieves is taken as the arithmetic or geometric mean of the openings of the sieves between which it is retained. The determination of shape factor can be made with a microscope. The shape factor of 0.7 has been found to be about average for natural sands (Interagency Committee, 1957).

13. Effect of Size of Settling Column on Fall Velocity.—Fall velocities are usually measured by observing individual grains settling in a column of still fluid. In making such measurements, care must be taken to use containers that are large enough so that the confining effect of the walls on the flow around the grain is negligible. The size of cylinder to use to accomplish this can be judged from Fig. 2.11, which shows the effect of d/D and the Reynolds number of the sphere on the drag coefficient, C_D, in which D = the diameter of the cylindrical settling column; and d = the diameter of the sphere (McNown and Newlin, 1951).

For values of d/D less than 0.3 and for Reynolds numbers less than 0.1,

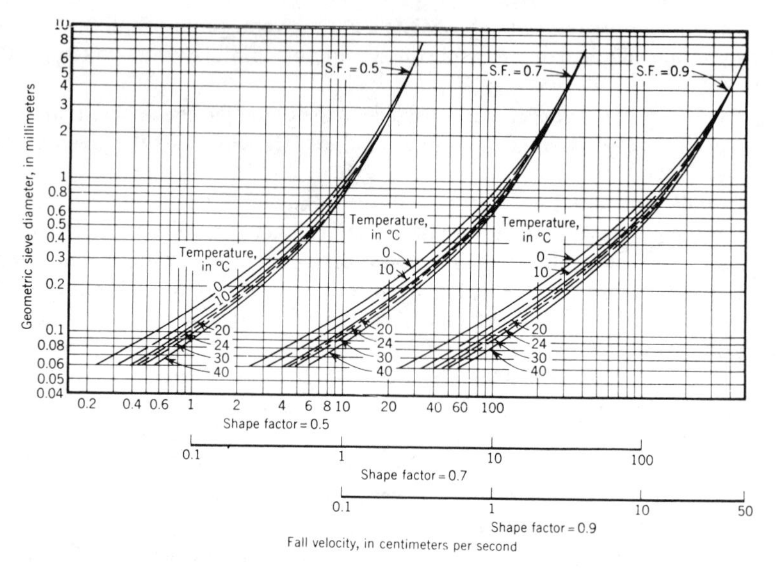

FIG. 2.9.—Relation between Sieve Diameter, Fall Velocity, and Shape Factor for Naturally Worn Quartz Sand Particles Falling in Distilled Water (Interagency Committee, 1957)

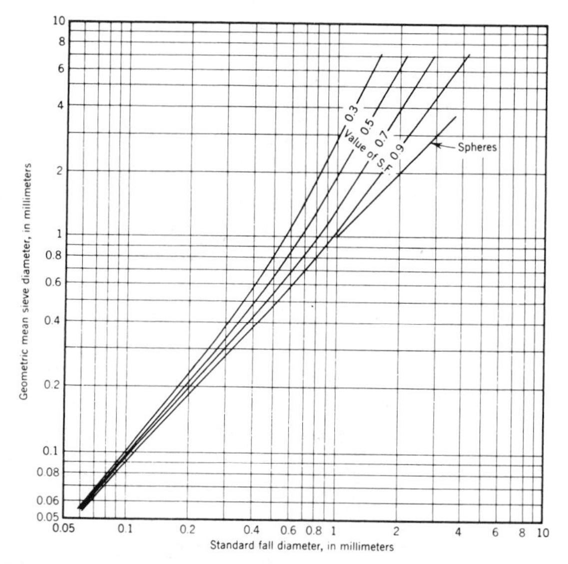

FIG. 2.10.—Relation between Sieve Diameter, Standard Fall Diameter, and Shape Factor for Naturally Worn Sand Particles (Interagency Committee, 1957)

McNown, et al. (1948) gave the following relation between d/D and w/w_D, the ratio of fall velocity of a sphere in a fluid of infinite extent and in one confined in a cylinder of diameter D:

$$\frac{w}{w_D} = 1 + \frac{9}{4}\frac{d}{D} + \left(\frac{9}{4}\frac{d}{D}\right)^2 \qquad (2.8)$$

From Fig. 2.11 it can be seen that, for the lower values of d/D, the retarding effect of the cylinder walls diminishes as the Reynolds number increases. This relation is illustrated by the following data which give values for $d/D = 0.2$ calculated from Fig. 2.11

R	0.1	1	10	100	1,000
$\frac{w_D}{w}$	0.77	0.79	0.90	0.94	0.95

McNown, et al. (1948) also found that the values of C_D for spheres usually found in textbooks and elsewhere tended to be high because boundary effects in the apparatus used to determine the values were not always negligible. In the range between R = 10 to 100, C_D was found to be 5% high. The values of C_D in Fig. 2.1 that were used to calculate Fig. 2.2, are the accepted values usually found in textbooks and are higher than those given (McNown and Newlin, 1951) in Fig. 2.11 for $d/D = 0$.

14. Effect of Turbulence on Fall Velocity.—In attempts to represent particles falling in a turbulent fluid, workers have made theoretical and experimental studies of spheres in oscillating fluids. Field (1968) and Houghton (1968) showed theoretically that spherical particles would settle more slowly in a fluid oscillating in the vertical direction than in one at rest. Field (1968) also confirmed his result by experiments. Both authors indicated that the reduction in fall velocity resulted

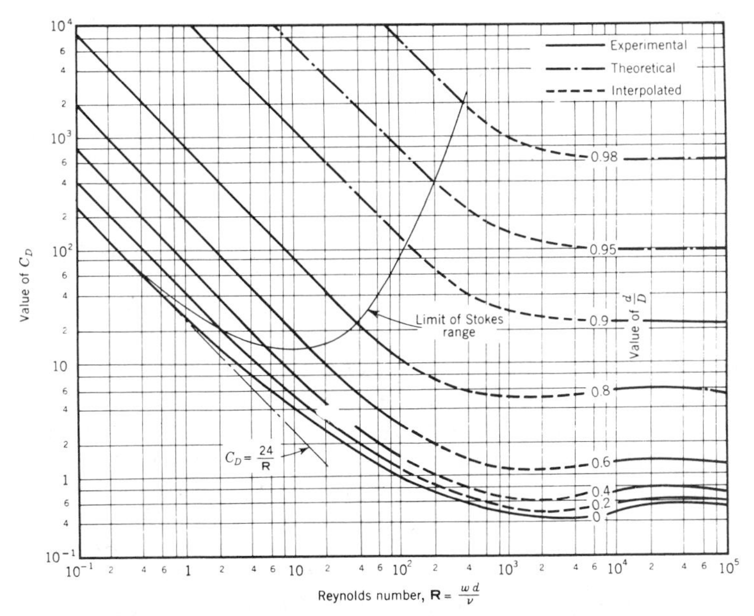

FIG. 2.11.—Variation of Drag Coefficient of Spheres with Reynolds Number and d/D (McNown, et al., 1948)

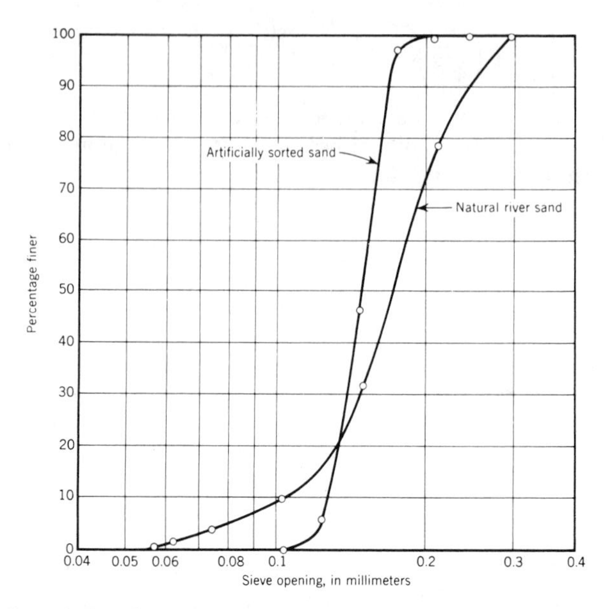

FIG. 2.12.—Cumulative Semilogarithmic Size-Frequency Graphs for Two Sands

from the nonlinear relation between drag on the particles and their velocity relative to the fluid. The oscillating fluid is not a realistic model of a turbulent flow. However, it seems reasonable to expect that because the relative velocity between the fluid and a particle in a turbulent flow has an unsteady component the tendency is for the fall velocity of the particle to be less than in a quiescent fluid.

15. Size-Frequency Distribution.—Because natural sediments are made up of grains with wide ranges of size, shape, and other characteristics, it is natural to resort to statistical methods to describe these characteristics (Otto, 1939). The discussion that follows deals only with size distribution, although the techniques apply equally well to fall velocity and other properties.

The process of obtaining the size distribution, which is essentially the separation of a sample into a number of size classes, is known as the mechanical analysis. The results of such analyses of sediment are usually presented as cumulative size-frequency curves, where the fraction or percentage by weight of a sediment that is smaller or larger than a given size is plotted against the size.

Figs. 2.12 and 2.13 show such cumulative size-frequency curves for sieve diameter of a river sand and an artificially sorted sand. In Fig. 2.12 the data have been plotted on a semilogarithmic graph and in Fig. 2.13 a logarithmic probability graph has been used. The median size, d_{50}, i.e., the size for which 50% of the material is finer or coarser, can be read from each of the curves in the two figures, and it is seen that the values for each sand are the same on Figs. 2.12 and 2.13. From Fig. 2.13 one can also read the geometric mean size. As shown by Otto (1939), this size is read at the intersection of the 50% line and a line passing through the cumulative curve at 15.9% and 84.1% or $d_g = \sqrt{d_{84.1} d_{15.9}}$, in which $d_{84.1}$ and $d_{15.9}$ are the grain sizes for which 84.1% and 15.9% by weight, respectively, of the sediment is finer.

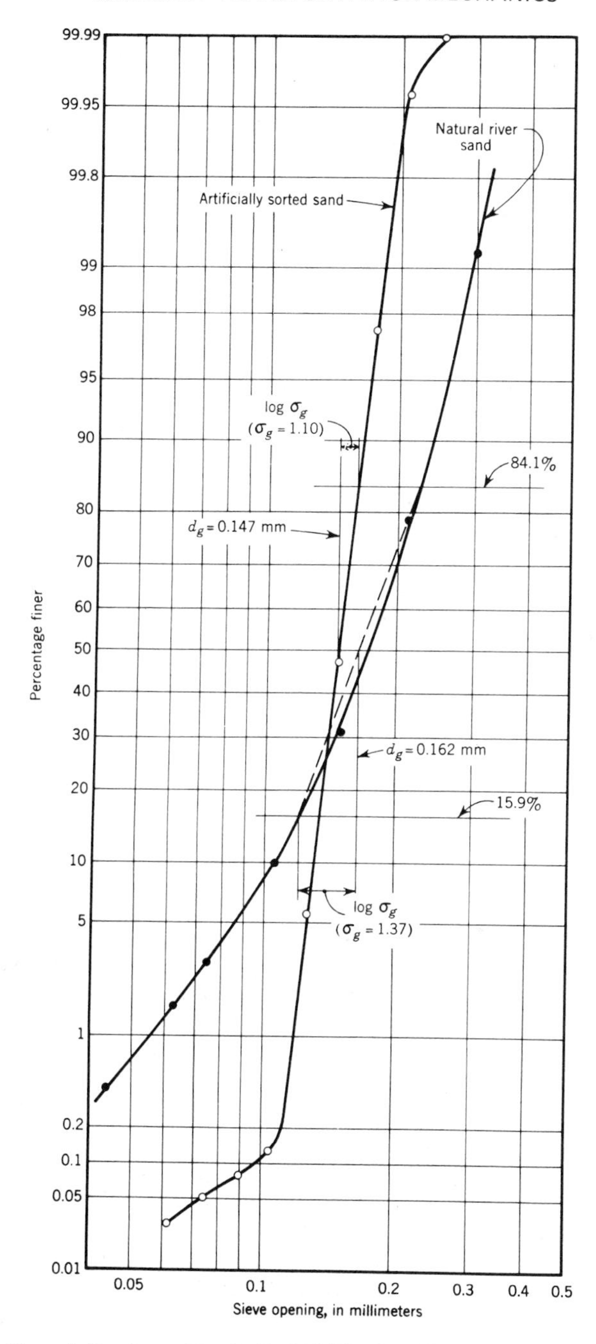

FIG. 2.13.—Cumulative Logarithmic Probability Size-Frequency Graphs for Sands of Fig. 2.12

In the case of the artifically sorted sand, the cumulative curve is a straight line over 99.75% of the material so that the logarithms of the sizes are normally distributed or follow the normal error law. In this case, the distribution is symmetrical, and the geometric mean and median sizes are identical. For the river sand, the median size is larger than the geometric mean, and the size distribution is said to be skewed to the left. Since, in Fig. 2.13, the sizes are plotted on a logarithmic scale, the mean size is the one corresponding to the mean logarithm, which is the geometric mean size and will be denoted by d_g.

The geometric standard deviation, σ_g, of the sizes also can be read from Fig. 2.13 as the ratio of the size $d_{84.1}$ to d_g or the ratio of d_g to $d_{15.9}$ (Otto, 1939) or $\sigma_g = \sqrt{d_{84.1}/d_{15.9}}$.

The geometric mean size and geometric standard deviation obtained in the manner described previously herein are correct only if the logarithms of the grain sizes are normally distributed, i.e., if the distribution is log-normal. However, the practice is to determine d_g and σ_g as described previously herein even when the size distribution is not log-normal. The practice is also to take σ_g as $\sqrt{d_{84}/d_{16}}$ instead of $\sqrt{d_{84.1}/d_{15.9}}$ and d_g as $\sqrt{d_{84}d_{16}}$, in which d_{84} and d_{16} are the grain sizes for which 84% and 16% by weight, respectively, of the sediment is finer. Inman (1952) also proposed the following quantities in addition to d_g and σ_g to further describe the size distribution of a sediment, i.e.

$$\text{Skewness} = \frac{\log\left(\dfrac{d_g}{d_{50}}\right)}{\log \sigma_g}$$

$$\text{2nd skewness} = \frac{\log\left(\dfrac{\sqrt{d_{95}d_5}}{d_{50}}\right)}{\log \sigma_g} \qquad \text{Kurtosis} = \frac{\log\sqrt{\dfrac{d_{16}}{d_5}\dfrac{d_{95}}{d_{84}}}}{\log \sigma_g}$$

In these expressions d_5 and d_{95} are the grain sizes for which 5% and 95% by weight, respectively, of the sediment is finer.

Kennedy and Koh (1961, 1963) have presented a method of determining the distribution of fall velocities of a sediment from the distribution of sieve diameters for the case of log-normal distribution of sieve diameter.

16. Specific Weight of Sediment Deposits.—The term specific weight of a sediment deposit, as used herein, is defined as the dry weight of the sedimentary material within a unit volume, which in the foot-pound system of units is expressed in pounds per cubic feet. Because there are voids between the grains of sediment, the specific weight of the deposit is always less than that of the grains themselves. One of the main interests of those working with the specific weight of sediments arises in connection with predicting the depletion of reservoir storage by deposits of river sediments. Measurements are usually made of the sediment transported by the river prior to the construction of a dam, so that a reliable estimate can be made of the weight of sediment that will be brought into the proposed reservoir in a given period of time. To obtain the volume occupied by the sediment, it is necessary to have its specific weight. Experience has shown that the specific weight of deposits varies over a range of severalfold and that, therefore, it has an important bearing on estimates of rates of depletion of reservoir capacity. Such variation in specific weight is shown by Table 2.4 presented by Koelzer and Lara

TABLE 2.4.–Specific Weights of Sediments Showing Extreme Variation

Location (1)	Predominant class of sediment (2)	Specific weight, in pounds per cubic foot (3)
Lake Niedersonthofen, Bavaria, upper layer	marl[a]	21.6
Lake Niedersonthofen, 20 m depth	marl[a]	89.6
Lake Arthur, South Africa	clay	38
Iowa River at Iowa City, Iowa	silt	52
Missouri River near Kansas City, Mo.	silt	74
Lake Claremore, Oklahoma	silt	54
Lake McBride, Iowa	silt	60
Powder River, Wyoming	silt	81
Castlewood Reservoir, Colorado	sand	92
Cedar River near Cedar Valley, Iowa	sand	109
Lake Arthur, South Africa	sand	100

[a]As used herein, marl is a mixture of calcium carbonate or dolomite and clay.

(1958) from data summarized by Hembree, et al. (1952) and Lane and Koelzer (1953).

Factors influencing the specific weight of a deposit are: (1) Its mechanical composition; (2) the environment in which deposits are formed; and (3) time. Coarse materials, e.g., gravel and coarse sand, are laid down with specific weights very nearly equal to their ultimate value and change little with time. However, fine materials, e.g., clay and silt, may settle in thin "soupy" masses that have an initial specific weight of only a fraction of their ultimate value and will reach the ultimate weight only after many years or decades. This is illustrated by Table 2.5, which shows the specific weights of sediments of various sizes that have been in deposits for 1 yr or less (Koelzer and Lara, 1958; Hembree, et. al., 1952). Fig. 2.14, taken from the work of Lane and Koelzer (1953), shows the specific weight of a large number of samples plotted against the percentage by weight of material in the deposit that is coarser than 0.05 mm. All of the deposits sampled were laid down recently. Some were in reservoirs and others were in streambanks and beds. The average curve follows the formula

$$W = 51\,(P + 2)^{0.13} \qquad (2.9)$$

in which W = the specific weight of the deposit; and P = the weight percentage of the material coarser than 0.05 mm. (The numbers next to the points refer to the percentage of moisture in the sample as taken.) It will be seen that most of the points group around the average line, although a number of points definitely do not. These are believed (Hembree, et al., 1952) to have come from types of deposits that are unimportant in reservoirs and, therefore, of little interest in this analysis. Lines are also drawn on Fig. 2.14 showing 5% and 10% deviation from Eq. 2.9. It will be seen that 50% of the data fall within 5% of the mean line and that 75% fall within 10%. Fig. 2.14 shows that, despite the wide variation in conditions under which the deposits occurred, the initial specific weight can still be expressed fairly well as a function of the sand content alone, and indicates the importance of this factor.

TABLE 2.5.—Initial Specific Weights

California Division of Water Resources		Trask (Laboratory)	
d_{90}, in millimeters (1)	Specific weight, in pounds per cubic foot (2)	Size range, in millimeters (3)	Specific weight, in pounds per cubic foot (4)
256	140		
128	138	0.5-0.25	89
64	132	0.25-0.125	89
32	124	0.125-0.064	86
16	116	0.064-0.016	79
8	109	0.016-0.004	55
4	103	0.004-0.001	23
2	98	0.001-0	3
1	95		
0.5	93		
0.25	92		
0.125	92		

Komura (1963a) proposed the following formula for the initial specific weight of sediment deposits based on the analysis of extensive data from a wide range of sources:

$$W = 125 - 7d_{50}^{0.21} \qquad (2.10)$$

in which d_{50} = the median sediment size in feet; and W = the specific weight in pounds per cubic feet. Data on specific weight have been compared with Eq. 2.10 by Komura (1963).

Another important factor in determining the specific weight of deposits is the environment or, more specifically, whether or not the material becomes exposed to the air and is allowed to dry. Drying accelerates the compaction process, and material thus compacted remains so even after reinundation. Because the amount of deposits in a given reservoir that become exposed to the air depends on the variation of water level, it is seen that the method of operating a reservoir will influence the specific weight of the deposits. Obviously, the effect of drying is greatest on fine sediments (i.e., clays and silts) and for reasons already outlined has almost no effect on coarse sands.

When the consolidation of a deposit of sediment occurs, the grains get closer together and are compressed by the pressures applied. At the same time, the water is squeezed out of the pores between grains. The analytic treatment of the problem considers the relation between the loads on a deposit due to its weight, the pressure of the water in the pores between the grains, the resistance of the water to being squeezed from the pores, and the stress in the grains themselves.

This problem has been treated by Terzaghi (1943) for the case of a load applied to a deposit that was already well compacted. The resulting reduction in volume is small compared to the original volume. This theory is not applicable to the case in which the compaction is large, e.g., for clays in reservoirs, which may have an

of Sediments of Various Grain Size

Hembree, Colby, Swenson, and Davis (Countrywide)		Happ (Middle Rio Grande)	
Median size, in millimeters (5)	Specific weight, in pounds per cubic foot (6)	Median size, in millimeters (7)	Specific weight, in pounds per cubic foot (8)
1.0	120	–	–
0.5	104	–	–
0.25	89	–	–
0.1	77	0.1	88
0.05	70	0.05	78
0.01	57	0.01	73
0.005	52	0.005	68
0.001	42	0.0012	48

initial specific weight of only a fraction of its ultimate value. The problem of large compaction in sediments has been studied theoretically and experimentally by Long (1961). He derived a nonlinear partial differential equation describing the void ratio of the deposit in depth and time in terms of the relations between void ratio and permeability and void ratio and intergranular stress.

The results of one set of Long's experiments on the compaction of clay slurries are reproduced in Fig. 2.15. The clay slurry that consisted of 10.4 g of Montmorillonite clay and 1.79 g of common salt per liter of water was introduced into the top of a column 4¼ in. in diameter at a uniform rate of 1.04 g of clay per minute. At intervals, a small quantity of red clay was introduced to form the marker lines noted in Fig. 2.15. The common salt was added to produce flocculation and thus approximate natural conditions. The observations of the position of the surface of the slurry deposit and the red markers in time are shown by the data points in Fig. 2.15 and the curves shown were fitted to the points by eye. Long solved the differential equation for large compaction for the special case of uniform rise of the surface of the deposit corresponding to his experiment. This solution agreed well with the data of Fig. 2.15.

Although the theoretical work previously outlined is useful, it has not progressed to the point at which it can be used to predict specific weights of reservoir deposits. Therefore, the engineer is forced to rely on empirical relations for this purpose. Lane and Koelzer (1953) have presented such a relation for estimating the specific weight of deposits in reservoirs, taking into account the grain size of the sediment, the method of operating the reservoir, and time:

$$W = W_1 + B \log T \qquad (2.11)$$

in which W = the specific weight in pounds per cubic foot of a deposit with an age of T years; W_1 = its initial weight, usually taken to be the value after 1 yr of

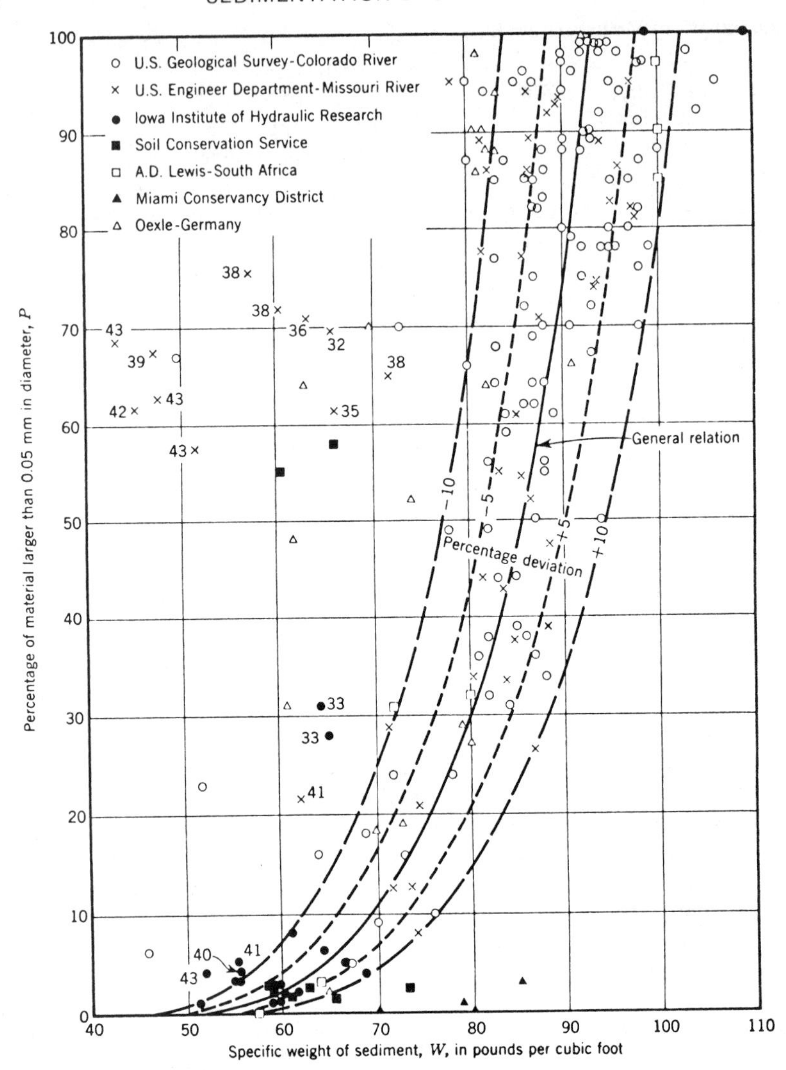

FIG. 2.14.—Relation of Specific Weight of Sediment Deposited to Percentage of Sand (Lane and Koelzer, 1953)

consolidation; and B = a constant with dimensions of pounds per cubic foot. Both W_1 and B are functions of sediment size and method of operating the reservoir and are given in Table 2.6. Eq. 2.11 and the quantities in Table 2.6 are based on measurements of the weights of reservoir sediments and, therefore, will give average results that will not apply in specific cases where conditions are abnormal.

Where the sediment contains material in more than one size class, the weight for each class given by Eq. 2.11 should be combined in proportion to their relative weights. For example, suppose it is desired to calculate the average specific weight of sediment in a reservoir that is 100 yr old, where the sediment is composed of

TABLE 2.6.—Constants in Eq. 2.11 for Estimating Specific Weight of Reservoir Sediments

Reservoir operation	Sand		Silt		Clay	
	W_1	B	W_1	B	W_1	B
(1)	(2)	(3)	(4)	(5)	(6)	(7)
Sediment always submerged or nearly submerged	93	0	65	5.7	30	16.0
Normally a moderate reservoir drawdown	93	0	74	2.7	46	10.7
Normally considerable reservoir drawdown	93	0	79	1.0	60	6.0
Reservoir normally empty	93	0	82	0.0	78	0.0

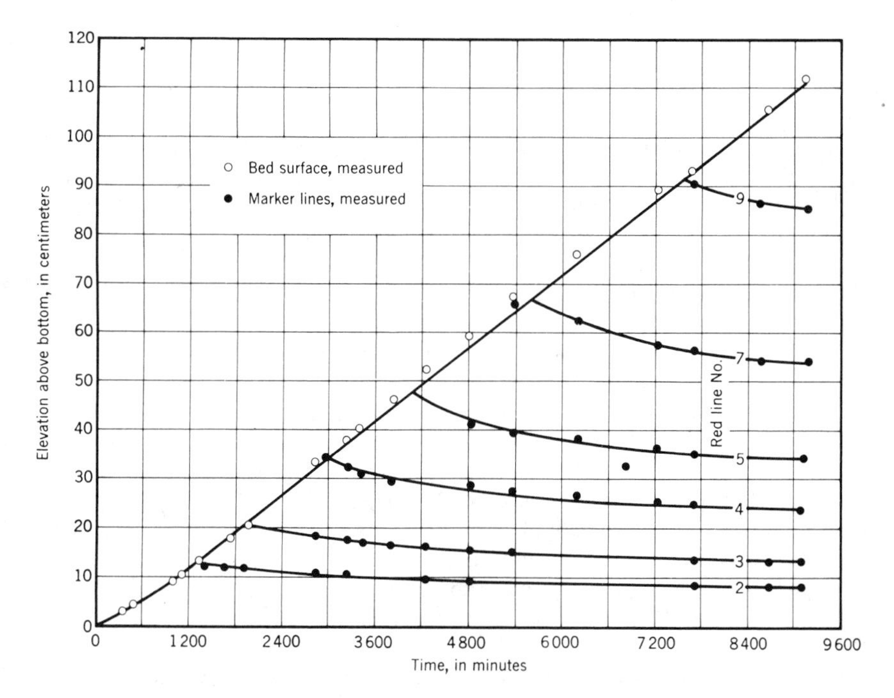

FIG. 2.15.—Observed Compaction of Montmorillonite Clay Bed while It Was Being Deposited at Uniform Rate (Long, 1961)

20% sand, 40% silt, and 40% clay, and the sediment has always been submerged. The specific weights given by Eq. 2.11 are 93.0 pcf, 76.4 pcf, and 62.0 pcf for sand, silt, and clay, respectively. The average specific weight of the entire deposit is then $93.0 \times 0.20 + 76.4 \times 0.40 + 62.0 \times 0.40 = 74.0$ pcf.

Colby (1963) proposed that the specific weight of a sediment deposit containing several size classes should be calculated by combining the various fractions according to their relative volumes instead of weights. Calculating the aforementioned example by this method gives

$$\frac{1}{\frac{0.20}{93.0} + \frac{0.40}{76.4} + \frac{0.40}{62.0}} = 72.3 \text{ pcf}$$

a value not significantly different from the previous one. However, as pointed out by Colby (1963) if it is assumed that the clay and silt are less compact and have specific weights of 65.0 pcf and 30 pcf, respectively, the resulting specific weights will be 56.6 pcf when the size fractions are weighted according to weight and 46.2 pcf when weighted according to volume.

Eq. 2.11 gives the specific weight of material that has consolidated for a period of T years after having reached its initial specific weight in a short period of about 1 yr. The average specific weight, W_{ave}, of the sediments in a reservoir after T years of operation during which deposits accumulated at a uniform rate is obtained by integration of Eq. 2.11 with respect to time. Performing the integration from one to T years and dividing by $(T - 1)$ years gives

$$W_{ave} = W_1 + B\frac{T}{T - 1}\log T - 0.434\,B \qquad (2.12)$$

as obtained by Miller (1953).

Based on extensive studies, Miller (1953) was led to believe that the initial specific weights proposed by Lane and Koelzer (1953) in Table 2.6 were too large for the finer sediments and that the values in Table 2.5 compiled by Trask (1931) were more appropriate, particularly for the finer sediments.

C. Erosion of Sediment—Local Scour

17. Concepts.—Erosion is the removal of soil particles from their environment. Within the context of this section, it refers to the removal of soil particles by water. It constitutes the beginning of motion of particles that were previously at rest and their removal from the region under consideration. Where erosion exists, important changes in the configuration and character of soil mass may occur, giving rise to many problems of considerable magnitude. The loss of productivity, or even destruction of agricultural land due to erosion, is a problem of worldwide concern. Large expenditures have been made to develop means of preventing or lessening the erosion that is inherent in most agricultural processes. The prevention of damaging scour at the base of bridge piers and spillways is an important part of the design of hydraulic structures.

On the basis of areal extent and, to a somewhat lesser degree on erosional intensity, erosion may be divided into land or soil erosion and local scour. Land erosion as an engineering problem applies primarily to the accelerated erosion of agricultural lands as distinguished from normal or geologic erosion. Local scour, on the other hand, is used to describe those erosion phenomena involving a single unified flow pattern as, e.g., the local scour around bridge piers or at the base of outlet structures. Land erosion combines the effects of innumerable local scours of varying intensities and patterns. When combined they form a typical land erosion pattern covering a large area.

Soil materials, from the point of view of erosivity, may be loosely classified as: (1) Noncohesive sediments; and (2) cohesive sediments. As the term implies, noncohesive sediments are those consisting of discrete particles, the movement of which, for given erosive forces, depends only on particle properties, e.g., shape, size, and density, and on the relative position of the particle with respect to surrounding particles. Sands and gravels of stream beds or beaches that have been previously transported, or deposits laid down by wind or water, may be included in this class.

Cohesive sediments on the other hand are those for which the resistance to initial movement or erosion depends also on the strength of the cohesive bond between particles. This resisting force may far outweigh the influence of the characteristics of the individual particles. This class includes soils containing clay, as may be found in many residual soils. Reservoir deposits exposed to the drying action of the sun and wind may fall into this classification, even though they acted as noncohesive particles during the process of deposition. In the broad sense, cohesive soils also include soils bound by a root network and protected by a vegetative cover, either natural or cultivated. In general, cohesive sediments are considerably more resistant to erosion than are corresponding noncohesive sediments. In fact, many of the soil conservation practices designed to lessen the rate of erosion have the purpose of increasing the erosion-resisting forces, where necessary, by the promotion of vegetative growths or other artificial cover during periods when the soil is particularly subject to erosive forces.

18. Land Erosion.—The quantitative aspects of land erosion and sediment yield are covered in detail in Chapter IV. In this section land erosion is discussed in general terms and in relation to the principles of local scour. The characteristics of land erosion on the broad scale are similar for both cohesive and noncohesive soils, differing primarily in degree or rate of erosion. Land erosion or soil erosion may be divided into two general types: (1) Sheet erosion, which includes the so-called rill erosion; and (2) gully erosion (Bennett, 1939). Sheet erosion and gully erosion represent, in a sense, the two extremes of the same process. Rill erosion, being gully erosion on a miniature scale, lies between these extremes. Sheet erosion occurs particularly in cultivated areas of mild slope when the runoff is not concentrated in well-defined channels. The erosion thus takes place gradually over large areas, as though the soil were removed in sheets. Erosion normally proceeds at an increasing rate as the upper layers of soil are successively removed. The more absorptive humus-charged topsoil is generally more resistant to erosion than the less absorptive, less stable layers beneath. The process involves the complex action of both running water and raindrop impact. Unprotected land varies widely in its susceptibility to sheet erosion, the differences depending on topographic features, climatic environment, and the character of the soil. The most important factor appears to be the inherent erodibility of the soil (Bennett, 1939). Quantitative data regarding the erosional characteristics of soils must, of necessity, be obtained by experiment on various types of soils and for various treatments under controlled conditions. Numerous experiments have been performed to establish the erodibility of the soil and the effect of various treatments in reducing the rate of erosion.

Table 2.7 gives typical data on actual amounts of erosion and the percentage of total precipitation lost as runoff under natural conditions of precipitation from five widely separated, extensive, and important types of farm land, as collected by the United States Department of Agriculture (1938). On steeper slopes, where the precipitation is intense and the soils have a low absorptive capacity, the runoff tends to concentrate in streamlets with the formation of small, but well-defined, incisions in the land surface. Considered over the eroding area as a whole, rill

TABLE 2.7.—Annual Soil and Water Losses per acre from Five Widely Separated Types of Land under Conditions of Clean Tillage and Dense Cover of Vegetation[a]

			Clean-Tilled Crop		Dense Cover—Thick-Growing Crop		Approximate Number of Years To Remove 7 in. of Soil[b]	
Soil, location, and years of measurements (1)	Average annual precipitation, in inches (2)	Slope, as a percentage (3)	Annual soil loss, in tons (4)	Annual water loss, as a percentage[c] (5)	Annual soil loss, in tons (6)	Annual water loss, as a percentage[c] (7)	Clean tillage (8)	Dense cover (9)
Shelby silt loam, Bethany, Mo., 1931-1935	34.79	8.0	68.78	28.31	0.29	9.30	16	3,900
Kirvin fine sandy loam, Tyler, Tex., 1931-1936	40.82	8.75	27.95	20.92	0.124	1.15	49	11,100
Vernon fine sandy laom, Guthrie, Okla., 1930-1935	33.01	7.7	24.29	14.22	0.032	1.23	50	33,200
Marshall silt loam, Clarinda, Iowa, 1933-1935	26.82	9.0	18.82	8.64	0.06	0.97	48	15,200
Cecil clay loam, Statesville, N.C., 1931-1935	45.22	10.0	22.58	10.21	0.012	0.33	51	95,800

[a]Measurements at the soil and water conservation experiment stations of the Soil Conservation Service.
[b]Based on actual volume-weight determinations of the several soils as follows: Shelby silt loam 1.43 g/ml; Kirvin fine sandy loam 1.73 g/ml; Vernon fine sandy loam 1.54 g/ml; Marshall silt loam 1.15 g/ml; and Cecil clay loam 1.45 g/ml.
[c]Of total precipitation.

erosion is a more intense form of sheet erosion. Considered separately, they are essentially miniature gullies, in that the erosion is caused by flowing water in a defined channel. From a practical standpoint, rill erosion forms small channels that can be obliterated by ordinary methods of tillage, whereas gullies are too large to be crossed with farm equipment.

Control measures for the prevention or lessening of sheet erosion in areas used for agricultural purposes include construction of terraces, strip-cropping, and contour farming. The purposes of these measures are the reduction in the length of slope, the reduction of water concentration in channels, and the storage of runoff. Terraces are shallow channels constructed on the contour or with very flat slopes that serve to store the runoff or to drain it off very slowly. In contour farming, row crops are planted on the contour so that each row serves as a miniature terrace. Strip-cropping involves alternate strips along the contour of close-growing, erosion-resistant crops and open, nonerosion-resisting crops. It may be used with or without terraces, depending on local conditions (Bennett, 1939). In the worst cases of susceptibility to erosion, it is often necessary to keep the soil in permanent vegetative cover (which has the effect of increasing the cohesiveness of the soil through its root system while, at the same time, the foliage serves to reduce the raindrop impact). Numerous experiments using these various practices have demonstrated a very appreciable reduction in soil loss (Bennett, 1939).

Gullies are examples of local scour and are formed when the runoff from an area is concentrated in well-defined channels. The concentration of the flow permits a more concerted local attack on the sediment. Usually the development of gullies follows severe sheet erosion, but often gullies have their origin in slight depressions of the land surface where in time the flow may cut a considerable channel. The shape of the incision that is formed is generally influenced by the relative resistivity (which includes cohesiveness) of the soil strata and the underlying rock. Where the subsoil is resistant to rapid cutting because of its texture, or where the incision is all in the same horizon, gullies develop sloping banks and take a V-shaped cross section. This form is usually found in humid areas where the surface soil is underlain by a stiff clay and usually develops less rapidly than other types.

In loessal regions and alluvial valley fills of the western United States, both surface soil and subsoil are commonly friable and are easily cut by flowing water. Under such conditions, gullies tend to develop vertical walls that result from undermining and collapse of the banks and tend to assume a U-shape. In certain localities, a combination of the V-shaped and U-shaped gullies prevails. It develops first as a V-shaped gully, but after the water has cut through the resistant subsoil, it strikes an underlying stratum of loose or soft rock material. Undercutting and caving then occur and the incision changes to the U-shaped channel. In regions where decomposed granite underlies a subsoil of brittle clay, gullies of this type have developed in a half century to depths of 50 ft–100 ft or more and cover large areas.

19. Local Scour.—The property of cohesiveness as related to the erosion-resisting characteristics of a soil mass introduces a factor into the relationship between the erosive forces inherent in the flow and the physical properties of the soil particles that has not been evaluated. Because the forces required to erode or scour cohesive sediments are generally greater than those for noncohesive sediments with approximately the same grain size, the rate and extent of the scour depends on this property rather than on the properties of the particles. Once the bond has been broken, the individual particles become a part of the noncohesive population. Further deposition, scour, or transport become a function of the properties of these separate particles.

Aside from numerous specific experiments relating to land erosion, the generalized study of scour by flowing water or wind has been limited to studies involving noncohesive sediments. Although the property of cohesiveness is of paramount importance in such scour phenomena as that occurring below conservation control structures, in bank caving of streams, and in the design of artificial channels constructed in cohesive materials, there are also many problems involving scour of noncohesive materials, e.g., bed scour in natural streams and scour around bridge piers and downstream of outlet works. In addition, considerable progress has been made in the design of structures discharging on cohesive soils through model studies using noncohesive sediments.

Work by Laursen (1952) on the nature of scour has crystallized many of the scattered notions of scour into some general principles: (1) The rate of scour will equal the difference between the capacity for transport out of the scoured area and the rate of supply to the area; (2) the rate of scour will decrease as the flow section is enlarged by erosion; (3) there will be a limiting extent of scour for given initial conditions; and (4) the limit will be approached asymptotically with respect to time.

The first principle is expressed symbolically by

$$\frac{df(B)}{dt} = g(B) - g(S) \qquad (2.13)$$

in which $f(B)$ = the mathematical description of the boundary; t = time; $g(B)$ = the sediment discharge or transport rate out of the scour zone as a function of the boundary shape and position; and $g(S)$ = the rate of supply to the scour zone. The remaining three principles serve as reasonable boundary conditions to be satisfied by the solution of Eq. 2.13 that describes not only the case of scour, but that of equilibrium transport and deposition as well. If the local rate of transport is greater than the rate of supply, $df(B)/dt$ is positive and scour results. If the local rate of transport is less than the rate of supply, deposition occurs and $df(B)/dt$ is negative. When the local rate of transport is equal to the supply including the case in which they are both identically zero, $df(B)/dt$ is equal to zero and the bed is stable.

Some success has been attained in establishing these functions in the important case of the equilibrium transport of noncohesive sediment in alluvial channels where $df(B)/dt = 0$ and the rate of transport, $g(B) = g(S)$, becomes a function of the flow conditions and sediment properties only. By means of analysis and experiment, useful relationships between sediment characteristics and the flow conditions have been found. This phase will be considered in detail in Sections G and H. The application of Eq. 2.13, in general, depends on the establishment of mathematical functions to describe the bed geometry, the capacity, and supply as functions of the flow conditions and time. Generally, in the absence of such descriptions, recourse must be had to experiments to determine these functions empirically or to obtain an integral solution directly. In any case, the use of Eq. 2.13 and the other characteristics of local scour provides a means for interpreting the results of experiment or of field observations that involve changes in bed configuration.

The integral form of Eq. 2.13 may be written functionally as

$$f(B) = \phi_1[g(B),g(S),t] \qquad (2.14)$$

i.e., the bed configuration is a function of the local rate of transport, rate of supply, and time. It may be assumed that the rate of transport depends on the flow pattern and the properties of the sediment. The flow pattern, in turn, will depend on the boundary geometry, including the bed configuration, which is a function of time, the characteristic velocities, and the fluid properties. Utilizing the procedures of dimensional analysis, Eq. 2.14 might be written as

$$f(B) = \phi_3\left[\frac{V}{\sqrt{\frac{\tau_c}{\rho}}}, \sigma, \frac{Vt}{a}, \frac{Va}{\nu}, \frac{V^2}{ga}, \frac{b}{a}, g(S)\right] \qquad (2.15)$$

in which $f(B)$ = a dimensionless function of the bed configuration. In Eq. 2.15, V, a typical velocity, is taken as characterizing the kinematic properties of the flow pattern; a = a characteristic length describing the size of the system; $g(S)$ denotes the supply rate which may be a constant or a function of time; τ_c = the critical tractive force for the sediment composing the bed; σ = the standard deviation of the particle size distribution; t denotes time; ν refers to the kinematic viscosity of the fluid; g = the acceleration due to gravity; ρ = the fluid density; and b describes a typical length of the flow pattern or scour geometry. Additional ratios of dimensions may be needed to describe the geometry of the system, or additional dimensionless kinematic variables may be needed to describe the flow pattern.

The fluid properties enter in a characteristic Reynolds number and a characteristic Froude number, and the sediment properties for noncohesive sediments are characterized by the critical tractive force and the standard deviation. For cohesive sediments, an additional term, probably a critical scouring velocity, should be added. For highly cohesive sediment, the critical tractive force becomes less significant and may be neglected as compared to the scouring velocity. Little is known at the present (1972) regarding the scouring characteristics of cohesive soils.

The supply, $g(S)$, is a dimensionless parameter that depends on the rate at which sediment, if any, enters the scour area under consideration. It should be realized that the terms descriptive of the flow pattern may also be functions of time, and that the particular meaning assigned to each independent variable must be determined for each situation. Particular solutions have been obtained analytically in some cases, while experimental research has yielded valuable results in others. Some of these will be described briefly in the following articles.

20. Scour by Jets.—Understanding of the scour problem has been aided considerably by investigations of scour by jets. Such an investigation permits the initial flow pattern to be more simply described and, thus, permits the other basic variables to be systematically studied. The pioneering investigation by Rouse (1940) of the scour of a bed of noncohesive sediment by a vertical jet provided an impetus for similar investigations. Johnson (1950) examined the scour pattern caused by the flow over an idealized spillway in which both the size of the spillway and sediment were varied. Experiments on scour by vertical solid and hollow jets and scour below a free overfall have been reported by Doddiah, et al. (1953).

Laursen (1952) and Tarapore (1956) have described results of experiments on scour of a sand bed by a horizontal jet. In all of these investigations, the scouring agency, the jet, was idealized so that the resulting flow pattern could be described by two parameters, the velocity and size of the jet. The data in each case provided a particular solution of Eq. 2.15. Laursen used his data to establish the transport functions in Eq. 2.13 and obtained a solution by integration.

In experiments of this type, a somewhat simpler form of Eq. 2.15 may be used to provide the necessary parameters. The supply rate of sediment is zero. The flow pattern may be characterized by the jet velocity and size or, in the case of the free overfall, by the unit discharge and the height of fall. Because noncohesive sediment composed the scoured medium, the critical tractive force, τ_c, serves to describe the sediment. Assuming that the velocities are high and that the jet is submerged, the flow pattern would be relatively independent of Reynolds and Froude number influence. Eq. 2.15 can then be simply written as

$$f_1(B) = \phi\left[\frac{V}{\sqrt{\frac{\tau_c}{\rho}}}, \frac{Vt}{a}\right] \quad \text{(2.16)}$$

Laursen (1952) showed that the scour patterns caused by jets were similar in shape as the scour progressed and, thus, could be described by a single parameter, e.g., depth or lateral extent. Fig. 2.16 shows the dimensionless shape of the scour hole with the passage of time. As a result of this, the parameter, X_D, the distance from the inlet to the crest of the dune, can be used to describe the bed configuration. Consequently, Eq. 2.16 can be written as

$$\frac{X_D}{a} = \phi\left[\frac{V}{U_{*c}} \frac{Vt}{a}\right] \quad \text{(2.17)}$$

in which $U_{*c} = \sqrt{\tau_c/\rho}$.

The results of experiments by Laursen and Tarapore showing the extent of scour caused by a horizontal jet as a function of time and the characteristics of the flow and the sediment have been plotted in Fig. 2.17. A separate curve is obtained for each value of the parameter, V/U_{*c}. The sand diameters and values of V/U_{*c} for each curve are listed in Table 2.8. The value of τ_c for each size of sediment was determined by the relationship of Shields (Fig. 2.43) for critical tractive force. In

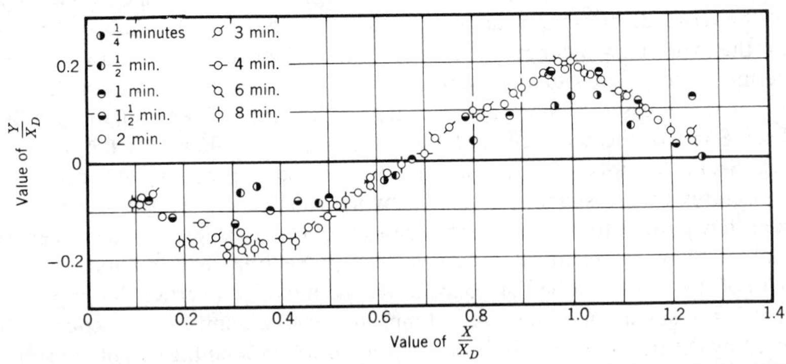

FIG. 2.16.—Dimensionless Scour Profiles Showing Similarity Independent of Time and Sediment Size (Laursen, 1952)

TABLE 2.8.—Data for Curves of Fig. 2.17

Curve number (1)	Sand (2)	Sand diameter, in millimeters (3)	V/U_{*c} (4)
1	B	0.24	160.0
2	A	0.70	132.4
3	B	0.24	118.2
4	M	1.60	90.5
5	A	0.70	96.0
6	M	1.60	82.2
7	B	0.24	83.6
8	X	0.70	89.2
9	M	1.60	67.0
10	Y	1.40	62.7
11	A	0.70	65.5
12	B	0.24	58.5
13	M	1.60	51.6
14	X	0.70	57.2
15	M	1.60	46.0
16	Z	4.4	29.2
17	X	0.70	43.3
18	A	0.70	47.4
19	X	0.70	48.8
20	B	0.24	42.5
21	B	0.24	29.2
22	M	1.60	27.8
23	A	0.70	23.7

Note: Sands M, A, and B from Laursen (1952); and sands X, Y, and Z from Tarapore (1956).

Fig. 2.17 the data points have been omitted and only the trend lines are shown. The curves show that the extent of the scour increases linearly over the range shown with the logarithm of the time and that the greater the ratio, V/U_{*c}, the steeper is the curve and the greater is the rate of increase. The linear relationship implies that the extent of the scour increased indefinitely, contrary to the preceding hypothesis stated by Laursen. It must be remembered, however, that such a logarithmic function is empirical and is based on fitting the relationship to experimental points. In fitting such a curve, it was often necessary to omit points near the origin in Fig. 2.17, because it is obvious that the logarithmic function cannot hold for small values of time. At the beginning of the experiment, when t is zero, X_D/a is also equal to zero rather than minus infinity as required by a logarithmic function. An exponential function as suggested by Breusers (1963) would satisfy this condition. For larger values of t, it appears that the logarithmic function fits the data better, while for an intermediate range, either a logarithmic or an exponential function might be suitable. As a case in point, the experiments of Ghetti and Zanovello cited by Breusers show a tendency toward a logarithmic relationship for the larger values of t, even though it is clearly exponential for the smaller time values. Although experiments do not indicate the condition clearly, it is plausible that for large values of time, when the local shear stress is approaching the critical value for the sediment, the scour-time relation may again depart from the logarithmic form. In the absence of an analytical solution that describes the

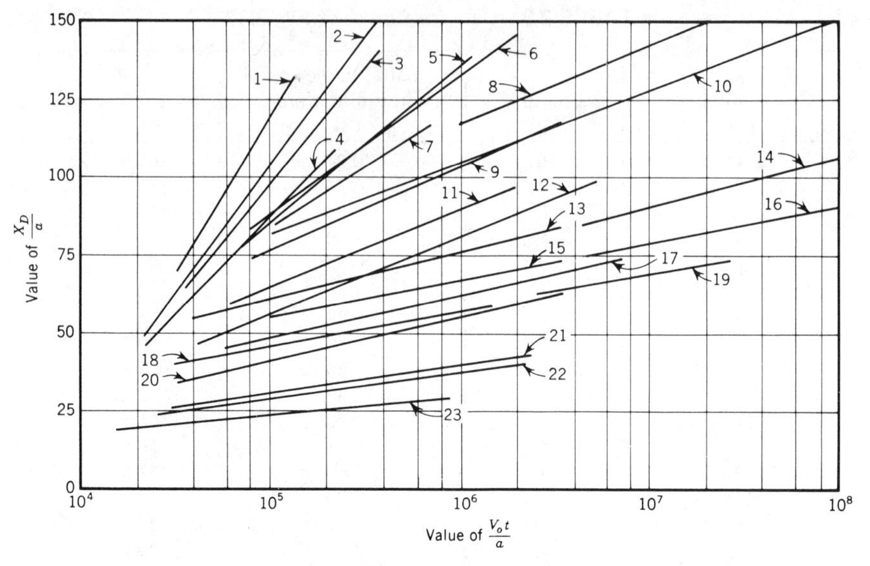

FIG. 2.17.—Increase in Extent of Scour as Function of Time

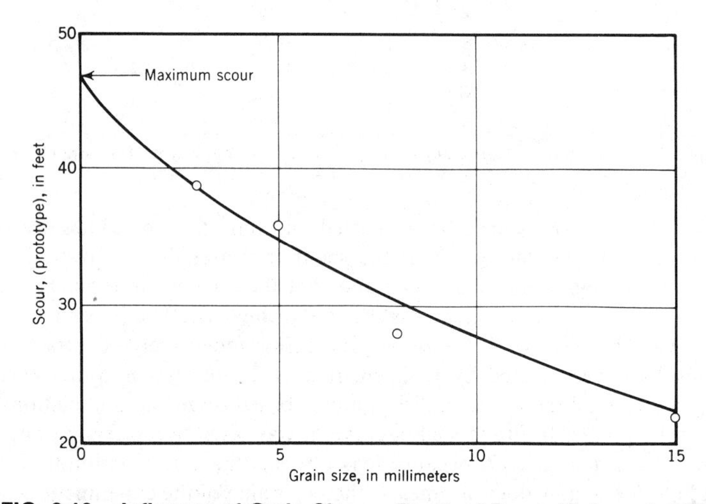

FIG. 2.18.—Influence of Grain Size on Depth of Scour (Scimemi, 1947)

entire range, recourse must be had to empirical descriptions of the various significant parts of the entire curve.

Note that, by virtue of the logarithmic character of the development of the scour region with time, a practical equilibrium is reached after a relatively short time, after which the increase in the depth and extent or scour becomes virtually imperceptible. The curves in Fig. 2.17 suggest that the greater the ratio, V/U_{*c}, the more rapidly a practical equilibrium would be reached.

The limiting extent of scour appears to be a function of the ratio of the dynamic characteristics of the flow pattern to the resistive properties of the sediment. Fig. 2.17 suggests that, although the experiments were not carried to the point of establishing the limit in each case, the limiting extent of scour will be larger for larger values of V/U_{*c}, i.e., for a given flow pattern, the limiting size will be larger as the size of sediment forming the bed becomes smaller. This effect was suggested in experiments performed (Scimemi, 1947) in the course of a model study of a spillway for a debris barrier. The jet from the free overfall impinged on a bed of noncohesive sediment that was made successively finer in consecutive tests. In each case, the depth of scour, presumably representing a practical scour limit, was measured. These data are given in Fig. 2.18 as a function of grain size of the bed sediment. The plot shows that the limiting scour depth increases as the size of the sediment decreases or as the critical tractive force decreases. The extrapolation to

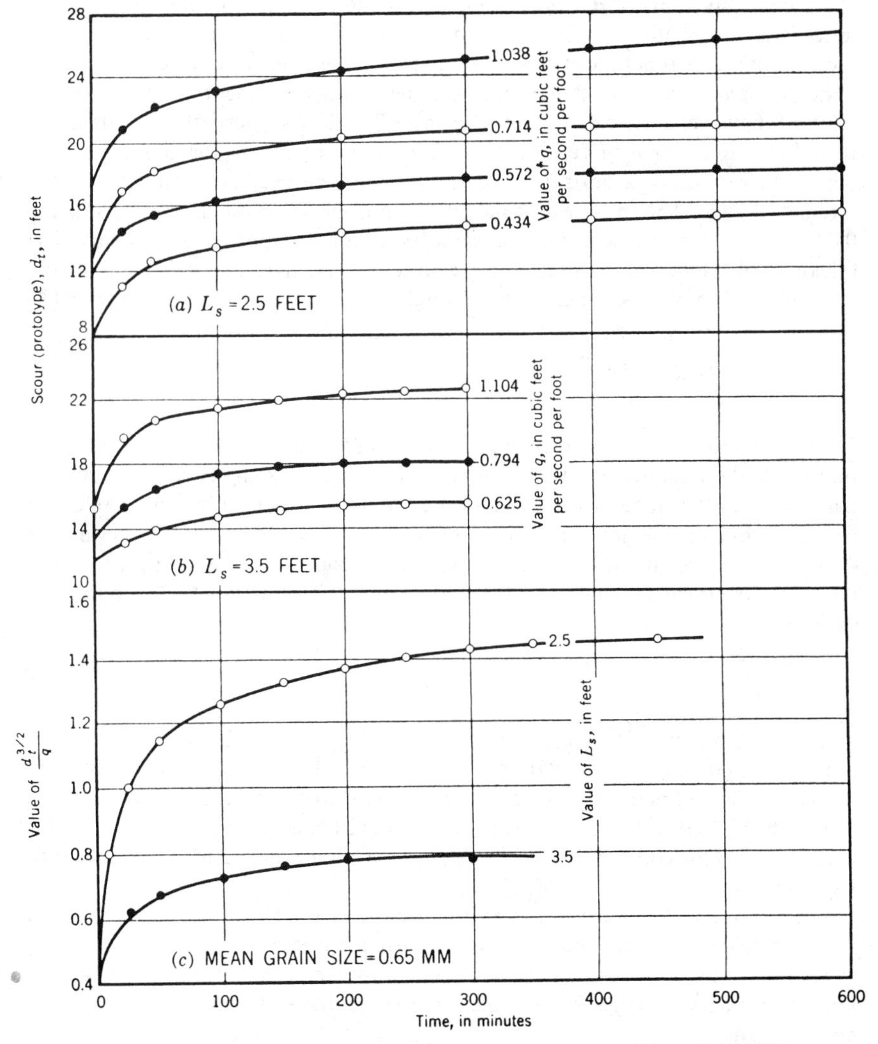

FIG. 2.19.—Effect of Discharge on Extent of Scour (Ahmad, 1953b)

a zero grain size presumably corresponded to the condition for which all of the jet energy was consumed in the pool of water into which it fell and, thus, represents the maximum possible scour. In Scimemi's experiments, the ratio, V/U_{*c}, was in effect, increased by decreasing the size of the sediment and, thus, the critical tractive force.

Qualitatively similar results were obtained by Ahmad (1953b) in a model study of a river barrage or weir. In this case, however, the sediment size was held constant and the discharge or jet energy was increased. For a given sediment size, the depth of scour as a function of time increased with the discharge or, in effect, an increase in the ratio, V/U_{*c}. Fig. 2.19 (a) and 2.19 (b) shows these results for two lengths of apron, L_s. It is apparent that the increased length of apron had the effect of reducing the effective jet velocity. Ahmad also found that plotting the scour depth in terms of $d_t^{3/2}/q$, in which d_t = the maximum depth; and q = the unit discharge, results, for the various discharges in a single curve [Fig. 2.19(c)] of scour depth as a function of time, and clearly shows the effect of apron length.

For the particular flow pattern and geometrical arrangement, the nature of the curves for the separate experiments shown in Fig. 2.17 suggests that the data can be reduced to a single curve. The data were all obtained by subjecting a granular bed to the attack of a horizontal two-dimensional submerged jet located at the initial bed level. The thickness of the jet was not the same for the two sets, being 0.025 ft for Laursens experiment and 0.0104 ft for Tarapore's experiments. The thickness of the jet can be used as the characteristic length and a measure of the scale of each set of experiments. Each straight line in Fig. 2.17 can be expressed as

$$\frac{X_D}{(X_D)_1} = \frac{\tan\theta}{\left(\frac{X_D}{a}\right)_1}\log\frac{t}{t_1} + 1 \qquad (2.18)$$

in which X_D/a = the relative extent of the scoured area eroded in time, t; $(X_D/a)_1$ represents the relative extent of scoured area eroded in some particular time, t_1; $\tan\theta$ denotes the slope of the curve in the semilogarithmic form; and a refers to the thickness of the jet. Eq. 2.18 represents a single curve if the term, $\tan\theta/(X_D/a)_1$, is a constant. It will pass through the coordinates, $X_D/(X_D)_1 = 1$ and $t/t_1 = 1$. Arbitrarily taking $\tan\theta/(X_D/a)_1 = 1.25$, $(X_D/a)_1$ can be computed for each curve as

$$\left(\frac{X_D}{a}\right)_1 = \frac{1}{1.25}\tan\theta \qquad (2.19)$$

and the value of $V_o t_1/a$ corresponding to $(X_D/a)_1$ can be taken from the appropriate curve (by extrapolation, if necessary). Since individual data points are not given, the end points of each curve can be used to compute $X_D/(X_D)_1$ and t/t_1. These are plotted in Fig. 2.20 along with the data taken from Breusers (1963) for comparison. This comparison is the basis for the choice of 1.25 for the constant in Eq. 2.18.

The superposition of the two curves shows the character of a transition from a logarithmic curve to an exponential curve near the origin. In experiments carried to values of t in the logarithmic range, the data for small values of t show the same kind of departure and go in the same direction as the exponential portion of the curve.

The constant 1.25 for the slope of the dimensionless curve was chosen

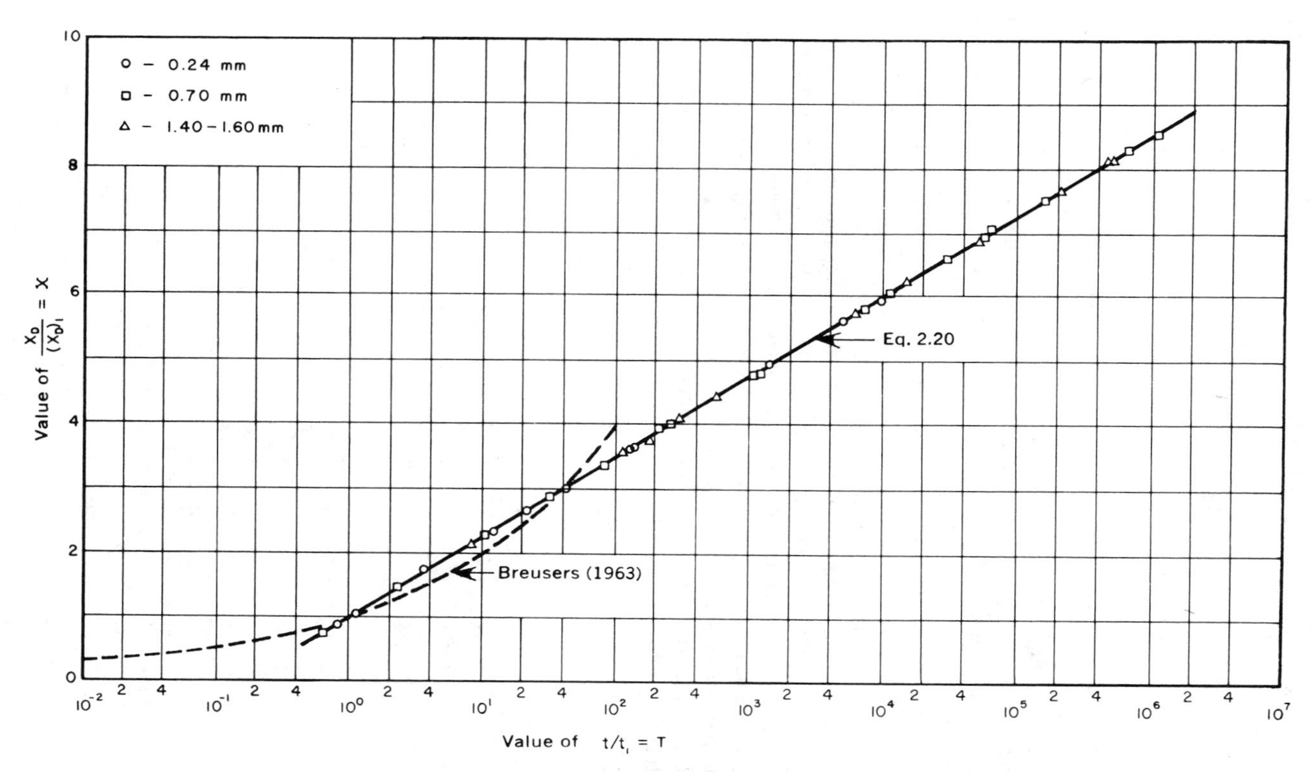

FIG. 2.20.—Relative Extent of Scoured Area as Function of Time

specifically to make this comparison. At this time, it is not apparent what the basis for a more appropriate choice should be. However, using this value, Eq. 2.18 can be written as

$$X = 1.25 \log T + 1 \quad (2.20)$$

in which
$$X = \frac{X_D}{(X_D)_1} \quad (2.21)$$

and
$$T = \frac{t}{t_1} \quad (2.22)$$

In this form, Eq. 2.20 is a kind of universal curve that will apply to all curves having a logarithmic scour-time relationship. The question then arises as to how the scour-time function for different flow patterns can be delineated. It appears that this is possible through the magnitude of tan θ. According to Eq. 2.19, if tan θ is large, $(X_D/a)_1$ is large, and, consequently, the points will fall at the lower end of the curve. For small values of tan θ, the points will fall on the upper part of the curve. Tan θ, in turn, appears to be a function of V_o/U_{*c}, as suggested by Fig. 2.17 and Table 2.8. The scour at any point is caused by the local shear stress generated by the velocity near that point in relation to the sediment critical shear stress. This local velocity is a function of the inlet velocity and the boundary geometry, so that, for a given boundary geometry, the local shear stress depends on V_o, and the scour pattern depends on V_o/U_{*c}. Therefore, tan θ is dependent not only on V_o/U_{*c}, but also on the boundary geometry of the system. In Fig. 2.21, tan θ is plotted as a function of the ratio V_o/U_{*c} for the particular experiments shown in

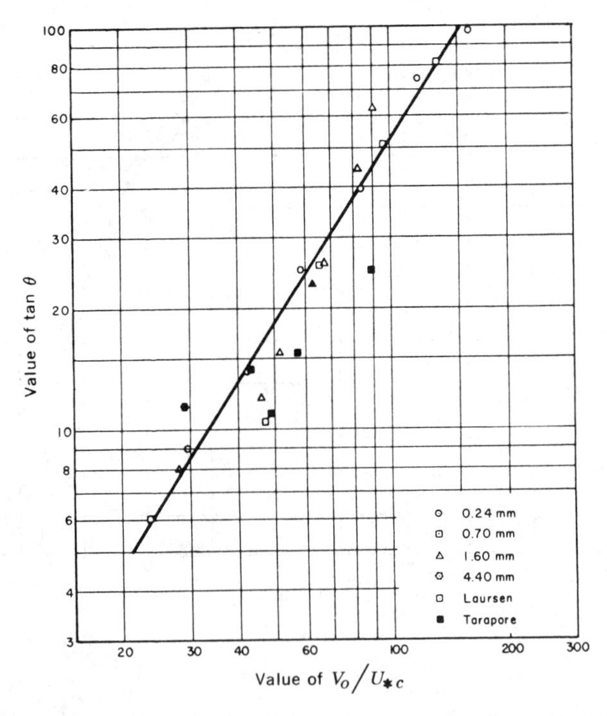

FIG. 2.21.—Relation of Slope of Scour-Time Curve to Scour-Velocity Ratio

Fig. 2.17. A single curve describes the relationship reasonably well, and presumably represents the functional relationship between tan θ and V_o/U_{*c} for this particular flow pattern; i.e., a horizontal jet discharging at the initial level of the bed. An examination of Fig. 2.21, however, shows that there is a tendency for the points obtained in Tarapore's experiments to fall to the right of those obtained by Laursen. Conceivably, this shift could represent the slight difference in flow pattern created by the difference in jet thickness. It would be expected that the form of the function would also be different for other kinds of flow patterns, e.g., those created by weirs, free falling jets, spillways, and other structures. If this is so, the functional relationship between $(X_D/a)_1$ (or tan θ) and the ratio V_o/U_{*c} serves as a criterion to classify or identify the various flow patterns involved in scour problems.

An application of the relationship given by Eq. 2.20 lies in the determination of the "practical limit." In model studies, observation indicates that after the passage of a certain time, a practical limit is reached where a neglibible quantity of sediment is being eroded. Such a limit is suggested by the arithmetically plotted curves in Fig. 2.19, where the curves become increasingly flat. The choice of the practical limit is arbitrary and represents some point, on the curve of Fig. 2.19, where the slope has become sufficiently flat. The choice based on a "negligible" rate of scour implies that the local bed shear stress is approaching the critical value for the sediment. The bed shear stress, however, depends on the flow geometry and the duration of the erosion process. The slope of the scour-time curve at the practical limit may then be taken as proportional to the ratio τ_o/τ_c or $(V_o/U_{*c})^2$. Then

$$\left(\frac{dX}{dT}\right)_p = \frac{0.434 \times 1.25}{T_p} = k\left(\frac{V_o}{U_{*c}}\right)^2 \qquad (2.23)$$

or

$$T_p = \frac{0.542}{k\left(\frac{V_o}{U_{*c}}\right)^2} \qquad (2.24)$$

in which T_p = the dimensionless time to the practical limit. Using this value of T_p in Eq. 2.20, the dimensionless practical limit becomes

$$X_p = \frac{\left(\frac{X_D}{a}\right)_p}{\left(\frac{X_D}{a}\right)_1} = 1.25\left[\log 0.542 - \log k\left(\frac{V_o}{U_{*c}}\right)^2\right] + 1 \qquad (2.25)$$

Using the relationship between tan θ and V_o/U_{*c} given in Fig. 2.21 for this particular flow geometry

$$\left(\frac{X_D}{a}\right)_1 = 0.8 \tan\theta = 0.0424\left(\frac{V_o}{U_{*c}}\right)^{3/2} \qquad (2.26)$$

and

$$\left(\frac{X_D}{a}\right)_p = 0.053\left(\frac{V_o}{U_{*c}}\right)^{3/2}\left[\log 0.542 - \log k\left(\frac{V_o}{U_{*c}}\right)^2\right] + 1 \qquad (2.27)$$

becomes the "practical limit." Again, in this application, the practical limit depends on the functional relationship between tan θ and V_o/U_{*c}. Its value also

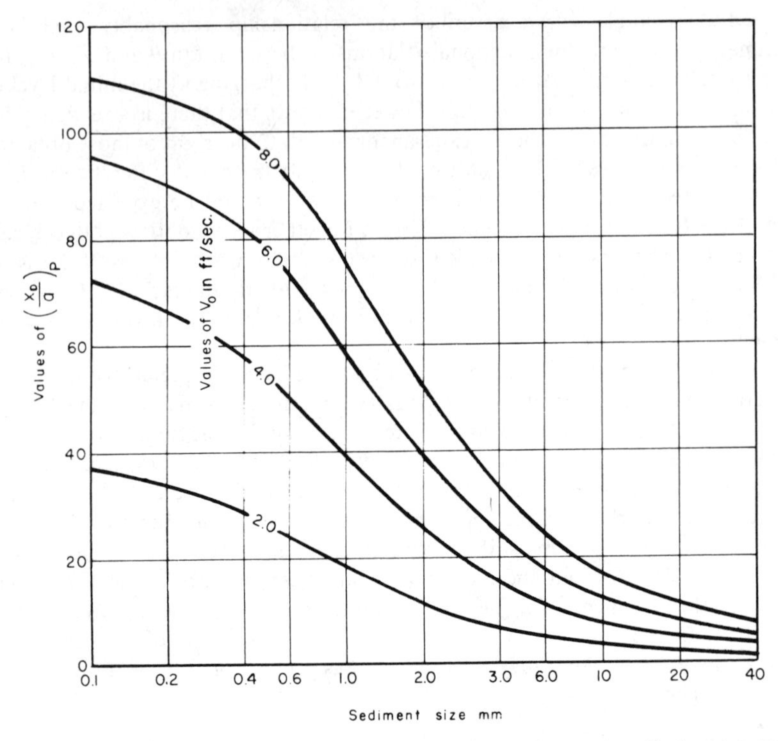

FIG. 2.22.—Practical Extent of Scour for Various Sizes for Horizontal Two-Dimensional Submerged Jet [Based on Experiments by Laursen (1952) and Tarapore (1956)]

depends on the choice of k to represent the slope of the scour-time curve assumed to represent this limit. Using the experimental results presented in Fig. 2.17 as a guide, a value of $k = 1.5 \times 10^{-5}$ was adopted, and the value of $(X_D/a)_p$ was computed for various values of V_o/U_{*c}. These have been plotted in Fig. 2.22, in terms of the grain size for various values of V_o. The results show that the extent or depth of scour at the practical limit increases with decreasing grain size (see Fig. 2.22). There appears to be an inflection point in the neighborhood of 0.5 mm, so that for grain sizes less than this, the curves tend to flatten out. This is somewhat in accord with the conclusion of Breusers (1963) that, for sediment in this range, the practical limit is almost independent of the diameter. The upper part of the curves for larger grain sizes bears out the results of Scimemi (1947) that showed that the depth of scour at the practical limit increased as the grain size decreased.

21.Scour in Channel Constrictions.—For the special case of scour in constrictions that are long enough so that it can be assumed that uniform flow exists in both the normal and constricted reaches, the equations for flow of water and sediment developed for such uniform flows may be used. Straub (1935) treated the problem by using the DuBoys equation (Eq. 2.224) for sediment transport at the equilibrium condition. Then for $dt(B)/dt = 0$

$$g(B) - g(S) = \psi_D b_2 \tau_2(\tau_2 - \tau_c) - \psi_D b_1 \tau_1(\tau_1 - \tau_c) = 0 \qquad (2.28)$$

in which for wide channels

$$\tau_2 = \gamma d_2 S_2 \qquad (2.29)$$

and $\tau_1 = \gamma d_1 S_1 \qquad (2.30)$

Continuity of discharge in the two sections is satisfied by the Manning formula for wide channels for which

$$Q = \frac{1.49}{n} b_2 d_2^{5/3} S_2^{1/2} = \frac{1.49}{n} b_1 d_1^{5/3} S_1^{1/2} \qquad (2.31)$$

In Eqs. 2.28 and 2.31, τ_1 and τ_2 = the tractive forces or boundary shear stresses; d_1 and d_2 = the depths; S_1 and S_2 = the slopes; b_1 and b_2 = the breadths in the normal and constricted section, respectively; τ_c = the critical tractive force for the sediment; n = the Manning roughness coefficient, which is assumed to be the same in both sections; and ψ_D = the DuBoys transport parameter, the magnitude of which depends on the sediment size, and is the same in both sections.

Combining Eqs. 2.28 and 2.31 and the definition of the boundary shear stresses, τ_1 and τ_2, the relative depths can be written as

$$\frac{d_2}{d_1} = \left(\frac{b_1}{b_2}\right)^{3/7} \left\{ \frac{-\frac{\tau_c}{\tau_1} + \left[\left(\frac{\tau_c}{\tau_1}\right)^2 + 4\left(1 - \frac{\tau_c}{\tau_1}\right)\frac{b_1}{b_2}\right]^{1/2}}{2\left(1 - \frac{\tau_c}{\tau_1}\right)} \right\}^{3/7} \qquad (2.32)$$

Eq. 2.32 is plotted in Fig. 2.23 with d_2/d_1 as a function of τ_c/τ_1 for various values of the parameter, b_1/b_2.

For normal values of the parameter, the relative depths are essentially independent of the ratio, τ_c/τ_1, and, even for a contraction ratio, b_1/b_2, as great as four, the effect of τ_c/τ_1 on d_2/d_1 is appreciable only for the larger values of τ_c/τ_1. If τ_c is small compared to the mean boundary shear stress in the normal section, τ_1, Eq. 2.32 reduces to the following simple form:

$$\frac{d_2}{d_1} = \left(\frac{b_1}{b_2}\right)^{9/14} = \left(\frac{b_1}{b_2}\right)^{0.642} \qquad (2.33)$$

A special case might exist if the supply were cut off by a hydraulic structure, and the normal section were eroded until the boundary shear stress had been reduced to the sediment critical. The critical boundary shear stress would also exist in the constricted section. This situation corresponds to $\tau_c/\tau_1 = 1$. Solving Eq. 2.32 with $\tau_c/\tau_1 = 1$

$$\frac{d_2}{d_1} = \left(\frac{b_1}{b_2}\right)^{6/7} = \left(\frac{b_1}{b_2}\right)^{0.857} \qquad (2.34)$$

This treatment can be carried a step further by prescribing that the shear stress in the normal section be less than the critical boundary shear stress, τ_c, while, in the constricted section, the boundary shear stress is initially greater than the critical. The bed in the constricted section will be scoured until the shear stress here is reduced to the critical. Then for equilibrium

$$\tau_c > \tau_1 = \gamma d_1 s_1 \text{ and } \tau_c = \gamma d_2 s_2 \qquad (2.35)$$

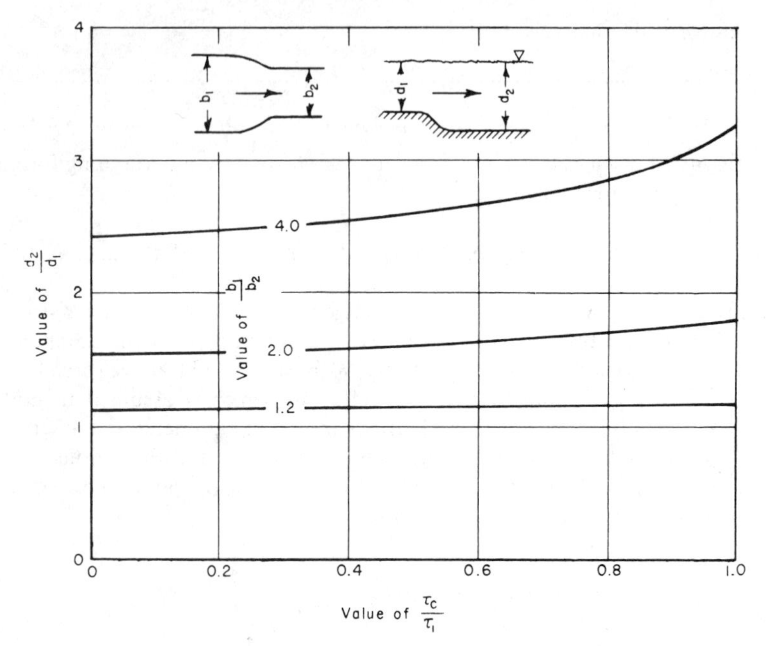

FIG. 2.23.—Relative Depth in Long Constriction

The use of Eqs. 2.35 in Eq. 2.32 shows the relative depths to be

$$\frac{d_2}{d_1} = \frac{\left(\frac{b_1}{b_2}\right)^{6/7}}{\left(\frac{\tau_c}{\tau_1}\right)^{3/7}} \qquad (2.36)$$

and shows that the relative depths are reduced for a given contraction ratio by an amount depending on the ratio of the critical boundary shear stress to the actual boundary shear stress in the normal section.

Since the exponent in the equation relating the relative depths to the relative widths is 9/14 for $\tau_c/\tau_1 = 0$ and 6/7 for $\tau_c/\tau = 1$, it would appear that Eq. 2.32 may be closely approximated over the entire range, $0 < \tau_c/\tau_1 < 1$ by

$$\frac{d_2}{d_1} = \left(\frac{b_1}{b_2}\right)^{\theta} \qquad (2.37)$$

in which $9/14 < \theta < 6/7$. The variation of θ as a function of τ_c/τ_1 was established empirically and is shown in Fig. 2.24.

These relationships show that for a given contraction ratio the depth in the constricted reach will be essentially proportional to the depth in the normal reach but that as the depth increases with increasing discharges the depth in the constricted reach will also increase but at a somewhat faster rate.

One of the principal functions of regulatory works in natural rivers is to establish and maintain depths sufficiently large during low-water periods so that navigation can be continued during these periods. During high-water periods such training works are not necessary. The application of the foregoing equations

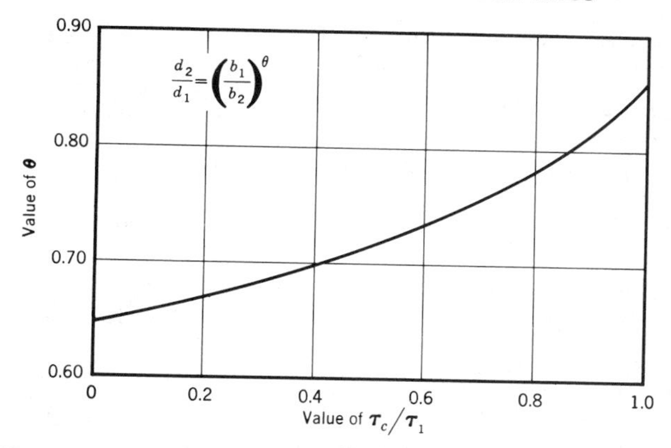

FIG. 2.24.—Variation of Exponent θ with Relative Transport Rate

(Anderson, 1968) to the situation when constricting works are submerged during high flows shows that the depth in the constriction can be increased during low-water periods and decreased relatively to the depth in the normal section during high-water periods. These considerations lead to an equation for the relative depths as

$$\frac{d_2}{d_1} = \left(\frac{Q_2}{Q_1}\right)^{6/7} \left(\frac{b_1}{b_2}\right)^{\theta}; \ \frac{9}{14} < \theta < \frac{6}{7} \quad \text{.......... (2.38)}$$

in which Q_2 = the discharge transporting sediment in the constricted reach; and Q_1 = the total river discharge, part of which takes place over the constricting grains. Eq. 2.38 shows that as the ratio, Q_2/Q_1, decreases the ratio, d_2/d_1, also decreases.

Komura (1963b) treated the same problem but used more recent results to describe the transport rate and the flow over movable beds. The basic concepts of continuity of sediment transport and water discharge for a stable bed are the same. He described the fluid velocity by

$$\frac{U}{U_*} = E\left(\frac{r}{k_s}\right)^q \quad \text{.......... (2.39)}$$

in which U_* = the friction velocity; r represents the hydraulic radius; k_s = the equivalent roughness; q = a dimensionless exponent; and E = a constant. The sediment transport was defined by the Kalinske-Brown equation as

$$\frac{q_s}{U_* d_s} = a_s \left\{ \frac{U_*^2}{\left[\left(\frac{\rho_s}{\rho}\right) - 1\right] g d_s} \right\}^P \quad \text{.......... (2.40)}$$

in which q_s = the transport rate; a_s and P = constants; ρ_s and ρ = the densities of the sediment particles and the water, respectively; g denotes the acceleration due to gravity; and d_s refers to the mean diameter of the bed material.

By applying the equilibrium condition (Eq. 2.13) relative depths in the constricted reach to the normal reach were obtained for so-called "dynamic equilibrium" as

$$\frac{d_2}{d_1} = \left(\frac{b_1}{b_2}\right)^{0.686} \quad \text{.......... (2.41)}$$

which corresponds to an equilibrium for conditions of sediment transport through both the normal and constricted reaches. For the so-called "static equilibrium" case of no transport in either the normal and constricted reaches

$$\frac{d_2}{d_1} = \left(\frac{b_1}{b_2}\right)^{0.857} \qquad (2.42)$$

These results are in very good agreement with those obtained by the use of simpler expressions.

By an analysis based upon regime equations and supported by some experimental data, Griffith (1939) proposed a similar equation given as

$$\frac{d_2}{d_1} = \left(\frac{b_1}{b_2}\right)^{1/(n+1)} = \left(\frac{b_1}{b_2}\right)^{0.637} \qquad (2.43)$$

in which n = the exponent for the Kennedy regime velocity and is taken as 0.57.

22. Scour around Bridge Piers.—In the preceding Article 20, the local scour caused by jets was concerned with the solution of the scour equation (Eq. 2.13) for the special case of zero sediment supply. It examined both the change in bed configuration as a function of time and the limiting extent or "practical limit" of the scour as $g(B)$ decreases and $df(B)/dt$ approaches zero. This condition developed as the consequence of the decreased local shear as the scour hole was enlarged. This situation is characteristic of the scour pattern downstream from spillways and other outlet works where the sediment supply is trapped in the upstream impoundment. In Article 21 a solution of the scour equation for the final equilibrium or limiting configuration when the sediment supply is not zero was attempted. Here equilibrium is reached when the rate at which sediment is tranported out of the scour area is equal to the rate of supply. In the case of long channel constrictions this condition could be approximated by assuming that the constricted reach was long enough for normal flow to develop so that the transport rate could be described by established equations. With this assumption, the equilibrium configuration could be estimated and related to the upstream flow conditions.

The scour around bridge piers and other local obstructions is another example of such scour in which the limiting configuration is reached when the transport rate through the scour region is equal to the supply rate, i.e., $g(B) = g(S)$. In this case, however, the scour is essentially localized and is caused by a very complex flow pattern around the obstruction so that the transport rate and thus the depth of scour for equilibrium state cannot be determined. Because of the complexity in both the flow pattern and the transport function, it is often necessary to utilize model studies of specific structures so that the equilibrium scour depth can be measured directly. For this purpose a section of the stream including the piers and abutment is modeled and the transport of sediment through the section is simulated. The resulting scour pattern can be measured as it becomes stable and estimates made of the expected prototype scour. It is sometimes difficult to appropriately simulate all of the many variables that play a role in such a process, but reasonably good results can be obtained.

The nature of the flow pattern upstream and around a bridge pier has been described by Tison (1961), who showed that a downward flow in front of the pier should exist as a result of the horizontal curvature of the streamlines in front of the pier and the fact that the velocity near the bed is reduced by friction. This downward velocity component gives rise to the so-called "horseshoe vortex" that is wrapped around the pier near the junction with the bed. The vortex as well as the downward velocity component itself is effective in initiating and extending the scour around and close to the pier. Tison showed that the magnitude of the vertical component and thus the strength of the vortex depended to a high degree on the radius of curvature of the streamlines as they were deflected by the pier. Since the radius of curvature of the streamline about a pier with a rectangular nose is considerably less than that about a pier with a streamlined or lenticular nose, the scour generated by rectangular piers is considerably greater in general than that for lenticular piers. His experiments substantiated this view and showed a very significant reduction in scour when the pier noses were pointed.

Because of the difficulty of evaluating the flow pattern and shear forces generated by the flow pattern, most of the data upon which estimates of depth of scour are made have been obtained by experiments. A careful investigation of this phenomenon by means of hydraulic models was carried out by Laursen and Toch (1953). They related the depth of scour around model bridge piers to the flow parameters and pier geometry. The depth of scour was measured for a rather wide range of velocities and sediment sizes. One significant general result was that the equilibrium depth appears to depend only on the initial depth of flow and to be independent of both the mean velocity and the sediment characteristics. These results are shown in Fig. 2.25. The explanation given by Laursen and Toch is that at the equilibrium scour the velocity of the currents formed in the scour hole and the velocity near the bed of upstream section, which is the transporting agency, are both similar functions of the mean velocity. Consequently, a change in the mean velocity will increase the bed velocity of the incoming stream as well as the local velocity in the scour hole. This will result in approximately equal increases in the sediment discharge into and out of the scour area, thus maintaining the existing depth. The lack of dependence of the depth of scour upon sediment size was explained on a similar basis, since for uniform sediment the size will have no effect on any existing balance of transport capacities. These conclusions are also implied by Eq. 2.32 and Fig. 2.23, which shows that d_2/d_1 is essentially independent of the ratio, τ_c/τ_1, and the flow velocities. Consequently, the scour depth, d_2, is dependent only upon the antecedent depth of flow, d_1, and the constriction geometry, b_1/b_2, which would be equivalent to the geometry of the bridge pier or abutment as the case might be. These concepts led to further analysis by Laursen (1962) who attempted to derive a relationship for the scour that was an adaptation of those developed for long constrictions to the case of localized scour at bridge piers and bridge abutments. The resulting relationship was given as

$$\frac{b}{d} = 5.5\frac{d_t}{d}\left(\frac{d_t}{ad} + 1\right)^{1.7} - 1 \qquad (2.44)$$

in which b = the width of the pier; d_t = the depth of scour from the mean bed elevation; d = the depth of flow; and a = a factor to represent the scour in the hypothetical long contraction.

Because of the importance of scour around bridge piers and abutments many experiments and model studies have been made with respect to specific structures. These have lead to empirical relationships between the scour and the various flow parameters. However, when these are applied to a particular structure the

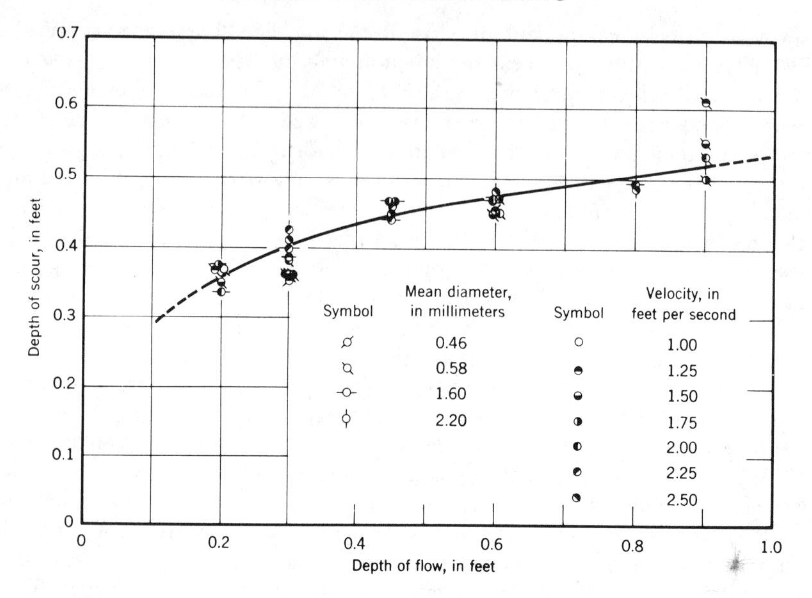

FIG. 2.25.—Depth of Scour around Bridge Piers (Laursen, 1952)

predicted depth of scour may vary widely. The reasons, therefore, are probably due to different experimental conditions that may not be appropriate to the particular hydraulic geometry to which they were applied. A short resume of the state of scour experimentation as applied to the scour problems of bridge waterways has been given in a report of the Highway Research Board (1970). References to other studies on scour are also given in that report.

Similar results were obtained by Ahmad (1953a) in model studies of spur dikes. In a model of the Sutlej River, the spur projected into the channel one-third of the channel width. Two sands (mean diameter 0.354 mm and 0.695 mm) were used. The results in Fig. 2.26(a) show that the ultimate or limiting scour was independent of the sediment size, but the rate of development of scour was greater for the finer size. He also found that plotting the data in terms of $d_t/q^{2/3}$ (in which d_t = the depth of scour, and q = the unit discharge) as a function of time resulted in a single curve for each sediment [Fig. 2.26(b)].

The results of experiments on local scour indicate that when the rate of supply into the scoured region is zero, the extent of scour depends on the relative sediment size as well as the flow energy and pattern; whereas, if the rate of supply is not zero, the extent and pattern of scour is independent of the sediment size and depends only on the flow pattern as governed by boundary geometry.

23. Degradation and Aggradation.—The scour equation (Eq. 2.13) is also fundamental to studies of the process and extent of degradation or aggradation of streams. For various reasons, the sediment supply of a stream may be increased or decreased beyond the local transporting capacity with the consequent change in bed elevation. In many cases, dams or other hydraulic structures may decrease the rate of supply to zero permanently or for considerable periods of time. Downstream of the structure, the bed will be degraded because the clear stream will pick up a new equilibrium load. Stated in another way, the rate of supply at

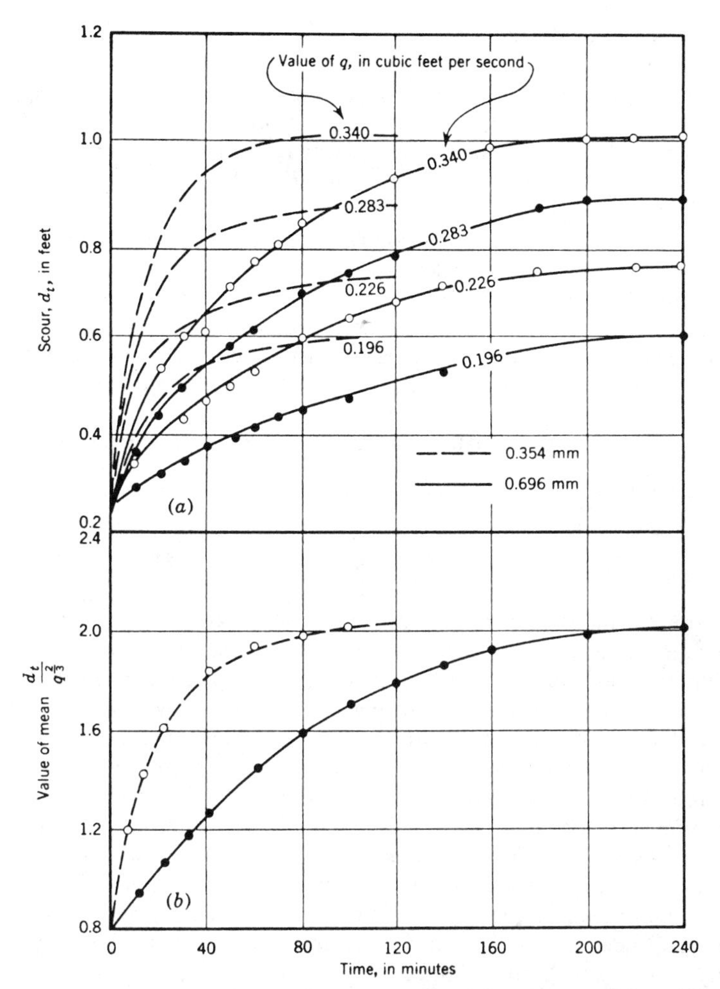

FIG. 2.26.—Effect of Sediment Size on Depth of Scour around Spur Dike

the structure is zero but the transport rate is appreciable; therefore, the bed scours. Beyond this point, the load that the stream has picked up becomes the supply of the succeeding section. The rate of change of the bed elevation then becomes proportional to the rate of change of the sediment discharge, G_s, in the downstream direction, or

$$\frac{\partial[f(B)]}{\partial t} = \frac{\partial(G_s)}{\partial x} \qquad (2.45)$$

The solution of Eq. 2.45 is the solution of the degradation or aggradation problem. Because of the complexity of boundary conditions in these problems, approximations must be made and solutions found, taking into account the characteristics of the stream obtained by finite steps (United States Army Corps of Engineers, 1949). Tinney (1955) solved Eq. 2.45 for a particular case of a straight rectangular channel and found good agreement with laboratory experiments.

D. Suspension of Sediment

24. Turbulence.—Because turbulence is the most important factor in the suspension of sediment, it is pertinent to describe it briefly at this point. Turbulence is the irregular motion of a flowing fluid that one observes commonly in streams and in the atmosphere. The turbulent motion results from eddies that are swirling in an irregular manner as they are carried along by the flow. The eddies are being formed continuously by the shearing action of the fluid while eddies already in existence are being dissipated into heat by viscous friction.

If a steady turbulent flow is observed at a point, the velocity will be observed to fluctuate with time, both in magnitude and direction. To describe a turbulent flow quantitatively, let u, v, and w be the instantaneous velocity components and $\bar{u}$, $\bar{v}$, and $\bar{w}$ be the mean velocity components in the Cartesian coordinate directions, x, y, and z, respectively. One can then define components of the turbulent fluctuations, u', v', and w', of the velocities in the three directions by

$$u = \bar{u} + u' \quad \text{(2.46a)}$$

$$v = \bar{v} + v' \quad \text{(2.46b)}$$

and $$w = \bar{w} + w' \quad \text{(2.46c)}$$

In these equations, the mean value of a velocity component is obtained by averaging many instantaneous values. Because the mean velocities are defined in this way, the mean value of the turbulent fluctuations are zero. In the example previously assumed, the flow was taken as steady, so that by definition the mean values of the velocities at a point are constant. If, in this example, the x-direction is taken in the direction of the mean flow, the mean velocities in the other coordinate directions are zero or $\bar{v} = \bar{w} = 0$ and $v = v'$ and $w = w'$. Flows in which the mean velocities, $\bar{u}$, $\bar{v}$, and $\bar{w}$,at a point do not change with time are said to be steady, although, strictly speaking, such flows are not steady because the turbulent fluctuations vary with time. If $\bar{u}$, $\bar{v}$, and $\bar{w}$ do not change as the x-coordinate of the position in the stream is varied, keeping y and z constant, the flow is said to be uniform. In uniform flow in a stream, e.g., as river or canal, the velocity, $\bar{u}$, varies with both y and z.

The root-mean-square (rms) values $(\overline{u'^2})^{1/2}$, $(\overline{v'^2})^{1/2}$, and $(\overline{w'^2})^{1/2}$ are taken as measures of the intensity of turbulence. In the preceding expressions the bar denotes mean value, for instance, $\overline{u'^2}$ denotes the mean value of u'^2. By far, most measurements of turbulence have been made in air, using the hot-wire anemometer.

However, since the hot-film anemometer was developed, turbulence has been measured in water in flumes by several workers. Fig. 2.27 is a plot of relative longitudinal turbulence intensity, $(\overline{u'^2})^{1/2}/U_*$ against y/d for both air and water, in which $U_* = \sqrt{\tau_o/\rho}$ = shear velocity; τ_o = bed shear stress; ρ = density of fluid; y = distance from the wall or bed; and d = water depth or half thickness of the two-dimensional air flow studied by Laufer (1950). The Reynolds numbers, R, in Laufer's experiments are defined as $4\bar{u}_{max}d/\nu$ and in the experiments with water R = $4Vr/\nu$ in which $\bar{u}_{max}$ = center line velocity of the wind channel; ν = kinematic viscosity of the fluid; V = mean velocity in the cross section of the flume flow; and r = hydraulic radius of the flume flow. The data in Fig. 2.27(a) are for flows in smooth channels and those in Fig. 2.27(b) are for flows in a rectangular flume with smooth sides and beds roughened with 2.45 mm sediment.

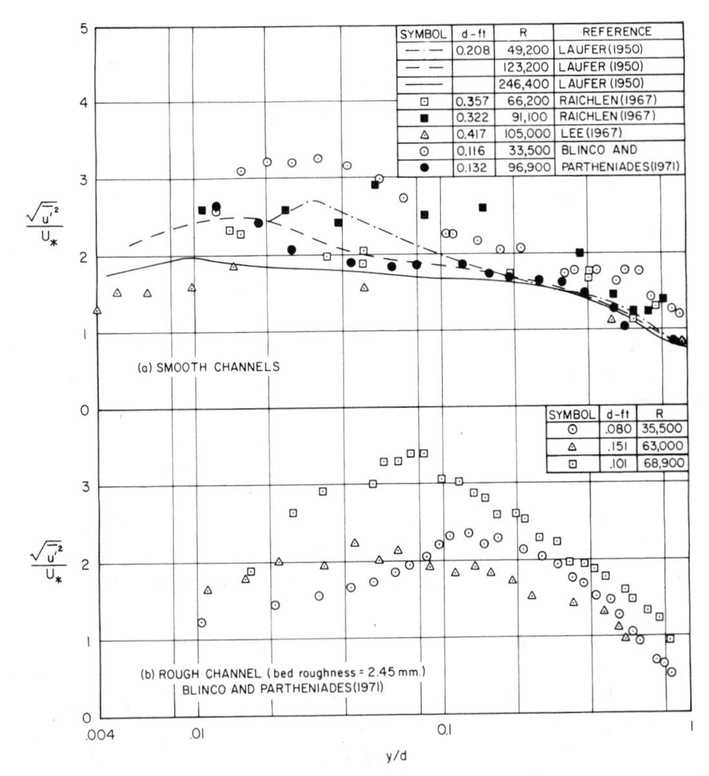

FIG. 2.27.—Distribution of Relative Longitudinal Turbulence Intensity in Smooth Wind Channel and Flume and in Flume with Roughened Bed

All of the data of Fig. 2.27 show that the longitudinal turbulence intensity, $(\overline{u'^2})^{1/2}$, reaches a maximum a short distance from the wall or bed and diminishes each way from this point, reaching its minimum value at $y/d = 1.0$. The Laufer (1950) data indicate that $(\overline{u'^2})^{1/2}/U_*$ tends to decrease as Reynolds number increases. The measurements in water do not show a consistent trend with Reynolds number. The values of $(\overline{u'^2})^{1/2}/U_*$ in the rough channels are not significantly different from those in the smooth ones. However, because U_* in a rough channel is larger than in a smooth one with the same depth and velocity, the turbulence intensity $(\overline{u'^2})^{1/2}$ will also then be larger in the rough channel.

Laufer (1950) also measured the turbulence intensities in the transverse directions. He found that both $(\overline{v'^2})^{1/2}/U_*$ and $(\overline{w'^2})^{1/2}/U_*$ varied approximately linearly from about 0.9 at $y/d = 0.1$ to 0.6 at $y/d = 1.0$ and were considerably smaller than the corresponding longitudinal intensities.

Another property of turbulence of importance to sediment transportation is its scale or the mean size of the eddies forming the turbulence. Several expressions have been developed for the scale. One of these is given by

$$l = \int_0^\infty R(\xi)\, d\xi \qquad (2.47)$$

in which $$R(\xi) = \frac{\overline{u'(x)\, u'(x + \xi)}}{\overline{u'^2}} \qquad (2.48)$$

and $u'(x)$ and $u'(x + \xi)$ are simultaneous values of the longitudinal turbulence fluctuations at position x and $(x + \xi)$, respectively, and the bar denotes mean values of the products of many simultaneous measurements.

The function $R(\xi)$ is a correlation coefficient with a value of unity when $\xi = 0$, because at this point $u'(x) = u'(x + \xi)$. At large values of ξ, $u'(x + \xi)$, and $u'(x)$ are not related so that there is equal chance that a simultaneous pair of velocity fluctuations will have the same sign as there is that they will have opposite signs, and one would expect the correlation coefficient to fall to zero. Measurements (Laufer, 1950) of turbulence show that this is indeed what happens, so that the expressions for the scale l, which is merely the area under the correlation curve, has a finite value.

The quantity, l, may be thought of as a measure of the size of the eddies that make up the turbulence. Measurements of $R(\xi)$ made in wind tunnels with turbulence produced by grids show that the resulting value of l is proportional to the size of the units making up the grid or of other devices that produce the turbulence. Because apparatus for making such measurements in rivers is not perfected, no quantitative information is available regarding l in such flows. However, it appears reasonable to expect from visual observations of rivers that the maximum size of the eddies will vary with the size or, e.g., depth, of the stream. That is, other conditions being equal, the larger, or the deeper flow will have the larger eddies.

25. Diffusion in Turbulent Flow.—If a small amount of dye is injected instantaneously into a steady uniform turbulent flow, it will be seen to spread as it is carried downstream with the mean velocity, $\bar{u}$, as indicated in Fig. 2.28. At time, $t = 0$, the instant of injection, the dye will occupy a relatively small volume as shown at point "a". At a later time, $t = t$, the dye will have moved downstream an average distance equal to $\bar{u}t$ and the volume occupied by the dye will have increased and changed shape. The increase in size and change in shape is due to the turbulence components of the velocities. Because these components have zero mean values, the increase in volume cannot be due to simple transport or flow, but must result from the more involved diffusion mechanism. The diffusion mechanism involves two essential features. One is the simple transport that results from the turbulent-velocity fluctuations, and the other is a mixing of the transported fluid with its new surroundings. Without this mixing, the volume of fluid that

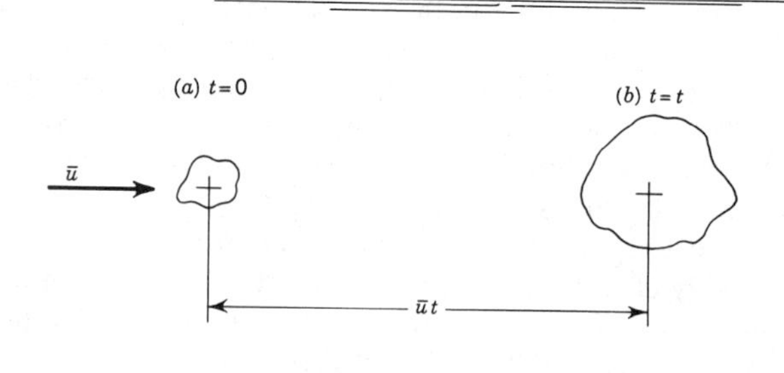

FIG. 2.28.—Diffusion in Turbulent Flow

contains the dye could be broken up and dispersed but it could not be increased. The result of transport without mixing would be a series of small regions of dyed fluid interspersed with clear fluid.

At the periphery of the dyed volume where dyed fluid is transported into the clear fluid, it is simple to see that dye is carried outward away from the center of the dyed volume. Outward transport of dye also occurs within the dyed volume when a small amount of dyed fluid is transported to a region with lower dye concentration and mixed, thus raising the average local concentration. In the example previously considered, the core of the volume will, on the average, have higher dye concentration than the fringes, and the flow of dye due to diffusion will be outward. Thus, it is seen that in the diffusion process, the flow of dye is in the direction of decreasing concentration, or opposite to the concentration gradient, and that the tendency is to equalize the concentration. In case the concentration is uniform throughout the entire fluid, it is clear that there will be no net transport of dye even though vigorous transport of fluid and mixing are occuring.

Transport of dye was selected as a convenient example to illustrate the diffusion mechanism, but any other substance or property that can be carried by the fluid also will be diffused in the fluid by the same process that transports the dye. For instance, suspended sediment, heat, and momentum are all transported in turbulent flows by this mechanism.

26. Differential Equation for Suspension of Sediment.—Consider diffusion of suspended-sediment particles of uniform size, shape, and density in a two-dimensional, uniform, turbulent flow. Let the mean flow velocity, $\bar{u}$, be horizontal in the x-direction, and let y be vertical, as shown in Fig. 2.29. Because the flow is two-dimensional and uniform, the average concentration of sediment in the flow does not change in the z and x-directions and no diffusion will occur in these directions. The only diffusion will be in the y-direction, or vertically. An expression can now be written for the rate of upward transport of sediment through an elementary horizontal area, $dx\, dz$. At any instant of time, the upward velocity through this element of area will be v', which as explained previously, fluctuates with time in magnitude and sign and has the average value zero. The flow rate through this area will be $v'\, dx\, dz$, and the instantaneous rate of transport of sediment will be $v'c\, dx\, dz$, in which c is the instantaneous value of the sediment concentration that like the instantaneous velocities also will fluctuate with time. To get the transport per unit area, divide by $dx\, dz$ which gives $v'c$, and to get the average transport, g_1, per unit area the time average of $v'c$ is taken. The result can be written

$$g_1 = \overline{v'c} \qquad (2.49)$$

in which the bar denotes mean value or time average. The instantaneous concentration can be written

$$c = \bar{c} + c' \qquad (2.50)$$

in which $\bar{c}$ denotes the mean value of c; and c' = the fluctuation from the mean value. Substituting for c in Eq. 2.49 yields

$$g_1 = \overline{v'(\bar{c} + c')} = \overline{v'\bar{c}} + \overline{v'c'} \qquad (2.51)$$

in which, again, the bar denotes mean value. But the term, $\overline{v'\bar{c}}$, at the right can be written $\overline{v'}\,\bar{c}$ and is zero because $\bar{v}' = 0$. The equation then becomes

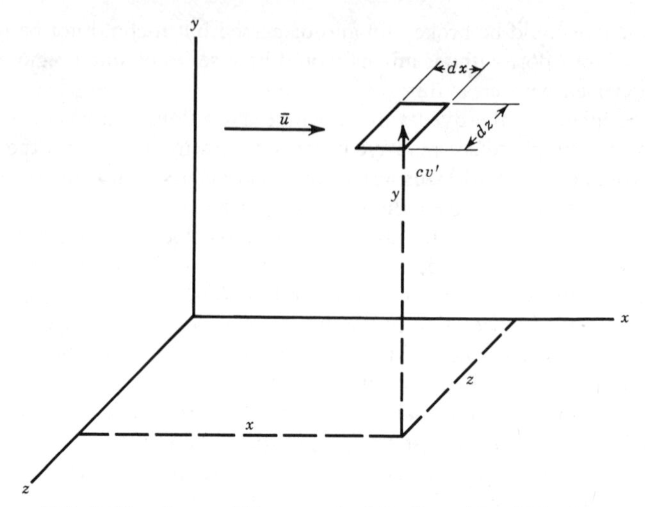

FIG. 2.29.—Upward Transport of Sediment by Diffusion

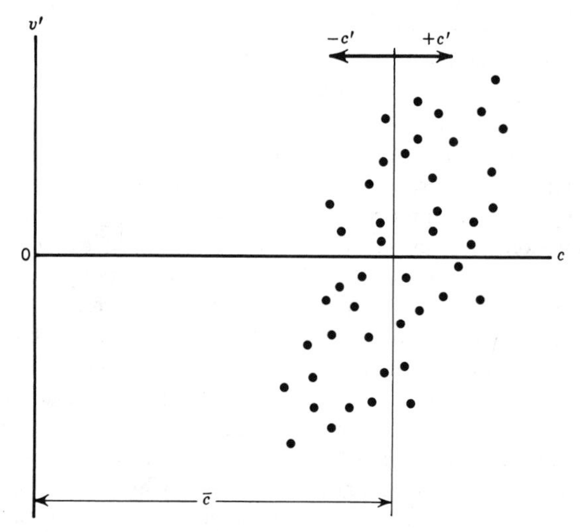

FIG. 2.30.—Schematic Scatter Diagram of Vertical Velocity Fluctuation v' and Instantaneous Sediment Concentration c

$$g_1 = \overline{v'c'} \qquad (2.52)$$

It is clear from the definitions of v' and c' that they can vary in absolute magnitude as well as in sign so that products $v'c'$ of instantaneous values also can vary in magnitude and sign.

In streams carrying suspended sediment, $\bar{c}$ is a function of the distance up from the bed and, because of the gravitational force on the grains, the average concentration decreases with distance up from the bed. In this case, fluid moving upward through the area, $dx\ dz$, will have come from a region in which the average concentration is higher than locally and chances are that it will have a concentration, $\bar{c} + c'$, that is higher than the local average value. The opposite would be true of downward flow. This means that there will be a preponderance of upward or positive v' associated with $+c'$ and of $-v'$ associated with $-c'$. Both of these kinds of events produce positive products $v'c'$ and contribute to the upward transport of sediment. Even when the average concentration, $\bar{c}$, has a definite gradient, as it has in natural streams, it is possible to have some negative values of v' associated with positive values of c' and vice versa, so that there can be some negative values of the product, $c'v'$.

The relation of c' and v' is shown schematically by the scatter diagram in Fig. 2.30. Here each point represents simultaneous values of c' and v' or of c and v', and it can be seen that although the mean values of c' and v' are both zero, their mean product need not be zero.

If, as in Fig. 2.30, the mean product, $\overline{c'v'}$, has a value other than zero, the quantities, c' and v', are correlated and the degree of correlation can be expressed by a correlation coefficient, β_1, which is defined by

$$\beta_1 = \frac{\overline{c'v'}}{\sqrt{\overline{c'^2}}\sqrt{\overline{v'^2}}} \qquad (2.53)$$

in which $\overline{c'^2}$ = the mean of c'^2; and the other symbols are as defined previously. To put Eq. 2.53 into a more convenient form, assume that

$$\sqrt{\overline{c'^2}} = l_1 \left| \frac{d\bar{c}}{dy} \right| \qquad (2.54)$$

in which l_1 = a factor with the dimensions of length that can be thought of as analogous to the so-called mixing length, l_2, which will be defined (Prandtl, 1952); and y = the distance up from the bed. Introducing this assumption into Eqs. 2.52 and 2.53 and eliminating $\overline{v'c'}$ gives

$$g_1 = -\,|\beta_1|\sqrt{\overline{v'^2}}\,l_1\frac{d\bar{c}}{dy} \qquad (2.55)$$

in which the minus sign has been introduced to express the fact, stated previously, that the transport is in the direction of decreasing concentration.

The product, $\beta_1(\overline{v'^2})^{1/2}l_1$, is known as the diffusion coefficient for sediment and is given the symbol, ϵ_s, so that Eq. 2.55 can be written

$$g_1 = -\epsilon_s \frac{dC}{dy} \qquad (2.56)$$

in which $\bar{c}$ has been replaced by C to simplify the notation.

An equation for the shear stress in a two-dimensional flow can be derived that is similar to Eq. 2.55. The local shear stress, τ, in a two-dimensional turbulent flow with local mean velocity $\bar{u}$ in the x-direction is given by the Reynolds equation

$$\tau = -\rho\,\overline{u'v'} \qquad (2.57)$$

in which ρ = the mass density of the fluid and the bar over the product denotes mean value. A correlation coefficient, β_2, between u' and v' can now be defined by

$$\beta_2 = \frac{\overline{u'v'}}{\sqrt{\overline{u'^2}}\sqrt{\overline{v'^2}}} \qquad (2.58)$$

Following a method similar to the one used in deriving Eq. 2.55, it can be assumed that

$$\sqrt{\overline{u'^2}} = l_2 \left| \frac{d\bar{u}}{dy} \right| \qquad (2.59)$$

which, when introduced into Eq. 2.57 for the shear stress along with Eq. 2.58 gives

$$\tau = -\rho\, \beta_2 \sqrt{\overline{v'^2}}\, l_2 \frac{d\bar{u}}{dy} \qquad (2.60)$$

The quantity, l_2, is the Prandtl mixinglength that is defined as the average distance that an element of fluid moves transverse to the flow before it mixes with its new surroundings. Eq. 2.60 may be written

$$\tau = \epsilon_m \frac{d(\rho\bar{u})}{dy} \qquad (2.61)$$

in which $\epsilon_m = |\beta_2| \sqrt{\overline{v'^2}}\, l_2$ .... (2.62)

is the kinematic eddy viscosity or the diffusion coefficient for momentum, the derivative is the momentum gradient; and the sign of the right side of the equation has been changed to agree with the convention that positive shear stress is produced by positive velocity gradient. As considered at greater length subsequently, ϵ_s and ϵ_m do not appear to have the same value. This may also be interpreted to mean that the correlation coefficients, β_1 and β_2, and the mixing lengths, l_1 and l_2, also are different. This appears reasonable, because it is difficult to imagine that the mechanisms of transport of sediment and momentum are identical.

Because in deriving Eq. 2.56, the flow is assumed to be steady, the average concentration at any level will be constant and the net average flow of sediment through a horizontal area will be zero. Therefore, the upward flow due to diffusion will be balanced by settling of the sediment due to its weight. The settling rate per unit area is Cw in which w = the terminal settling velocity of the individual sediment particles. Equating the upward transport due to diffusion to the downward transport due to settling yields

$$Cw + \epsilon_s \frac{dC}{dy} = 0 \qquad (2.63)$$

Eq. 2.63 holds for two-dimensional steady uniform flow and applies to sediment with settling velocity w. The diffusion coefficient, ϵ_s, is generally a function of y that must be known before Eq. 2.63 can be solved for C. Eq. 2.63 was developed by Wilhelm Schmidt (1925) in connection with studies of dust in the atmosphere and by M. P. O'Brien (1933) in studies of suspended sediment in streams.

To derive the equation for unsteady, nonuniform distribution of sediment in a two-dimensional steady uniform flow, an expression is developed which states that in a small time, Δt, the flow of sediment into an element of volume minus the flow out is equal to the change in concentration in the volume. Fig. 2.31 indicates the flow of sediment in time Δt into and out of the element of volume in the x and y-directions due to flow of the water and to diffusion. The width of the element normal to the xy-plane is taken to be unity and the x-axis is taken to be horizontal. In Fig. 2.31 $\bar{u}$ and $\bar{v}$ are the components of the flow velocity parallel to the x-axis and y-axis, respectively, and ϵ_x and ϵ_y are the diffusion coefficients for sediment in these same directions.

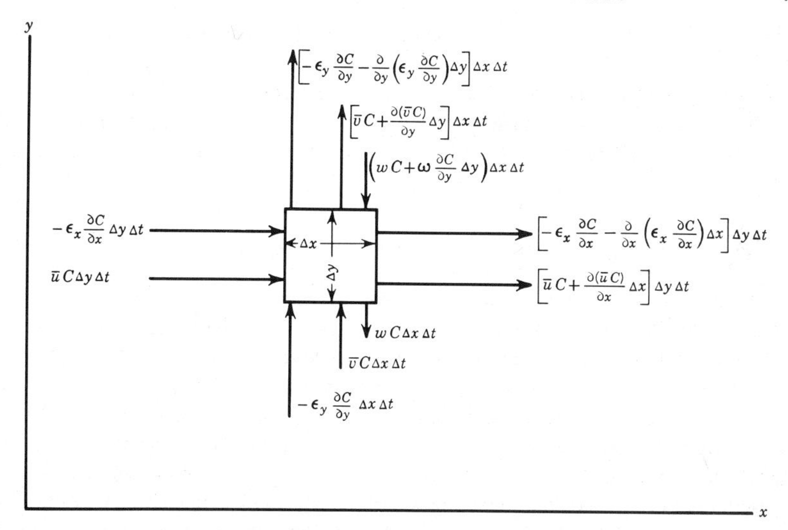

FIG. 2.31.—Transport of Sediment into and out of Element of Volume in Two-Dimensional Flow for Horizontal X-Axis

The flow of sediment due to settling under gravitational attraction is denoted by the two terms containing the settling velocity, w. There is no contribution of sediment through the faces parallel to the xy plane because the mean velocity and the concentration gradient normal to the faces are both zero. The differential equation for the concentration can be written by equating the contributions of sediment from the x and y-directions to the increase in concentration within the elementary volume in time Δt, as follows

$$\left[-\frac{\partial}{\partial x}(\bar{u}C) + \frac{\partial}{\partial x}\left(\epsilon_x \frac{\partial C}{\partial x}\right) - \frac{\partial}{\partial y}(\bar{v}C) + \frac{\partial}{\partial y}\left(\epsilon_y \frac{\partial C}{\partial y}\right) + w\frac{\partial C}{\partial y}\right]\Delta y\,\Delta x\,\Delta t$$

$$= \frac{\partial C}{\partial t}\Delta x\,\Delta y\,\Delta t \quad \text{.......} \quad (2.64)$$

Dividing through by $\Delta x\,\Delta y\,\Delta t$, expanding, and noting that $\partial u/\partial y + \partial v/\partial y = 0$, gives

$$-\bar{u}\frac{\partial C}{\partial x} - \bar{v}\frac{\partial C}{\partial y} + \epsilon_x \frac{\partial^2 C}{\partial x^2} + \frac{\partial \epsilon_x}{\partial x}\frac{\partial C}{\partial x} + \epsilon_x \frac{\partial^2 C}{\partial y^2}$$

$$+ \frac{\partial \epsilon_y}{\partial y}\frac{\partial C}{\partial y} + w\frac{\partial C}{\partial y} = \frac{\partial C}{\partial t} \quad \text{.......} \quad (2.65)$$

If the sediment distribution is steady and uniform, and the mean flow is horizontal, i.e., in the x-direction, $\partial C/\partial t = 0$; $\bar{v} = 0$ and all derivatives with respect to x are also zero. Putting these conditions into Eq. 2.65 letting $\epsilon_y = \epsilon_s$ and integrating with respect to y yields Eq. 2.63.

It will be noted that in deriving the differential equations for the distribution of suspended sediment, no forces were considered, but only kinematic considerations were involved. Therefore, it must be expected that dynamic effects of the sediment that must be present will not be accounted for by the theory as represented by Eqs. 2.63 and 2.65.

27. Equations for Distribution of Suspended Sediment in Turbulent Flow.—If the turbulence in a two-dimensional uniform flow is uniform from top to bottom the diffusion coefficient, ϵ_s, would be constant and Eq. 2.63 can be integrated to yield

$$\frac{C}{C_a} = \exp\left[-\frac{w}{\epsilon_s}(y - a)\right] \qquad (2.66)$$

in which C_a = the concentration at the level, $y = a$. The turbulence, and thus the diffusion coefficient in streams is not constant over the depth, and Eq. 2.66 would not be expected to apply. However, Hurst (1929) and Rouse (1938) achieved a uniform distribution of turbulence by mechanically agitating water in a small cylindrical tank, called a turbulence jar by Rouse. Hurst found that the distribution of sand in his experiments followed an exponential relation with elevation above the bed, and varied with sand size and rotation speed of the propellers that he used to agitate the water. Hunter Rouse used a coarse mesh grid reciprocating in the vertical direction to produce the uniform turbulence. His results agreed with Eq. 2.66, including the explicit dependence of the process on the fall velocity, w, of the grains, and thus established the validity of the basic theory. This work also furnished experimental evidence that the mixing of the fluid and sediment were not the same, a question that is not yet resolved, as will be seen from the discussion of the value of β in Eq. 2.69.

Dobbins (1944) made experiments in an apparatus like that of Rouse, except that Dobbins studied the transient condition in going from one steady-state condition to another. A transient was begun once a steady state was established, by dropping the cylindrical tank containing the water a small distance with respect to the agitator while the experiment was in progress. This increases the distance from the sand bed in the tank to the lowest member of the agitator, thus reducing the rate of pickup of sediment, and causing a gradual decrease in the amount of sand in suspension. This transient condition is described by the one-dimensional diffusion equation for unsteady conditions. This is obtained from Eq. 2.65 by noting that in the turbulence jar, $\bar{u}$, $\bar{v}$, and all derivatives with respect to x are zero, and that $\epsilon_y = \epsilon_s$ = constant. The resulting equation is

$$\epsilon_s \frac{\partial^2 C}{\partial y^2} + w\frac{\partial C}{\partial y} = \frac{\partial C}{\partial t} \qquad (2.67)$$

Dobbins obtained the solution of Eq. 2.67 by conventional methods and found that it agreed well with his measurements of concentration profiles at various times. In this solution, the known concentration profile at the beginning of the experiment was taken as the initial condition. The two boundary conditions were that there was no net transfer of sediment through the water surface and that the rate of pickup from the bottom was constant. Camp (1946) applied Dobbins' solution of Eq. 2.67 to the problem of the retardation of the settling of particles in sedimentation basins due to turbulence resulting from fluid friction on the boundaries of basins. To do this, Camp made the transformation

$$x = Vt \qquad (2.68)$$

in which x = the distance from the inlet of the basin; and V = the mean forward velocity of flow in the basin.

In a stream, the coefficient, ϵ_s, is an unknown function of y, the distance up from the bottom. Kalinske and Pien (1943) have measured the diffusion coefficient in a water flow by applying the theory of Taylor (1921) to analyze observations of the spread of droplets of a liquid mixture that had the same density as the water. Using these measured values of the diffusion coefficient, Kalinske was able to predict the observed distribution of sediment concentration in the same flow.

To obtain an expression for ϵ_s the assumption is made that

$$\epsilon_s = \beta\,\epsilon_m \qquad (2.69)$$

in which β = a numerical constant; and ϵ_m = the coefficient for momentum exchange defined in Eq. 2.62. The reciprocal of β is often called the turbulent Schmidt number by workers concerned with diffusion of matter in turbulent flows (Daily and Harleman, 1966).

For a two-dimensional, steady, uniform flow in an open channel, τ can be expressed by

$$\tau = \gamma\,S\,(d - y) \qquad (2.70)$$

in which γ = the specific weight of the liquid; S = the slope of the channel; and d = the depth of the flow. Setting $y = 0$, the preceding expression gives the shear stress at the bed

$$\tau_o = \gamma\,d\,S \qquad (2.71)$$

and τ can be expressed in terms of τ_o by

$$\tau = \tau_o\left(\frac{d - y}{d}\right) \qquad (2.72)$$

If the velocity distribution, $U(y)$, is known, ϵ_s can be expressed as a function of y by means of Eqs. 2.61, 2.72, and 2.69, and Eq. 2.63 can then be solved for C. (For convenience of notation, $\bar{u}$ has been replaced by U.) A convenient expression for U is the well-known Prandtl-von Karman velocity defect law

$$\frac{U - U_{max}}{\sqrt{\frac{\tau_o}{\rho}}} = \frac{2.3}{k}\log\frac{y}{d} \qquad (2.73)$$

in which U_{max} = the maximum velocity over the depth, i.e., the surface velocity; and k = von Kármán's universal constant having a mean value of 0.4 for clear fluids. From Eq. 2.73

$$\frac{dU}{dy} = \frac{\sqrt{\frac{\tau_o}{\rho}}}{ky} = \frac{U_*}{ky} \qquad (2.74)$$

in which $U_* = \sqrt{\tau_o/\rho}$ is known as the shear velocity. By means of Eqs. 2.61, 2.72, and 2.74, there follows

$$\epsilon_m = k\,U_*\,\frac{y}{d}\,(d - y) \qquad (2.75)$$

and from Eq. 2.69

$$\epsilon_s = \beta k U_* \frac{y}{d}(d - y) \quad \text{(2.76)}$$

Substituting Eq. 2.76 into Eq. 2.63 and separating the variables and integrating gives

$$\frac{C}{C_a} = \left(\frac{d - y}{y} \frac{a}{d - a}\right)^z \quad \text{(2.77)}$$

in which, as in Eq. 2.66, C_a denotes the concentration of sediment with settling velocity w at the level, $y = a$, and

$$z = \frac{w}{\beta k U_*} \quad \text{(2.78)}$$

Eq. 2.77 was developed by Rouse (1937a) and will be referred to as the Rouse equation. The exponent, z, in Eq. 2.77 has been called the Rouse number in some French literature in recognition of this important contribution. This equation gives the concentration, C, at any level, y, in terms of C_a, the concentration at some arbitrary level $y = a$, and, therefore, does not give an absolute value of C.

In presenting Eq. 2.77 Rouse (1937a) refers to a similar development that was being carried out concurrently by A. T. Ippen but that has been published only recently (Ippen, 1971). It is based on the Krey (1927) equation for the velocity profile which is

$$\frac{U}{U_{max}} = \frac{\ell n\left(1 + \frac{y}{a}\right)}{\ell n\left(1 + \frac{d}{a}\right)} \quad \text{(2.79)}$$

in which a, a small distance from the bed, is defined by

$$\frac{U_{max}}{U_*} = \frac{U_* a}{\nu} \ell n\left(1 + \frac{d}{a}\right) \quad \text{(2.80)}$$

From these equations

$$\frac{dU}{dy} = \frac{U_* a}{\nu} \frac{U_*}{(y + a)} \quad \text{(2.81)}$$

and by comparison with Eq. 2.74 it is seen that for a, much smaller than y, $U_* a/\nu$ is equivalent to $1/k$. The solution of Eq. 2.63 with the aforementioned value of dU/dy gives

$$\frac{C}{C_c} = \left(\frac{d - y}{y + a} \frac{2a}{d}\right)^{z_1} \quad \text{(2.82)}$$

in which

$$z_1 = \left(\frac{w}{\beta U_*}\right)\frac{U_* a}{\nu} \quad \text{(2.83)}$$

and C_a = concentration at distance, a, from the bed; and a is given by Eq. 2.80 instead of being chosen arbitrarily as in Eq. 2.77.

Fig. 2.32 shows a graph of the Rouse equation for several values of the exponent, z. It is seen that for low values of z the concentration tends towards becoming uniform over the depth, and that for large z values, the concentration is

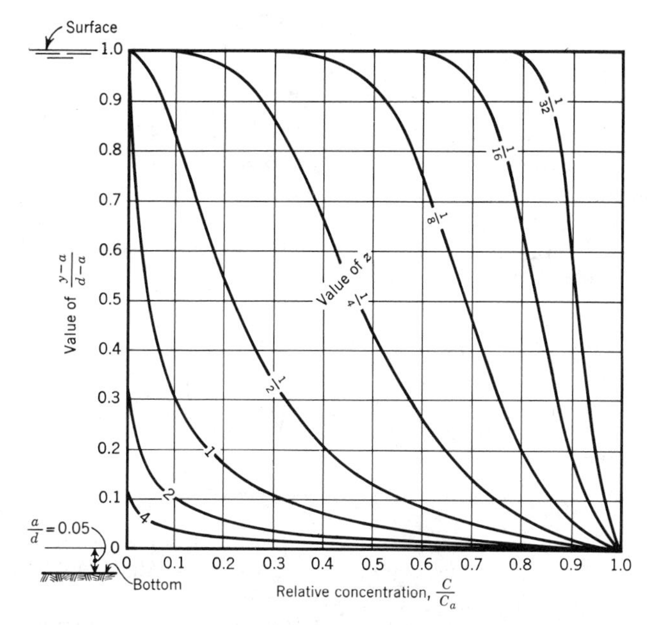

FIG. 2.32.—Graph of Rouse Suspended Load Distribution Equation for $a/d = 0.05$ and Several Values of z

small near the surface and relatively high near the bottom. Referring to Eq. 2.78, it is seen that for a given shear stress, i.e., in a given flow at a given time, z is proportional to the settling velocity, w, so that the small grains will have small z and be relatively uniformly distributed over the depth of the stream and the coarse grains with large z will be concentrated near the bottom.

Hunt (1954) has developed the following differential equation for the distribution of suspended sediment in two-dimensional steady uniform flow

$$\epsilon_s \frac{dc_v}{dy} + c_v \frac{dc_v}{dy}(\epsilon_w - \epsilon_s) + (1 - c_v)c_v w = 0 \quad \text{(2.84)}$$

in which ϵ_w = the diffusion coefficient for water such that the rate of flow of water into a unit area normal to the velocity U is $U(1 - c_v) - \epsilon_w(\partial/\partial x)(1 - c_v)$; c_v = the sediment concentration by volume; and the other terms are as defined previously. When $\epsilon_w = \epsilon_s$, Eq. 2.84 becomes

$$\epsilon_s \frac{dc_v}{dy} + (1 - c_v)c_v w = 0 \quad \text{(2.85)}$$

that was also derived by Halbronn (1949) and which differs from Eq. 2.63 only in that the second member contains the quantity $(1 - c_v)$. When c_v is negligible compared with unity, Eq. 2.85 becomes the same as Eq. 2.63. The quantity $(1 - c_v)$ results by taking into account the volume of the sediment in setting up expressions for flow of sediment similar to those shown in Fig. 2.29. The diffusion coefficient, ϵ_s, used by Hunt was derived as was done for Eq. 2.76, except that dU/dy was obtained from

$$\frac{U - U_{max}}{U_*} = \frac{1}{k}\left(\sqrt{1 - \frac{y}{d}} + B \ln \frac{B - \sqrt{1 - \frac{y}{d}}}{B}\right) \quad (2.86)$$

in which B = a constant to be determined from experimental data. If it is assumed that the velocity gradient dU/dy is infinite at the bed, the constant $B = 1$ and Eq. 286 become the von Kármán equation (1934) resulting from his well-known similarity hypothesis. Introducing the expression for ϵ_s derived from Eq. 2.86 and integrating gives Hunt's equation for the distribution of suspended sediment

$$\left(\frac{c_v}{1 - c_v}\right)\left(\frac{1 - c_a}{c_a}\right) = \left[\sqrt{\frac{1 - \frac{y}{d}}{1 - \frac{a}{d}}}\left(\frac{B_s - \sqrt{1 - \frac{a}{d}}}{B_s - \sqrt{1 - \frac{y}{d}}}\right)\right]^q \quad (2.87)$$

$$\text{in which } q = \frac{w}{k_s B_s U_*} \quad (2.88)$$

and B_s and k_s are constants similar to B and k, which are to be determined from measurements of sediment distribution. Using Vanoni's data (1946), Hunt showed that Eq. 2.87, with B_s and k_s determined for best fit agreed more closely with measured concentration than did Eq. 2.77. Values of B_s were found to be between 0.990 and 1.000 and k_s was between 0.31 and 0.44. The fact that Eq. 2.87 has two arbitrary constants whereas Eq. 2.77 has only one will make possible a better fit of data with Eq. 2.87. Eq. 2.87 has not come into use, probably because of its complicated form and the added difficulty of having two instead of one constant. Paintal and Garde (1964) plotted flume data for a flat bed and a dune bed according to Eq. 2.77 and the Hunt Eq. 2.87. They concluded that the two equations fitted the data equally well and, therefore, that Eq. 2.77 was preferable because it is simpler.

Einstein and Chien (1952) proposed several modifications of the theory on which Eq. 2.77 is based, in an attempt to explain discrepancies between river data and the theory. The modifications all involved changes in the concepts of the details of the turbulent exchange process. One of these, outlined briefly herein, illustrates some of the concepts used by these authors. This theory is based on the idea of the mixing length. In this concept it is assumed that the fluid and sediment move vertically through a horizontal area at level y; a distance that, on the average, is equal to the mixing length and then mixing occurs with the new surroundings. In addition, the present theory of Einstein and Chien is based on the following model or assumptions: (1) Upward flow through a unit horizontal area at level y begins at a level, $y - A_1 \ell$, and downward flow starts at level $y + (1 - A_1)\ell$ in which ℓ = the mixing length; and A_1 = a numerical factor with value less than unity; (2) the mean upward and downward flow velocity of suspended sediment grains is, respectively, $v' - w$ and $v' + w$, in which w = the fall velocity of the grains; and v' = the vertical water velocity due to turbulence; and (3) the average concentration of sediment in the upward flow is $C - A_1 \ell (dC/dy)$ and in the downward flow is $C + (1 - A_1)\ell (dC/dy)$ in which C = the concentration at level y. Using these relations, it is possible to write $C - A_1\ell(dC/dy)(v' - w)$ for the rate of upward flow of grains per unit area and

$[C + (1 - A_1)\,\ell\,(dC/dy)](v' + w)$ for the rate of downward flow. Setting these two rates equal to each other for steady-state conditions, taking mean value, and introducing the following from Prandtl's theory of turbulence

$$\overline{v'\ell} = 2kU_*\left(\frac{d-y}{d}\right)y \qquad (2.89)$$

and
$$\ell = B_1 ky\sqrt{\frac{d-y}{d}} \qquad (2.90)$$

in which B_1 = a dimensionless factor, gives Einstein's and Chien's differential equation for distribution of suspended sediment

$$wC + kyU_*\left(\frac{d-y}{d}\right)\frac{dC}{dy} + Nwky\left(\frac{d-y}{d}\right)^{1/2}\frac{dC}{dy} = 0 \qquad (2.91)$$

in which
$$N = B_1\left(\frac{1}{2} - A_1\right) \qquad (2.92)$$

Solution of Eq. 2.102 yields

$$\frac{C}{C_a} = \left[\frac{1-\sqrt{\frac{d-a}{d}}}{1-\sqrt{\frac{d-y}{d}}}\right]^{\frac{z_1}{1+Nkz_1}}\left[\frac{1+\sqrt{\frac{d-a}{d}}}{1+\sqrt{\frac{d-y}{d}}}\right]^{\frac{z_1}{1-Nkz_1}}\left[\frac{\sqrt{\frac{d-a}{d}}+Nkz_1}{\sqrt{\frac{d-y}{d}}+Nkz_1}\right]^{\frac{2z_1}{N^2k^2z_1^2-1}} \qquad (2.93)$$

in which
$$z_1 = \frac{w}{kU_*} \qquad (2.94)$$

For very small z, Eq. 2.93 reduces to Eq. 2.77. To make Eq. 2.93 useful, further studies are needed to establish its validity, and the values of the constants N and k to be used in it.

Tanaka and Fugimoto (1958) have also developed a theory for the distribution of suspended sediment. Their relation appears in Paintal and Garde (1964).

Bagnold (1954) has advanced the idea that sediment grains can be suspended in a flow by intergranular collisions alone without the aid of turbulence. Because the grains are carried along by the fluid, their forward velocity will increase with distance from the bed. Due to this variation in forward velocity, collisions of grains at a particular level with those above will impart forward and downward

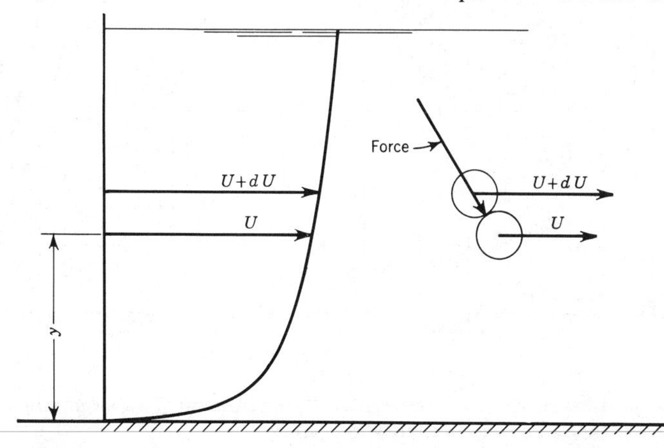

FIG. 2.33.—Force on Suspended Grains due to Collisions

TABLE 2.9.—Data for Curves of Fig. 2.34

Exponent, z (1)	Flow depth, d, in feet (2)	Slope, in feet per foot (3)	Size of suspended sand, in millimeters (4)	Concentration at reference level y = 0.05d, in grams per liter (5)	Site of measurements (6)
0.16	9.9	0.000125	0.044-0.062	0.134	Missouri River at Omaha 10/18/51
0.32	0.590	0.00125	0.10	6.95	Laboratory
0.43	7.7	0.000121	0.062-0.074	0.240	Missouri River at Omaha 10/17/51
0.56	0.295	0.00125	0.10	3.20	Laboratory
0.81	0.295	0.00125	0.10	17.0	Laboratory
1.12	7.7	0.000121	0.149-0.210	2.53	Missouri River at Omaha 10/17/51
1.93	0.600	0.00125	0.208-0.295	0.56	Laboratory

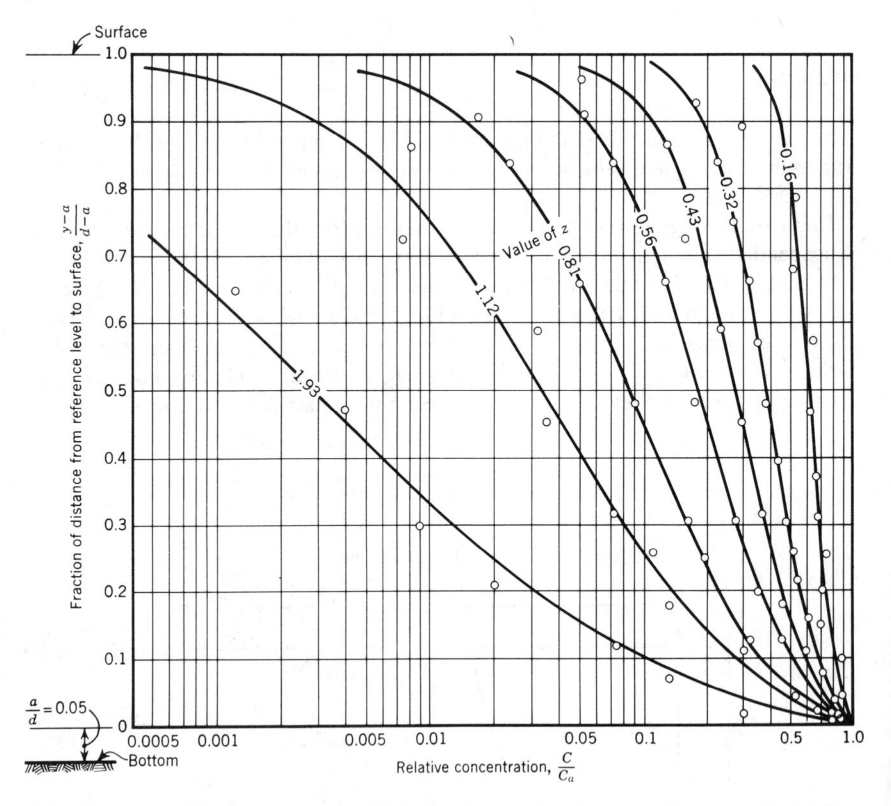

FIG. 2.34.—Vertical Distribution of Relative Concentration C/C_a Compared with Eq. 2.77 for Wide Range of Stream Size and z Values

momentum to the grains, and collisions with grains below will impart backward and upward momentum, as shown in Fig. 2.33. Continuous impacts of this kind will develop stresses with vertical and horizontal components that will be transmitted to the bed. The horizontal stress will be observed as a shear stress.

The existence of these stresses was demonstrated by Bagnold (1954) in experiments with grains with the same specific weight as the water in the annular space between two concentric cylinders in which the outer cylinder rotated and the inner one was fixed. By measuring the torque on the fixed cylinder and the pressure on the wall of the fixed cylinder, the normal and tangential components of the stresses due to grain collision were observed. These stresses were found to bear a constant ratio to each other and to become very small when the grain concentration fell to approx 25% by volume, which is extremely high for streams. Bagnold (1955) also made experiments with grains only slightly denser than water in a small flume, and observed effects that he interpreted to confirm the existence of stresses due to grain collisions.

Because grain collisions are important only at very high concentrations, the effects of such collisions advanced by Bagnold can be appreciable only very near the bed and normally should not be of any consequence in the suspension process in the main body of streams.

28. Agreement between Suspended Load Distribution Equation and Measurements.—Measurements of the distribution of suspended sediment have been made in streams by Anderson (1942) and United States Army Corps of Engineers (1951); in canals by Bender (1956); and in flumes by Vanoni (1946), Brooks (1954), and Barton and Lin (1955). Fig. 2.34 is a graph of Eq. 2.77 for several values of z showing the measured data from experiments in a flume by Vanoni (1953b), and in the Missouri River as indicated in Table 2.9. The z values used are those that give the best fit of Eq. 2.77, and thus the graph indicates only how well data fit the form of the equation. It is seen that the data well fit the form of the theoretical curves despite the large variation in size of the flows in which the measurements were made.

A convenient way to present sediment distribution data for a vertical in a flow is to plot the concentration against the ratio, $(d-y)/y$, on a logarithmic graph in which d is the flow depth and y is the distance up from the stream bed to the sampling point. Fig. 2.35 is such a graph for three of the sets of flume measurements that are plotted on Fig. 2.34. The fact that the data plot as straight lines on Fig. 2.35 shows that they can be represented by an equation of the form of Eq. 2.77. From Eq. 2.77, it can be seen that the value of z that fits any set of data is merely the reciprocal of the slope of the straight line that fits the data on a graph, e.g., Fig. 2.35, where the unit of length is the logarithmic cycle on the graph.

The exponents, z, used to determine the curves of Fig. 2.34 were obtained from graphs like Fig. 2.35, and not from theory. Thus it is seen that the theory does not predict the distribution of relative concentration. However, it is clear from Figs. 2.34 and 2.35 that the form of the theoretical distribution equation agrees closely with that of the observed distributions. To predict the distribution of relative concentration, one has to know w, β, and k in Eq. 2.78 for the exponent, z. The value of k is approx 0.4 for clear fluids, but it has been observed to diminish to as low as 0.2 in flows with high concentrations of suspended material. The value of β has been reported to vary in laboratory experiments from approx 1.0–1.5. However, careful laboratory determinations (Brooks, 1954) of this factor in which

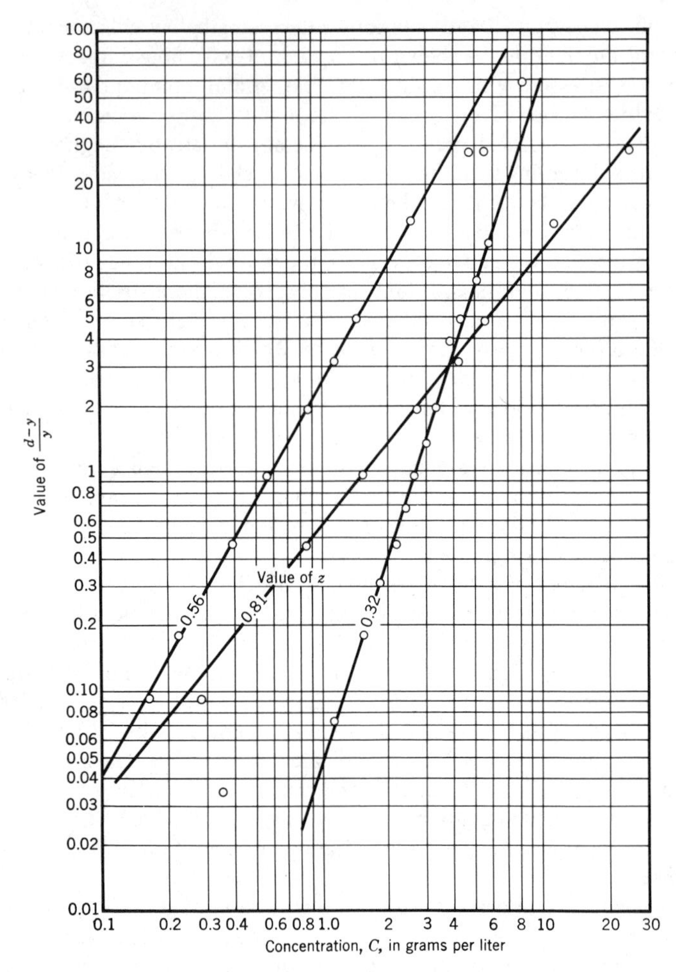

FIG. 2.35.—Logarithmic Graph of Concentration against $(d - y)/y$ for Three of Flume Measurements Plotted on Fig. 2.34

corrections have been made for side-wall effect on the bed shear stress, τ_o, and for concentration on the settling velocity, w, of the sediment, indicate that β is close to unity. Both of these quantities will be examined at greater length in the next article.

Einstein and Chien (1954) determined values of the exponent, z, for the Missouri River at Omaha, Neb., and for the Atchafalaya River, in Mississippi, between Simmesport and Melville by plotting concentration data from point samples as shown in Fig. 2.35. It was found that for low z the values determined by the preceding graphical procedure agreed closely with the quantity, w/ku_*, in which k was determined from measured velocity profiles. However, as z increased, w/ku_* exceeded z by a continuously increasing fraction indicating a continuous increase in β. Their surmise that this discrepancy was due to defects in the theory led them to the study of the theory and the development of Eq. 2.93. This equation gives relations between z and w/ku_* similar to those found by Einstein and Chien for rivers.

29. Effect of Suspended Sediment on Flow Characteristics.—Fig. 2.36 shows semilogarithmic graphs of typical velocity profiles in the Missouri River at Omaha (United States Army Corps of Engineers, 1951) and in a laboratory flume. It is seen that these measurements can be fitted well by a straight line of the form of Eq. 2.73, which has been used in developing the theory of sediment suspension. It can also be seen from Eq. 2.73 that the reciprocal, N, of the slope of the straight line of the graph is

$$N = \frac{k}{2.3\sqrt{\frac{\tau_o}{\rho}}} \qquad (2.95)$$

in which the increment of the ordinate is measured in cycles of the logarithmic scale. To calculate k, one needs in addition to N (which is determined from graphs, e.g., Fig. 2.36) the value of τ_o which in two-dimensional uniform flow is given by

$$\tau_o = \gamma d S \qquad (2.96)$$

This requires the measurement of the slope, which is difficult to do with precision because this involves the determination of a small difference in elevation between two water surfaces where waves and surges are usually occurring. The combination of the difficulties of measuring the slope with those of fitting the curve to the velocity profile graph, required to determine N, results in appreciable scatter in k values.

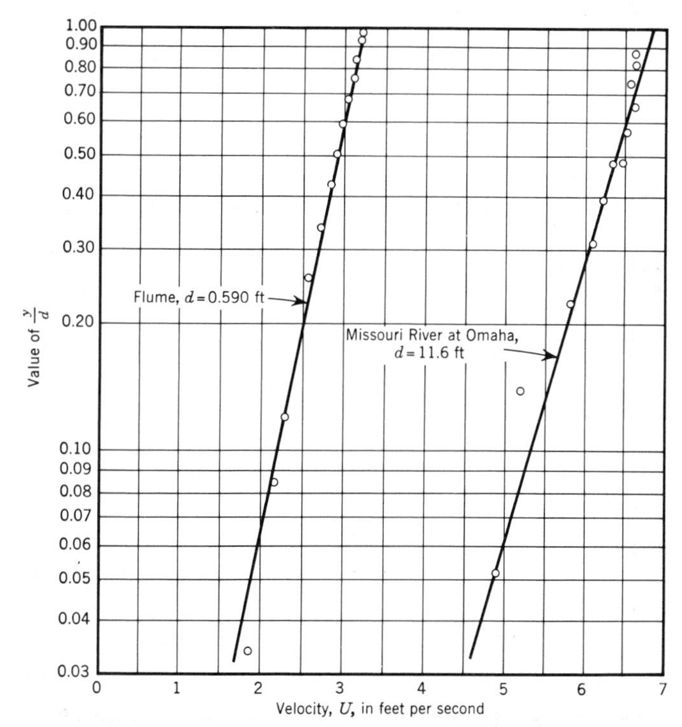

FIG. 2.36.—Semilogarithmic Graphs of Velocity Profiles of Missouri River at Omaha and of Flow in Flume

Fig. 2.37 shows graphs of k against $\overline{C}$, the mean concentration of sediment over the depth in a flume 33-in. wide for four sets of experiments (Vanoni, 1953b). In each set of experiments the slope and depth were kept constant and the sediment load, most of which was carried in suspension, was increased in increments by adding well-sorted sand of 0.10 mm grain size to the flow system. Fig. 2.37 shows clearly that k diminishes from about 0.4 for clear water to as little as 0.2 for the highest concentration. This trend in k to decrease as the concentration of suspended sediment increases has been observed in numerous flume experiments (Vanoni, 1946, 1953b; Brooks, 1954) and in the experiments carried out by Ismail (1952) in a rectangular pipe. A very interesting analysis has been made by Fowler (1953) of velocity profile measurements on the Missouri River that carries a heavy load of suspended sediment, and on the St. Clair and St. Mary's Rivers in the Great Lakes chain, which carry negligible sediment. This analysis showed very definitely that the k values for the sediment-laden Missouri River are appreciably less than for the clear streams, thus confirming the results obtained in laboratory flumes shown in Fig. 2.37. Because all of the evidence is in agreement, it can be concluded that the effect of suspended sediment in a flow is to reduce the value of the von Kármán k below its value for clear fluids.

Referring to Eq. 2.74 for the slope of the velocity profile or velocity gradient, it can be seen that for a given U_* and y, a decrease in k will result in an increase in the velocity gradient, dU/dy. From Eqs. 2.61 and 2.72, it can also be seen that for a given τ_o, d, and y, an increase in dU/dy will result in a decrease in ϵ_m because τ_o remains constant. A similar effect on ϵ_m by a decrease in k is shown by Eq. 2.75. Because in each set of experiments from which the data of Fig. 2.37 were obtained,

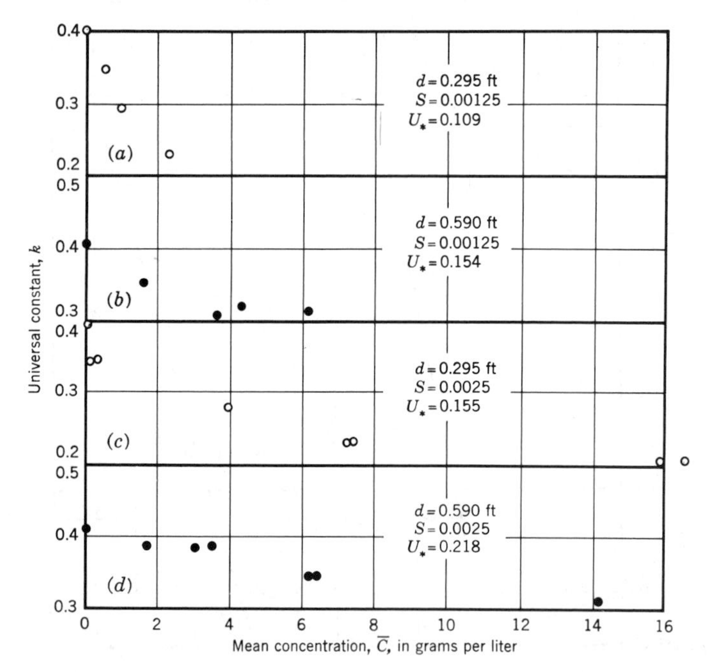

FIG. 2.37.—Variation of Von Karman k with Mean Concentration of Suspended Sand in Flume

the depth and slope, and thus τ_o, U_*, and τ for any y remain constant, a graph of the velocity profiles for different k values in a series will show directly the effects outlined. Fig. 2.38 shows such a graph in rectangular and semilogarithmic coordinates for two of the experiments of the series from which data for Fig. 2.37(c) were obtained. The values of k for each of the profiles and of depth and slope for the series are shown on Fig. 2.38. The quantities, C_m, on the face of the figure are the sediment discharge concentrations of the flow, i.e., the concentration which, when multiplied by the flow rate, will give the total sediment discharge of the flow.

To explain the observed decrease in k and the corresponding decrease in the momentum transfer coefficient, ϵ_m, when sediment is in suspension, Vanoni (1946) has hypothesized that the turbulence is damped by the sediment. It was shown previously that the coefficient, ϵ_m, is of the same form as ϵ_s and therefore that it depends on the root-mean-square of the vertical turbulence velocity, $(\overline{v'^2})^{1/2}$. However, the sediment is actually kept in suspension by the vertical velocity, and the energy to do this must therefore come from the turbulence. This is the same as saying that the turbulence is damped.

The amount of energy per unit time, or the power to hold a grain of sand in suspension is the submerged weight of the grain times the fall velocity of the grain. The power, P_{sw}, to support the sediment of a given size or settling velocity in a column of fluid of unit cross-sectional area and height equal to the depth, d, of the stream is

$$P_{sw} = \left(1 - \frac{\gamma}{\gamma_s}\right) wg \int_0^d C\, dy \qquad (2.97)$$

in which γ and γ_s = the specific weights of the water and sediment, respectively, and, as before, C is the concentration of sediment at the level, y, expressed in mass per unit volume. This can be written

$$P_{sw} = \left(1 - \frac{\gamma}{\gamma_s}\right) wg\overline{C}d \qquad (2.98)$$

in which $\overline{C}$ = the mean concentration over the depth; and thus $\overline{C}d$ = the value of the integral in Eq. 2.97. This expression gives the power required to support the fraction of sediment with mean settling velocity, w. If the suspended load is made up of several size fractions, the total power, P_s, to support this sediment in the column of water is

$$P_s = \Sigma P_{sw} = \left(1 - \frac{\gamma}{\gamma_s}\right) dg\Sigma\overline{C}w \qquad (2.99)$$

in which the summation is performed over all values of w. The power, P_f, to overcome the friction on the column of water is

$$P_f = \gamma d V S \qquad (2.100)$$

in which V = the mean velocity at the vertical; and S = the slope of the stream. The ratio, P_s/P_f, is

$$\frac{P_s}{P_f} = \left(1 - \frac{\gamma}{\gamma_s}\right) \frac{\Sigma\, \overline{C}\, w}{\rho\, V\, S} \qquad (2.101)$$

Einstein and Chien (1952, 1955) have correlated k against P_s/P_f using data of

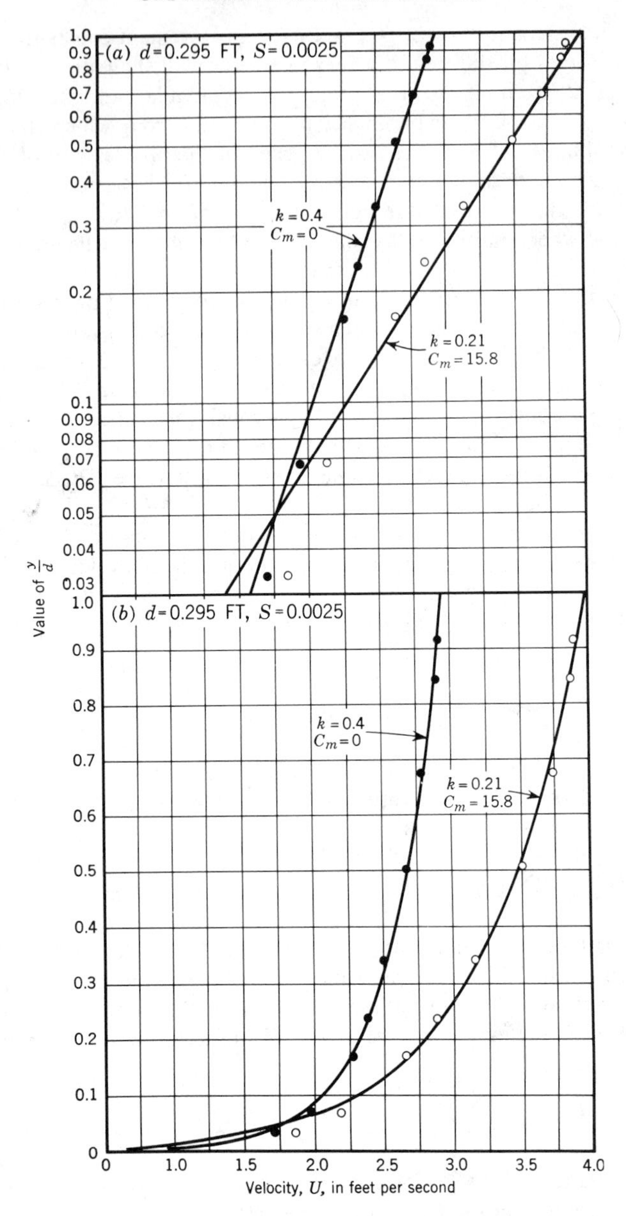

FIG. 2.38.—Semilogarithmic and Linear Graphs of Velocity Profiles in Flow 0.295 ft Deep and 33 in. Wide with Clear Water and with Heavy Suspended Load of 0.1 mm Sand

several investigators. Their results presented in Fig. 2.39 show reasonable correlation and can be used to estimate k. Fig. 2.40(a) is a graph similar to Fig. 2.39 of data obtained from experiments with well-sorted 0.1 mm sand that yielded the data of Fig. 2.37. In Fig. 2.40 (b) the same k values are plotted against P_s'/P_f in which P_s' is the power to support the sediment contained in a column of unit area lying between levels y_1 and y_2 near the bed. The power, P_s', to support sediment of uniform velocity w is given by

$$P_s' = \left(\frac{\gamma_s - \gamma}{\gamma_s}\right) \bar{c}_1 w(y_2 - y_1) \quad \text{.......} \quad (2.102)$$

in which $\bar{c}_1$ = the mean sediment concentration between levels $y = y_1$ and $y = y_2$. In the caluculations for Fig. 2.40 (b), these levels were taken as 0.001 and 0.01 times the depth, d, respectively. The correlation in Fig. 2.40 (b) is better than in Fig. 2.40 (a) and tends to support the ideas of Buckley (1922) and Chien (1955) that the major effect of the sediment on the flow occurs near the bed.

Laursen (1953) has obtained results from flume experiments, with sediment completely covering the bed, that are similar to those described previously, but his explanation of the mechanism by which the observed effects are brought about differs from that of Vanoni. Laursen advances the idea that there are two distinct relationships operating simultaneously. One is that the concentration is correlated with the roughness of the bed, and the other is that k also varies with roughness independently of the sediment concentration. This, then, leads him to the conclusion that k and the concentration are correlated only because both of these quantities are correlated with roughness. In presenting his arguments, Laursen

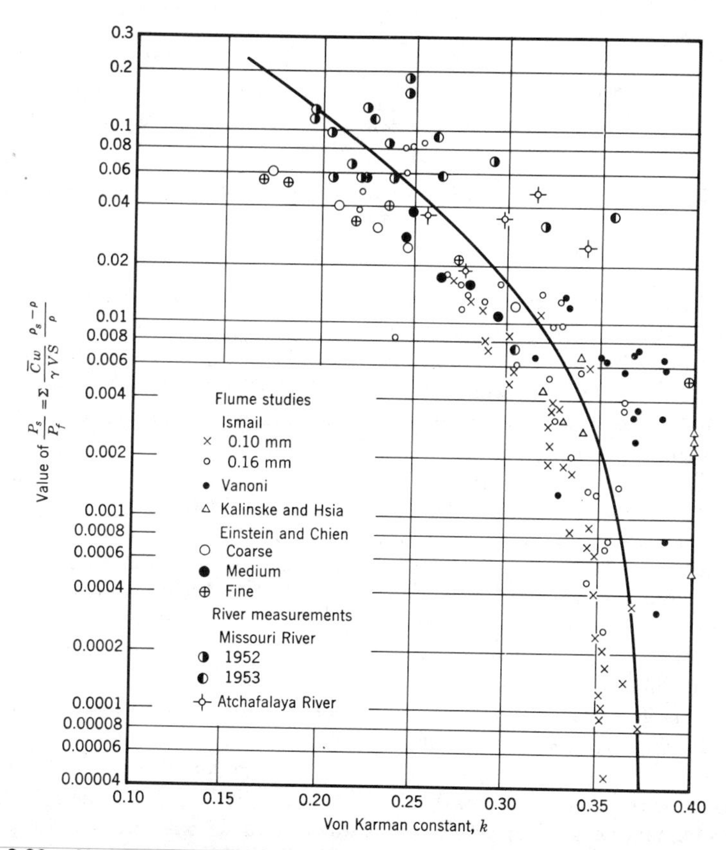

FIG. 2.39.—Variation of Von Karman k with Suspended-Sediment Concentration (Einstein and Chien, 1955; Fig. 15)

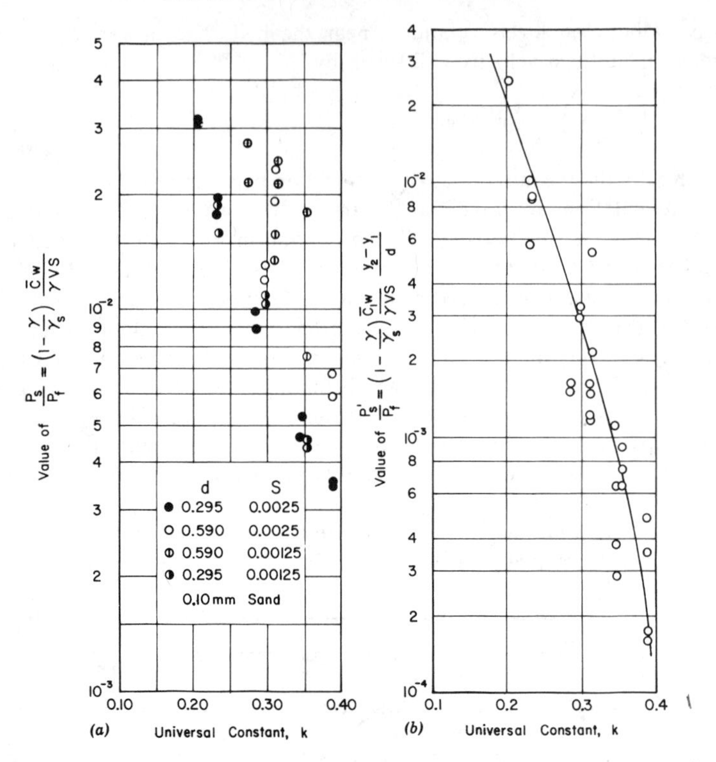

FIG. 2.40.—Variation of Von Karman *k* with Suspended-Sediment Concentration in Flume

refers to the flume experiments of Rand (1953) with clear water and artificial roughness. Rand reported that with roughness on the flume bottom composed of 1/4 in. square bars spaced at 1 in. intervals, the k values were 0.30 and 0.37, respectively, for relative roughnesses of 0.3 and 0.003 where the relative roughness is the ratio of the linear dimension of the roughness element, i.e., 1/4 in., and the water depth.

Paintal and Garde (1964) plotted values for k for flows with suspended load against τ_*, the dimensionless shear stress defined by $\tau_* = \tau_o/(\gamma_s - \gamma)d_{50}$ and found that k tended to diminish with increase in τ_*, although the dispersion in the data was large. The relation between k and τ_* is in agreement with those shown in Figs. 2.39 and 2.40 since the sediment concentration and sediment discharge tend to increase as τ_* increases.

Elata and Ippen (1961) have observed the variation in k and in the mean-square turbulence intensity, $\overline{u'^2}$, in flows of water in a horizontal flume containing relatively large concentrations of plastic spheres with median diameters of 0.12 mm, specific gravity of 1.05, and mean settling velocity in water in excess of 0.1 cm/s. It was found that k diminished as the concentration of the plastic sediment increased, as reported by others. Their analysis also indicated that the main effect of the sediment occurred near the bed as was concluded by Vanoni and Nomicos (1960). On the other hand, it was observed that the presence of sediment tended to increase turbulence rather than to damp it, as postulated by Vanoni (1946). Experimental studies by Daily and Chu (1961) and Daily and Hardison (1964) of water flows in a pipe 2 in. in diameter transporting fine plastic spheres only slightly more dense than water gave the same trends for k and $\overline{u'^2}$ with increase in concentration as reported by Elata and Ippen.

Bohlen (1969) measured turbulence in a free-surface flow of silicone oil 20 cm wide by 1.75 cm deep with a mean velocity of 28.15 cm/s. The measurements were made with and without suspended particles that had a diameter of 0.595 mm and a fall velocity of 0.37 cm/s in the oil. He found that for low particle concentrations (Conc $\leq$ 0.8% by volume) the turbulence level near the bed was less than with clear fluid and that it increased with increase in mean concentration. In the upper parts of the flow the turbulence level tended to increase with increase in concentration for all concentrations. For the highest concentration (Conc = 3.4% by volume) the turbulence intensity exceeded that for clear flow over the entire depth. Bohlen also observed that the gradient of the velocity profiles increased with increase in particle concentration and was smallest for the clear flow, thus indicating that the von Karman constant, k, decreased as concentration increased.

Ippen (1971) developed the following expression for the effect of suspended sediment on the flow

$$\frac{1}{k'} = \frac{1}{k}\left[\frac{1 + 2.5\,C_o}{1 + \overline{C}(s-1)}\right] \qquad (2.103)$$

in which C_o = maximum sediment concentration near the bed in fraction by volume; $\overline{C}$ = mean value of volume concentration in the entire flow; $s = \rho_s/\rho$ = specific gravity of the sediment; k = the von Karman constant for clear fluids; and k' = a factor analogous to the von Karman k. Since the term in parentheses exceeds unity, k' is always less than k and decreases with an increase in C_o. With k = 0.385 the foregoing equation agreed well with flume data. The numerator of the term in parentheses is recognized as the ratio of fluid viscosity with sediment concentration C_o to that of clear fluid according to the classical expression of Einstein. The denominator expresses the increase in bed shear stress due to the presence of suspended sediment. Ippen also employed k' in developing an expression for the velocity profile for flow with suspended sediment and from this derived another equation for distribution of suspended sediment that agreed well with observations.

A complete theoretical treatment of the effect of particles suspended in turbulent flows was presented by Hino (1963). The theory predicts that the von Karman constant diminishes as concentration of suspended particles increases for neutrally buoyant as well as for heavier materials. The theory agrees remarkably well with the k-value observed by Elata and Ippen (1961) with suspensions of neutrally buoyant particles as well as with those with suspended sands by Vanoni (1946), Ismail (1952), and Vanoni and Nomicos (1960). Hino's theory also predicted that the longitudinal turbulence intensity would increase with concentration of suspended neutrally buoyant particles and is in agreement with the experimental results of Elata and Ippen (1961). The theory indicates that suspended particles that are denser than the fluid cause only slight damping of the turbulence and is not in agreement with the results of Bohlen's (1969) experiments.

Ample experimental evidence has been presented to show that the turbulence intensity of a flow containing essentially neutrally buoyant particles tends to increase as concentration of particles increases. The evidence on the effect of particles that are denser than the fluid is scarcer and not so decisive. The one set of measurements on flows with dense sediment indicates that the turbulence level increases over the entire flow only for appreciable concentrations. Further study and more measurements are needed to clarify the effect of dense sediment on the turbulence of streams.

The fact that k varies with amount of suspended sediment means that ϵ_m also varies because, according to Eq. 2.75, ϵ_m is proportional to k. It is now of interest to inquire regarding the effect of concentration on ϵ_s. Eliminating $\beta\, k\, U_*$ between Eq. 2.76 and 2.78 yields

$$\epsilon_s = \frac{w}{z}\frac{y}{d}(d - y) \qquad (2.104)$$

Fig. 2.41 shows z plotted against C_m, the sediment discharge concentration in the flow for the four sets of experiments that yielded the data of Fig. 2.37. The exponent, z, was determined graphically as the slope of a curve of C against $(d - y)/y$ plotted on a logarithmic graph. In three of the experiments, z is seen to increase consistently with C_m; in the fourth series the result is not conclusive. Graphs similar to Fig. 2.41 for experiments in a rectangular pipe (Ismail, 1952) show that z increases with C_m in most cases. This seems to justify the conclusion that z tends to increase as the concentration increases, although the law of increase cannot be established from the data at hand.

Referring to Eq. 2.104, it is seen that an increase in z tends to decrease ϵ_s, the sediment transfer coefficient. However, the settling velocity, w, decreases, and it follows that ϵ_s is decreased by an increase in concentration. As has already been noted, an increase in z will tend to make the distribution of suspended load less nearly uniform. Referring to Eq. 2.78 for z, it is seen that z can be increased by increasing w or, e.g., by increasing the size of the sediment. It can be said then that increasing the concentration (i.e., the sediment discharge) tends to have the same effect on z as increasing the size of the sediment. Actually, the settling velocity of the grains tends to decrease as the concentration increases, so to increase z, the product, βk, must also decrease. As was shown in Fig. 2.37, k decreases as the concentration increases, thus accounting at least for a part of the increase in z. Analysis of results from experiments that gave the data on Figs. 2.40 and 2.41 showed no marked trend in β as the concentration was changed, keeping U_* constant. This finding suggests that if there is a change in β due to a change in the suspended-sediment concentration, it is small in comparison with the relative change observed in k. Therefore, it appears that the increase in z accompanying the increase in the suspended-sediment concentration occurs largely because of the decrease in k.

Eq. 2.77, giving distribution of concentration of suspended load, applies for sediment with a settling velocity, w. If the sediment in a flow is well sorted, the variation in settling velocity of the particles will be small, and it can be expected that the concentration of all particles taken together will be distributed over the depth according to Eq. 2.77. However, in streams the variation in particle size and settling velocity is appreciable, and the distribution of total concentration does not agree with Eq. 2.77. If point samples of suspended load at a vertical of a stream are separated into several size classes, it has been found that the concentration distribution for each class does follow the theory of Eq. 2.77. Reference to Eq. 2.78 for z indicates that the z-values for each size fraction should be proportional

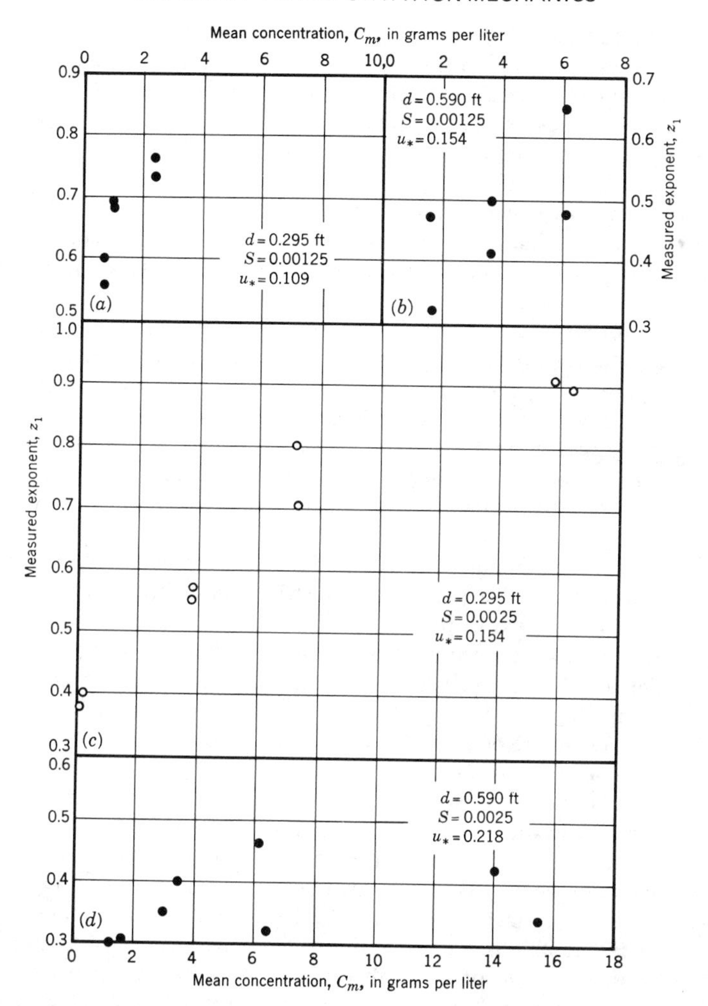

FIG. 2.41.—Variation of Exponent z with Concentration of Sediment in Four Series of Experiments in Flume 33 in. Wide

to the settling velocity, w, of the fraction. Analysis of such suspended-sediment measurements made by the United States Army Corps of Engineers on the Missouri River at Omaha (1951; Vanoni, 1953a), and by Anderson (1942) on the Enoree River show that z-values for the different size fractions at a vertical tend to be proportional to w. This aso shows that β at a vertical is constant or changes little with the size of sediment particles. This kind of study does not shed any light on the effect of concentration on β.

E. Initiation of Motion

30. General.—Water flowing over a bed of sediment exerts forces on the grains that tend to move or entrain them. The forces that resist the entraining action of the flowing water differ according to the grain size and grain size distribution of the sediment. For coarse sediments, e.g., sands and gravels, the forces resisting motion are caused mainly by the weight of the particles. Finer sediments that contain appreciable fractions of silt or clay, or both, tend to be cohesive and resist entrainment mainly by cohesion rather than by the weight of the individual grains. Also, in fine sediments groups of grains are entrained as units whereas coarse noncohesive sediments are moved as individual grains. Cohesion of the fine sediments is a complicated phenomenon that appears to vary with mineral composition and environment. Although the problem of entrainment resistance of cohesive sediments is important, it has received less attention than the similar problem of entrainment of noncohesive sediments. This situation is reflected in the fact that most of this section concerns noncohesive sediments.

When the hydrodynamic force acting on a grain of sediment or an aggregate of particles of a cohesive sediment has reached a value that, if increased even slightly will put the grain or aggregate into motion, critical or threshold conditions are said to have been reached. When critical conditions obtain values of such quantities as the bed shear stress, the stage of a stream or its mean velocity are said to have their critical or threshold values. Under critical conditions, a stream is also said to be competent to move its sediment, i.e., it has reached its competent shear stress, stage, or velocity.

The problem of determining critical conditions for entrainment of sediment has long been considered. Lelliavsky (1955) who gives an excellent review of early work, reports that A. Brahms presented a formula for critical velocity in 1753. This problem is still of significant practical importance. For example, one important application is in the design of earth channels to convey clear water without eroding the boundaries. From a scientific viewpoint, the problem of impending motion is one of the simpler of the many complicated sedimentation problems, and therefore one that might be attacked with hope of success.

31. Analysis, Noncohesive Sediment.—The forces acting on a grain of noncohesive sediment lying in a bed of similar grains over which a fluid is flowing are the gravity forces of weight and buoyancy, hydrodynamic lift normal to the bed, and drag parallel to the bed. The lift is often neglected without proper justification because both analytical (Jeffries, 1929) and experimental studies (Nemenyi, 1940; Einstein and El-Samni, 1949) have established its presence. Most treatments of forces on a grain on a bed consider only drag; lift does not appear explicitly. But, because the constants in the resulting theoretical equations are determined experimentally and because lift depends on the same variables as drag, the effect of lift regardless of its importance is automatically considered.

The forces on a grain on the bed of a stream are depicted in Fig. 2.42, in which ϕ = the slope angle of the bed; and θ = the angle of repose of the sediment submerged in the fluid, and intergranular forces are ignored. When motion is impending, the bed shear stress attains the critical or competent value, τ_c, which is also often termed the critical tractive force. Under critical conditions, also, the particle is about to move by rolling about its point of support. The gravity or weight force is given by $c_1(\gamma_s - \gamma)d_s^3$, in which $c_1 d_s^3$ = the volume of the particle; d_s = its size, usually taken as its mean sieve size; and γ and γ_s are specific weights of fluid and sediment, respectively. The critical drag force is $c_2 \tau_c d_s^2$ in which $c_2 d_s^2$ = the effective surface area of the particle exposed to the critical shear stress, τ_c. Equating moments of the gravity and drag forces about the support point yields

$$c_1 (\gamma_s - \gamma) d_s^3 a_1 \sin(\theta - \phi) = c_2 \tau_c d_s^2 a_2 \cos\theta \qquad (2.105a)$$

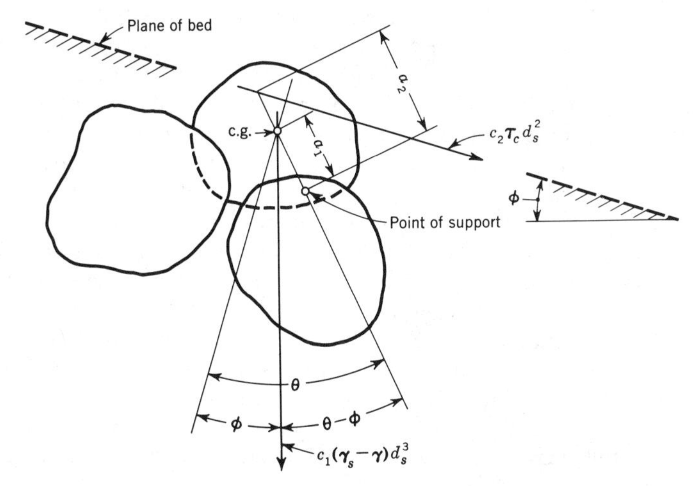

FIG. 2.42.—Forces on Sediment Grain in Bed of Sloping Stream

which, when rearranged, yields

$$\tau_c = \frac{c_1 a_1}{c_2 a_2} (\gamma_s - \gamma)\, d_s \cos \phi\, (\tan \theta - \tan \phi) \qquad (2.105b)$$

which is similar to an equation by White (1940). For a horizontal bed, $\phi = 0$, and Eq. 2.105b becomes

$$\tau_c = \frac{c_1 a_1}{c_2 a_2} (\gamma_s - \gamma)\, d_s \tan \theta \qquad (2.106)$$

When a_1 and a_2 are equal the forces on the grain act through its center of gravity and the fluid forces are caused predominantly by pressure. Also, when a_1 and a_2 are equal it will be seen that the ratio of the forces on the grain acting parallel to the bed to those acting normal to the bed is equal to $\tan \theta$. It will also be seen that the critical shear stress on a horizontal bed is greater than on a downward sloping bed (positive ϕ) but less than on an upward sloping bed. White (1940) has observed that when a particle is in turbulent flow, the resultant fluid forces are caused mostly by pressure and tend to pass through the center of gravity of the grain. When viscous effects are important, the particle is subject to skin friction and the resultant fluid forces tend to fall above the center of the particle, i.e., a_2 exceeds a_1.

Taking $\tau_c \sim u_{oc}^2$, in which u_{oc} is the fluid velocity near the bed under critical conditions and substituting this relation into Eq. 2.106 gives us $u_{oc}^2 \sim d_s$. Cubing both sides of the relation gives $d_s^3 \sim u_{oc}^6$ which is the well-known sixth power law attributed to Brahms (1753) by Lelliavsky (1955) and to Leslie (1829) by Rubey (1948). Because the volume or weight of a grain is proportional to d_s^3, the law states that the weight of the largest particle that a flow will move is proportional to the sixth power of the velocity in the neighborhood of the particle. Rubey (1948) found that this law applied only when d_s is large compared with the thickness of the laminar sublayer and flow about the grain is turbulent. This finding agrees with those of White (1940), Shields (1936), and Tison (1953).

32. Observed Critical Shear Stress.—Most data on critical shear stress for noncohesive sediments have been observed in flume experiments. Such experiments show that the motion of sediment grains at the bed of a stream is highly unsteady and nonuniformly distributed over the bed area. Near critical conditions the motion of grains in any small area of bed occurs in gusts whose incidence increases as the mean shear stress increases. Observation of a large area of a sediment bed when the shear stress is near the critical value will show that the incidence of gusts of sediment motion appears to be random in both time and space. This suggests, as observed by Shields (1936), that the process of initiation of motion is statistical in nature. Einstein (1942) was the first to develop a transport relation based on statistical concepts.

Because of the statistical nature of the entrainment process, there is no truly critical condition for initiation of motion for which motion begins suddenly as the condition is reached. Data available on critical shear stress are based on more or less arbitrary definitions of critical conditions. As an example, Kramer (1935) defined the following three intensities of motion near the critical or threshold condition:

1. Weak movement indicates that a few or several of the smallest sand particles are in motion in isolated spots in small enough quantities so that those moving on 1 cm^2 of the bed can be counted.
2. Medium movement indicates the condition in which grains of mean diameter are in motion in numbers too large to be countable. Such movement is no longer local in character. It is not yet strong enough to affect bed configuration and does not result in appreciable sediment discharge.
3. General movement indicates the condition in which sand grains up to and including the largest are in motion and movement is occurring in all parts of the bed at all times.

The fact that the definition of critical conditions is rather indefinite can explain the variation in results of different workers. However, despite this nebulous

TABLE 2.10.—Data on Critical Shear

Experiment number (1)	Bed sediment (2)	Mean size of sediment, d_s, in millimeters (3)	Fluid (4)	Critical shear stress, τ_c, in dynes per square centimeter (5)
1	sand	0.21	oil	9.5
2	sand	0.90	oil	26
3	sand	0.122	water	3.56
4	sand	0.90	water	16.9
5	steel shot	0.71	water	30.8
6	sand	0.90	air	22.5
7	sand	5.6	water	50.0
8	sand	5.6	water	89.0
9	sand	5.6	water	133.0
10	sand	5.6	air	146.0

definition there seems to be reasonable agreement among published results from several sources. Weak movement, as previously defined, seems to agree well with workers' idea of critical movement. For example, it agrees with Shields' critical condition. However, the critical shear stress was determined by Shields as the value of the stress for zero sediment discharge obtained by extrapolating a graph of observed sediment discharge versus shear stress and does not depend on a qualitative criterion.

Experiments by White (1940), whose analysis was followed in deriving Eqs. 2.105 and 2.106, yielded the results in Table 2.10. All of the sediment, except that for experiment No. 5, was well-sorted sand prepared by sieving. The sands in experiments 1 and 2 were less well-sorted than the other sands. In experiment 5, steel shot was used as the sediment. Experiments 1 and 2 were made in a flume 5.08 cm wide, 100 cm long, with lubricating oil as the fluid. The flow in these two experiments was entirely laminar. Experiment 3 was made with water flowing in a nozzle contracting from the sides in such a manner that the shear stress on the sand bed that formed the bottom of the nozzle was uniformly distributed. The boundary layer in this experiment was laminar. Experiments 4, 5, 6, 8, and 10 were made with air or water in a nozzle contracting from the top where the bottom was formed by the sediment bed. The flows in these experiments were always turbulent. Experiments 7 and 9 were performed with water flowing over a sand bank. As will be noted in experiment 7, the bank sloped downward (positive ϕ) and in experiment 10, the slope was upward. In experiment 3, the dimensionless quantity, $U_* d_s/\nu$, was less than 3.5, and the flow around the grains was laminar. For the high speed experiments nos. 4–10, the boundary Reynolds number, $U_* d_s/\nu$, was much larger than 3.5 and the flow in the boundary layer and around the grains was turbulent. In the boundary Reynolds number U_* = the shear velocity defined by $U_* = \sqrt{\tau_o/\rho}$; and ν = the kinematic viscosity of the fluid.

Column 6 of Table 2.10 shows the values of the constant, c_1a_1/c_2a_2, in Eq. 2.105 calculated from observed value of τ_c. It will be seen that the constant for the flows with turbulent boundary layers, experiments 4–10, is only from 50%–65% of that in experiments 1–3 where the flow was laminar. White attributed this difference to

Stress Obtained by White (1940)

c_1a_1/c_2a_2 (6)	$\tau_c/(\gamma_s - \gamma)d_s$ (7)	$(U_* d_s)/\nu$ (8)	tan ϕ (9)	tan θ (10)	Description of flow (11)
0.19	0.27	0.04	0.0	1.4	laminar
0.16	0.17	0.28	0.0	1.05	laminar
0.20	0.20	2.1	0.0	1.0	laminar boundary layer
0.119	0.119	33	0.0	1.0	turbulent boundary layer
0.094	0.064	35	0.0	0.685	laminar boundary layer
0.098	0.098	80	0.0	1.0	turbulent boundary layer
0.113	—	360	0.45	1.0	laminar boundary layer
0.101	0.101	480	0.0	1.0	turbulent boundary layer
0.112	—	590	-0.50	1.0	laminar boundary layer
0.102	0.102	1,280	0.0	1.0	turbulent boundary layer

velocity fluctuations in the turbulent flows that cause fluctuations in the bed shear stress and in the forces on the grains. It is clear that in turbulent flows the observed mean critical shear stress, τ_c, may not be high enough to entrain the sediment and that, most, if not all of the motion results from pulsating values of the shear stress that exceed the mean value, τ_c. Because fluctuations of 50% have been observed in the mean velocity of turbulent flows near the bed and because the shear stress is approximately proportional to the local velocity squared, pulses of shear stress that have more than twice the average intensity are to be expected. White concluded that the true critical shear stress required to move a particular grain has a fixed value and that this value is given by the experiments with laminar flow. His equation for τ_c for sediment in a horizontal bed is

$$\tau_c = 0.18\,(\gamma_s - \gamma)\,d_s \tan\theta \qquad (2.107)$$

in which the constant is obtained from the experiments with laminar flow.

If it is assumed that initiation of motion is determined by τ_c, $\gamma_s - \gamma$, d_s and the density ρ and viscosity $\mu \sim$ of the fluid, dimensional analysis immediately yields

$$\frac{\tau_c}{(\gamma_s - \gamma)d_s} = f\left(\frac{U_{*c}d_s}{\nu}\right) \qquad (2.108)$$

in which $U_{*c} = \sqrt{\tau_c/\rho}$; $\nu = \mu/\rho$; and f denotes function of. This result was obtained by Shields (1936) by a much less direct method. The left-hand member of Eq. 2.108 will be called the dimensionless critical shear stress and denoted by τ_{*c} and the variable on the right is called the critical boundary Reynolds number and denoted by R_{*c}. When values of bed shear stress τ_o, other than τ_c are used in the two quantities they are termed dimensionless shear stress and boundary Reynolds number, and are denoted by τ_* and R_*, respectively.

Fig. 2.43 is a graph of the data used by Shields to determine the function of Eq.

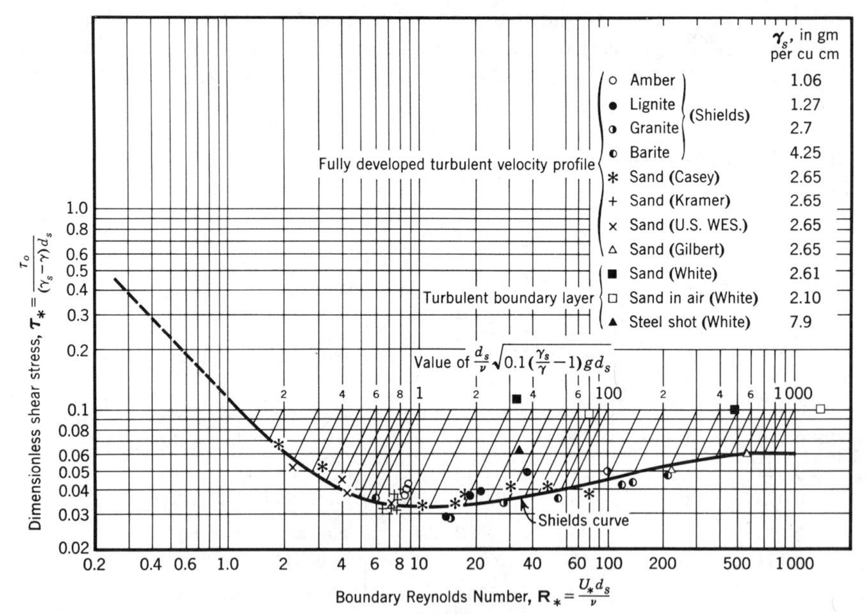

FIG. 2.43.—Shields Diagram with White Data Added

2.108. The data delineating the curve were obtained by Shields and several other workers from experiments in flumes with fully-developed turbulent flows and artificially flattened beds of noncohesive sediments. In this graph, the quantity, d_s, is taken as the mean size of the sediment. The data by White in experiments, 4, 5, 6, 8, and 10 are also plotted on Fig. 2.43. These data differ from those of Shields not only because τ_{*c} is larger than given by Shields but also because τ_{*c} appears to be independent of R_*. It is believed that these differences result from the differences between the structures of flows with turbulent boundary layers as in White's case and fully-developed turbulent flows as found in flumes. The fact that White's value of τ_{*c} for steel shot falls below that for sands is probably caused by the difference in shapes of the particles. The extrapolation of the Shields' curve for R_* less than two was suggested by Shields and is not based on any data. Shields did not fit a curve to the data but indicated the relationship between τ_{*c} and R_{*c} by a band of appreciable width. The curve in Fig. 2.43 termed the Shields curve herein was first proposed by Rouse (1939).

Shields' results shown in Fig. 2.43 have been widely accepted although some workers have reported somewhat different values for the parameters. Tison (1953) reported slightly lower values of τ_{*c} than Shields. Egiazaroff (1950) reported values of τ_{*c} that fell above the Shields curve in the range of R_* from approx 1.5–15 and then fell slightly below the curve for higher R_* but also reached the constant value of $\tau_{*c} = 0.06$ at R_{*c} of approx 1,000. Egiazaroff's curve reached a minimum value of $\tau_{*c} = 0.03$ at R_{*c} in the neighborhood of 30 while the Shields curve Fig. 2.43 reaches a minimum of $\tau_{*c} = 0.033$ at R_{*c} near 10. Conversely, Matsunashi's (1957) results agreed closely with those of Shields. All of these workers obtain curves having the general shape shown in Fig. 2.43.

Egiazaroff (1957, 1965) proposed a theory for the function expressed in Eq. 2.108 and delineated in Fig. 2.43. To derive this theory, one begins with the expression defining τ_{*c} in Eq. 2.108 and then uses the equations

$$\bar{u}_c = \sqrt{\frac{8}{f}}\sqrt{g r_c S} = \sqrt{\frac{8}{f}}\,U_{*c} = \sqrt{\frac{8}{f}}\sqrt{\frac{\tau_c}{\rho}} \qquad (2.109)$$

$$(\gamma_s - \gamma)\frac{\pi}{6}d_s^3 = C_D\frac{\pi}{4}d_s^2\frac{1}{2}\rho w^2 \qquad (2.110)$$

to eliminate τ_c, $(\gamma_s - \gamma)$, and d_s from the expression. Eq. 2.109 is recognized as the Chezy equation with the coefficient expressed in terms of the Darcy-Weisbach friction factor, f. The quantity $\bar{u}_c$ = the mean velocity in the cross section; g = the gravitational constant; r_c = the hydraulic radius; S = the channel slope; and the subscript, c, denotes threshold values of the quantities. Eq. 2.110 equates the submerged weight of a sphere of diameter d_s and specific weight γ_s in a fluid of specific weight, γ, to the fluid drag force experienced by the sphere in free fall. The quantity C_D = the drag coefficient; w = the terminal fall velocity; and ρ = the fluid density. The expression that results from eliminating τ_c, γ_s, γ, and d_s from the expression for τ_{*c} is

$$\tau_{*c} = \frac{1}{6}\frac{f}{C_D}\left(\frac{\bar{u}_c}{w}\right)^2 \qquad (2.111)$$

The assumption is now made that at threshold conditions the velocity that the particle on the bed experiences is equal to its free fall velocity or

$$u_{oc} = w \qquad (2.112)$$

which yields Egiazaroff's equation

$$\tau_{*c} = \frac{1}{6}\frac{f}{C_D}\left(\frac{\bar{u}_c}{u_{oc}}\right)^2 \qquad (2.113)$$

The friction factor, f, can now be eliminated by means of Eq. 2.109 yielding

$$\tau_{*c} = \frac{4}{3C_D}\left(\frac{U_{*c}}{u_{oc}}\right)^2 \qquad (2.114)$$

The quantity, u_{oc}/U_{*c}, is calculated from Eq. 2.119 given as follows, assuming that $k_s = d_s$ and that the velocity, u_o, occurs at $y = 0.63\, d_s$. Substituting this result into Eq. 2.114 yields

$$\tau_{*c} = \frac{4}{3C_D(a_r + 5.75 \log 0.63)^2} \qquad (2.115)$$

The drag coefficient, C_D, for spheres determined experimentally as a function of the Reynolds number, $w\, d_s/\nu$, is given in Fig. 2.1. If the sediment is coarse, the bed will be hydrodynamically rough and $a_r = 8.5$. Also the value of C_D for quartz spheres settling in water with sizes ranging from approx 3 mm–12 mm is approx 0.4. With $a_r = 8.5$ and $C_D = 0.4$, Eq. 2.115 yields $\tau_{*c} = 0.06$ which is the value given by Shields for large R_{*c}. As R_{*c} diminishes, a_r and C_D increase, thus decreasing τ_{*c} and agreeing qualitatively with the Shields curve.

The auxiliary scale on Fig. 2.43 with lines with positive slopes of 2.0 is included to facilitate the calculation of τ_c when one has the sediment and fluid properties. To find τ_c the value of $(d_s/\nu)\sqrt{0.1[(\gamma_s/\gamma) - 1]gd_s}$ is calculated for the case of interest using a consistent set of units, e.g., feet, pounds, and seconds; or meters, kilograms, and seconds and this value is located on the auxiliary scale. A line is then drawn through this point parallel to the inclined lines and the value of τ_* is read at the intersection with the Shields curve from which τ_c can be calculated.

Fig. 2.44 is a graph of τ_c against mean sediment size, d_s calculated from Shields' curve of Fig. 2.43 for quartz sand ($\rho_s = 2.65$ g/cm³) and water at several temperatures. Several values of the boundary Reynolds number, R_*, have been shown on the graph. Note that the largest temperature effect occurs for sands with sizes from 0.1mm–0.5mm. In this range the value of R_* is less than approx five. It will also be seen that for the finest sediment, τ_c becomes independent of sediment size. Egiazaroff (1950) showed that this relationship follows from the fact that for R_{*c} less than about two, the Shields curve has a negative slope of unity.

Fig. 2.44 also shows values of τ_c recommended by Lane (1955) for design of irrigation canals with noncohesive beds that are to convey clear water. For the coarse material, Lane uses d_{75} as the representative diameter of the sediment, in which d_{75} is the size for which 75% of the bed material by weight is finer. The value of τ_c recommended by Lane for the coarse material and shown in Fig. 2.44 is given by

$$\tau_c = 0.0164\, d_{75} \qquad (2.116)$$

in which d_{75} is in millimeters and τ_c is in pounds per square foot. This value was taken by Lane as 20% less than the values actually observed in a system of stable canals to allow a factor of safety. Assuming that the bed sediment has the common value of 1.4 for the geometric standard deviation of sizes and that the size distribution is log-normal d_{75}/d_g is found to be 1.25 (d_g is the geometric mean size). This means that Lane's recommended values of τ_c for coarse sediment agree closely with those of Shields as may be seen from Fig. 2.44, but it also means that the observed values are more than those of Shields.

As shown in Fig. 2.44, Lane's recommended values of τ_c for canals with sand beds are from 1.5 to almost 10 times as large as those given by Shields. The reasons for this are not clear, and it is only possible to speculate about them. Two possible reasons may be considered. One is that stable canals from which Lane's data were obtained are actually discharging sediment and, thus, are operating with bed shear stresses well above the critical. A second possibility is that dunes exist in these canals and, as explained in Article 35, act to increase the critical shear stress although dunes are formed only when sediment is moved.

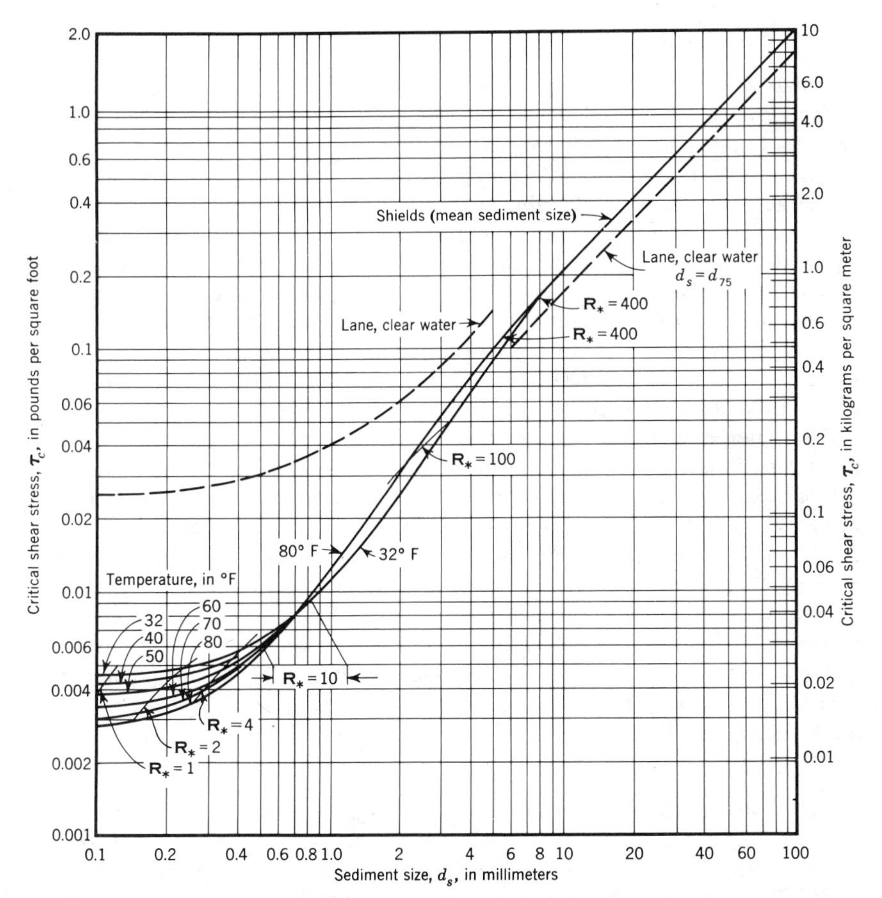

FIG. 2.44.—Critical Shear Stress for Quartz Sediment in Water as Function of Grain Size; Shields (1936) and Lane (1955)

The fact that critical conditions are usually identified by observing motion of sediment grains implies that they are not conditions for impending motion, but for small but finite sediment transport rate as pointed out by Einstein (1966). Willis (1967) observed in degradation experiments that sediment movement did not cease until the shear stress was approx 10% less than for the Kramer "weak movement" that has been adopted by many workers to indicate critical conditions. Gessler (1965, 1970) also determined τ_{*c} from degradation experiments with well-graded bed sediment. He found that τ_{*c} for the condition that there was a 50% probability of having no motion, varied from 20% less than the Shields value of 0.06 for large R_* to about 5% less at $\mathsf{R}_* = 20$. For a 95% probability of no motion, he determined that the τ_{*c} for large R_* would be reduced to 0.024 compared to 0.06 obtained by Shields. Gessler assumed that the fluctuations in shear stress around the mean value are normally distributed and that for $\tau_o = \tau_c$ there is a 50% probability of motion. The probability, p_i, that no sediment of mean size, d_{si}, would be moved was determined from the degradation experiments from the relation

$$p_i = \frac{W_{Di}}{W_{Di} + W_{Ei}} \qquad (2.117)$$

in which W_{Di} = weight per unit bed area of particles of mean size, d_{si}, in the surface layer of the armored bed; and W_{Ei} = the weight per unit bed area of particles of mean size d_{si} eroded from the flume.

Taylor (1971) made some definitive experiments with very low sediment discharge over flat beds of sediment that further clarified the ideas regarding critical conditions for initiation of sediment motion. In these experiments also reported by Taylor and Vanoni (1972a), the sediment discharge was measured by trapping all of the sediment moved off the end of the flume. Taylor's data are plotted in Fig. 2.45 as dimensionless shear stress τ_* against boundary Reynolds number, R_*. Values of the dimensionless sediment discharge, $q_{s*} = q_s/(U_* d_g)$ in which q_s = sediment discharge in volume per unit width were determined for each data point and the contours of q_{s*} for values 10^{-3}, 10^{-4}, and 10^{-5} were drawn in Fig. 2.45 by interpolation. By extrapolation of these data the contour for $q_{s*} = 10^{-2}$ shown in Fig. 2.45 was also drawn in and the Shields curve was added for comparison. The curve $q_{s*} = 10^{-2}$, is seen to fall remarkably close to the Shields curve although it reaches a constant value of τ_* at a lower value of R_* than the Shields curve and reaches its minimum value of τ_* at a higher value of R_* than the Shields curve. The contours of q_{s*} show that the sediment discharge is extremely sensitive to changes in τ_* so that a reduction of τ_* of only a few percent will reduce q_{s*} by an order of magnitude.

Taylor's data as displayed in Fig. 2.45 indicate that for a given sediment (given values of sediment size, size-distribution, density, and grain shape) the sediment discharge, q_s, can be expressed by

$$q_s = f(\tau_o, \mu, \rho, d_g, \gamma_s - \gamma) \qquad (2.118)$$

Eq. 2.118 and Taylor's data are in agreement with the concept that the bed shear stress, τ_o, expresses the dynamic conditions at and near the bed. This concept is also the basis for the velocity profile, Eq. 2.119, and the Shields Eq. 2.108.

Fig. 2.45 also shows the effect of water temperature on bed load discharge when the transport rate is very low. Like symbols in each of the several series represent data from pairs of experiments with the same depth and mean velocity but

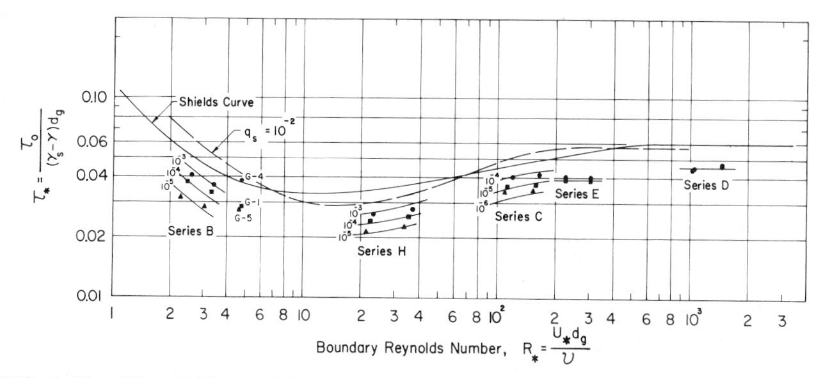

FIG. 2.45.—Plot of Lines of Constant Dimensionless Sediment Discharge q_{s*} on Shields Diagram; Taylor (1971)

different water temperatures. For low values of R_* (series B) the sediment discharge contours, q_{s*} = Const, have a negative slope. In this region an increase in temperature (decrease in ν and increase in R_*) results in an increase in q_{s*} and bed load discharge q_b. For intermediate values of R_* (Series H and C) temperature has the opposite effect, an increase in temperature results in a decrease in q_b. For large R_* (Series E and D) q_b is not affected by water temperature.

The intermittent nature of sediment motion under critical conditions and at very low transport rates indicates that under these circumstances the mean shear stress is not large enough to move sediment grains. The movement occurs when the instantaneous shear stress is increased above the mean value enough to overcome the stabilizing forces acting on the grains. To move grains, high velocity fluid must act on them and when the grains are fine the viscous layer of fluid at the bed, the so-called viscous sublayer, must be penetrated or disrupted or both. The thickness, δ', of the viscous sublayer may be written $\delta' = 11.6\, \nu/U_*$ although many workers prefer the factor 4 instead of 11.6. The ratio $d_s/\delta' = (1/11.6)\, R_*$. Thus when $R_* = 11.6$ the viscous sublayer is 1 grain diam in thickness and when $R_* = 2$ it is almost 6 diam thick and the grain is well submerged in the sublayer.

Einstein and Li (1958) studied the intermittency of the sublayer experimentally and theoretically. They visualized a cyclic behavior of the sublayer in which it develops and then is disrupted in a time that is small compared to the development time. Sutherland (1967b) made simultaneous observations of the entrainment of sediment grains and the disruption of lines of dyed fluid on the bed. He found that grain movement coincided with violent disruption of the dyed fluid. On the basis of his observations, Sutherland (1967b) visualized that turbulent bursts of high velocity fluid from the interior of the flow impinged on the bed, temporarily sweeping away the viscous sublayer and creating a pulse intense enough to set grains in motion. Apperley (1968) explained his detailed observation of turbulence fluctuations near a sediment bed in terms of a conceptual model much like Sutherland's.

33. Critical Velocity.—The earliest observations of critical or threshold conditions for initiation of motion of sediments were reported in terms of velocity. For example, Lelliavsky (1955) presents data that appeared in a book by du Buat in 1816. Another example is the classical work of Fortier and Scobey (1926) on permissible canal velocities that formed the basis for canal design for many years.

The trend in reporting work on critical conditions has been to abandon velocity in favor of bed shear stress because it is a more satisfactory quantity.

The velocity at a vertical, i.e., the velocity profile, for a two-dimensional, free-surface flow over a flat sediment bed is given by

$$\frac{u}{U_*} = a_r + 5.75 \log \frac{y}{k_s} \quad (2.119)$$

in which u = the velocity at distance y above the bed; k_s = the characteristic roughness size of the sediment; and a_r = a function of the boundary Reynolds number, $U_* k_s/\nu$, which has been obtained experimentally. (The quantity a_r is related to x given in Fig. 2.97 by $a_r = 8.5 + 5.75 \log x$.) For values of the boundary Reynolds number in excess of approx 90, a_r assumes the constant value of 8.5. When R_* is less than 3.5, the boundary is hydrodynamically smooth, and a_r is given by

$$a_r = 5.5 + 5.75 \log \frac{U_* k_s}{\nu}$$

and the equation for the velocity profile becomes

$$\frac{u}{U_*} = 5.5 + 5.75 \log \frac{U_* y}{\nu} \quad (2.120)$$

Eq. 2.119 shows that if two flows of different depth have flat beds of identical sediment and the same bed shear stress, the velocities at any distance y above the bed will also be the same in the two flows. However, because the mean velocity occurs at y equal to a constant fraction of the depth, the deeper flow will have the larger mean velocity. Thus, it is seen that mean velocity alone cannot express the scouring action of the water at the bed and that to completely specify conditions the depth must also be given. The bed conditions can also be specified by a velocity at a given value of y. The advantage of using shear stress to specify critical conditions is that the one quantity suffices whereas if velocity is used one must also report the depth or the position at which the velocity is observed. The fact

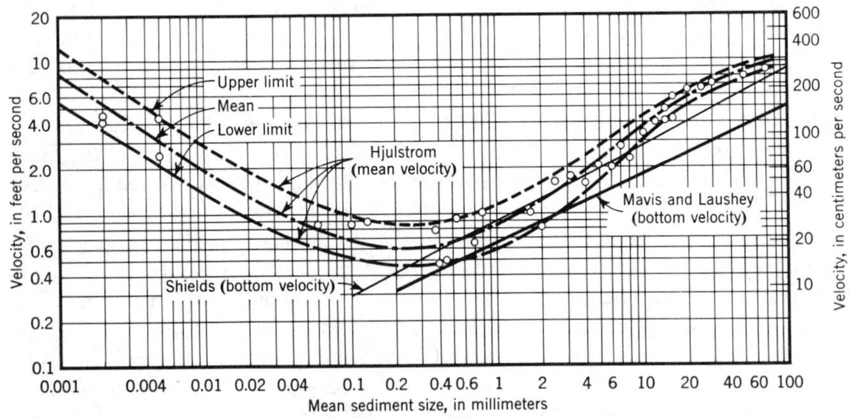

FIG. 2.46.—Critical Water Velocities for Quartz Sediment as Function of Mean Grain Size

that deep canals have higher critical mean velocities than shallow ones was known many years ago to irrigation engineers who considered this when planning canal systems. As shown by Lane (1955) this fact supports the idea that shear stress is the appropriate quantity to express critical conditions.

Fig. 2.46 shows data on critical velocity plotted against mean sediment size for quartz sediment in water ($\rho_s = 2.65\ g/cm^3$) obtained from three sources. The data points and the curves of the upper limit, mean, and lower limit of the critical mean velocity are taken from the work of Hjulström (1935) who prepared the curves based on data of several workers. The curves are for flows with depths of at least 1 m. The data for mean sediment size less than 0.01 mm were taken from Fortier and Scobey (1926). In such fine sediments, cohesion is an important factor in determining critical conditions. The curve of Mavis and Laushey (1949) yields critical bottom velocity u_{oc} for sands according to the relationship

$$u_{oc} = 0.5\left(\frac{\gamma_s}{\gamma} - 1\right)^{1/2} d_s^{4/9} \qquad (2.121)$$

in which d_s = the mean size of the sediment in millimeters; and u_{oc} is in feet per second. The relationship was developed by fitting a curve to observed data most of which were obtained under Mavis's supervision. In these cases, the bottom velocity was obtained by extrapolating velocity profile measurements to the plane of the bed. The curve in Fig. 2.46, labeled Shields, was calculated from Shields' diagram for quartz sand in water at 65° F by means of Eq. 2.119. In the calculations, the roughness size, k_s, of the sediment was taken equal to the mean size and the bottom velocity was assumed to occur at $y = d_s$ or $y = 11.6\nu/U_*$, whichever was larger. The curve marked "Shields" in Fig. 2.46 has a slope of 0.5 indicating that the bottom velocity, u_{oc}, is proportional to $d_s^{1/2}$ which agrees with the sixth power law (Sutherland, 1966).

The curve for Shields' data gives substantially higher critical bottom velocities than that of Mavis and Laushey. However, Mavis and Laushey (1966) have called attention to the fact that Eq. 2.121 represents a lower envelope of the data on which it was based. Because of this, the data of Mavis and Laushey are in better agreement with the Shields results than is indicated on Fig. 2.46. Bottom velocities were calculated from the Hjulström data by the same relationships used in calculating u_o from Shields' data assuming a flow depth of 1 m. These results did not agree with either of the values given by the other two curves of Fig. 2.46. Thus, the data for critical velocity are considerably less consistent than those for critical shear stress. Therefore, it is recommended that data on critical shear stress be used wherever possible.

The expression of critical conditions for initiation of sediment motion in terms of the shear stress as in Eq. 2.108 and in the Shields curve is not accepted by a number of workers. Blench (1966b) argues that Eq. 2.108 should contain two more terms, d/d_s and $(\gamma_s/\gamma) - 1$. Barr and Herbertson (1966) are of the opinion that the terms, d/d_s, and the ratio, ρ_s/ρ, of sediment density to fluid density should appear in Eq. 2.108. The main arguments in favor of omitting these terms [ASCE (1967)] are that the variables in Eq. 2.108 give good correlation of data from a number of sources and that use of these variables is consistent with the basic developments in turbulent flow that produced Eq. 2.119.

34. Lift on Particles.—Most works on initiation of motion consider only shear stress and completely ignore lift on particles, despite the fact that lift must be present. The only quantitative observations of lift on sediment in a bed were made by Einstein and El-Samni (1949) and Apperley (1968). Einstein and El-Samni measured the difference in mean static pressure in sediment beds at the level of the bottom of the top layer of sediment and at the wall of the channel at the level of the top of the top layer of sediment.

In these experiments the velocity was less than the critical value and none of the sediment moved. Two sediment beds were used, one made of hemispheres of 0.225 ft diam cemented to the bed in a hexagonal pattern and the other gravel with mean size approx 0.225 ft. These measurements yielded a pressure difference or lift pressure on the grains of

$$\Delta p = 0.178 \frac{1}{2} \rho u_o^2 \qquad (2.122)$$

in which u_o = the velocity at a distance of 0.35 d_{35} above the theoretical bed; and d_{35} = the size of grains for which 35% by weight of the bed material is finer. For the uniform hemispheres this size was taken as the diameter, i.e., 0.225 ft. The theoretical bed was defined as the position of the origin of y for which the measured velocity profiles conformed to Eq. 2.119 with a_r equal to 8.5. This position was found to be below the tops of the uppermost grains, a distance equal to 0.2 times the sphere diameter for the bed with hemispheres and 0.2 d_{67} for the gravel bed, in which d_{67} = the size of sediment for which 67% by weight is finer. The values of R_* in the experiments used to obtain Eq. 2.122 were high; of approx 50,000. Therefore, the lift pressure given by Eq. 2.122 should be valid only for rough boundaries.

It is interesting to compare Δp given by Eq. 2.122 and the shear stress, τ_o, existing concurrently with Δp. This can be done by calculating the ratio, $\Delta p/\tau_o$, in which τ_o can be calculated in terms of u_o from the velocity profile, Eq. 2.119, with a_r = 8.5. This kind of calculation gives values of $\Delta p/\tau_o$ of approx 2.5 for sediment with σ_g of approx 1.4. This tends to indicate that the lift is of considerable importance in entraining sediment.

Apperley (1968) measured the lift and drag forces on a sphere 1/4 in. in diameter in a bed of well-rounded 1/4 in. gravel in a flow 9.25 in. deep with a mean velocity of 3.40 fps. When the sphere was in the bed of gravel the mean drag force and mean lift force on it were 0.024 g and 0.012 g, respectively, giving lift/drag = 0.5. When the sphere was raised a distance equal to 1/4 of its diameter the drag and lift increased to 0.063 g and 0.049 g, respectively, giving lift/drag = 0.78. When the sphere was raised further the drag continued to increase but the lift decreased sharply and became negative at a distance of 1.5 diam above the bed. The mean drag force with the sphere imbedded with the other particles was very nearly equal to the product of the bed shear stress and the cross-sectional area of the sphere.

The distribution of instaneous lift force observed by Apperley indicated that there was a predominance of negative values, but that there were infrequent bursts of large positive lift forces. These large lift forces are apparently the ones that entrain particles.

35. Critical Conditions for Dune Beds.—The data on critical shear stress and velocity presented thus far are for artificially flattened beds. When the sediment is in the silt or sand size, dunes or ripples form as soon as transport is established. The critical shear stress or velocity for initiating motion on a dune bed is different than for a flat bed. The flow resistance of a dune bed is composed of a part resulting from the form drag caused by nonuniform pressure over the dune profile and another part resulting from tangential or skin friction on the dune surface. Although the shear stress, τ_o, as normally calculated includes both form drag and skin friction resistance, only the skin friction resistance acts to move sediment grains. This means that the critical shear stress for a given sediment is larger when the bed is dune-covered than when it is flat. This was shown by several workers including Shields (1936), White (1940), Einstein and Barbarossa (1952), and Sundborg (1956).

Quantitative data on critical velocity for flat and ripple beds were presented by Menard (1950), Rathbun and Goswami (1966), and Willis (1967). Menard found that the critical velocity for dune beds of sand ranging in size from 0.062 mm–0.7 mm was from 75%–88% of that for flat beds. Rathbun and Goswami observed that the critical velocity for 0.3 mm sand was 23.0 cm/s when the bed was flat and only 10.5 cm/s when the bed was rippled. The critical shear stresses were 0.00224 g/cm^2 and 0.00302 g/cm^2, respectively, for flat and rippled beds. The Darcy-Weisbach friction factors of the channel observed by Rathbun and Goswami were 0.035 and 0.215, respectively, for the flows with flat and rippled beds.

The incipient motion on rippled beds occurs near the center of the upstream face of ripples (Menard, 1950). This is the region where the flow that separates at the ripple crest reattaches to the ripple and thus is near a stagnation point (Raudkivi, 1963; Vanoni and Hwang, 1967) where the velocity and shear stress are low as shown in Fig. 2.55. Raudkivi (1966) also reported that despite the fact that the mean shear stress and velocity are small near the stagnation point of a ripple the longitudinal turbulence intensity is much larger than elsewhere on the ripple. This intense turbulence must cause the initiation of motion on rippled beds.

36. Critical Velocity for Prisms.—Novak (1948) made flume experiments to determine the critical mean velocity, V_c, required to overturn a cube with sides of length h with a face normal to the flow, and resting on a solid channel bottom with flow depth d. The cubes in the experiments were held against sliding by resting against pins projecting only slightly from the bed. The experimental results were corrected to yield V_c for overturning about the downstream edge. The data were also extended by calculation to the case of a rectangular prism of length L in the flow direction with a square face of width and height h exposed to the flow. In making these calculations, it was assumed that the overturning moment exerted by the flow on the prism depends on h but is independent of L. The experiments covered cube sizes h, ranging from approx 0.7 cm–10 cm and ratios h/d ranging from 1/37–1/3.

The critical velocity for overturning a rectangular prism about its downstream edge is given by

$$V_c = \frac{L}{h}\left[1.59 \; \frac{1.28}{\left(5.8 + \frac{d}{h}\right)^{1/3}}\right]\sqrt{\frac{\rho_s - \rho}{\rho} gh} \qquad (2.123)$$

in which L = the length of the prism parallel to the flow direction; ρ_s and ρ = densities of cube and fluid respectively; g = the acceleration of gravity; and h and d are as previously defined. The equation is dimensionally homogeneous, and can be used with any consistent set of units.

When the cube was rotated so the diagonal of a horizontal face was parallel to

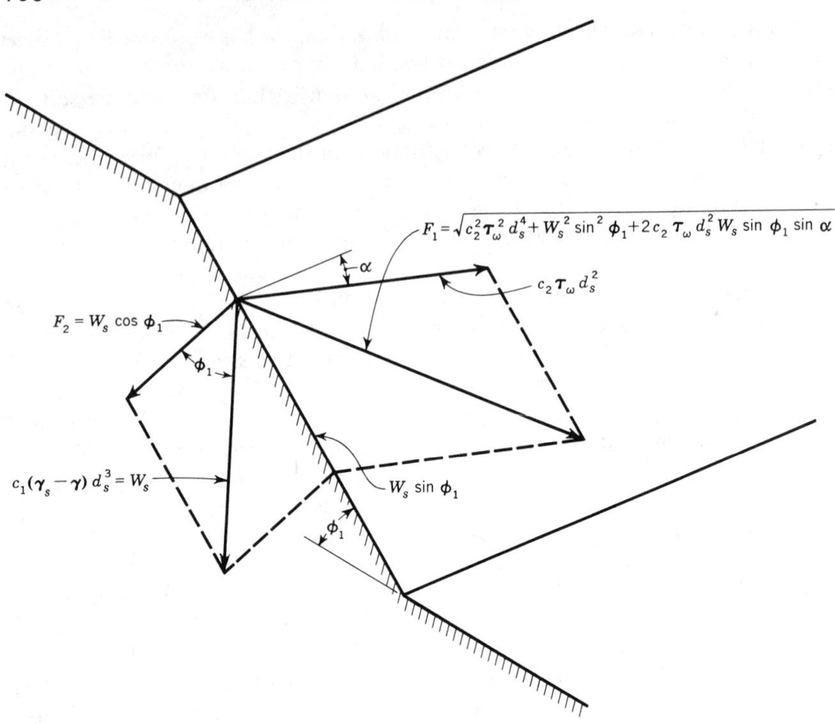

FIG. 2.47.—Forces on Sediment Particle on Bank of Stream

the flow velocity, the critical velocity increased approx 4%. Experiments were also made with a row of cubes across the flume with gaps between them, and with a flow depth of 3.9 h. The critical velocity for these experiments varied from 69% of that for the single cube when there was no gap, to 100% for a gap of 1.7 h.

37. Critical Shear Stress for Particle on Bank.—The forces on a sediment particle on the bank of a stream or canal differ from those on particles in the bed because of the relatively large gravity force tending to detach the particle. The forces on such a particle are shown in Fig. 2.47 for the case where the channel slope is small and intergranular forces are negligible. The hydrodynamic force on the particle is given by $c_2 \tau_w d_s^2$ in which τ_w = the shear stress on the bank; d_s = the representative size of the bank sediment; and c_2 = a constant defined previously.

As seen in Fig. 2.47, the hydrodynamic force and, thus, the velocity at the bank surface is inclined to the axis of the channel. This is the case that obtains when secondary circulation occurs as in channel curves. The other force on the particle is its submerged weight, W_s, given in terms of quantities already defined. Those forces have been resolved into a force, F_1, tending to move the particle, and F_2 tending to inhibit motion. The assumption is now made that when motion is impending, the ratio, F_1/F_2, is equal to tan θ, which as explained previously is the same as assuming that the hydrodynamic force applied to the particle passes through the center of gravity of the particle. Applying this assumption yields

$$\frac{F_1}{F_2} = \frac{\sqrt{c_2^2 \tau_{wc}^2 d_s^4 + W_s \sin^2 \phi_1 + 2c_2 \tau_{wc} d_s^2 W_s \sin \phi_1 \sin \alpha}}{W_s \cos \phi_1} = \tan \theta \quad \text{..........} (2.124)$$

in which τ_{wc} = the critical value of the shear stress on the bank. For a particle on the bed, i.e., for $\phi_1 = \alpha = 0$, Eq. 2.124 yields

$$\tau_c = \frac{c_1}{c_2}(\gamma_s - \gamma)\, d_s \tan\theta \qquad (2.125)$$

and from these equations it is possible to obtain

$$\frac{\tau_{wc}}{\tau_c} = -\frac{\sin\phi_1 \sin\alpha}{\tan\theta} + \sqrt{\left(\frac{\sin\phi_1 \sin\alpha}{\tan\theta}\right)^2 + \cos^2\phi_1\left[1 - \left(\frac{\tan\phi_1}{\tan\theta}\right)^2\right]} \qquad (2.126)$$

the relationship derived by Brooks (1963). When $\alpha = 0$ the relationship becomes

$$\left.\frac{\tau_{wc}}{\tau_c}\right|_{\alpha=0} = \cos\phi_1\sqrt{1 - \left(\frac{\tan\phi_1}{\tan\theta}\right)^2} = K \qquad (2.127)$$

first presented by Lane (1955). As will be noted, Eq. 2.125 is equivalent to Eq. 2.105 for the case of $a_1 = a_2$.

Eq. 2.127 shows that when $\phi_1 = \theta$, i.e., when the slope of the bank is at the angle of repose of the material, $\tau_{wc} = 0$. This is to be expected because, under these conditions without a shear force on the grain, motion of bank particles is impending under gravity forces alone. To be stable, the slope angle of the bank, ϕ_1, must be less than the angle of repose, θ. Lane (1955) has presented values of the angle of repose, θ, for sediment in water. Eqs. 2.126 and 2.127 when thoroughly verified, can be used to design channels. The procedure is to make values of τ_{wc} given by the equations, the same as or greater than values of the shear stress to which the bank is subject. In the calculations one must use a value of τ_c that will give negligible transport because particles eroded from the bank cannot be replaced and even a very small rate of erosion of the bank can in time cause failure or serious damage. In contrast, the movement of particles on the bed is much less serious because there is a good chance that they will be replaced by others moved in from upstream. Fig. 2.45 is helpful in judging the value of τ_c or τ_{*c} to be used in stability calculations. It must be emphasized that the foregoing equations are for straight or gently curving uniform channels. In highly sinuous channels and or in those that may carry large floating debris, e.g., logs or ice, the stability of particles on a bank is determined by the collision of these objects with the bank and not by the fluid forces alone as assumed in deriving Eq. 2.124.

38. Critical Conditions for Cohesive Sediments.—Sediments composed of or containing significant fractions of fine-grain material in the silt and clay sizes have greater resistance to entrainment than coarser sediments consisting only of sands. This is shown by the Hjulström curve of Fig. 2.46 for sediments finer than approx 0.1 mm. The fact that these fine sediments have so high a critical velocity is ascribed to cohesion that acts with the weight of the sediments to inhibit entrainment. The behavior of fine sediments under the attack of flows is complex and depends on many factors including the electrochemical environment of the sediment. Few studies have been made of this problem and knowledge of this phase of sedimentation is therefore in a primitive state.

The oldest information on the erosion resistance of cohesive sediments is that obtained by engineers through experience in design and operation of canals. An example of this is the permissible velocity data of Fortier and Scobey (1926) on which the part of the curve of Fig. 2.46 for fine sediment is based. Although this information is useful, it does not clarify the mechanics of the entrainment process for fine sediments. Sundborg (1956) suggested that the cohesive force resisting entrainment of a grain is proportional to the shearing strength of the sediment as determined in standard soil tests, and that it acts in a direction opposite to the fluid force. Based on this idea, he obtained the following relationship for the critical shear stress for cohesive sediment, on a horizontal bed

$$\tau_c = \frac{c_1 a_1}{c_2 a_2} (\gamma_s - \gamma) d_s \tan\theta + c_3 S_v \qquad (2.128)$$

in which S_v = the shearing strength of the sediment expressed as a stress with the same units as τ_c; c_3 = a constant, the value of which is not known; and the rest of the symbols are as in Eq. 2.105. Sundborg expected that for fine sediments, the first term on the right of Eq. 2.128 would be negligible compared to the other, and that the inverse would be true for sands and coarser sediments.

Since 1956, several studies have been made in which the critical shear stress for cohesive sediments has been correlated with such quantities as shear strength, S_v, plasticity index I_p, and the percentage by weight of fine material, e.g., clay, or silt and clay. The shearing strength of the sediment is usually determined by some standard laboratory tests. The plasticity index is the difference between the liquid limit and the plastic limit of the sediment. The liquid limit is the water content in percentage by weight of the dry sediment at which the sediment exhibits a small shearing strength, and the plastic limit is the water content in percentage by weight of dry sediment at which the sediment begins to crumble when rolled into thin cylinders. For further information on these quantities and how they are measured, the reader is referred to a textbook on soil mechanics.

Dunn (1959) determined τ_c for sediments ranging from sand to silty clay taken from several channels in Nebraska, Wyoming, and Colorado. His apparatus consisted of a tank in which a submerged jet of water was directed vertically downward onto a sample of the sediment. The shear stress exerted by the jet on the sediment was determined by actually measuring the fluid force on a plate 1 in. square placed at the level of the sample in the zone where sediment was first eroded by the jet. Table 2.11 shows values of critical shear stress and other data obtained by Dunn for cohesive materials. The samples tested were first passed through a sieve with 2-mm square openings, oven dried, and then consolidated by pressing them between porous plates while in a saturated state. Two or more samples of each material were subjected to different consolidation pressures before testing. This accounts for the fact that a particular sediment has maximum and minimum values of S_v and τ_c. The consolidation pressures ranged from 25 psf–300 psf.

For cases where the plasticity index fell between 5 and 16, Dunn's data fitted

$$\tau_c = 0.001\,(S_v + 180)\tan(30 + 1.73\,I_p) \qquad (2.129)$$

in which τ_c and S_v are in pounds per square foot and the argument of the tangent is in degrees. The shear strength, S_v, was determined from measurements of the maximum torque required to rotate a vane that is pressed into the sample. The vane was made of two metal plates 2 in. wide by 4 in. long intersecting at right

TABLE 2.11.—Data on Critical Shear Stress of Channel Sediments Observed by Dunn

Sediment	Geometric mean grain size d_g, in millimeters	Geometric standard deviation, σ_g	Percentage of silt and clay	Plasticity index, I_p	Vane Shear Strength of Sediment S_v, in pounds per square foot		Observed Critical Shear Stress τ_c, in pounds per square foot	
					Min	Max	Min	Max
(1)	(2)	(3)	(4)	(5)	(6)	(7)	(8)	(9)
1	0.022	6.08	69.0	11.1	77	167	0.48	0.49
2	0.072	6.00	35.0	0.0	268	520	0.18	0.33
3a	0.328	—	12.5	—	96	138	0.057	0.063
3b	0.319	—	18.0	—	96	118	0.057	0.083
3c	0.308	—	26.0	—	110	110	0.11	0.12
3d	0.250	—	44.0	—	128	140	0.14	0.15
4	0.078	3.25	41.0	0.0	75	233	0.19	0.28
5	0.081	2.79	31.0	0.0	102	243	0.11	0.15
6	0.038	3.25	46.0	2.5	173	242	0.30	0.33
7	0.016	4.83	78.0	8.8	105	250	0.30	0.45
8	0.015	4.66	81.0	13.3	137	163	0.40	0.41
9	0.026	3.48	56.0	3.5	100	300	0.19	0.32
10	0.014	3.48	88.0	11.2	144	187	0.43	0.48
11	0.014	3.05	95.0	15.6	137	143	0.48	0.49
12	0.139	1.57	10.0	0.0	43	155	0.053	0.053
13	0.173	1.29	5.0	0.0	95	110	0.043	0.033

angles along their center lines. Correlation of τ_c with the fraction of silt and clay were possible only when the sediment contained appreciable fractions of sand.

The data of Table 2.11 indicate at least qualitatively the effect on τ_c of adding varying amounts of silt and clay to an otherwise sandy sediment. This can be seen by comparing τ_c for sediments 3a–3d with values obtained from Fig. 2.44 for noncohesive sediments. For sediments with the same mean size, the maximum values of τ_c in Table 2.11 are from 15 times those in Fig. 2.44 for sediment No. 3a to almost 40 times as much for sediment No. 3d.

Smerdon and Beasley (1961) determined τ_c for 11 cohesive farm soils from Missouri ranging from silty loam to clay by observing them in a tilting flume. The flume was 60 ft long, 2.51 ft wide and the soil to be tested was placed in a depression in the floor 2.5 in. deep by 18 ft long, with its upstream end 30 ft from the flume inlet. The soil was thoroughly mixed, lumps were broken, and foreign matter was removed as the soil was placed in the flume. The soil was then leveled to the level of the solid floor of the flume, the soil was wetted by flooding, the water was drained, and the soil allowed to compact and dry for approx 20 hr at which time the erosion test was begun. Fig. 2.48 shows τ_c plotted against the plasticity index, I_p, for the 11 soils tested. Fig. 2.49 shows the same data plotted against percentage clay in the soils. The lines shown in these figures were fitted by the method of least squares. From inspection, it appears that the scatter about the curves of best fit in the two figures are approximately the same. The values of τ_c in Figs. 2.48 and 2.49 were calculated from Eq. 2.119, and measured velocities at two distances y, from the bed at a vertical. These values of τ_c were as much as 50% less than those calculated from the formulas, $\tau_c = \gamma dS$ or $\tau_c = \gamma d S_f$, in which S and S_f = the slope of the channel and the energy grade line, respectively.

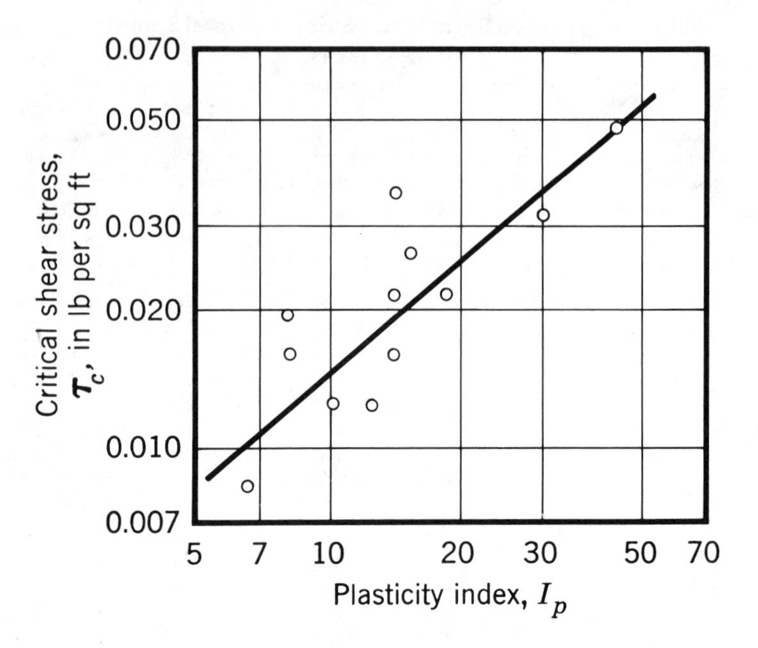

FIG. 2.48.—Critical Shear Stress for Several Farm Soils as Function of Plasticity Index; Smerdon and Beasley (1961)

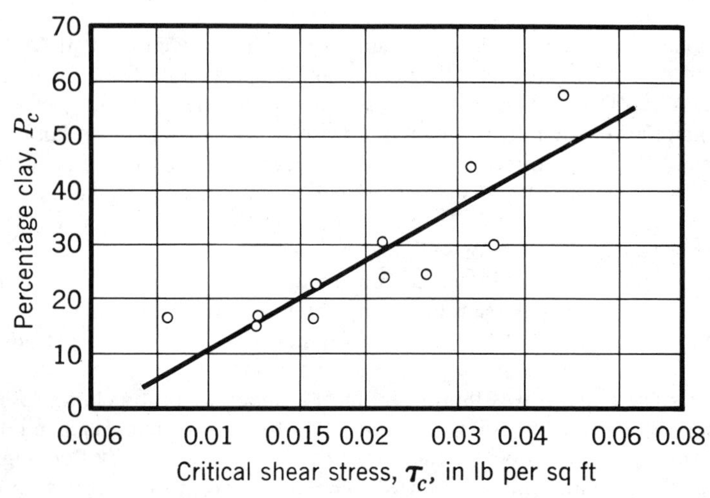

FIG. 2.49.—Critical Shear Stress for Several Farm Soils as Function of Percentage of Clay; Smerdon and Beasley (1961)

From the investigation of sediments in 12 natural perennial and ephemeral channels in the western United States, Flaxman (1963) found that the compressive strength of unconfined saturated and undisturbed samples of the sediment was a good indication of the shear stress it would withstand without excessive erosion. The undisturbed samples were 1.9 in. in diameter and 4 in. long. Fig. 2.50 is a graph of unconfined compressive strength of the sediment against the calculated maximum shear stress to which it was subjected. The maximum shear stress was calculated from the expression $\tau_c = \gamma r S$ in which r = the hydraulic radius of the channel calculated for the flow at the high water mark; and S = the channel slope. The symbols on the graph show whether the channel was stable or eroding and the numbers opposite the points give the hydraulic radius and mean velocity at maximum stage. Flaxman (1963) presented a graph of the unconfined compressive strength of his bed sediment against the product, $\tau_o V$, that gave a much sharper separation between stable and eroding channels than that of Fig. 2.50. The product, $\tau_o V$, of shear stress and mean flow velocity is sometimes referred to as stream power or tractive power.

Of the five channels that were eroding, only the one with the material with the lowest compressive strength was eroding rapidly, indicating that even the eroding channels were not subjected to stresses greatly in excess of the limiting value. Although the graph does not clearly separate the stable and eroding channels, it indicates a definite trend. Flaxman also observed that some channels in sediment with small or negligible plasticity index were stable and therefore resistant to

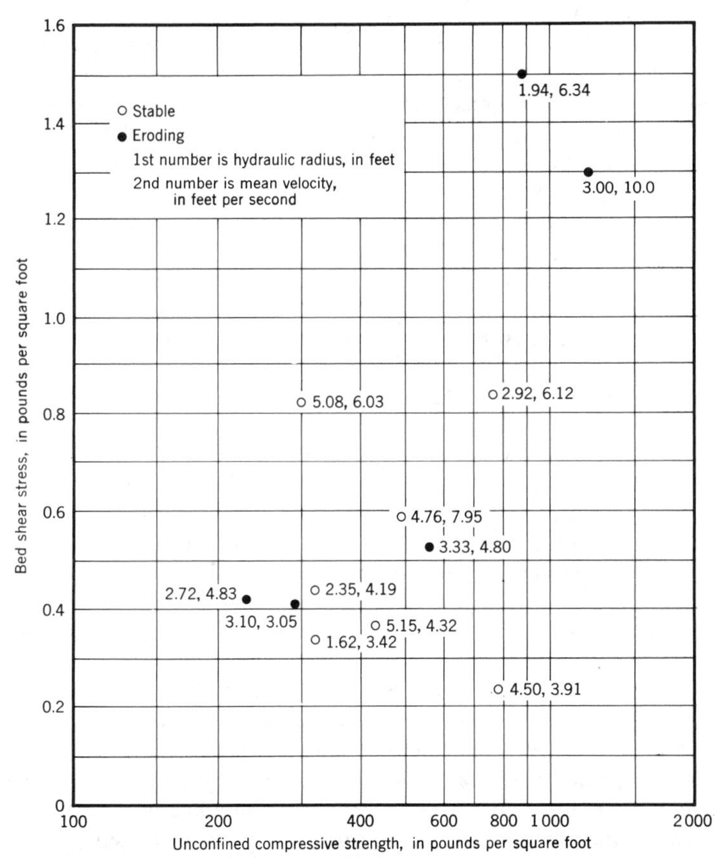

FIG. 2.50.—Stability of Natural Channels Related to Bed Shear Stress and Unconfined Compressive Strength of Sediment; Flaxman (1963)

erosion indicating that plasticity index alone was not an adequate indicator of erosion resistance or critical shear stress. The data of Dunn in Table 2.11 also agree with this. Flaxman also determined the permeability of the undisturbed samples and was of the opinion that it was a factor in determining the amount of erosion that would occur once the critical shear stress was exceeded.

Abdel-Rahmann (1964) studied the erosion resistance of a clayey sediment termed opalinuston in Switzerland. This sediment had a median size of approx 0.009 mm. Also, 22%–25% by weight of the material was finer than 0.002 mm and 90%–96% was finer than 0.2 mm. It had a plasticity index of 23. The silicate content of the sediment was more than 90% by weight. A bed of the material 6 m long and 5 cm thick was laid down in a flume 90 cm wide, 10.5 m long, on a fixed slope of 0.05.

Two series of tests were made: (1) The vane shear strength of the bed sediment, S_v, was kept the same at 1,267 kg/m^2 and the mean bed shear stress, τ_o, varied from 0.0745 kg/m^2–0.440 kg/m^2; and (2) the shear stress was kept constant at 0.261 kg/m^2 and the compaction and water content of the sediment were varied to yield vane shear strengths at the end of the tests ranging from 900 kg/m^2–2,810 kg/m^2. The vane used to measure the shear strength had vanes 2 cm wide and 3.5 cm long.

Fig. 2.51 shows this author's results as mean depth of erosion plotted against time for each of the two series of five experiments. As may be seen, erosion always occurred in his experiments regardless of the values of τ_o and S_v. As shown by Fig. 2.51, most of the erosion occurred in the first 80 hr and the time to reach maximum erosion depth in the first series of tests (Fig. 2.51a) varied little suggesting that it was independent of τ_o.

As the bed was eroded, holes developed in it and its hydrodynamic roughness tended to increase. The clay used in these experiments was high in silicate content and is of the kind that swells when it absorbs water. The erosion process as visualized by the author was intimately related to the swelling of the clay. When erosion ceased, the bed surface was observed to be covered by a thin layer of sticky material.

Moore and Masch (1962) made scour tests on samples of natural and composite cohesive sediments in an apparatus similar to the one used by Dunn (1959) in which a submerged jet of water impinges vertically downward on a sample of the sediment. In these experiments, the quantity of eroded material was determined as a function of time for sets of fixed values of jet diameter, velocity, and distance from nozzle to sample. An interesting result was that for any experiment the depth of erosion was proportional to the logarithm of duration of the experiment that is the same relationship obtained in experiments on scour of noncohesive sediments shown in Fig. 2.17. The results shown on Fig. 2.51 also follow roughly a relationship of this kind.

Partheniades (1965) made flume experiments with San Francisco Bay mud using salt water with the same salinity as sea water. The mud was a blue-grey plastic clay composed predominantly of montmorillonite with some illite. Sixty percent by weight of the sediment was finer than 2 μ, and a trace of very fine sand was present. Two series of tests were made; one with a bed of the mud placed at field moisture, and the other with a bed formed of the same material deposited in the flume from suspension in the flow. The mud in the first series of tests had a vane shear strength, S_v, of approx 40 psf; and S_v for the mud in the second series

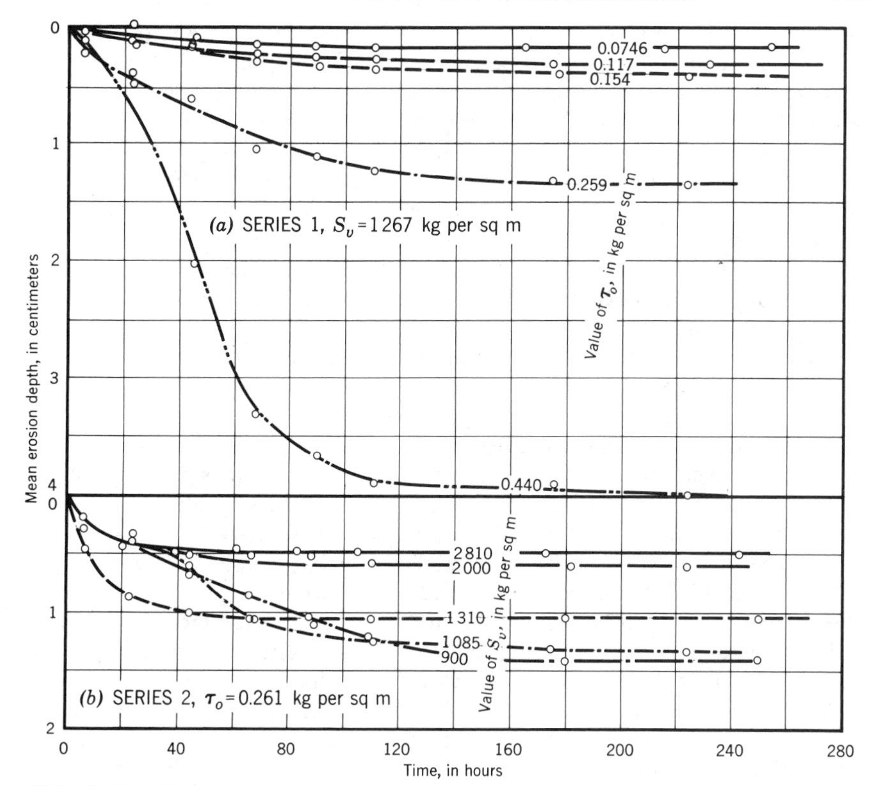

FIG. 2.51.—Erosion Depth in Clay Soil in Flume as Function of Time; Abdel-Rahmann (1964)

was only approx 1% of that in the first. The values of τ_c for these two beds were found to be approximately the same, i.e., from 0.001 psf–0.002 psf. From this, Partheniades concluded that τ_c was independent of S_v. He postulated that τ_c depends on the bond strength of clay flocs that does not depend on the degree of consolidation.

The investigations of the behavior of cohesive sediments referred to previously have been concerned mainly with mechanical properties and have not dealt to any extent with the effect of electrochemical factors. That these are important has been noted by several workers. Grissinger and Asmussen (1963) found that the erosion resistance of clay soils increased with the time they were kept in a wetted state. The explanation offered for this behavior was that when clay is first wetted the free water releases the bonds between particles, but as free water is absorbed and the clay minerals hydrate, the bond is strengthened. Because hydration is a slow process, a clay sediment can be eroded for some time before it reaches its full strength. In the tests performed by these authors, erosion rates did not tend to decrease to their ultimate value until materials had been in wetted states for approx 8 hr. Martin (1962) has noted the importance of such factors as water content, salt content, and ion exchange on the strength of clay. For example, he reports that by merely leaching the salt from a clay, its strength can be reduced from 1/100 to 1/1,000 of its value with the original salt content. He also showed that identical treatments of two different clays, e.g., illite and montmorillonite clays, had totally different effects on their strength. Altschaeffl (1965) reported that the erodibility of a clay can be reduced by adding a dispersing agent to the water, thus, again indicating the importance of the chemical environment on clay behavior.

The results of several workers have been presented mainly to illustrate the state of knowledge of erosion of cohesive sediments. It is clear from this work that the properties of the sediment that determine its resistance to erosion are not completely defined. Shear strength and plasticity index and, perhaps, clay content have an important bearing on the phenomenon but they apparently do not describe it completely. The chemical and environmental factors outlined by Grissinger and Asmussen (1963), Martin (1962), and Altschaeffl (1965) must also be considered. The effect of these have not been studied systematically in the laboratory investigations referred to herein. The work of Abdel-Rahmann indicates that erosion will occur over a range of fluid shear stresses and sediment properties if given sufficient time. This result suggest that there is no critical shear stress for clays in the sense that one exists for noncohesive sediments. Abdel-Rahmann's observation that when his clay bed reached stability it was coated with a sticky substance, suggests that the stabilizing process involves some chemical action of the kind suggested by Grissinger and Asmussen (1963) because, as shown by Fig. 2.51, a critical time seems to be involved. For further discussion of the erosion of cohesive sediments, the reader is referred to an ASCE Task Committee report (1968) and a paper by Partheniades and Paaswell (1970).

The fact that the properties affecting initiation of motion of cohesive sediments are not delineated, and the fact also that there is no precise manner in which to tell when critical conditions exist, can account for much of the disparity in results of different workers. These conditions also make it difficult to compare results of different studies. Flaxman calculated shear stresses in stable natural channels ranging from 0.24 psf–1.5 psf. The next highest values ranging from 0.12 psf–0.49 psf were found for plastic sediment by Dunn in his jet testing apparatus and values found in a flume by Smerdon and Beasley ranged from 0.008 psf–0.050 psf. The shear stresses in the experiments of Abdel-Rahmann ranged from 0.015–0.090 psf although some erosion occurred even for the lowest shear stress. Thus, it is seen that the measured shear stresses vary almost 200 fold in these four studies. Whether this is a true variation or is caused by errors or differences in techniques cannot be said. However, it does suggest that the resistance to erosion of cohesive sediments varies greatly.

F. Hydraulic Relations for Alluvial Streams

39. General.—A flow confined by boundaries composed of noncohesive granular material that can be transported by the flow has the curious property that the flow itself, or to be more precise, the interaction between the flow and the boundary, molds the geometry of the channel and casts its hydraulic roughness. Thus, alluvial channel flows are simultaneously sculptor and sculpture. In effect, all of the flow boundaries and not just the water surface, are free surfaces in the sense that their profiles are not fixed or known *a priori*, and that at least two mathematical boundary conditions (a kinematical condition, a dynamical condition or a sediment transport condition, or both) must be fulfilled on these unknown boundaries. Therein lies the critical distinction between single-phase flows in rigid channels and alluvial channel flows. The former have their geometry fixed, by man or nature, and the flows within their boundaries are incapable of changing either the large-scale (cross-sectional and plan form) or small-scale (boundary roughness) configuration of the channel. The states of knowledge of fluid mechanics and engineering hydraulics have progressed to the point that it is now possible to predict with accuracy, adequate for most practical requirements, the hydraulic roughness of fixed-geometry channels. For alluvial channel flows the picture is by no means so bright. The fundamental difficulty is that the channel geometrical characteristics, both large and small-scale, and thus also the hydraulic roughness depend on the depth, velocity, and sediment transport rate of the flow; but these flow properties are in turn strongly dependent on the channel configuration and its hydraulic roughness. Thus, to make depth-discharge predictions for alluvial streams, an additional item of information is required: a relationship between the flow parameters, fluid and sediment properties, and the hydraulic roughness of the channel. There arises the problem, one of central importance to the subject of fluvial hydraulics and one that to date has been solved only imperfectly.

It is useful to make a distinction between the effects on hydraulic relations of the large-scale or gross geometric characteristics and the small-scale features. Into the former category falls the channel pattern (Leopold and Wolman, 1957), braided, meandering, straight, etc., which, to be sure, has an effect on the depth-discharge relation of a reach of a stream. However, the understanding and formulation of this effect is still so primitive that at present little of a meaningful quantitative nature can be stated about it. For example, prediction of the wavelength of meanders has only recently received analytical treatment (Langbein and Leopold, 1966; Anderson, 1967; Hansen, 1967; Callander, 1969), and it was not demonstrated until the mid-1960's that the energy gradient of flow in a meandering channels is heavily Froude-number dependent (Rouse, 1965). The shape of the channel cross section also plays a role in the hydraulic relations, and varies with changing discharges of water and sediment; however it appears possible to take account of its effect, at least approximately, by means of the hydraulic-radius characterization of the section.

By far the most nettling aspect of the problem surrounds the role of the small-scale features that constitute the bed configurations. Because of the changes the bed configuration undergoes with varying flow conditions, from very rugged arrays of ripples and dunes that are generated by low discharges, to flat beds with their irreducible roughness that accompany larger discharges and flow velocities, the friction factor of flows occurring in a given river reach can vary by a factor of 10 or more during, e.g., the passage of a spring flood. The wide variability the friction factor can undergo is nicely shown by the carefully measured field data obtained by Nordin (1964) on the Rio Grande and summarized in Fig. 2.52(a) [after Alam and Kennedy (1969)] and Fig. 2.52(b). Measured sediment discharges in Section F are presented as functions of water discharge and mean velocity in Figs. 2.70 and 2.71, respectively. The field data of Colby (1960) and Coleman (1962) also reveal this phenomenon.

Indeed, the variability of alluvial channel roughness is so great that, as was first pointed out by Brooks (1958), it is possible for several discharges to occur at some depth-slope combinations. This is seen to be the case for the flows in the Rio

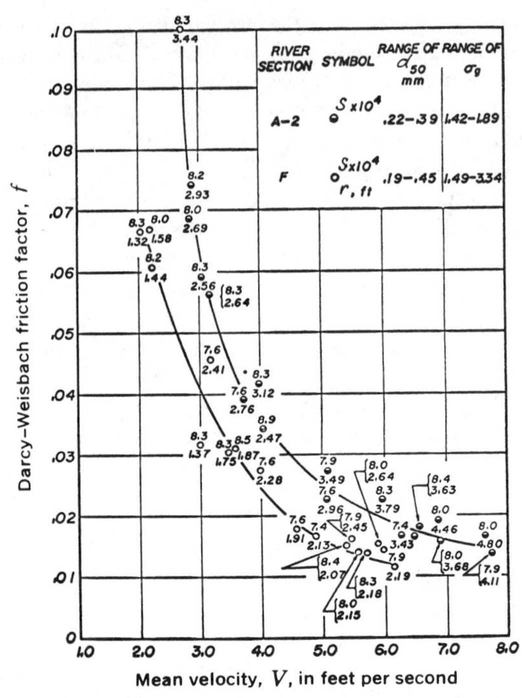

FIG. 2.52(a).—Variation of Darcy-Weisbach Friction Factor with Velocity for Two Sections of Rio Grande in New Mexico [Data Obtained by Nordin (1964); Figure prepared by Alam and Kennedy (1969)]

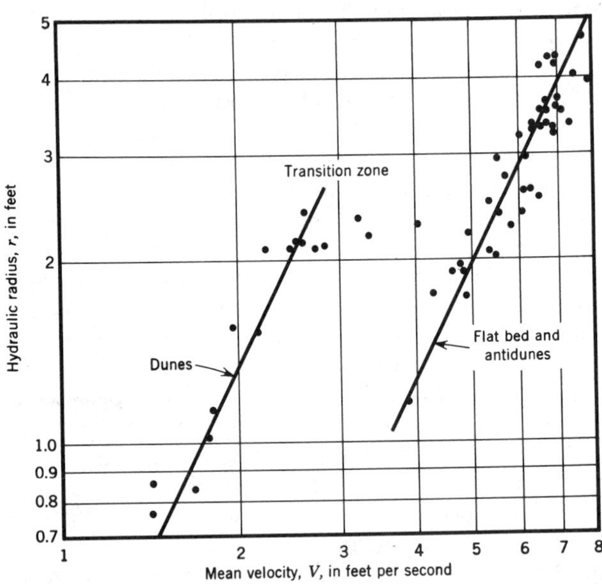

FIG. 2.52(b).—Discontinuous Rating Curve for Rio Grande near Bernalillo, N.M. (Nordin, 1964, Fig. 3)

Grande (Nordin, 1964) summarized in Fig. 2.52(b), and also for the laboratory data presented by Vanoni and Brooks (1957) and depicted in Fig. 2.53. Alternately, at some slopes it is possible for a given water discharge to occur at two or more different depths: a large depth and low velocity accompanying a ripple or dune-covered bed; a small depth and high velocity with a flat bed; and possibly one or more other depths and velocities with intermediate bed configurations. This peculiar aspect of river behavior, albeit a thorn in the side of the river engineer, is another of those subtle miracles of nature that works most decidedly in man's favor. For if flat beds did not occur at large discharges, and instead ripples and dunes with their high rugosity persisted, floods would give rise to much higher river stages. On the other hand, the rugged arrays of ripples and dunes produce larger depths and lower velocities at low discharges than would a flat bed, with the obvious benefits to navigation, diversion works, etc. Thus, the variability of the bed configuration and channel roughness to reduce the variations of depth and stage accompanying changes in discharges of water and sediment.

There are yet two more physical complications that arise in hydraulic

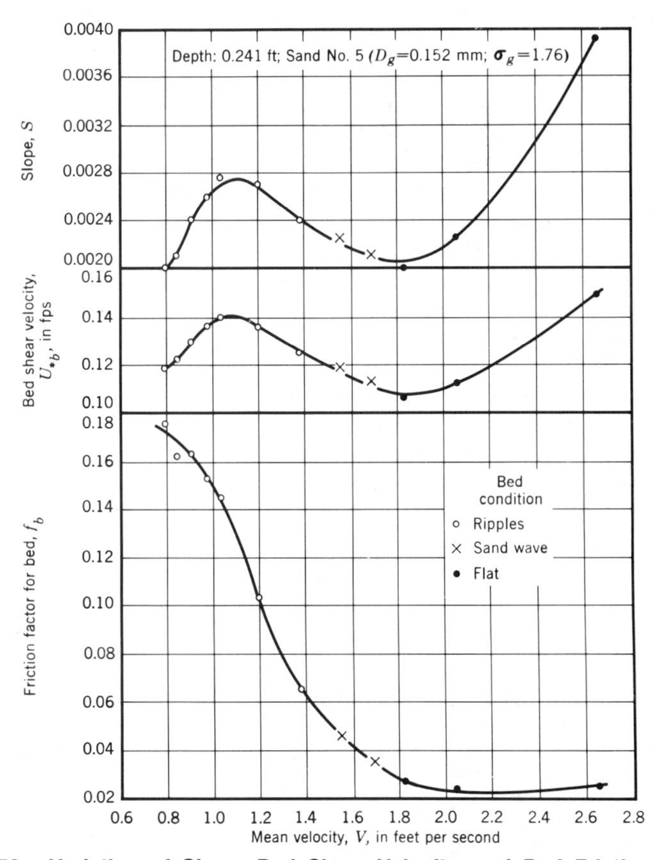

FIG. 2.53.—Variation of Slope, Bed Shear Velocity, and Bed Friction Factor in Constant-Depth Flume Experiments Reported by Vanoni and Brooks (1957, Fig. 21) (Flume Width = 10.5 in.)

calculations for alluvial channel flows. Firstly, the energy dissipation characteristics of turbulent flows of particle suspensions have been found to be dependent on the particle characteristics and concentration. In experimental comparisons of pairs of single-phase and sand laden flume flows with the same mean depths and velocities, Vanoni (1946) discovered that suspended sand concentrations of 1.2 g/ℓ-3.3 g/ℓ can reduce the Darcy-Weisbach friction factor for flat beds by as much as 18%. In similar flume experiments with flows over solidified sediment ripples, Vanoni and Nomicos (1960) found that suspended sand reduced the friction factor by up to 25%. The suspended sediment modifies the form of the velocity profile, increasing the velocity at every elevation above the bed and increasing the discharge accompanying a given depth and slope. The details of the interaction between the flow and the sediment suspension are discussed in Section D of this chapter. Secondly, temperature changes have been found to have a marked effect on alluvial channel roughness, both in laboratory flumes (Franco, 1968; Hubbel and Al-Shaikh, 1961) and natural streams (United States Army Corps of Engineers, 1968). To the present time, this effect has not been well elucidated, even quantitatively. Franco's (1968) flume data, summarized in Fig. 2.94, e.g., indicate that the friction factor decreases with increasing temperature, while data from the Missouri River near Omaha (United States Army Corps of Engineers, 1968) point to the opposite trend (see Fig. 2.91). It is not known whether this apparent discrepancy is due to differences in the ranges of depth and velocity involved, or in the functioning of the flume and river; in Franco's flume experiments water and sediment discharges were held constant and the slope was varied to maintain uniform flow as the temperature was changed, while in the field case the discharge and slope remained nearly invariant. In Article 55 of this chapter the temperature effect is discussed at some length, and the seeming paradoxes previously outlined are interpreted in the light of the results of laboratory experiments reported by Taylor and Vanoni (1972b).

This section will be concerned primarily with prediction of depth-discharge relations for steady uniform flows in straight alluvial channels. The emphasis will be on the role of the small-scale geometric characteristics, although the effects of channel meandering and other nonuniformities, suspended sediment, temperature, etc., will enter indirectly into some of the considerations. Because of the central role bed forms play in determining alluvial channel roughness, Article 40 which follows is devoted to a brief description of their properties and occurrence. The subsequent articles take up, in turn, alternate approaches to obtaining hydraulic relations for alluvial channels, a presentation of some of the techniques that have been developed for calculating stage-discharge relations; a comparison of depth-discharge relations obtained from some of the predictors; a critique of various approaches to the problem; and a summary discussion of the present status of the problem and the prospects and promising avenues for future developments. Finally, a calculational procedure is presented for determining the friction factor for just the bed section, f_b, in cases where the roughnesses of the channel bed and sides are significantly different.

40. Occurence and Properties of Sedimentary Bed Forms.—A free surface flow over an erodible sand bed generates a variety of different bed forms and bed configurations (arrays of bed forms). The type of bed configuration and the dimensions of the bed forms are dependent on the properties of the flow, fluid, and bed material. Descriptions of and definitions for the principal types of bed forms and configurations have been presented by the ASCE (1966) Task Force on Bed Forms and Alluvial Channels, Committee on Sedimentation. Table 2.12 summarizes from the Task Force report the salient characteristics of the various bed forms in the general order of their occurrence with increasing velocity and Froude number. The report also presents a collection of photographs illustrating the various bed forms and configurations. In Article 47 of this chapter, photographs and sketches of the various types of bed forms are presented,

TABLE 2.12.—Summary Description of Bed Forms and Configurations Affecting Alluvial Channel Roughness [Material Taken from ASCE (1966) Task Force on Bed Forms]

Bed form or configuration (1)	Dimensions (2)	Shape (3)	Behavior and occurrence (4)
Ripples	Wavelength less than approx 1 ft; height less than approx 0.1 ft.	Roughly triangular in profile, with gentle, slightly convex upstream slopes and downstream slopes nearly equal to the angle of repose. Generally short-crested and three-dimensional	Move downstream with velocity much less than that of the flow. Generally do not occur in sediments coarser than about 0.6 mm
Bars	Lengths comparable to the channel width. Height comparable to mean flow depth	Profile similar to ripples. Plan form variable	Four types of bars are distinguished: (1) Point; (2) alternating; (3) Transverse; and (4) Tributary. Ripples may occur on upstream slopes
Dunes	Wavelength and height greater than ripples but less than bars	Similar to ripples	Upstream slopes of dunes may be covered with ripples. Dunes migrate downstream in manner similar to ripples
Transition	Vary widely	Vary widely	A configuration consisting of a heterogeneous array of bed forms, primarily low-amplitude ripples and dunes interspersed with flat regions
Flat bed	—	—	A bed surface devoid of bed forms. May not occur for some ranges of depth and sand size
Antidunes	Wave length = $2\pi V^2/g$ (approx)[a] Height depends on depth and velocity of flow	Nearly sinusoidal in profile. Crest length comparable to wavelength	In phase with and strongly interact with gravity water-surface waves. May move upstream, downstream, or remain stationary, depending on properties of flow and sediment.

[a] Reported by Kennedy (1969).

together with further description of field and laboratory observations on their properties and behavior and a brief survey of theoretical models that have been developed to explain their occurrence.

The mechanisms responsible for the formation of the various bed forms are by no means fully understood as yet. However, as Kennedy (1969) points out in his survey paper on the formation of ripples, dunes, and antidunes, there are several general features of their occurrence and behavior that are self-evident and common to all of them, as follows. Firstly, all sedimentary bed forms are the result of an orderly pattern of scour and deposition. Their growth results from material being scoured from the trough regions and deposited over the crests, and continues until there is no further net scour from the bottoms of the troughs or net deposition at the peaks of the crests. This mechanism is shown in Fig. 2.54 [adapted from Kennedy (1969)], which shows for idealized bed-form geometry how ripples and dunes increase in amplitude and migrate downstream, and how antidunes grow and move upstream or downstream. In this figure, u is the local perturbation velocity near the bed in the streamwise direction induced by the waviness of the lower boundary, $\eta(x,t)$ is the bed profile, and δ is the distance by which changes in the local sediment transport rate lag changes in the local velocity near the bed. The transport rate has been assumed to vary as a power of the velocity, and thus increases (decreases) in velocity produce corresponding increases (decreases) in the local transport rate, but shifted a distance, δ, downstream. Fig. 2.54 also shows how under conditions in which scour occurs over the crests and deposition in the troughs any chance bed disturbance is diminished, leading to the flat bed. Note that regardless of the details of the responsible mechanisms, growth, or diminution of the bed forms requires a phase shift, δ in Fig. 2.54, between the bed profile and pattern of scour and deposition, and thus also between the bed profile and streamwise distribution of the local sediment transport rate. Fig. 2.54 also makes it evident that as bed forms approach their equilibrium amplitude and their growth ceases, δ must tend to zero for antidunes moving upstream, $L/2$ for antidunes moving downstream, and L for ripples and dunes. Many factors, most of them as yet not formulated, probably contribute to δ; the dominant constituent phenomena appear to be the phase shift between the local bed elevation and local flow properties (boundary shear stress, mean velocity, turbulence intensity, etc.) and the phase shift between the local flow properties and local transport rate. But regardless of the origin of δ, purely kinematical considerations dictate that the bed waves cannot form unless the sediment transport rate is diminishing in the streamwise direction over the crest regions and increasing in the streamwise direction over the troughs. The converse situation, increasing transport rate over the crests and decreasing transport over the troughs, holds for flat beds, which accompany flows such that any disturbance is spontaneously reduced in amplitude until it is obliterated.

The second general observation is that bed forms are the consequence of an instability phenomenon. A small disturbance on an initially flat bed will, under some conditions, affect the flow and the local sediment transport rate in such a way that deposition and scour occur over the crests and troughs, respectively, and increase the amplitude of the initial bed undulation, as shown in Figs. 2.54(a), 2.54(b), and 2.54(d). As the bed profile disturbance increases in amplitude it further perturbs the local sediment transport rate that in turn increases the rates of local scour and deposition and thereby further enhances the growth rate of the

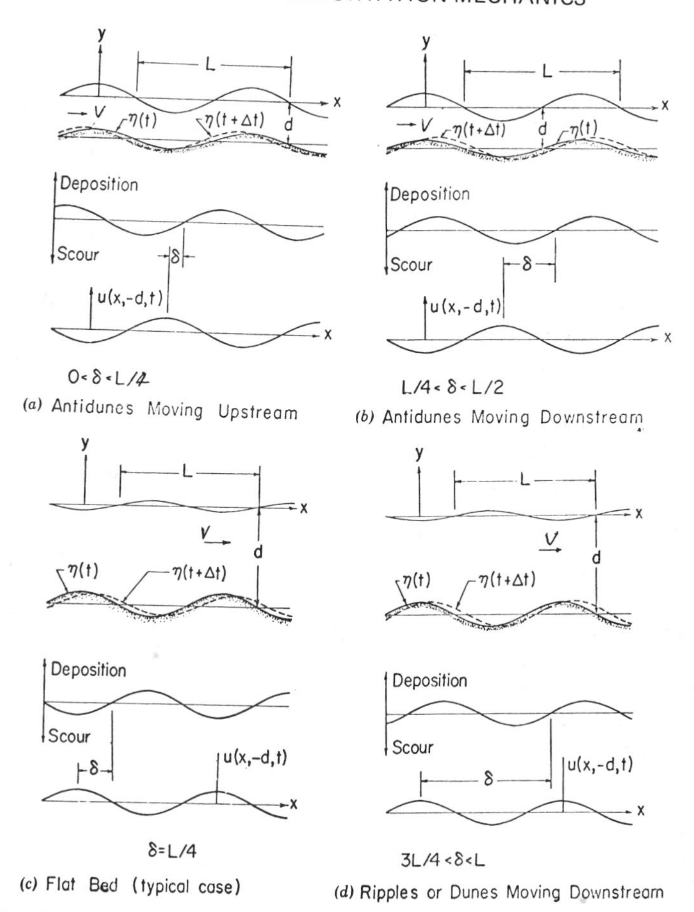

FIG. 2.54.—Schematic Representation of Mechanism of Formation of Antidunes, Flat Beds, and Ripples and Dunes (Kennedy, 1969)

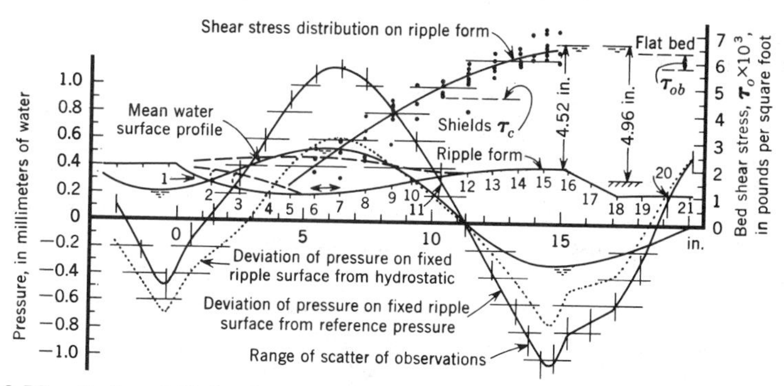

FIG. 2.55.—Bed and Water Surface Profiles and Distributions of Bed Pressure and Shear Stress, Measures by Raudkivi (1963, Fig. 4) Using Fixed, Smooth Ripple Form (d = 4.96 in., V = 0.98 fps; Flume Width = 3 in.)

bed forms, and so on, until factors associated with the increased amplitude of the bed waves intervene and limit their growth and fix their equilibrium amplitude. Under other flow conditions an initial disturbance of any wavelength will affect the distribution of the local sediment transport capacity in such a way that the amplitude of the bed perturbation is diminished; whence the flat bed configuration [see Fig. 2.54(c)]. These general features, an initial perturbation disturbing the flow so as to increase (instability) or decrease (stability) the magnitude of the disturbance, are common to many flow phenomena, e.g., the origin of turbulence, hydroelastic vibrations, surface-tension breakup of laminar jets, etc.

A final observation of sufficient generality to warrant mention herein is that each bed configuration possesses a geometrical characterization that is dependent upon the flow parameters and bed-sediment properties. For simple, orderly bed forms, e.g., antidunes or wave-generated and aeolian ripples, the wavelength and height of the features usually suffice to give a virtually complete description of the configuration. For the less regular bed configurations, e.g., aqueous current ripples and dunes, topographic regularity may be present only in the statistical sense, and it may be necessary to use the spectral density function, which was first applied to sedimentary bed forms by Nordin and Algert (1966), of the three-dimensional bed profile, or another statistical description, to obtain a meaningful representation of its geometry. The significant point is that the bed topography is dependent in an orderly way on the properties of the generating flow, including its sediment transport rate, and on the bed material in which it is formed.

Ripples and dunes play a particularly significant roles in the makeup of the hydraulic roughness of an alluvial channel. The lee slopes of these bed forms are regions of strong adverse pressure gradients that invariably produce flow separation that gives rise on the downstream ripple slopes to pressures that are lower than those at the same elevations on the upstream slopes. This is shown in Fig. 2.55 by the experimental data of Raudkivi (1963), obtained with flow over a rigid bed whose profile duplicated that of a two-dimensional ripple configuration; similar pressure distributions were measured by Vanoni and Hwang (1967) on solidified sediment ripples. Thus, these bed forms behave much as bluff bodies do, and can exert significant upstream drag forces on the flow. For example, Nordin (1964) reports that the Darcy-Weisbach friction factors for the ripple and dune-bed flows represented in Fig. 2.52(a) were 4.5–8.7 times as great as a pipe friction diagram would predict for the observed flow over a flat bed with immobile sand-grain roughness. [Pipe friction diagrams based on Nikuradse's sand-grain roughened pipes or data from commercial pipes are included in most texts on fluid mechanics and hydraulics; see, e.g., Streeter (1962), pp. 218 and 220.] As was previously discussed, the geometry and thus also the hydraulic roughness of the bed configuration depend upon, among other factors, the depth and velocity of flow. But for a given discharge of water and sediment the depth and velocity of flow are regulated in part by the roughness of the bed. Since the analyst is presented with no obvious point of departure in the problem, it is worthwhile to consider the alternate approaches possible.

41. Approaches to Problem.—River systems and river processes are so complex that there is not even general accord on which aspects of their behavior are causes and which are effects. Apparently much of the confusion stems from not making adequate distinctions between long and short-term behavior of rivers, between the functioning of natural rivers and laboratory flumes, and between single-valued and multiple-valued relationships. It is necessary that these distinctions be made and their implications be understood before any meaningful discussion of river hydraulics can be undertaken.

Consider first the question of single and multiple-valued relations. A dependent variable is said to be a single-valued function of a group of independent variables if for each set of values of the independent variable the dependent variable takes on one and only one value. Thus in Fig. 2.53, the friction factor for the bed, f_b, is a single-valued function of the mean velocity, V; for each value of V, there is only one value of f_b. It so happens in this case that V is also a single-valued function of f_b. For certain values of slope S, on the other hand, e.g., $S = 0.0024$, there are three possible different values of V; thus V is a multiple-valued function of S. However, S is a single-valued function of V since for each velocity there is only one slope. Likewise, the bed shear velocity, U_{*b}, is a single-valued function of V, but V is a multiple-valued function of U_{*b}.

An independent variable is one whose value is imposed externally on the system, and a dependent variable is one whose value is determined by the values specified for the independent variables. In designating the independent variables for a system one must take care not to specify too many, which can give rise to an unrealizable situation, or too few, which may not determine a unique situation. For example, for the flows summarized in Fig. 2.53, specification of the depth, d, and S (in addition, of course, to the sediment and fluid properties and channel width) does not determine a unique situation; as many as three different values of V and sediment discharge, G_s, may result. If, however, the values of d and V are specified, then a specific flow is described in which S and G_s take on unique values. This would be the case also if d and G_s were chosen as the independent variables. An overconstrained, and thus unattainable situation could result if d, S, and G_s were specified; there may be no value of V possible with the given d and S that could transport the specified G_s.

In determining how many and which independent variables specify a unique, realizable physical situation it is often helpful to consider what the inputs and responses are in the system under consideration as it occurs naturally. For example, the inputs to a river reach are the water and sediment discharges, and the primary responses are the width, depth, and velocity of flow, sediment discharge through the reach, and rates of water and sediment storage (plus or minus) in the reach. The bed roughness and friction factor may be regarded as secondary responses; their values are interrelated with the depth and velocity of flow, sediment transport rate, and probably also the rate of scour or deposition.

To generalize on the question of how many independent variables one can specify, consider, following Kennedy and Brooks (1963), the elementary case of an alluvial stream flowing with steady uniform depth and velocity. The relevant quantities are: water discharge, Q; sediment discharge (of sizes found in the bed material), G_s; concentration of wash load (clay-sized particles and other fine material not present in significant quantities in the bed), C_w; width, b; mean depth, d; hydraulic radius, r; mean velocity, V; energy gradient, S; Darcy-Weisbach friction factor, f; fluid kinematic viscosity, ν; fluid density, ρ; particle density, ρ_s; geometric mean size of bed material, d_g; geometric standard deviation of particle sizes in the bed material, σ_g; mean settling velocity of particles, w; gravitational constant, g; and plan-form geometry (meander dimensions, etc.).

The relationships that are known (in principle at least), or are known to exist in the stream, between the foregoing variables are:

1. Continuity

$$Q = Vbd \qquad (2.130)$$

2. Definition of f

$$V = \sqrt{\frac{8g}{f}}\sqrt{rS} \qquad (2.131)$$

3. Roughness or friction-factor relation; not concisely formulated.
4. Sediment transport relation: not concisely formulated.
5. Width-depth relation: not concisely formulated; b = constant for flumes.
6. Hydraulic radius-depth relation (a channel shape factor): not concisely formulated.
7. Relation between channel plan-form and flow and sediment properties: not formulated.

Since there are 17 basic variables possessing seven interdependencies, there are 10 independent and seven dependent variables. For a laboratory flume with fixed walls, or a river channel or canal with fixed banks, the width and plan-form are predetermined and thus relations 5 and 7 are deleted from the foregoing list. Therefore, if 10 variables (12 in flumes) are selected or specified, seven variables (five in flumes) will be physically determined by the flow. In flumes it has been found that not all selections of independent variables will specify unique flows, as previously discussed. For rivers, however, the situation is not so clear-cut because it is not possible to control the variables as needed to make the definitive experiments.

Finally, it is necessary to differentiate between long-term and short-term behavior of natural alluvial streams. Over the periods of geologic time a river evolves in such a way that, over the long run, it can transport the sediment delivered to it with the available water runoff. The adjustment occurs in the slope, width, and plan-form of the stream. If, e.g., the sediment brought into the upper reaches of a stream over a period of many years is too great to be carried with the available water discharge, deposition will occur, the stream will steepen, and the velocity will increase until the sediment transport capacity of the stream equals the imposed sediment discharge. Adjustments may also occur in the width-depth ratio or degree of meandering of the stream, both of which also affect its sediment and water transport capacities. Thus, for very long periods it appears logical to treat the water and sediment discharges as the primary independent variables, and the channel slope and other geometrical characteristics as dependent variables. To be sure, the water and sediment discharges imposed on a stream by a watershed are not wholly independent; indeed they generally take on related sequences of values, often with identifiable statistical characteristics. But regardless of their independence or interdependence, the stream must evolve so that over the long term each reach transmits as much water and sediment as are delivered to its upstream end; continuity must prevail! This is the concept of the graded stream proposed by Mackin (1948). Considering still a longer time scale, the evolution of a stream is governed by the geology and climate of the region, the properties of matter, etc. From this point of view, every property of a river is a dependent variable governed by external factors.

Viewed against a shorter time scale (e.g., days, weeks, or months), streams seldom if ever achieve a strictly steady state, because of changes in the water and sediment discharges and the finite time required for adjustments to occur in the channel roughness (bed forms), flow depth, etc. Short-term transitory accommodation of the differences between the water and sediment discharges into and out of a reach is accomplished through storage or depletion of water by a rise or fall in river stage or overbank flooding, and of sediment by deposition or erosion. The depth and velocity of flow will adjust as required to produce continuity of water and sediment motion and continuous bed, water surface, and energy profiles throughout the reach at any moment. Over the periods of time under consideration, the average channel slope must remain practically constant over reaches of appreciable length, although it can vary significantly over short reaches. On the one hand, the local sediment transport rate (at a river section) must depend on the local flow properties, and sediment will be scoured or deposited as required to simultaneously satisfy the requirements of sediment continuity and the physical relationship between the fluid motion and sediment transport rate. On the other hand, the sediment discharge is instrumental in setting the depth and velocity for a given water discharge, or the velocity for a given depth and slope; e.g., it is the sediment discharge that determines which of the three alternate velocities possible for some slopes in Fig. 2.53 will occur.

Therefore, over the short term the slope and water discharge are logically viewed as independent variables in the hydraulic relations at a section. The local rate of sediment transport at a river section is dependent upon the depth and velocity of flow, and thus from this perspective, G_s is a dependent variable. The sediment discharge performs as an independent variable in that it is one of the factors determining which of a variety or range of values of possible depths and velocities will occur. However, there may be no depth-velocity combination that is compatible with the water discharge and channel slope and that has a transport capacity just equal to the sediment discharge imposed on the reach, in which case bed erosion or deposition will occur.

In Table 2.13, adapted from Kennedy and Brooks (1963), an attempt has been made to summarize the foregoing discussion of the hydraulic behavior of alluvial streams. In this table various selections of dependent and independent variables are included to indicate which choices yield single-valued functional relations. Plan-form geometry and wash-load concentration have not been included in this table.

The subsequent articles will focus on the problem of calculating depth-discharge relations for alluvial streams. At this point it is in order to examine in the light of the foregoing discussion the possible approaches to the problem. Firstly, it must be borne in mind that the bed configuration plays a major role in fixing the channel roughness, and thus it is necessary to take into account the bed-form geometry. This can be done indirectly by assuming that the bed geometry and its roughness depend on the same factors as does the energy gradient, and attempting to relate directly V, d, and S or U_*, and also the fluid and sediment properties. The alternate approach is to examine first the relationship between the flow characteristics and the bed configuration to determine the geometry and hydraulic roughness of the bed, and then calculate the friction factor and the energy gradient. Secondly, care should be exercised in the selection of the independent variables. Table 2.13 shows that any calculation that uses slope (or shear stress or shear velocity) as an independent variable cannot be expected

TABLE 2.13.—Choices of Independent and Dependent Variables for Flow in Alluvial Streams (Adapted from Kennedy and Brooks, 1963)

	Independent Variables[a]			Functional Relations	
System (1)	Properties of fluid, sediment, gravity, etc. (2)	Characteristics of flow systems (not all combinations listed) (3)	Dependent[a] variables (not all combinations listed) (4)	Single-valued (5)	Multiple-valued for some ranges (6)
Flumes	$\nu, \rho, \rho_s, d_g, \sigma_g, w, g$	Q, d, b	G_s, V, r, S, f	x	
		Q, G_s, b	d, r, U, S, f	x	
		V, d, b	Q, G_s, r, S, f	x	
		d, S, b	Q, G_s, r, V, f		x
		r, S, b	Q, G_s, d, V, f		x
		Q, S, b	G_s, d, r, V, f		x
Natural streams Short term	$\nu, \rho, \rho_s, d_g, \sigma_g, w, g$	Q, d	G_s, b, r, V, S, f	x	
		d, S	Q, G_s, b, r, V, f		x
		r, S	Q, G_s, b, d, V, f		x
		Q, S	G_s, b, d, r, V, f		x
Long term (graded stream)	ν, ρ, ρ_s, g	Q, G_s	$b, d, r, V, S, f, d_g, \sigma_g, w$	x	
Very long term	ν, ρ, ρ_s, g, geology	climate, man-made works	$Q, G_s, b, d, r, V, S, f, d_g, \sigma_g, w$	x	

[a]Note that plan-form geometry and wash-load concentration are not considered.

to yield unique relations among the variables of interest for the full range of flows and bed configurations occurring in alluvial streams. On the other hand, analyses that treat slope as a dependent variable can produce single-valued relationships. It must also be pointed out (see, e.g., Fig. 2.53) that shear velocity and thus also the related variables, shear stress and slope, undergo relatively small variations in alluvial channel flows. This also diminishes their usefulness compared to velocity and depth, or discharge and depth, as a starting point for any calculation.

Many aspects of flow in rigid and mobile boundary channels are quite similar. For example, the mean velocity is distributed according to the logarithmic law (ASCE, 1963), the shear stress is linearly distributed in two-dimensional flows, etc. A thoroughgoing treatment of friction factors in rigid open channels will not be included herein; instead, the reader is referred to recent survey papers (Rouse, 1965; ASCE, 1963) and textbooks (Chow, 1959; Henderson, 1966) on this subject.

42. Stage-Discharge Predictors for Alluvial Channels.—No attempt will be made herein to review all of the predictors that have been developed for alluvial channel depth-discharge relations. Many of the techniques that have been proposed have been found not to have adequate reliability or sufficiently wide applicability to be of lasting value. Others that presented significant advances when first presented have been supplanted by later developments, while still others are not sufficiently unlike similar techniques to warrant separate attention. The analyses and investigations summarized herein were selected because of their historical significance, unique features, or present value; they are presented in roughly chronological order of their development, and the date each was published is noted. The limitations of the various methods and a general evaluation of them will be given.

Regime Formulation (1895-1970).—The regime formulation traces its origin to British engineers who were charged late in the 19th century with designing and operating extensive irrigation systems in India. They recognized that in channels that performed satisfactorily the depth and velocity of flow were such that at or near the design flow, the water and sediment discharges were in equilibrium and produced no objectionable scour or deposition. Such channels were said to be "in regime." Not all canals constructed exhibited this happy balance, and so it was natural for the engineers to seek to determine what was unique about those channels that did function well. This was done by obtaining data from "regime" channels, plotting the variables judged pertinent in various combinations, and trying thereby to discover what relations hold among the hydraulic and sediment characteristics obtained in these channels. More recent researchers in this field have also examined the interrelations existing in natural streams, which, as previously discussed, generally evolve over geologic time to be "in regime." The forerunner of all regime relations is the famous Kennedy (1895) formula for the nonsilting, nonscouring velocity

$$V = 0.84\, d^{0.64} \text{ (ft-sec units)} \qquad (2.132)$$

the many deficiencies of which are apparent. Kennedy's equation was followed by a host of other regime formulas, which are concisely summarized in the report of the ASCE Task Force on Friction Factors in Open Channels (1963), and graphical relations. Recent contributions include the graphical relations presented by Simons and Albertson (1963) and by Blench (1969).

Most regime formulations include three relations that yield values of the channel width, depth, and slope as functions of the water discharge and bed material size. Some of the more refined ones also take into account bank cohesiveness, sediment discharge or concentration, and fluid viscosity. A good example of a regime formulation is that of Blench (1970). The width is given by

$$b = \sqrt{\frac{F_b Q}{F_s}} \qquad (2.133)$$

in which F_b, the so-called bed factor, is g times the Froude number squared:

$$F_b = \frac{V^2}{d} \qquad (2.134)$$

and the side factor, F_s, is defined as

$$F_s = \frac{V^3}{b} \qquad (2.135)$$

The depth equation is

$$d = \sqrt[3]{\frac{F_s Q}{F_b^2}} \qquad (2.136)$$

while the slope is expressed as

$$S = \frac{F_b^{7/8}}{K b^{1/4} d^{1/8}\left(1 + \dfrac{C}{2{,}330}\right)} \qquad (2.137)$$

in which C = the concentration, in parts per million, of bed material transported by the flow, and

$$K = \frac{3.63\, g}{\nu^{1/4}} \qquad (2.138)$$

Foot-pound-second units are to be used in these relations. The bed and side factors, F_b and F_s, are generally determined largely on the basis of the engineer's experience with flows similar to the one being analyzed or designed for. In the absence of better information, Blench suggests that the following values be used:

$$F_b = 1.9\sqrt{d_g} \qquad (2.139)$$

in which d_g is in millimeters; and F_b is in feet per second squared

$$\left.\begin{aligned} F_s &= 0.10 \text{ for friable banks} \\ F_s &= 0.20 \text{ for silty, clay, loam banks} \\ F_s &= 0.30 \text{ for tough clay banks} \end{aligned}\right\} \qquad (2.140)$$

More complete guidelines for determining F_b and F_s are given by Blench (1966a).

One of the major deficiencies of the regime approach is that the relations for the depth, width, and slope generally involve one or more loosely defined quantities, e.g., the bed and side factors in Blench's formulation, which must be estimated largely on the judgment and experience of the engineer. Regime relations should never, of course, be applied in cases in which the flow, sediment transport, and channel characteristics differ widely from those from which the particular formulation was derived. In general, they are applicable only to flows at low Froude numbers, in the ripple-dune regime. Blench's monograph (1966a) gives a rather thorough exposition of the "regime" philosophy, and his paper (Blench, 1969) on "Coordination in Mobile-Bed Hydraulics" provides further guidance to the application of the regime approach; one should study these and related publications before undertaking use of the regime method.

Chien (1957), Henderson (1961), and Gill (1968) have presented examinations of the physical basis for the regime relations.

Einstein-Barbarossa Analysis (1952).—Einstein and Barbarossa (1952) were the first to develop a depth-discharge predictor taking any formal account of the contribution the bed forms make to the channel roughness. They proposed that the cross-sectional area, A, and hydraulic radius, r, of the channel each be treated as consisting of two additive parts: one section, with area A' and corresponding hydraulic radius r', in which the component of the gravitation force along the channel is balanced by the hydrodynamic force exerted on the grain roughness; and a second section, of area A'' and hydraulic radius r'', for which the gravity force is balanced by the drag exerted on the bed forms and other channel irregularities. This division of the cross section is a useful conceptual tool; in fact, however, there is no such strict geometric division, the streamwise gravitational force on each fluid element actually being balanced by turbulent momentum exchange that is generated by the composite grain and bed form roughness.

Relations developed for flow in rigid-boundary channels and conduits were used to calculate A' and r'. In particular, Einstein and Barbarossa recommended use of the Manning-Strickler equation:

$$\frac{V}{U'_*} = 7.66\left(\frac{r'}{k_s}\right)^{1/6} \qquad (2.141)$$

in which $U'_* = \sqrt{gr'S}$ .... (2.142)

is the shear velocity corresponding to the grain roughness; and k_s, the equivalent fixed sand grain roughness diameter, was taken to be d_{65}, the bed sediment size such that 65% of the material is finer. For cases in which the grain roughness does not produce a hydraulically rough surface [i.e., when $k_s U'_*/(11.6\mu)$ is less than about five] a logarithmic type formula was recommended:

$$\frac{V}{U'_*} = 5.75 \log\left(12.2\,\frac{r'}{k_s}\,x\right) \qquad (2.143)$$

in which x, a function of $k_s U'_*/(11.6\nu)$ given in Fig. 2.97, is a correction factor accounting for the effects of viscosity.

Einstein and Barbarossa then argued that the bed form contribution to the friction factor

$$f'' = 8\left(\frac{U''_*}{V}\right)^2 = 8\,\frac{gr''S}{V^2} \qquad (2.144)$$

must be dependent on the topography of the bed configuration, which in turn is a function of the sediment transport rate along the channel bed. But according to Einstein's (1942) concept, the bed load transport depends solely upon the dimensionless variable, ψ':

$$\psi' = \frac{\rho_s - \rho}{\rho}\,\frac{d_{35}}{r'S} \qquad (2.145)$$

in which ρ_s and ρ = the densities of the sediment and fluid, respectively. Thus they sought among data obtained from natural streams in California and the Missouri River basin a relation between ψ' and V/U''_*, and obtained the result

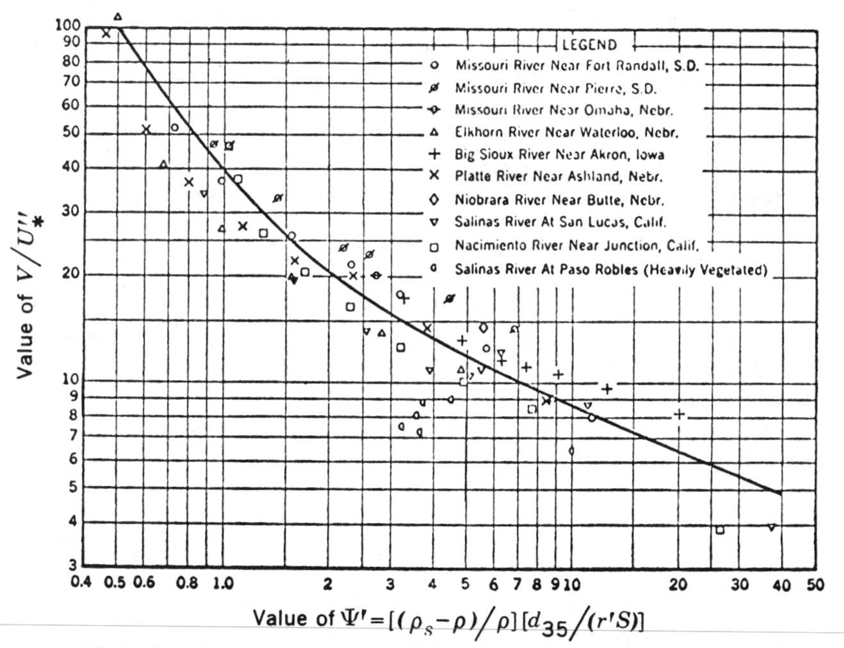

FIG. 2.56.—Einstein-Barbarossa (1952, Fig. 3) Bar Resistance Graph

shown in Fig. 2.56. This graphical relation has come to be known as the bar resistance curve.

Determination of a depth-discharge relation using the Einstein-Barbarossa method for a channel with a known cross section, slope, and bed sediment size distribution proceeds as follows:

1. Select a value of r', calculate $U'_* = \sqrt{gr'S}$ and ψ', and determine V from Eq. 2.141 or Eq. 2.143.
2. Obtain V/U''_* from Fig. 2.56, and calculate r''.
3. Calculate $r = r' + r''$.
4. Determine A and d from curves of r versus A and d, based on the known channel section dimensions.
5. Calculate $Q = VA$.

Einstein and Barbarossa indicated that in the case of steep banks the friction loss on the banks should be calculated separately according to the side-wall correction procedure outlined in Article 44 and their procedure applied only to the area, A_b, and hydraulic radius, r_b, of the bed section. They also pointed out that in applying their method, an average slope and average channel cross section dimensions obtained along a reach of the river should be used instead of the slope and cross section at one station. The irregularities and nonuniformities that invariably characterize river channels make this very sound advice, which all to often has been overlooked or ignored, as applicable to other analyses and methods as it is to theirs.

Shen (1962) undertook to improve the Einstein-Barbarossa method, and in particular to extend it to materials other than sand and to eliminate the systematic deviation that had been found to exist between the Einstein-Barbarossa curve (Fig. 2.56) and flume data. He found that for $wd_{50}/\nu > 100$ (approx), U''_*/V may be expressed as a function of ψ' alone, but for $wd_{50}/\nu < 100$ (approx), U''_*/V is a function of ψ' and wd_{50}/ν. His principal results are presented in two graphs: one depicting a correction factor, α, as a function of wd_{50}/ν, and a second giving U''_*/V as a multiple-valued function of ψ'/α.

Veiga da Cunha (1967) analyzed data from a Portugese river with a bed composed of relatively coarse material. He concluded that for ψ' greater than about five, U''_*/V is a double-valued function of ψ', and presented a graphical relation between these two quantities.

Garde and Raju's Analysis (1966).—Garde and Raju (1966) collected a large body of laboratory and field data obtained by other investigators and used it to evaluate the depth-discharge predictors proposed by Einstein and Barbarossa (1952), Shen (1962), and Liu and Hwang (1961). They concluded that none of these methods is particularly reliable and then proceeded to analyze the data to develop new relations. Several graphical renderings of the data led them to conclude that an equation of the Manning form with a variable coefficient should be adopted:

$$\frac{V}{\sqrt{\frac{\rho_s - \rho}{\rho} g d_{50}}} = K\left(\frac{r}{d_{50}}\right)^{2/3} \sqrt{S \frac{\rho}{\rho_s - \rho}} \qquad (2.146)$$

in which K = a function of $V/\sqrt{[(\rho_s - \rho)/\rho]gr}$ defined in graphs they developed from field and laboratory data. For sand bed channels, $K = 3.2$ for ripples and 6.0 for the antidune and transition regimes. The scatter exhibited in their key diagram (Fig. 8), does not lend support to the validity of their analysis, and the poor depth-discharge predictions calculated by Alam (1967) on the basis of their proposed technique and presented in his discussion of their paper casts further doubt on its usefulness.

Simons and Richardson's and Haynie and Simons' Analyses (1966–1968).—The United States Geological Survey conducted an extended laboratory study of alluvial channel flow, using the 8-ft wide, 150-ft long and 2-ft wide, 60-ft long laboratory flumes at Colorado State University. A summary of the data obtained in this investigation has been prepared by Guy, Simons, and Richardson (1966). Simons and Richardson (1966) analyzed these data together with those obtained in natural streams and irrigation canals by several other investigators with the goal of obtaining an improved friction factor predictor for alluvial channels. They proposed two different but closely related approaches to the problem: a slope-adjustment method and a depth-adjustment method.

In the first of these, the slope is regarded as consisting of two components and the Darcy-Weisbach friction factor is cast in the form

$$f = \frac{8}{\left(\frac{C}{\sqrt{g}}\right)^2} = \frac{8grS}{V^2} = \frac{8gr}{V^2}(S' + \Delta S) \qquad (2.147)$$

in which $C/\sqrt{g}$ = the nondimensional Chezy coefficient. The quantity S' = the slope that would be required for the specified discharge to flow at the given hydraulic radius if the bed were flat, nonerodible, and composed of the alluvial channel bed material; S = the actual slope for the alluvial channel flow with hydraulic radius r; and $\Delta S = S - S'$. The slope corresponding to just the grain roughness, S', is estimated from a logarithmic relation they deduced from an analysis of data they obtained with flat-bed flows in the laboratory flumes

$$\frac{C'}{\sqrt{g}} = \frac{V}{\sqrt{grS'}} = 7.4 \log \frac{d}{d_{85}} \qquad (2.148)$$

with C', the Chezy coefficient, corresponding to S'. Eqs. 2.147 and 2.148 can be combined to yield

$$\frac{C}{\sqrt{g}} = \left(\frac{C'}{\sqrt{g}}\right)\sqrt{\frac{S'}{S}} = \left(7.4 \log \frac{d}{d_{85}}\right)\sqrt{\frac{S'}{S}} \qquad (2.149)$$

The investigators obtained individual empirical relations for $\sqrt{S'/S}$ as functions of S, d, and U_* for the different sand sizes used in their laboratory study and the various bed configurations. They did not present a generally applicable means for estimating $\sqrt{S'/S}$.

In their second method, Simons and Richardson proposed that the depth be adjusted to account for the separation regions downstream from ripples and dunes, since these wake zones do not contribute to the cross-sectional area actually available for conveying fluid. The modified depth, d', and corresponding modified velocity, V', are related by the continuity equation

$$q = Vd = V'd' \qquad (2.150)$$

in which $d' = d - \Delta d$; and Δd = the depth adjustment. The velocity, V', was obtained from the Chezy equation using a constant average value of $C'/\sqrt{g}$ determined from laboratory flows in the flat bed regime for each individual sand size investigated. The depth adjustment was presented as a function of d in empirical graphical relations, a different graph being required for each bed configuration and each sand size. In similar plots for canal data it was found that Δd also depends on S. Graphs were presented for estimating Δd for various other types of roughness, e.g., rigid cubes and battens, rocks, cobbles, etc. As with the slope-adjustment method, no procedure generally applicable to a range of sediment sizes and all bed configurations was forthcoming for the prediction of Δd.

Haynie and Simons (1968) abandoned the slope and depth-correction methods in the development of their technique for designing stable channels in alluvium, and instead based their analysis of flows in the ripple and dune regimes on a velocity correction computed for a large number of river, canal, and laboratory flows as follows. For each flow, the velocity correction, ΔV, was calculated as the difference between the actual velocity and that predicted by Tracey and Lester's (1961) friction-factor relation for smooth rectangular channels:

$$-\frac{\Delta V}{U_*} = \frac{V}{U_*} - 5.75 \log \frac{U_* d}{\nu} + 2.5 \qquad (2.151)$$

The values of ΔV so determined were found to be related to r and S, as shown in Fig. 2.57. Application of the results of this analysis in the calculation of depth-discharge relations proceeds according to the following steps:

1. Select values of r, d, and S, and obtain ΔV from Fig. 2.57.
2. Compute $U_* = \sqrt{grS}$, $\Delta V/U_*$, and $U_* d/\nu$.
3. Compute the velocity, V^*, that would obtain if the boundary were smooth, $V^*/U_* = 5.75 \log [(U_* d)/\nu] + 2.5$.
4. Calculate the velocity to be expected in the alluvial channel, $V = V^* + \Delta V$.
5. Calculate the stream power, $\rho g V d S$, and for the known value of d_g ascertain from Simons and Richardson's (1966) empirical graph (derived from field and laboratory data) if the flow is expected to be in the ripple or dune regimes. If not, the estimate of V should be disallowed, since Fig. 2.57 is valid only for these bed configurations.

The investigators pointed out that the data used in preparing this graph had values of d_{50} ranging from 0.12 mm–0.82 mm, and thus Fig. 2.57 should not be applied to channels with d_{50} outside this range. The absence of any dependency of V on bed material size would appear to be a serious deficiency since it is well known [see Simons and Richardson (1966)] that f is heavily dependent on the size of the bed sediment.

Engelund's Analysis (1966–1967).—Engelund (1966) proposed a new line of analysis based on similarity considerations. In the closing discussion to his paper (Engelund, 1967) he introduced some modifications of his proposed model. The monograph authored by Engelund and Hansen (1967) presents a more complete exposition of the underlying similarity hypotheses utilized. Devlopment of the friction factor predictor he evolved proceeds as follows.

Let the energy gradient, S, again be treated as consisting of two components

$$S = S' + S'' \qquad (2.152)$$

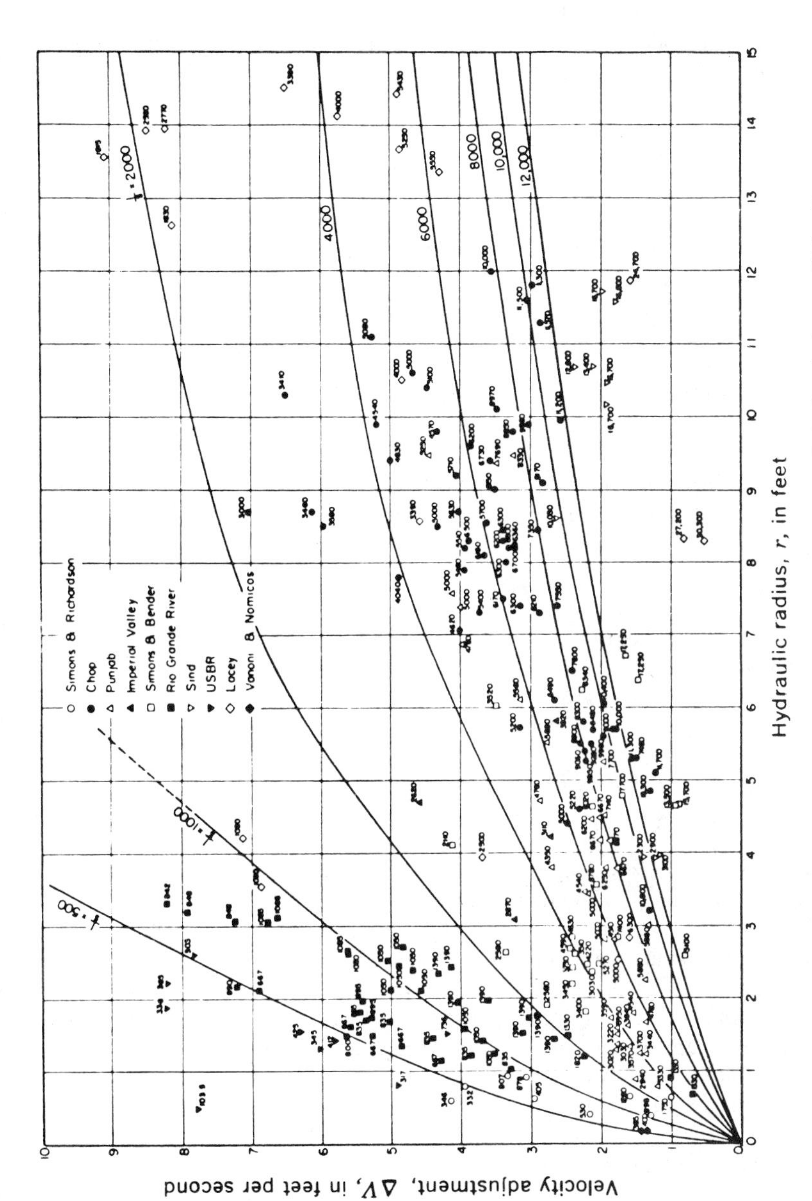

FIG. 2.57.—Haynie and Simons' (1966, Fig. 3) Empirical Relation between Velocity Adjustment ΔV, Hydraulic Radius, and Slope

in which S' = the grain-roughness slope discussed in conjunction with Eq. 2.147; and S'' = the additional slope engendered by the hydrodynamic drag on the bed forms. According to Meyer-Peter and Mueller (1948), the idea of dividing S into constituent components traces its origin to H. A. Einstein in 1939. Engelund reasons that S'' is due primarily to the expansion losses in the separation zones downstream from the ripples and dunes, and utilizes Carnot's formula for expansion losses in closed conduits to obtain for an estimate of S''

$$S'' = \alpha \frac{V^2}{8gL}\left(\frac{h}{d}\right)^2 \qquad (2.153)$$

in which h and L = the height (trough to crest) and wavelength, respectively, of the bed forms; and α = a geometric factor dependent on L, h, and d. Introduction of Eq. 2.153 into Eq. 2.152 expressed in terms of the friction factor defined by Eq. 2.131 yields

$$f = f' + f'' = f' + \alpha \frac{h^2}{dL} \qquad (2.154)$$

in which f' and f'' correspond to S' and S'', respectively, calculated on the basis of r and V. Engelund then introduced into his analyses the dimensionless shear stress, τ_*, and its components, τ'_* and τ''_*

$$\tau_* = \tau'_* + \tau''_* \qquad (2.155a)$$

in which $$\tau_* = \frac{\tau_o}{\gamma(s-1)d_s} = \frac{\gamma d S}{\gamma(s-1)d_s} \qquad (2.155b)$$

$$\tau'_* = \frac{d S'}{(s-1)d_s} \qquad (2.155c)$$

$$\tau''_* = \frac{\mathsf{F}^2}{8}\frac{\alpha h^2}{(s-1)d_s L} \qquad (2.155d)$$

in which γ = the specific weight of the fluid; s = specific gravity of the sediment; d_s = a representative grain diameter (usually d_g or d_{50}); and F = the Froude number. The channel has been regarded as sufficiently wide that the depth is an adequate approximation for the hydraulic radius.

The essence of Engelund's similarity hypothesis is: (1) In two dynamically similar streams τ'_* takes on equal values; and (2) in two dynamically similar streams the expansion loss (due to form roughness) is the same fraction of the total energy loss. Application of Eq. 2.154 to two different streams, identified by the subscripts 1 and 2, yields

$$f_1 = f'_1 + \frac{\alpha_1 h_1^2}{d_1 L_1};\ f_2 = f'_2 + \frac{\alpha_2 h_2^2}{d_2 L_2} \qquad (2.156)$$

After normalizing each of these equations by its respective value of f and setting the resulting expressions equal to each other, it is seen that the second similarity condition can be expressed as

$$\frac{f_1}{f_2} = \frac{f'_1}{f'_2} \qquad (2.157)$$

From the definitions of f and τ_*, Eq. 2.157 can be expressed as

$$\frac{\tau_{*1}}{\tau_{*2}} = \frac{\tau'_{*1}}{\tau'_{*2}} \qquad (2.158)$$

The first similarity hypothesis dictates that τ'_* must take on the same value in both flows, and therefore τ_* must also be equal in both. But if Eq. 2.158 is to be satisfied, this can be true in general only if τ_* is a function of just τ'_*.

To verify this conclusion Engelund used flume data reported by Guy, Simons, and Richardson (1966). The quantity, τ'_*, was calculated from

$$\tau'_* = \frac{\tau'}{\gamma(s-1)d_s} = \frac{dS'}{(s-1)d_s} = \frac{d'S}{(s-1)d_s} \qquad (2.159)$$

with d' obtained from a logarithmic resistance formula Engelund obtained from flume experiments

$$\frac{V}{\sqrt{gd'S}} = 6 + 2.5 \ln \frac{d'}{2d_{65}} \qquad (2.160)$$

The result is presented in Fig. 2.58. It is particularly noteworthy that τ'_* is a multiple-valued function of τ_*, and thus this graphical relation used as a depth-discharge predictor has the potential of producing discontinuous rating curves of the type shown in Fig. 2.52(b). In fact, Engelund (1967) demonstrated that his predictor yields a depth-velocity relation that is in very good agreement with Fig. 2.52(b).

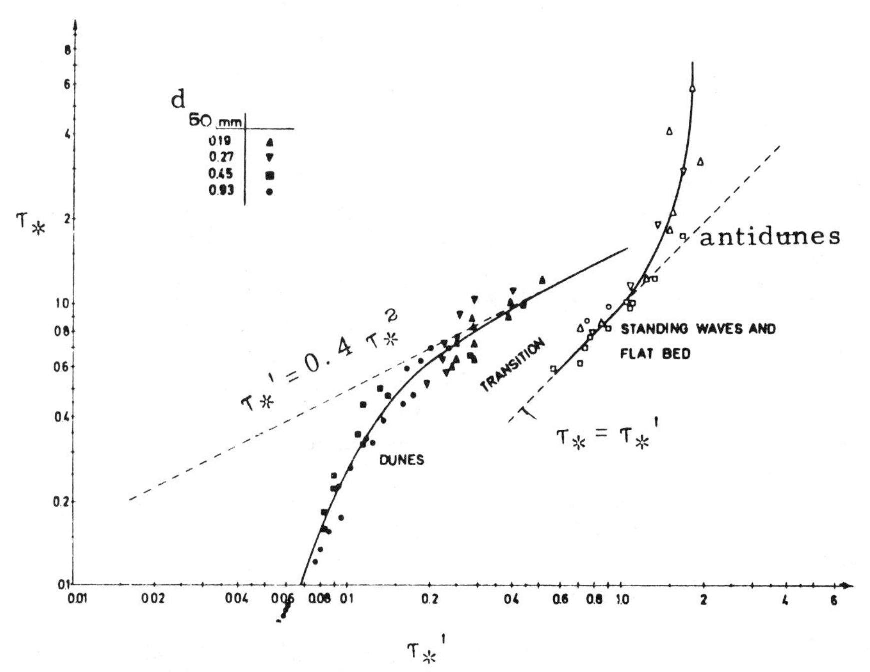

FIG. 2.58.—Engelund's (1967, Fig. 12) Universal Relation between Normalized Grain-Roughness Shear Stress $\tau'_* = \tau'_o/[\gamma(s-1)d_s]$ and Normalized Total Shear Stress $\tau_* = \tau_o/[\gamma(s-1)d_s]$

Calculation of the depth-velocity relation for a stream with known slope and distribution of bed particle sizes proceeds as follows:

1. Select a value of d' and calculate the corresponding values of τ'_* (Eq. 2.159) using $d_s = d_g$, and V (Eq. 2.160).
2. Obtain τ_* from Fig. 2.58.
3. Compute $d = [\tau_*(s - 1)d_s]/S$, again taking $d_s = d_g$.
4. Calculate the unit discharge (discharge per unit width), $q = Vd$.

A direct solution of these equations is given by Fig. 2.107.

The validity of Engelund's analysis rests heavily upon his similarity hypothesis, which has not yet been verified. The depth-discharge relations presented in his closing discussion (Engelund, 1967) tend to corroborate the reliability of his approach. The method does not, of course, take account explicitly of the effects of temperature and suspended-sediment concentration on the channel roughness; whether these effects are covered by the similarity hypothesis remains to be seen.

Vanoni and Hwang's Investigation (1967).—As previously noted, Einstein and Barbarossa (1952) were the first to attempt to take a formal accounting of the role of the bed forms in the makeup of alluvial channel roughness. Subsequently, investigations of rigid, geometrically regular arrays of roughness elements (generally cubes) on open channel boundaries were conducted by Koloseus and Davidian (1961), Sayre and Albertson (1963), O'Loughlin and MacDonald (1964), and Rouse (1965), as well as others. These investigations revealed that the size (height) and areal concentration of the elements are dominant and comparable in importance in determining the hydraulic roughness of a surface, and that there is an optimum concentration of 15% to 25%, depending on element shape and arrangement, that produces maximum surface roughness. Additionally, it was found that the semilogarithmic velocity distribution and friction factor laws are a valid basis for analysis of flow in channels with rough boundaries, provided an adequate accounting is taken of the roughness geometry.

Vanoni and Hwang (1967) undertook to determine experimentally what geometric characteristics of alluvial channel bed forms determine their hydraulic roughness. To this end, ripples generated by flows in laboratory flumes were solidified chemically in such a way that their shape and surface texture remained unaltered. The roughness characteristics of the stabilized bed forms were investigated at various velocities, and detailed observations and measurements were made on the flow field around the bed forms. The mean length, L, and mean trough-to-crest height, h, of the bed forms were measured directly. A measure of the areal concentration was obtained by outlining on plan-view photographs of the bed, the crests, and bases of the lee slopes of the ripples, and then measuring with a planimeter the areas so delineated. The upstream lighting used for the photographs made the lee slopes readily identifiable. The ratio of this area to the total area was the areal concentration, e, they utilized.

Vanoni and Hwang's principal findings may be summarized as follows:

1. Ripple-covered beds are hydrodynamically rough in the sense that the friction factor is independent of Reynolds number. The normalized pressure distribution on the ripples was also found to be Reynolds-number independent.
2. The pressure on a ripple attains its minimum value at the crest and its

maximum value on the upstream face near the point of reattachment of the separation streamline.

3. The quantity, eh, is an adequate length measure of the hydraulic roughness of a rippled bed, and the friction factor, f'', given by

$$f'' = f - f' \qquad (2.161)$$

in which f' = the friction factor obtained from the Nikuradse pipe friction diagram using r (rather than r' or d' as in the Einstein-Barbarossa and Engelund procedures, respectively) can be expressed as

$$\frac{1}{\sqrt{f''}} = 3.5 \log \frac{r_b}{eh} - 2.3 \qquad (2.162)$$

as is demonstrated in Fig. 2.59.

4. The ripple steepness, h/L, is inferior to e as a parameter describing the ripple roughness characteristics, and eh is a more consistent measure of roughness length than an equivalent sand grain roughness. Examination of data obtained by several researchers indicated that f'' can also be roughly estimated from

$$\frac{1}{\sqrt{f''}} = 3.3 \log \frac{Lr}{h^2} - 2.3 \qquad (2.163)$$

This equation is much less accurate than Eq. 2.162.

One generally has no estimate of e, and thus cannot use Eq. 2.162 as the basis for a depth-discharge calculation. The results of this investigation are nevertheless valuable because of the insight they give into the hydraulic roughness of bed forms.

Znamenskaya's Analysis (1967):—As part of a continuing research program in the Soviet Union on the mechanics of alluvial streams, Znamenskaya (1967) made extensive measurements of distributions of instantaneous velocities in flows in the ripple-dune regime. She distinguished three different zones on the basis of their

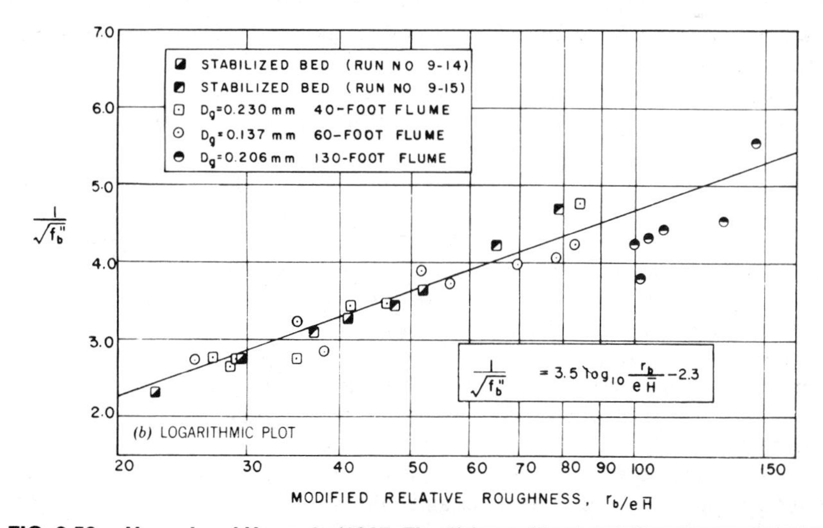

FIG. 2.59.—Vanoni and Hwang's (1967, Fig. 5) Logarithmic Friction Factor Relation

energy dissipation characteristics; the region above the bed where the streamlines are sinuous; the captive-eddy-regions in the lees of the bed forms; and the regions of intense shearing and micro-eddies just above the upstream slopes of the ripples and dunes. The rates of energy dissipation in the separation eddies and in the regions on the upstream slopes of the bed forms were found to be dominant by an order of magnitude or more over the dissipation rate in the zone of sinuous streamlines. Some success was enjoyed in obtaining empirical graphical relations between a geometric parameter consisting of the product of the bed form height and wavelength, the ratio of the energy dissipation rate in the intense eddies to the total dissipation rate, and the ratio of mean velocity to the critical velocity for incipient motion. A different relationship was obtained for each of several types of bed configurations. The relations so obtained were used with considerable success to make slope predictions for flows with known bed form dimensions reported by other investigators.

Raudkivi's Graphical Relations (1967).—Raudkivi (1967) conducted a set of laboratory flume experiments using 0.4 mm diam sand, and analyzed the data obtained in his investigation together with those reported by others for field and laboratory streams. Some of the laboratory data he utilized were obtained in experiments with lightweight sediments. Dimensional analysis and physical reasoning, guided largely by his experiments on the distribution of shear stress and pressure over ripples and dunes (see Fig. 2.55), led him to a graphical relation that he later modified slightly (Raudkivi, 1971), with the result shown in Fig. 2.60, in which U_{*c} is the shear velocity at incipient sediment motion. If $U_{*c} \ll U_*$, Eq. 2.131 shows that the ordinate of Fig. 2.60 is proportional to $1/\sqrt{f}$. The left side of the solid curve in this figure corresponds to flows with velocities just above the critical value for initiation of motion, where f increases with V or U_*, a flow regime occurring at lower velocities than included in the experiments summarized in Fig. 2.53. Over a significant range of abcissa, $V/\sqrt{U_*^2 - U_{*c}^2}$ is seen to be a multiple-valued function of $U_*^2/[(s - 1)\, gd_{50}]$; in this region of the graph the bed configuration is undergoing transition from ripples and dunes to flat bed. Most of the points to the right of the multi-valued range correspond to the flat bed or antidune regimes. The ability of Raudkivi's relation to predict multiple-valued depth-velocity relations [of the type shown in Figs. 2.52(b) and 2.53] and the relation the diagram includes for gravel rivers are two of its attractive features.

The application of Raudkivi's graphical relation, Fig. 2.60, to calculate depth-velocity relations is quite straightforward in principle. However, it is not always altogether clear which of the various curves through the points should be used in a particular case. If a family of curves, a different one for each constant value of, e.g., a sediment-size parameter, could be drawn through the points, this difficulty might be minimized.

Flat Bed Friction Factors (1969).—Several of the analyses previously described used as a starting point the friction factor corresponding to the sand grain roughness of a flat bed, and assume that this can be predicted from one of the formulas developed for flow in rigid-boundary conduits if use is made of a suitable measure of the equivalent rigid-boundary sand grain roughness, k_s, for the erodible bed material. Einstein and Barbarossa used $k_s = d_{65}$; Engelund analyzed data from some experiments and concluded $k_s = 2d_{65}$; Simons and Richardson used $k_s = d_{85}$ in one of their analyses and a different constant value of Chezy's C for each sand size in the other. Haynie and Simons used a smooth-boundary

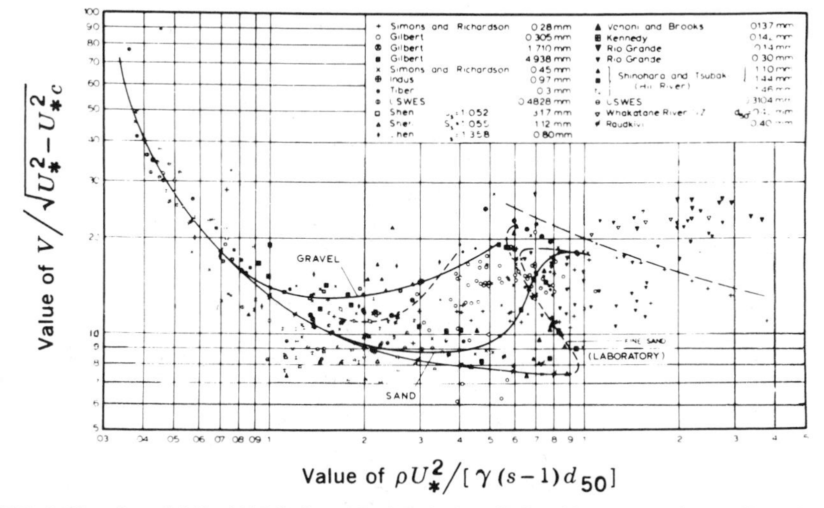

FIG. 2.60.—Raudkivi's (1971) Graphical Relation Giving Mean Velocity as Function of Shear Velocity

relation. The numerous obvious objections that can be raised to calculation of friction factors for sediment laden flows over mobile beds on the basis of formulas derived for sediment-free flows past fixed boundaries prompted Lovera and Kennedy (1969) to analyze data obtained from field and laboratory alluvial channel flows that are known, or can reasonably be assumed, to have produced flat beds and an active state of sediment transport.

Dimensional analysis and imposition of some only moderately restrictive assumptions (bed sediment characterized by d_{50} only; analysis limited to water and sand; no free-surface-wave effects, and thus gravity disregarded) led them to seek a relation of the form

$$f_f = f_f\left(\mathsf{R} = \frac{Vr_b}{\nu}, \frac{r_b}{d_{50}}\right) \quad \text{(2.164)}$$

in which f_f = the Darcy-Weisbach friction factor for a flat, erodible bed; and r_b = the hydraulic radius of just the bed section and is obtained from a side-wall correction calculation. The graphical relation they obtained is presented in Fig. 2.61. The variation of the friction factor with R for constant values of r_b/d_{50} is seen to be very dissimilar in flat bed alluvial channels and rigid conduits. In the latter case, the contours of constant r_b/d_{50}, or r/k_s, are nearly horizontal for the ranges of r_b/d_{50} and R shown, whereas for the mobile-bed case the friction factor is seen to be strongly dependent on both R and r_b/d_{50}. Superposition of the r/k_s contours obtained from either the Moody or Nikuradse pipe friction diagrams or one of their equaivalent logarithmic formulations shows that at the points of intersection, $r/k_s > r_b/d_{50}$, which explains why earlier investigators were led to adopt for k_s particle sizes larger than d_{50}. But as Fig. 2.61 reveals, the forms of the $f - r/k_s -$ R relations are fundamentally different in the cases of rigid-boundary and alluvial channel flows, and moreover they cannot be made the same by using for k_s some particular particle size. It is also noteworthy that no flows were found to have friction factors less than those of smooth boundary flows at the same Reynolds numbers.

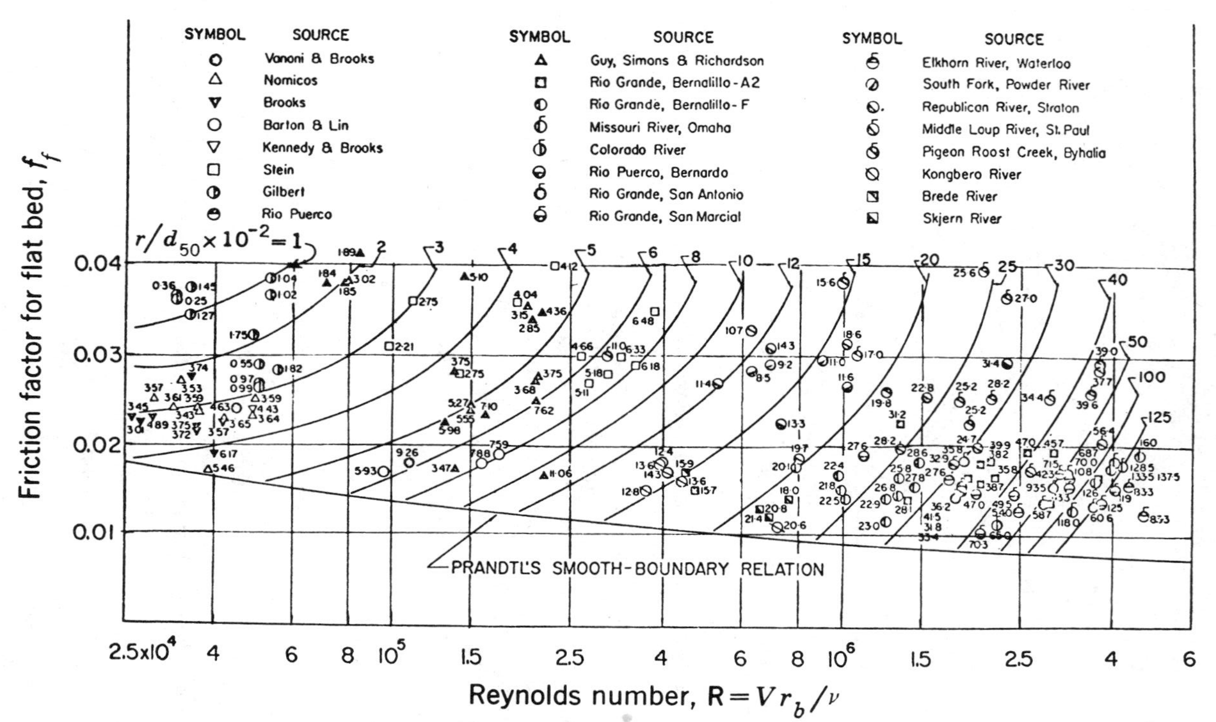

FIG. 2.61.—Lovera and Kennedy's (1969, Fig. 1) Flat Bed Friction Factor Diagram for Alluvial Streams

Alam and Kennedy's Friction Factor Chart (1969).—The results of investigations such as those previously discussed that showed a strong dependence of f'' the friction factor contribution resulting from the bed forms and other effects of boundary mobility, on the bed topography led Alam and Kennedy (1969) to attempt to obtain a generalized predictor for f'' as a function of those quantities that govern the occurrence and dimension of the bed forms. Following several unsuccessful attempts (Alam, et al., 1966) to obtain for f a predictor, that does not entail division of the slope or friction factor into components, they adopted as the basis for their predictor Eq. 2.152, with S' calculated using the flat bed friction factor graph proposed by Lovera and Kennedy and presented in Fig. 2.61:

$$S' = f_f \frac{V^2}{8gr_b} \qquad (2.165)$$

and $f = f_f + f''$ (2.166)

Dimensional analysis was used to identify the nondimensional groupings of terms governing the hydrodynamic drag exerted on a two-dimensional bed profile that can be characterized by two length parameters, L and h. It was assumed that the effects on f'' of the distribution of bed particle sizes and fluid viscosity are of secondary importance, and the analysis was limited to sand and water. Kennedy's (1963, 1969) theory of the mechanics of ripple and dune formation was used in determining the functional relationships between L and h and the independent variables utilized: V, d, g, r_b, and d_{50}. Incorporation of the expressions obtained for L and h into the relation for f'' resulting from the dimensional analysis yielded

$$f'' = f''\left(\frac{V}{\sqrt{gd_{50}}}, \frac{d_{50}}{r_b}\right) \qquad (2.167)$$

If a side-wall correction procedure (see Article 44) is utilized, f'' is replaced by f_b''; likewise, f is replaced by f_b in the calculation. Alam and Kennedy plotted the data presented by several investigators in the format of Eq. 2.167, with the result shown in Fig. 2.62.

Depth-discharge relations are calculated from Figs. 2.61 and 2.62 in the following way:

1. From known values of ν and d_{50}, a selected value of V, and an assumed value of r_b, calculate $V/\sqrt{gd_{50}}$, r_b/d_{50}, and Vr_b/ν.
2. Obtain f_f from Fig. 2.61 and f'' from Fig. 2.62. Use the value of f_f corresponding to smooth boundaries if the $\mathsf{R} - r_b/d_{50}$ intersection would fall below the smooth boundary relation.
3. Calculate $f = f_f + f''$.
4. Calculate the corresponding hydraulic radius using f and the known value of S

$$r_b = f \frac{V^2}{8gS} \qquad (2.168)$$

5. Compare the calculated and assumed values of r_b. If they are not in satisfactory agreement, use the value of r_b calculated from Eq. 2.168 and repeat the procedure. Iterate until the assumed and calculated values of r_b are equal.

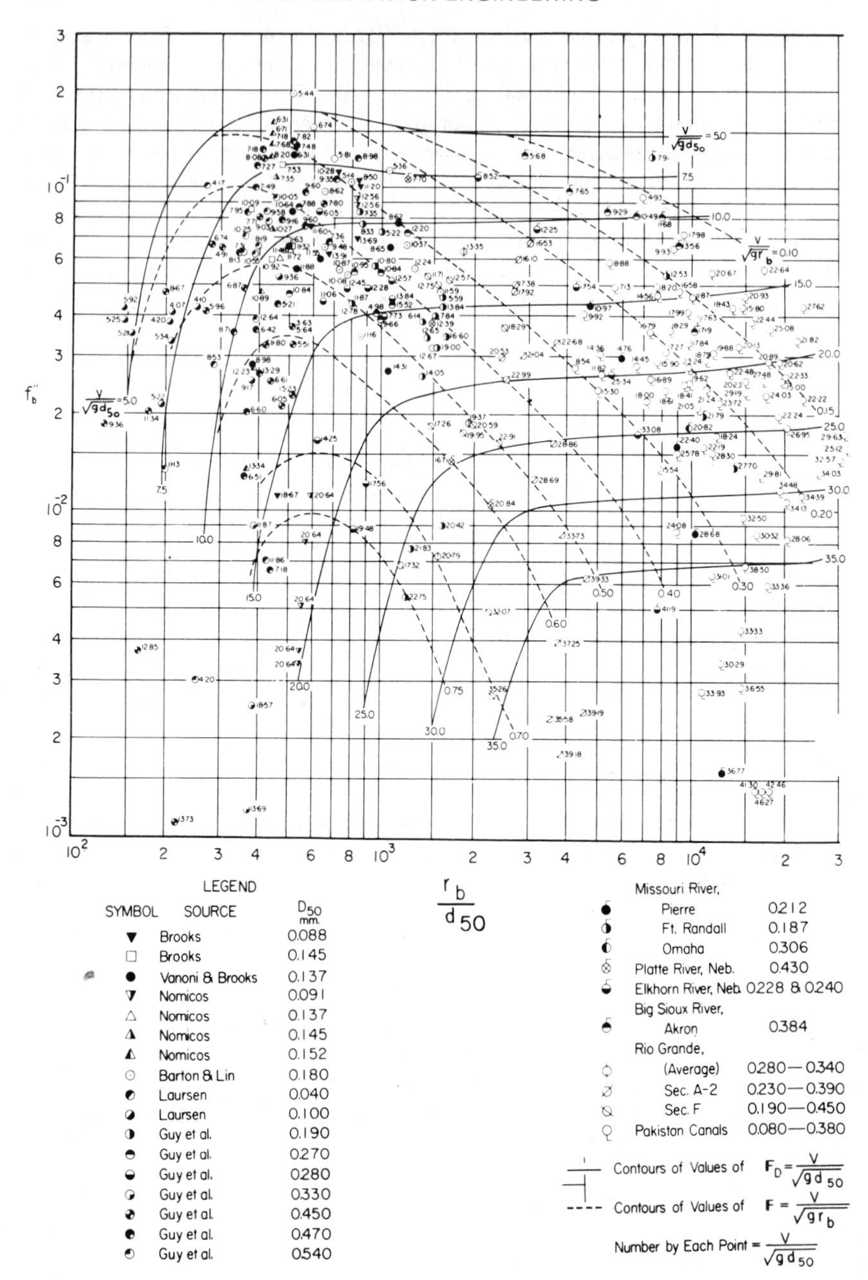

FIG. 2.62.—Alam and Kennedy's (1964, Fig. 3) Graphical Expression of f_b'' as Function of Froude Number and r_b/d_{50}

Alam and Kennedy pointed out that in the region of Fig. 2.62 where the $V/\sqrt{gd_{50}}$ contours are practically horizontal, i.e., where f'' is independent of r_b/d_{50}, their graphical relation is equivalent to the Einstein-Barbarossa (1952) bar resistance curve, Fig. 2.56. It follows from the definition of the Darcy-Weisbach f (Eq. 2.131) that the ordinate of Fig. 2.56 can be written

$$\frac{V}{U''_*} = \sqrt{\frac{8}{f''}} \qquad (2.169a)$$

Moreover, ψ', given by Eq. 2.145, is for sand and water:

$$\psi' = \frac{\rho_s - \rho}{\rho}\frac{d_{35}}{r'S} = 1.68\,\frac{8gd_{35}}{f'V^2} \qquad (2.169b)$$

Now f' does not vary widely among different alluvial channels; this is clearly indicated by Fig. 2.61. Furthermore, most natural sands have nearly the same geometric standard deviation of particle sizes, and thus d_{35} is roughly proportional to d_{50}; therefore

$$\psi' \sim \frac{gd_{50}}{V^2} \qquad (2.170)$$

Thus the Einstein-Barbarossa bar resistance curve is in effect a relation between f'' and $V/\sqrt{gd_{50}}$. At small values of r/d_{50}, f'' is dependent on both parameters appearing in Fig. 2.62, and for these conditions the Einstein-Barbarossa relation cannot be expected to be valid.

Maddock's Analysis (1969).—A rather unique conception of the mechanics of alluvial channel flows was put forth by Maddock (1969). According to his view, the only determinate relationship among the quantities appearing in Table 2.13 is that between mean velocity, water discharge, and sediment discharge; moreover, this relationship is different for each of the three velocity ranges (low-velocity, midvelocity, and high-velocity) he distinguished. Later (Maddock, 1971) he declared that not even this relation is deterministic.

He argues further that relationships between mean velocity, depth, and slope are generally indeterminate because of the variability in the channel roughness resulting from changes in the bed configuration. Maddock's contention is that consistent d-V-S relations result only if certain constraints are imposed upon the flow, and even these relations are not unique but only the most probable. The forms of the relations are then such that the minimization criteria set forth by Langbein (1964) are satisfied. For example, if the slope is constrained to be constant, the sum of the variances of shear stress and friction factor are minimized with the result

$$V = 5.5\left[\frac{\gamma V d}{\sqrt{\rho}}\sqrt{\frac{w}{g(\rho_s - \rho)d_{50}}}\right]^{0.4} \text{(ft-lb-sec units)} \qquad (2.171)$$

in which the numerical constant has been determined from experimental data. On the other hand, if the variances of V and f are minimized, there results

$$V = 14.4(VdS)^{0.3}\text{(ft-lb-sec units)} \qquad (2.172)$$

Several other relations are presented for cases of different constraints and minimization of the variances of various quantities.

Maddock presents no concise guidelines for determining what variances should be minimized in a given situation. This deficiency, together with the uncertain physical or mathematical basis for the minimization hypothesis he employs and the dimensional nonhomogeneity of some of the resulting equations (e.g., Eq. 2.172) tend to distract from his analysis.

Mostafa and McDermid's Manning Coefficient Graph (1971).—Mostafa and McDermid (1971) sought to avoid the questionable step of dividing the hydraulic radius, slope or friction factor, or both, into components associated with the grain roughness and bed form resistance. They expressed the Manning equation in dimensionally homogeneous form as

$$V = \frac{\sqrt{g}}{C_m d_{50}^{1/6}} r^{2/3} S^{1/2} \qquad (2.173)$$

in which C_m = the nondimensional Manning coefficient that is related to the Darcy-Weisbach friction factor, appearing in Eq. 2.131, by

$$f = 8\, C_m^2 \left(\frac{d_{50}}{r}\right)^{1/3} \qquad (2.174)$$

They concluded, on the basis of physical and dimensional considerations, that

$$C_m = C_m\left(\frac{V}{\sqrt{\dfrac{gA}{T}}}\ \frac{d_{50}}{\delta}\right) \qquad (2.175a)$$

in which T = width of the channel at the level of the free surface; and δ = thickness of the viscous sublayer. They used published field data to give quantitative expression to Eq. 2.175, with the result shown in Fig. 2.63, which displays rather good correlation. However, it should be borne in mind that since

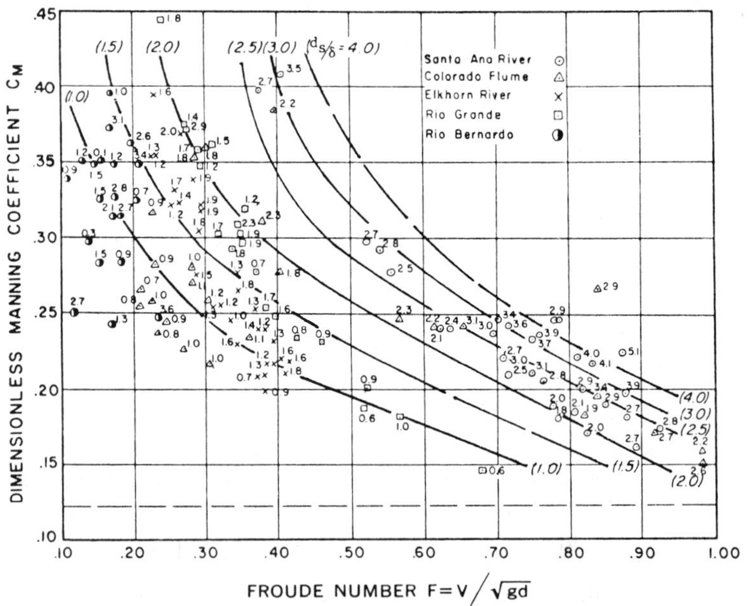

FIG. 2.63.—Mostafa and McDermid's (1971) Chart for Determination of Non-dimensional Manning Coefficient

C_m varies as $\sqrt{f}$, use of C_m instead of f as the dependent variable reduces the apparent scatter of the data included in the graph without improving the accuracy of the predictor.

To apply their method to a wide river for which the depth and hydraulic radius are nearly equal, one proceeds as follows. Eq. 2.173 is rewritten as

$$\frac{V}{\sqrt{gd}} C_m = FC_m = \left(\frac{d}{k_s}\right)^{1/6} S^{1/2} \quad \text{(2.175b)}$$

For selected values of $k_s = d_{50}$, S, and d_{50}/δ, FC_m can be calculated. Eq. 2.175b can be superimposed on Fig. 2.63 for the conditions chosen, and the intersection of this curve with the d_{50}/δ curve for the same flow gives C_m and F, from which V can be calculated. However, note that Eq. 2.175b gives hyperbolas in this case, which do not have sharp intersections with the d_{50}/δ curves that are also somewhat hyperbolic in shape. This suggests that Fig. 2.63 includes some degree of spurious correlation.

43. Criticism of Depth-Discharge Predictors.—Several deficiencies are common to all of the depth-discharge predictors previously considered. Firstly, none takes adequate account of the seemingly disproportionate effect even small water temperature changes can have on the bed configuration and thus also on channel roughness (Franco, 1968; Taylor and Vanoni, 1969, 1972 a–b). Under certain flow conditions a temperature change of only a few degrees can cause the bed configuration to change between the flat bed and ripple-dune regimes. It would appear that temperature variations by affecting the fluid viscosity and particle fall velocity alter the mobility of the particles in the transporting fluid and thereby modify the stability behavior of the bed-fluid interface. It seems not unreasonable to conjecture that this is accomplished through the medium of changes in the transport relaxation distance, δ, shown in Fig. 2.54. But until the details of this temperature effect are better understood and formulated, further progress in eliminating this shortcoming of alluvial channel hydraulic relations will not be forthcoming.

A second and more far reaching objection is that none of the analyses is wholly logical in its development or philosophically satisfying in its composition. A complete formulation of the hydraulic functioning of an alluvial stream would consist of a system of seven equations expressing the interdependencies enumerated in conjunction with the development of Table 2.13. These equations would have to be solved simultaneously for the seven dependent variables. In particular, the friction factor would be expressed as a function of the depth and velocity of flow (or depth and slope, or discharge and slope, etc.; see Table 2.13), sediment discharge or sediment discharge concentration, and channel-plan form geometry and cross section shape (in addition, of course, to the fluid and sediment properties). All of the predictors considered characterize the cross section by either the mean depth or hydraulic radius, a measure that has so much successful precedent that it is difficult to lodge substantive objection against it. None of the techniques described includes the channel pattern, except as it enters willy-nilly through the field data used to give quantitative expression to the functional relations. Given the present state of knowledge of open channel flow and the inherent quasi-randomness of river-pattern geometry, this deficiency is probably unavoidable, and perhaps, at least in the case of meandering channels, not a severe limitation in the light of laboratory results reported by Onishi, et al. (1972).

These investigators performed triplets of experiments in a straight and two different meandering channels and found the effect of meandering on the friction factor is relatively minor. In their results, the bend-loss coefficient, K_L, defined as

$$K_L = \frac{L}{4r_b}(f_{bc} - f_{bs}) \qquad (2.176)$$

in which L = the channel length along one-half of a complete meander cycle; and f_{bc} and f_{bs} = the bed friction factors for the curved and straight channels, respectively, varied from about -0.5 to 1.3. It was found that K_L increases with increasing depth of flow and Froude number. The friction factor reduction found in some of the meandering channel experiments was attributed by them to the very large effect of channel sinuosity on the transverse profiles of the bed, and thus also on the channel cross section shape and velocity distribution. The effect of meandering on the sediment transport rate was found to be much more significant, and strongly dependent on channel width. In their experiments the sediment transport rate per unit width was typically about 50% greater in the 8-ft wide meandering channel than in the 3-ft wide straight flume for a given set of mean-flow conditions, while the unit transport rate on the 4-ft wide meandering flume was only about 65% of that on the straight channel.

None of the analyses incorporates explicitly the effect of sediment discharge or concentration on the friction factor or bed roughness. It can be argued that if the friction factor is expressed as a function of those quantities that govern the sediment transport rate, the role of the transport will be included implicitly. This implies that the sediment transport relation is uncoupled from the friction factor relation (i.e., G_s is not a function of f, except as f is interrelated with d, V, S, etc.), which indeed appears to be the case and is probably an adequate measure provided the friction factor relation includes all the quantities that affect G_s, and provided further that the independent variables are chosen judiciously. The first of these two provisions is not met by any of the formulations; it would require introduction of several additional variables, principally distribution of particle sizes and fall velocities, fluid viscosity, and wash-load concentration. The resolution of the available data is simply not adequate to permit precise determination of these effects, nor is there an analytical framework to guide the design of meaningful experiments to define and quantify their roles. The second provision has several important ramifications. The quantities selected to serve as independent variables should produce unique values of the dependent variables (i.e., the resulting relations should be single-valued). Additionally, the dependent variables should be neither overly sensitive nor extremely insensitive to the independent variables; i.e., an incremental change in an independent variable should produce changes that are neither too large nor too small in the dependent variables. In the former case the inherent inaccuracies in the data used to obtain quantitative relations can produce such large corresponding errors in the dependent variables that the formulation has no practical value, while in the latter case the interdependency is too weak to permit definition of the interrelation sought. Thus the independent variables should be quantities whose values vary widely for the range of conditions under consideration, and that markedly affect the dependent variables. These considerations abase slope or any of its combinational forms (shear stress, shear velocity, stream power, etc.) as an independent variable. As Table 2.13 indicates, its use as such invariably produces multiple-valued relations. Moreover,

Figs. 2.52(b) and 2.53 indicate that slope, shear velocity, etc., are insensitive measures of the hydraulic changes streams undergo. For example, in the midrange of Fig. 2.52(b), the velocity varies some two-fold while the shear stress remains practically unchanged (assuming constant slope), and an examination of the data (Nordin, 1964) indicates the sediment discharge changed by a factor of five or more. In Fig. 2.53 it is seen that the shear velocity varied by only some 30% while the mean velocity changed by a factor of about three; the corresponding change in the sediment discharge was almost two orders of magnitude. To be sure, in these two examples the flow was in the range in which the bed configuration was undergoing radical changes that tended to keep the shear velocity, etc., nearly constant. But natural streams appear to evolve such that they do in fact operate in and about this range of conditions, for they then have maximum latitude in adjusting to convey the imposed water and sediment discharges.

The foregoing considerations indicate that the sediment transport rate and flow velocity are not particularly sensitive to slope and shear stress, and thus those depth-discharge predictors that use slope or any quantity including slope as an independent variable are inherently inferior in at least two respects (uniqueness and sensitivity) to those that treat S as a dependent variable. Into the former category fall the techniques proposed by Garde and Raju, Simons and Richardson, and Haynie and Simons. As discussed earlier, Einstein and Barbarossa's independent variable, ψ', while appearing to be a shear-stress parameter is in reality heavily velocity dependent; the same is true of Engelund's τ'_*.

In evaluating the various predictors it is also necesssary to ascertain if the presentation is such that the dependent variable is somewhat insensitive to the independent variables, or if the apparent correlation is in fact only spurious. For example, in the Einstein-Barbarossa bar-resistance curve (Fig. 2.56) the apparent verticle scatter of the points have been reduced by a factor of two in the logarithmic plot by using V/U''_* rather than f'' in the ordinate scale, since V/U''_* varies as $1/\sqrt{f''}$. As previously noted, this is also the case in the graphical relations of Mostafa and McDermid given in Fig. 2.63. It has also been observed (Nordin, 1964, p. 34) that the bar-resistance curve is not sensitive to the mean velocity; i.e., velocities predicted using it can vary widely from the measured values even though the corresponding values of ψ' and V/U'_* would fall very near the curve. For further discussion of the matter of spurious correlation the reader is referred to the paper by Benson (1965). It is also in order to record a criticism of those methods [e.g., the methods proposed by Simons and Richardson (1966)] that have been verified using the same data that were utilized in determining the relationships between the variables involved. Finally, it is interesting to consider the formats of the various predictors and to compare them with the forms of a friction factor relation for the relatively simple case of uniform, single-phase flow in rigid conduits, e.g., the Moody or Nikuradse pipe friction diagram. In the latter case, f is a function of Reynolds number and relative roughness, and is portrayed as a family of curves of constant r/k_s in the f-R plane. This suggests that it is overly optimistic to expect in the far more complex case of an alluvial channel flow that f, or f'', can be expressed as a function of just one parameter. Simple dimensional analysis of even an idealized alluvial channel flow will suggest that f must be a function of at least three dimensionless quantities, and the experimental evidence is that no one of them is generally dominant. Accordingly, one should endeavor to relate f or f'' to as many of the pertinent quantities as the available data permit. One is also skeptical of analyses which demonstrate no unifying characteristics; e.g., the methods of Simons and Richardson that require separate graphs for each sand size and each bed configuration, and Maddock's analysis in which the variances of apparently arbitrarily selected quantities are minimized in some loosely-defined sense. Natural processes are generally not so disordered or devoid of unifying tendencies, and formulations demonstrating such derangement are suspect of being ill-founded.

The question of whether an alluvial stream functions in a stochastic or a deterministic manner appears to be more one of philosophy of approach than of the physics involved, as is the case in most mechanics problems. The outcome of an event may depend in such a complex way on so many variables and the degree of refinement required in the solution may be such that it simply is not worthwhile or practical to undertake to calculate the dependent variable as a deterministic function of the independent variables. Such is the case, e.g., in predicting the outcome of a coin toss, a problem in which, given modern computing techniques, the result of an individual trial could be predicted exactly if the initial and boundary conditions were sufficiently well prescribed. Likewise, the many variables involved and the many degrees-of-freedom that an alluvial stream has makes the prediction of its behavior very difficult. Relations of the kind presently available among the variables will probably predict average behavior but not the deviations therefrom. These deviations might be viewed as random, in which case the behavior of streams would be considered as nondeterministic. However, it seems evident that if the input of water and sediment discharges to a reach of a stream were prescribed functions of time, and if the initial channel planform geometry and the bed sediment properties were given, then the mean velocity, water-surface and bed profiles, and bed form characteristics in the reach would be completely set as functions of time and, at least in principle, predictable. Such a system is considered to be deterministic.

Comparison of Depth-Discharge Predictors.—The value of any predictor for depth-discharge relations in alluvial streams must ultimately be judged on the basis of its accuracy when applied to natural streams. To aid in the evaluation of the methods previously reviewed, seven of them were used to calculate depth-velocity relations for four natural streams, and the forecast relations were compared with measured results. The data used in preparing the calculated

TABLE 2.14.–Summary of Data Used in Calculation of

River (1)	Reference (2)	Slope, in feet per foot times 10^3 (3)	d_{35}, in millimeters (4)
Niobrara	Colby and Hembree (1955)	1.3	0.23
Colorado	United States Bureau of Reclamation (1955)	0.22	0.29
Republican	Dawdy (1961)	1.7	0.26
Middle Loup	Dawdy (1961)	1.0	0.22

[a]In calculations of the depth-velocity relations, the following values were adopted: γ =

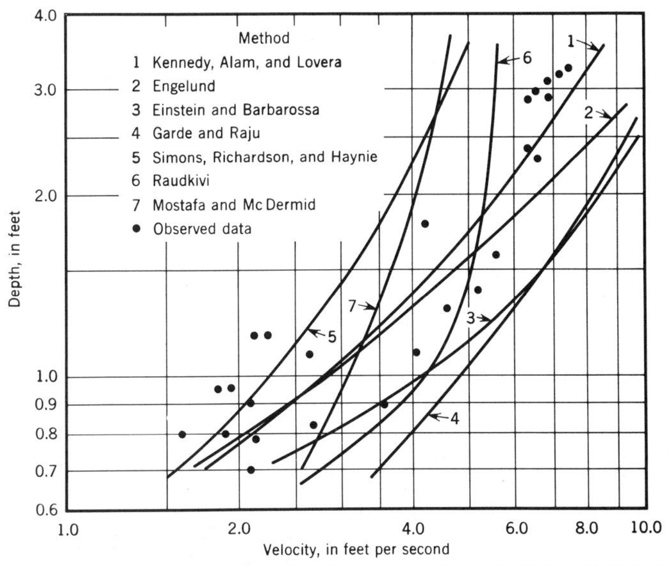

FIG. 2.64.—Comparison of Predicted and Observed Depth-Velocity Relations for Republican River

relations are summarized in Table 2.14, and the outcome of the evaluations is presented in Figs. 2.64, 2.65, 2.66, and 2.67. In applying Raudkivi's method, the curve labeled "sand" was utilized. Perhaps the most encouraging aspect of these comparisons is that the more recently developed predictors appear generally to be more reliable.

Suggestions to the Engineer.—What guidelines can be provided for the engineer who must calculate depth-discharge relations for alluvial streams? Firstly, he should not hesitate to use and base his decision on the results obtained from several different predictors; indeed, he should make use of all of the potentially useful tools at his disposal. Secondly, he should be guided to the extent possible by the available experience with similar streams, especially those in the same locale and preferably the stream being analyzed. In this regard, the regime approach,

Depth-Velocity Relations Given in Figs. 2.71-2.74[a]

d_{50}, in millimeters (5)	d_{65}, in millimeters (6)	Range of depth, in feet (7)	Range of velocity, in feet per second (8)
0.28	0.33	0.7-3	1.7-5
0.33	0.38	4-12	2-3.2
0.32	0.39	0.3-3	1-8
0.26	0.32	0.6-8	1-10

62.4 pcf; s = 2.65; and $\upsilon = 1.21 \times 10^{-5}$ sq ft/sec.

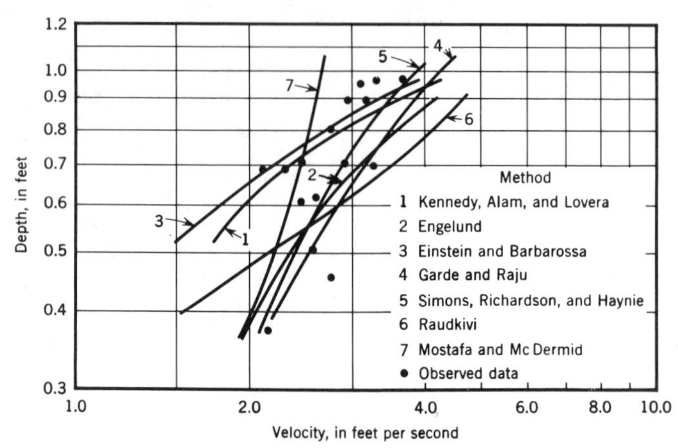

FIG. 2.65.—Comparison of Predicted and Observed Depth-Velocity Relations for Colorado River

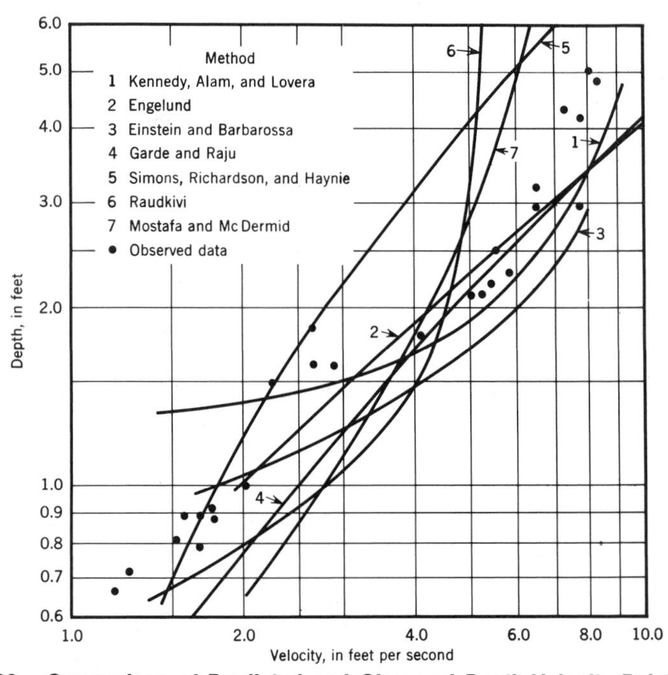

FIG. 2.66.—Comparison of Predicted and Observed Depth-Velocity Relations for Middle Loup River

despite its many obvious deficiences, should not be overlooked, provided the regime method used was derived from data for conditions similar to those being considered. The regime formulations are attempts to systematize favorable experience with alluvial channel flows, and as such should not be discounted. No method should be applied to conditions differing from those for which it was derived. For example, the regime formulations and the techniques of Einstein and

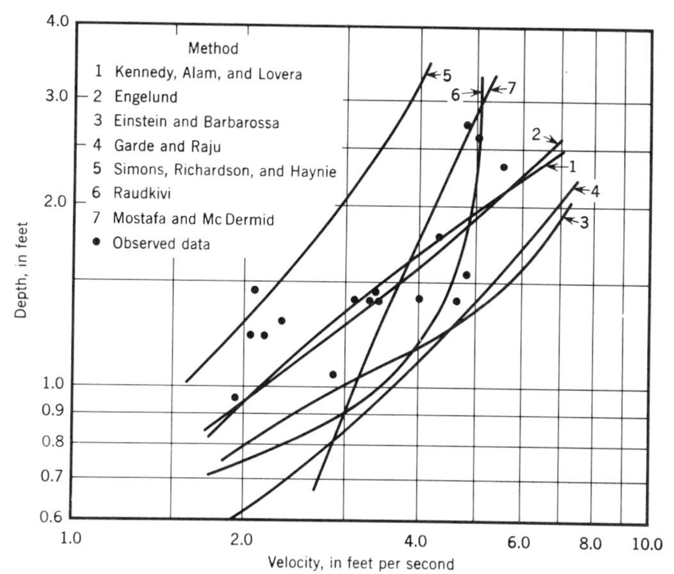

FIG. 2.67.—Comparison of Predicted and Observed Depth-Velocity Relations for Niobrara River

Barbarossa, Simons and Richardson, and Haynie and Simons, were obtained for the ripple-dune regime and should not be applied to other bed conditions. The charts of Lovera and Kennedy, and Alam and Kennedy, Figs. 2.61 and 2.62, are valid only for sand and water. This is true also of the Einstein-Barbarossa curve, Fig. 2.56, and Engelund's graph, Fig. 2.58, which although including the densities of the fluid and solid, were derived from data on sand-transporting flows of water. In short, before applying any predictor one should study its derivation and understand its limitations. Finally, in evaluating the results of the various predictions and arriving at a decision, one should give greater weight to those obtained from methods consistent with the principles of uniqueness, sensitivity, and logical structure set forth earlier in this section.

The foregoing remarks should make it clearly evident that the formulation of hydraulic relations for alluvial streams is far from complete. In the matter of formulating the friction factor the dominant stumbling block is the incomplete understanding of the mechanics of the bed forms and the relationship between their geometrical characteristics and hydraulic roughness. Included in this problem is the ancillary question of the temperature effect on bed stability and roughness. As long as the bed form problem remains incompletely solved, so will the friction-factor problem. For just as one cannot reasonably expect fully to understand an organism without first understanding its constituent cells, one cannot anticipate a fully satisfactory formulation for alluvial channel roughness factors that takes no account of the most variable and generally dominant component of channel roughness, the bed forms.

There are also, of course, other factors that affect the friction factor in ways that have not yet been fully elucidated. Probably the most important among these is the effect of sediment discharge or concentration on the bed roughness and friction factor. It is a challenging intellectual exercise to attempt to reason through the process by which a change in sediment discharge affects the bed configuration and roughness so as to make it possible for some unit water discharges to occur at two or more different depth-velocity combinations as is the case for some pairs of points in Fig. 2.52(b). Clarification of the mechanics of the interaction between the sediment transport, flow, and bed stability is essential to further progress. Other unresolved questions center around the effects of the wash load, particle size and fall velocity distributions, rate of scour and deposition, and channel geometry.

It would appear that flume experiments of the type pioneered by Gilbert (1914), in which attention is focused on the gross parameters, e.g., slope, depth, water and sediment discharges, etc., and since performed by a host of other investigators have reached the point of diminishing returns. They have, to be sure, been invaluable in demonstrating and permitting formulation of many aspects of stream behavior. But the present need is for cleverly conceived and adroitly executed experiments designed to clarify the specific aspects of stream behavior previously discussed.

44. Side-Wall Correction Procedure.—In flume experiments the sand-covered bed will generally be much rougher than the flume walls, and thus will be subjected to higher shear stress. In natural streams the banks may be either more or less rough than the bed. The problem considered herein is development of a calculational procedure for determining the average shear stress on the bed and the related values of shear velocity, friction factor, etc., from known values of V, S, r, etc., measured in a laboratory flume experiment or in a natural channel flow. Separation of the shear force exerted on the bed from that on the lateral boundaries was first proposed by Einstein (1942). The line of analysis pursued as follows is that proposed by Johnson (1942) and modified by Vanoni and Brooks (1957).

The principal assumption is that the cross-sectional area can be divided into two parts, A_b and A_w, in which the streamwise component of the gravity force is resisted by the shear force exerted in the bed and walls, respectively. It is further assumed that the mean velocity and energy gradient are the same for A_b and A_w, and that the Darcy-Weisbach relation can be applied to each part of the cross section as well as to the whole, i.e.

$$\frac{V^2}{S} = \frac{8gA}{fp} = \frac{8gA_b}{f_b p_b} = \frac{8gA_w}{f_w p_w} \qquad (2.177)$$

in which p = the wetter perimeter; and the subscripts b and w refer to the bed and wall sections, respectively. Introducing the geometrical requirement

$$A = A_b + A_w \qquad (2.178)$$

into Eq. 2.177 results in

$$pf = p_b f_b + p_w f_w \qquad (2.179)$$

whence $$f_b = \frac{p}{p_b} f - \frac{p_w}{p_b} f_w \qquad (2.180)$$

For a rectangular channel, $p = 2d + b$; $p_w = 2d$; and $p_b = b$, for which Eq. 2.180 can be expressed as

$$f_b = f + \frac{2d}{b}(f - f_w) \qquad (2.181)$$

The remaining problem is the determination of f_w. To this end consider the Reynolds number for each section

$$\mathsf{R}_b = \frac{4r_bV}{\nu}; \quad \mathsf{R}_w = \frac{4r_wV}{\nu}; \quad \mathsf{R} = \frac{4rV}{\nu} \qquad (2.182)$$

Substituting for the values of r from Eq. 2.177 and recalling that V and S are common to all sections, Eq. 2.182 becomes

$$\frac{\mathsf{R}_b}{f_b} = \frac{\mathsf{R}_w}{f_w} = \frac{\mathsf{R}}{f} \qquad (2.183)$$

The quantities, R and f, will be known from the experimental data, and can be used to compute R/f and thus also R_w/f_w. Then f_w can be obtained from the Moody or Nikuradse pipe friction diagrams or a similar friction factor predictor if the wall roughness is known. Fig. 2.68, prepared by Vanoni and Brooks (1957), permits direct determination of f_w for the case of smooth walls.

Calculation of the friction factor, shear velocity, and hydraulic radius for the bed and wall sections proceeds as follows:

1. Calculate R and f from the experimental data, and compute R/f, which from Eq. 2.183 equals R_w/f_w.
2. Obtain f_w from Fig. 2.68.
3. Calculate f_b from Eq. 2.181.
4. Calculate $r_b = (A_b/p_b)$ from Eq. 2.177.
5. Calculate the bed shear velocity, $U_{*b} = \sqrt{gr_bS}$.
6. The bed shear stress is given by $\tau_b = \rho U_{*b}^2$.

Despite its several obvious deficiencies (division of the cross section into two noninteracting parts, determination of friction factors for section components on

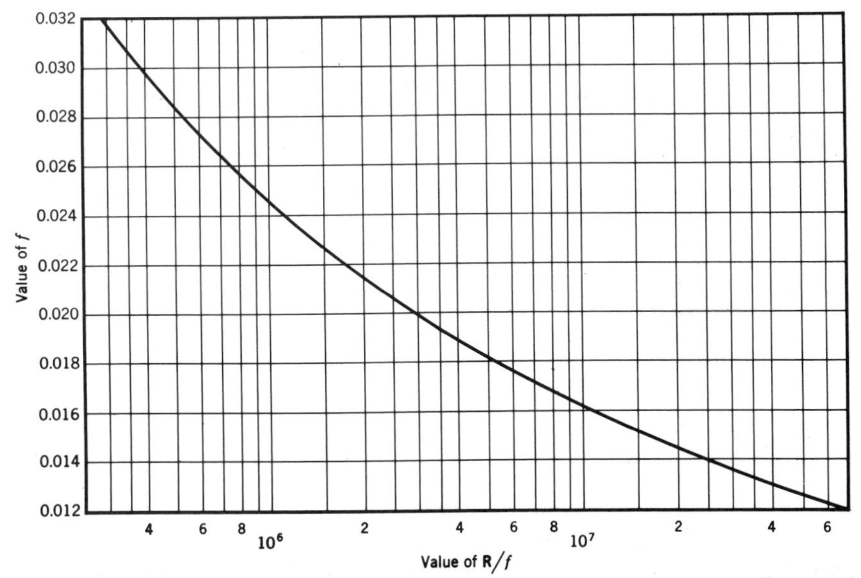

FIG. 2.68.—Friction Factor as Function of R/f for Smooth Boundary Channels and Conduits, for Use in Side-Wall Correction Calculation (Vanoni and Brooks, 1957)

the basis of a pipe friction diagram, use of the same mean velocity for each subsection, etc.), the side-wall correction procedure appears to yield fairly reliable estimates of the friction factors for flow over sand beds with no flume walls present. Nevertheless, a preferable procedure is to measure the side-wall shear stress directly with a Preston tube or other shear stress transducer. The side-wall correction calculation described previously does not, of course, introduce any correction for the effects of the walls on the velocity distribution and sediment transport characteristics of the flow, which, as noted previously, affect the bed configuration and thus also the bed roughness.

G. Fundamentals of Sediment Transportation

45. General.—In this section some of the basic ideas and findings upon which knowledge of sedimentation is based will be discussed briefly. In so doing an attempt will be made to emphasize the truly basic points and some developments may be presented only in sufficient detail to bring out those points. Considerable reliance will be placed directly on results of field and laboratory investigations because theoretical approaches have had only limited success in dealing with many of the complex phenomena involved. Frequent reference will be made to previous sections of this work and some duplication of material in these sections will occur where it is found necessary for convenience of the reader. This is especially so in the case of general relationships that follow immediately.

46. General Relationships.—Table 2.13, first presented by Kennedy and Brooks (1963), lists the variables involved in determining the behavior of alluvial channels and classifies them into several sets of independent and dependent groups. Each of the dependent variables can be determined as a function of the independent variables. In some cases the functions are known and dependent variables can be determined easily. Perhaps the simplest such function is the continuity equation stating that discharge Q is equal to the product of stream width b, depth d, and mean velocity V or $Q = bd\,V$. In the first line of Table 2.13 for the case of flumes, the independent variables are fluid properties of kinematic viscosity, ν; mass density, ρ; sediment properties of density, ρ_s; geometric mean size, d_g; geometric standard deviation of sizes, σ_g; fall velocity, w; the acceleration of gravity, g; and flow system characteristics Q, b, and d. The dependent variables are sediment discharge, G_s; mean velocity, V; hydraulic radius of cross section, r; energy gradient, S; and Darcy-Weisbach friction factor, f.

The sediment discharge, G_s, can be expressed as

$$G_s = f(Q, d, b, \nu, \rho, \rho_s, d_g, \sigma_g, w, g) \qquad (2.184)$$

In a particular flume of a given width the fluid and sediment properties can be kept constant and b and g are constant and the discharge can be replaced by V by means of the continuity equation. When the depth and sediment and fluid properties are kept constant the relation reduces to

$$G_s = f(V) \qquad (2.185)$$

Such a relation from experiments by Vanoni and Brooks (1957) is shown in Fig. 2.69. Other data from this set of experiments are shown in Fig. 2.53 in which slope, bed shear velocity, U_{*b}, and bed friction factor, f_b, are plotted against V.

In the preceding example (Eq. 2.184) the independent variable, V, is determined

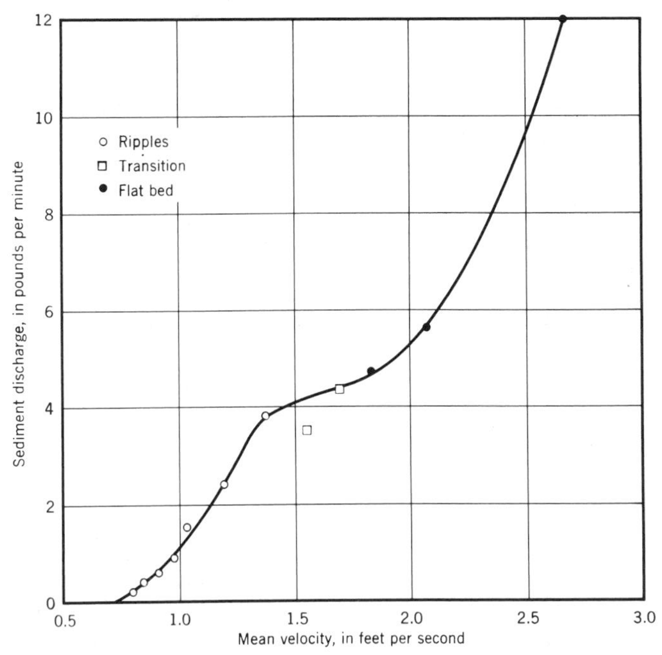

FIG. 2.69.—Sediment Discharge as Function of Mean Velocity for Flow 0.241 ft Deep in Bed of Fine Sand (Flume Width = 10.5 in., Bed Sediment Size d_{50} = 0.152 mm, σ_g = 1.76 (For Plot of Other Data for This Flow, See Fig. 2.53)

immediately from the continuity equation because V is given in terms of Q, d, and b which are known. In cases where this is not possible as in lines 2 and 6 of Table 2.13 and in all cases of natural streams, the variables in the continuity equation that are dependent must be obtained from one or more of the several relationships listed in Chapter II, Section F. As noted by Kennedy and Brooks (1963), the existence of several of the unformulated relations has been inferred mainly from laboratory experiments. One such relationship is expressed, at least approximately, by several flow resistance relations in the literature; e.g., the Alam and Kennedy (1969) friction factor chart relating f, V, r, g, and fluid and sediment properties presented in Fig. 2.62. The relation for sediment discharge of alluvial streams is even less well defined than the friction factor relation. It can be expressed according to line 7 of Table 2.13 as

$$G_s = f(Q, d, \nu, \rho, \rho_s, d_g, \sigma_g, w, g) \qquad (2.186)$$

In principle one can also express the other five dependent variables, b, r, V, S, and f in terms of the nine independent variables at the right of Eq. 2.186. As one now has a complete solution for the flow system one can obtain a rating curve for the stream that is merely a curve relating depth and discharge or $d = f(Q)$. Through this relation d can be eliminated from Eq. 2.186. Assuming that the fluid and sediment properties and g remain constant at a station on the stream under discussion, one obtains the bed sediment discharge curve for the station

$$G_s = f(Q) \qquad (2.187)$$

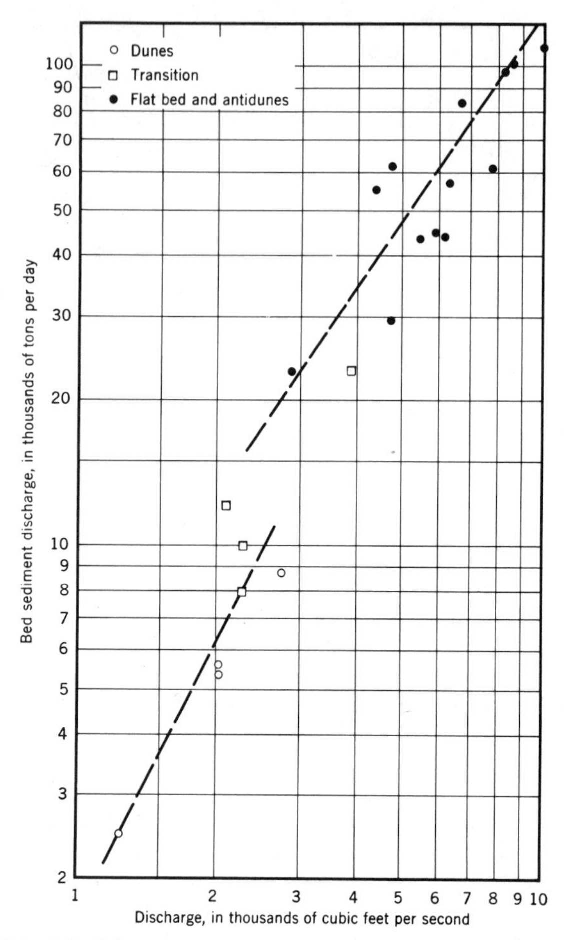

FIG. 2.70.—Plot of Bed Sediment Discharge against Water Discharge at Section F **of Rio Grande, near Bernalillo, N.M.; Nordin (1964)**

Such a curve for Section F of the Rio Grande near Bernalillo (Nordin, 1964) is shown in Fig. 2.70 in which the bed form of the stream is indicated. As reported by Nordin (1964) this section is in a truly alluvial reach of the river in which the width, b, is self adjusting. The large scatter of data in Fig. 2.70 can be attributed, at least in part, to the fact that the fluid and sediment properties were not constant during the observations that were made in the spring and summer of four different years. For example, the water temperature varied from 61° F–81° F and the median size of the bed sediment was observed to vary appreciably even considering sampling errors inherent in such determinations. Other factors contributing to data scatter were variation of wash load concentration and the unavoidable errors in determining the total bed sediment discharge from suspended-sediment concentration measurements by the modified Einstein method of Colby and Hembree (1955) (also see Article 58).

Despite the scatter of the data, Fig. 2.70 may be taken as evidence that the

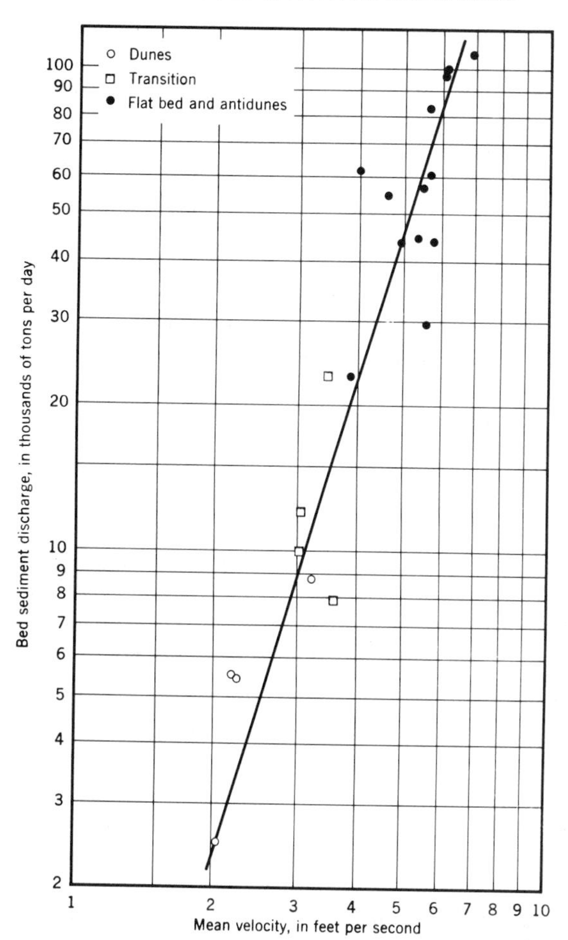

FIG. 2.71.—Plot of Bed Sediment Discharge against Mean Velocity for Section F of Rio Grande, near Bernalillo, N.M.

function defined by Eq. 2.187 exists. The figure also indicates that for this section the relation is discontinuous and multivalued for a certain range of Q. The branch of the relation for the low range of Q is for a dune-covered bed and the other branch is for flat and antidune beds. The discontinuity in the function was to be expected because the relation between Q and d used to eliminate d from Eq. 2.184 is also known to be double valued for a range of Q-values. Multivalued functions occur because the bed form changes drastically. All streams are not expected to have discontinuous relations between discharge and sediment discharge.

Fig. 2.71 shows bed sediment discharge G_s of Fig. 2.70 plotted against mean velocity, V. Again the scatter of the data is large but nevertheless the correlation suggests that the relation, $G_s = f(V)$, does exist. A significant feature of Fig. 2.71 is that the function, $G_s = f(V)$, is continuous and single valued while $G_s = f(Q)$ is discontinuous, as shown in Fig. 2.70. The fact that the relation between G_s and V

is a continuous one is demonstrated clearly in Fig. 2.72 by Colby (1964) based on data from the Mississippi River at St. Louis reported by Jordan (1965). The bed sediment discharge was determined by the Modified Einstein Method (see Article 58) as for the Rio Grande, and errors in these determinations contribute to the scatter of the data in Fig. 2.72. The substantial variation in fluid and sediment properties over the period of the investigation also contribute to the scatter. The sizes of the Rio Grande and Mississippi Rivers are of a different order of magnitude. The depth range for the Rio Grande data in Fig. 2.70 is 2.5 ft–5 ft and in Fig. 2.72 for the Mississippi the range is 15 ft to almost 60 ft.

Table 2.13 indicates that for flumes the relations are unique, i.e., single valued, when Q, d, and b, or V, d, and b are independent variables and thus known. The continuity equation enables one to determine the velocity, V, from Q, d, and b so that the velocity is known in both cases. The fact that functions with V as an independent variable are unique is shown by Figs. 2.71 and 2.72. This fact is also shown in Fig. 2.52(b) in which V is plotted against r. In this case the relation is discontinuous but r is still a unique function of V.

Brooks (1958) pointed out that when Q and G_s are taken as independent variables all of the dependent variables are uniquely determined by orderly and logical relationships among the variables. This is indicated in Table 2.13. Selecting Q and G_s as independent variables is equivalent to saying that if water with a certain temperature flows into a reach of stream at a given rate and carries sediment of a given size composition, density, and grain shape at a given rate, the stream will adjust itself to a unique velocity, depth, bed form, etc. to accommodate the imposed Q and G_s. Suppose now that G_s is a dependent variable and say Q and d are taken as independent variables with the same values as before and that the fluid and sediment remain the same. Table 2.13 indicates that all quantities will be the same in the two systems. That this is true was confirmed by flume experiments of Rathbun, et al. (1969). When Q and G_s were independent variables a sediment feed system was used and the sediment leaving the flume was accumulated in a settling basin during any one experiment. When G_s was a dependent variable the sediment and water discharging from the flume were recirculated. These two kinds of experiments gave essentially the same values of depth, slope, etc. for the same Q, sediment, and fluid regardless of whether G_s was an independent variable and sediment was fed to the flume or a dependent variable as in recirculating flumes. The conclusion of Guy, et al. (1969) that

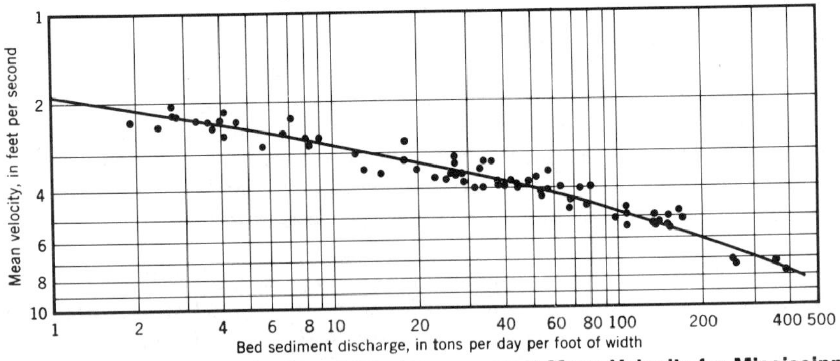

FIG. 2.72.—Plot of Bed Sediment Discharge against Mean Velocity for Mississippi River at St. Louis, Mo.; Colby (1964)

recirculating and feed-type flumes give identical results was challenged by Maddock (1968). This challenge was on the grounds that the data of Guy, et al. obtained with the recirculating flume follow a relation between velocity, and depth, and slope that is characteristically different than that of other data from recirculating flumes. This objection was subsequently refuted by Guy, et al. in the concluding discussion of their original paper.

In rivers, Q and G_s are independent variables. The water and sediment flow at the head of reach of river are actually imposed on the reach. There is some evidence to indicate that a stream does adjust itself to transport water and sediment flow delivered to it. Maddock (1970) called attention to a spectacular example of such an adjustment in the Maraloa-Ravi Link Canal in West Pakistan (1962). About 4 miles downstream from the intake the slope is approx 1:5000 and the depth is about 12 ft when the discharge is 15,000 cfs. Thirty miles downstream from the intake the channel width, discharge sediment size and concentration are the same as upstream, the slope is approx 1:9,000, and the depth is about 8 ft and thus the velocity is 1.5 times that in the upstream reach. The important adjustment made to achieve this condition was in the bed form. In the upper reach the bed was dune covered and in the lower reach it was flat.

A stream can also make adjustments in other characteristics, e.g., bed sediment size and slope that do not affect the stream as much as changes in bed form but still may cause significant changes in equilibrium conditions. Nordin (1964) reported that the median bed sediment size of the Rio Grande near Bernalillo was about 0.3 mm but that it varied from as little as 0.2 mm to more than 0.4 mm during the observations that were made in 4 yr of a 10-yr interval. Jordan (1965) found that the median bed material size of the Mississippi River at St. Louis was between 0.2 mm and 1.1 mm and that it tended to increase with mean annual discharge taken over periods of 1 yr or 2 yr as shown by Fig. 2.116. This large variation in sediment size is not thought to be a common one. He also observed that bed elevation decreased as sediment size increased. These data suggest that during periods of abnormally high runoff when sediment transport capacity is high the bed degrades and coarsens. This coarsening could occur by selective removal of the finer particles or exposure of coarse material left by previous high flows, or both.

The average slope of long reaches of rivers must be constant over periods of a few years or decades. However, as has been observed, over short reaches the slope can vary appreciably. On the Rio Grande near Bernalillo the water surface slope measured over a reach 8,240 ft long (Nordin, 1964) was essentially constant over a 10-yr period. However, on the Mississippi River at St. Louis the slope measured over a reach 6.5 miles long varied as much as four-fold (see Table 2.19). Such large variations in slope can cause appreciable adjustments in the local characteristics of a stream.

Eqs. 2.184 and 2.186 can be expressed in dimensionless form by applying the π theorem. Several different dimensionless relations have been developed by this means based on most of the variables listed in these equations or their equivalent. Barr and Herbertson (1968) presented the equation

$$\phi\left(C_m, \frac{Q g^{1/3}}{\nu^{5/3}}, \frac{d_{50}\sqrt{g d_{50}}}{\nu}, \frac{d_{50}}{r}, \frac{d}{r}, \frac{b}{d}\right) = 0 \quad \text{.......} \quad (2.188)$$

in which $C_m = G_s/Q$. These authors verified the validity of this equation by testing it against data from a given flume with a given sediment and flows of constant depth that means that the only variables are C_m and Q. Cooper and Peterson (1968) and Blench (1969) correlated most of the available flume data according to the equation

$$C_m = \phi\left(F^2, \frac{d}{d_{50}}, g^{1/3}\frac{d_{50}}{\nu^{2/3}}\right) \quad (2.189)$$

These workers found that the correlation was good and also that these variables indicated the bed form. Willis and Coleman (1969) developed the relation

$$\phi\left(\frac{\rho_s - \rho}{\rho} C_m, F, \frac{Vd}{\nu}, g^{1/3}\frac{d_{50}}{\nu^{2/3}}\right) = 0 \quad (2.190)$$

by normalizing the Navier-Stokes equations and including the effect of sediment in transit on the density and viscosity of the water-sediment mixture. By analyzing flume data they found that C_m depended on F and $(d_{50}g^{1/3})/\nu^{2/3}$, but not on the

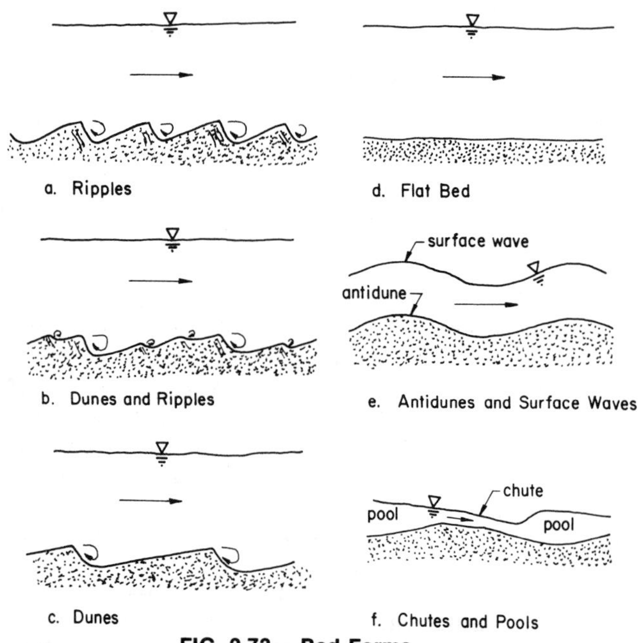

FIG. 2.73.—Bed Forms

FIG. 2.74.—Ripples in Bed of Fine Sand in Flume 10.5 in. Wide; Hwang (1965) (Flow was to Left; Flow Depth = 0.241 ft, Mean Velocity = 1.25 fps, d_{50} = 0.230 mm, σ_g = 1.43)

FIG. 2.75.—Dunes in Coarse Sand in Flume 8 ft Wide; Simons (ASCE, 1966) (Flow Depth = 1.02 ft; Mean Velocity = 2.78 fps, d_{50} = 0.93 mm)

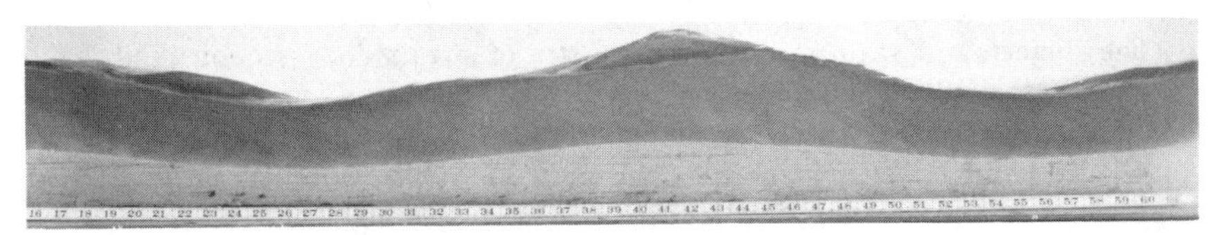

FIG. 2.76.—Antidunes in Flume 10.5 in. Wide; Kennedy (1960) (Flow was to Left; Scale in Figure is in inches; Mean Flow Depth = 0.248 ft, Mean Velocity = 3.30 fps, Froude Number = 1.17, d_{50} = 0.233 mm, σ_g = 1.47)

flow Reynolds number, Vd/ν. The equation of Blench and Cooper and Peterson can be obtained from Eq. 2.184 by neglecting the effect of σ_g and grain shape and keeping the sediment density, ρ_s, constant.

47. Bed Forms.—The definition of common bed forms have been given in Table 2.12 according to the ASCE (1966) Task Committee on Bed Forms that also presents photographs of bed forms. A comprehensive set of photographs and description of such forms is given by Pettijohn and Potter (1964). Fig. 2.73 shows bed forms arranged in increasing order of sediment transport rate. Ripples sketched in Fig. 2.73(a) occur at transport rates or flow velocities that are less than for the flat bed and rarely occur in sediments coarser than approx 0.6 mm (ASCE, 1966). Fig. 2.74 shows a ripple-covered bed of fine sand in a flume 10.5 in. wide. The mean length and height of these ripples are 0.53 ft and 0.046 ft, respectively.

Dunes are longer than ripples and occur at transport rates that are larger than for ripples but smaller than for flat beds. Fig. 2.75 shows dunes in coarse sand in a flume 8 ft wide. The mean length and height of these dunes are 6.3 ft and 0.3 ft, respectively. For some sediments and flows the flat bed does not occur and the bed form goes directly from dunes to antidunes. The antidune is characterized by a surface wave that is sometimes called the sandwave (Pierce, 1917). Fig. 2.76 is a photograph by Kennedy (1960) of an antidune in fine sand in a flume 10.5 in. wide. These antidunes moved upstream rather rapidly. The commonest bed form in all but very small streams is the dune. At relatively high stages in large rivers the bed form is believed to tend to change to flat bed and dunes or to flat bed. When the bed is flattened the friction factor is observed to fall to a very low value. Antidunes are much less common than the dune and flat bed forms and occur only in steep streams. The ripple seems to occur at very shallow flow depths and is not important except in streams with small depths and low velocities (Nordin, 1964) but may be of major importance in movable bed hydraulic models. The bed form known as chutes and pools sketched in Fig. 2.73(f) occurs in very steep streams, e.g., mountain torrents.

48. Stream Forms.—Self-formed river channels tend to be sinuous, and straight reaches with lengths equal to more than a few times the stream width are a rarity (Leopold and Wolman, 1957; Lane, 1957). The sinuosity or sinuosity ratio of a stream is expressed as the ratio of the length along the center line of the stream to the length along the valley. Leopold and Wolman (1957) observed that even in reaches that were almost straight, with a sinuosity ratio approaching unity, the thalweg or the line locating the lowest point of the stream bed, was sinuous, i.e., it wandered within the channel. This was also observed by Brooks (1958) in a straight flume. As the thalweg weaves back and forth across the bed, bars are formed in a staggered pattern at the banks of the channels. These have been called alternate bars (ASCE, 1966). A striking picture of such bars in a straight reach of the Rio Grande appears in a paper by Maddock (Maddock, 1969, Fig. 50).

The subjects of stream forms and geometry of stream cross sections and their relation to hydraulic and sediment characteristics of streams have been studied by Leopold and Wolman (1964), Schumm (1960, 1963), and Leopold and Maddock (1953). A summary of this subject appears in a book by Leopold, et al. (1964). Stream forms and stream geometry are discussed further in Chapter V, Section B.

49. Prediction of Bed Forms.—Work on prediction of bed forms has involved both theoretical and experimental approaches. In the former category use is made of the continuity equation for sediment. This can be applied to flow over a bed form sketched in Fig. 2.77 to yield

$$W\frac{\partial \eta}{\partial t} + \frac{\partial g_s}{\partial x} = 0 \qquad (2.191)$$

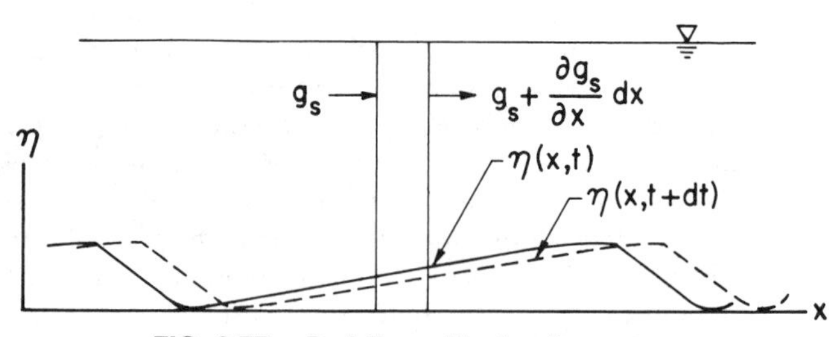

FIG. 2.77.—Bed Form Moving Downstream

in which W = specific weight of the sediment deposit in the bed; η = the height of the bed form at position x along the stream and time t; and g_s = the local sediment discharge in weight per unit width and time. The first term expresses the rate of deposition of sediment on the bed and the second one gives the change in sediment discharge with change in distance x along the stream. Obviously the two terms are always of opposite sign, i.e., when the bed is being built up, $\partial\eta/\partial t$ is positive and $\partial g_s/\partial x$ is negative and vice versa.

Fig. 2.77 indicates the profiles at times t and $t + dt$ of a bed form that is moving downstream. On the upstream or stoss side of the form the bed lowers with time, i.e., $\partial\eta/\partial t$ is negative, and from Eq. 2.191 it is seen that $\partial g_s/\partial x$ is positive so that g_s increases continuously up to the crest of the form. On the lee side the reverse is true. Introducing into Eq. 2.191 the simplifying assumption that $g_s = A_o u_o$, in which u_o = the flow velocity near the bed and A_o = a constant, results in

$$W\frac{\partial\eta}{\partial t} + A_o\frac{\partial u_o}{\partial x} = 0 \qquad (2.192)$$

presented by Exner (1925). This and other early work on bed forms is summarized concisely by Lelliavsky (1955). Other theoretical treatments of this subject were presented by von Karman (1947), Anderson (1953), Kennedy (1963), Reynolds (1965), Engelund (1966), Gradowczyk (1968), and Hayashi (1970). For a brief discussion of the Kennedy theory the reader is referred to Chapter II, Section F and Kennedy (1969).

Kennedy's result for the dominant wave length of bed forms is

$$\mathsf{F}^2 = \frac{V^2}{gd} = \frac{1 + kd\ \tanh kd\ + jkd\cot jkd}{(kd)^2 + (2 + jkd\cot jkd)kd\ \tanh kd} \qquad (2.193)$$

in which F = Froude number of the flow; d = flow depth; $k = 2\pi/L$ = wave number; L = wave length; $j = \delta/d$ = lag ratio; and δ = distance by which changes in the local sediment discharge lag changes in local velocity near the bed. The concept of the lag distance, δ, was first introduced by Kennedy (1963) and is an essential feature of his analysis.

Fig. 2.78, taken from Kennedy (1963), shows theoretical curves derived from Eq. 2.193 and data plotted as F against kd. The theory is in remarkably good agreement with the data. Antidunes are seen to occur at F^2, greater than $(1/kd)$ tanh kd, and at kd less than two. The curve $\mathsf{F}^2 = (1/kd)\tanh kd$ gives the upper limit of F for ripples and dunes.

Kennedy proposed the following simple relation between antidune wave length, L, and mean flow velocity, V:

$$V^2 = \frac{gL}{2\pi} \qquad (2.194)$$

which is recognized as the equation for the celerity of a deep-water wave. Observed antidune wave lengths (Kennedy, 1960; Nordin, 1964) agreed roughly with this relation.

Kennedy (1963) found experimentally that surface waves over antidunes broke when their steepness (ratio of wave height to wave length) fell between 0.13 and 0.16 which bracket the accepted value of 0.14 for the steepness of deep-water waves at incipient breaking.

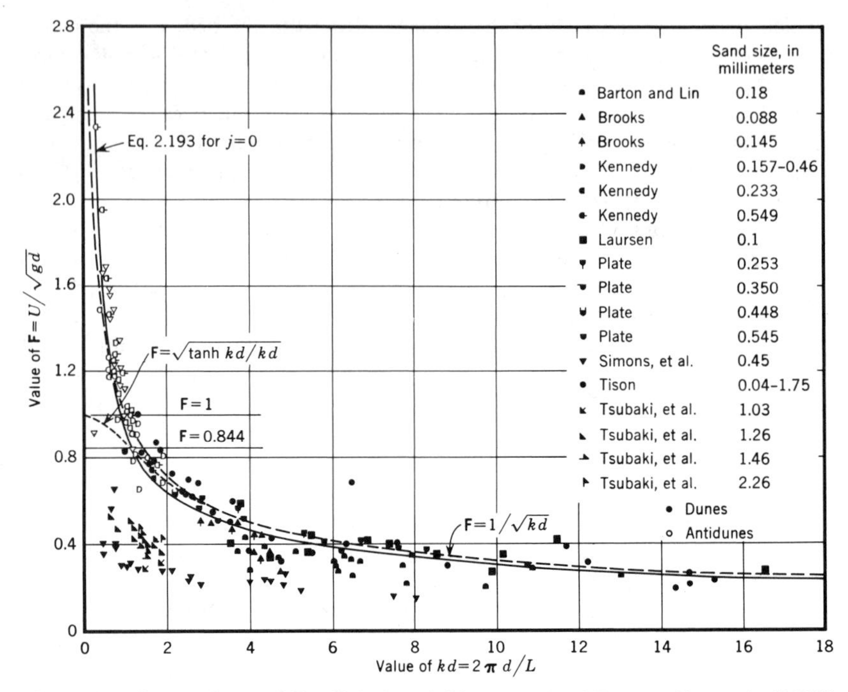

FIG. 2.78.—Comparison of Predicted and Observed Bed Forms; Kennedy (1963)

Fig. 2.79 is a chart prepared by Simons and Richardson (1966) for predicting bed forms. It is based on extensive data from flumes and several rivers and canals. The ordinate of the chart represents the product of bed shear stress, and mean velocity. This quantity has been called stream power by Bagnold (1960). The bed forms given by Fig. 2.79 are in good agreement with those observed by Nordin (1964) in the Rio Grande in which the depths were less than 5 ft and the velocities were relatively high. However, in the case of the Mississippi River (Jordan, 1965), with velocities in the same range as for the Rio Grande, but with depths up to 50 ft, the figure predicts flat beds in some cases where evidence indicates that the beds are probably dune covered.

Taylor and Brooks (1962) related flow depth, mean velocity, and bed form for a given bed sediment and the ratio of total friction factor, f, to sand grain friction factor f'. They calculated f' from the Moody pipe friction diagram (Moody, 1944) using $4r$ instead of pipe diameter and d_{50} for the roughness length. The graphs of depth, velocity, bed form, and f/f' prepared from flume data for two sands showed that dune and ripple beds had f/f' ratios in excess of two, and that flat and antidune beds had ratios less than two. Nordin (1964) found that this relation also held for the Rio Grande that has flow depths up to about 5 ft and median sediment size of 0.3 mm. This simple method can be used to infer the bed form from observed data but cannot predict the bed form.

50. Observed Bed Forms.—Much of what is known about bed forms comes from observations in flumes. Such observations show that ripples and dunes move downstream by erosion of their stoss or exposed faces and deposition on their lee faces. As already noted this means that the sediment transport rate increases with

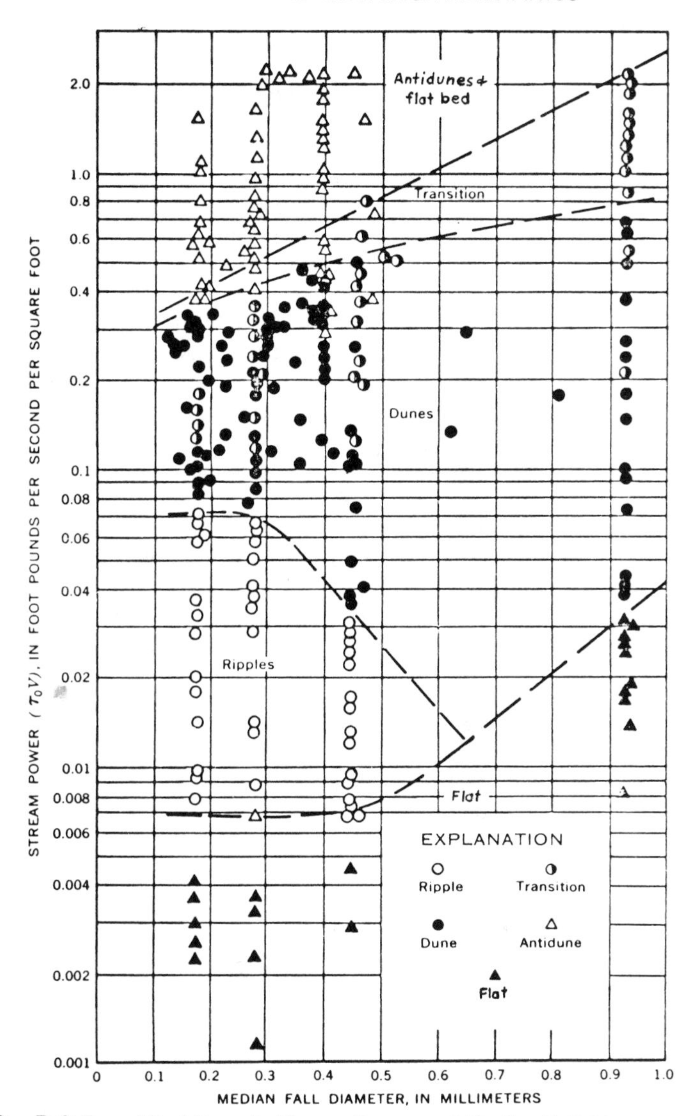

FIG. 2.79.—Relation of Bed Form to Stream Power and Median Fall Diameter of Bed Sediment Proposed by Simons and Richardson (1966)

distance up the exposed face and then diminishes suddenly as sediment slides down the steep lee face of the dune or ripple. In sliding down the steep lee face the grains orient themselves in such a way that the sloping bedding planes parallel to the face are visible in the profile of the form. Observations of the movement of rippled beds by eye and with the aid of time lapse motion pictures shows that the movement of individual ripples is variable in both time and space. Some are overtaken and combined with others, new ones may appear and disappear giving a feeling of great variability. The velocity or celerity of ripples and dunes as reported by Liu (Brooks, 1958), is only a small fraction of the flow velocity and seems to increase as sediment discharge increases.

Observations of bed forms in the Lower Mississippi River as early as 1879 have been summarized by Lane and Eden (1940). These observations, which consisted of soundings at frequent intervals at a series of longitudinal ranges, showed a great variation in size, celerity, and general behavior of the bed forms much like those in laboratory studies. The celerity varied from a few feet per day up to a maximum of 81 ft/day, the average heights ranged from a few inches up to 8 ft with maximums of 22 ft and average lengths varied between 100 ft and 3,000 ft. Typical values of the steepness, ratio of height to wave length, varied between 0.0077 and 0.033. Water depths ranged from about 20 ft–100 ft, and the median size of the bed sand varied from 0.3 mm–0.5 mm. The behavior of the bed forms at different stages and stations was very variable. At some stations, the maximum height for a given stage was found in the swiftest current, and the highest dunes occurred when high stage persisted for some time. At one station, a sudden rise in stage of 18 ft above low water caused the dunes to grow to their maximum height, and these dunes persisted for 5 months, and then disappeared on a falling stage of 20 ft above low water. On the other hand, at another station the dunes were destroyed by a sudden rise in stage, and a new series was formed when the crest of the flood was finally reached. At this same station, dunes were also observed to be destroyed on a sudden drop in stage. Some of the bed forms had crests normal to the direction of flow, and others had crests inclined to the normal. In some instances, most of the bed was covered with bed forms, and in others half or less was covered.

Fig. 2.80 shows longitudinal echo sounder profiles of the bed of the Lower

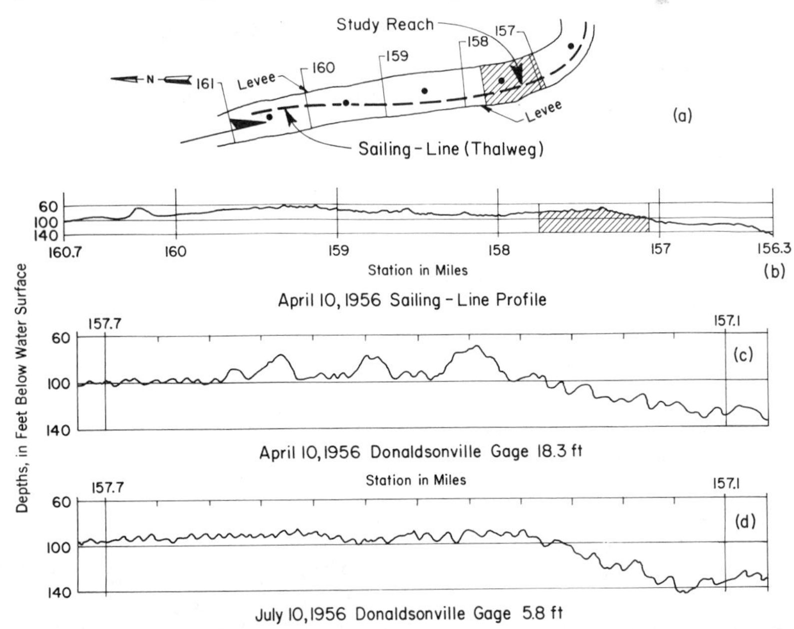

FIG. 2.80.—Bed Profiles of Lower Mississippi River near Donaldsonville 57 miles Upstream of New Orleans; Carey and Keller (1957); (a) Plan Showing Study Reach and Sailing Line along which Profiles were Taken; (b) Small Scale Profile on April 10, 1956; (c) Profile of about 0.6 miles of Bed on April 10, 1956; (d) Profile of Same Reach as in c **on July 10, 1956**

Mississippi River near a bend for two river stages taken by Carey and Keller (1957). From such profiles of a number of reaches, it was found that the bed forms were highest at the higher stage as shown in Fig. 2.80. In some cases, the bed at crossings was observed to become flat at the high stage. The maximum height of bed forms observed was about 30 ft. The high stage was still below bank-full stage and no observations were available for higher stages. Detailed observations of the bed in a rather straight reach of the Missouri River at Omaha (Einstein and Chien discussion of Brooks, 1958) showed that on one side of the river the bed was covered with dunes, and on the other it was flat. Where dunes existed, the water was deeper, the velocity was lower, and the bed material was finer than where the bed was flat.

The preceding data have some similarity to those for laboratory streams, and may be interpreted in the light of laboratory experience. First of all, it is clear that bed forms in rivers change greatly with discharge, as they do in the laboratory. The fact that the bed of the Lower Mississippi flattens at crossings at high stage is in agreement with laboratory observation. On the other hand, the behavior of the bed in becoming flat during a sudden rise, and then developing bed forms at the high stage bears no similarity to any laboratory observations. However, it might be explainable in terms of a change in sediment discharge imposed on the stream and causing a change in bed form to accommodate to the changed set of conditions. A slight tendency for wave length of bed forms to increase with sediment discharge has been observed in the laboratory (Vanoni and Brooks, 1957), but no such large changes as observed in the Mississippi have been reported. Theories do not predict such increases in length and the simultaneous occurrence of bed forms with greatly different lengths, e.g., those observed on the Mississippi, although spectral analysis (Nordin and Algert, 1966) of flume bed profiles show that a broad range of wave length exists.

Colby (1960) was the first to document the existence of discontinuities in stage-discharge relations of streams. The two small streams that he studied were straight with relatively stable banks and median size of bed sediment of 0.4 mm. The transition from dune bed to flat bed was observed to occur approximately between 3 fps–5 fps. Manning's roughness factor n was about 0.035 for the dune beds and fell to as low as 0.011 for the flat beds. The equivalent Darcy-Weisbach friction factors, f, were 0.1 and 0.02, respectively, for dunes and flat bed. In the transition zone where the stage-discharge curve had two branches, the discharge for a flat bed was as much as twice that for a dune bed at the same stage. Nordin [1964, Fig. 2.52(b)] found that in the Rio Grande transition from dunes to flat bed also occurred in the range of 3 fps–5 fps. The transition occurred with a mean depth of about 2 ft and the sediment had a median size of 0.3 mm.

Colby (1960) observed that the straight regular streams that he studied were probably not typical and he did not expect all sand streams to have discontinuous stage-discharge relations. He expected discontinuities in the stage-discharge relation in streams that have uniform sand, uniform lateral and longitudinal distribution of flow, and slopes and velocities in the range that will produce both dunes and flat bed. In many natural streams the bed form is usually not uniform either across or along the stream and changes in bed form do not occur simultaneously over the entire stream (Colby, 1964). For this reason the stage-discharge relations for many natural streams are continuous. The stage-discharge and sediment discharge-water discharge relations for the Rio Grande are discontinuous as shown by Fig. 2.52(b) and Fig. 2.70, respectively. This suggests that if one of these relations is discontinuous the other will also be discontinuous, although sufficient evidence to substantiate or disprove this statement has not been assembled.

Colby and Scott (1965) reported that dunes in natural streams are soft and that a stream gager sinks into them while wading the stream. Ripples on beds of laboratory flumes are also soft. Experience on the Rio Grande (Nordin, 1964) and in flumes indicates that flat beds and antidune beds are firm.

51. Sediment Discharge Theories.—Work on sediment transportation prior to about 1930 dealt almost exclusively with bed load. Then for almost two decades the study of the problem concentrated towards the suspended load and the application of the theory of turbulent flow developed largely by Prandtl and von Karman. As predicted by Lelliavsky (1955) bed phenomena are again (1971) receiving the attention their importance requires. Despite this attention, only modest progress has been made in the basic understanding of the bed phenomena and their relation to sediment movement.

Bed Load Discharge.—Much of the early development in analysis of bed load was influenced by the work of DuBoys (1879), who proposed the idea of the bed shear stress or tractive force, and presented the transport formula that bears his name. He imagined the bed material to move in layers of thickness d' and mean velocity increasing linearly toward the bed surface as indicated in Fig. 2.81. If there are $n - 1$ layers in motion the surface layer will have a velocity $(1/2)(n - 1)\Delta u$ in which Δu is the velocity increment between adjacent layers. The sediment discharge, g_s, per unit width and time is

$$g_s = nd'\frac{1}{2}(n - 1)\Delta u W \qquad (2.195)$$

in which W = the weight of sediment deposit in a unit volume of the bed. Assuming that the friction between layers is proportional to the submerged weight of the overlying grains, the following expression is obtained for the distance, nd', below the surface of the bed at which the frictional resistance between grains is just equal to the fluid shear stress, τ_o, at the bed and no motion occurs:

$$\tau_o = c_f\, nd'W\left(1 - \frac{\gamma}{\gamma_s}\right) \qquad (2.196)$$

in which c_f = the coefficient of friction between grains; and γ and γ_s = the specific weights of the fluid and the sediment grains, respectively. When $n = 1$ in Eq. 2.196 the surface layer is about to move and τ_o has the critical or threshold value, τ_c. Using this relation in Eq. 2.196 gives $\tau_o = n\tau_c$. Eliminating n from Eq. 2.195 gives the DuBoys formula

$$g_s = \Psi_d \tau_o(\tau_o - \tau_c) \qquad (2.197)$$

in which $\Psi_d = (1/2)d'\ \Delta u\ W\ \tau_c^2$. Values of Ψ_d and τ_c determined experimentally by Straub (1935) are given for river sands in Fig. 2.95 as functions of sediment size only.

Einstein (1942) based a bed load formula on probability concepts. He first introduced the idea that grains move in steps of average length, L_1, proportional to the sediment size, d_s. The number of grains per unit width that pass a given cross section is then taken as the product of the number of grains in the surface

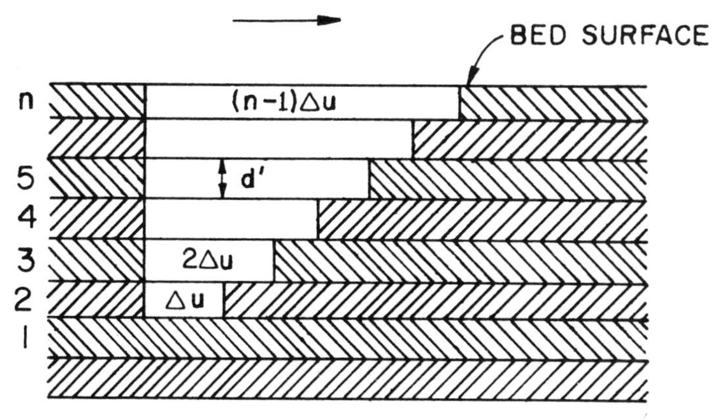

FIG. 2.81.—Layers of Moving Sediment on Bed as Conceived by DuBoys (1879)

layer of a bed area of unit width and length, L_1, and the probability, p_s, that in any second the drag force is great enough to set a particle in motion. The sediment discharge, g_s, in weight per unit width and time is given by

$$g_s = \frac{L_1}{A_1 d_s^2} p_s \gamma_s A_2 d_s^3 = \frac{A_2}{A_1} \gamma_s \lambda_o d_s^2 p_s \quad \text{(2.198)}$$

in which $A_1 d_s^2$ = the bed area occupied by a single grain; $A_2 d_s^3$ = volume of a grain; λ_o = the constant of proportionality in the relation; $L_1 = \lambda_o d_s$; and γ_s = the specific weight of the grains. The probability, p_s, has the dimensions of seconds to the minus one power, and it is made dimensionless by multiplying by a characteristic time that was taken as d_s/w in which w = the settling velocity of a grain in water. Eq. 2.198 can now be written as

$$g_s = \frac{A_2}{A_1} \gamma_s \lambda_o d_s w p \quad \text{(2.199)}$$

in which $p = p_s d_s/w$ = the number of steps in time d_s/w that will start from any point. The quantity, p, was also considered to be the probability that the lift force acting on a particle exceeds its submerged weight. The submerged weight of a particle is $(\gamma_s - \gamma) A_2 d_s^3$ and the average lift was taken proportional to $\tau_o d_s^2$. The probability, p, was then taken as a function of the ratio of the submerged weight to lift or

$$p = f\left[\frac{(\gamma_s - \gamma)\, d_s}{\tau_o}\right] \quad \text{(2.200)}$$

Einstein employed the Rubey equation (1933) for w:

$$w = F_1 \sqrt{\frac{\gamma_s - \gamma}{\gamma} g d_s} \quad \text{(2.201)}$$

$$\text{in which } F_1 = \sqrt{\frac{2}{3} + \frac{36\,\nu^2}{g d_s^3\left(\frac{\gamma_s}{\gamma} - 1\right)}} - \sqrt{\frac{36\,\nu^2}{g d_s^3\left(\frac{\gamma_s}{\gamma} - 1\right)}} \quad \text{(2.202)}$$

and ν = kinematic viscosity of water. Eliminating p and w from Eq. 2.199 and rearranging gives the Einstein equation

$$\phi = f(\Psi) \qquad (2.203)$$

in which
$$\phi = \frac{g_s}{F_1 \gamma_s d_s \sqrt{\left(\frac{\gamma_s}{\gamma} - 1\right) g d_s}} \qquad (2.204)$$

$$\Psi = \frac{(\gamma_s - \gamma) d_s}{\tau_o} \qquad (2.205)$$

The quantity, Ψ, is recognized as the reciprocal of τ_*, the Shields parameter or dimensionless shear stress.

Fig. 2.82 is a plot of ϕ against Ψ from flume experiments with well-sorted coal, sand, and gravel ranging in median size from 0.3 mm–28 mm. The straight line on the graph has the equation

$$0.465\phi = e^{-0.391\Psi} \qquad (2.206)$$

It fits the data moderately well for Ψ in excess of about six but deviates markedly from the data for small Ψ. Curve 2 on the graph is based on a refinement of the Eq. 2.206. It fits data for Ψ larger than five but still deviates from the data for small Ψ. The curve

$$\phi = 40\left(\frac{1}{\Psi}\right)^3 \qquad (2.207)$$

which was fitted to the data by Brown (1950), is seen to fit data for low values of Ψ, i.e., high values of ϕ, τ_o, and g_s. A curve which follows Eq. 2.207 up to Ψ approximately equal to 5.5 and follows Eq. 2.206 for higher Ψ is sometimes referred to as the Einstein-Brown relation.

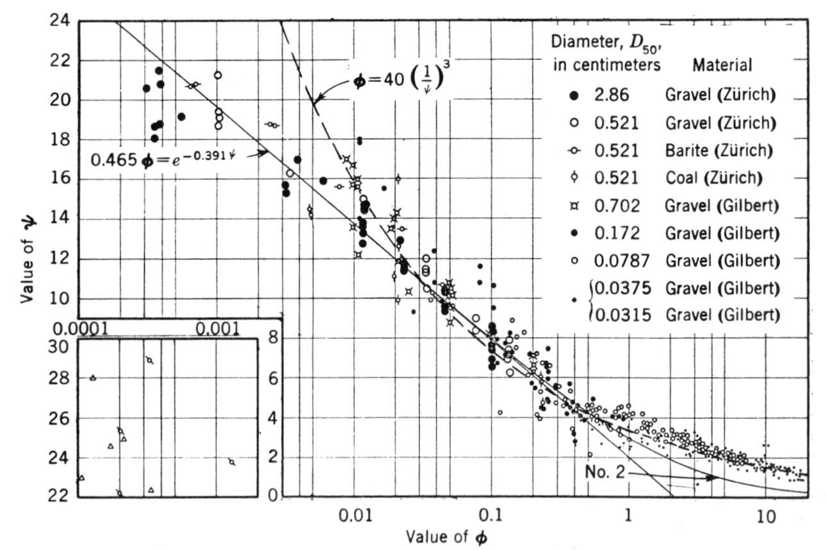

FIG. 2.82.—Graph of Einstein (1942) and Einstein-Brown (1950), Ψ-ϕ Relations for Bed Load Discharge

The Einstein relation Eq. 2.206 indicates that some transport will occur even at very large Ψ, i.e., very small τ_o and in this sense is more realistic than equations of the DuBoys type that give no transport for $\tau_o \leq \tau_c$. The critical value of the Shields parameter, $\tau_* = 1/\Psi$, for initiation of motion on a flat bed (Fig. 2.43) ranges between 0.033–0.060 so that corresponding values of Ψ are 30 and 16, respectively. Eq. 2.206 indicates that in this range of Ψ the transport rate is small but not zero.

Flume data for graded sediments plotted on a ϕ-Ψ graph gave a poor correlation and departed appreciably from the relations indicated in Fig. 2.82. One of the factors contributing to these departures was the large increase in friction factor due, no doubt, to the formation of ripples and dunes. The important contribution of this work (Einstein, 1942) was in the introduction of the probability concept in bed load movement. The writer was aware of the limitations of his relation and did not intend that it should be considered as a universal one.

Kalinske (1942, 1947) based a relation for bed load discharge on the local velocity and shear stress at the bed including fluctuations in these quantities due to turbulence. He first assumed that the velocity, u_g, of a sediment grain moving on the bed is given by

$$u_g = m(u_o - u_c) \qquad (2.208)$$

in which m = a constant of proportionality; u_o = the instantaneous water velocity near the grain; and u_c = the critical velocity for initiation motion of the grain. The instantaneous velocity can be expressed by $u_o = \bar{u}_o + u'_o$ in which $\bar{u}_o$ = mean value of u_o; and u'_o = the fluctuating or turbulence component of the velocity. Letting $P/(\pi/4)d_s^2$ be the number of grains per unit area of the surface layer, the sediment discharge per unit width is given by

$$g_s = \frac{2}{3} P \gamma_s d_s \bar{u}_g \qquad (2.209)$$

in which $\bar{u}_g$ = mean velocity of the grains.

The mean grain velocity, $\bar{u}_g$, can be expressed by

$$\bar{u}_g = m \int_{u_c}^{\infty} (u_o - u_c) f(u_o)\, du_o \qquad (2.210)$$

in which $f(u_o)$ is the frequency distribution of u_o so that $f(u_o)du_o$ is the fraction of the time that the value of the instantaneous fluid velocity will be between u_o and $u_o + du_o$. Assuming that the velocity fluctuations, u'_o, are normally distributed Eq. 2.210 can be integrated giving

$$\frac{\bar{u}_g}{\bar{u}_o} = m \frac{\sqrt{\overline{u_o'^2}}}{\bar{u}_o} \left[\frac{1}{\sqrt{2\pi}} e^{-t_c^2/2} - \frac{t_c}{\sqrt{2}} + \frac{\tau_c}{\sqrt{2\pi}} \int_o^{t_c} e^{-t^2/2} dt \right] \qquad (2.211)$$

in which $t_c = (u_c - \bar{u}_o)/[\overline{u_o'^2}]^{1/2}$ and $t = u'_o/[\overline{u_o'^2}]^{1/2}$. Assuming that bed shear stress is proportional to the square of the bed velocity one gets $\sqrt{\tau_c/\tau_o} = u_c/\bar{u}_o$ by means of which Eq. 2.211 can be expressed in terms of τ_c/τ_o and $[\overline{u_o'^2}]^{1/2}/\bar{u}_o$. The ratio, $\bar{u}_g/\bar{u}_o$, is shown graphically in Fig. 2.83 in terms of τ_c/τ_o and $[\overline{u_o'^2}]^{1/2}/\bar{u}_o$. For zero turbulence level, $[\overline{u_o'^2}]^{1/2}/\bar{u}_o = 0$, $\bar{u}_g$ goes to zero as τ_o approaches τ_c. But for a nonzero turbulence level transport continues even when τ_o is less than τ_c. In this regard the Kalinske and Einstein formulations are similar.

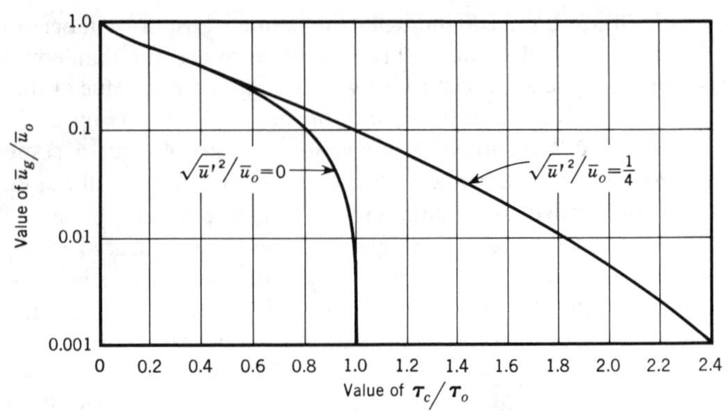

FIG. 2.83.—Kalinske's (1947) Relation for Ratio of Mean Grain Velocity, $\bar{u}_g$, to Mean Fluid Velocity, $\bar{u}_o$, near Bed in Terms of Ratio of Critical Shear Stress, τ_c to Bed Shear Stress τ_o for Two Values of Turbulence Intensity, $(\overline{u'^2})^{1/2}/\bar{u}_o$

Eq. 2.211 can be substituted into Eq. 2.209 to give the Kalinske bed load equation. This is not done since the objective is only to present the concepts on which the relation is based.

Suspended-Sediment Discharge.—The discharge, g_{ss}, of suspended sediment per unit width of stream is given by

$$g_{ss} = \int_{y_o}^{d} CUdy \qquad (2.212)$$

in which C and U are, respectively, the sediment concentration and mean velocity at distance y above the bed; d = flow depth; and y_o = some small value of y taken as the lower limit of integration. If C and U are given as functions of y, g_{ss} can be found by integration.

Brooks (1965) integrated Eq. 2.212, substituting the Rouse equation, i.e., Eq. 2.77 for C and the following relation of U:

$$U = V + \frac{U_*}{k}\left(1 + \ln\frac{y}{d}\right) \qquad (2.213)$$

in which k = von Karman constant; and U_* = shear velocity. Substituting the relations for C and U into Eq. 2.212, taking $a = 0.5\,d$ and noting that the discharge per unit width, $q = Vd$, gives the Brooks relation

$$\frac{g_{ss}}{qC_{md}} = \left(1 + \frac{U_*}{kV}\right)\int_{\eta_o}^{1}\left(\frac{1-\eta}{\eta}\right)^z d\eta$$

$$+ \frac{U_*}{kV}\int_{\eta_o}^{1}\left(\frac{1-\eta}{\eta}\right)^z \ln \eta d\,\eta \qquad (2.214)$$

in which C_{md} = the sediment concentration at mid-depth; $\eta = y/d$; and $\eta_o = y_o/d$. The relation of Eq. 2.214 obtained by numerical integration is shown in Fig. 2.84 for values of z of 2 or less. In obtaining the curves, Brooks took y_o such that $U = 0$, giving $\eta_o = e^{-(kV/U_*+1)}$. This results in the maximum value for g_{ss} since for η_o less than this value U becomes negative. The quantity $V/U_* = \sqrt{8/f}$ and

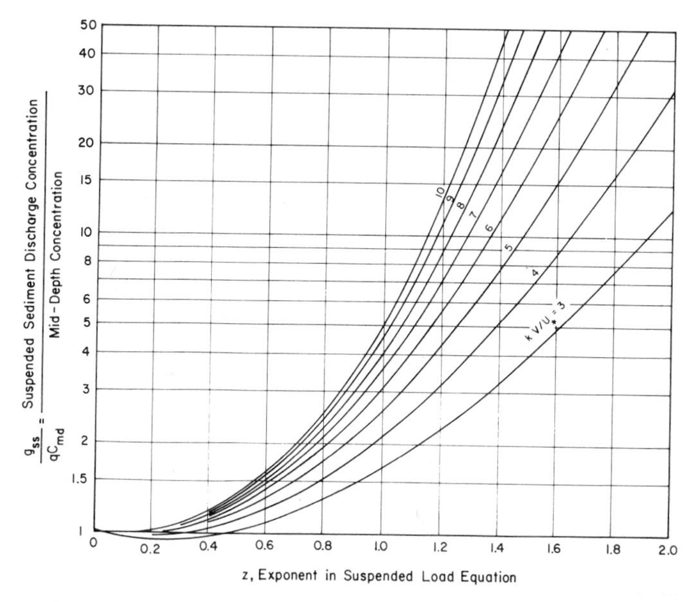

FIG. 2.84.—Chart for Determining Suspended-Sediment Discharge, g_{ss}, from Known Values of kV/u_*, Exponent z, and Mid-Depth Concentration C_{md}; Brooks (1965)

can often be estimated. The ratio, g_{ss}/q, is the suspended-sediment discharge concentration and tends to be less than C_{md} for certain low values of z as shown in Fig. 2.84. For very low values of z the concentration tends to become uniform so the discharge concentration and mid-depth concentration are essentially equal and the value of the ordinate becomes unity. To calculate the sediment discharge, g_{ss}, by Fig. 2.84 one needs to know z, q, C_{md}, and kV/U_*. The figure is not useful for predicting sediment discharge but it can be used to calculate suspended-sediment discharge from stream measurements. It is also helpful in displaying the relation between suspended-sediment discharge and the several variables, e.g., discharge, friction factor, and z.

The Rouse equation for distribution of concentration of suspended sediment is valid only for well-sorted sediment. The suspended sediment of natural streams is usually made up of grains with a broad spectrum of sizes and fall velocities. Eq. 2.212 and the Rouse equation can be applied to streams by dividing the sediment into a number of size fractions and applying the equations to each fraction taking w to be the mean fall velocity for the fraction.

Einstein (1950) applied Eq. 2.212 and the Rouse equation in developing his bed load function using the logarithmic equation for the point velocity in the following form

$$\frac{U}{U'_*} = a_r + 5.75 \log \frac{y}{d_{65}} \qquad (2.215)$$

in which $U'_* = \sqrt{g\, r'_b\, S}$ = grain roughness shear velocity; g = acceleration of gravity; r'_b = bed hydraulics radius for grain roughness (see Chapter II, Section F); S = the stream slope; d_{65} = the bed sediment size for which 65% by weight of the sediment is finer; and a_r is a constant taken by Einstein to be a function of the boundary Reynolds number, $U'_* d_{65}/\nu$. Einstein (1950) used Eq. 2.215 which is based on $k = 0.4$ and also assumed that $k = 0.4$ in the expression for z and replaced U_* by U'_* in the expression for z. Einstein was the first to take account of both bed load and suspended load in an expression for sediment discharge and to calculate discharge by size fractions. He took $y_o = 2d_{si}$ in which d_{si} is the mean grain size of the size fraction being considered and assumed that the sediment discharge, g_{si}, of sediment of size d_{si} was p_i times the discharge that would occur if all of the bed sediment were of uniform size d_{si}. The quantity, p_i, is the fraction by weight of the bed sediment with mean size, d_{si}.

Note that the Rouse equation gives distribution of concentration in terms of the concentration, C_a, at an arbitrary level $y = a$ and to calculate g_{ss} it is necessary to obtain this concentration. Several methods for calculating C_a have been developed, notable among which are those of Lane and Kalinske (1939) and Einstein (1950).

The Lane and Kalinske theory is based on the following assumptions: (1) The vertical turbulence fluctuations, v', near the bed are normally distributed; (2) the rate at which particles of any size are picked up is proportional to the relative amount by weight $p_i(w)$ of those particles present in the bed; the magnitude of the velocity, v', that is capable of picking up particles; and the fraction of the time that such velocities exist; and (3) only fluid currents with velocities v' in excess of the fall velocity, w, of the sediment particles can pick up particles. The normal distribution of v' is given by

$$f(v') = \frac{1}{\sqrt{2\pi}\sqrt{\overline{v'^2}}} e^{-(1/2)(v'^2/\overline{v'^2})} \qquad (2.216)$$

in which $\overline{v'^2}$ = mean square of the turbulence velocity and is assumed proportional to U_*^2. According to the assumptions, the rate at which grains with settling velocity w are picked up per unit area of bed is proportional to

$$p_i(w) \int_w^\infty v' f(v') dv' \qquad (2.217)$$

For equilibrium the rate of pickup is equal to the rate of settling $C_o w$ in which C_o = the sediment concentration near the bed so that the quantity given by Eq. 2.217 is proportional to $C_o w$. Substituting Eq. 2.216 into Eq. 2.217 and integrating gives

$$\frac{C_o}{p_i(w)} \sim \frac{1}{t_c} e^{-1(1/2)t_c^2} \qquad (2.218)$$

in which $t_c = w/U_*$. Lane and Kalinske (1939) plotted canal and river data according to Eq. 2.218 and found that $C_o/p_i(w)$ would be expressed roughly as a power function of the quantity at the right of Eq. 2.218. This particular development was not carried further by its authors or others. It is presented because it is one of the very few theories for predicting the concentration near the bed.

As already stated, Einstein (1950) selected the value $y_o = 2d_{si}$ for the lower limit of integration in Eq. 2.212. He also defined the bed load as material moving in a layer of thickness, $2d_{si}$, at the bed and assumed the following relation between the concentration, C_{ai}, at level $y_o = 2d_{si}$ and the bed load discharge per unit width, g_{sbi}:

$$g_{sbi} = C_{ai} 2 d_{si} \bar{u}_g \qquad (2.219)$$

in which $\bar{u}_g$ = mean grain velocity and the subscript, i, indicates that the quantities refer to a particular size fraction of the bed sediment. He further assumed that $\bar{u}_g = 11.6\ U'_*$, which is recognized as the velocity at the edge of the viscous sublayer and obtained

$$C_{ai} = \frac{g_{sbi}}{(2)11.6 d_{si} U'_*} \qquad (2.220)$$

Einstein's relation for bed load discharge (1950) is based on a model much like the one used to develop Eq. 2.203, and the assumption that the lift on particles on the bed depends on an unsteady velocity whose fluctuating component is normally distributed. Several empirical correction functions of ν, U'_*, d_{si}, and d_{65} are also applied to y_o and the lift on the particles.

Einstein's expression for sediment discharge can be written symbolically as

$$g_s = f(r', d, S, \nu, \rho, \rho_s, d_g, w, g) \qquad (2.221)$$

The Einstein-Barbarossa (1952) method of predicting flow velocity and depth, used in the Einstein bed load function, can be written as

$$V = f(r', S, \nu, d_g, \sigma_g, g) \qquad (2.222)$$

and $$d = f(r', S, \rho, \nu, \rho_s, d_g, \sigma_g, g) \qquad (2.223)$$

Using the latter two equations to eliminate r' and S from Eq. 2.221 and introducing the continuity equation, $Q = bdV$, gives Eq. 2.184. Of the many equations proposed, Einstein's is the only one that contains all of the pertinent variables.

Despite the fact that Einstein's relations contain all pertinent variables the functional forms used are apparently not adequate to give satisfactory accuracy for most engineering purposes. The relations have been criticized by Laursen (1957) and Brooks (1958). Among the items questioned was the use of d_s/w as a characteristic time of removal of grain from the bed in arriving at Eq. 2.200. It was argued, with justification, that this time should depend on flow properties and not on fall velocity w. The use of $k = 0.4$ and U'_* instead of U_* in the expression for the exponent, z, in the Rouse equation was also objected to. The use of U'_* instead of U_* tends to give too large a value of z and to exaggerate the difference in concentration of suspended sediment between the bed and surface. A value of k of 0.4 tends to be too large and tends to compensate for the effect of U'_*. The use of the Einstein bed load function to arrive at general relations was also challenged.

Although the use of U'_* has been criticized, as outlined in the preceding paragraphs, the concept of U'_* and of τ'_o, the bed shear stress due to grain roughness, introduced by Meyer-Peter and Muller (1948) is an interesting and useful one. The basic concept is that the total bed shear stress, τ_o, is the sum of two parts, a part τ'_o, due to a tangential or skin friction stress on the grains lying on the bed and a part, τ''_o, due to normal stresses on the bed and referred to as the shear stress due to bed forms. Relating bed load discharge to the tangential stress, τ'_o, as done by Einstein, seems appropriate since it is inconceivable that the normal stress is effective in moving sediment at the bed. However, to use U'_* instead of U_* in calculating z in the suspended load equation is inappropriate because the diffusion coefficient for sediment upon which the equation is based depends on the total shear stress, τ_o, and not only on τ'_o. In the Einstein-Barbarossa (1952) method of calculating velocity or friction factor of alluvial streams, mean velocity V is related to U'_*, r', d_{65}, and ν as may be seen from Eq. 2.143.

52. Calculation of Sediment Discharge from Stream Measurements.—The development of the depth-integrating sampler (see Chapter III, Section A) made it possible to sample the suspended load of streams on a routine basis. In the United States, Federal Agencies, e.g., the Geological Survey, Department of Agriculture, and Army Corps of Engineers are among those responsible for collecting data of this kind as well as hydraulic data on streams. Suspended load samplers cannot sample the flow within a layer a few inches thick at the bed and do not include any of the bed load. For fine sediments, e.g., those making up the wash load, which are nearly uniformly distributed over the depth, the concentrations measured are essentially equal to the discharge concentration. However for the bed sediment that is coarser than the wash load, the concentrations in the samples are usually considerably less than the true discharge concentration. To get the true sediment discharge it is necessary to add the contribution of the unsampled suspended load and the bed load to the measured sediment discharge.

Methods of estimating the suspended-sediment discharge in the unsampled layer near the bed and the bed load discharge have been proposed by Colby (1957) and Colby and Hembree (1955). The latter method, known as the Modified Einstein method, is based on observed suspended load samples, mean velocity, depth, cross section, and size composition of the bed sediment. Its name derives from the fact that it uses a modified version of the Einstein bed load function (1950) in estimating the unmeasured sediment discharge. The sediment discharge based on suspended-sediment samples and estimates of the unmeasured discharge are much more reliable than those based on formulas alone. The two methods of estimating unmeasured sediment discharge are presented in detail in Chapter II, Section H.

53. Observed Sediment Discharge Relations.—Because of the complexity of sediment transport phenomena, many of the theoretical relations do not completely describe the processes and their agreement with observations is often less than satisfactory. Many field and laboratory observations point up relations not given by theories that ultimately will form the basis for improved theories. Such observed information makes up much of the knowledge of sedimentation, yet it is difficult to present because it has not yet been incorporated into theories. In this article, some selected items of information are presented and discussed in the light of the theories already outlined.

As noted in the analyses already presented, sediment discharge has been related to the bed shear stress by a number of workers. Through a series of carefully performed experiments, Brooks (1958) showed that the relation between sediment discharge and shear stress is more complicated than had been generally visualized. These experiments were carried out with two well-sorted fine sands in a tilting flume 10.5 in. wide and 40 ft long in which the water and sediment were recirculated. Brooks found that it was possible to get more than one value of sediment discharge and water discharge for a given depth and slope, and thus shear stress, because of the large change in friction factor and velocity when the bed changed from ripple covered to flat. Results from two pairs of runs with the same depth and slope shown in Table 2.15 illustrate this fact. For runs 2a and C the beds were covered with ripples, which gave high friction factors and for runs 7 and A the beds were flat and had low friction factors. Despite the fact that in the

TABLE 2.15.–Data by Brooks (1958) from Two Pairs of Flume Experiments with Constant Depth and Slope Showing Effect of Bed Form on Velocity and Sediment Discharge

Run number (1)	Depth, in feet (2)	Slope, in feet per foot (3)	Bed shear stress, in pounds per square foot $\tau_b = \gamma r_b S$ (4)	Velocity, in feet per second (5)	Sediment discharge, in pounds per minute (6)	Darcy Weisbach friction factor, f (7)	Bed form (8)
(*a*) $d_g = 0.152$ mm, $\sigma_g = 1.76$, $\tau = 25°$ C							
2a	0.241	0.0021	0.029	0.85	0.40	0.115	ripples
7	0.241	0.0021	0.024	1.69	4.4	0.029	flat
(*b*) $d_g = 0.145$, $\sigma_g = 1.30$, $\tau = 25°$ C							
C	0.241	0.0021	0.029	0.91	0.17	0.101	ripples
A	0.241	0.0021	0.022	2.06	3.0	0.020	flat

runs with rippled beds the bed shear stresses were somewhat higher than in the flat-bed runs, the velocity and sediment discharge were much lower. Brooks' finding that for a given shear stress more than one value of velocity and sediment discharge may result is also shown by Fig. 2.53 which is a plot of S, U_{*b}, and f_b against V for flows with constant depth and water temperature. In a certain range of slopes, flows with a given slope and thus shear stress could occur at three different mean velocities. The friction factor decreased and sediment discharge increased as the velocity increased thus making it impossible to express sediment discharge as a unique function of shear stress. Fig. 2.52(b) and Fig. 2.70 that are plots of hydraulic radius against velocity and sediment discharge against water discharge, respectively, for the Rio Grande show discontinuous relations. Fig. 2.52(b) shows that for some values of hydraulic radius, and thus shear stress, two different velocities can occur. As shown by Fig. 2.69 sediment discharge is a unique function of mean velocity so that the two flows with the same shear stress but different velocities will have different sediment discharges.

Experience indicates that sediment discharge is strongly correlated with mean velocity. This is shown by Fig. 2.69 for flume data and by Figs. 2.71 and 2.72 for river data. The relations are seen to be continuous and single-valued functions of the velocity. Colby (1964) developed the relation (Fig. 2.104) of sediment discharge as a function of mean velocity, depth, water temperature, and concentration of fine suspended sediment. This relation shows that the velocity is the predominant variable.

Bagnold (1960) has correlated sediment discharge with stream power, $\tau_o V$, the product of bed shear stress and mean velocity. The fact that sediment discharge is not generally a single-valued function of shear stress is reason to believe that the relation between sediment discharge and stream power is not a very useful one. The sediment discharge may be a unique function of $\tau_o V$ in cases where the bed form and thus friction factor does change drastically. However, when the bed changes from dunes to flat and the depth-discharge relation is discontinuous the relation between sediment discharge and stream power will not be unique.

Some streams carry very high concentrations of fine suspended sediment that appears to enhance the capacity of the stream to transport coarser sediments.

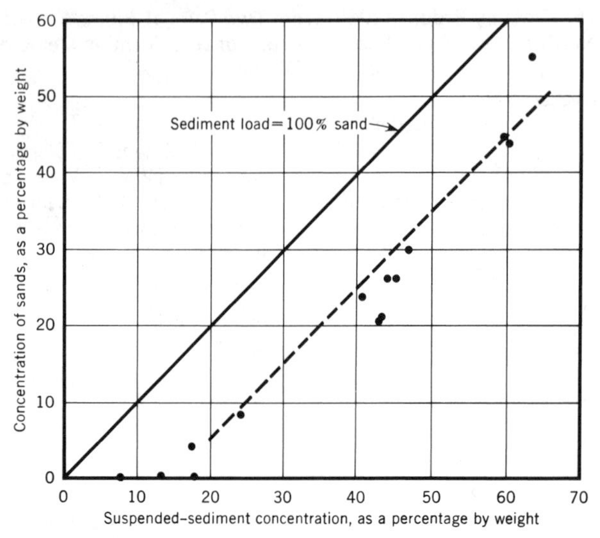

FIG. 2.85.—Relation of Concentration of Suspended Sands to Total Suspended-Sediment Concentration; Beverage and Culbertson (1964)

Lane (1940) reported observed concentrations of fine sediment as high as 55.79% by weight where the concentration is calculated as the dry weight of sediment in a given volume of flow as a percentage of the weight of sediment and water in the volume. Beverage and Culbertson (1964) have reported a number of concentrations of suspended sediment in four rivers in the United States ranging from 40% by weight to as high as 63% by weight. The suspended sediments consisted of sand, silt, and clay. They referred to concentrations in this range as hyperconcentrations.

Fig. 2.85 is a plot by Beverage and Culbertson (1964) of total suspended-sediment concentration against concentration of suspended sands for several rivers in the Southwestern United States. The concentration of sands is seen to increase as total concentration of sediment increases despite the fact that the discharge of the streams had a variation as high as tenfold and total sediment concentration was not correlated with discharge.

The effect of suspended bentonite clay on discharge of bed sediment was investigated in flumes 2 ft wide and 8 ft wide by Simons, et al. (1965). The concentration of clay ranged from a fraction of 1% up to 6.37%, a concentration well below the hyperconcentration range. The bed sediment had median sizes of 0.47 mm and 0.54 mm, respectively, in the 8-ft and 2-ft flumes and geometric standard deviations, σ_g of 1.5.

In the 8-ft flume, the addition of as little as 0.48% of clay to a clear water flow over a ripple bed tended to stabilize the bed and to reduce the sediment discharge and friction factor. For flows over dune beds in this flume, the addition of clay in concentrations from 0.5%–2.83% resulted in a slight increase in bed sediment discharge and a substantial decrease in the friction factor. In the 2-ft flume with dune beds, concentrations of clay as high as 2.4% had little or no effect on bed sediment discharge and friction factor. The addition of clay to flows with dune beds was observed to reduce the heights and increase the lengths of the dunes, which probably accounts for much of the reduction in friction factor. The effect of clay on flows over antidune beds was also investigated by Simons, et al. (1963). As long as the antidunes did not move, additions of clay had little effect on either sediment discharge or friction factor. However, when antidunes moved upstream the addition of clay increased the frequency with which the surface waves broke and also tended to increase the sediment discharge and friction factor. Antidunes were observed to form at lower Froude numbers when clay was added to flows.

Suspensions of fine sediment, e.g., clay in water tend to behave as liquids with higher density and viscosity than clear water. This means that the fall velocity of sand particles in such suspensions will be less than in clear water. Simons, et al. (1963) measured the viscosity of suspensions of clay with a stormer viscometer and verified that the viscosity increases with concentration. Simons, et al. (1963) and Beverage and Culbertson (1964) also measured fall velocity of sand particles in clay suspensions using a visual accumulation tube. The latter authors found that the fall velocity of fine sand in a clay suspension of 10% concentration was only about 0.4 times that in clear water. The increase in sediment discharge resulting from the addition of clay to flows, as observed by Simons, et al. (1963), can be explained in part in terms of the reductions in fall velocity of the bed sediment in clay suspensions. No explanation can be offered for the fact that the addition of fine sediment to a flow affects the bed form.

54. Size Distribution of Bed Sediment and Bed Sediment Load.—Many observations in flumes and some observations in rivers show that with few exceptions the bed sediment load is finer than the bed sediment over which it is carried. Exceptions to this relation have been reported by Kennedy (1960) and Guy, et al. (1966). Kennedy found that with antidune beds in coarse sand the load always had about the same or slightly larger median size than the bed sediment. With antidune beds in very fine sand however, he found that the load was always finer than the bed sediment. In the extensive experiments reported by Guy, et al. it was found that with bed sediments of median grain size of 0.28 mm or finer the load was always finer than the bed sediment for all bed forms including antidunes and chutes and pools. For bed sediments with median sizes of 0.3 mm–0.94 mm the load was coarser than the bed sediment in some but not all runs with antidunes and in an insignificant number of runs with ripple, dune, or flat beds. These results suggest that the load tends to be coarser than the bed sediment when the transport rate is extremely high and that in most cases the load is finer than the bed sediment.

Table 2.16 gives data on hydraulic characteristics and size distribution of bed sediment and bed sediment load in three series of flume experiments by Brooks (1958) in which the bed form was either ripples or flat. The first and second series were made in a flume 33.5 in. wide and the other series was made in a flume 10.5 in wide. The mean size of the load tends to be considerably less than that of the bed sediment for all flows but those with the highest velocities and sediment discharges. In the experiments with the 0.137-mm bed sediment with $\sigma_g = 1.38$ the value of σ_g for the load exceeds 1.38. In the series of experiments with the well-graded sediment ($\sigma_g = 1.76$) the load tended to have about the same value of σ_g as the bed sediment. Fig. 2.86 shows the size distribution of the bed sediment and the load in run 2-1 for which data are shown in Table 2.16. It is seen that the load contains small amounts of the coarsest fractions of the bed sediment and relatively large amounts of the finest fractions. In the sediment load, 28.9% by

TABLE 2.16.—Comparison of Size of Bed Sediment and Bed Sediment Load in Some Flume Experiments by Brooks (1958)

				COMPOSITION OF SEDIMENT				
				Bed Sediment		Bed Sediment Load		
Run number (1)	Depth, in feet (2)	Mean velocity, in feet per second (3)	Sediment discharge, in pounds per minute (4)	Geometric mean size, in millimeters (5)	Geometric standard deviation, σ_g (6)	Geometric mean size, in millimeters (7)	Geometric standard deviation, σ_g (8)	Bed form (9)
(a) Flume 33.5 in. wide								
2-3	0.243	0.90	0.54	0.137	1.38	0.094	1.53	ripples
2-1	0.240	1.28	6.0	0.137	1.38	0.086	1.56	ripples
2-6	0.249	1.44	5.3	0.137	1.38	0.086	1.64	ripples
2-2	0.233	2.13	13	0.137	1.38	0.095	1.60	flat
2-5	0.528	1.04	0.39	0.137	1.38	0.085	1.60	ripples
2-11	0.536	1.49	5.6	0.137	1.38	0.078	1.52	ripples
2-4	0.544	2.53	16.5	0.137	1.38	0.124	1.45	flat
(b) Flume 10-1/2 in. wide								
2b	0.241	0.80	0.19	0.152	1.76	0.070	1.7	ripples
3	0.241	1.04	1.5	0.152	1.76	0.061	1.9	ripples
4	0.241	1.20	2.4	0.152	1.76	0.062	1.6	ripples
5	0.241	1.38	3.8	0.152	1.76	0.064	1.7	ripples
8	0.241	1.83	4.7	0.152	1.76	0.073	1.8	flat
1	0.241	2.06	5.6	0.152	1.76	0.085	2.3	flat
9	0.241	2.66	12	0.152	1.76	0.163	1.7	flat

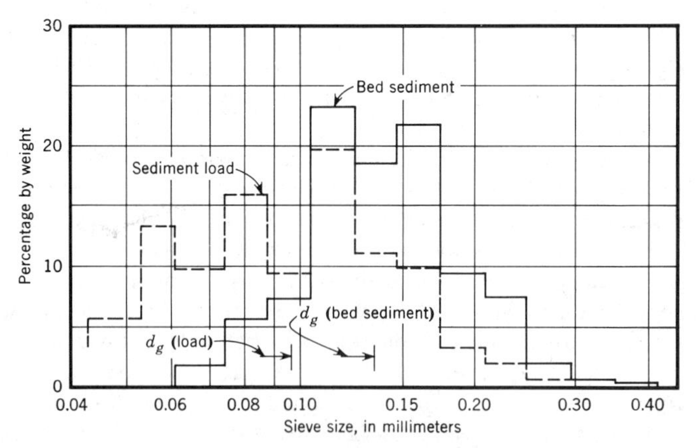

FIG. 2.86.—Size-Frequency Distribution of Bed Sediment and Bed Sediment Load in Flume Experiment 2-1 of Table 2.16

weight of the material is finer than 0.074 mm but the bed sediment contains only 3.5% of material finer than 0.074 mm. The large fraction of fine material in the load can be obtained by extracting all of the grains finer than 0.074 mm from a layer of the bed 1.5 mm in thickness.

Data on sizes of bed sediment and load are available from two small sand bed rivers in Nebraska, the Niobrara (Colby and Hembree, 1955) and the Middle Loup (Hubbell and Matejka, 1959). In the Niobrara River the sediment discharge was measured at a contracted section where the velocity was high and most of the sediment load was in suspension. A special flume was built on the Middle Loup River to facilitate the sediment discharge measurement. During the observations in these rivers the mean flow depths in the normal sections did not exceed 2 ft. The bed sediment in normal sections of the Niobrara River had a median size of about 0.3 mm and the total bed sediment load had a median size of approx 0.2 mm. In the Middle Loup River the median size of the bed sediment was between 0.3 mm–0.4 mm and the median size of the bed sediment load was between 0.2 mm–0.3 mm.

The many data available on suspended load of rivers taken with depth-integrating samplers show that the bed sediment load measured in this manner is considerably finer than the bed sediment. For example, on the Colorado River at Taylor's Ferry (United States Bureau of Reclamation, 1958) during the winter 1955–1956 with mean flow depth from 4.5 ft–10 ft the median size of the bed sediment was 0.3 mm and the median size of the suspended bed sediment collected with an integrating sampler was about 0.15 mm. The size of the suspended bed sediment collected with a depth-integrating sampler is finer than that of the total bed sediment load partly because the sampler does not collect any bed load or sediment in a layer near the bed where the mean size of the load is the coarsest. However, even when the size of the entire load is determined, as for the Niobrara and Middle Loup Rivers, the load was found to be finer than the bed sediment.

The fact that the bed sediment load of streams is finer than the bed sediment from which it derives, leads to the conclusion that as a stream degrades its bed will coarsen (Vanoni, 1962). Conversely, the fact that beds of degrading streams do coarsen may be taken as evidence that the load is finer than the bed sediment. Stream beds downstream from large reservoirs are known to coarsen as they degrade. This was observed to occur on the Colorado River downstream from Hoover Dam. Lake Meade, the reservoir formed by the dam, has a capacity of twice the mean annual runoff so that the river is completely controlled. Also, no significant flow is contributed to the river by the arid area downstream of the dam. Lane, et al. (1949) found that after closure of the dam the discharge of sediment decreased consistently as shown by Fig. 2.88. This decrease was explained by the observed coarsening of the bed sediment.

Prolonged degradation and coarsening of the bed sediment of streams can lead to armoring of the bed and to drastic reduction in the rate of degradation and sediment discharge. A spectacular case of this kind on the Missouri River downstream from Fort Randall Dam was reported by Livesey (1965). The tail water level downstream from the dam was expected to lower at the rate of about 1 ft/yr but after 10 yr the level had lowered less than 3 ft. Inspection of the bed at low water revealed that the bed surface was armored with one layer of gravel as shown in Fig. 2.87. The median size of the sediment in the bed surface before closure of the dam was about 0.17 mm and tended to coarsen after closure.

Harrison (1950) studied bed armoring in a flume by first recirculating the sediment load and then trapping it at the discharge end of the flume and introducing clear water flow into the flume. When sediment was recirculated the coarse sediment particles that moved very slowly or not at all tended to collect at the base of the dunes thus forming a lens of large particles. When clear water was introduced into the flume the dunes moved through the system in unison leaving behind a flat armored bed. The armor particles were always only one particle in thickness and covered less than one half of the bed surface. Despite this incomplete coverage the sediment discharge with the armored beds was 1% or less of that of the same water flow with recirculated sediment. Fig. 2.87 shows that the armor in the Missouri River also does not cover the entire bed area although it appears to cover more than half the bed. This figure also shows that the armored bed is flat as Harrison found in his experiments.

Harrison observed that the particles armoring the bed were arranged in a shingle pattern, i.e., they were tilted with the downstream end resting on an adjacent particle and raised higher than the upstream end.

Einstein and Chien (1953a) outlined the transport process for wash load. They advance the idea that the rate of transport of a given size-fraction of bed material of a stream depends on: (1) The ability of the stream to entrain the grains of a given size; and (2) the availability of grains of this size on the bed. If the material is available in large quantities, as in the case of bed sediment, the availability remains essentially constant and the sediment discharge depends only on the ability of the flow to entrain the grains. For this case a unique relation exists between velocity and bed sediment discharge. On the other hand, if, as in the case of wash load, the material is easily entrained and very little of it is present in the bed, so that entrainment of some of it will materially reduce the supply, the discharge of this material will depend mainly on its supply. The experiments of Einstein and Chien showed that although the amount of fine sediment found in

FIG. 2.87.—Armored Bed of Missouri River below Fort Randall Dam in 1962 (Livsey, 1965) (Scale in Picture is 6 in. Long; Flow is toward Lower Left Corner of Picture)

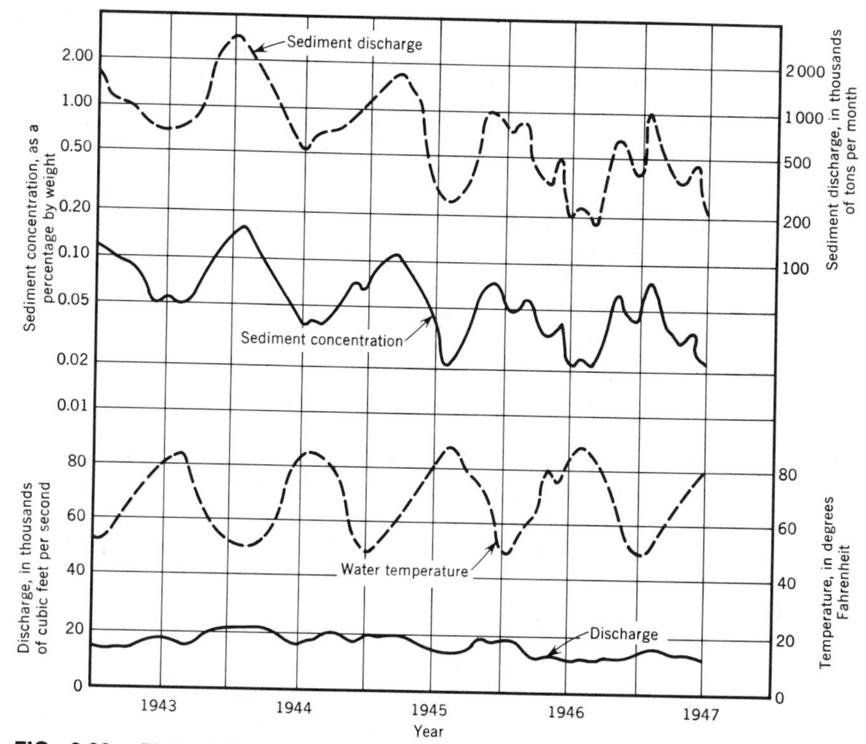

FIG. 2.88.—Plot of Suspended-Sediment Discharge and Concentration, Water Temperature, and Water Discharge against Time for Colorado River at Taylor's Ferry; Lane, Carlson, and Hanson (1949)

the bed increased as wash load discharge increased, the relative amounts present were small. Thus with a relatively large wash load discharge a relatively small amount of fine sediment is found in the bed. Conversely much greater quantities of the coarser fractions are found in the bed when the bed material discharge is relatively small.

55. Effect of Water Temperature.—According to the Rouse equation the concentration of suspended sediment will tend to become more nearly uniform over the depth of a given flow as the fall velocity of the sediment grain diminishes. Because fall velocity decreases as the temperature drops, a reduction in water temperature alone also will cause the concentration of suspended sediment to tend toward uniformity. If one assumes that the sediment concentration near the bed and the velocity and bed shear stress do not change with temperature, the Rouse equation shows that the discharge of suspended sediment will tend to increase as temperature falls, although there is no sound basis for these assumptions.

Lane, et al. (1949) found that on the Lower Colorado River the sediment discharge for a given water discharge was in fact larger in winter, when the temperature was low, than in summer. Results of their observations at Taylor's Ferry, given in Fig. 2.88, show that in winter when the water temperature dropped to 50°F the sediment discharge was as much as 2-½ times as great as in summer when the water temperature was about 85°F. As already stated, the gradual reduction in sediment discharge with time shown in Fig. 2.88 is due to coarsening of the bed sediment that had a median size of about 0.32 mm. Since the Lower Colorado River is completely controlled by dams and negligible flow is contributed to it by tributaries, all of the load is derived from the bed. Lane, et al. found that the fractions of the sediment finer than 0.295 mm exhibited large changes in transport rate with changes in temperature and that little change occurred in the transport rate of the coarser fractions. This agrees with the theory of suspended load since the fall velocities of sand particles in this size range are sensitive to water temperature.

Fig. 2.89 by Colby and Scott (1965) shows that Manning's friction factor, *n*, increased over 50% as the water temperature of the Middle Loup River increased from 32°F to 80°F. The median size of the bed sediment in the measuring reach was about 0.35 mm. Fig. 2.90, by Colby and Scott (1965) shows two sets of three bed profiles each of the Middle Loup River taken at high and low-water temperatures with essentially the same discharge. The profiles marked *a, b,* and *c* at the left third, middle third, and right third of the stream, respectively, were taken in June, 1959 when the water temperature was 83°F. The set marked *d*, *e*, and *f* was taken in December, 1959 when the water temperature was 39°F. It is seen that the height and steepness (ratio of height to length) of the bed forms was much smaller when the water was cold than when it was warm. This accounts for the lower friction factor of the cold water flow shown in Fig. 2.89.

A very illuminating and complete set of observations over a 7-mile reach of the Missouri River near Omaha, Neb., was made by the United States Army Corps of Engineers (1969). The flow of the river is controlled by releases of water from Gavins Point Dam about 150 miles upstream from Omaha, which keep the discharge between 30,000 cfs and 35,000 cfs during the navigation season, March 15 to November 15, and then reduce it to approx 10,000 cfs for the rest of the year. Observations during the fall of 1966 showed that the stage or water surface elevation for a given discharge diminished appreciably. For the regulated discharge of about 33,000 cfs the stage dropped about 2 ft. For this discharge the Missouri River at Omaha is about 700 ft wide and 10 ft deep. Fig. 2.91 is a plot of water temperature, discharge, average velocity, and the Manning roughness coefficient, *n*, against time for the Missouri River during the fall of 1966. From

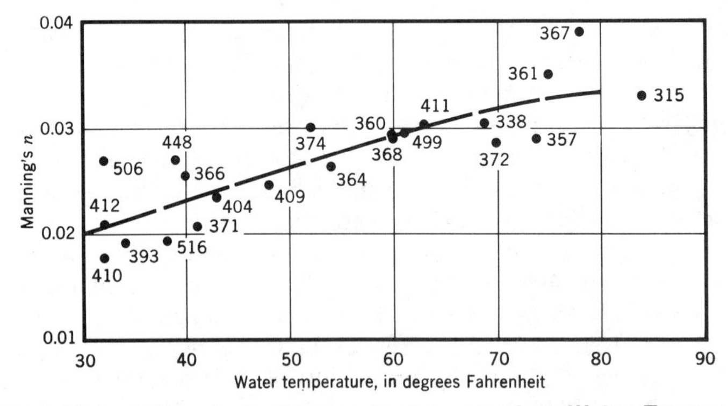

FIG. 2.89.—Plot of Manning's Friction Factor *n* against Water Temperature for Middle Loup River at Dunning, Neb.; Colby and Scott (1965) (Numbers beside Points Indicate Discharge in cubic feet per second)

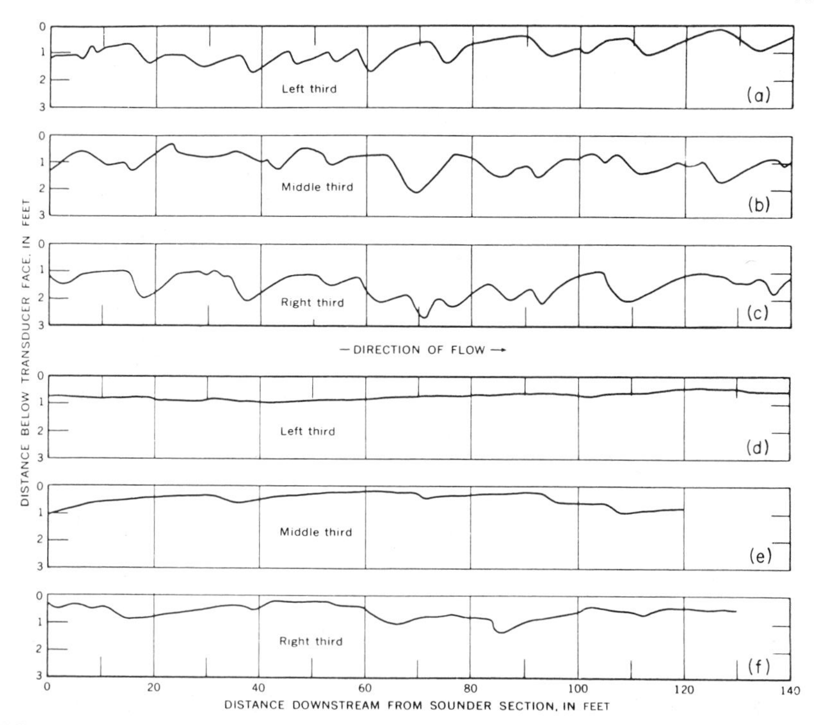

FIG. 2.90.—Two Sets of Three Longitudinal Bed Profiles Each of Middle Loup River at Dunning, Neb., at High and Low Water Temperature; Colby and Scott (1965) (Profiles a, b, c Were Taken on June 25, 1959, with Discharge = 350 cfs and Water Temperature = 83°F; Profiles d, e, f Were Taken on December 5, 1959 with Discharge = 360 cfs and Water Temperature = 39°F

early September to early November the water temperature dropped from 76°F to 42°F, the average velocity increased from 4.5 fps to 5.1 fps, and n diminished from 0.019 to 0.015. Extensive surveys of the river bed with echo sounding equipment yielded the bed-form data plotted in Fig. 2.92. It is seen that as n diminished the fraction of the bed that was flat increased, the number of dunes decreased, the dune length increased, the mean dune height decreased, and thus the dune steepness decreased. It is clear from these remarkable data that the reduction in friction factor is the result of the modification in bed forms. The fact that a part of the bed was flat and part dune-covered illustrates a characteristic of streams discussed by Colby and Scott (1965) that is different from that of flume flows. In flumes, bed forms tend to be much more uniform and patches of flat bed interspersed with patches of dunes or ripples do not occur as in natural streams.

Investigations of the Missouri River (United States Corps of Engineers, 1969) included depth-integrated suspended load samples and bed sediment samples. From early September to early November the measured suspended sand discharge concentration increased from about 500 ppm to 740 ppm. During the same period the median size of the bed sediment increased from 0.20 mm to 0.24 mm showing a slight tendency to armor.

Carey (1965) and Burke (1965) made observations of bed profiles of the

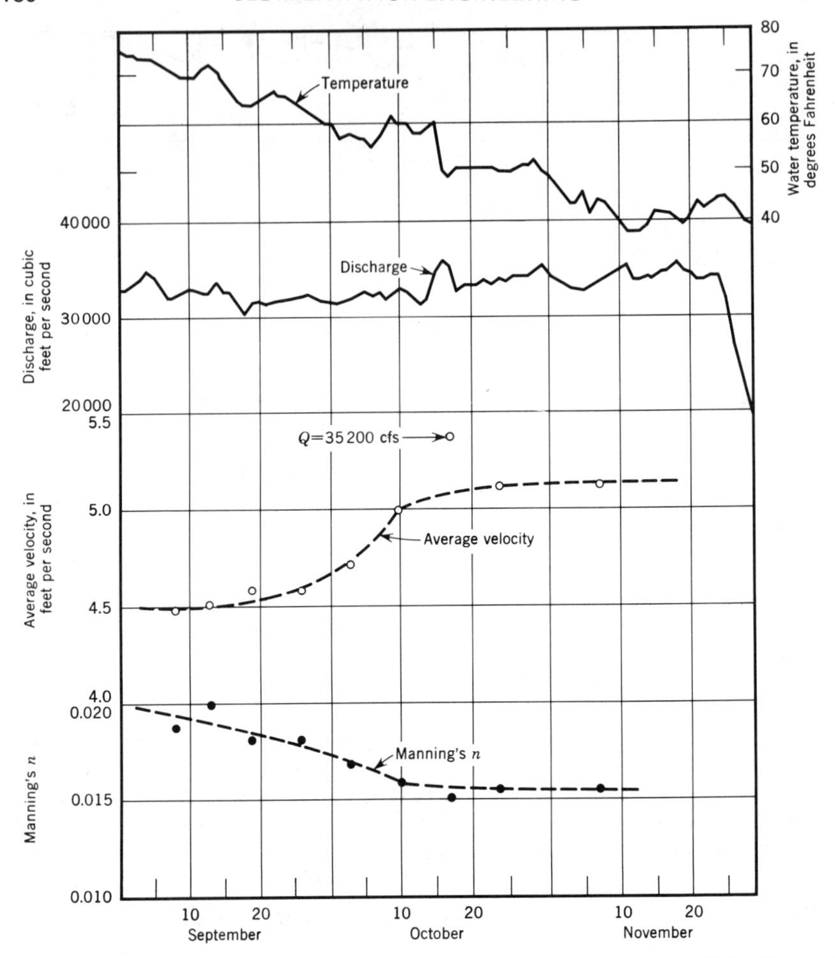

FIG. 2.91.—Variation of Water Temperature, Discharge, Average Velocity, and Manning's *n* for Missouri River at Omaha, Neb., during Fall of 1966 (United States Army Corps of Engineers, 1969)

Mississippi River over a 200-mile reach upstream of New Orleans. In this reach the water temperature normally varies from 80°F in summer to 40°F in winter and the median size of the bed sediment is about 0.4 mm. Carey (1965) observed that as the temperature lowered there was a general tendency for the bed at crossings to lower and for the height of the highest dunes to reduce. It was also observed that for a given gage height the discharge tended to increase as the water temperature decreased. This tendency is shown in Fig. 2.93 in which Carey's observations of discharge and temperature for two gage heights are plotted. The fact, that for a given gage height the discharge tends to increase as the temperature falls, indicates that the tendency is also for the velocity to increase and friction factor to decrease and agrees with observations on the Missouri and Middle Loup Rivers presented previously.

Laboratory investigations of the effect of water temperature on flow over

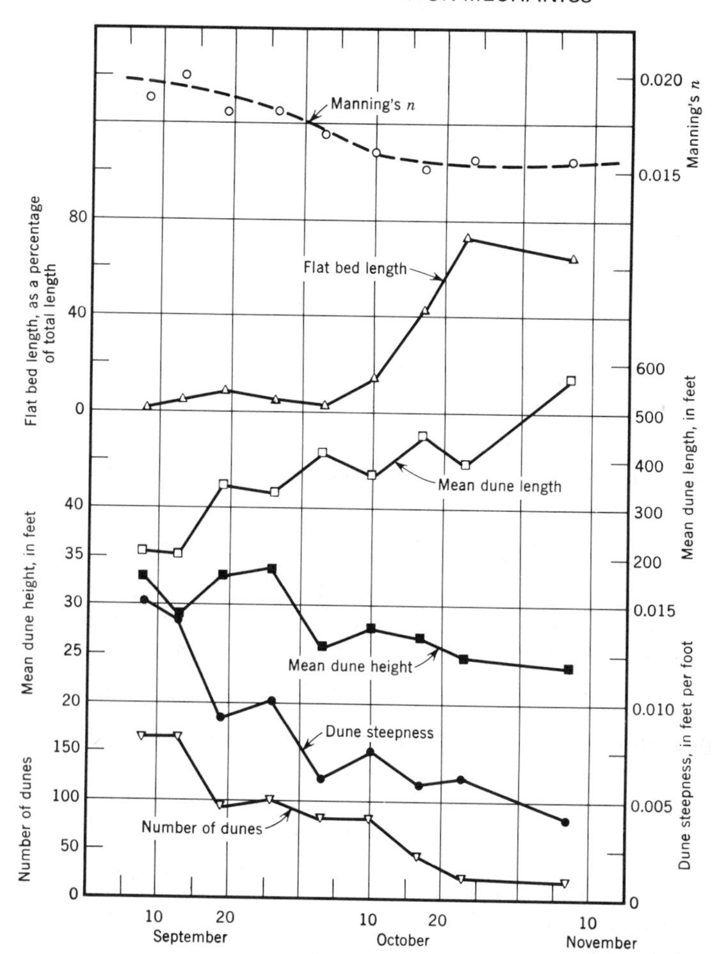

FIG. 2.92.—Data on Manning's n and Bed Form Characteristics of Missouri River at Omaha, Neb., during Fall of 1966 (United States Army Corps of Engineers, 1969)

movable beds have not shed much light on the phenomena involved. Systematic experiments on temperature effects were carried out by Franco (1968) in a flume 3 ft wide by 75 ft long charged with fine sand. The results of his experiments with a constant discharge and three water temperatures are shown in Fig. 2.94 (Taylor and Vanoni, 1969). The bed in these experiments was always rippled and most of the sediment moved as bed load. It is seen that for a given velocity a drop in temperature results in an increase in sediment discharge, slope, and friction factor. For a given slope, which corresponds to the case of a natural stream, a drop in temperature results in a decrease in velocity and sediment discharge and an increase in friction factor. Obviously, this behavior is opposite to that of the rivers previously described in which a drop in temperature was accompanied by an increase in velocity and sediment discharge and a decrease in friction factor.

The temperature effect was clarified for the case of heavily laden flows over flat beds by Taylor (1971) and Taylor and Vanoni (1972b). It was shown experi-

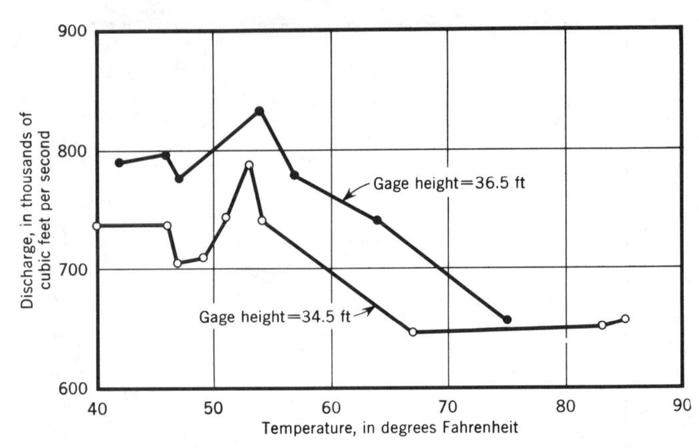

FIG. 2.93.—Variation of Discharge with Water Temperature for Constant Gage Readings of 36.5 ft and 34.5 ft for Mississippi River at Red River Landing for Period 1958-1962

mentally that the effect of water temperature on the sediment discharge could be expressed in terms of the boundary Reynolds number, R_*. When R_* was less than about 13 a decrease in water temperature caused a decrease in sediment discharge and when R_* was between 20 and 200 a decrease in water temperature caused an increase in sediment discharge. Since in flumes with flows over fine sand beds the value of R_* tends to be less than 13 and in rivers it tends to be considerably higher than this value, Taylor's results on temperature effect agree qualitatively with both field and laboratory experience. Taylor also found that the effect of temperature on the discharge of particles of different size in a given sediment could be expressed by the value of R_* based on the size of the grains. In some experiments the value of R_* for the different size fractions of the sediment extended over the range from less than 13 to in excess of 20 so that a decrease in water temperature would cause a decrease in the discharge of the finer fractions and an increase in that of the coarser ones.

From the data presented it is clear that there are at least two sets of phenomena governing the behavior of sediment-laden flows with changing water temperatures One of these involves the change in fall velocity of the sediment grains with change in water temperature and thus viscosity. A drop in temperature and thus fall velocity tends to increase the suspended-sediment concentration in the upper levels of the flow relative to those near the bed. Since the flow velocity in the upper levels is higher than near the bed an increase in concentration in these levels will also tend to increase the suspended-sediment discharge. This tendency to increase sediment discharge with a drop in temperature is observed in streams and flumes. The other phenomena governing the effect of water temperature on stream behavior operates at the bed and is still not clarified especially for ripple and dune beds. It appears that a drop in temperature tends to modify conditions at the bed. In rivers this causes the bed forms to reduce in height and to diminish the resistance to flow. In flumes with ripple beds of fine sand a drop in temperature causes the opposite effect, i.e., the friction factor tends to increase as water temperature fall, e.g., see Fig. 2.94. One can hypothesize that a drop in temperature (increase in viscosity) tends to protect the bed by increasing the thickness

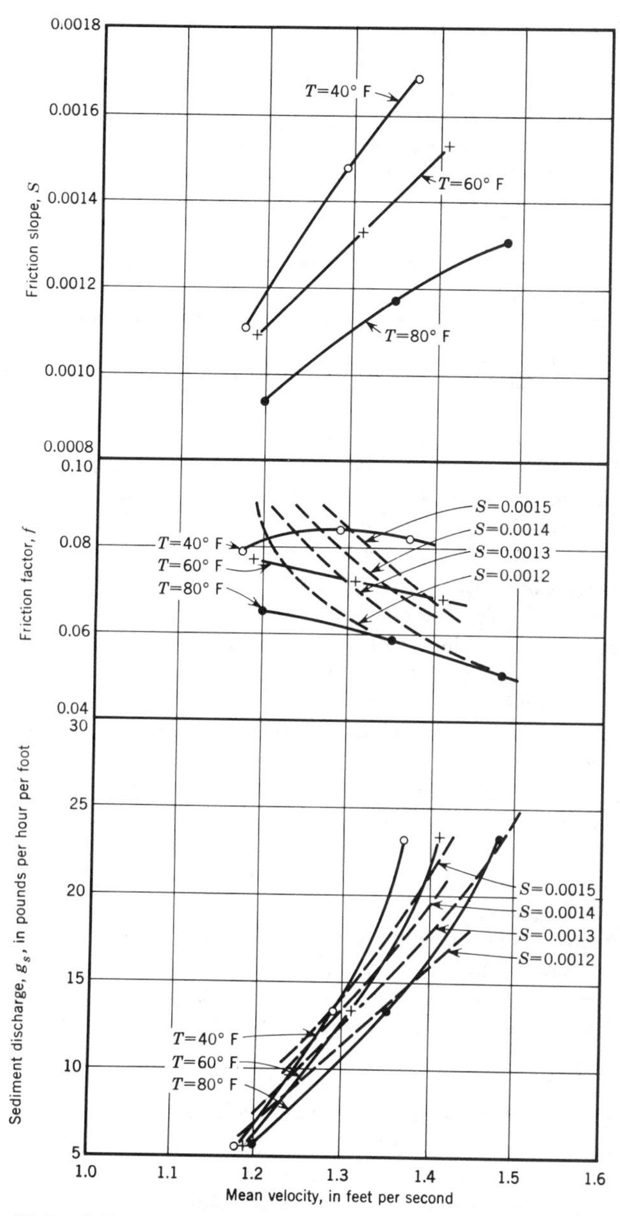

FIG. 2.94.—Plot of Data from Experiments by Franco (1968) with Constant Discharge of 0.62 cfs /ft of Width over Bed of Fine Sand Showing Effects of Water Temperature (Flume Width = 3 ft, Median Size of Bed Sediment = 0.23 mm)

of the laminar sublayer while also increasing the effectiveness of the currents to attack the bed. Based on this hypothesis, it would appear that in rivers the increase in current effectiveness with drop in temperature exceeds the increase in the protection of the laminar sublayer. In flumes the balance is apparently reversed.

H. Sediment Discharge Formulas

56. General.—Engineers engaged in river regulation and design and operation of canal systems have great need for methods of computing sediment discharge. In fact, obtaining such methods is probably the most important practical objective of research in sedimentation. Unfortunately, available methods or relations for computing sediment discharge are far from completely satisfactory with the result that plans for works involving sediment movement by water cannot be based strongly on such relations. At best these relations serve as guides to planning and usually the engineer is forced to rely strongly on experience and judgment in such work. The guiding provided turns out to be important especially when the conditions in the problem area differ from those in the experience of the planning engineers. For this reason it seems pertinent to present some of the available relations in this section.

Relations for calculating sediment discharge will be referred to as formulas even though parts or all of some relations are presented in graphical form and, strictly speaking, are not formulas. The objective of presenting them is to make them conveniently available to those who may need to employ them and to give some information that may help in evaluating them.

Many formulas have appeared in the literature since DuBoys (1879) presented his tractive force relation. The problem of the engineer is to select one or more of these for use in solving his particular problem. This selection is not straightforward since the results of different formulas often differ drastically and it is not possible to determine positively which one gives the most realistic result. To help engineers select formulas, a brief outline of the data on which each formula is based will be given and an attempt will be made to evaluate the formulas by comparing observed sediment discharges in rivers with values calculated by the formulas. Only a few of the many formulas available will be presented and discussed. These were selected because they are used by many engineers or because it may appear that they show promise of being adopted in the future. Some formulas considered important by many experts in sedimentation have probably been omitted because of this subjective method of selection. No derivations will be attempted in this section because the fundamental ideas upon which the derivations are based are discussed in Chapter II, Section G.

In addition to the formulas, procedures will be presented for estimating the discharge of bed sediment from suspended load samples and normal stream flow measurements. This approach to obtaining sediment discharge is one of the most important developments in river sedimentation in recent years. By making use of observed quantities it gives results that are very much more reliable than those given by the formulas. A compilation of sediment discharge formulas by Shulits and Hill (1968) gives computer programs for a number of the formulas listed herein.

57. Formulas.—All of the formulas presented in this section are for discharge of bed sediment under conditions of uniform steady flow and do not include the wash load. The formulas are as follows:

DuBoys (Brown, 1950), 1879	Eq. 2.224
Meyer-Peter (Meyer-Peter and Muller, 1948)	Eq. 2.225
Schoklitsch (Shulits, 1935)	Eq. 2.226
Shields (1936)	Eq. 2.227

Meyer-Peter and Muller (1948) Eq. 2.228
Einstein-Brown (Brown, 1950) Eq. 2.229
Einstein Bed Load Function (Einstein, 1950) Eq. 2.230
Laursen (1958) Eq. 2.231
Blench Regime Formula (Blench, 1966a) Eq. 2.232
Colby (1964a-b) Eq. 2.233
Engelund-Hansen (Engelund, 1966; Engelund and Hansen, 1967) Eq. 2.234
Inglis-Lacey (Inglis, 1968) Eq. 2.235
Toffaleti (1969) Eq. 2.236

To apply these formulas the flow depth, d, and slope, S, or mean velocity, depth, and slope must be given.

DuBoys Formula (Brown, 1950)

$$g_s = \Psi_D \tau_o (\tau_o - \tau_c) \qquad (2.224)$$

in which g_s = sediment discharge, in pounds per second per foot of width; Ψ_D = coefficient with dimensions of cubic feet per pound per second; $\tau_o = \gamma r_b S$ bed shear stress, in pounds per square foot; γ = specific weight of water, in pounds per cubic foot; τ_c = critical bed shear stress at which sediment movement begins; r_b = bed hydraulic radius, in feet (determined by the Side-wall Correction method outlined in Chapter II, Section F); and S = slope of stream, in feet per foot. Values of Ψ_D and τ_c obtained by Straub and reported in Brown, 1950 are given as functions of median size of the bed sediment, d_{50}, in Fig. 2.95. These quantities were based mainly on data from experiments by Gilbert (1914; Johnson, 1943) in small flumes. Eq. 2.224 as presented herein is valid only for the foot-pound-second system of units.

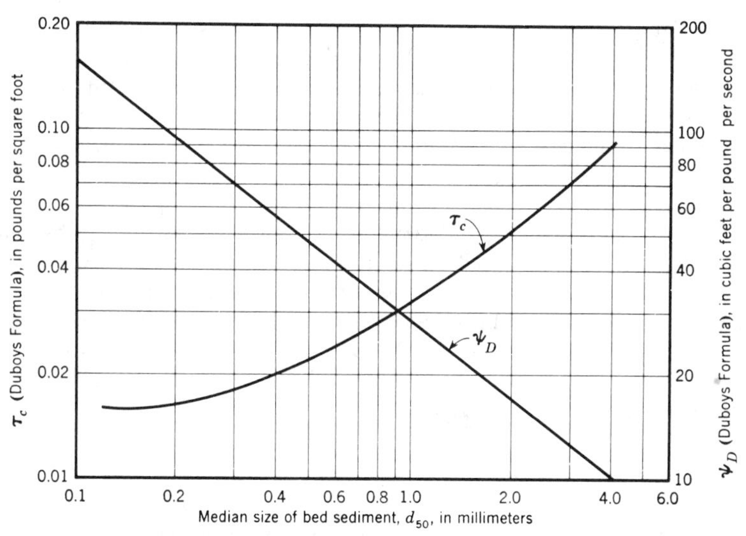

FIG. 2.95.—Coefficient Ψ_D and Critical Shear Stress τ_c for DuBoys Eq. 2.227 as Functions of Median Size of Bed Sediment

Meyer-Peter Formula (Meyer-Peter and Muller, 1948)

$$g_s^{2/3} = 39.25\, q^{2/3}\, S - 9.95\, d_{50} \qquad (2.225)$$

in which q = water discharge, in cubic feet per second per foot of width; d_{50} = median size of bed sediment, in feet; and g_s and S are as defined previously. Eq. 2.225 is valid only for the foot-pound-second system of units. For the meter-kilogram-second units the numbers 39.25 and 9.95 in Eq. 2.225 are replaced by 250 and 42.5., respectively.

The constants in Eq. 2.225 were determined by fitting the equation to data obtained in experiments with five well-sorted river sediments ranging in median size from 3.1 mm–28.6 mm. The experiments with three of these sediments were made by Gilbert (1914) in flumes 8 in., 12 in., and 16 in. wide with flow depths ranging from 1 in.–6 in. The experiments with the 28.6-mm sediment and one other were made in a flume 2 m wide. These coarse sediments do not produce rugged bed forms so that in these experiments the flow resistance was due mainly to grain roughness. Therefore Eq. 2.225 is valid only for beds of relatively coarse sediments for which the flow resistance due to bed forms is a small part of the total resistance.

Schoklitsch Formula (Shulits, 1935)

$$g_s = \sum_i p_i \frac{25.3}{\sqrt{d_{si}}} S^{3/2} (q - q_{ci}) \qquad (2.226a)$$

$$q_{ci} = 0.638 \frac{d_{si}}{S^{4/3}} \qquad (2.226b)$$

in which q_{ci} = critical value of q for initiating motion of sediment of mean size, d_{si} as given by Eq. 2.226b; p_i = fraction by weight of that fraction of the bed sediment with mean size, d_{si}; the symbol, Σ_i, denotes summation for all sets of values of p_i, d_{si}, and q_{ci}; and other symbols are as defined previously. All quantities in Eq. 2.226b are expressed in the foot-pound-second system of units.

To determine sets of values of p_i and d_{si} a mechanical analysis of a representative sample of the bed sediment is made and a size distribution curve prepared. A set of size grades is then selected and the corresponding p_i values can be determined from the size distribution curve. The size grades frequently used are those shown in Table 2.1. The mean size, d_{si}, of a fraction is often taken as the geometric mean of the extreme sizes in the fraction.

The Schoklitsch formula was based mainly on data from experiments by Gilbert (1914) in small flumes with well-sorted and also graded sediments with median sizes ranging from 0.3 mm–5 mm. Sediment discharges calculated with the formula also agreed well (Shulits, 1935) with bed load discharges measured with samplers in two European rivers that have gravel beds. This suggests that it is a bed load formula that should not be applied to sand bed streams that carry considerable bed sediment in suspension.

Shields Formula (Shields, 1936)

$$g_s = 10 q\, S \frac{(\tau_o - \tau_c)}{\left(\dfrac{\gamma_s}{\gamma} - 1\right)^2 d_{50}} \qquad (2.227)$$

in which γ_s = specific weight of the sediment grains; τ_c = critical bed shear stress for sediment of size d_{50} given by Shields graph, Chapter II, Section E, Fig. 2.43; γ_s = the specific weight of the sediment; and all other quantities in Eq. 2.227 are already defined except that since the equation is dimensionally homogeneous, the quantities can be expressed in any consistent set of units.

(In the formulas presented in this section, sediment discharge and concentration are usually given in weight units even though they should be given in mass units. In the traditional systems of units, i.e., foot-pound-second and meters-kilogram-second this confusion gives no difficulty since pounds and kilograms of mass are numerically equal to those of weight. However, in the SI system of units the formulas as presented will give sediment discharge and concentration in terms of Newtons that numerically are 9.8 times as large as the same quantities expressed in kilograms of mass.)

The Shields formula is based mainly on data from two flumes with widths of 40 cm and 80 cm, respectively, with five sediments of specific gravities ranging from 1.06–4.2. The lightest sediment was made of amber particles with a median size of 1.56 mm. The other sediments were well sorted with median sizes ranging from 1.7 mm–2.5 mm. Ripples were produced on the bed but none of them was very high or steep. Because the sediments in the experiments were coarse and the shear stresses low, essentially all of the sediment moved was bed load.

Meyer-Peter and Muller Formula (1948)

$$\left(\frac{k_r}{k'_r}\right)^{3/2} \gamma r_b S = 0.047\,(\gamma_s - \gamma)\, d_m + 0.25\left(\frac{\gamma}{g}\right)^{1/3}\left(\frac{\gamma_s - \gamma}{\gamma_s}\right)^{2/3} g_s^{2/3} \qquad (2.228a)$$

$$\frac{k_r}{k'_r} = \sqrt{\frac{f'_b}{8}}\,\frac{V}{\sqrt{g r_b S}} \qquad (2.228b)$$

in which g = acceleration of gravity; f'_b = Darcy-Weisbach bed friction factor for the sand grain roughness defined in Chapter II, Section F; and V = mean flow velocity of the stream. The quantities, k_r and k'_r, are defined by

$$V = k_r\, r_b^{2/3}\, S^{1/2} \qquad (2.228c)$$

and $V = k'_r r_b^{2/3} S'^{1/2}$ $\qquad$ (2.228d)

in which S' is that part of the total slope, S, required to overcome the grain resistance and is defined in terms of f'_b as

$$V = \sqrt{\frac{8}{f'_b}}\,\sqrt{g r_b S'} \qquad (2.228e)$$

The friction factor, f'_b, is obtained from the well-known pipe friction graph of the Nikuradse pipe friction data in which the friction factor, f, is expressed as a function of Reynolds number VD/ν and relative roughness D/k_s in which D = pipe diameter; ν = kinematic viscosity of the water; and k_s = the grain size of the sand forming the roughness at the pipe wall. To obtain f'_b from the pipe friction graph, the diameter, D, is replaced by $4r_b$ and k_s is replaced by d_{90}, the grain size of the bed sediment for which 90% is finer. When the boundary

Reynolds number, $\sqrt{(f_b'/8)}\, V(d_{90}/\nu)$, equals or exceeds a value of approx 100, the boundary will be hydrodynamically rough and k_r' is given by

$$k_r' = \frac{26}{d_{90}^{1/6}} \qquad (2.228f)$$

in which d_{90} is in meters and k_r' is in meters to the one-third power per second. The quantity, d_m, is the effective diameter of the sediment given by

$$d_m = \sum_i p_i d_{si} \qquad (2.228g)$$

in which p_i, d_{si}, and the summation sign are as in the Schoklitsch formula.

Eqs. 2.228a, 2.228b, and 2.228e are dimensionally homogeneous so that any consistent set of units may be used with them. On the other hand, Eq. 2.228f is valid only when d_{90} is expressed in meters and time is in seconds. When k_r' is obtained from Eq. 2.228f the quantity, k_r, is to be calculated from Eq. 2.228c in which V and r_b are expressed, respectively, in meters per second and meters. Once k_r and k_r' are obtained in meter-second units any other consistent set of units may be used for all other quantities in Eqs. 2.228a, 2.228b, and 2.228e.

The Meyer-Peter and Muller formula is based on data from experiments in flumes ranging in width from 15 cm–2 m with slopes varying from 0.0004–0.02 and water depths ranging from 1 cm–120 cm. The sediments used in the experiments ranged from coal with a small specific gravity, γ_s/γ, = 1.25, to river sediment to barite with a specific gravity in excess of four. Some of the sediments were graded and others were sorted. The mean sizes and effective diameters, d_m, of the sediments ranged from 0.4 mm–30 mm. The advantage of this formula over the older Meyer-Peter formula, Eq. 2.225, is that it can be used for graded sediments under flow conditions that give rise to dunes and other bed forms. Most of the data upon which the formula is based were obtained in flows with little or no suspended load that suggests that the formula is not valid for flows with appreciable suspended loads.

Einstein-Brown Formula (Brown, 1950)

This formula was presented in Chapter XII of Rouse, 1950. It is a modification developed by Hunter Rouse, M. C. Boyer, and E. M. Laursen of a formula by Einstein (1942). Its name derives from the name of the original author and the author of the chapter where the formula first appeared. The formula is

$$\Phi = f\left(\frac{1}{\Psi}\right) \qquad (2.229a)$$

in which the function, $f(1/\Psi)$, as given in Rouse (1950) is shown in Fig. 2.96

and $$\Phi = \frac{g_s}{\gamma_s F_1 \sqrt{g\left(\frac{\gamma_s}{\gamma} - 1\right) d_s^3}} \qquad (2.229b)$$

$$\frac{1}{\Psi} = \frac{\tau_o}{(\gamma_s - \gamma) d_s} = \tau_* \qquad (2.229c)$$

$$F_1 = \sqrt{\frac{2}{3} + \frac{36\nu^2}{g d_s^3\left(\frac{\gamma_s}{\gamma} - 1\right)}} - \sqrt{\frac{36\nu^2}{g d_s^3\left(\frac{\gamma_s}{\gamma} - 1\right)}} \qquad (2.229d)$$

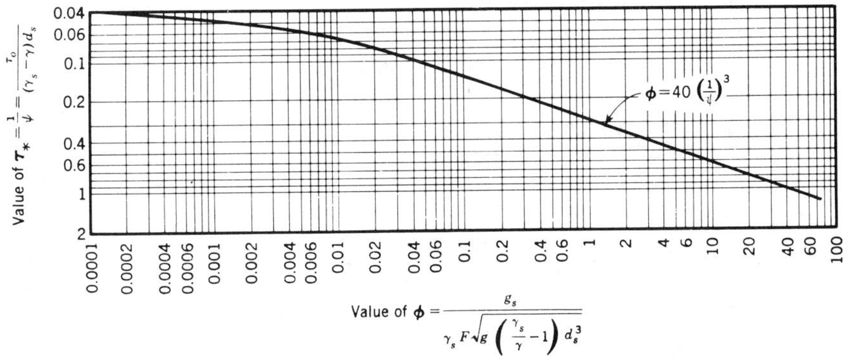

FIG. 2.96.—Function $\phi = f(1/\Psi)$ for Einstein Brown Eq. 2.229

As shown by Fig. 2.96, Eq. 2.229a becomes $\Phi = 40(1/\Psi)^3$ for $1/\Psi$ in excess of 0.09.

The quantity, d_s, is the representative size of bed sediment and is usually taken as the median size, d_{50}, or geometric mean size, d_g. The bed shear stress, τ_o, is usually taken as $\gamma r_b S$. The quantity, F_1, appears in the Rubey (1933) formula for fall velocity w of sediment of size d_s:

$$w = F_1 \sqrt{\left(\frac{\gamma_s}{\gamma} - 1\right) g d_s} \qquad (2.229e)$$

Note that $1/\Psi$ is the same as the dimensionless shear stress, τ_*, introduced by Shields (1936). Since all of Eqs. 2.231 are dimensionally homogeneous any consistent set of units may be used in them.

The Einstein-Brown formula was based on flume data by Gilbert (1914) and Meyer-Peter and Muller with well-sorted sediments. The Gilbert data were obtained in small flumes with river sediment with median sizes from 0.3 mm–7 mm. The other data used were obtained with a flume 2 m wide with 28.6-mm gravel and with a smaller flume with 5.21-mm gravel, barite, and coal. The specific gravities of the barite and coal were 4.2 and 1.25, respectively.

Einstein Bed Load Function (Einstein, 1950)

In this method the sediment discharge is computed for individual size fractions of the bed material. This means that one also obtains the size distribution of the sediment load. The equations and relations used in the calculations follow:

$$g_s = \sum_i g_{si} \qquad (2.230a)$$

$$G_s = b g_s \qquad (2.230b)$$

$$g_{si} = g_{sbi} [P_r I_1(\eta_{oi}, z_i) + I_2(\eta_{oi}, z_i) + 1] \qquad (2.230c)$$

In these equations g_s = discharge of bed sediment in weight per unit width and time as defined previously; G_s = the total bed sediment discharge of the stream, in weight per unit time; b = the bed width of the stream; g_{si} = discharge of bed sediment of mean size d_{si}, i.e., the ith size fraction, per unit width; the summation sign, $\sum_i$, indicates the sum of g_{si} for all size fractions; and g_{sbi} = discharge of bed load of mean size d_{si}, in weight per unit width. The product of g_{sbi} and the first two terms in the parentheses at the right of Eq. 2.230c gives g_{ssi}, the discharge of suspended load of mean size, d_{si}, per unit width. From this it follows that the total sediment discharge per unit width is the sum of g_{ssi} and g_{sbi}. The terms, $I_1(\eta_{oi}, z_i)$, $I_2(\eta_{oi}, z_i)$, and P_r are

$$I_1(\eta_{oi}, z_i) = 0.216 \frac{\eta_{oi}^{z_i-1}}{(1-\eta_{oi})^{z_i}} \int_{\eta_{oi}}^{1} \left(\frac{1-\eta}{\eta}\right)^{z_i} d\eta \quad \text{(2.230d)}$$

$$I_2(\eta_{oi}, z_i) = 0.216 \frac{\eta_{oi}^{z_i-1}}{(1-\eta_{oi})^{z_i}} \int_{\eta_{oi}}^{1} \left(\frac{1-\eta}{\eta}\right)^{z_i} \ln\eta \, d\eta \quad \text{(2.230e)}$$

$$P_r = 2.3 \log \frac{30.2\, x r_b}{d_{65}} \quad \text{(2.230f)}$$

$$z_i = \frac{w_i}{0.4\, U'_*} \quad \text{(2.230g)}$$

in which $\eta_{oi} = 2\, d_{si}/r_b$; $\eta = y/r_b$ = a dimensionless variable of integration; x = a dimensionless quantity in the logarithmic velocity distribution law defined by Eq. 2.143 and given in Fig. 2.97 in terms of $U'_* d_{65}/11.6\nu$; d_{65} = size of bed sediment for which 65% of the sediment by weight is finer; w_i = the fall velocity of a grain of bed sediment of size d_{si} given by the Rubey Eq. 2.229e; and $U'_* = \sqrt{g r'_b S}$ is the shear velocity based on the shear stress due to grain roughness as defined by Einstein and Barbarossa (1952) and presented in Chapter II, Section F. The functions, I_1 and I_2, are given in Figs. 2.98 and 2.99, respectively, as functions of z_i and $\eta_{oi} = 2d_{si}/r_b$. These functions derive from the expression

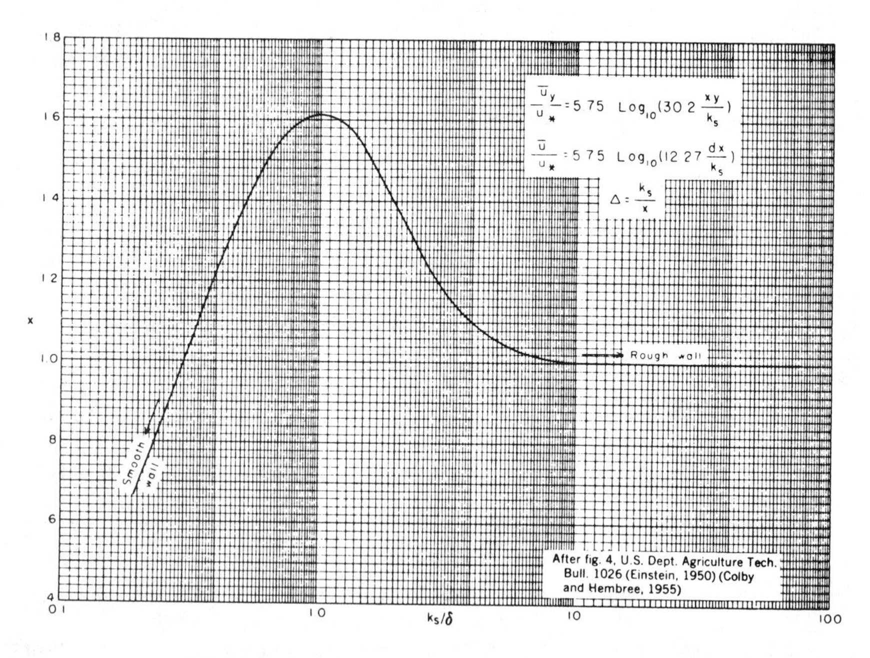

FIG. 2.97.—Factor x in Velocity Distribution Equation

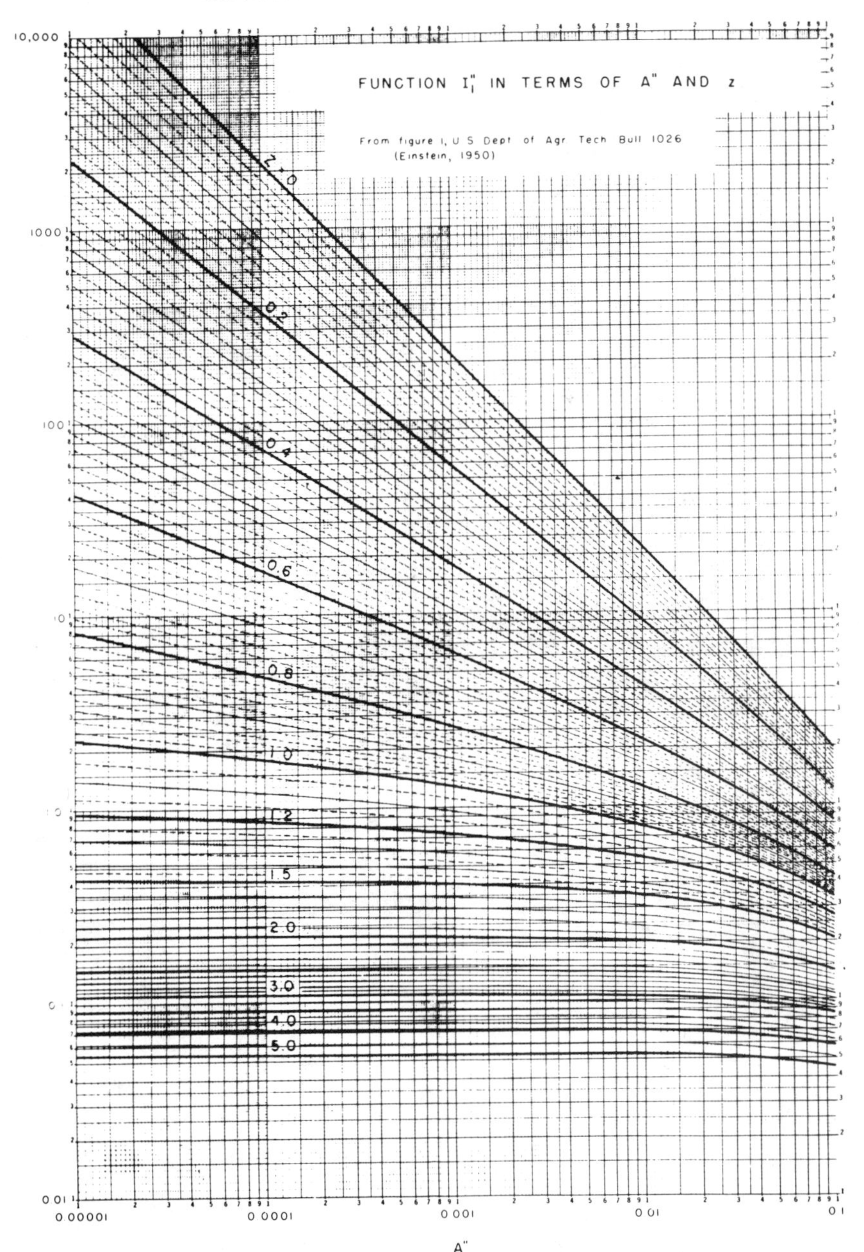

FIG. 2.98.—Integral I_1 in Terms of Exponent z and Lower Limit η_o; Einstein (1950) ($A'' \equiv 2\, d_{si}/r_b = \eta_{oi}$)

$$g_{ssi} = \int_{a_i}^{r_b} C_i U\, dy \quad \text{(2.230h)}$$

in which C_i = concentration of suspended sediment of mean size d_{si} at a distance y above the bed in weight per unit volume given by Eq. 2.77; U = velocity of flow at distance y above the bed; and $a_i = 2\, d_{si}$ = the lower limit of integration at which level the concentration, C_{ai}, of sediment of mean size d_{si} is related to g_{sbi} by

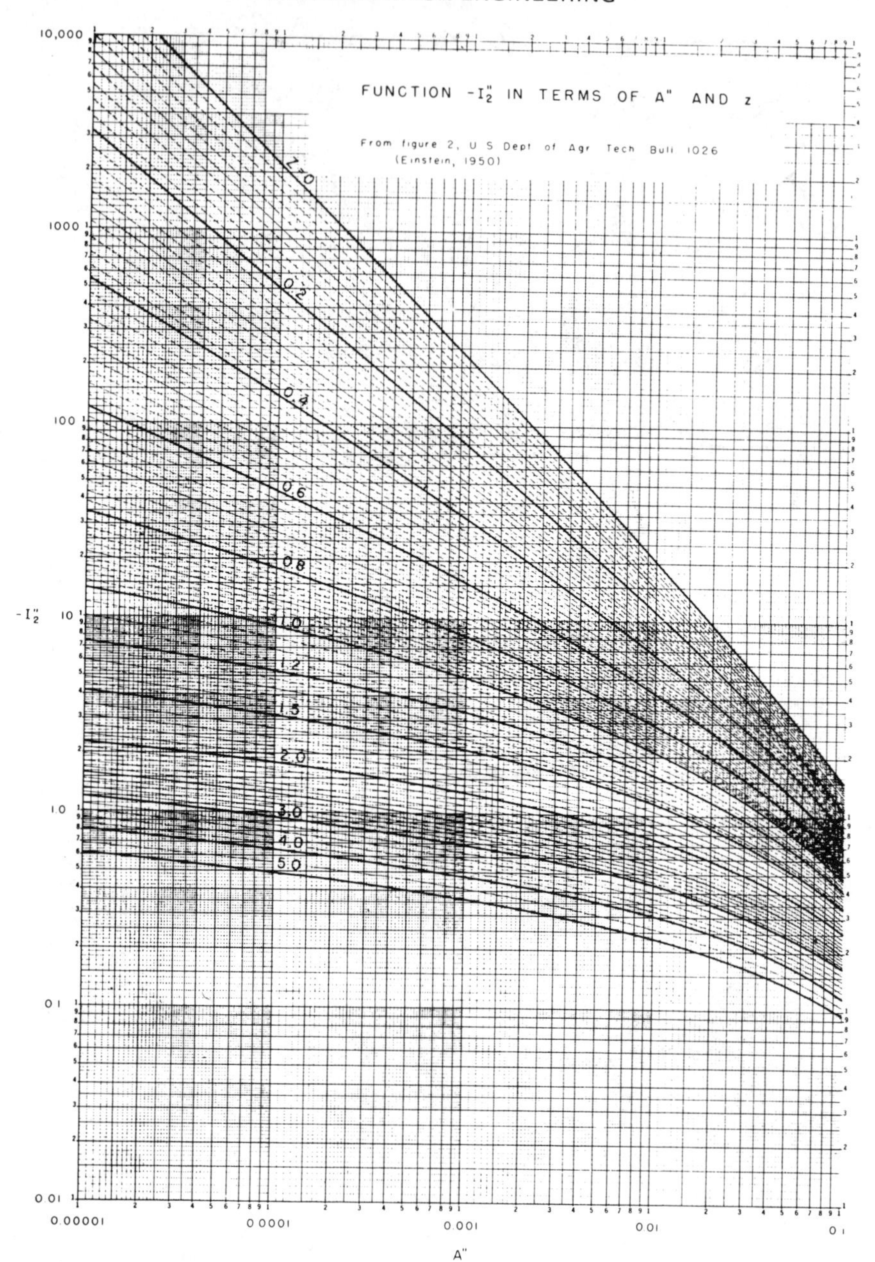

FIG. 2.99.—Integral I_2 in Terms of Exponent z and Lower Limit η_o; Einstein (1950) ($A'' \equiv 2d_{si}/r_b = \eta_{oi}$)

$$g_{sbi} = 11.6\, C_{ai}\, 2\, d_{si} U'_* \quad \text{(2.230i)}$$

Ordinarily one uses the flow depth, d, as the upper limit of integration in Eq. 2.230h. Einstein uses r_b instead of d, based on the idea, inherent in the side-wall correction method, that r_b is the depth of a two-dimensional stream equivalent to the real stream. For this same reason the variable, η, in Eq. 2.230d is taken as y/r_b instead of y/d.

The relation for g_{sbi} remaining to complete the Einstein theory is the actual Einstein Bed Load Function

$$1 - \frac{1}{\sqrt{\pi}} \int_{-(1/7)\Psi_{*i}-2}^{(1/7)\Psi_{*i}-2} e^{-t^2} dt = \frac{43.5\Phi_{*i}}{1 + 43.5\Phi_{*i}} \qquad (2.230j)$$

$$\text{in which } \Psi_{*i} = \xi_i Y \left(\frac{\log 10.6}{\log \dfrac{10.6x\,X}{d_{65}}} \right)^2 \frac{(\gamma_s - \gamma) d_{si}}{\gamma r'_b S} \qquad (2.230k)$$

$$\Phi_{*i} = \frac{1}{p_i} \frac{g_{sbi}}{\gamma_s} \sqrt{\left(\frac{\gamma}{\gamma_s - \gamma} \right) \frac{1}{g d_{si}^3}} \qquad (2.230l)$$

In these equations ξ_i = a function of d_{si}/X given in Fig. 2.100; Y = a function of d_{65}/δ given in Fig. 2.101

$$\left. \begin{aligned} X &= 0.77 \frac{d_{65}}{x} \text{ when } \frac{d_{65}}{x\delta} > 1.80 \\ X &= 1.398\,\delta \text{ when } \frac{d_{65}}{x\delta} < 1.80 \end{aligned} \right\} \qquad (2.230m)$$

$$\delta = 11.6 \frac{\nu}{U'_*} \qquad (2.230n)$$

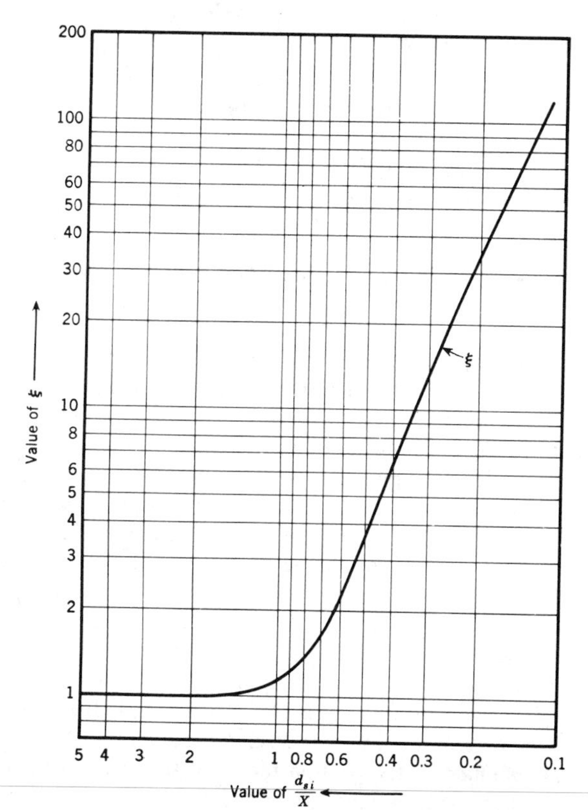

FIG. 2.100.—Factor ξ in Einstein's Bed Load Function (Einstein, 1950) in Terms of d_{si}/X

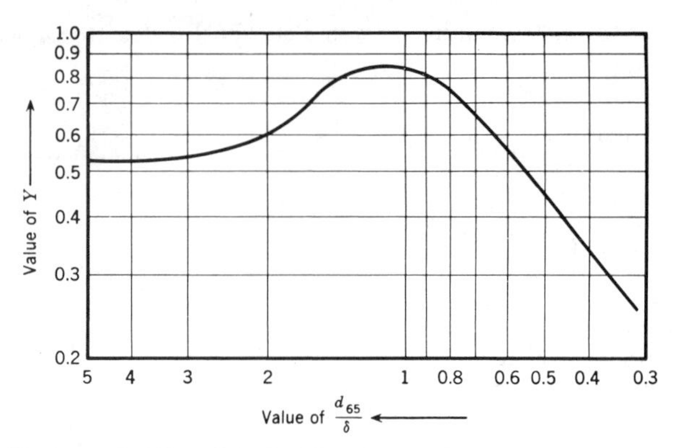

FIG. 2.101.—Factor Y in Einstein's Bed Load Function (Einstein, 1950) in Terms of d_{65}/δ

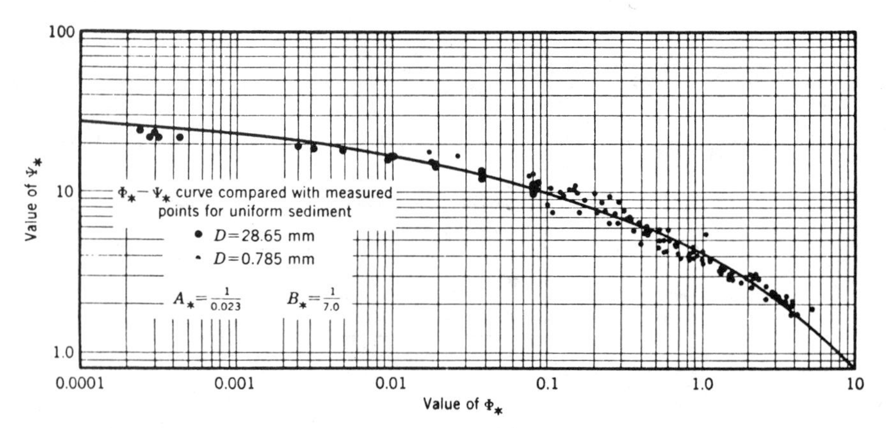

FIG. 2.102.—Einstein's Φ_* – Ψ_* Bed Load Function (Einstein, 1950)

Term r'_b is the Einstein-Barbarossa bed hydraulic radius corresponding to the sand grain resistance of the bed and is determined as described in Chapter II, Section F. For bed sediment of uniform size, $X = 1.39\delta$ and $Y = 1.0$.

As already stated Eq. 2.230j is the Einstein bed load function that gives the bed load discharge, g_{sbi}. Fig. 2.102 gives the bed load function, i.e., Φ_* as a function of Ψ_*. This result then enables one to calculate the suspended-sediment concentration, C_{ai}, at a distance, 2 d_{si}, above the bed by means of Eq. 2.230i.

Calculation of the sediment discharge of a stream by the Einstein bed load function can be carried out in three parts. The first part involves collecting field data that include measurements of: (1) Slope; (2) data on cross section from which one can determine the bed width and the cross-sectional area and wetted perimeter of the banks as functions of depth; (3) a representative sample of the bed sediment; (4) a mechanical analysis of the bed sample from which the weight fraction, p_i, can be determined for each size, d_{si}, and values of d_{35} and d_{65} can be read; and (5) an estimate of the friction factor of the channel banks.

In the second part, calculations are made to determine values of the bed hydraulic radius, r'_b, due to sand grain roughness for a range of values of water discharge Q. Such calculations are described in Chapter II, Section F.

The third part of the calculation uses the results of the first two to finally determine the total bed sediment discharge, G_s, for several values of Q. A method of step by step calculation of G_s is outlined in Table 2.17. In this table the foot-pound-second system of units is used but since the relations are dimensionally homogeneous, any consistent set of units could have been used.

The steps in the calculation of sediment discharge by the Einstein bed load function are:

1. Field data required for calculation are: (*a*) S = channel slope; (*b*) data on bed width and cross-sectional area and wetted perimeter of banks as a function of water depth; (*c*) friction factor of banks of channel (estimated); (*d*) sample of bed sediment from which values of weight fractions p_i can be determined for several mean sizes d_{si} along with the specific weight; γ_s, d_{35}, and d_{65}; and (*e*) water temperature.
2. Hydraulic calculations—these are described in Chapter II, Section F and will give r_b' and r_b for a range of values of Q.
3. Sediment discharge calculation (see Table 2.17).

The coefficients, 1/7 and 43.5, of Ψ_* and ϕ_*, respectively, in Einstein's bed load function, Eq. 2.230j, were determined by fitting the function to flume data. Eq. 2.230j is plotted in Fig. 2.102. The data plotted in this figure were obtained in flume experiments with two well-sorted sediments of mean size 28.65 mm and 0.785 mm, respectively. These are shown in this figure to demonstrate the fact that the theory agrees with data from experiments with sorted sediments. The data used to determine the coefficients in Eq. 2.230k were from experiments in a flume 10.5 in. wide and 40 ft long with high transport rates of graded fine sands.

Laursen Formula (1958)

$$C_m = 0.01\gamma \sum_i p_i \left(\frac{d_{si}}{d}\right)^{7/6} \left(\frac{\tau_o'}{\tau_{ci}} - 1\right) f\left(\frac{U_*}{w_i}\right) \quad \text{(2.231a)}$$

$$\tau_o' = \frac{\rho V^2}{58}\left(\frac{d_{50}}{d}\right)^{1/3} \quad \text{(2.231b)}$$

$$\tau_{ci} = \tau_{*c}(\gamma_s - \gamma)d_{si} \quad \text{(2.231c)}$$

$$g_s = C_m q \quad \text{(2.231d)}$$

in which C_m = sediment discharge concentration, in weight per unit volume; d = depth of flow; τ_o' = Laursen's bed shear stress due to grain resistance defined by Eq. 2.231b; τ_{ci} = critical shear stress for particles of size d_{si}, $U_* = \sqrt{gdS}$ = total bed shear velocity; w_i = fall velocity of particles of mean size d_{si} in water; $f(U_*/w_i)$ = the function shown in Fig. 2.103; and τ_{*c} = dimensionless critical shear stress that is defined by Shields (see Fig. 2.43). Laursen used $\tau_{*c} = 0.039$ for sediments ranging in median size from 0.088 mm–4.08 mm. If this value of τ_{*c} is substituted into Eq. 2.231c with $\gamma = 62.4$ pcf and $\gamma_s/\gamma = 2.65$ (quartz) one gets Laursen's equation for sands, $\tau_{ci} = 4d_{si}$ psf in which d_{si} is in feet. The value $\tau_{*c} = 0.039$ should be used in Eq. 2.231c. Eq. 2.231c was presented instead of Laursen's equation to bring out the nature of the equation and to achieve dimensional homogeneity. For the same reason Eq. 2.231b differs from Laursen's expression for τ_o' in that ρ has been added and the numerical constant changed. With these changes in Eqs. 2.231b and 2.231c all of the Laursen

TABLE 2.17.—Calculation of Sediment Discharge by Einstein Bed Load Function

(1)	(2)	(3)	(4)	(5)	(6)	(7)	(8)	(9)	(10)	(11)	(12)
(*a*) Calculations Not Involving d_{si}											
r'_b, in feet assume	$U'_* = \sqrt{g r'_b S}$, in feet per second	$\delta = 11.6(\nu/U'_*)$, in feet Eq. 2.220*n*	d_{65}/δ	x from Fig. 2.97	X Eq. 2.230*m*	Y Fig. 2.101	$10.6Xx/d_{65}$	log (Col. 8)	[log 10.6/log (Col. 8)]^2	$30.2 r_b x/d_{65}$ r_b from (b)	$P_r = 2.3$ log (Col. 11)
(*b*) Calculations Involving d_{si}											
(13)	(14)	(15)	(16)	(17)	(18)	(19)	(20)	(21)	(22)	(23)	(24)
d_{si}, in feet	p_i	r'_b	d_{si}/X, in feet	ξ_i Fig. 2.100	$(\gamma_s - \gamma)d_{si}/(\gamma r'_b S)$	Ψ_{*i} Product of Cols. 17, 18, 10, and 7 (Eq. 2.230*k*)	ϕ_{*i} Fig. 2.102	g_{sbi}, in pounds per second per foot Eq. 2.230*l*	$\eta_{oi} = 2d_{si}/r_b$	w_i, in feet per second Eq. 2.229*e*	$z_i = w_i/(0.4U'_*)$
(*b*) Continued											
(25)	(26)	(27)	(28)	(29)	(30)						
I_{1i} Fig. 2.98	I_{2i} Fig. 2.99	$P_r I_{1i} + I_{2i} + 1$	g_{si} = (Col. 27) × (Col. 21), in pounds per second per foot Eq. 2.230*c*	$g_s = \sum_i g_{si}$, in pounds per second per foot Eq. 2.230*a*	$G_s = b g_s$, in pounds per second Eq. 2.230*b*						

Note: Include entire set of d_{si}, p_i values with each set of values of r'_b and r_b.

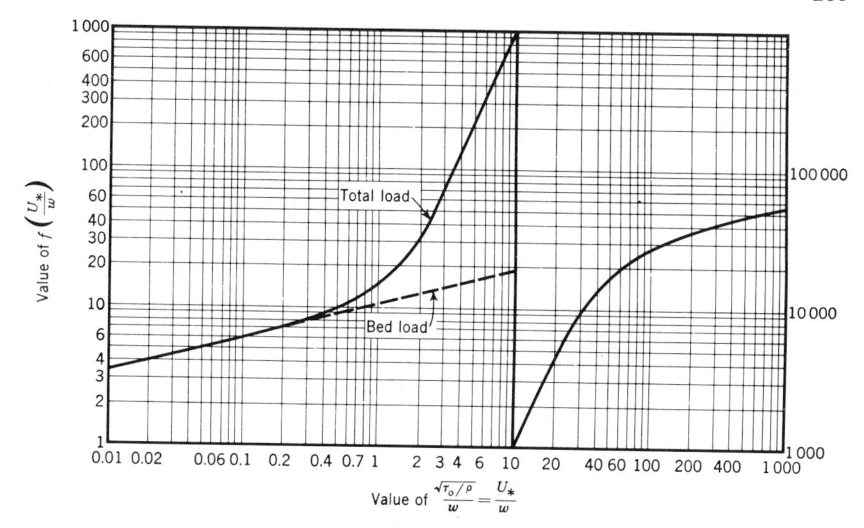

FIG. 2.103.—Function $f(U_*/w)$ for Laursen Formula, Eq. 2.231a

formulas are dimensionally homogeneous and can be used with any consistent set of units. The Laursen formula is intended to apply only to natural sediments with specific gravity of 2.65.

Laursen determined the function in Eq. 2.231a and in Fig. 2.103 by correlating values of the function obtained from flume data of several investigators with U_*/w from the data. The flumes used by the investigators ranged in size from one 10.5 in. wide and 40 ft long to the one used by Laursen that was 3 ft wide and 90 ft long. The sediments all had specific gravities of close to 2.65 and varied in median sizes from 0.011 mm–4.08 mm with grain size distributions from well-sorted to well-graded.

Laursen also compared values of sediment discharge calculated by Eqs. 2.231 with values observed on three small streams: the Niobrara River near Cody, Neb. (Colby and Hembree, 1955); Mountain Creek in South Carolina (Einstein, 1944), and West Goose Creek in Mississippi (Einstein, 1944). These streams had flow depths in the range from 0.12 ft–1.3 ft and bed sediment with median sizes 0.277 mm, 0.86 mm, and 0.287 mm, respectively. The agreement between observed and calculated sediment discharge was good for the Niobrara River, but only fair for the other two streams.

Blench Regime Formula (1966a)

$$\frac{\left(1 + 0.12 \times 10^5 \frac{C_m}{\gamma}\right)^{11/12}}{\left(1 + \frac{1}{233} \times 10^5 \frac{C_m}{\gamma}\right)} = \frac{3.63\, g b^{1/4} q^{1/12} S}{k_m \nu^{1/4} \left[1.9\, \sqrt{(d_{50})\ \mathrm{mm}}\right]^{11/12}} \qquad (2.232)$$

in which C_m = sediment discharge concentration, in pounds per cubic foot; (d_{50}) mm = median size of bed sediment, in millimeters; b = width of stream, in feet; and k_m = a meander coefficient with values of 1.25 for straight reaches, 2.0 for streams with well-developed meanders, and 2.75 for very sinuous streams. The

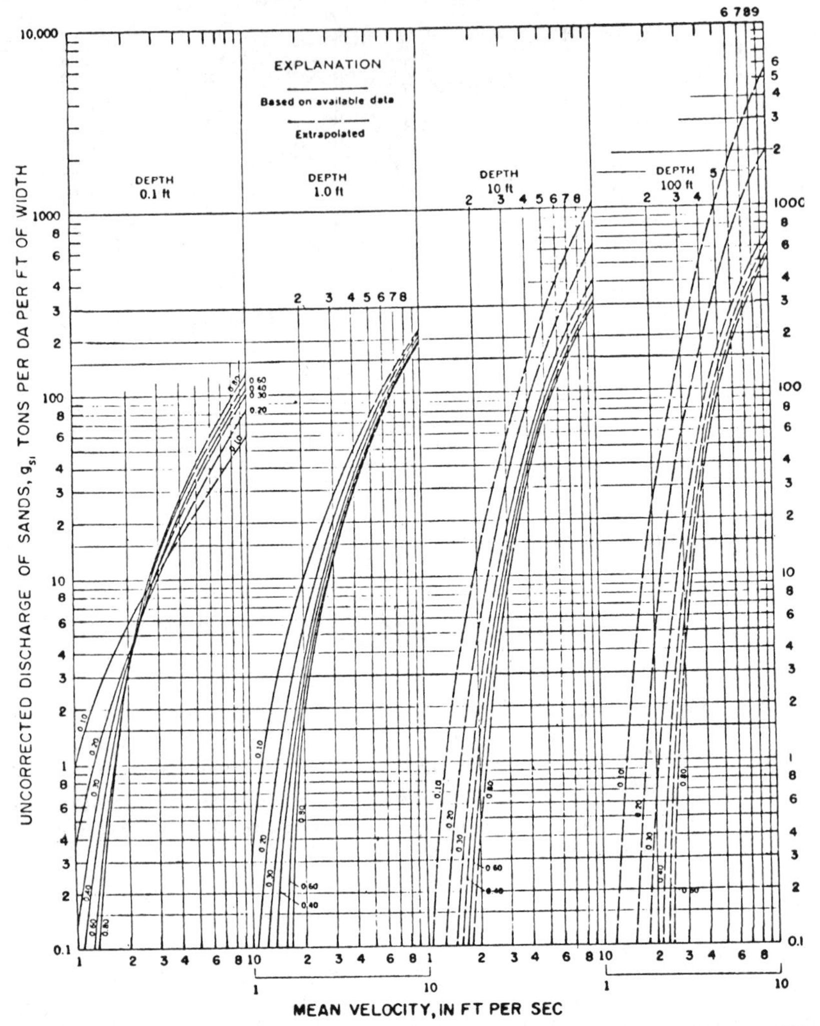

FIG. 2.104.—Colby's (1964b) Relationship for Discharge of Sands in Terms of Mean Velocity for Six Median Sizes of Bed Sands, Four Depths of Flow, and Water Temperature of 60° F

other quantities are as defined previously. Eq. 2.232 is intended to apply only to sand bed streams that are in regime, i.e., in equilibrium and have dune-covered beds. Blench (1966a) uses a slightly different equation for streams with gravel beds. Once C_m is obtained from the preceding equation the sediment discharge, g_s, is calculated by Eq. 2.231d. In Eq. 2.231, d_{50} is in millimeters and all other quantities are in foot-pound-second units.

The form of Eq. 2.232 and the constants in it were derived from regime relations and by correlating the relations mainly with data of Gilbert (Gilbert, 1914 and Johnson, 1943) obtained in small flumes with well-sorted sands ranging in size from 0.3 mm–7 mm.

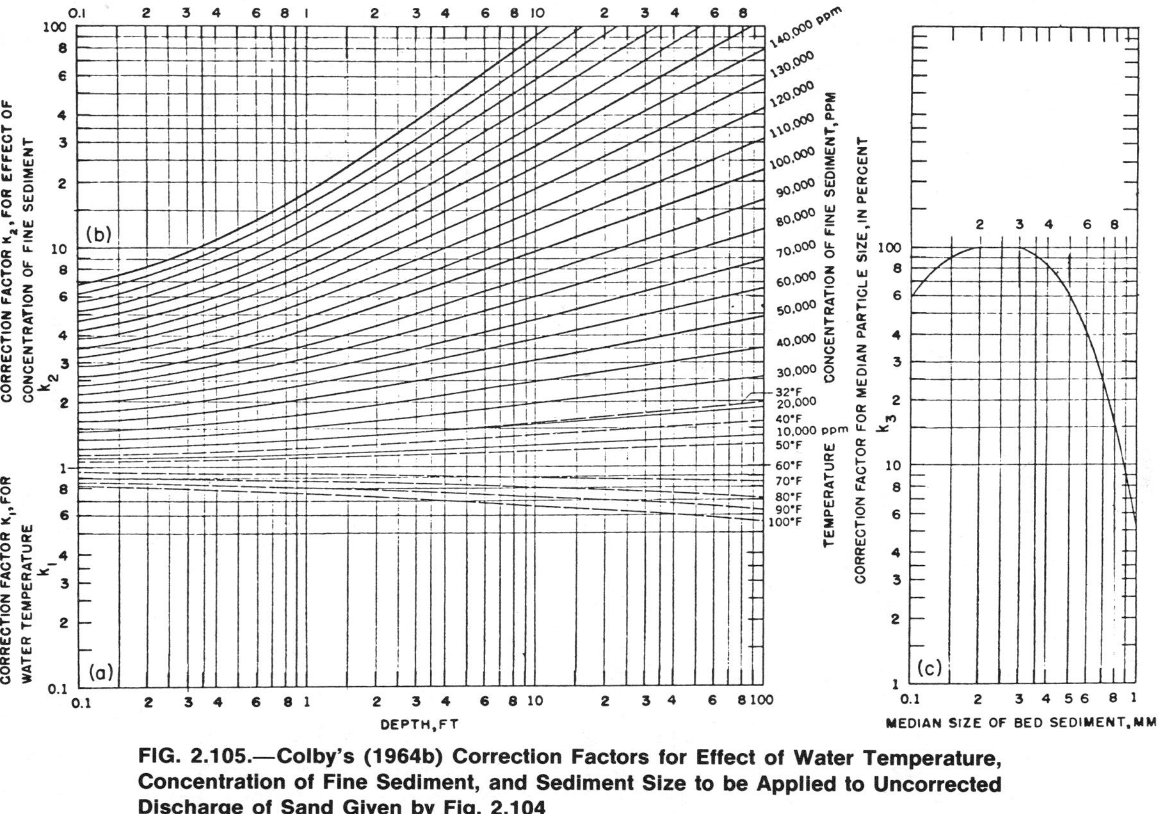

FIG. 2.105.—Colby's (1964b) Correction Factors for Effect of Water Temperature, Concentration of Fine Sediment, and Sediment Size to be Applied to Uncorrected Discharge of Sand Given by Fig. 2.104

Colby Relations (Colby, 1964a and 1964b)

Colby's relations for sediment discharge that apply only for sands are presented in four graphs. The first of these shown in Fig. 2.104 gives the uncorrected sediment discharge, g_{s1}, as a function of V, d, and d_{50}:

$$g_{s1} = f(V, d, d_{50}) \quad (2.233a)$$

and actually consists of four sets of curves for water depths, d, of 0.1 ft, 1 ft, 10 ft, and 100 ft, respectively. Each curve in each set is for a water temperature of 60° F, a given d_{50}, and curves are given for d_{50} of 0.10 mm, 0.20 mm, 0.30 mm, 0.40 mm, 0.60 mm, and 0.80 mm, respectively. Before the graphs of Fig. 2.104 can be used one must obtain the velocity and depth either by observation or calculation. To obtain the sediment discharge for flows with depths other than the four values for which curves are given, one first reads g_{s1} for the known velocity for two of the depths indicated in Fig. 2.104 which bracket the desired depth. Then one interpolates on a logarithmic graph of d versus g_{s1}, to get the sediment discharge per unit width for the desired depth and velocity.

The sediment discharge, g_{s1}, obtained in this manner is correct only for sediments with median sizes ranging from 0.2 mm–0.3 mm in water at 60°F containing negligible amounts of fine sediment, i.e., silt and clay, in suspension. The two correction factors, k_1, and k_2, shown in Figs. 2.104(*a*) and 2.104(*b*), respectively, account for the effect of water temperature and concentration of fine suspended sediment on the sediment discharge when the median size of the bed sediment ranges from 0.2 mm–0.3 mm. If the bed sediment size falls outside this range, the factor, k_3, shown in Fig. 2.105(*c*) is applied to correct for the effect of sediment size. The true sediment discharge, g_s, corrected for the effect of water temperature, presence of fine suspended sediment, and sediment size is given by

$$g_s = [1 + (k_1 k_2 - 1)0.01\, k_3]\, g_{s1} \quad (2.233b)$$

As Fig. 2.105 shows, $k_1 = 1$ when the temperature is 60°F; $k_2 = 1$ when the concentration of fine sediment is negligible; and $k_3 = 100$ when d_{50} lies between 0.2 mm–0.3 mm. Under these conditions, i.e., $k_1 = k_2 = 1$ and $k_3 = 100$, Eq. 2 233b shows that $g_s = g_{s1}$. When d_{50} falls in the range of 0.2 mm–0.3 mm, Fig. 2.105 gives $k_3 = 100$ and Eq. 2.233b shows that $g_s = k_1 k_2 g_{s1}$.

Fig. 2.106 shows Colby's curves of observed bed sediment discharge per foot of width against mean velocity for five rivers adjusted to a water temperature of 60° F. Colby recommends that Fig. 2.106 be used to calculate sediment discharge whenever the characteristics of the stream being studied are even approximately the same as those of the streams listed in the figure. Colby also recommended that Fig. 2.106 be used to roughly check all calculations.

In arriving at his curves Colby was guided by the Einstein Bed Load Function and an immense amount of data from streams and flumes. Data were used from at least a score of streams including those from the Niobrara River by Colby and Hembree (1955), the Middle Loup River by Hubbell and Matejka (1959), the Colorado River at Taylor's Ferry (United States Bureau of Reclamation, 1958), and the Mississippi River at St. Louis (Jordan, 1965).

During the measurements, the Niobrara and Middle Loup Rivers had beds of medium sand and depths that did not exceed 2 ft. Measurements of sediment discharge in these two streams are noteworthy in that they are the first direct measurements of total sediment discharge to be made on natural streams of any

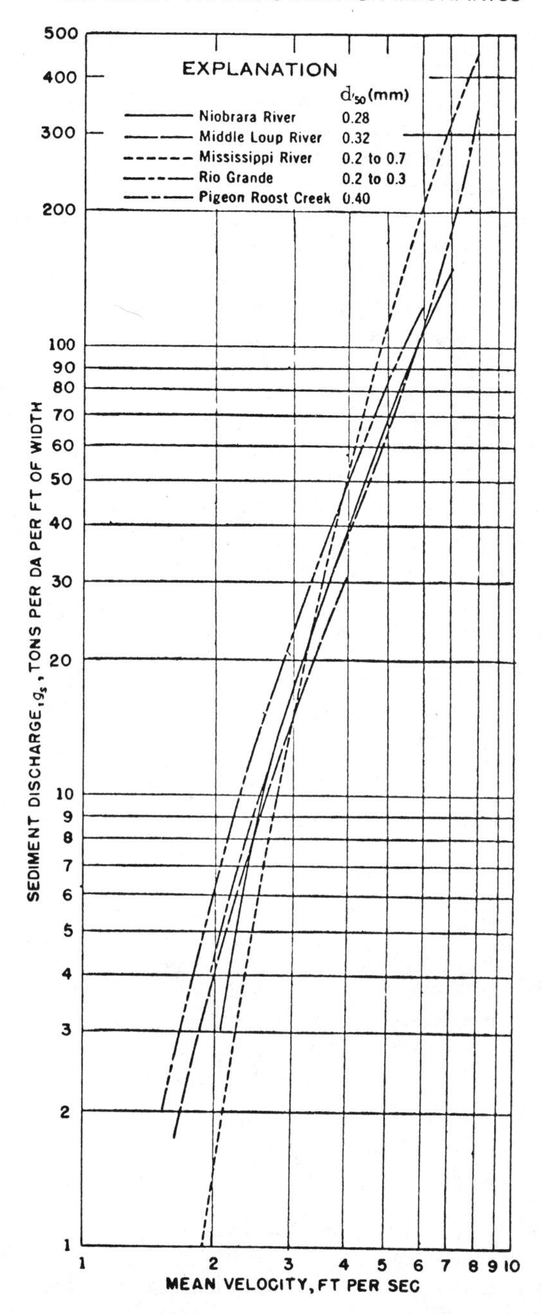

FIG. 2.106.—Relationship between Observed Discharge of Sands and Mean Velocity for Five Sand-Bed Streams at Average Temperatures of about 60° F; Colby (1964b)

size. In the Niobrara River the measurements were made at a natural narrow section where the velocity was high and virtually all of the sediment load was in suspension and thus could be measured with a suspended load sampler. In the Middle Loup River the sediment discharge was also measured with suspended load samplers but in a flume section constructed for this purpose.

The sediment discharge measurement in the Colorado and the Mississippi Rivers and in other rivers (aside from the Niobrara and Middle Loup) used in determining the Colby relations, were based on suspended load samples in normal cross sections. In such measurements the bed load discharge and that part of the suspended load discharge in a thin layer near the bed are not measured. To get the total sediment discharge in such cases the unmeasured sediment discharge is estimated, by methods presented later, and added to the measured discharge. At Taylor's Ferry on the Colorado River where sediment discharge was determined the bed was of medium sand and flow depths ranged from 4 ft–12 ft. In the Mississippi River the bed was also of medium sand but flow depths ranged as high as 57 ft.

The many flume data used by Colby in determining his relations included those of Guy, et al. (1966) in a flume 8 ft wide and 150 ft long and of Brooks (1958) in flumes, 10.5 in. wide by 40 ft long and 33.5 in. wide by 60 ft long, respectively.

Engelund-Hansen Formula (Engelund, 1966 and Engelund and Hansen, 1967)

$$f_e \phi_e = 0.1\, \tau_*^{5/2} \qquad (2.234a)$$

$$f_e = 2\frac{\tau_o}{\rho V^2} = \frac{2gdS}{V^2} \qquad (2.234b)$$

$$\phi_e = \frac{g_s}{\gamma_s \sqrt{\left(\frac{\gamma_s}{\gamma} - 1\right) g d_{50}^3}} \qquad (2.234c)$$

$$\tau_* = \frac{\tau_o}{(\gamma_s - \gamma) d_{50}} \qquad (2.234d)$$

in which f_e = friction factor defined by Eq. 2.234b; d_{50} = median fall diameter of bed sediment; ϕ_e = a dimensionless sediment discharge defined by Eq. 2.234c; τ_* = dimensionless shear stress defined by Eq. 2.234d; and other quantities are as defined previously. Eliminating f_e, ϕ_e, and τ_* from Eq. 2.234a gives

$$g_s = 0.05\, \gamma_s V^2 \sqrt{\frac{d_{50}}{g\left(\frac{\gamma_s}{\gamma} - 1\right)}} \left[\frac{\tau_o}{(\gamma_s - \gamma) d_{50}}\right]^{3/2} \qquad (2.234e)$$

Fig. 2.107 is a chart presented by Engelund and Hansen (1967) that combines the Engelund flow formula (presented in Chapter II, Section F) and the preceding sediment discharge formula thus solving for the stage-discharge relation and water discharge-sediment discharge relation in one graph. This chart is valid only for dune-covered beds for which the boundary Reynolds number, $U_* d_{50}/\nu$, exceeds 12, in which $U_* = \sqrt{\tau_o/\rho}$; τ_o = bed shear stress; and ρ = mass density of the water. The chart does not apply to rippled beds. Its authors do not recommend it for cases in which the median size of the sediment is less than 0.15 mm and the geometric standards deviation of the grain size of the sediment is greater than approximately two. Since Eq. 2.234 is dimensionally homogeneous it can be used with any consistent set of units.

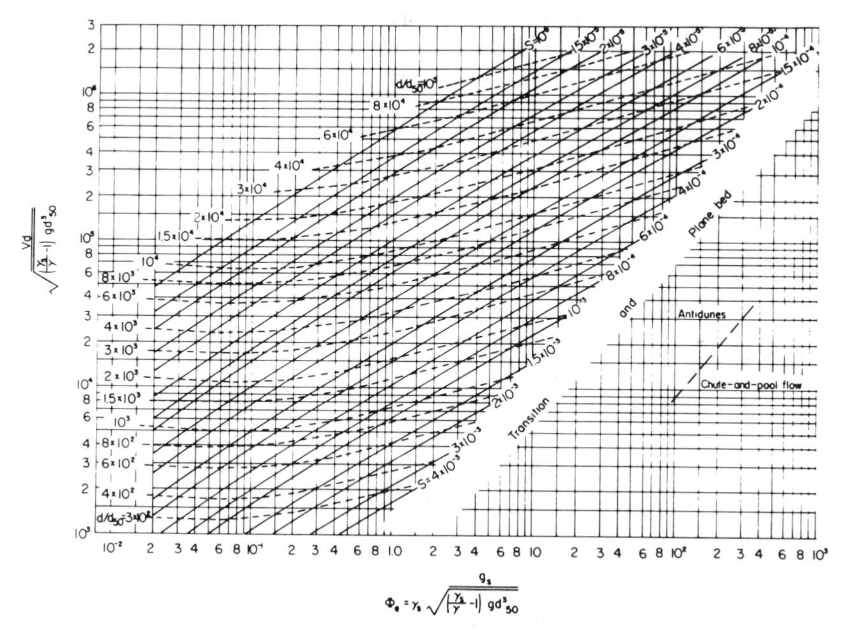

FIG. 2.107.—Engelund and Hansen Chart for Calculating Velocity and Sediment Discharge

In developing Eqs. 2.234, Engelund and Hansen relied heavily on data from four sets of experiments in a large flume 8 ft wide and 150 ft long reported in Guy, et al. (1966). The sediments in these experiments had median fall diameters of 0.19 mm, 0.27 mm, 0.45 mm, and 0.93 mm, respectively, and geometric standard deviations of particle sizes of 1.3 for the finest sediment and 1.6 for the others.

Inglis-Lacey Formula (Inglis, 1968)

$$g_s = 0.562 \frac{(vg)^{1/3}}{w} \frac{V^2}{gd} \frac{\gamma V^3}{g} \qquad (2.235)$$

In Eq. 2.235, w = the fall velocity of a characteristic sediment particle that is assumed to be the particle having the median size of the bed material. The other quantities are as defined previously. Since the equation is dimensionally homogeneous it can be used with any consistent set of units. The Inglis-Lacey Formula was developed by Inglis (1968, 1947) by introducing into the Lacey (1929) regime relations the mean size and fall velocity of the bed sediment and the sediment discharge concentration. The exact form in which these three quantities were introduced was determined from data from model experiments. The Lacey regime relations were based on data from large stable irrigation canals.

Toffaleti Formula (Toffaleti, 1968, 1969)

In this formula the actual stream for which the sediment discharge is to be calculated is assumed to be equivalent to a two-dimensional stream of width, b, equal to that of the real stream and of depth, r, equal to the hydraulic radius of the real stream. The bed sediment is divided into the standard size fractions listed in

Table 2.1 in which the maximum size in each fraction is twice the minimum size. The mean size, d_{si}, of each fraction is taken as the geometric mean of the maximum and minimum sizes in the fraction. The sediment discharge is taken as the sum of that for each of the fractions. The assumption is made, as by Einstein (1950), that the discharge of material in a size fraction of the bed sediment is proportional to the weight fraction, p_i, of the fraction.

For purposes of calculation, the depth, r, of the hypothetical stream is divided into the four zones shown in Fig. 2.108. These are: (1) The bed zone of relative thickness $y/r = 2\,d_{si}/r$; (2) the lower zone extending from $y/r = 2\,d_{si}/r$ to $y/r = 1/11.24$; (3) the middle zone extending from $y/r = 1/11.24$ to $y/r = 1/2.5$; and (4) the upper zone extending from $y/r = 1/2.5$ to the surface. The velocity profile is represented by the power relation

$$U = (1 + n_v)\,V\left(\frac{y}{r}\right)^{n_v} \quad \text{(2.236a)}$$

in which U = the flow velocity at the distance y above the bed; V = the mean velocity, in feet per second of the real stream; and the exponent, n_v, is given by the empirical relation

$$n_v = 0.1198 + 0.00048T \quad \text{(2.236e)}$$

in which T = the water temperature, in degrees Fahrenheit. The concentration distribution of each size fraction is given by a power relation for each of the three upper zones, i.e., by Eqs. 2.236b–d as shown in Fig. 2.108. The equations closely approximate the theoretical distribution of concentration given by Eq. 2.77. The exponent, z_i, in Eqs. 2.236b–d is given by

$$z_i = \frac{w_i V}{c_z r S} \quad \text{(2.236f)}$$

in which w_i = the fall velocity, in feet per second, of the sediment of size, d_{si}, in water at temperature T; S = the slope of the real stream; and c_z is given by the empirical equation

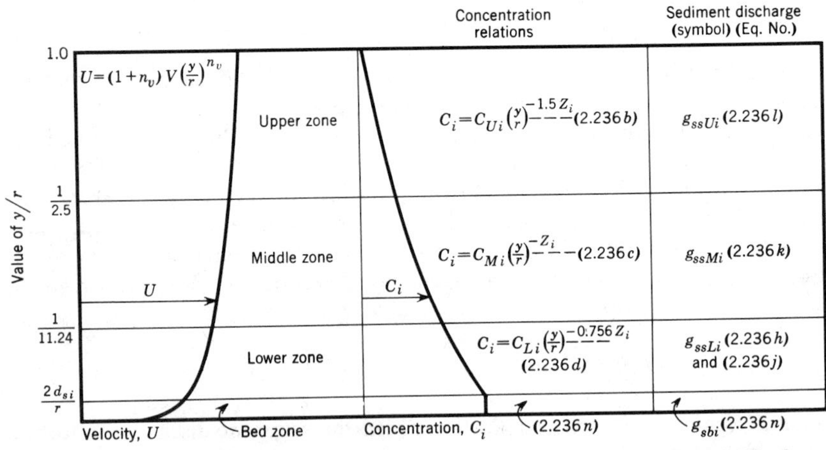

FIG. 2.108.—Toffaletti's (1969) Velocity, Concentration, and Sediment Discharge Relations

$$c_z = 260.67 - 0.667T \qquad (2.236g)$$

in which T = water temperature, in degrees Fahrenheit. When the value of z_i is less than n_v, z_i shall be arbitrarily taken equal to 1.5 n_v. The quantities $n_v - 1.5\,z_i$ and $n_v - z_i$ appear as the exponent of y/r in the product, CU, for the upper and middle zones, respectively. This arbitrary procedure assures that this exponent is never positive and that the product, CU, decreases as y/r increases as it does in streams.

Expressions for the suspended load discharges, g_{ssLi}, g_{ssMi}, and g_{gssUi} in the lower, middle, and upper zones, respectively, are obtained by substituting U from Eq. 2.236a and the appropriate relations for C_i into Eq. 2.230h and integrating between the appropriate limits. Since C_{Ui} and C_{Mi} in Eqs. 2.236b and 2.236c can be expressed in terms of C_{Li}, the sediment discharges in the three zones will all be expressed in terms of C_{Li}, which is still unknown. This is determined from Eq. 2.236h, an empirical equation for g_{ssLi}:

$$g_{ssLi} = \frac{0.600\,p_i}{\left(\dfrac{T_T A\,k_4}{V^2}\right)^{5/3}\left(\dfrac{d_{si}}{0.00058}\right)^{5/3}}\ \text{(tons/day/ft)} \qquad (2.236h)$$

in which A = a function of $(10^5\nu)^{1/3}/10U'_*$ given in Fig. 2.109(a); ν = the kinematic viscosity of the water, in square feet per second; U'_* = shear velocity, in feet per second based on the bed shear stress due to sand grain roughness, defined previously; and T_T is given by

$$T_T = 1.10\,(0.051 + 0.00009T) \qquad (2.236i)$$

in which T = water temperature, in degrees Fahrenheit. The number 0.00058 in Eq. 2.236h is the geometric mean size of fine sand expressed in feet. Therefore the sand size, d_{si}, in Eq. 2.236h must also be expressed in feet. The dimensions of A are such that g_{ssLi} in Eq. 2.236h is given in tons per day per foot of width. The factor,

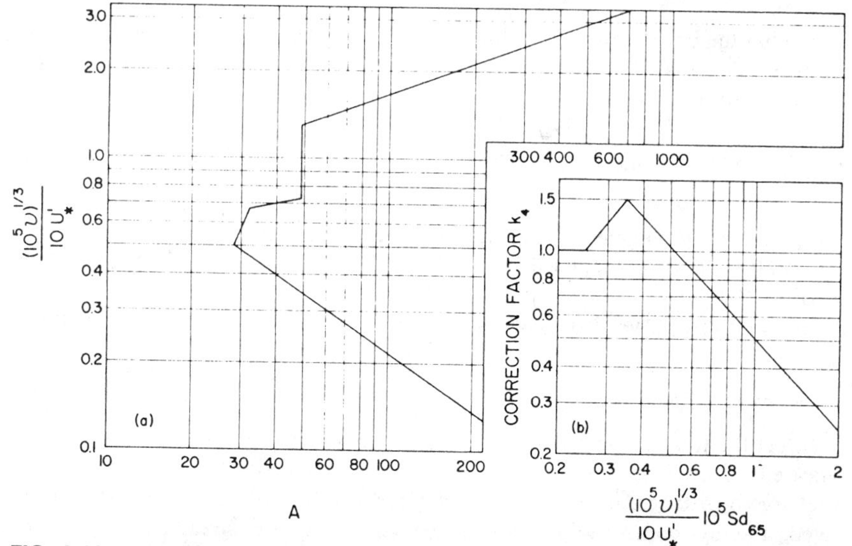

FIG. 2.109.—Factors in Toffaletti Relations: (a) Factor A in Eq. 2.236h; (b) Correction Factor k_4 in Eq. 2.236h

k_4, is given in Fig. 2.109(*b*). Note that k_4 has a value of 1.00 for values of the abscissa of Fig. 2.109(*b*) less than 0.25, a value exceeded only in cases of some flume experiments. When the value of k_4 results in Ak_4 less than 16.0 the value of 16.0 is arbitrarily used for this quantity.

The suspended-sediment discharges in the three upper zones obtained by integration according to Eq. 2.230h with the appropriate limits are:

$$g_{ssLi} = M_i \frac{\left(\frac{r}{11.24}\right)^{1+n_v-0.758z_i} - (2d_{si})^{1+n_v-0.756z_i}}{1 + n_v - 0.756\, z_i} \quad \text{(2.236j)}$$

$$g_{ssMi} = M_i \frac{\left(\frac{r}{11.24}\right)^{0.244z_i} \left[\left(\frac{r}{2.5}\right)^{1+n_v-z_i} - \left(\frac{r}{11.24}\right)^{1+n_v-z_i}\right]}{1 + n_v - z_i} \quad \text{(2.236k)}$$

$$g_{ssUi} = M_i \frac{\left(\frac{r}{11.24}\right)^{0.244z_i} \left(\frac{r}{2.5}\right)^{0.5z_i} \left[r^{1+n_v-1.5z_i} - \left(\frac{r}{2.5}\right)^{1+n_v-1.5z_i}\right]}{1 + n_v - 1.5\, z_i} \quad \text{(2.236l)}$$

in which

$$M_i = 43.2\, p_i C_{Li}(1 + n_v)\, V r^{0.758z_i - n_v} \quad \text{(2.236m)}$$

To obtain C_{Li} one sets g_{ssLi} from Eq. 2.236j equal to the value given by Eq. 2.236h and solves for C_{Li} since it is the only unknown in the resulting equation.

The bed load discharge is assumed to be given by the product of p_i, the concentration and velocity at $y = 2d_{si}$ and the distance, $2d_{si}$, which gives

$$g_{sbi} = M_i\, (2d_{si})^{1+n_v-0.758z_i} \text{ (tons/day/ft)} \quad \text{(2.236n)}$$

In making the calculations according to Eq. 2.236n, the concentration at $y = 2d_{si}$ must be calculated to be sure that it is not unrealistically high. From Eq. 2.236d this concentration is

$$(C_i)_{y=2d_{si}} = C_{Li}\left(\frac{2d_{si}}{r}\right)^{-0.736z_i} \text{ (pcf)} \quad \text{(2.236o)}$$

If the concentration given by Eq. 2.236o is in excess of 100 pcf, the concentration, C_{Li}, in all equations is reduced so that Eq. 2.236o gives a value of concentration equal to 100 pcf.

The total discharge of sediment g_{si} in the *i*th size fraction per foot of width is given by

$$g_{si} = g_{sbi} + g_{ssLi} + g_{ssMi} = g_{ssUi} \text{ (tons/day/ft)} \quad \text{(2.236p)}$$

To get the total sediment discharge, g_s, per foot of width these calculations are repeated for each size fraction and the resulting g_{si} values are summed. The total discharge of bed sediment in tons per day is given by the product bg_s. As will be noted, the hydraulic radius, r; sediment size, d_{si}; and stream width, b, are expressed in feet in the Toffaleti formula. Velocity is in feet per second, concentration is in pounds per cubic foot and sediment discharge is in tons per day per foot of width or tons per day.

The Toffaleti formula is based on extensive data from seven rivers and flume data from four investigators. The rivers are the Mississippi at St. Louis (Jordan 1966), Rio Grande at Bernalillo (Nordin, 1964), Middle Loup (Hubbell and Matejka, 1959), Niobrara (Colby and Hembree, 1955), and three rivers in the lower Mississippi Basin, the data for which are not published. These seven rivers had depths ranging from less than 1 ft to over 50 ft and bed sediments in the fine and medium sand ranges. The flume data were by Kennedy (1961), Vanoni and Brooks (1957), Einstein and Chien (1953b), Guy, et al. (1966), and Waterways Experiment Station of the United States Corps of Engineers. The flumes ranged in width from 10.5 in.–8 ft, the flow depths ranged from as little as 2 in.–2 ft and the bed sediment had median grain sizes ranging from 0.3mm–0.93 mm.

The formulas presented may be classified into groups according to the form of the equations. Four of the 13 formulas are of the DuBoys type, i.e., Eqs. 2.224, 2.226, 2.227, and 2.231. These formulas contain terms of the form $\tau_o - \tau_c$ or $q - q_c$ that go to zero along with g_s when $\tau_c = \tau_o$ or $q_c = q$ and indicate that movement of sediment ceases abruptly when these threshold conditions obtain. Negative sediment discharge is indicated by the formulas when $\tau_c > \tau_o$ or $q_c > q$. However, this is meaningless and these formulas should not be used for such values of τ_o and q. Formulas in which the sediment discharge is expressed in terms of bed shear stress (Eqs. 2.224, 2.227, 2.228, and 2.229) are sometimes called bed load formulas and it might be expected that they would not apply to sand bed streams that carry apprreciable suspended load. It appears that such an expectation is not always realized and some engineers have applied bed load equations to sand bed streams with satisfactory results. For this reason the curves for bed load equations are shown in Figs. 2.113 and 2.114 for the Colorado and Niobrara Rivers even though these streams transport considerable suspended bed sediment. Formulas of the bed load type indicate that sediment discharge increases as bed shear stress increases. Therefore these formulas will show a decrease in sediment discharge for a given water discharge as the friction factor of the stream decreases and velocity increases. This is a questionable result since experience shows that sediment discharge increases with velocity as illustrated by Colby's graphs, Figs. 2.104 and 2.106.

The two Meyer-Peter formulas, Eqs. 2.225 and 2.228, exhibit behavior similar to those of the DuBoys type. The term, 9.94 d_{50}, in Eq. 2.225 and the term, 0.047 $(\gamma_s - \gamma)d_m$, in Eq. 2.228 express critical shear stress. Therefore those formulas also indicate that sediment motion ceases when the shear stress is approximately equal to the critical value. For low transport rates of well-sorted sediments that are coarse enough to produce an hydrodynamically rough bed the Meyer-Peter and Muller equation and the Einstein bed load function agree.

The Einstein-Brown formula and the Einstein bed load function, Eqs. 2.229 and 2.230, respectively, give continuous values of g_s at low transport rates and differ from those of the DuBoys type in this regard. They are based on the concept that due to the turbulent structure of the flow the fluid forces on the bed sediment fluctuate in a random fashion. This causes a small amount of sediment to be moved randomly in jumps or steps even when the mean hydraulic forces at the bed are very small.

In the Inglis-Lacey formula and the Colby relations the sediment discharge is expressed in terms of the mean velocity. In the Inglis-Lacey equation the sediment discharge varies as the fifth power of the velocity. In the Colby relation the exponent of the velocity varies, as may be seen from Fig. 2.104. For a given sediment size and water depth the exponent decreases continuously as velocity and sediment discharge increase. For the curves shown in Fig. 2.104 the exponents vary from a maximum of about nine at low velocities and a depth of 0.1 ft to a minimum of about 1.4. The velocity at which the exponent of the curves reaches a value of five increases as the stream depth increases. The exponent of five occurs at less than 2 fps for a depth of 0.1 ft and between 2.5 fps and 4.5 fps for a depth of 100 ft.

58. Calculation of Sediment Discharge from Stream Measurements.—Normal measurements made on many streams include suspended load samples as well as discharge measurements. The latter measurements consist of depth and mean velocity determinations at a number of selected verticals from which one can get the total discharge, Q, and the cross-sectional area, A. The suspended load samples give the mean measured sediment discharge concentration, C_s', at each of the verticals in which the mean velocity is measured. The concentration C_s' is defined by

$$C_s' \int_{a'}^{d} U dy = \int_{a'}^{d} C U dy \qquad (2.237)$$

in which a' = distance from sampler inlet tube to the stream bed when the sampler is in its lowest position in the vertical; d = stream depth at the vertical; U = water velocity at a point a distance y above the bed; and C = sediment concentration at this same point. From the preceding data at the several selected verticals one can get the mean suspended-sediment discharge concentration, $\overline{C_s'}$, in the sampled part of the stream cross section, i.e., in that part of the cross section lying above a layer of thickness, a', on the bed. This layer forms what is called the unsampled zone of the cross section and the part of the cross section lying above the layer is called the sampled zone. The details of the measurement techniques and calculation procedures for obtaining $\overline{C_s'}$ are outlined in Chapter III, Section A. The measured sediment discharge is usually reported as the product of $\overline{C_s'}$ and Q. This includes sediment of all grain sizes, wash load as well as bed sediment load.

The suspended-sediment measurements just described do not sample the bed load or the suspended load in the layer of water of thickness a' near the bed. The modified Einstein procedure developed by Colby and Hembree (1955) and the Colby method (1957), to be described, are methods for estimating the total bed sediment discharge once suspended load samples and other data at a station on a stream have been observed. The data needed for the modified Einstein procedure are stream discharge Q, mean velocity V, cross-sectional area A, stream width b, mean value d_v of the depths at verticals where samples were taken, the measured sediment discharge concentration $\overline{C_s'}$, size distributions of the measured load, the size distribution of the bed sediment at the cross section and water temperature. Given these data one can separate $\overline{C_s'}$ into wash load discharge concentration, C_w, and bed sediment discharge concentration, $\overline{C_m'}$, and determine the value of concentration $\overline{C_{mi}'}$ for each of several selected size fractions of the mean size, d_{si}, and also determine the weight fraction, p_i, of the bed sediment in each of these size fractions. The size distribution is usually determined by sieving or by one of several fall velocity methods and the size fractions are commonly taken to conform to the classes in the grade scale shown in Table 2.1. In this grade scale the upper limit of grain size is twice the lower limit and the mean size d_{si}, of the fraction is taken as the geometric mean of the two limiting sizes.

The next step is to calculate $g'_{ssi} = G'_{ssi}/b$, the suspended-sediment discharge in a given size fraction per unit width of the stream in the sampled zone of the cross section. To obtain this quantity Colby and Hembree (1955) assume the relation

$$g'_{ssi} = \overline{C'_{mi}} \int_{a'}^{d_v} U dy = \frac{\overline{C'_{mi}}\, Q'}{b} \qquad (2.238)$$

in which d_v = average of depths at verticals where suspended sediment samples were taken. To evaluate the integral, i.e., to determine Q' in Eq. 2.238 the point velocity, U, is taken to be

$$U = 5.75\, U'_* \log \frac{30.2\, xy}{d_{65}} \qquad (2.239)$$

Introducing U from Eq. 2.239 into Eq. 2.238 and integrating yields

$$Q' = 2.50 b\, U'_* d_v [(1 - \eta_v)(P_m - 1) - 2.3 \eta_v \log \eta_v] \qquad (2.240)$$

in which, $\eta_v = a'/d_v$ and

$$P_m = 2.3 \log \frac{30.2\, x d_n}{d_{65}} \qquad (2.241)$$

in which $d_n = A/b$; A = area of stream cross section; and x is given in Fig. 2.97. The total discharge, Q, is obtained from Eq. 2.240 by letting η_v go to zero giving

$$Q = 2.50\, b U'_* d_v (P_m - 1) \qquad (2.242)$$

From Eqs. 2.240 and 2.242 one obtains

$$\frac{Q'}{Q} = (1 - \eta_v) - 2.3 \frac{\eta_v \log \eta_v}{P_m - 1} \qquad (2.243)$$

One can now write

$$G'_{ssi} = b g'_{ssi} = \overline{C'_m}\, Q \left[(1 - \eta_v) - 2.3 \frac{\eta_v \log \eta_v}{P_m - 1} \right] \qquad (2.244)$$

The function of Eq. 2.243 is available in graph form in Colby and Hembree (1955), Hubbell and Matejka (1959), and Colby and Hubbell (1961).[2]

The suspended-sediment discharge, g'_{ssi}, of sediment of mean size, d_{si}, per unit width in the sampled zone is taken as

$$g'_{ssi} = \int_{a'}^{d_v} C_i U dy \qquad (2.245)$$

The integral is evaluated substituting C_i from the Rouse suspended load distributions equation (Eq. 2.77), U from Eq. 2.239 and getting the reference

[2] The graphs in these three references give values of Q/Q' that differ slightly from those of Eq. 2.243. To be consistent, use values from the graphs.

concentration, C_{ai}, from Einstein's Eq. 2.230i. This calculation yields

$$g'_{ssi} = \left(\frac{\eta'_{oi}}{\eta_v}\right)^{z'_i-1}\left(\frac{1-\eta_v}{1-\eta'_{oi}}\right)^{z'_i} g_{sbi}\,[P_m I_1(\eta_v, z'_i) + I_2(\eta_v, z'_i)] \qquad (2.246)$$

in which $\eta'_{oi} = 2d_{si}/d_n$; $\eta_v = a'/d_v$; d_v = average of depths at verticals sampled; and the functions $I_1(\eta_v, z'_i)$ and $I_2(\eta_v, z'_i)$ differ from those defined by Eqs. 2.230d and 2.230e in that η_v and z'_i replace η_{oi} and z_i, respectively. The quantity, x, is given in Fig. 2.97.

$$g'_{ssi} = 0.216\,\frac{(\eta'_{oi})^{z'_i-1}}{(1-\eta'_{oi})^{z'_i}}\,g_{sbi}[P_m J_1(\eta_v, z_i) + J_2(\eta_v, z'_i)] \qquad (2.248)^3$$

in which the functions, J_1 and J_2, are

$$J_1(\eta_v, z'_i) = \int_{\eta_v}^{1}\left(\frac{1-\eta}{\eta}\right)^{z'_i} d\eta \qquad (2.249)$$

and $$J_2(\eta_v, z'_i) = \int_{\eta_v}^{1}\left(\frac{1-\eta}{\eta}\right)^{z'_i} \ln \eta\, d\eta \qquad (2.250)$$

The quantities, z'_i and g_{sbi}, in Eq. 2.246 and 2.248 are still unknown and need to be determined. To obtain g_{sbi} one replaces Ψ_* in Einstein's bed load function, Eq. 2.230j by Ψ_m which is taken as the larger of the values given by Eq. 2.251a and 2.251b:

$$\Psi_m = (0.4)(1.65)\frac{d_{si}}{(rS)_m} \qquad (2.251a)$$

$$\Psi_m = 1.65\frac{d_{35}}{(rS)_m} \qquad (2.251b)$$

The quantity $(rS)_m$ is defined by Eq. 2.252

$$V = 5.75\sqrt{g(rS)_m}\,\log\frac{12.2\,xd_n}{d_{65}} \qquad (2.252)$$

in which x = a function of d_{65}/δ given in Fig. 2.97; in which $\delta = 11.6\nu/\sqrt{g(rS)_m}$ and d_n, d_{65}, and V are defined previously. The quantity $(rS)_m$ is obtained by a simultaneous trial and error solution of Eq. 2.252 and the graphical function, x. Once $(rS)_m$ is available, Ψ_m is calculated and ϕ_* is read from Fig. 2.102 using Ψ_m in place of Ψ_*. The bed load discharge, g_{sbi}, is then calculated from

$$g_{sbi} = \frac{1}{2}\phi_* p_i \gamma_s \sqrt{\frac{\gamma_s}{\gamma} - 1}\,\sqrt{g d_{si}^3} \qquad (2.253)$$

With g_{sbi} known, Eqs. 2.246 and 2.248 can be solved for z'_i by trial and error. Eq. 2.246 can be solved with the aid of Figs. 2.98 and 2.99 in evaluating $I_1(\eta_v, z'_i)$ and

---

[3] Note that there is no Eq. 2.247.

$I_2(\eta_v, z_i')$. These figures extend only to $\eta = 0.1$ which is often less than η_v especially for shallow streams. For cases in which η_v exceeds 0.1, Eq. 2.248 can be solved for z_i' employing graphs of $J_1(\eta_v, z_i')$ and $J_2(\eta_v, z_i')$ that extend to η values of 0.5 and that are available in Colby and Hembree (1955) and Hubbell and Matejka (1959). In this way one can solve for the z_i' value for each size fraction. However, Colby and Hembree (1955) discovered that the values of z_i' in Eqs. 2.246 and 2.248 were related to the fall velocities of sediment by

$$\frac{z_i'}{z_1'} = \left(\frac{w_i}{w_1}\right)^{0.7} \qquad (2.254)$$

in which z_1' is obtained by solving Eq. 2.246 or Eq. 2.248 for the size fraction of the suspended bed sediment of mean size d_{s1}; w_1 = the fall velocity of a sediment grain of size d_{s1} calculated from the Rubey Eq. 2.229e; and z_i' and w_i = values corresponding to any other size fraction of mean size d_{si}.

It was recommended that the value of z_1' be determined by solution of Eq. 2.246 or Eq. 2.248 only for the size fraction of measured sediment discharge containing the largest relative amount of bed sediment and that the z' values for other sizes be calculated by Eq. 2.254. Colby and Hubbell (1961) have presented graphs that greatly facilitate solution of Eqs. 2.246 and 2.248 for z' and Eq. 2.254.

Eq. 2.254 appears to be in conflict with the Rouse suspended load theory, Eqs. 2.77 and 2.78, which show that at a given vertical at a given time the exponent, z, is proportional to the fall velocity of the sediment grains. The differences between the theories of the modified Einstein procedure (Eq. 2.246) and of suspended load (Eq. 2.77) arises from the way in which the theory of suspension is applied in developing the Einstein bed load function that is also applied in the modified Einstein method. Einstein's definition of z in Eq. 2.230g contains U_*' where the suspension theory contains the larger total shear velocity, U_*, and thus gives a smaller z and relatively larger suspended-sediment discharge. Einstein's bed load discharge, g_{sbi}, appearing in Eq. 2.246, reflects the effect of defining z in terms U_*'. Another disagreement between the two theories arises from using average values of quantities, e.g., stream depth sediment discharge and point velocities in a real stream in Eq. 2.245 that is valid only for the two-dimensional case. In view of these approximations, dictated by practical considerations, it is not surprising that disagreements exist. Since z defined in terms of fall velocity and shear velocity is fundamentally different from the one obtained solving Eq. 2.246 or Eq. 2.248; the latter kind of z has been denoted by z'.

The total sediment discharge per unit width g_{si} of the fraction of mean size, d_{si}, is assumed to be given by Einstein's Eq. 2.230c in the form

$$g_{si} = g_{sbi}[P_v I_1(\eta_{oi}', z_i') + I_2(\eta_{oi}', z_i') + 1] \qquad (2.255)$$

This enables one to calculate the several values of g_{si} corresponding to each size fraction of bed sediment. The sum of all g_{si} gives the total discharge of bed sediment per unit width, g_s, as indicated by Eq. 2.230a. The total bed sediment discharge, G_s, is the product bg_s.

The Colby method (1957) of estimating total bed sediment discharge from stream measurements consists of several semi-empirical relationships based on some of the same stream measurements employed in developing the modified Einstein method. It is much simpler to apply than the modified Einstein method.

However, it does not give a breakdown of the unmeasured sediment discharge into size fractions and it requires a precise determination of mean velocity since the unmeasured bed sediment discharge is very sensitive to changes in velocity.

To apply the Colby method one needs the mean stream velocity, V, the stream width, b, the mean depth, d_n, the measured mean suspended sediment discharge concentration, $\overline{C}'_s$, which includes wash load and bed sediment load and the measured discharge concentration of bed sediment, $\overline{C}'_m$. The size distribution of the bed sediment is not needed. In the absence of bed sediment size-distribution the limiting size between the wash load and bed sediment is taken at the lower limit of the sand size, i.e., d_s = 0.062 mm.

The procedures in the calculations are as follows:

1. From Fig. 2.110 read an estimate of the unmeasured bed sediment discharge per foot of width for the given mean velocity.
2. From Fig. 2.111 read the relative concentration of suspended sands in parts per million for the given depth and velocity.
3. Calculate the ratio of the measured discharge concentration of suspended sands in parts per million to the relative concentration obtained in the previous step 2. This is also known as the availability ratio.

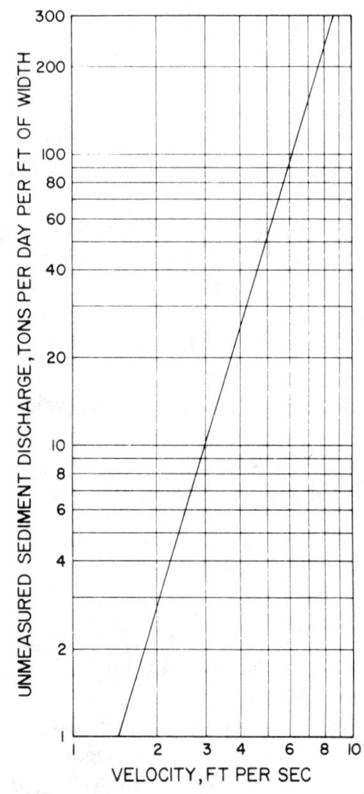

FIG. 2.110.—Colby's (1957) Graph of Unmeasured Sediment Discharge against Mean Velocity

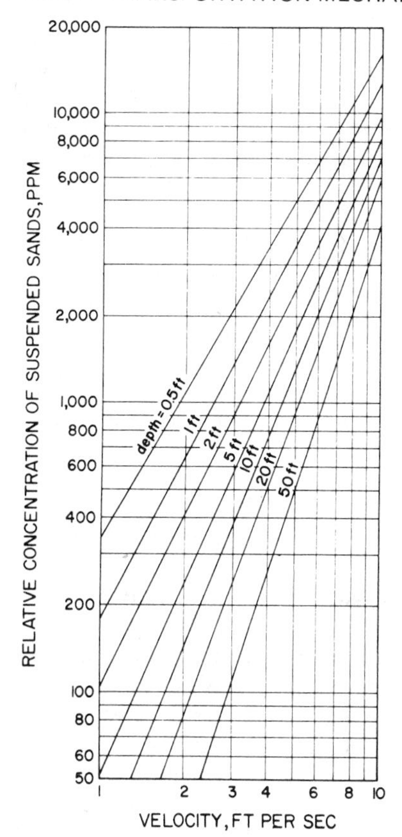

FIG. 2.111.—Colby's (1957) Chart of Relative Concentration against Mean Velocity

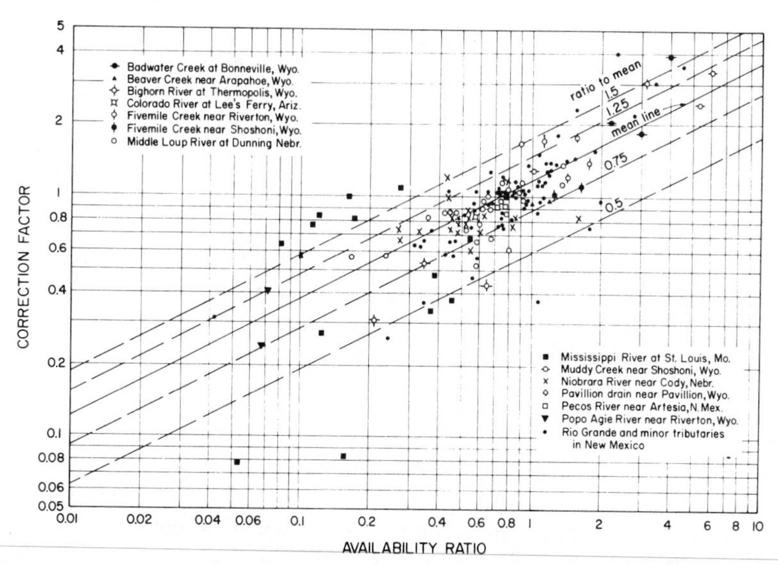

FIG. 2.112.—Colby's (1957) Chart of Availability Ratio against Correction Factor

4. Enter Fig. 2.112 with the availability ratio obtained in step 3 and read the correction factor to be applied to the unmeasured sediment discharge obtained in step 1.

5 The unmeasured bed sediment discharge is then the product of the unmeasured sediment discharge in step 1, the correction factor in step 4, and the stream width.

6. The total bed sediment discharge is the sum of the measured bed sediment discharge and the unmeasured discharge in step 5. The measured bed sediment discharge is defined as the product of Q in cubic feet per second, the measured bed sediment discharge concentration in pounds per cubic foot, and the factor 43.2 that gives the result in tons per day.

7. If the concentration, $\overline{C}'_s$ is especially high or low, Colby (1957) has developed another correction factor, k_5, which when multiplied by the result in step 5 gives a better estimate of the unmeasured bed material discharge. A good approximation of k_5 is given by

$$k_5 = 0.18\,(\overline{C}'_s)^{0.23} \qquad (2.256)$$

in which $(\overline{C}'_s)$ = measured mean sediment concentration, in parts per million.

59. Evaluation of Formulas.—A straightforward way to evaluate a formula is to compare the results of calculations with it with bed sediment discharge actually measured on natural streams. In making the calculations one needs to know hydraulic quantities, e.g., mean velocity, depth and, hydraulic radius. These quantities can be observed when the stream under study already exists or they can be predicted by the methods outlined in Chapter II, Section F when observed quantities are not available or the channel under consideration is yet to be built. Sediment discharges calculated with observed hydraulic quantities are expected to be more reliable than those obtained with predicted quantities.

The comparisons of calculated and observed sediment discharge can be made in two ways. One of these involves plotting both the observed and calculated sediment discharge against water discharge for a given stream, to obtain a sediment transport curve for the stream. On such a graph calculated values by several formulas can be displayed and compared directly with each other and the observed sediment discharge. The other way to make comparisons is by plotting the observed values against the calculated ones. In this method, results from only one formula can be shown in any one graph but the observed results can be obtained from any number of streams including laboratory flumes. Both methods of comparison have been used herein.

One of the problems of developing and evaluating formulas is that very few direct measurements of sediment discharge of rivers have been made. Most of the growing body of data on sediment discharge of rivers is obtained by adding an estimate of the unmeasured sediment discharge to that measured with suspended load samplers. The modified Einstein method and the Colby methods presented previously are those commonly used in estimating the unmeasured sediment discharge.

Bed sediment transport curves are shown in Figs. 2.113 and 2.114, respectively, for the Colorado River at Taylor's Ferry and Niobrara River near Cody, Neb. These are modifications of curves presented by Vanoni, Brooks, and Kennedy (1960). As observed previously, the total bed sediment discharge for the Colorado

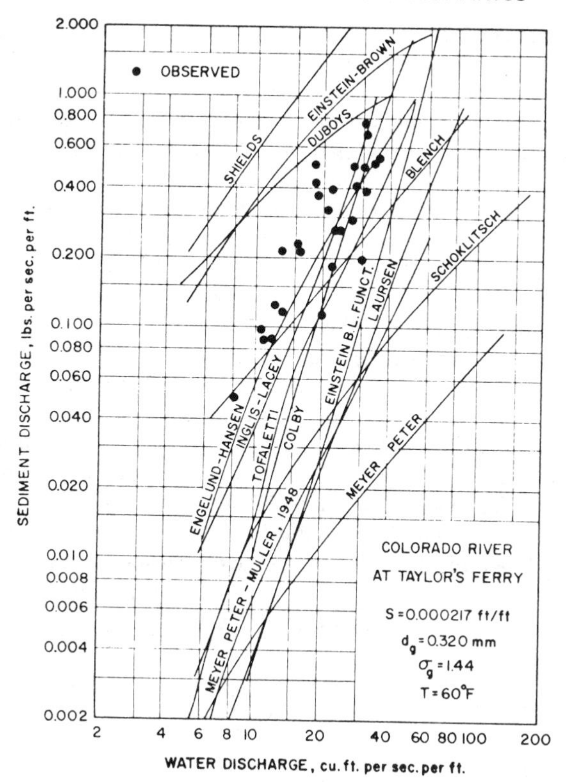

FIG. 2.113.—Sediment Discharge as Function of Water Discharge for Colorado River at Taylor's Ferry Obtained from Observations and Calculations by Several Formulas

River (United States Bureau of Reclamation, 1958) was obtained from suspended sediment samples by the modified Einstein method and the sediment discharge for the Niobrara River (Colby and Hembree, 1955) was measured directly. The figures also show curves of sediment discharge against water discharge for each of the 13 formulas presented previously. Some characteristics of the Colorado and Niobrara Rivers are listed in Table 2.18. As will be noted the bed sediment of these streams is medium sand and the stream depths range from less than 1 ft–12 ft. Of particular interest are the substantial ranges in observed slope and water temperature. In the calculations the mean values of slope and temperature shown in Table 2.18, and Figs. 2.113 and 2.114 were used. Also in the calculations, the depth and velocity for a given discharge were determined by the method of Einstein and Barbarossa (1952).

The points shown in Figs. 2.113 and 2.114 represent the observed data for the streams. Neither set of data show a definite relation although the scatter of the points about a mean line is obviously greater in the case of the Colorado River than in the Niobrara River. The sediment discharges for the Niobrara River are among the most reliable available since they were measured directly. Some likely factors contributing to the scatter of these data are variations in water temperature, stream slope, bed sediment size, and size distribution and measurement errors. These factors also contribute to the scatter in the data for the Colorado River but in this case there is also a contribution resulting from the error in calculating the unmeasured sediment discharge by the Modified Einstein procedure. The ratio of the unmeasured to the measured bed sediment discharge in the Colorado River varied between 0.46 and 1.29 and had a mean value of 0.65.

To obtain a rough estimate of the reliability of the modified Einstein method one can compare the sediment discharges computed by this method with those measured directly in the Niobrara River (Colby and Hembree, 1955) and the Middle Loup River (Hubbell and Matejka, 1959). However, this is not entirely satisfactory because these data were used to develop the method. About one-half of the 24 calculated total bed sediment discharges of the Niobrara River fell within 20% of the measured values and only three values deviated more than 50% from the measured values. For the Middle Loup River over 50% of the 63 calculated values of bed sediment discharge were within 20% of the measured values and only seven values deviated more than 50% from the measured values. These results serve to establish confidence in the modified Einstein procedure for use on streams similar to the Niobrara and Middle Loup Rivers. In applying it to rivers, e.g., the Colorado, with somewhat different characteristics one must rely on the theoretical basis of the relation to give realistic results. Much of the

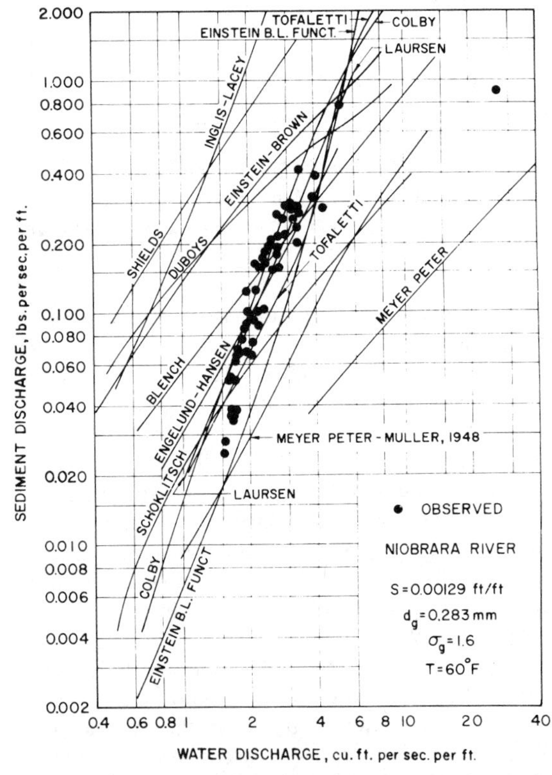

FIG. 2.114.—Sediment Discharge as Function of Water Discharge for Niobrara River near Cody, Neb., Obtained from Observations and Calculations by Several Formulas

uncertainty in the results is removed because the calculations are based on suspended sediment samples and observed mean velocity. This method is the best one available and workers in sedimentation engineering will be relying on it until a better one is developed. Such improvements in sediment discharge relations will require further direct measurements of sediment discharge in natural streams.

The data in Fig. 2.113 and 2.114 can be fitted roughly by power relations of the form

$$g_s = Bq^{n_s} \qquad (2.257)$$

that are straight lines on logarithmic graphs. Straight lines were fitted to the data of these figures in the range of q for which observed data were available. The exponents, n_s, are approx 2.0 and 2.8, for the Colorado and Niobrara Rivers, respectively.

In using Figs. 2.113 and 2.114 to evaluate various formulas it must be kept in mind that in the calculations, average values of slope and water temperature were used and the relation between depth and discharge or discharge and mean velocity was calculated. The velocity-depth relations calculated by Einstein-Barbarossa and other methods for the Colorado and Niobrara Rivers are shown in Figs. 2.65 and 2.67, respectively. In Fig. 2.65 for the Colorado River it is seen that for low depths the velocities predicted by the Einstein-Barbarossa method and used in calculating the sediment discharge curves of Fig. 2.113, are appreciably higher than those observed. For the higher depths the predicted and observed velocities agree. The velocities of the Niobrara River predicted by the Einstein-Barbarossa method shown in Fig. 2.67 are all lower than the observed ones. Calculations were

TABLE 2.18.—Properties of Colorado and Niobrara Rivers at Sites of Measurement of Sediment Discharge

Data (1)	Stream: Colorado River (2)	Niobrara River (3)
Depth range, d, in feet	4-12	0.7-1.3
Range in q, in cubic feet per second per foot	8-35	1.7-5
Mean width, in feet	350	110
Slope, in feet per foot		
Minimum value	0.000147	0.00116
Maximum value	0.000333	0.00126
Value used in calculations	0.000217	0.00129
Water temperature, in degrees Fahrenheit		
Minimum value	48	33
Maximum value	81	86
Value used in calculations	60	60
Geometric mean sediment size, in millimeters	0.320	0.283
Geometric standard deviation, σ_g	1.44	1.60
d_{35}, in millimeters	0.287	0.233
d_{50}, in millimeters	0.330	0.277
d_{65}, in millimeters	0.378	0.335
d_{90}, in millimeters	0.530	0.530
Mean size, d_m, in millimeters	0.396	0.342

made with some of the formulas using mean values of the observed depth and discharge. Most of the sediment discharges obtained in this way were in closer agreement with the observed values than those obtained with calculated $q - d$ values. The formulas of Meyer-Peter, Schoklitsch, and Blench are independent of the $q - d$ relation and therefore showed no change in g_s when the observed values were used. This was true also of the Einstein bed load function and the Engelund-Hansen relations because these include methods of calculating the $q - d$ relation.

From Figs. 2.113 and 2.114 it appears that in these two cases the sediment discharges calculated by the Shields, Einstein-Brown and DuBoys formulas tend to be high and those by the Meyer-Peter formula tend to be low. These formulas and the Schoklitsch formula give curves with considerably less slope than straight lines fitted to the data in Figs. 2.113 and 2.114. The formulas that give best agreement with the observed sediment discharge in these two cases are those of Colby, Tofaletti, and Engelund-Hansen. The curves for these formulas have slopes that are close to those of lines fitted to the data. The slopes of the curves for the Einstein bed load function, and the Laursen and Inglis-Lacey formulas are also close to those of a mean line for the data but the curves do not fit the data as well as several others. The curve for the Blench formula intersects the data but has two small a slope.

Since the sediment transport curves calculated by formulas and displayed on the logarithmic graphs, Figs. 2.113 and 2.114, have little curvature they can be fitted satisfactorily by power relations of the form of Eq. 2.258 especially over the narrow ranges of discharge for which data are available. Calculated curves with slopes, i.e., with exponents n_s, close to those of the line fitting the data are useful even if they do not give the correct values of sediment discharge. Such curves will give correct relative values of g_s, i.e., the ratios of sediment discharges at any two values of water discharge given by formula will be the same as observed in the stream.

The sediment discharge concentration, C_m, is given by the quotient of sediment discharge and water discharge. From Eq. 2.257 this concentration is

$$C_m = Bq^{n_s - 1} \qquad (2.258)$$

If the exponent, n_s, is equal to unity, Eq. 2.258 gives the result that C_m is constant and independent of q and if n_s exceeds unity C_m increases as q increases. Since it is reasonable to expect that the concentration will increase as discharge increases it is also expected that realistic values of n_s should exceed unity. This criterion can be applied in judging the reliability of formulas. Those that give values of n_s near unity are to be discounted. It also appears from experience that commonly n_s should lie between 2 and 3 for sand bed streams as is the case for the Colorado and Niobrara Rivers as shown in Figs. 2.113 and 2.114. However, Swaminathan and Dakshinamurti (1972) have observed values of n_s of 0.9 in two Indian rivers. Such low values of n_s must be due to special circumstances and cannot be viewed as typical. The reader is reminded that this discussion and all the material in this section applies to bed sediment discharge and does apply to wash load.

As mentioned previously, a second method of judging the reliability of a formula is to plot values of g_s given by formula against observed values. Several such graphs have appeared in the literature. Notable among these is Fig. 2.115 prepared by Colby (1964) for the Colby relation with data from 11 streams in the

United States. These streams ranged in size from the Niobrara and Middle Loup Rivers with depths of from about 1 ft–2 ft; the Colorado and Missouri Rivers with depths of up to about 10 ft to the Mississippi River at St. Louis with depths varying from about 15 ft to over 50 ft. The median size of bed material for these streams varied from about 0.2 mm–0.4 mm.

All of the data in Fig. 2.115 were used to establish the Colby relation and thus indicate the best correlation that can be expected between observations and calculations with this relation. About 45% of the calculated values of sediment discharge deviate only 20% from the observed amounts and about 75% fall between 0.5 and 2.0 times the observed amounts. To further test his relations Colby (1961) prepared a correlation graph, similar to Fig. 2.115, with data not used to establish the relation. These data were taken from flume experiments by Gilbert (1914) and from observations in medium to small streams. The correlation obtained was essentially the same as displayed in Fig. 2.115.

A comprehensive set of graphs comparing sediment discharge calculated by the Toffaleti method with values observed in flumes and rivers was presented by Toffaleti (1969). The comparisons showed roughly the kind of correlation displayed in Fig. 2.115.

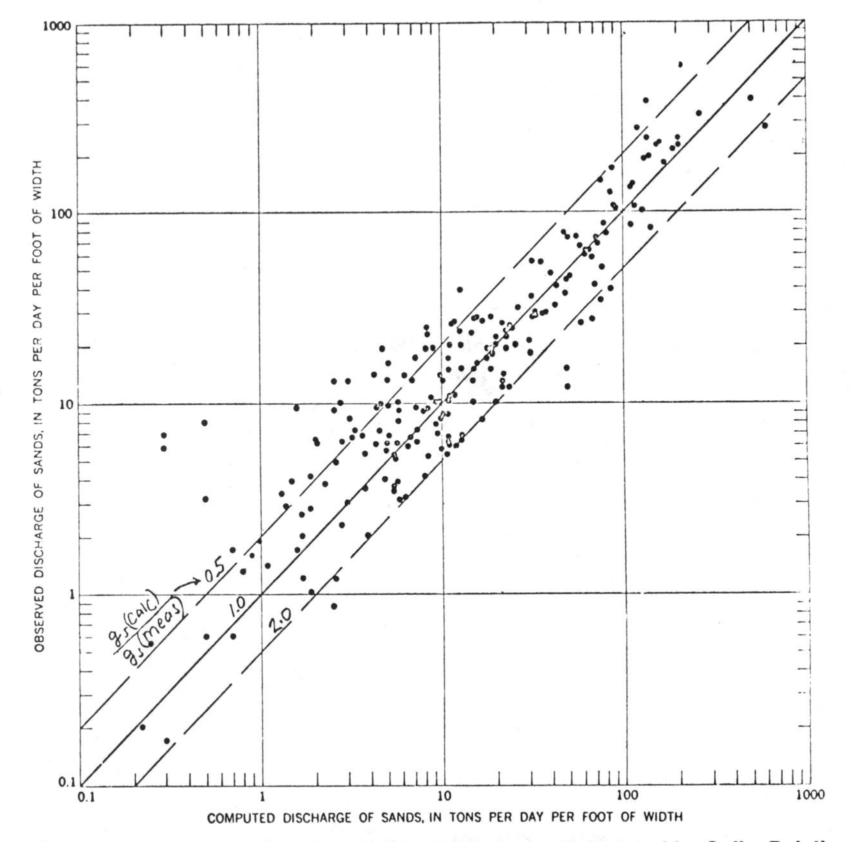

FIG. 2.115.—Comparison of Bed Sediment Discharge Calculated by Colby Relation and Observed on Several Rivers in United States; Colby (1961) (Observed Data Were Used to Establish Colby Relation)

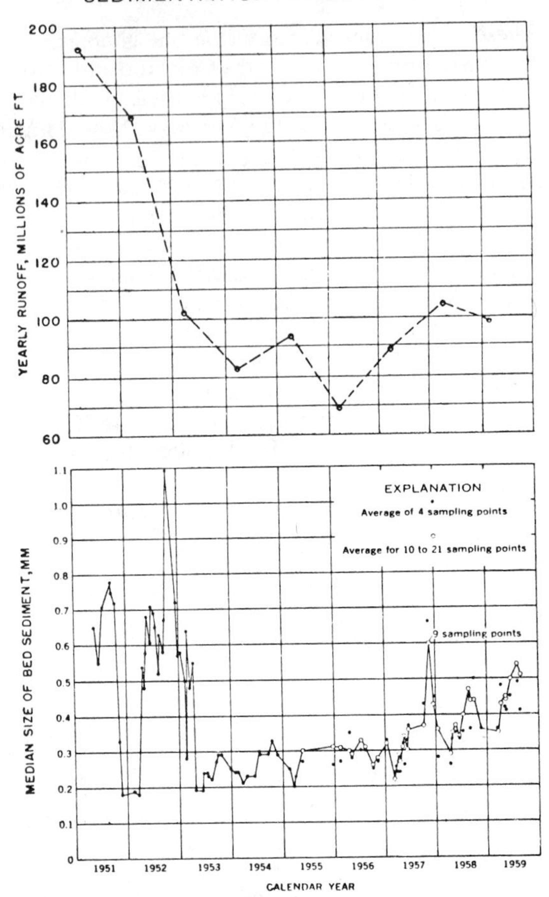

FIG. 2.116.—Variation of Yearly Runoff and Median Bed Sediment Size with Time, for Mississippi River at St. Louis; Jordan (1965)

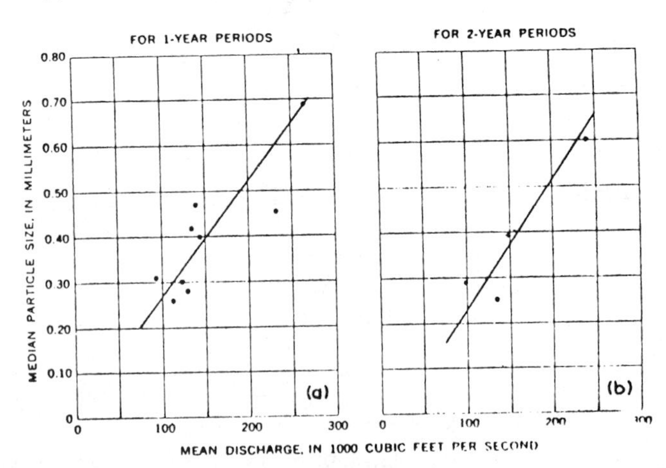

FIG. 2.117.—Relation of Bed Material Size to Discharge for Mississippi River at St. Louis; Jordan (1965)

Zernial and Laursen (1963) followed a different method of comparing sediment discharge data from the Middle Loup Rivers (Hubbel and Matejka, 1959), Five Mile Creek (Colby, et al. 1956), and some unpublished data from the Rio Grande with calculations by the Laursen formula (Eq. 2.231). They showed that calculated sediment discharges with the observed extreme values of water temperature, mean sediment size, and depth for a given discharge bracketed observed values of sediment discharge for the three rivers previously listed. These authors also clearly called attention to the great variability of characteristics of natural streams, e.g., size and size distribution of bed sediment, relation between water depth and velocity and water temperature. These are characteristics that strongly affect sediment transportation and that must be predicted before the sediment discharge can be calculated with greater reliability than is now possible.

The variability of stream characteristics of the kind listed previously has apparently come to the attention mainly of those intimately involved in stream investigations. Since this is of importance to the subject of this section some data are presented which illustrate concretely the surprising extent of these variations. Data of this kind are presented in Table 2.19 for four rivers and in Figs. 2.116–2.118 for the Mississippi River.

The large variation in size of bed sediment and in slope shown in Table 2.19 is remarkable. The median sediment size varied from about 1.5 fold for the Colorado River to over 5 fold for the Mississippi. Even allowing for large sampling errors the size variation is still large and will result in large fluctuations in transport rate as indicated by the Colby relation, Fig. 2.104. The variation in surface slope, which is essentially equal to the energy slope, ranges from about 10% for the Niobrara River to in excess of an astounding four fold in the Mississippi. This later is all the more remarkable when it is observed that the

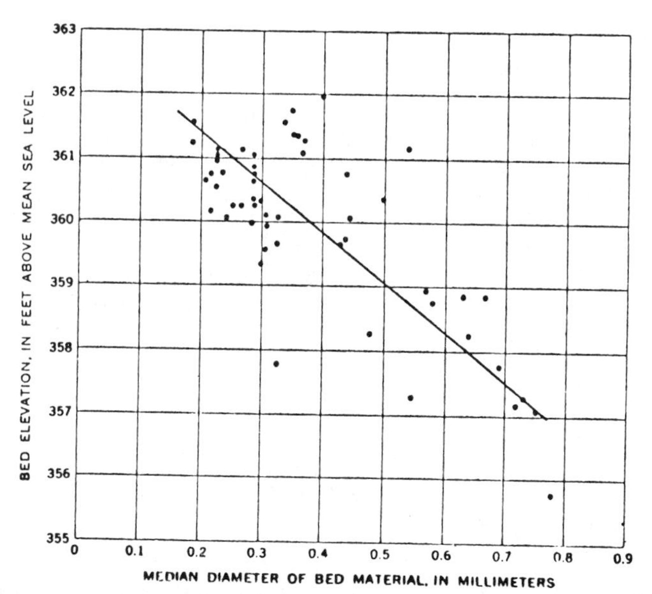

FIG. 2.118.—Relation of Median Size of Bed Sediment against Bed Elevation for Mississippi River at St. Louis; Jordan (1965)

slopes were determined from simultaneous readings of staff gages 6.5 miles apart so that the error in slope reading cannot be vary large. The variation in temperatures shown in Table 2.19 result in viscosity changes of from 1.5 to 2 fold, which will have an appreciable effect on sediment transportation. For example, for a stream 10 ft deep flowing over a bed of medium sand, Colby's relations, Eq. 2.233, indicates that by lowering the temperature from 80° F to 50° F the sediment discharge will almost double. Studies by Lane, et al. (1949) on the

TABLE 2.19.—Ranges in Some Characteristics of Four Rivers

River characteristic (1)	Niobrara near Cody, Neb.[a] (2)	Middle Loup at Dunning, Neb.[b] (3)	Colorado at Taylor's Ferry[c] (4)	Mississippi at St. Louis[d] (5)
Water discharge, in cubic feet per second				
Minimum	208	319	3,720	46,000
Maximum	668	482	11,000	763,000
Width, in feet				
Minimum	104[e]	123[g]	300	1,490
Maximum	178	154	354	1,780
Mean depth, in feet				
Minimum	0.97[e]	0.8[g]	3.8	15
Maximum	2.42	1.4	11.0	57
Velocity, in feet per second				
Minimum	1.8[e]	1.9[g]	2.3	2.0
Maximum	4.5	3.7	3.3	7.5
Median bed sediment size, in millimeters				
Minimum	0.20[e]	0.23[g]	0.27	0.18[h]
Maximum	0.40	1.0	0.40	1.1
Water surface slope, in feet per foot				
Minimum	0.00116[f]	0.00102[g]	0.000147	0.0000312
Maximum	0.00125	0.00150	0.000333	0.000139
Temperature, in degrees Fahrenheit				
Minimum	32	32	48	33
Maximum	78	88	80	84

[a]Colby and Hembree, 1955.
[b]Hubbell and Matejka, 1959.
[c]United States Bureau of Reclamation, 1958.
[d]Jordan, 1965.
[e]at Sec. C6.
[f]Mean surface slope Sec. C1 to C6.
[g]Sec. E.
[h]Over 9-yr period.

Colorado River, which has the characteristics assumed in the foregoing calculation, showed that in winter when the water was cold the bed sediment discharge was about 2.5 times that in summer for about the same water discharge.

Fig. 2.116 is a graph of median bed sediment size and mean annual discharge of the Mississippi River at St. Louis against time over a period of 9 yr. As observed by Jordan (1965) the similarity of the curves for sediment size and water discharge suggests that these two characteristics are correlated. Fig. 2.117 showing plots of mean discharge against sediment size indicates that there is indeed a relation between these two quantities. The discharge plotted in Fig. 2.117(*a*) is calculated for 1-yr periods and in Fig. 2.117(*b*) it is calculated over 2-yr periods. Jordan (1965) reported that plots of sediment size against mean discharge over periods of less than 1 yr showed much poorer correlation than in Fig. 2.117. Fig. 2.118 shows Jordan's graph of bed elevation against median bed sediment size for the Mississippi River at St. Louis. Jordan found that bed elevation did not respond to short-duration discharges thus suggesting that it was a cumulative effect that responded to events that occur over an extended period.

The information presented in Table 2.19 and Figs. 2.116 to 2.118 serve to emphasize the complexities of some alluvial streams. It is clear that slope and bed material of alluvial streams can vary with time and are not always constant as assumed in determining the curves of Figs. 2.113 and 2.114. An example that sediment size is not always variable was shown in observations on the Clyde River in Scotland by Fleming and Poodle (1970). Water temperature is also an important variable in determining sediment discharge. These large variations in important properties help to explain the poor agreement between calculated and observed sediment discharge in Figs. 2.113 and 2.114. They also help to explain the much better agreement between calculations and observations shown in Fig. 2.115. Results of using observed data in calculations were reported by Tarapore and Dixit (1972). They found that sediment discharge for the Colorado and Niobrara Rivers calculated by the DuBoys formula agreed much more closely with observed values than shown by Figs. 2.113 and 2.114 when observed flow depths and other observed data were used in the calculations. This result would not be expected because the DuBoys Formula is of the bed load type that is not normally considered to be applicable to sand bed streams carrying suspended load. A probable reason for this is that the values given by a sediment discharge formula depend not only on the form of the equation but also on the characteristics of the flow systems that gave the data used to evaluate the constants in the equation. The quantities, ψ_D and τ_c, in the DuBoys formula were determined from experiments with flows in which there was both suspended load and bed load. The calculated sediment discharges shown in Fig. 2.115 were determined using observed slope, velocity, water depth, temperature, and bed sediment characteristics. Despite this, some calculated values still fall outside the range of 0.5–2.0 times the observed values. In view of results of the kind shown in Fig. 2.115 one must assume that the probable error in sediment discharge calculations under the most favorable circumstances is large. Errors as large as 50%–100% can be expected. When calculations are based on average values of slope, bed material characteristics, temperature and calculated flow depth, and velocity as in Figs. 2.113 and 2.114 larger errors can be expected.

60. Summary and Recommendations.—Based on the material presented it is clear that sediment discharge formulas, at best, can be expected to give only

TABLE 2.20.—Variables in Wind Erosion Problem

Wind (1)	Surface (2)	Topography (3)	Soil (4)	Surface effects (5)
Speed Direction Structure Temperature Humidity Burden carried	Roughness Cover Obstructions Temperatures	Flat Undulating Broken	Texture Structure Organic content Moisture content Soil binders	Removal Deposition Surface markings Dune formation

estimates. It appears that sediment transport curves for sand bed streams can be approximated by power relations having the form of Eq. 2.257. Based on experience it also appears that common values of the exponent, n_s, for sand bed streams lie between two and three. Formulas that give transport curves with exponents very far outside this range are probably not applicable and should be used only if it is possible to justify the results, e.g., by showing that they are consistent with data from other similar streams.

It is very desirable to give guide lines that will enable the engineer to select an appropriate formula or relation that is suitable for use under any specific set of conditions. Unfortunately, this cannot be done except in a very general way. A rough way to judge the suitability of a formula is to compare the characteristics of the flows in the problem at hand with those of the flows that gave the data for determining the formula. Data for such flows are given for each formula as far as they are available. If it appears that the problem flow has appreciable suspended load then the formula selected should be based on data from flows that also carried suspended load, etc. It must be emphasized that the foregoing method of judging the suitability of a formula is a rough one and may not give satisfactory results. For this reason it is recommended that the selection of a formula or formulas be based on checking calculated sediment discharge against any observed values of the stream under consideration or on similar ones. Finally, after reasonable checks have been made one can then make some kind of judgment of the reliability of the calculations that should be kept in mind in using them in planning works.

I. Wind Erosion and Transportation

61. Concepts.—The wind is an important geological agent. It removes, deposits, and mixes soil, and to some degree it is a factor in soil formation. However, it is also one of the great destructive agents of soils.

The mechanics of wind erosion is a broad, complex subject. The transport of soil material by wind is a special case of the broader field of the transport of solids by fluids. Certain fundamental and relatively great differences exist, however, between the transport of material by wind in comparison to its transport by water. For example, soil, as it is exposed to wind erosion, is composed of solid, liquid, and gaseous components while soil exposed to water erosion is made up principally of solid and liquid components. The greatest fundamental difference between air and water transportation, however, is associated with the difference in the density of the fluid media. Malina (1941) has listed some of the variables in wind erosion, as shown in Table 2.20. Others can be added to this list.

62. Velocity Distribution and Wind Shear over Stable Surfaces.—The most useful relation used in the study of the phenomena of erosion of soil by wind has been the von Kármán-Prandtl logarithmic velocity distribution law in the form

$$\frac{\bar{u}}{U_*} = \frac{2.3}{k} \log \frac{y}{y_1} \qquad (2.261)$$

in which $\bar{u}$ = the average velocity of the wind at a distance y from a rough boundary surface; $U_* = (\tau_o/\rho)^{1/2}$ is the friction velocity over surfaces with an equivalent sand roughness height, k_s, not exceeding 1.5 mm, in which τ_o is the tractive force or, more precisely, the shear stress at the bed and ρ is the mass density of the fluid; and k = the von Kármán universal constant for turbulent flow. The term, y_1, denotes a reference distance equal to that value of y at which $\bar{u}$ in Eq. 2.261 is equal to zero. For clear fluids the value of k is approx 0.40. The value of y_1 has been considered to be 1/30 the height of the equivalent sand roughness height. Different investigators (Goldstein, 1950; White, 1940) have found that the surface is hydrodynamically smooth when the boundary Reynolds number

$$\mathsf{R}_* = \frac{U_* k_s}{\nu} \leqq 3.5 \text{ to } 4 \qquad (2.262)$$

indicating that under these conditions the surface roughness elements are within the laminar sublayer. In Eq. 2.262, k_s = the roughness height and ν = the kinematic viscosity of the fluid. When the boundary Reynolds number reaches a value of approx 90, the laminar layer appears to be disrupted completely and the boundary becomes hydrodynamically rough.

The von Kármán-Prandtl Eq. 2.261 is applicable to flows with hydrodynamically rough boundaries. However, sedimentation studies in water channels and wind tunnels have shown that the relationship of y_1 to height of roughness elements varies. R. A. Bagnold (1943) found the value of y_1 approximately equal to 1/30 times the diameter of sand grains on the bed. White (1940) obtained values of approx 1/9 times the grain diameter.

Zingg (1953b) found y_1 to vary as the logarithm of grain diameter, d_s, according to the relationship

$$y_1 = 0.081 \log \frac{d_s}{0.18} \qquad (2.263)$$

in which d_s and y_1, in Eq. 2.263 are in millimeters.

The von Kármán-Prandtl Eq. 2.261 applies reasonably well to the atmospheric wind profile to 5 ft above field surfaces if the origin of y is displaced slightly downward from the mean ground surface level (Chepil and Woodruff, 1963; Geiger, 1957; Sheppard, 1947; Woodruff, 1952). Modifications of the logarithmic law have also been used to describe the surface wind (Rossby and Montgomery, 1935; Sutton, 1949; Thornthwaite and Koser, 1943); however, all these modifications retain the universal constant, k. Power laws for the velocity profile have been used by several investigators (Geiger, 1957; Sutton, 1949; Kepner, et al., 1942; Sverdrup, 1936). Monin and Obukhov (1954), Ellison (1957), and H. A. Panofsky, et al. (1960) have developed relationships for wind profiles from similarity theory. The wind profile implied by the relationship first given by Ellison, and which is approximated by the Monin-Obukhov log-linear profile, fits data well in neutral and unstable conditions. However, another important result of these theories is that the logarithmic law is not valid under unstable conditions but is approached only as a limiting relationship very close to the ground. Measurements made in fully turbulent flow when wind erosion is in progress have confirmed this and have shown that the logarithmic law with k equal to 0.40 is valid if applied to profile data measured below 5 ft (Chepil and Woodruff, 1963).

Sheppard (1947) has suggested that τ_o might be expressed as a function of a drag coefficient, C_D, similar to that used in aerodynamics. This provides a useful index of surface roughness, particularly for vegetative-covered surfaces, where measurements have indicated that τ_o for a given natural wind varies significantly with aerodynamic surface roughness heights greater than about 1.5 mm (Chepil and Woodruff, 1963; Woodruff, 1952).

In describing atmospheric wind movement near a field surface, the level of atmospheric turbulence is of major importance. Kalinske (1943a) has demonstrated that the velocity fluctuations in rivers are distributed according to the normal error law, i.e.

$$f(u') = \frac{1}{\sqrt{\overline{u'^2}}\,\sqrt{2\pi}}\, e^{-u'^2/2\overline{u'^2}} \qquad (2.264)$$

in which $f(u')$ = the frequency function; $[\overline{u'^2}]^{1/2}$ = the standard deviation of u'; u' = the fluctuation of the velocity about the mean; and e = the base of Naperian logarithms. Investigation of velocity fluctuations of atmospheric wind movement above different surfaces has shown that they are normally distributed some distance above the ground during relatively stable atmospheric conditions, but have a slightly skewed distribution close to the ground when the atmosphere is unstable (Chepil and Siddoway, 1959; Sutton, 1955; Woodruff, 1952).

Four turbulence characteristics can be used to describe the atmospheric wind (Chepil and Siddoway, 1959): (1) Relative intensity; (2) relative magnitude; (3) scale; and (4) a turbulence factor. The intensity of turbulence at a particular height is the ratio of standard deviation of turbulence fluctuation to the mean velocity at that height and is equal to $[\overline{u'^2}]^{1/2}/\overline{u}$. It varies with height and between windstorms (Woodruff, 1952; Zingg and Chepil, 1950). An example of the distribution of $(\overline{u'^2})^{1/2}/\overline{u}$ with height is shown in Fig. 2.119(b). Magnitude of turbulence, a measure of fluctuations of velocity at a particular height in proportion to the shear velocity, is defined by $6[\overline{u'^2}]^{1/2}/U_*$. It increases directly, though not proportionately, with the roughness of the surface. Fig. 2.119(a) shows a distribution of magnitude of turbulence with elevation. The scale of turbulence is a measure of the size of eddies and is equal to $\overline{u}/n$, in which n is the number of velocity fluctuation cycles in a second. As shown in Fig. 2.119(c), it increases as the logarithm of height above the surface.

Many problems of transport of soil materials by wind and water are dependent on maximum impulses or pressures in turbulent flow. Chepil and Siddoway (1959) used the expression $\overline{P} + 3(\overline{p^2})^{1/2}$ as the maximum pressure and ratio $\overline{P} + 3(\overline{p^2})^{1/2}/\overline{P}$ was termed the turbulence factor, in which $\overline{P}$ = mean pressure; and $(\overline{p^2})^{1/2}$ is standard deviation of p. It was found to have a value of approx 2.7 at the surface near the topmost grains on the soil bed and to decrease to nearly 1.0 for nearly turbulence-free flow, at a height that varied with surface roughness.

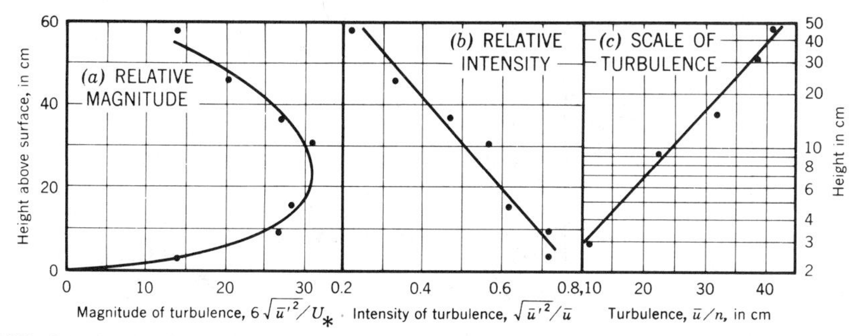

FIG. 2.119.—Variation of Turbulence Characteristics with Height above Surface Composed of Gravel Ridges 10 cm High and 40 cm Wide (Chepil and Siddoway, 1959)

These very major effects of wind turbulence on wind forces demonstrate the limitations of using average velocity or shear stress obtained in wind tunnels, where the turbulence factor is very low, to directly depict soil movement by wind. This limitation has been recognized and consideration has been given to the occurrence-frequency of atmospheric wind velocity and the time duration of the force associated with it (Zingg, 1953b; Chepil, 1960; Chepil, et al. 1962; Zingg, 1950; Zingg, 1951).

63. Velocity Distribution and Wind Stress over Drifting Surfaces.—Sketches showing the forms of velocity profiles obtained by different investigators above drifting surfaces are shown in Fig. 2.120. Figs. 2.120(a), 2.120(b) and 2.120(c) represent the results of Bagnold (1943), Chepil (1945b), and Zingg (1953b), respectively. The solid lines show the approximate ranges of height traversed whereas the broken lines are straight-line projections made for various purposes. Bagnold (1943) and Chepil (1945b) concluded that velocity distribution curves for different wind velocities converged to a focal point and conform to the equation

$$\bar{u} = 5.75\ U_* \log \frac{y}{y_t} + u_t \qquad (2.265)$$

in which the value $5.75 = 2.3/k$ when $k = 0.40$; $\bar{u}$ = the mean velocity at any height y; and y_t = the height at which the velocity distribution curves project to the focal point at a velocity, u_t. Zingg (1953b) found that k should be less than 0.40 and suggested an approximate value of 0.375 for drifting surfaces. This modifies the velocity distribution equation to

$$\bar{u} = 6.13\ U_* \log \frac{y}{y_t} + u_t \qquad (2.266)$$

Zingg further concluded that the velocity profiles over a given sand surface for varying wind forces do not follow the straight-line semilogarithmic relationship for values of y less than approx 0.05 ft and that they showed a curved convergence that appeared to approach a constant velocity near the bed as indicated by the solid lines of Fig. 2.120(c).

There appears to be a rather important and critical difference in these two interpretations of velocity profiles. With the straight-line projection, the stronger the wind the lower the velocity below the height y_t, but with the curved convergence where the line does not intersect the y-axis, the stronger the wind the

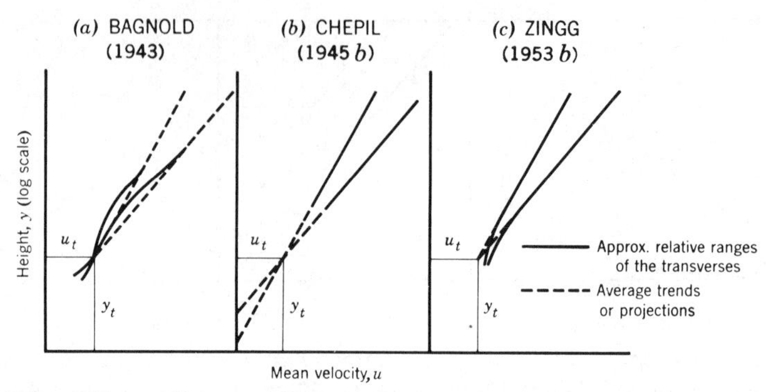

FIG. 2.120.—Different Forms of Velocity Profiles Obtained above Drifting 0.25 mm diam Sand Surfaces

higher the velocity near the surface. One explanation for the former is that with strong winds there is a greater concentration of saltating grains to take up the energy and lower the wind velocity (Chepil and Woodruff, 1963). The explanation for the latter interpretation is that the energy of the wind is transmitted to the grain above the elevation of the projected focal point. The grains in turn transmit a portion of the gained energy to the bed. The velocity obtained by the grains propelled from the bed to the upper portion of the sand cloud is apparently greater than the velocity of the wind near the bed and, as these faster moving grains descend to the bed, they tend to speed up the air near the bed (Zingg, 1953b).

64. Initiation of Particle Movement.—Various attempts have been made to describe a threshold wind velocity or shear stress at which beds of soil grains start to be moved by the wind. Einstein and El-Samni (1949), Ippen and Verma (1953), and Chepil (1959a) have recognized forces resulting from impact or velocity, viscosity, and static or internal pressures as the three forces exerted on a soil grain by a moving fluid. These forces may be resolved into a drag force acting horizontally in the direction of the wind and a lift force acting vertically. The drag force is visualized as being composed of two parts. The first results from normal wind pressure on the grain and is known as form drag; the second is the result from tangential stresses on the grain and is known as skin friction drag, and the sum of these is the total drag. The lift is merely the vertical component of the force on the grain as a result of the tangential and normal stresses applied at the grain surface. This has been recognized by Einstein and El-Samni (1949) and Chepil (1959a). Other investigators (White, 1940; Einstein and El-Samni, 1949; Ippen and Verma, 1953; Chepil, 1958; and Jeffreys, 1929) have determined that diameter, shape, immersed density of the grain, angle of repose, closeness of packing, and impulses of wind turbulence all influence the initiation of movement of a grain. By applying equilibrium of forces acting on a grain, as shown in Fig. 2.121, Chepil (1959a) derived the equation

$$F_c = (0.52\ g d_s^3 \rho' - L_c) \tan \phi \qquad (2.267)$$

in which F_c = the threshold drag force acting on the top grain; g = acceleration of gravity; d_s = diameter of grain; ρ' = the difference in density between the grain and the fluid; L_c = lift on top of grain; and ϕ = the angle of repose of the grain with respect to direction of gravitational forces acting through center of gravity of grain. Chepil also determined from wind tunnel experiments that L_c was equal to approx 0.85 F_c and by mathematical deduction obtained the relationship

$$\tau_c = \frac{0.66\, g d_s \rho' \eta \tan\phi}{1 + 0.85 \tan\phi} \qquad (2.268)$$

for winds of uniform velocity, and for turbulent flow

$$\bar{\tau}_c = \frac{0.66\, g d_s \rho' \eta \tan\phi}{(1 + 0.85 \tan\phi)\, T} \qquad (2.269)$$

in which τ_c and $\bar{\tau}_c$ = threshold shear stress and mean threshold shear stress on the whole bed, respectively; η = ratio of drag and lift on the whole bed to drag and lift on the topmost grain moved by the fluid; and T = the turbulence factor expressed as ratio of maximum to mean lift and drag on the soil grain. Chepil also evaluated the several variables in Eq. 2.269 and found ϕ equal to approx 24°; η equal to 0.2; and T equal approximately to 2.5. Some representative values of $\bar{\tau}_c$ determined from wind tunnel tests and computed by Eq. 2.269 are shown in Table 2.21.

Zingg (1953b), using a wind tunnel, graphically determined a "saltation threshold" from the points of intersection of force trend lines obtained for stabilized and drifting surfaces. He expressed his results by the empirical equation

$$\tau_s = 0.007\, d_s \qquad (2.270)$$

in which d_s = grain diameter in millimeters; and τ_s = saltation threshold shear stress in pounds per square foot of bed area. For grain sizes ranging from 0.15 mm–0.84 mm in diameter, Eq. 2.270 gives values very close to those of Table 2.21.

Bagnold (1943) used an experimental coefficient in a dimensionless formula to describe a threshold velocity. From equilibrium of forces acting on the grain under threshold conditions, the following expressions were derived:

$$U_{*t} = A\sqrt{\frac{\rho_s - \rho}{\rho}\, g d_s} \qquad (2.271)$$

and $$u_t = 5.75\, A\sqrt{\frac{\rho_s - \rho}{\rho}\, g d_s}\, \log\frac{y}{y_1} \qquad (2.272)$$

TABLE 2.21.—Threshold Drag Values for Different Grain Sizes Determined by Chepil (1959a)

Minimum grain diameter, d_s, in millimeters (1)	Density difference $\rho' = \rho_s - \rho$, in grams per cubic centimeter (2)	Threshold Shear Stress $\bar{\tau}_c$, in dynes per square centimeter	
		Computed (3)	Measured (4)
0.25	2.65	1.16	0.85
	1.91	0.84	0.69
0.42	2.65	1.95	1.60
	1.91	1.40	1.08
2.00	1.65	5.78	5.15
4.75	1.55	12.90	14.00

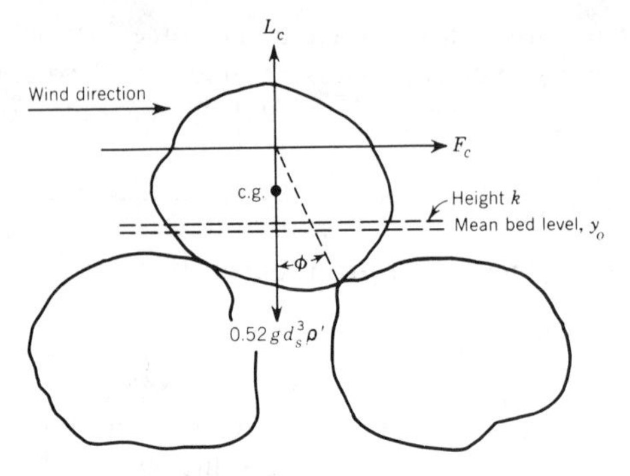

FIG. 2.121.—Forces Acting on Soil Grain in Windstream at Threshold of Movement of Grain (Chepil, 1959a)

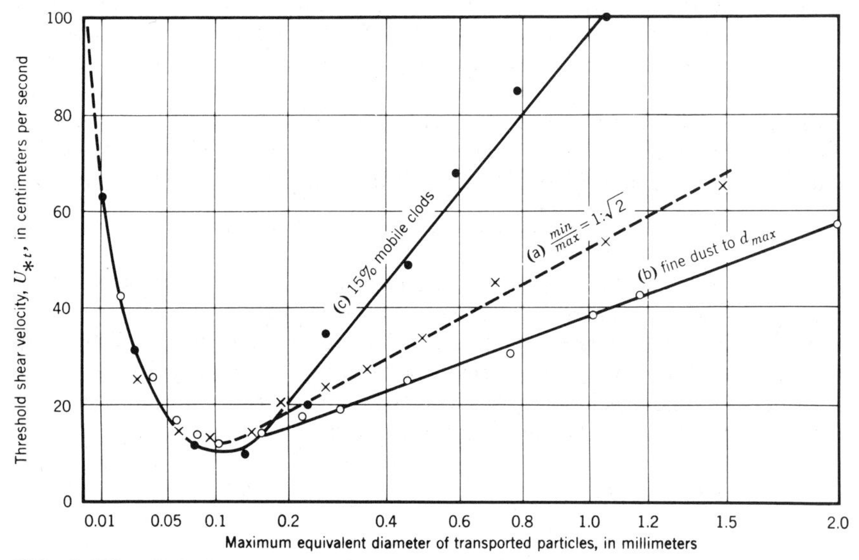

FIG. 2.122.—Relation of Fluid Threshold Shear Velocity of Wind to Maximum Equivalent Diameter of Transported Particles

in which U_{*t} = threshold shear velocity; ρ_s = the grain density; ρ = the air density; and A = an experimental coefficient. Bagnold found A = 0.1 for nearly uniform sand grains of diameters larger than 0.2 mm. Chepil (1945b) recognized a minimal and maximal fluid threshold. He found A equal to approximately 0.1 for beds composed of only erodible fractions but indicated that its value depended on the range of the equivalent size of the particles present on the eroding surface. Equivalent size or equivalent diameter is defined as $\rho_g d_s/\rho_s$ in which ρ_s is the density of the material composing the grain; ρ_g is the density of the grain itself which is different than ρ_s because of the voids in the aggregate; and d_s is the size of the grain determined by sieving.

Working with various ratios of minimum to maximum equivalent diameters and mixtures of erodible and nonerodible fractions, Chepil obtained the curves shown in Fig. 2.122 for threshold drag velocity as a function of maximum equivalent diameter. Each of the data of Fig. 2.122 was obtained with different sediment samples placed in trays and exposed in a wind tunnel. The most erodible discrete soil particles are approx 0.1 mm in diameter. The relationship between threshold velocity and equivalent diameter for particles above and below 0.1 mm in diameter follows the square root law indicated as Eq. 2.271. Curve (a) of Fig. 2.122 represents the case of a soil composed of erodible fractions of a limited size range, and the coefficient A in centimeter-gram-second units is equal to approx 0.1. The coefficient for curve (b), in which the soil is composed of erodible fractions ranging from the largest down to the smallest erodible particles is equal to about 0.85. The square root law holds for curve (c) in which 15% of the clods are nonerodible but the drag velocity is increased considerably because much of it is dissipated against the nonerodible fractions.

For field soils in their natural state there is no definite threshold shear velocity; rather, there is a range that varies with many factors including the previous erosional history. The lowest threshold velocity, measured at 1-ft height, for dry dune material is approx 13 mph and the highest for a previously noneroded soil is variable (Chepil, 1945b).

65. Nature of Particle Movement.—Once a soil grain is entrained in the airstream, the forces acting on it change rapidly. Chepil (1961) found that both lift and drag forces act on the grains; however, lift decreases with height and becomes hardly detectable at a distance of a few grain diameters above the ground surface. Drag forces, however, increase with height because of the direct pressure of the wind. The changes in force application as a spherical particle moves up into the airstream are shown in Fig. 2.123. The measuring sphere was surrounded by gravel mounds of the same diameter as the sphere, arranged three diameters apart in a hexagonal pattern. Pressure measurements were made at 15 points around the surface of the sphere, and direction of the resultant is that of the total pressure.

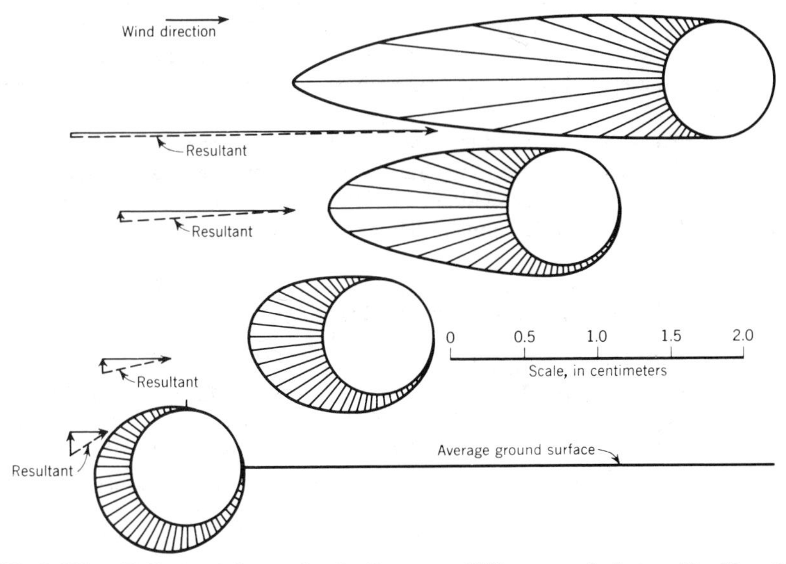

FIG. 2.123.—Pattern of Approximate Pressure Differences between Position 1 on Top of Sphere and Other Positions of Sphere at Various Heights in Windstream

The experiments were conducted in a wind tunnel at a location 50 ft downwind from the fan. Surface roughness was 0.4 cm and depth of turbulent boundary layer was approx 30 cm. Wind velocity at 2 cm above the aerodynamic surface was about 770 cm/s. The length of lines in the shaded areas outside the circular line (sphere) denote the relative differences in air pressures. The sphere is 0.8 cm in diameter and the drag velocity is 98 cm/s.

When a particle is dislodged from the surface it moves downstream by suspension, surface creep, or saltation (Bagnold, 1943). Those moving in suspension must have a terminal fall velocity less than the velocity of at least a small fraction of the upward eddy currents of the average surface wind. The size of the particle carried thus depends on shape and density but, in general, is smaller than 0.1 mm in diameter (Free, 1911). Von Kármán, as reported by Malina (1941), made an estimate of the length of time and distance traveled by soil particles when lifted from the surface. By using Stokes' law and the assumption that the rms of the height, h, that a particle reaches in arbitrary time, t, because of turbulence is given by the exchange-coefficient, E, the following expressions are obtained:

$$t = \frac{81\,E\,\mu^2}{2\rho'^2 g^2 d_s^4} \qquad (2.273)$$

and
$$L = \frac{40\,E\,\mu^2 \bar{u}}{\rho'^2 g^2 d_s^4} \qquad (2.274)$$

in which t = time; L = distance of travel; d_s = the diameter of particles; μ denotes fluid viscosity; ρ' = the density difference of fluid and particle; g = the gravitational constant; and $\bar{u}$ = the mean wind velocity. Assuming that the value of E for moderately strong winds lies between 10^4 and 10^5 cm²/s, the results given in Table 2.3 were obtained (Malina, 1941). Sand, silt, and clay have diameter ranges of 1.0 mm–0.05 mm, 0.05 mm–0.002 mm, and less than 0.001 mm, respectively. From Table 2.22 it is readily seen that discrete soil particles of the size of silt and clay are transported easily for great distances and to great heights by atmospheric winds once they are entrained in the wind. However, clay particles are generally aggregated or cling to larger particles. Silt is the most mobile soil constituent (Chepil, 1958).

The differences between surface creep and saltation are gradual and it is difficult to draw a fixed dividing line. Surface creep comprises the material driven along the immediate surface. Udden (1894) has indicated that surface creep involves particles 0.5 mm–1.0 mm in diameter. While grains moving in saltation

TABLE 2.22.—Time of Flight and Maximum Height of Rise of Particles in Atmosphere According to Eqs. 2.273 and 2.274

Diameter of particles, in millimeters (1)	Velocity of fall, in centimeters per second (2)	Time of flight (3)	Range for 15-m/s wind (4)	Maximum height (5)
0.001	0.00824	9 yr-90 yr	2.5×10^6 miles -25×10^6 miles	3.8 miles-38 miles
0.01	0.824	8 yr-80 yr	250 miles-2,500 miles	200 ft-2,000 ft
0.1	82.4	0.3 sec-3 sec	150 ft-1,500 ft	2 ft-20 ft

receive most of their forward momentum directly from the pressure of the wind, those in surface creep received nearly all their momentum by impact from the saltating grains returning to the bed.

The characteristics of saltating grains in wind have been described by several research workers (White, 1940; Bagnold, 1943; Bagnold, 1951; Chepil, 1945a; Kalinske, 1943b; von Karman, 1947; Zingg, 1953a). In general, the grains rise nearly vertically from the bed, gain momentum from the airstream, and return to the bed on paths at angles of 6°–12° from the horizontal. Bagnold (1943, 1951) made calculations of characteristic paths of saltating grains and noted the connection between rise of grains into the air and impact of grains returning to the bed surface. He also indicated that the violent impacts of saltating grains form and control surface ripples and showed that ripple wavelength corresponded to mean distance between takeoff and landing of saltating grains. Kalinske (1943b) has demonstrated that the height attained by given saltating particles is approx 1/800 as great in water as in air for conditions of the same value of shear stress. Chepil (1945a) found saltating grains to have appreciable rotation and determined the speed of rotation to be of the order of 200 rps–1,000 rps. He also indicated the ratio of height of rise to horizontal distance of grain leap to be on the average 1:7, 1:8, 1:9, and 1:10 for heights of rise up to 2 in., 2 in.–4 in., 4 in.–5 in., and above 6 in., respectively, for wind speeds of 17 mph measured at a height of 1 ft above the surface.

Zingg (1953a) used high-speed photography techniques to clarify the motion of saltating grains. He found that saltating grains have components of motion in the vertical direction, transverse to the wind and with the wind, as shown in Fig. 2.124. He also found that a variable angular velocity gained from impacts either on the bed or by collision in the air associated with movement of grains by saltation process. The orientation of axes of rotation are variable, and rotation may be in either direction with respect to such axes. Momentum conservation equations were also presented to describe the aspects of grain collision. Von Kármán (1947) analyzed the formation of ripples and for wavelength, L, of the ripples obtained the equation

$$L = 2\pi U\left(\frac{h}{g}\right)^{1/2} \qquad (2.275)$$

in which U = the wind velocity near the ground; h = the height of the layer near the bed in which most of the sediment is carried; and g = the gravitational constant. Values of L given by Eq. 2.275 have not been compared with observed values.

66. Distribution of Suspended Soil Particles above Bed Surface.—The first attempt to show the distribution of suspended sand over a surface moving by wind was made by Bagnold (1943). He reported the relative concentration of a uniform 0.25-mm sand to decrease markedly with height above the surface and to approach zero at an elevation of approx 4 in. (10 cm) for wind speeds ranging from 300 cm/s–450 cm/s measured at the 1-cm height. However, he noted that a few particles jumped at least 5 ft high. Surface creep at the bed level comprised about one-quarter of the total sand discharge.

In a study of eroding soils, Chepil (1945a) made determinations of the distribution of soil movement in suspension, saltation, and surface creep. The proportion was found to vary widely on different soils. In the cases examined,

FIG. 2.124.—Saltation Movement of Quartz Sand of Range 0.59 mm–0.84 mm in Wind with Shear Velocity of 92 cm/s (Zingg, 1953a)

between 50% and 75% by weight of the soil was carried in saltation, 3%–40% in suspension, and 5%–25% in surface creep. The relative concentrations of eroding material at different heights varied with different soils and conditions of the surface.

Chepil and Woodruff (1957) found, from measurements in dust storms in the Great Plains of Central United States, that the concentration of dust varied with height in accordance with an empirical power equation of the form

$$C_y = \frac{b}{y^n} \qquad (2.276)$$

in which C_y = concentration of dust in milligrams per cubic foot at height y, expressed in feet. Constant b and exponent n varied slightly with intensity of erosion but averaged 12.4 and 0.28, respectively, for 2 yr of measurement. Eq. 2.276 is the equivalent, for the vicinity of the surface, to the basic formula of Schmidt, as reported by Vanoni (1946). The data of Chepil and Woodruff (1957) and Langham, et al. (1938) were also used to determine the relationship between visibility, L_v, in miles and concentration, C_a at a 6-ft height in milligrams per cubic foot and C_m, total dust concentration, in tons per cubic mile. The relationships that are shown in Fig. 2.125 are $L_v = 0.9/C_a^{0.8}$ and $L_v = 15.0/C_m^{0.8}$.

Zingg (1953b), working with graded sizes of dune sand in a wind tunnel, determined a functional relationship for distribution of rate of sand transport or discharge at different heights above the bed. The general equation describing the distribution was of the type

$$\Delta g_s = \left(\frac{b}{y + a}\right)^{1/n} \qquad (2.277)$$

in which Δg_s = weight of sand per unit time and area at height y above the average elevation of the surface; b = a constant that varied with sediment size and shear stress τ_o; n = the exponent of the sand transport function Eq. 2.277; and a = a reference height. The total discharge of sand for each wind speed was obtained by the integration of Eq. 2.277. The mean height, y_a, above the bed, above and below which equal amounts of sand were transported, is shown in Fig. 2.126 for several ranges of sand size. An approximate empirical equation for mean height of saltation is

$$y_a = 7.7\, d_s^{3/2} \tau_o^{1/4} \qquad (2.278)$$

in which y_a is in inches; d_s = grain size, in millimeters; and τ_o is in units of pounds per square foot of bed area.

67. Rates of Soil Transport in Relation to Wind Shear Stress.—A number of investigators have measured the rate of soil transported by wind over sand and soil surfaces of different roughness.

O'Brien and Rindlaub (1936) published data on the relationship between the rate of sand transportation or discharge and wind velocity. For a beach sand with a mean diameter of approx 0.2 mm, they secured the relationship

$$g_s = 0.036\, u^3 \qquad (2.279)$$

in which the sediment discharge, g_s, is in pounds per foot of width per day; and u = the wind velocity in feet per second at a height of 5 ft above the surface.

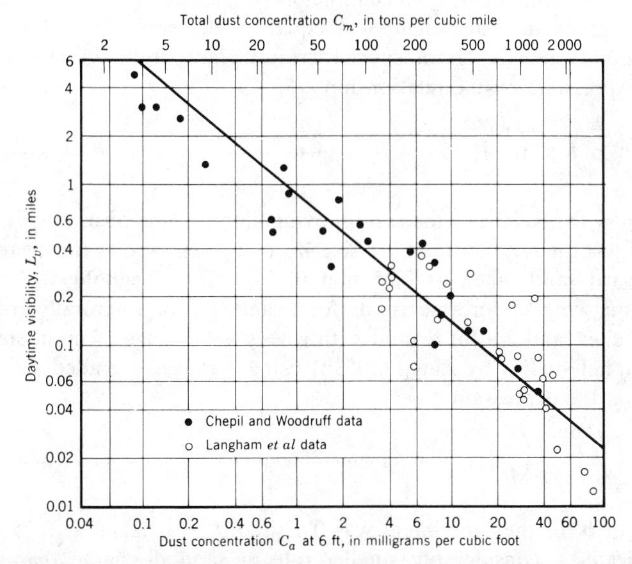

FIG. 2.125.—Relation between Daytime Visibility and Dust Concentration near Surface (Chepil and Woodruff, 1957)

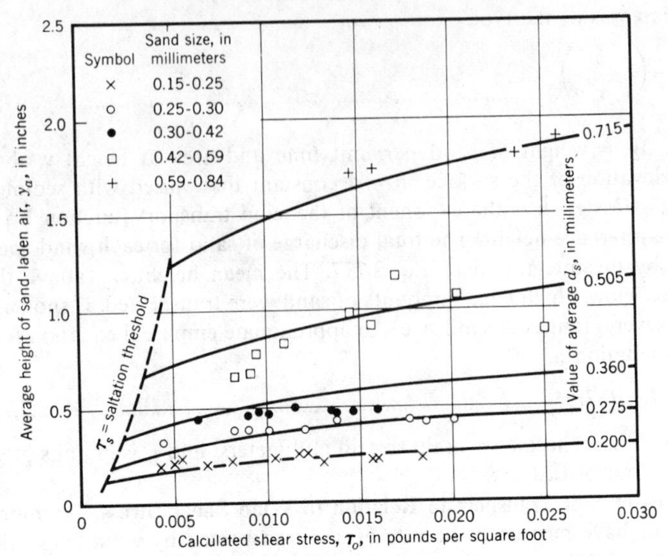

FIG. 2.126.—Average Height of Saltation Sand Movement in Relation to Calculated Shear Stress and Sand Size (Zingg, 1953b)

Bagnold (1943, 1938) made measurements of the discharge of sand both in sand storms in the Egyptian desert and in a wind tunnel. From the measurements in the desert, for sand having a mean diameter of 0.25 mm, the sand discharge in metric tons per meter width per hour was given by

$$g_s = 5.2 \times 10^{-4}(u - u_t)^3 \qquad (2.280)$$

in which u = the wind velocity in centimeters per second at 1 m height; and u_t, the threshold velocity at that height, was equal to 400 cm/s. In the wind tunnel Bagnold found the rate of transport for sand grains from 0.1 mm–1.0 mm in diameter expressible by the relationship

$$g_s = b\left(\frac{d_s}{d_{so}}\right)^{1/2} \frac{\rho}{g}\left(\frac{\tau_o}{\rho}\right)^{3/2} \qquad (2.281)$$

in which g_s = the weight of the sand moving along a lane of unit width per unit time; d_s/d_{so} = the ratio of the mean size, d_s, of a given sand to the mean size, d_{so}, of a standard sand taken as 0.25 mm in Eq. 2.281. Bagnold's values of the coefficient, b, were 1.5 for a nearly uniform sand, 1.8 for a naturally graded sand found in dunes, and 2.8 for a sand with a very wide range of grain sizes.

Wind tunnel studies by Zingg (1953b) using a range of graded sizes of dune sand yielded the expression

$$g_s = b\left(\frac{d_s}{d_{so}}\right)^{3/4} \frac{\rho}{g}\left(\frac{\tau_o}{\rho}\right)^{3/2} \qquad (2.282)$$

The value of b for the experiment was 0.83 and d_{so} was taken as 0.25 mm. This result indicates a considerably smaller rate of sand discharge than has been obtained from previous experiments. Zingg indicated that the differences could be because of: (1) The fact that his experiments considered saltation movement only;

(2) a different interpretation of τ_o, so that greater values were obtained for a given condition; (3) differences in size of wind tunnel used that affects development of a turbulent boundary layer.

Chepil (1945c), in research on rate of movement of dry field soils, has indicated that the general form of Eq. 2.281 proposed by Bagnold adequately expresses the rate of movement in relation to shear velocity. However, he has also shown that the coefficient, b, varies with the proportion of fine dust particles present in a mixture, the proportion and size of nonerodible fractions, roughness of the field, and amount of moisture in the soil (Chepil, 1941; Chepil, 1950; Chepil, 1956; Chepil and Milne, 1941).

Because of these factors and because rate of movement is seldom constant under some conditions the total weight of soil material removable from the surface by wind is a more accurate measure of erodibility than the rate of removal. The weight of soil material W that is removable from a given area may be expressed in terms of shear velocity, U_*, by

$$W = b\,U_*^5 \qquad (2.283)$$

in which b = a coefficient that varies with many factors. Because the quantity of erodible soil material also varies with degree of soil abrasion as influenced by the characteristic length of eroded area, it is better to express erodibility in dimensionless form applicable to any size of eroding area, direction of wind, or units of measure by

$$I_w = b\,U_*^5 \qquad (2.284)$$

in which I_w = soil erodibility index that is equal to W/W_o in which W_o is the weight of soil material removable per unit area from a "small" area (wind tunnel area with length of exposed area not exceeding 30 ft) when the soil contains 60% by weight of clods larger than 0.84 mm in diameter; and W is weight removable under the same set of conditions from soil containing any other proportion of clods larger than 0.84 mm (Chepil, et al. 1955; Chepil and Woodruff, 1959).

68. Avalanching.—The increase of soil discharge with distance downwind over an unprotected eroding area is known as soil avalanching (Chepil, 1957b). It is associated with the following processes: (1) Erodible soil fractions tend to accumulate on the surface with distance downwind, thereby creating a progressively more erodible soil condition to leeward; (2) progressively greater concentration of moving grains downwind increases the frequency of impacts and degrees of breakdown or abrasion of clods and surface crust fragments that, in turn, are moved by wind; (3) erodible soil fractions are usually dislodged from projections and trapped in depressions. This latter process, known as detrusion, tends to produce a progressively smoother surface that causes a progressively increasing rate of soil flow with distance downwind (Chepil, 1946).

The rate of soil discharge increases from zero on the windward edge to a maximum that a given wind velocity can sustain, provided the eroding area is of sufficient length. Maximum rate of soil discharge in a given wind is approximately the same for all field soils having a range of erodible and nonerodible soil fractions and is about equal to that of dune sand. The distance required for soil discharge to reach a maximum on a given soil is the same for any level of erosive wind. It depends only on erodibility of the surface and tends to decrease as erodibility increases, as shown in Table 2.23.

TABLE 2.23.—Relation between Wind Tunnel Erodibility and Distance to Maximum Soil Discharge

Wind tunnel erodibility, I_w (dimensionless) (1)	Distance (in field) for soil discharge to reach a maximum, in feet (2)
920	180
300	300
50	1,100
39	1,600
19	2,200
7.5	3,900
6.1	4,100
5.1	5,200

Note: Conditions equivalent to a 40-mph wind measured at 50 ft above a smooth, level, unsheltered terrain according to Chepil (1959b).

69. Sorting.—The two types of soil removal associated with wind erosion are: (1) Nonselective; and (2) selective. The former occurs primarily on loessal soils where wind removes virtually all of the surface soil, thus reversing the order in which it was laid down in past geologic eras. The latter occurs primarily on soils developed from glacial till, residual material, mountain outwash, and sandy soils of various origins. Wind removes, primarily, the silt and clay from these soils leaving the coarser fractions behind. Selective removal and the natural tendency of the wind to move finer, lighter particles faster than coarser and denser particles leads to the classification of soils into the following four distinct grades arranged according to increasing erodibility:

1. Residual soil materials—nonerodible clods and massive rocks that remain in place.
2. Lag sands, gravels, and soil aggregates—semi-erodible grains moved slowly and deposited randomly on the surface.
3. Sand and clay dunes—highly erodible grains deposited only short distances from source.
4. Loess—dust carried great distances by suspension and deposited in uniform layers.

There are no distinct demarcations of size between the various grades of wind-sorted materials. The size distribution of one grade overlaps considerably that of another grade. However, the size distribution of particles contained in any grade conforms to a logarithmic relationship discovered by Bagnold (1943) for desert sand. This relationship indicates the existence of a remarkable phenomenon of sorting of soil materials by the wind; i.e., sorting material transported at any height or deposited anywhere by any single windstorm is characterized by a predominant or peak diameter of the discrete particles and by arms on each side of the peak falling off each at its own uniform rate. The peak diameter varies from one graded material to another, depending on the physical composition of the soil and distance and height of transport. Materials deposited at a location by winds of different velocities are composed of mixtures of different grades that may show no resemblance to any single grade.

70. Abrasion.—Abrasion, the wearing away of solid materials by impacts of transported particles, is an important phase of the wind erosion process (Chepil, 1946). The thin surface crust covering most soils disintegrates because the abrasive action of a few grains moving in surface creep and saltation. Nonerodible clods are also broken down by impacts of saltating grains. The longer erosion continues, the greater is the quantity of erodible material formed by abrasion and the higher the rate of soil discharge.

Abrasion also abets wind erosion because it is extremely injurious to plants and tends to destroy them. The more vegetation destroyed, the more source material exposed to the wind action and, consequently, the higher the rate of soil discharge.

The amount of breakdown of soil structure by abrasion depends on the mechanical stability of the structural units. The mechanical stability, or resistance to disintegration by mechanical forces, varies with soil texture and with various phases of soil structure. Chepil (1955) has developed the equation

$$A = W_v\left(\frac{25}{U}\right)^2 \qquad (2.285)$$

in which A = a coefficient of abrasion; W_v = weight of soil abraded per unit weight of abrader blown at wind velocity U, expressed in miles per hour.

71. Deposition.—Soil particles once lifted in the air eventually return to the surface and come to rest when the wind subsides or when surface obstructions alter the velocity distributions and turbulent structure. Fine particles carried in suspension travel great distances from the source and are deposited in a more or less uniform mantle over extensive areas by settling when wind velocity falls below a sustaining velocity. Larger particles moving in saltation and surface creep are deposited by accretion or trapping by tall, closely spaced obstructions and by encroachment or trapping in depressions or sudden drops in surface elevation as on the lee side of a dune or ridge (Bagnold, 1943).

Because the wind is variable in direction, considerable mixing of soils, particularly of fine material moving in suspension, occurs. Movement in one direction is not always equal to that in another, and under some conditions considerable accretion of soil material occurs.

Considerable information is available on depth, composition, and distance from source of loess deposited in past geologic eras (Hanna and Bidwell, 1955; Waggoner and Bingham, 1961). Some information on the composition of dust presently deposited from the atmosphere is also available (Chepil, 1957a; Pewe, 1951; Swineford and Frye, 1945; Warn and Cox, 1951). All of these studies indicate that deposition of wind-transported sediment is a major factor in geological changes constantly occurring over the land surface. Man has greatly accelerated these changes since he began cultivating the soil.

J. Transportation of Sediment in Pipes

72. General

Scope.—The main objectives of this section are: (1) To present a survey of existing knowledge on the transportation of sediment in pipes; (2) to examine the similarities and differences of previous findings; and (3) to suggest design criteria based on the present status of knowledge. This presentation is limited to transportation of cohesionless sediment (sand, coal, and so forth) as Newtonian solid-liquid mixtures. Chemical reaction and coagulation of material are excluded.

The term sediment as used in this section may also refer to solids other than sand particles. It has been adopted primarily because of the wording of this manual and especially this section. The reader will surely find no ambiguity or difficulty in interpreting its meaning wherever it is used.

General Applications.—Transportation of sediment by fluids in pipelines has a wide variety of applications in industry. Some of the more familiar applications of pipeline transportation include dredging; conveyance of coal and ores; disposal of tailing, ashes, and other waste products; solid waste and cement slurries; and pneumatic conveyance of grains. Pipeline transportation systems are also used to handle raw materials, materials in process, and finished products in various segments of different industries including those involving agriculture, paper, oil, food, chemicals and mining. Ocean mining is another potential industry receiving increased attention, and the transport of these minerals by pipelines is being strongly considered.

Pneumatic solid waste removal by pipeline has been successfully and economically carried out in Sundeberg, a suburb of Stockholm, Sweden [Zandi and Hayden (1971)]. Zandi and Hayden also indicate that solid waste disposal by pipe conveyance in the downtown area of Philadelphia is economically competitive with present methods of disposal by trucks, even without the consideration of the benefits derived with respect to traffic obstructions, health hazards, and so forth. Transportation of sewage solids by pipeline for deposit outside the city in land reclamation has been tested in Morgan County, W. Va., Chicago, Ill., Long Island, N.Y., and so on (e.g., see Engineering News Record, May 23, 1968).

Regimes of Flow.—Researchers not only use various names for flow regimes, but also have classified flow regimes differently. For simplicity, the following divisions are proposed. Four modes or regimes of transport of solid-fluid mixtures through pipelines for a given fluid, sediment material, and pipe size, can be qualitatively described as shown in Fig. 2.127. These are homogeneous flow, heterogeneous flow, saltation, and stationary bed flow. [Durand (1953) and Carstens (1969) classify saltation as a part of heterogeneous mixtures; however, Newitt, et al. (1955), Babcock (1962), and Zandi (1971) tend to list moving bed and saltation as a separate regime from heterogeneous flow.] Each of these regimes has indistinct regions or subregimes, both narrow and wide, bordering the solid lines in the figure. There is no intent, of course, to imply that distinctive changes in physical phenomena occur at the lines, but the sediment-fluid flow interactive mechanisms are sufficiently different from regime to regime that the classification is justified. If there is more than one size of sediment in the flow, some smaller particles may be transported as homogeneous flow, and others may be transported as heterogeneous flow. This regime, where homogeneous and heterogeneous flows exist simultaneously, is referred to by Zandi (1971) as intermediate regime.

The region labeled "Flow as a homogeneous suspension" in the lower part of Fig. 2.127 represents the regime in which the particles being transported are so small that the fall velocities are insignificant compared with the vertical motion of fluid; thus, the vertical distribution of sediment particles is nearly uniform.

Heterogeneous flow shall be defined herein as the regime in which all solids are in suspension, but the vertical sediment concentration gradient is not uniform. This is probably the most important regime of sediment transportation in pipes

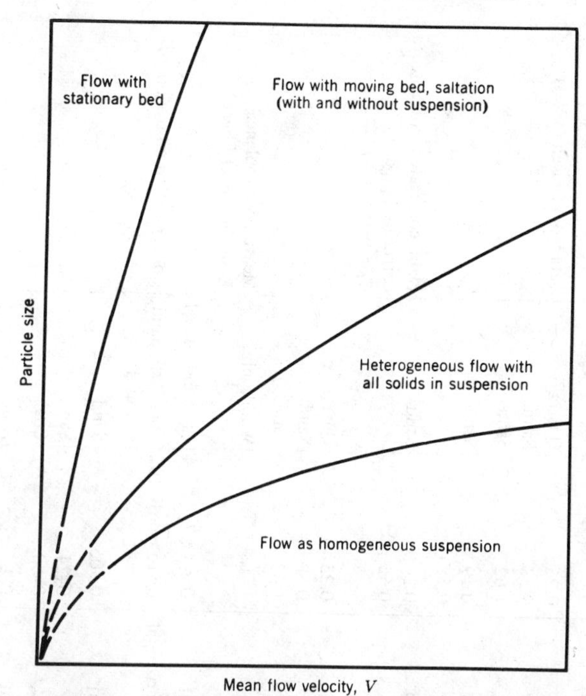

FIG. 2.127.—Flow Regimes for Given Fluid, Sediment, and Pipe Size, Qualitatively Only

because it is normally identified with economical operation, in the sense that the ratio of, e.g., the amount of material transported per unit power consumption is at a maximum. The regime of flow that involves a moving bed (including saltation) might normally be avoided in practice because the occurrence of irregularities (ripples and dunes) on the moving bed usually causes a distinctive increase of head loss. Finally, flow with a stationary bed involving no suspended material will not be analyzed in this article because it is essentially a rigid boundary problem not pertinent to sediment transportation in pipes.

General Considerations.—The important design considerations for a given pipeline system are: (1) Head (energy) losses; (2) sediment concentrations that can be transported including pipeline blockage and reliability of the entire system; (3) rate of pipeline abrasion; and (4) rate of particle attrition. Because the characteristics of flow in each regime are quite different, the prediction or identification of the probable flow regime for a given set of conditions is essential for the estimation of head loss and transport capability.

Typical head loss and sediment concentration curves for a given fluid, sediment characteristics, and pipe size are presented in Fig. 2.128. For pure fluid, the head loss should vary with flow velocity to the 1.75 and 2 powers for turbulent-smooth and turbulent-rough boundaries, respectively.

In general, head loss increases with increasing sediment concentration, C_v, by volume. The exception is in the homogeneous regime where the head loss for different sediment concentrations approaches that of pure fluid at large flow velocities. The region dividing heterogeneous flow and saltation is closely related to the region of minimum head loss for a given sediment concentration.

TABLE 2.24.—Summary of Test Conditions of Selected Investigators

Investigators		Pipe			Sediment		
Reference (1)	Fluids (2)	Diameter, in inches (3)	Material (4)	Slope (5)	Diameter, in millimeters (6)	Material (7)	Remarks (8)
Ambrose (1953)	Water	2 and 6	Lucite	Variable	0.25 0.58 1.62	Quartz Sand Sand	Data from Craven (1953)
Blatch (1906)	Water	1	Galvanized Iron, Brass	Horizontal	0.19 0.584	Sand	$C_v = 5.35\%$
Chamberlain (1955)	Water	12	Steel	Horizontal	0.2 0.65	Sand	Plain, corrugated and helicordal corrugated pipe. $C_v = 0.15$ (20%)
Craven (1953)	Water	2 and 6	Lucite	Variable	0.25 0.58 1.62	Quartz Sand Sand	
Daily and Bugliarello (1958)	Water					Wood pulp	Measured turbulence Long and short fibers
Durand (1953)	Water	1.5–28	Steel	Horizontal	0.2–24.7	Sand and Gravel	
Durand and Condolios (1952)	Water	1, 5, 4, 6, 10, 13, 23, 27.5	Steel	Mostly Horizontal	up to nearly 102	Coal, Ash, Sand, Gravel	Concentration up to 600 g/*l*
Durepaire[a] (1939)	Water	2.05	Steel	Horizontal	0.305	Sand	
Fontein (1958)	Water	6		Vertical	0–24 0–80	Sand Coal	
Garde[a] (1956)	Water	12	Steel	Horizontal	0.2 0.6	Sand	
Graf and Acaroglu (1967)	Water	3	Aluminum	0°, 11.25°, 22.5°, 45°, 90°	2.85	Sand	$C_v < 7\%$
Graf and Acaroglu (1968)	Water				0.091–2.0	Sand and Coke	Circular pipe, square pipe, flume, and natural river

Gregory (1966)	Water	4	Cast Iron	Horizontal	0.012	Clay Slurry	
Homayounfar (1965)	Water	1/2, 5/8, 3/4, 1, 1.5	Galvanized Iron	Horizontal	0.05–0.65	Coal	Flow of multicomponent slurries
Howard (1962b)	Water	2	Steel	Horizontal	0.01–0.4	Silt and Sand	Effect of fines on the pipe flow
Howard (1939)	Water	4	Steel	Horizontal	0.382 2.52	Sand Gravel	C_w = 10 (40%)
Howard (1941)	Water	2, 4	Steel	Horizontal	0.382 2.52 0.023	Sand Gravel Silt	Rifled and plain pipe, 4 lb/sec–30 lb/sec
Hughmark (1961)	Water	0.5–28		Horizontal	0.066–1.84	Boiler ash, Lime, Sand, etc.	Data from others
Ismail (1952)	Water	Rectangular 10.5 X 3		Horizontal	0.1 0.16	Sand	C_w = 0.015 (31%)
Newitt and others (1955)	Water	1	Brass	Horizontal	0.203–5.98	Perspex, Coal, Sand, Gravel, Manganese dioxide	
O'Brien and Folsom (1937)	Water	2, 3	Wrought Iron	Horizontal	0.178–1.7	Sand	C_v = 2 (26%)
Shih (1964)	Water	3	Lucite	0.00°, 8.73°, 17.71°	12.7	Presoaked wood	Velocity 10 fps–12 fps
Sinclair (1962)	Water Kerosene			Horizontal	d_{85} = 0.0006D to 1.5D		
Smith (1962)	Water	2 and 3		Horizontal	0.203, 0.305, 1.22	Sand	Mixed size
Spells (1955)	Water	3–11.8		Horizontal	0.08–0.82	Lime, Boiler ash, Sand	C_w = 2%–33%
Thomas (1962)	Air Water	0.62–1.75 0.496–32		Horizontal Horizontal	0.97–2 0.19–38		C_v = 0%–2% C_v = 0%–15%
Vogt and White (1948)	Air	1/2		Horizontal & Vertical	0.203–0.73, 0.42, 1.17, 4.02	Sand, Steel shot, Clover seeds, Wheat	Air speed 30 fps–120 fps

TABLE 2.24.—Continued

Investigators		Pipe			Sediment		
Reference (1)	Fluids (2)	Diameter, in inches (3)	Material (4)	Slope (5)	Diameter, in millimeters (6)	Material (7)	Remarks (8)
Weisman[a] (1963)	Water	0.5–24		Horizontal	0.0125–2	Sand, Steel, Glass, Lead, $BaSO_4$, Tungsten	C_v = 0.2%–33%
Wilson (1966)	Water	3.69 X 3.69 square pipe	Aluminum with 3.425 ID Perspex pipe for viewing	Horizontal	Average about 0.75	Sand	
Wilson (1942)	Water	18, 30	Wood stave	Horizontal		Ore tailings	Data from Howard (1939), Blatch (1906) and Climax Molybdenum Co.
Worster (1952)	Water	1		Horizontal	0.51–3.2	Perspex, Coal, Sand, MnO_2	Data from Turtle (1952)
Yotsukura (1961)	Water	4.0	Lucite	Horizontal	0.231–1.09 0.8–3.0	Bentonite clay, Sand, Poly-styrene, Pellets	
Yufin (1949)		0.4–18		Horizontal	0.25–7.36		$0\% < C_v < 30\%$
Zabrodsky[a] (1966)	Water Air	1–11.8 1.77		Horizontal Horizontal & Vertical	0.012–0.58 0.168–1.68	Sand	
Zandi and Govatos (1967)	Water	1–24		Horizontal	0.1–12.7		1,452 points of data
Zenz (1958)	Air	1.75	Lucite	22.5°, 45.0°, 67.5°	0.168–1.68	Rape seeds, Sand, Glass beads	Air speed 3 fps–40 fps

[a]These studies were not described in the main text of this manual; however, they provide valuable data and thus are listed herein.

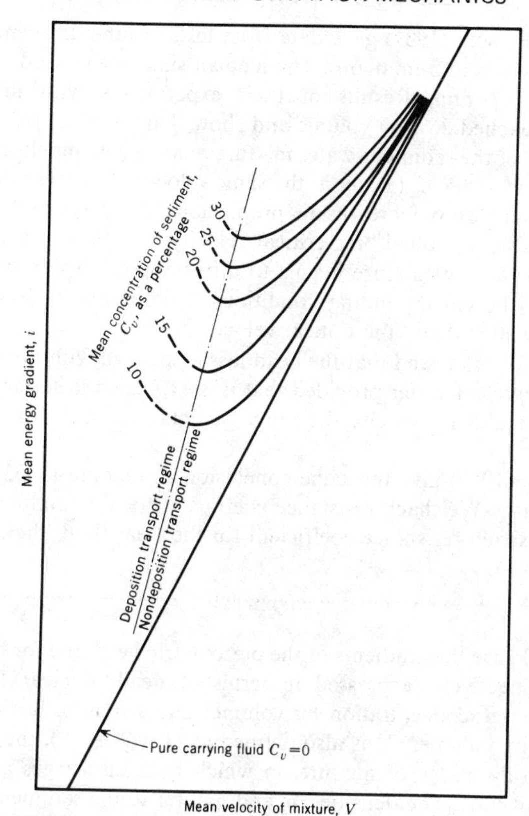

FIG. 2.128.—Typical Head Loss and Sediment Concentration Curves for Given Fluid, Sediment, and Pipe Size

Experimental Results.—Table 2.24 gives a summary of test conditions of some selected investigators.

73. Transportation of Uniform Material in Horizontal Pipes

Homogeneous Two-Phase Flow.—The basic mechanism by which sediment particles (assuming that solid is denser than fluid) may be maintained in suspension in a turbulent flow is the exchange of fluid containing greater sediment concentration at a lower level with fluid containing less sediment concentration at a higher level to offset the downward movement caused by gravity. This is done effectively by turbulent mixing. It is evident then that a vertical sediment concentration gradient may prevail while all the particles are permanently suspended. However, when the particles are so small that the fall velocities are insignificant, the sediment particles are uniformly distributed throughout the cross section of pipes, i.e., the flow is homogeneous.

From recent investigation by Jobson (1968), there is evidence that seems to indicate that turbulence near the boundary may diffuse solid particles more than fluid particles. If this is true, particles could be permanently suspended by this mechanism. However, without further positive evidence, the exchange mechanism as described in the previous paragraph is favored.

O'Brien and Folsom (1937) gave data from tests of nine different sands in two sizes of pipes (2-in. and 3-in. diam). The median sizes of the sand varied between 0.17 mm and 1.7 mm. Results of their experiments with maximum sand concentration reached 26% by volume and show that above a critical velocity the head loss in feet of the sediment-water mixture in any given pipeline is the same as that which occurs with clear water at the same velocity. It is clear then that in this range the pressure drop increases in proportion to the specific gravity of the mixture. The authors defined "the critical velocity" as the flow velocity at which the head loss in feet of mixture begins to differ appreciably from head loss for clear water for the corresponding conditions. Homogeneous flow occurs when velocities are greater than "the critical velocity."

Durand (1953) also found that the head loss associated with homogeneous flow is the same as in clear water provided that it is expressed in terms of the head of the mixture. He did not specify the range of data on which his conclusion was based.

Newitt, et al. (1955) give the same conclusion as Durand (1953). By assuming that f_m, the Darcy-Weisbach resistance coefficient for the mixture, is equal to f, the Darcy-Weisbach resistance coefficient for the clear fluid, they conclude that

$$\frac{i_m - i}{C_v i} = s - 1 \qquad (2.286)$$

in which i_m and i are the gradients of the piezometric head line for the mixture and pure fluid, respectively, expressed in terms of head of clear fluid; C_v = the sediment discharge concentration by volume; and $s = \rho_s/\rho$. Sediment discharge concentration by volume, C_v, is also defined as $Q_s/(Q + Q_s)$, the ratio discharge of sediment to discharge of mixture, in which both discharges are expressed in volume per unit time. The densities, ρ_s and ρ, refer to the sediment and the fluid, respectively. However, examination of their experimental results indicates that for Sand A (mean particle size 0.02 mm), $(i_m - i)/C_v i \approx 3$, which is greater than $s - 1$, and for Sand B (mean particle size 0.096 mm), $(i_m - i)/C_v i < s - 1$, for velocities greater than 5.5 fps.

Spells (1955) states in his introductory section that "As the rate of flow is increased, the pressure gradient ... approaches the equivalent true fluid value, eventually becoming identical with it." The "equivalent true fluid" is interpreted to mean a fluid with density equal to that of the slurry and with viscosity equal to that of water (clear fluid) flowing under the same conditions. The region where flow rate is large probably coincides with the two-phase homogeneous flow.

Howard (1962b) concludes from his tests that mixtures of water and fine sediments at small concentrations exhibit a similar hydraulic gradient characteristic to that of water. In fact, with the concentration of fine sediment (mean diameter 0.01 mm) up to 6% by volume, the test results are no different from those obtained in similar tests with water. No doubt, instrumentation inadequacies might account for the failure to measure the slight effect of the fine sediments at these low concentrations. As the concentration increases to a value greater than 6% by volume, the effect of sediment concentration on head loss becomes noticeable.

Ansley (1962) points out that the apparent kinematic viscosity increases quite rapidly as the concentration of silt is increased above 6%–8%. Therefore, in this range, the silt-water mixture should be treated as non-Newtonian.

Homayounfar (1965) found from his experiments that, provided the concentration of coal (finer than 0.254 mm diam) is less than 5% by weight, the head losses for slurries consisting of a mixture of water and coal are less than that of pure water.

Zandi (1967) tested coal (specific gravity 1.67, approximate size between 0.074 mm and 0.297 mm, concentration up to 5% by weight), fly ash (specific gravity 2.40, size smaller than 0.149 mm, concentration up to 0.5% by weight), clay (specific gravity 1.67, approximate size between 0.21 mm and 0.84 mm, concentration up to 7.5% by weight), and activated charcoal (specific gravity 1.56, 95% of particles smaller than 0.044 mm) in five different pipes (diameters ½ in., ¾ in., 1 in., 1 ½ in., and 2 in.). In the approx 800 data points collected, almost all head losses fell below corresponding points for clear water. Zandi describes several possible mechanisms for the suppressing action of fine particles on head losses, but did not succeed in establishing a criterion to predict the reduction of head loss under different conditions.

In summary, results from previous investigations appear to suggest that the homogeneous flow regime with very fine sediments may exhibit energy gradient suppressing action, but the homogeneous flow with slightly larger sediments can be treated as a homogeneous fluid, where the Darcy-Weisbach equation can be used to calculate the head loss in the pipeline, and the friction factor in the equation can be determined from the Moody diagram. In using this diagram, the Reynolds number can be calculated from $\mathsf{R} = VD/\nu$, in which V = the average flow velocity in the pipe; D = the pipe diameter; and ν = the kinematic viscosity of the pure fluid. The head loss would be expressed in terms of the specific weight of the liquid-sediment mixture.

Unfortunately, there is no known way at present to differentiate, a priori, between those liquid-sediment mixtures that may exhibit energy gradient suppressing action and those which would not.

Recommendations.—Based on previously mentioned investigations, some two-phase homogeneous flows may be treated as single fluids. Unfortunately, a criterion has not yet been found to determine when a solid-liquid mixture will exhibit energy gradient suppression behavior. However, previous investigations tend to show that, for sand coarser than, e.g., 0.15 mm (this rule is only a speculation and is by means established), the solid-liquid mixture may be treated as a single Newtonian fluid. A pilot model study to test the head loss curve of the solid-liquid mixture, under as close to a prototype operating condition as possible, is highly desirable. If the solid-liquid mixture can be treated as a single Newtonian fluid, the designer should consider using the kinematic viscosity of the mixture instead of that for pure fluid because it will result in a smaller Reynolds number of the flow and correspondingly, a greater friction factor. The effect may be noticeable only in a smooth pipe. For example, Elata and Ippen (1961) found that the change of kinematic viscosity, with a concentration of dylene polystyrene particles (mean diameter 0.12 mm and specific gravity 1.05) of 27% by volume, increases the kinematic viscosity of the fluid by 240%. The ratio of f_m/f would then be approx 1.24, which cannot be ignored. However, if the pipe is hydrodynamically rough under the oeprating conditions (normally galvanized pipes are rough), f_m is not a function of Reynolds number and therefore the effect of fine sediment on fluid viscosity does not enter the problem.

It is recommended that the following steps be taken in designing a pipeline system with a homogeneous two-phase flow that does not exhibit energy gradient suppression.

1. Measure the kinematic viscosity, ν_m, of the mixture with the desired sediment concentration. There are essentially four principal techniques of viscosity measurements. These comprise: (a) The capillary viscometer; (b) the rotational viscometer; (c) the Redwood type apparatus utilizing flow through a short nozzle; and (d) the calculation of mixture viscosity by measuring the mixture friction factor in either a small pipe or an open channel mixture flow with the same solid concentration. This is essentially a model study to obtain fluid viscosity. Colorado School of Mines (1963) and Bain and Bonnington (1970) give discussions on some of these.

2. Find f_m of the pipe wall from the Moody diagram with VD/ν_m and D/k_s in which k_s = the average roughness height.

3. Calculate head loss, h_L, in terms of head of the mixture as

$$h_L = f_m \frac{L}{D} \frac{V^2}{2g} \qquad (2.287)$$

in which L = the total length of pipe. Include bend losses, and other minor losses in the total head loss.

4. Estimate the power requirement to transport the solid by

$$hp = \frac{Q\gamma_m h_L}{550\eta} \qquad (2.288)$$

in which hp = the horsepower requirement; Q = the discharge of mixture, in cubic feet per second; γ_m = the specific weight of the mixture, in pounds per cubic foot; and η = the efficiency of the pump.

5. Some experimental data from laboratory studies are obtained from smooth pipes. The designer should consider the transferring of these results into rough pipes.

Since head loss may be reduced by the presence of fine sediment, the recommended procedures as previously stated may give a conservative estimate of power requirement (greater than necessary). However, with the present state of knowledge, the reduction of head loss due to fine sediment is not well defined. The designer should keep this in mind even if a small amount of fine sand with sizes less than 0.15 mm (this is only a rough criterion and by no means established), is present in the flow.

Criterion for Distinguishing between Homogeneous Flow and Heterogeneous Flow.—As mean flow velocity increases, the vertical sediment concentration distribution gradually approaches uniformity. It is difficult, therefore, to define a unique criterion to distinguish between homogeneous and heterogeneous flows. The following is a list of experimentally-determined criteria.

Durand (1953) states that turbulent water flow with clay, fine ash, and very fine powdered coal of less than 20μ–30μ in size (1μ = 0.001 mm) can be treated as homogeneous flow. He also indicates that silt, with sizes in the range of 25μ–50μ in water flow, can be treated as an intermediary mixture (between homogeneous and heterogeneous).

Newitt, et al. (1955) defined a critical velocity, V_H, to divide homogeneous and heterogeneous flows. They gave

$$V_H^3 = 1{,}800\, gDw \qquad (2.289)$$

in which w = the fall velocity of the sediment particle. Eq. 2.289 was derived on the basis of experimental evidence indicating that for the two sands tested, $(i_m - i)/C_v i$ approached 0.6 $(s - 1)$ in homogeneous flow.

Graf and Acaroglu (1967) state that if the transporting velocity is considerably larger than the settling velocity, the suspension is assumed to become homogeneous or pseudo-homogeneous. They do not define "considerably larger."

Spells (1955) defines "standard velocity" as the mean velocity of flow at which the pressure gradients for the slurry mixture and its equivalent true fluid become identical. The "equivalent true fluid" is defined as a fluid with a density equal to that of the slurry and a viscosity equal to that of water (pure fluid). Based on data collected by Blatch (1906), Howard (1939), and Worster (1952), Spells finds that

$$(\text{standard velocity})^{1.225} = 0.074\, g d_{85} \left(\frac{D \rho_m}{\mu_m}\right)^{0.775} (s - 1) \qquad (2.290)$$

in which d_{85} = the particle diameter such that 85% by weight of particles are smaller than d_{85}; μ_m = the dynamic viscosity of the slurry taken to be identical with the viscosity of the medium; ρ_m = the density of the mixture; and all dimensions are for consistent, absolute units. Spells also indicated that evidence suggested that Eq. 2.290 may be of little value when the particles are appreciably greater than those tested (820μ). Spells does not state that the "standard velocity" is the division between homogeneous and heterogeneous flows, but his criterion should be very close to it.

Comments.—There does not appear to be much uniformity in these criteria that separate homogeneous flows from heterogeneous flows. Let us, therefore consider a relationship for the concentration of the suspended particles for two-dimensional uniform free-surface flow [see Vanoni (1946)]

$$\frac{C}{C_a} = \left(\frac{d - y}{y}\,\frac{a}{d - a}\right)^z \qquad (2.77)$$

in which C and C_a = the concentrations at distances y and a, respectively, above the bed; d = the water depth, and

$$z = \frac{w}{kU_*} \qquad (2.291)$$

in which U_* = the shear velocity and k = Von Karman's universal constant.

In Eq. 2.77 homogeneous flow can be identified when $C = C_a$ for any y. Thus the value of z must approach zero. This means that either w must approach zero, or U_* must approach infinity. Although neither is strictly possible, various investigators conclude from their experiences that some flows can, nevertheless, be treated approximately as a homogeneous flow.

Let it be assumed that Eqs. 2.77 and 2.291 are applicable to pipe flow although such applicability has not been established. Since the main concern here is with nearly uniform sediment concentration distribution, the assumption of transferring results from two-dimensional flow to three-dimensional pipe flow should be reasonably valid. It would be interesting, therefore, to calculate the lower limit for a homogeneous suspension. Newitt, et al. (1955) found that when the average flow velocity was about 10 fps in a 1-in. smooth pipe, Sand B (with an average fall velocity of 0.032 fps) was conveyed as a homogeneous suspension. The hydraulic gradient for the flow was about 0.35 ft of water per foot. Based on these values

$$z = \frac{w}{k\sqrt{gi_m\frac{D}{4}}} = \frac{0.032}{0.35\sqrt{32 \times 0.35 \times \frac{1}{48}}} = 0.19 \quad (2.292)$$

The aforementioned point suggests that if an expression such as Eq. 2.77 can be derived for pipe flow, then a definitive criterion might be established that can be used to separate homogeneous from heterogeneous flows. The distinction between flows is important; for instance, if Eq. 2.287 is used for homogeneous flows, the calculated head loss may be considerably larger than that which is determined by Eq. 2.286 for the same velocity.

Recommendation.—Since the criterion between homogeneous and heterogeneous flow is not distinct, it is difficult to present a recommendation based on previous work. Engineering judgment for each application must be used to decide whether the flow will be homogeneous or heterogeneous using experience and any or all the results just considered.

Heterogeneous Flow.—Heterogeneous flow is the most important regime of sediment transport in pipes because it is normally the most economical regime in which to operate, i.e., it gives the most amount of sediment transport per unit of energy expended.

Due to its importance, great research effort has been concentrated in this regime, but unfortunately no generally accepted criterion to describe head loss under various flow conditions within this regime has yet been established as the following analysis will show.

Wilson (1942) divided the slope of piezometric head line into two parts. The first part is expressed by $f[V^2/(2gD)]$ representing the head loss due to pure liquid. The second portion is expressed as $K(w/V)C_v$ (in which K is a constant) representing the excess energy required to support the particles in suspension.

Einstein (see Wilson, 1942) points out that the two energy terms in Wilson's equation were added without regard to their interdependence. He further stated that the same energy may appear in the first term as mechanical energy being transformed into turbulence and again in the second term when this turbulence is dissipated in lifting the particles.

Wilson (1942) later modified his equation to

$$\frac{h_L}{L} = f\frac{V^2}{2gD} + KC_v\frac{w}{V}(1 + \delta) \quad (2.293)$$

in which δ can be either positive or negative to satisfy the requirement that under certain conditions the presence of solids may reduce head loss below that of pure liquid flow. This of course conflicts with his original assumption that excess energy is required to transport suspended sediment particles.

Durand (1953) and his co-workers at Sogreah Laboratory, Grenoble, France, contributed greatly to the understanding of sediment transport in pipes. They conducted 310 tests (reported in 1952) with sediment sizes ranging from 0.2 mm–25 mm, sediment concentrations ranging from 2%–23% by volume, and pipes ranging in size from 1.5 in.–28 in. in diameter. They conclude from their experiments that for heterogeneous flow

$$\phi_D = \frac{i_m - i}{iC_v} = K' \left(\frac{\sqrt{gD}}{V}\right)^3 \left(\frac{1}{\sqrt{C_D}}\right)^{1.5} \qquad (2.294)$$

in which C_D = the drag coefficient of the particle; and K' = a constant of proportionality. This functional relationship together with the data points are shown in Fig. 2.129.

Durand and Condolios in a later paper (1952) used the two terms $w/\sqrt{gd_s}$ and $1/\sqrt{C_D}$ interchangeably in their figures. According to Condolios and Chapus (1963a)

$$\frac{w}{\sqrt{gd_s}} = \sqrt{\frac{4}{3}(s - 1)\frac{\psi'}{C_D}} \qquad (2.295)$$

in which ψ' = particle form coefficient (ψ' = 1 for spheres) and thus for spherical sand particles, s = 2.65 and

$$\frac{w}{\sqrt{gd_s}} = 1.48\frac{1}{\sqrt{C_D}}$$

Therefore, for sand, $w/\sqrt{gd_s}$ is 50% greater than $1/\sqrt{C_D}$. For lightweight materials, the difference between these terms is smaller. Because of the great importance of Durand's work, these discrepancies are analyzed by many investigators (Zandi and Govatos, 1967; Homayounfar, 1965; Babcock, 1971).

Utilizing data collected by Turtle (1952) for coal and MnO_2 in water, Worster (1952) states that Eq. 2.294 is applicable to sand and gravel, but for other sediment densities the previous relationship should be modified to

$$\phi_D = K'\left[\frac{\dfrac{gD(s-1)}{1.65}}{V^2}\;\frac{w}{\dfrac{\sqrt{gd_s(s-1)}}{1.65}}\right]^{1.5} \qquad (2.296)$$

Worster also developed a new relation by assuming that in the equilibrium

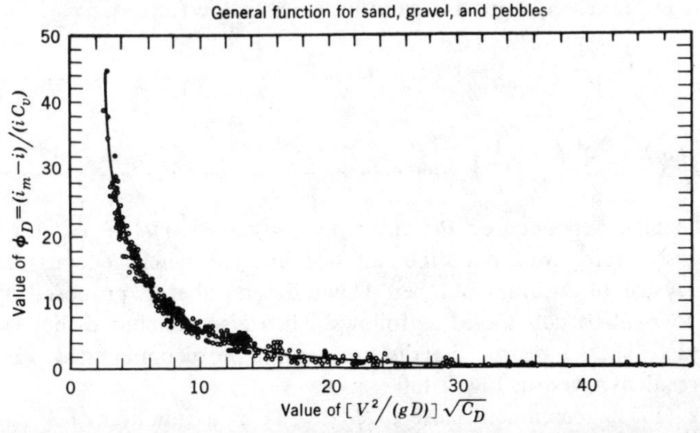

FIG. 2.129.—Head Losses in Pipes with Nondeposit Flow Regimes; Durand (1953)

condition, P, the probability of any one particle in the bed surface moving in one second is a function of

$$P = \frac{U_*}{l}\psi'' \qquad (2.297)$$

in which l = a characteristic length; U_* = the shear velocity; and ψ'' = a function of $U_*^2/[gd_s(s-1)]$.

Substituting Chezy's friction factor for $V\sqrt{g}/U_*$ and from experimental evidence, Worster obtained

$$\frac{C_{hw} - C_{hm}}{\sqrt{g}} = 4\left(\frac{w}{V}\right)^{0.173} \frac{C_{hw}}{\sqrt{g}}\left[\frac{C_v gD(s-1)}{V^2}\right]^{0.413} - 4 \qquad (2.298)$$

for $(C_{hw} - C_{hm})/\sqrt{g} < 13$. Here C_{hw} and C_{hm} are Chezy's friction factors for clear water and mixture, respectively. A major difference between Eqs. 2.294 and 2.298 is that the former includes particle size while the latter is nearly independent of it for particle size larger than, e.g., 1.5 mm.

Although Eq. 2.298 does not seem to apply strictly to heterogeneous flow where sediments are all moved in suspension and there is no bed surface, Turtle's data appear to justify his conclusion as given in Eq. 2.298. Actually, $U_*^2/[gd_s(s-1)]$ is probably a significant parameter for both, with or without bed load motion. A minor disadvantage of using C_{hw} is that this value varies with pipe size also.

Newitt, et al. (1955) used an energy approach very similar to that of Wilson (1942) and developed the following expression to describe the head loss for heterogeneous pipe flow

$$\frac{i_m - i}{C_v i} = 1{,}100\,\frac{gD}{V^2}\,\frac{w}{V}\,(s-1) \qquad (2.299)$$

They also point out that both Eq. 2.299 and Durand's Eq. 2.294 indicate that $(i_m - i)/C_v i$ is inversely proportional to the cube of the mean velocity, V; however, the two equations differ in the dependence on pipe diameter D. Since Eq. 2.299 was determinered with data from a 1-in. pipe, the effect of pipe diameter on head loss could not be tested. They also show that their data agree with Durand's equation almost as well as their own.

Zandi and Govatos (1967) collected all available data totaling 1,452 points and found that Durand's equation for sand in water in the form of

$$\phi_D = 176\left(\frac{\sqrt{gD}}{V}\right)^3\left(\frac{1}{\sqrt{C_D}}\right)^{1.5} \qquad (2.300)$$

$$\text{or } \phi_D = 81\left(\frac{\sqrt{gD}}{V}\right)^3\left(\frac{s-1}{\sqrt{C_D}}\right)^{1.5} \qquad (2.301)$$

the use of which depended on the value of $\psi = gD(s-1)/(V^2\sqrt{C_D})$, predicted the head losses fairly well once the saltation and the heterogeneous data were separated by use of an index number. This index number as proposed by Zandi and Govatos will be considered as follows. They also concluded that Eq. 2.298 was invalid to such a degree that its use cannot be recommended. Zandi and Govatos' result is given in Fig. 2.130.

Babcock (1971) recommends use of $\phi_D(s-1)$ as a function of V^2/gD until other parameters are proven to be superior. Flow in pipes with diameters larger than 1 in. behave only qualitatively the same as flow in a 1-in. pipe, but quantitatively the parameters $\phi_D(s - 1)$ and V^2/gD do not universally relate data for pipes of different sizes.

Comments and Recommendations.—Most investigators have attempted to determine the excess head loss, for flows transporting suspended sediment, over those with clear liquid. The relationship must clearly include the effects of sediment concentration, pipe roughness, and flow properties. The recommendation for analyzing heterogeneous flow is to use Eq. 2.294 by Durand and his co-workers because this equation seems to give the best agreement with observations. The value of the constant, K', in the equation is 176.

Criterion for Distinguishing between Heterogeneous Flow and Flow with Bed Load.—At some mean velocity less than that for heterogeneous flow, some of the suspended particles begin to settle and are moved along the bottom pipe boundary by the flow as bed load. At some velocity still less, an identifiable bed surface may be formed. The velocity at which the identifiable bed surface forms is not clearly distinguishable from the velocity at which sediment begins to settle out. For practical purposes, the need is to identify the latter. Depending on the prevailing flow conditions, the top of the deposited sediment bed layer may be either flat or dune shaped. If the bed surface is flat, the boundary may be treated approximately as a plane rough wall, and the increase of head loss will be relatively small. If dunes form on the surface of the bed layer, the boundary is considerably rougher hydraulically, and the increase of head loss will be larger than for the flat bed.

There is also a question of the uniqueness of this characteristic situation. The velocity which distinguishes heterogeneous flow from flow with bed load motion, sometimes referred to as critical velocity, is not unique. It depends on whether the velocity is decreasing or increasing in passing from one regime to the other. The critical velocity is less for decreasing flow velocity than for increasing velocity,

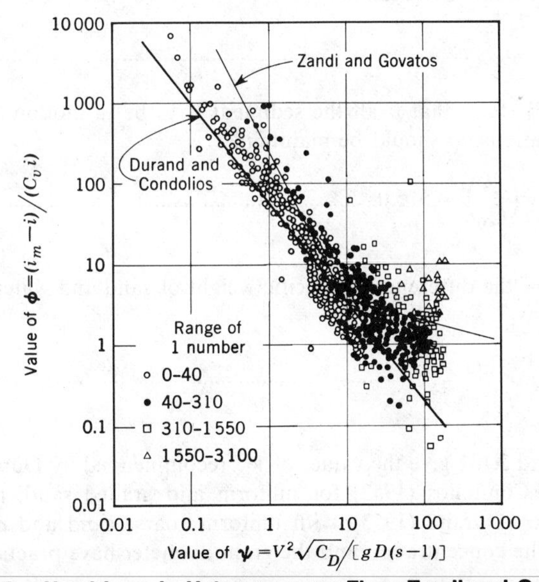

FIG. 2.130.—Head Loss in Heterogeneous Flow; Zandi and Govatos (1967)

thus strictly speaking, it is dependent upon past flow history. However, for practical purposes, one may assume that there is a single value for critical velocity.

Durand (1953) plotted the head loss (or energy gradient) as a function of the mean flow velocity and constant solid concentration as shown schematically in Fig. 2.128. A line drawn in dots and dashes delimits the zones of the regimes with and without deposits on the pipe bottom. The velocity corresponding to the passage from one regime to the other has been named by Durand as "limit deposit velocity." He also notes that it appears to correspond fairly accurately to the minimum of the head loss curves, and therefore to the most favorable operating conditions from the economical viewpoint. The value of this critical velocity is thus of great practical interest.

Durand's definition of the limiting deposit velocity has been used by many researchers and also will be used herein. However, from practical engineering design viewpoint, the most useful velocity would be the velocity associated with $\partial i/\partial V = 0$ of the function $i = \phi(Q_s, V)$ as pointed out by Carstens (1969). Engineers should investigate the minimum energy gradient for the transport of a certain Q_s and not for the transport of a certain solid concentration, C_v.

Yufin (1949) conducted experiments in pipelines ranging in diameter from 0.4 in.–1.5 ft with particle diameters from 0.25 mm–7.36 mm and concentrations from 0%–30% by volume. He found that the minimum head loss identifiable with a "limiting deposit velocity" could be expressed by

$$V_L = 14.23 d_s^{0.65} D^{0.54} \exp\{1.36[1 + C_v(s - 1)]^{0.5} d_s^{-0.13}\} \quad \text{(2.302)}$$

Wilson (1942) gives a criterion for the deposition of solids from suspension as

$$\frac{w}{\left(\frac{gD}{4} i_m\right)^{0.5}} = 1 \quad \text{(2.303)}$$

Eq. 2.303 can also be written as

$$\frac{w}{U_*} = 1 \quad \text{(2.304)}$$

Craven (1953) states that if all the sediment is to be in motion at all times, the following relationship should be maintained:

$$\frac{Q}{D^{2.5}\sqrt{\frac{\Delta\gamma}{\rho}}}\left(\frac{Q}{Q_s}\right)^{1/3} > 5.0 \quad \text{(2.305)}$$

in which $\Delta\gamma$ = the difference in specific weight of sand and water. Eq. 2.305 can be rewritten as

$$\frac{V_L}{\sqrt{Dg\frac{\Delta\rho}{\rho}}} > 6.4\, C_v^{1/3} \quad \text{(2.306)}$$

Figs. 2.131 and 2.132 give the values of V_L recommended by Durand (1953), and Durand and Condolios (1952) for uniform and graded sand, respectively. As pointed out by Durand (1953), with uniform coarse sand and pebbles (greater than 1 mm) the concentration and the grain diameter have practically no further

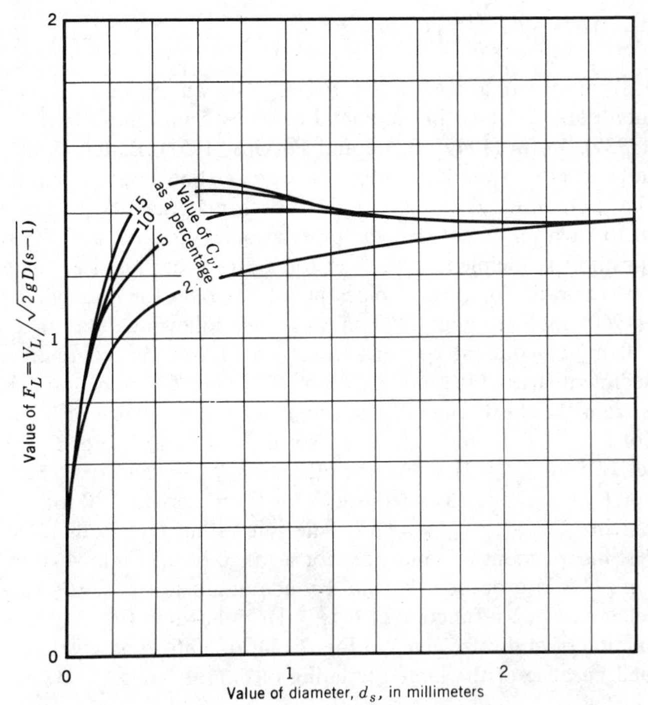

FIG. 2.131.—Limit Deposit Velocity for Uniform Material; Durand (1953)

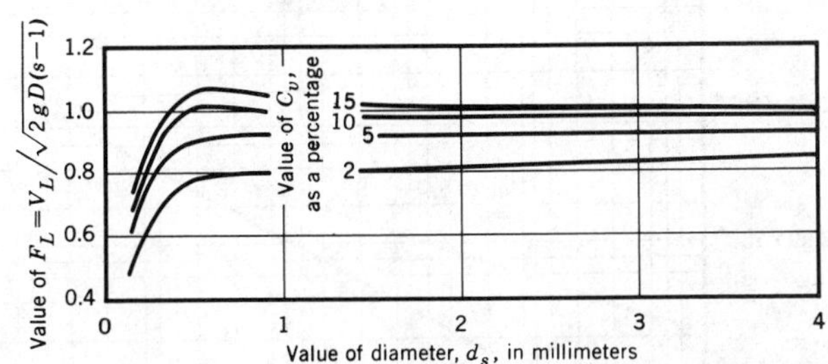

FIG. 2.132.—Limit Deposit Velocity for Nonuniform Material; Durand and Condolios (1952)

influence on the limit deposit velocity. However, this is certainly not true for nonuniform material as shown in Fig. 2.132.

Newitt, et al. (1955) equate their expression of head loss for flow with moving bed and heterogeneous flows to obtain

$$V_B = 17w \quad \text{(2.307)}$$

in which V_B = the mean flow velocity which separates the two regimes. They also state that Durand's limiting deposit velocity is the velocity below which a stationary bed exists.

Spells (1955) determined the following relationship from data collected by Howard (1939), Yufin (1949), Smith and Carruthers (1955), and Settle and Parkins (1951):

$$V_L^{1.225} = 0.025\, gd \left(\frac{D\rho_m}{\mu_m}\right)^{0.775} (s - 1) \quad \text{.......} \quad (2.308)$$

Zenz (1958) proposed a correlation for V_L, shown in Fig. 2.133, based on experimental data collected in horizontal pipes by Smith and Carruthers (1955), Howard (1939), Yufin (1949), Settle and Parkins (1951), Blatch (1906), Gregory (1966), Ambrose (1953) and Zenz (1958). The correlation assumes that the choking velocity that is defined as the velocity at which material being transported has settled out to such an extent that the pipe becomes clogged, and the velocity at which deposition is incipient, are equal for uniform particle size material. This assumption is correct for vertical pipes but is incorrect for horizontal pipes.

Gibert (1960) analyzed about 250 points for the following data range: Pipe size 40 mm–150 mm; sediment concentrations up to 20%, by volume; reasonably uniform sediment sizes of 0.2 mm, 0.37 mm, 0.89 mm, 2.05 mm, and 4.20 mm. He concludes from these experimental data that for sediment size of 0.2 mm (sand) the value of $V_L/\sqrt{gD}$ increases from 1.3 for C_v = 2.5% to 1.7 for C_v = 10%; and approaches 1.75 for $C_v >$ 10%. For the other four larger sand sizes, he found that the value of $V_L/\sqrt{gD}$ increases from 1.75 for C_v = 2.5% to 2.20 for C_v = 12.5%; and approaches 2.25 for $C_v >$ 12.5%. The interesting fact is that the value of $V_L/\sqrt{gD}$ is independent of sand sizes for a range of 0.37 mm–4.20 mm.

Hughmark (1961) proposed that the Froude number, $V_L/\sqrt{gD}$, based on the limiting velocity, V_L, is a function of $C_v(s-1)F_d$ shown in Fig. 2.134(*a*), in which F_d is a function of grain size given in Fig. 2.134(*b*). Unfortunately, his main curve is ill-defined because of the large scattering of data.

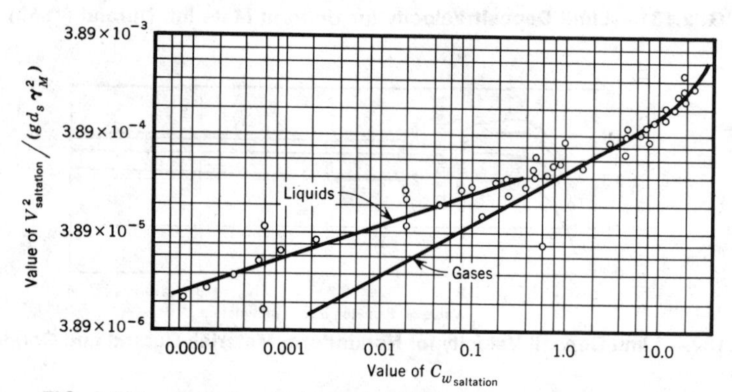

FIG. 2.133.—Saltation Velocity for Uniform Material; Zenz (1958)

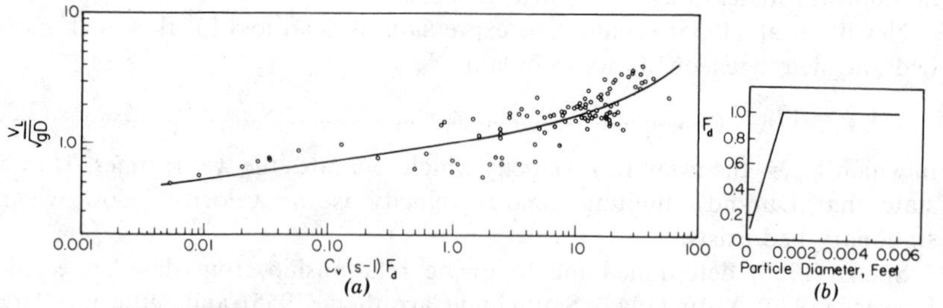

FIG. 2.134.—Limit Deposit Velocity; Hughmark (1961)

Sinclair (1962) presents data on the limit deposit velocity of heterogeneous suspensions of iron-kerosene, sand-water, and coal-water flowing in 1.0-in., 0.75-in., and 0.5-in. pipes at solid concentrations up to 20% by volume as shown in Fig. 2.135. It is shown that the limit deposit velocity depends on the particle-fluid system, particle-fluid density ratio, particle diameter, pipe diameter, and the transport concentration of the solids. The limit deposit velocity for any system is a maximum at some value of sediment concentration between 5% and 20%. As shown in Fig. 2.136, the maximum value is well correlated by a function of the form

$$\frac{V_{L\max}^2}{g d_{85}(s - 1)^{0.8}} = f\left(\frac{d_{85}}{D}\right) \qquad (2.309)$$

in which d_{85} is the sediment size of which 85% is finer; and f denotes "function of." For the sediment particles wholly immersed in a region where viscous forces predominate, $f(d_{85}/D) = 650$. When sediments are exposed in the turbulent region

$$f\left(\frac{d_{85}}{D}\right) = 0.19\left(\frac{d_{85}}{D}\right)^{-2} \qquad (2.310)$$

The normalized limit deposit velocity, $V_L/V_{L\max}$, has a unique relationship with concentration for each solid-liquid system as shown in Fig. 2.135. Sinclair also compared experimental values of V_L with those predicted by the correlations presented by Durand (1953), Spells (1955), Newitt, et al. (1955), Yufin (1949) and others, and found that only the correlation of Durand (1953) was acceptable, although the values were slightly overestimated.

Thomas (1962) divides flow into several regions. In his Flow Region, I, particles are smaller than the thickness of the viscous sublayer along the pipe wall and are transported predominantly in suspension. These fine particles also settle according to Stoke's law. With nonflocculated glass beads having a mean diameter of 0.078 mm in a 1-in. pipe, he found that the critical friction velocity, U_{*c} (defined as the friction velocity required to prevent the accumulation of a layer of stationary or sliding particles on the bottom of a horizontal conduit), can be expressed by

$$\frac{w}{U_{*c}} = 0.010\left(\frac{d_s U_{*c}}{\nu}\right)^{2.71} \qquad (2.311)$$

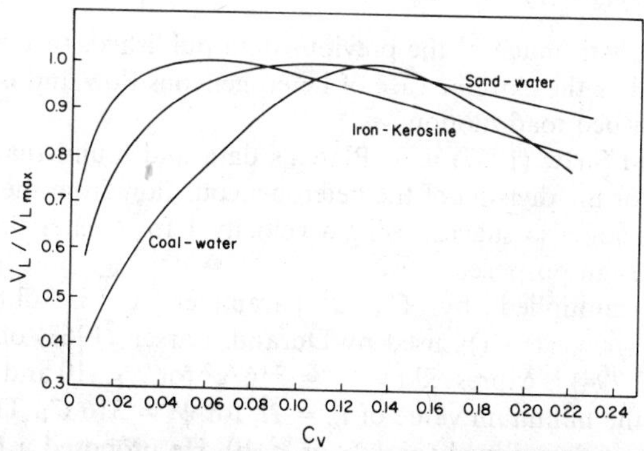

FIG. 2.135.—$V_L/V_{L\max}$ for Three Solids-Liquid Suspensions; Sinclair (1962)

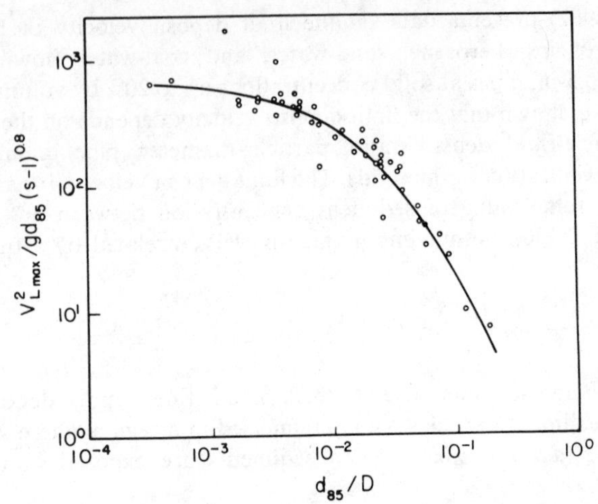

FIG. 2.136.—Proposed Correlation for Maximum Limit Deposit-Velocity; Sinclair (1962)

For larger particles (0.19 mm–38.1 mm), Thomas developed the following two relationships for U_{*c} using his data and those of previous investigators

$$\frac{w}{U_{*c}} = 4.90\left(\frac{d_s U_{*c}}{\nu}\right)\left(\frac{\nu}{DU_{*c}}\right)^{0.60}(s-1)^{0.23} \qquad (2.312)$$

and

$$\frac{U_{*cm}}{U_{*c}} = 1 + 2.8\left(\frac{w}{U_{*c}}\right)^{1/3} C_v^{1/2} \qquad (2.313)$$

in which U_{*c} = the critical frictional velocity in feet per second for infinite dilution (no suspension); and U_{*cm} = the critical friction velocity applicable to the concentration, C_v.

Zandi and Govatos (1967) used a dimensionless number, N_I, (labeled Index of the flow regime) for the upper limit of the transition between saltation to heterogeneous flow and found that a value

$$N_I = \frac{V^2\sqrt{C_D}}{C_v Dg(s-1)} = 40 \qquad (2.314)$$

was consistent with much of the previous data published. In their view, saltation may be treated as the extreme case of heterogeneous flow and is identifiable here with flow with bed load motion.

Babcock and Shaw (1967) used Blatch's data and found that the value of N_I should be 10 for the division of the heterogeneous flow from the flow regime with moving bed. They also suggest using a velocity 1 fps greater than the minimum velocity for design purposes.

If N_I is multiplied by C_v, a parameter, ψ, is obtained which is $\psi = (V^2/gD)[\sqrt{C_D}/(s-1)]$, used by Durand. Larsen (1968) observed that if ϕ_D (given in Eq. 2.294) is expressed by $\phi_D = 316/\psi^2$ for $\psi < 10$ and $\phi_D = 6.8/\psi^{1/3}$ for $\psi > 10$, then the minimum value of $i_m = 2i$, for $\psi^2 = 316\ C_v$. Then the minimum velocities can be determined for $\psi >$ or < 10. He proposed a modified index of $N_J = \psi^2/C_v$ and suggested that values of $N_J < 316$ identified flows with a moving bed (saltation in his term).

Wilson and Brebner (1971) suggested that the selection of the design limit deposit velocity be made from a ϕ_D versus V graph in a manner suggested in Fig. 2.137.

Summary of Previous Results.—The foregoing criteria for identifying the heterogeneous flow regime from flow with bed load should be examined to establish points of agreement as well as disagreement to form a reasonable guideline for design purposes.

The most important variables governing the criterion between heterogeneous flow and flow with bed load motion are flow velocity, fall velocity or size of sediment, pipe diameter, sediment concentration, sediment density, and fluid properties.

A limit deposit velocity, V_L, which is identified as the lower limit of the transition between heterogeneous flow and flow with moving bed, and which corresponds approximately to minimum head loss for a given sediment concentration, has been used to separate heterogeneous flow and flow with bed load. This limit velocity depends upon the sediment size in the following manner approximately

$$V_L \propto (d_{50})^{k_1} \qquad (2.315)$$

in which d_{50} = the median particle size. Thus: (1) $0.5 < k_1 < 1$ for $d_{50} > 0.5$ mm; (2) $0 < k_1 < 0.5$ for 0.5 mm $< d_{50} <$ 1.5 mm; and (3) $k_1 = 0$ for $d_{50} > 1.5$ mm.

Gibert (1960) concludes that the value of $V_L/\sqrt{gd}$ is independent of sand sizes for a range of sediment size between 0.37 mm–4.20 mm. However, without having his complete data, it is rather difficult to comment on his result. For instance, he might have used larger pipe sizes for larger sediment sizes.

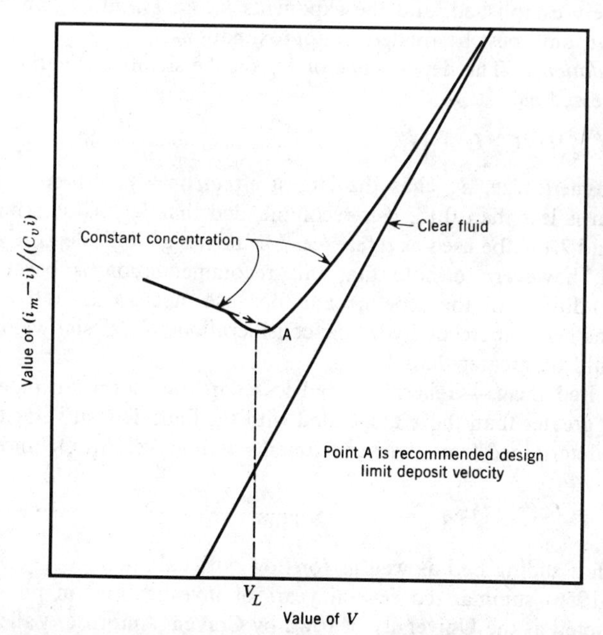

FIG. 2.137.—Suggested Method to Select Design Limit Deposit Velocity; Wilson and Brebner (1971), Qualitative Only

Near the upper limit of the transition a critical shear velocity is used to separate the heterogeneous flow from flow with bed load and can approximately be expressed as

$$U_{*c} \propto (d_{50})^{k_2} \qquad (2.316)$$

in which (1) $k_2 = -0.20$ for $d_{50} < 0.1$ mm; (2) $k = -0.33$ for $d_{50} > 1.5$ mm; and (3) $-0.33 < k_2 < -0.20$ for 0.1 mm $< d_{50} <$ 1.5 mm.

The effect of pipe size on V_L may be expressed as

$$V_L \propto D^{k_3} \qquad (2.317)$$

in which $k_3 \cong 0.5$. The value of k_3 may vary slightly (by about $\pm$ 0.1), the negative sign being associated with $d/D < 0.01$ and the positive sign with $d/D > 0.1$.

Although it might be thought that the critical shear velocity, U_{*c}, would be independent of pipe diameter, experimental results indicate that $U_{*c} \propto D^{0.13}$.

The limiting deposit velocity has been found to increase with sediment concentration but at a decreasing rate, so that in an expression

$$V_L \propto C_v^{k_4} \qquad (2.318)$$

the value of k_4 varies approximately from about 0.3–0.1 for sediment concentrations from 5%–15% respectively.

The effect of sediment and fluid densities on V_L may be expressed approximately as

$$V_L \propto (s - 1)^{k_5} \qquad (2.319)$$

with $k_5 \approx 0.5$. It must be strongly stressed here that Eqs. 2.315 through Eq. 2.319 are not entirely established, and the exponents k_1, k_2, k_3, and k_4 under various conditions can only best be treated as approximations.

Recommendations.—The dependence of V_L on the significant variable may be roughly expressed as

$$V_L \propto (d_{50})^{k_1} D^{k_3} C_v^{k_4} (s - 1)^{k_5} \qquad (2.320)$$

in which sediment size, d_{50}, has the largest effect on V_L. When the sediment concentration is less than 0.15, it is recommended that V_L, as determined from Figs. 2.131 and 2.136, be used as references for the design limit deposit velocity. It is important, however, to note that this recommendation is based only on hydraulic conditions in the pipeline and because mechanical and operational considerations may supercede hydraulic considerations, the design velocity in the pipeline should be greater than V_L.

Flow with Bed Load.—Generally, head losses in transportation pipelines with bed load are greater than those associated with the limit deposit velocity. Newitt, et al. (1955) determined from their experiments with gravel, MnO_2, and coal that

$$\frac{i_m - i}{C_v i} = 66(s - 1)\frac{D}{V^2} \qquad (2.321)$$

for flow with a sliding bed as well as for flow with saltation.

Laursen (1956) summarized several years of investigation on pipe transportation conducted at the University of Iowa by Craven, Ambrose, Vallentine, and Ince, the result of which is reproduced as Fig. 2.138. Most of the data were collected from a 6-in. Lucite pipe, although a limited number of Craven's

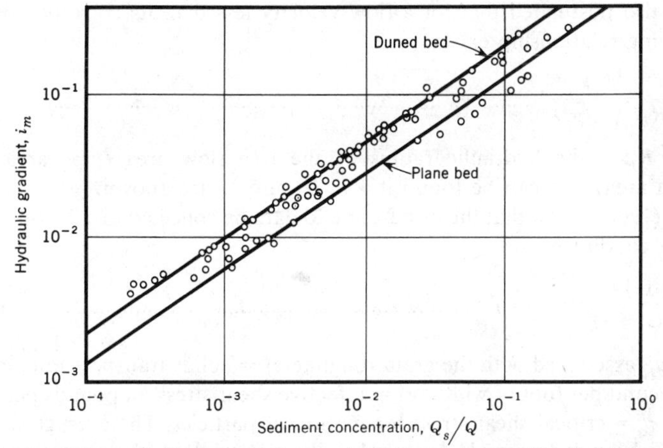

FIG. 2.138.—Effect of Sediment Concentration on Hydraulic Gradient (Uniform Sands); Laursen (1956)

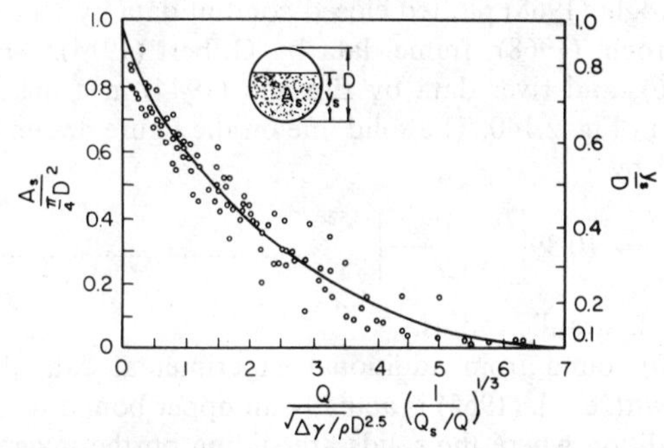

FIG. 2.139.—Blockage due to Sediment Deposition for Uniform Sand, Graded Sands, and Slag; Laursen (1956)

experiments were conducted with a 2-in. Lucite pipe. Three uniform sands with d_{50} of 0.25 mm, 0.58 mm, and 1.62 mm; two graded sands with sizes between 0.4 mm–2.1 mm and between 0.1 mm–1.5 mm, respectively, and a slag (density 2.80) with sizes between 0.3 mm–3.0 mm, were used. According to the studies, head loss is only a function of sediment concentration and is independent of flow velocity and pipe and sediment size, as noted in Fig. 2.138. Fig. 2.139 gives the blockage in the pipeline due to sediment deposition and is a useful chart for design engineers.

Ismail (1952) conducted experiments in a rectangular conduit (0.27 m wide by 0.076 m deep) with two different well-sorted sands of median sizes of 0.091 mm and 0.147 mm. His results indicated that the universal constant of turbulent exchange, k, decreased with the increase of the suspended material. He has also collected total sand load under various conditions but did not attempt to correlate load with flow conditions.

Gibert (1960) determined from additional data that Durand's (1953) curve as shown in Fig. 2.129 and also expressed as Eq. 2.300 is also applicable to flow with deposit. (Valid only for uniform sand in water in that form.) His data, for flow with deposit, covers (almost uniformly) a range of ϕ_D between 3.8–176.

Gibert also postulated that for a flow velocity less than its corresponding V_L, the following relationship exists:

$$\frac{V}{\sqrt{4gR_h}} = \frac{V_L}{\sqrt{gD}} \quad (2.322)$$

in which R_h is the hydraulic radius of the free flow area (pipe area minus deposition area) and can be found if V_L, D, and V are known.

Wilson (1966) found that the best fit line to data he collected in a 3.69-in. square aluminum conduit is

$$q_s = \frac{0.14}{(s-1)} (\tau' - \tau_c)^{1.5} \quad (2.323)$$

in which q_s (associated with the grain roughness) = solids transport rate, in cubic feet per second per foot of width; τ' = effective shear stress, in pounds per square foot; and τ_c = critical shear stress for motion of particles. The average sand size used was about 0.8 mm. However, he also found that this equation is not applicable to his results of tests with nylon particles (with specific gravity of 1.138 and sphere diameter of 3.88 mm).

Graf and Acaroglu (1968) plotted closed conduit data by Ismail (1952), Wilson (1966), and Acaroglu (1968); flume data by Gilbert (1914), Ansley (1963), and Guy, et al. (1966); and river data by Einstein (1944) in a massive plot of data points, as shown in Fig. 2.140. The solid line on the figure drawn through the data can be expressed by

$$\frac{C_v V r}{\sqrt{(s-1)gd_s^3}} = 10.39\left[\frac{(s-1)d_s}{i_m r}\right]^{-3.52} \quad (2.324)$$

in which r = the hydraulic radius.

Babcock (1970) found from additional experimental data that Eq. 2.321 as presented by Newitt, et al. (1955) is actually an upper bound to the head loss and represents a condition where the solids are sliding on the invert of the pipe.

Recommendations.—The variation of frictional resistance of a deposited sediment bed due to ripples, dunes, and so forth, has been extensively investigated in open channel flow. Nearly every investigator of the problem has developed his own formula. Unfortunately, no specific formula has generally been accepted. A formula for use in pipeline transportation is therefore not suggested, although experienced specialists may be able to recommend some values for friction coefficients that may be used in specific cases. The curve presented by Graf and Acaroglu (1968) in Fig. 2.140 may be used as a guide.

The amount of sediment deposition on the bottom of the pipe is certainly a function of the amount of sediment inflow. The same as for open channel flow, flow in a pipe will attempt (though sometimes may not be able) to adjust itself (by increasing deposit, decreasing flow area, and thus increasing flow velocity) so that the sediment inflow and outflow is balanced. Eq. 2.322 must be limited to a particular combination of sediment inflow and other flow conditions.

Flow with Stationary Bed and without Sediment Movement.—When flow velocity decreases below the corresponding incipient motion condition, no sediment movement occurs.

From a preliminary experimental study, Shen and Wang (1970) found that Shields' criterion for incipient sediment motion as obtained in open channel

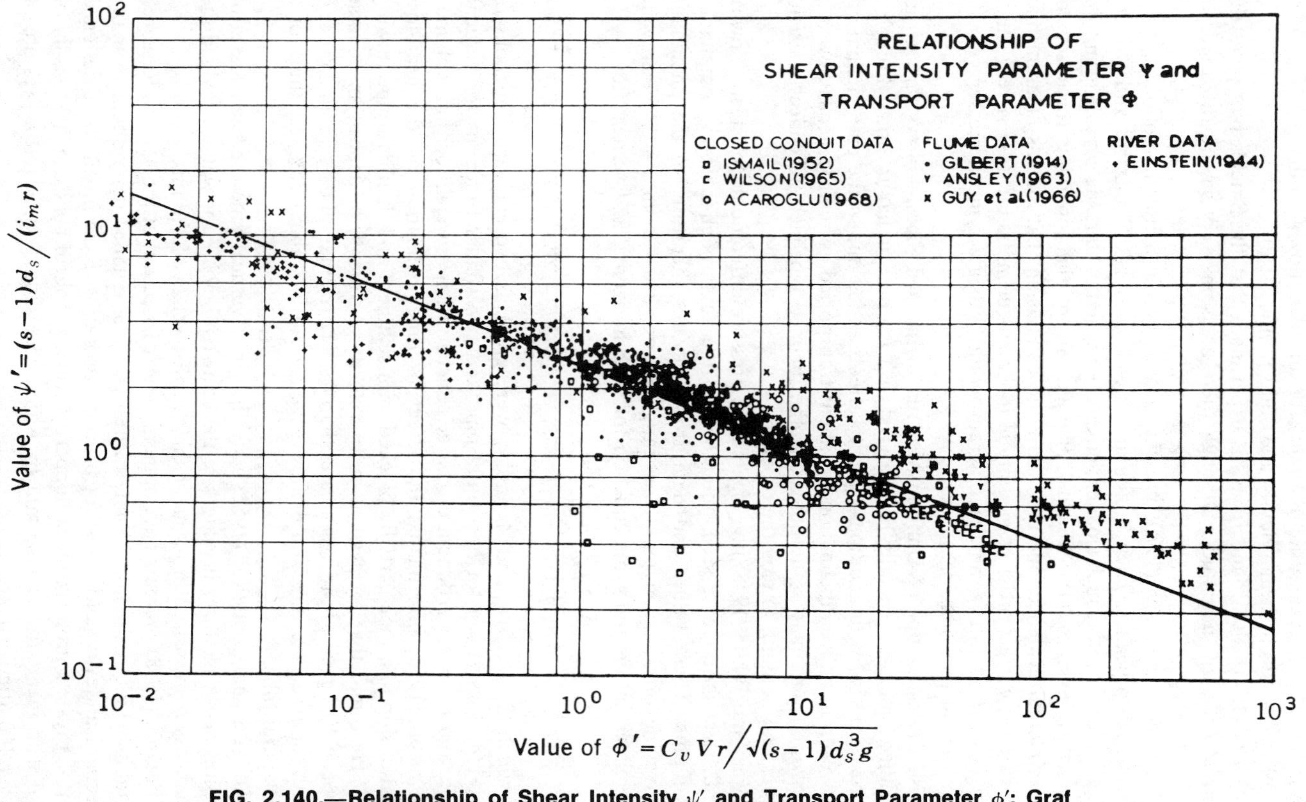

FIG. 2.140.—Relationship of Shear Intensity ψ' and Transport Parameter ϕ'; Graf and Acaroglu (1968)

turbulent flow is applicable to a rectangular closed conduit with turbulent flow. However for laminar flow of glycerine, a much greater Shields number is required for incipient sediment motion than that for turbulent flow with the same Reynolds number (based on shear velocity and solid particle size).

Wicks (1971) also found from his experiment that the Shields plot gives the correct order of magnitude, although there are some very significant trends of deviation.

Lazarus and Kilner (1970) present curves for the incipient motion of solid capsules in pipelines. Their minimum ratio of d/D is 0.5. Lazarus and Kilner's data indicate that a much smaller velocity is needed to start the capsule in motion than that given by the Shields' diagram. From the foregoing three sets of experimental evidence, one may draw the conclusion that Shields diagram may provide the correct order of magnitude for incipient motion if d/D is very small. For large d/D ratio's, Shields diagram would indicate a much greater velocity requirement than necessary.

74. Transportation of Uniform Material in Inclined Pipes.—Flows through an inclined pipe can generally be analyzed in a manner similar to flows through a horizontal pipe with additional gravity effects added due to the pipe inclination. However, at a given inclination, sediment transported as heterogeneous flow in a horizontal pipe may settle out along the inclined pipeline and collect near the lower end, tending to clog the pipeline.

Vogt and White (1948) investigated the pressure differentials for steady flow of sand, steel shot, clover seed, and wheat in the air through a 0.5-in. commercial iron pipe. These values and data from previous literature on the pneumatic conveying of wheat in pipe sizes ranging from 2 in.–16 in. in diameter were correlated by the following equation for both horizontal and vertical flow:

$$\frac{i_m}{i} - 1 = A\left(\frac{D}{d_{50}}\right)^2\left(\frac{\rho}{\rho_s} \times \frac{C_w}{R}\right)^{k_6} \qquad (2.325)$$

in which $R = VD/\nu$, and A and k_6 were given as two empirical functions of $\sqrt{1/3(\rho_s - \rho)\rho g d_{50}^3}/\mu$ for horizontal pipe and two similar functions for vertical pipe; and C_w = the sediment discharge concentration in weight per unit volume.

Craven (1953) investigated the effect of tube inclination on the head-loss relationship in a pipe partially blocked with sediment and then in a pipe flowing full. He found that for a given mean velocity and depth of deposit an adverse slope of 10% gave a head loss greater than for a horizontal tube by about 25%, and a 10% slope in the direction of flow gave a head loss about 25% less than for a horizontal tube.

Durand and Condolios (1952) state that in a vertical pipe the head losses of homogeneous mixtures in water are the same as they are for water, provided the head loss is expressed in terms of the specific weight of the mixture.

Fontein (1958) presents development work carried out at the Mining Research Establishment of the Dutch State Mines. He gives detailed descriptions of many mechanical devices of interest for practical operation. Experiments were designed to investigate a prototype lock hopper system, capable of handling 700 tons/hr of coal with mean size of 0.80 mm by water, that should withstand a pressure of about 100 atmospheres.

Newitt, et al. (1961) found from taking flash photographs in their vertical pipe flow experiments that "whilst such (solid) particles are randomly distributed at low velocities they tend to move towards the center of pipe (1 in. and 2 in. brass pipes) at high velocities leaving an annulus of almost clear water at the wall. At solid concentrations greater than about 10% by volume, velocity measurements show that the central core moves with a uniform velocity there being a steep gradient in the clear annulus near the wall. The presence of the annulus of clear water explains why the pressure gradient for these slurries is much the same as that for pure water ... " Testing sands, pebbles, zircon (specific gravity 4.56), Manganese dioxide (specific gravity 4.20) and perspex (specific gravity 1.19) they found that

$$\frac{i_m - i}{C_v i} = 0.0037 \left(\frac{gD}{V}\right)^{1/2} \left(\frac{D}{d}\right) S^2 \qquad (2.326)$$

The effects of heterogeneously suspended solids in a pressure flow of liquid on head losses or energy gradient for various pipe slopes, solid concentrations and flow rates were investigated by Shih (1964). Half-inch wooden balls were presoaked in water until the average specific gravity reached about 1.16. The test pipe was 3 in. in diameter, 25 ft, 8 in. long, with a 54-in. test section. The tests were divided into three series for the three slope angles of the pipe: horizontal, adverse slope of 8.73°, and adverse slope of 17.71°. The solids concentration reached approx 18% by weight for all three slopes. For each pipe slope Shih gives: (1) The energy slope as a function of the Froude number of the pipe, $V^2/2gD$, and the solid concentration; and (2) the Darcy-Weisbach friction factor as a function of solid concentration. Shih found that the effect of pipe slope on the head loss coefficient became more pronounced for higher solid concentrations in the flow mixture.

In the reply to Carsten's discussion, Shih (1967) developed a head loss equation for a homogeneous mixture by balancing all the forces in a uniform pipe flow. He believes that this equation cannot be applied to heterogeneous flow.

Homogeneous suspensions in circular inclined conduits were investigated by Graf and Acaroglu (1967). They tested two uniform sands (sizes $d_{50} = 2.85$ mm and $d_{50} = 1.15$ mm) in a 3-in. ID aluminum pipe and in smooth galvanized steel pipes, respectively. The concentration varied from 0%–7% by volume, and the water velocity varied from 16.1 fps–20.5 fps with the coarse sand (average diameter 2.85 mm). The pipes were set at angles of 11.25°, 22.5°, 45°, and 90°, and horizontally. They concluded that with a friction factor, f, determined with clear-water flow in a pipe, the head loss for homogeneous mixture transport is

$$h_L = L \sin\theta (s - 1) C_v + f \frac{L}{D} \frac{V^2}{2g} [1 + (s - 1) C_v] \qquad (2.327)$$

in which θ = the inclination angle of the pipe; and h_L is in terms of clear water.

Kawashima and Noda (1970) present data indicating that for greater flow velocity (presumably homogeneous mixture) Eq. 2.327 is reasonably good, but for smaller flow velocity (presumably heterogeneous mixture) Eq. 2.327 gives a much too low head loss.

Comments and Recommendations.—It appears that a homogeneous mixture flowing in liquid can be treated as a single fluid, according to the experiments of Durand and Condolios, Graf and Acaroglu. A simple equation as derived by Shih (1967) can be used to calculate the gradient dh_L/dx. This equation is

$$-\frac{dh_L}{dx} = f_m\left(\frac{\rho_m}{\rho}\right)\left(\frac{V^2}{2gD}\right) + (s - 1)C_v \sin\theta \quad \text{(2.328)}$$

in which positive θ indicates an adverse pipe slope; and x = distance along the pipe and increases in the downstream direction.

For a hydrodynamically rough pipe, it is agreed that $f = f_m$ as asserted by Graf and Acaroglu. But for a hydrodynamically smooth pipe in which f_m is a function of the Reynolds number (this depends on the viscosity), it is believed that $f \neq f_m$. For smooth pipes f_m should be determined from a Moody diagram using the kinematic viscosity of the mixture to calculate the Reynolds number.

For a heterogeneous mixture in inclined pipe flow, the information is extremely limited. There has been no general solution presented. At some inclination and sediment concentration, the pipe may tend to become clogged near the lower end of the incline because the sediment tends to settle from the inclined pipe above.

For verticle pipe flow, the flow velocity should be greater than the terminal settling velocity of the solids. Since in turbulent flow, velocity fluctuates with time and space, an average flow velocity should, perhaps, be at least twice the solid's terminal settling velocity.

75. Transportation of Material of Nonuniform Sizes.—Little is known about the transportation of graded sediments having a wide range of particle sizes. Previous investigators have concentrated their efforts on the determination of a characteristic size that will represent the entire range of sediments.

Gilbert (1914) observed from his open channel flow experiments with graded sediments that the total transport rate increases when the size gradation of sediment is broadened, and the mean size remains unchanged. Gilbert speculates that the reason for this increase of transport rate is the reduction of friction losses by the presence of fine suspended material in the flow.

Durand (1953), when experimenting with graded sediments, found qualitatively that fine particles have a significant influence in tending to reduce head losses. He also observed that by introducing a sufficient amount of clay in the system, coarse sands transported as bed load flow in the pipeline could be made to flow as a homogeneous suspension. The flow system could then be analyzed in the homogeneous regime using the mass density and viscosity of the entire mixture. This phenomenon has been of considerable interest to industry for it suggests a possible means of increasing the total sediment transport capability of the pipeline system.

Durand and Condolios (1952) present a design curve, shown in Fig. 2.129, for selecting the "limit deposit velocity" for graded material. The value of V_L in this figure is less than the V_L obtained from Fig. 2.131 for uniform material if all other parameters are constant.

Condolios and Chapus (1963a) concluded from further study of the results that material with a large size distribution is characterized neither by the mean particle size nor by the weighted average of the settling velocity but by the weighted average of the apparent drag coefficient. If the particle size distribution of the sediment can be weighted with factors $p_1, p_2, \ldots p_n$, for as narrow a size range as practicable, the apparent drag coefficient of the graded sediment can be represented by

$$\sqrt{C_D} = p_1\sqrt{C_{D_1}} + p_2\sqrt{C_{D_2}} + \ldots + p_n\sqrt{C_{D_n}} \quad \text{(2.329)}$$

This apparent drag coefficient can then be applied to previous equations applicable for uniform sediment as for instance in Eq. 2.294.

Condolios and Chapus also consider flows where the fines are conveyed as homogeneous suspensions, and the coarse material is transported in the heterogenous regime. This situation is more complex in analysis. They found that the introduction of an additional amount of fines in a coarse mixture reduced head losses significantly more than did the normal amount of fines contained in graded sand. This effect became more accentuated when there were sufficient amounts of fine material added so that the slurry behaved as a non-Newtonian fluid. In these cases, the apparent drag coefficients, C_D, and the apparent viscosity were rather difficult to determine because of the variation that occurred with the flow. Condolios and Chapus found that near the limit deposit velocity, the reduction of head loss due to the presence of fine material amounted to about 45%.

Newitt, et al. (1955) found from their experiments conducted with a mixture of two uniform sizes of sand that when the smaller sizes are transported in suspension, the total transporting capacity of the pipeline is greater than would be expected if considering each material separately. This effect was observed over their entire composition range and, in many cases, there was an optimum composition that gave a maximum capacity of the pipeline. They suggested that the suspension of the fine sediments increased the effective density of the medium so that it maintained the coarse material in suspension more effectively by reducing its rate of settling. The buoyancy effect of the fine fractions was assumed in deriving an equation for the transport of mixed sizes. This equation is

$$\frac{i_m - i}{C_v i} = k_1 (s - 1) \frac{gD}{V^2} \frac{B}{V} \qquad (2.330)$$

in which $B = (1 - X)k_f w_f + X[1 - (1 - X)C_v]k_c w_c$; w_f and w_c = fall velocities for fine and coarse particles, respectively; and X = the fraction by weight of coarse material in the mixture. They state that since the values of k_f and k_c, which account for the effect of suspension on the fall velocities, were unknown and were likely to vary with concentration and composition, quantitative predictions of capacity could not be made.

Howard (1962b) found from his experiments with sand and with a mixture of sand and much finer material in a 2-in. pipe, that there is no noticeable reduction in hydraulic gradient due to the presence of fine slurry with concentrations up to 6% by volume. However, as the concentration of fines increases above 6% by volume, the increase of hydraulic gradient becomes quite noticeable. The median size of fine material was slightly more than 0.01 mm. He also concludes that the addition of fine material reduces the limiting velocity.

Smith (1962) considered three alternate methods to determine an equivalent diameter of graded sediment. These methods considered: (1) The particle whose volume was equal to the average volume of all the particles; (2) the particle whose surface area was equal to the average surface area of all the particles; and (3) the particle whose surface area-volume ratio was equal to the surface area-volume ratio of all the particles. He found that the best approximation was obtained by the third method based on his own data with different mixtures of two different sizes (fine sand and coarse sand with mean diameter of about 0.18 mm and 1.2 mm, respectively), and the head loss equation (Eq. 2.296). His actual approach to obtain the equivalent diameter, d_s^*, is given by

$$d_s^* = \frac{\Sigma p}{\Sigma p d_s} \qquad (2.331)$$

in which p = the fraction by weight of particles of diameter d_s. He also suggested that further work is required on mixed sizes in the range of 0.1 mm–1 mm.

Homayounfar (1965) concludes from his experiments with coal in ¾-in. and 1-in. pipes that the head losses for slurries are less than that of pure water, provided the concentration of fine coal (finer than 0.254 mm diam) is below 5% by weight. The same was also true for slurries containing two sizes of solids, the smaller of which was finer than 0.127 mm diam. He confirmed the existence of an optimum composition resulting in minimum head losses for slurries having mixtures of two coal sizes in water. Homayounfar also concludes that Durand's empirical formula for heterogeneous flow will not consistently predict the head losses for flows consisting of two fractional sizes. However, predictions closer to Durand's equation are possible if the weighted average diameter of the graded sediment is used to obtain the drag coefficient rather than the weighted average drag coefficient, as suggested by Condolios and Chapus (1963b).

Wasp, et al. (1971) present a model to determine the pressure drop in an intermediate flow regime where homogeneous and heterogeneous flows exist simultaneously. First the average sediment discharge concentration (dividing the total sediment discharge by total mixture discharge) is assumed to occur at 0.3 diam from the bottom of the pipe. With the concentration at this level known, concentration $C_{0.8}$ at 0.8 depth from the bottom of the pipe is calculated using Ismail's equation (1952). Since 0.8 depth is relatively far from the bottom, they assume that the concentration must be uniformly distributed in the vertical direction throughout the flow, and therefore that it is the sediment discharge concentration for the homogeneous part of the sediment load. The difference between the concentration at 0.3 depth from the bottom and the concentration for homogeneous part of the load is the discharge concentration for the heterogeneous part of the load. They then suggest determination of the pressure drops due to the homogeneous part of the load and the heterogeneous part of the load separately. The total pressure drop is the sum of the homogeneous part of the load and the heterogeneous part of the load.

Danel (1948), Kada (1959), Daily and Bugliarelio (1958), Yotsukura (1961), Vanoni (1946), Metzner (1961), and Zandi (1967) have investigated the reduction of energy loss due to the presence of fine sediments.

Comments and Recommendations.—There is very little knowledge concerning transportation of graded sediments having a wide range of particle sizes. The sediments may be divided into the following three categories:

1. If even a small amount of very fine sediment (perhaps finer than 0.15 mm) is present, the mixture may exhibit non-Newtonian behavior, and the rheological properties of the fine sediment-water mixture must be understood. Because this manual is limited to Newtonian fluid and since non-Newtonian fluid is an involved subject by itself, no recommendation is suggested. Perhaps after a careful review of literature on non-Newtonian fluid flow, a pilot study should be considered prior to design of the pipeline.

2. The second category includes fine sediments that are transported as a homogeneous mixture and do not exhibit energy-gradient, suppression behavior.

The fall velocity, w, should be calculated as suggested by Du Plessis, et al. (1967). In this case, the head loss in terms of the specific weight of the mixture in any given pipeline is the same as would occur with clear water at the same velocity. The pressure drop increases in relation to the specific gravity and viscosity of the mixture. In other words, flows with fine sediments in this range can be considered as fluid flows with increased specific weight and viscosity.

3. The graded coarse sediment can be assumed to be transported as a heterogeneous mixture. If Durand's equation (Eq. 2.294) is used for calculating energy losses, the increase of energy loss due to the presence of graded sediment in heterogeneous flow is a function of $(1/C_D)^{0.75}$, and thus the apparent drag coefficient of the graded sediment for the calculation of pressure drops should be represented by

$$C_D^{0.75} = p_1(C_{D_1})^{0.75} + p_2(C_{D_2})^{0.75} + \ldots + p_n(C_{D_n})^{0.75} \qquad (2.332)$$

It is difficult to choose a representative sediment size from a graded sediment mixture for the calculation of a limiting deposit velocity. The size most likely depends upon the concentration of the sediment in the flow. If very large concentrations are involved, perhaps d_{90} might be selected as the representative size with which the limiting deposit velocity may be established. If only moderate concentrations are involved, a less conservative choice of, e.g., d_{75}, may be called representative.

A pilot plant or a model study is strongly suggested if transportation of graded sediments is of concern. Certainly for this kind of material, particle attrition in transport may be of greater concern than in the transport of uniform material because small increases in fine sediment concentration may lead to a non-Newtonian flow regime.

76. Effect of Pipe Shape on Transportation of Sediment.—This article summarizes results of studies made to determine the effect of pipe shapes on sediment transportation. It also includes the results from several studies on artificially roughened pipelines, particularly those with spiral roughness introduced for the specific purpose of increasing sediment transport in horizontal circular pipes.

Zenz and Othmer (1960) considered the possibility of increasing the carrying capacity of pipelines by altering the shape of the conduit cross section. They investigated two elliptical pipes (with the major axis of the ellipse either vertical or horizontal) and two divided circular pipes (with a dividing plate placed either vertically or horizontally). They found that the limiting deposit velocities within all these shapes were greater than that within the round tube. Thus, they concluded that round pipes were the optimum shape.

Chiu and Seman (1971) studied the head loss of flows transporting sediment in circular and square pipes with spiraling motion. They conclude that even slight spiraling motion in a square pipe increases the tendency of particles to be maintained in suspension and thus decreases the axial limit deposit velocity.

Studies by the United States Corps of Engineers, summarized by Howard (1941), indicate that spiraling flow induced by a spiral corrugation or rifling of the discharge lines of dredges increases the transport capacity for coarse sands and gravels. The material which normally (no spiral flow), would be transported as bed load motion in plain pipes are transported in suspension as a heterogeneous mixture in rifled pipe.

Chamberlain (1955) compares the flow and transport characteristics of 12-in. diam helical-corrugated pipe, smooth circular pipe, and standard corrugated pipe. From the results of several experiments, he indicates that when 0.2 mm sand is being transported at flow velocities less than 10 fps, helical-corrugated pipe can deliver more sediment with a given power input than either standard corrugated or plain pipe. Apparently, secondary circulation induced by a continuous helical corrugation, thus spiraling flow, is more effective in keeping the sediment suspended than corrugations placed normally to the direction of the mean flow. The head loss in helical-corrugated pipe is less than it is in standard corrugated pipe. Chamberlain also investigated the influence of the different boundary corrugations on the form of the concentration distribution in a cross section of the pipe.

Comments.—If a homogeneous fluid is involved in closed conduit flow, circular pipe is the best shape because it has the least wetted perimeter for the given flow area. When a material having a density greater than that of the supporting fluid is to be transported, circular pipe may not be the best shape. The flow may be separated in three different regimes. In a homogeneous mixture regime where no limit deposit velocity occurs, the circular pipe is the best shape because the entire mixture can be treated as a homogeneous fluid. Rifled or other pipe wall corrugation may change the flow from a heterogeneous to a homogeneous regime by introducing a secondary circulation. In the heterogeneous flow regime where the limit deposit velocity is of concern, spiral rifling of a pipe wall may well proved to be beneficial in terms of transporting more material at slower velocities than can be accomplished in a plain circular pipe. The use of rifled pipe (or other shapes) has two opposing effects: it increases flow resistance and decreases the limit deposit velocity. The design depends then on comparative power requirements to deliver a given sediment discharge.

77. Considerations in Design of Pipeline Systems.—Some thoughts are presented herein, and Article 78 gives several references for more detail discussions.

The first group of factors can perhaps affect the possibility of even being able to transport solids by liquid pipe flows. The factors are: (1) The availability of sufficient quantity of usable fluids; (2) the rate of sediment particle attrition; (3) the chemical stability of sediments in air or in the available fluids; (4) the availability of pipes, pumps, power supply, etc.; and (5) the topographical conditions, particularly the length and gradient of the adverse slopes.

The second group of factors can possibly be classified as the basic hydraulic factors, e.g., the determination of: (1) Pipe flow velocity with consideration of incipient motion and limit deposit velocity; (2) pipe diameter; (3) sediment concentration including pipe blockage problem; and (4) power requirements including the possibility of adding fine sediment to reduce head loss. Minor losses are normally unimportant for long pipelines and information on this subject is very limited. Graf (1971) stated that they have tested a 90° bend and found that the head loss over the pipe bend for mixture flow was the sum of a head loss due to clear water and a head loss due to the presence of solids. The head loss due to solids was found to be a linear function of the solid concentration. Iwanami and Suu (1970) tested head losses for fly ash and sand slurries through 90° T-sections. Data from their Fig. 8 for a particular flow condition indicates that the proportional increase of pressure loss through the main straight section is almost the same as the solid concentration by weight. In other words, the pressure loss through the main straight section for slurry with a solid concentration (by weight) of 50% would be 50% above the pressure loss for clear water.

The third group of factors are related to equipment design and these include the selection of: (1) Pipe material and shape; (2) pipeline connections; (3) valves; (4) feeder systems; (5) pumps and pumping stations; and (6) system to remove solid from the pipe flow etc. Consideration must be given to the maintenance problem that includes the reliability of the entire system. According to Fraddick and Babcock (1971) the past long experience at the Colorado School of Mines Research Institute indicates that the majority of the clients rank the design considerations in the following order of importance for short distance slurry pipelines: (1) Reliability; (2) rate of pipe abrasion; (3) rate of particle attrition; (4) throughput; and (5) energy requirements.

The fourth group of factors for design consideration is the optimization of the entire system design. All equipment chosen and all design procedures must be compatible with each other. Many pipeline installations, although technically feasible, may not be economically justified when compared to other methods of transporting the solids. It is expected, however, that as technology and experience with constructed systems increase, pipeline transportation will assume greater importance even than it has today. Transportation of material by pipeline affords low operational costs that certainly will encourage its use. When information is meager for a particular design problem, it will be necessary to build a pilot plant for further investigation, supplemented perhaps by special laboratory studies on specific aspects of the problem whether they be mechanical or hydraulic in nature.

As stated by the Mining Magazine (1972) "As a result of world wide pollution, control of the environment even in the remote areas of the world is becoming a factor of major importance. From this respect, pipelining has everything to commend it. There is a complete absence of atmospheric pollution due to exhaust gases, the pipeline is aesthetically pleasing since it can be buried underground or at least be quite inconspicuous, it causes no noise and last, by no means least important, it is safe in the extreme, no-one having even been killed or injured by the flow of slurry through a pipe ... "

78. Books and Pamphlets Giving Comprehensive Analysis of Transporting of Solids in Pipe Flows

1. Basic theoretical analysis.—Soo (1969).

2. Discussion of pipeline system, including pumps, pipes, valves, feeders, etc. —Tek (1961), Colorado School of Mines (1963), Condolios and Chapus (1963a and 1963b), Stepanoff (1965), Bain and Bonnington (1970), British Hydromechanics Research Association (1970a), Zandi (1971), Graf (1971), Mining Magazine (1972). Hunt and Hoffman (1968) on optimization.

3. Measuring techniques.—Fortino (1966), Brooks (1962), ASME (1963), Ackerman and Niyomthai (1964), Graf (1967, 1971).

4. Bibliography.—British Hydromechanics Research Association (1970a), Mih, et al. (1971).

79. Summary and Conclusions.—The scope of this section is as stated in the opening paragraph. The tremendous number of investigations made in this area clearly demonstrates the importance to different industries of sediment transport in pipes. An attempt has been made to examine the similarities and differences of previous findings, and views have been given on each subject covered. From these analyses, the reader will undoubtedly find that many critical issues are far from settled. One of the major shortcomings is the lack of data for large size pipes. Data from smaller pipe systems scatter widely so that one cannot define design criteria precisely. This can be done only with more data from large pipes (e.g., 12 in.–18 in. in size). The fact that such data are not available may be partially attributed to the policy of guarded secrecy of many competitive industries.

The writers researched the available knowledge in this area from Eastern European Journals. Unfortunately, most of their equations involve too many "to be determined coefficients," thus restricting their usefulness.

K. Density Currents

80. General.—In a gravitational field, fluid motions that are originated or influenced by variations in density within the fluid are characterized by the term stratified flow. In general, all free-surface liquid motions are stratified flows in the sense that the lower liquid is overlain by a lighter fluid that is the earth's atmosphere. In this case, the density difference is so large that the density, and therefore the inertial effects of the upper fluid, may be neglected in comparison with the lower. In conventional usage, stratified flow refers to motions involving fluid masses of the same phase. For purposes of analysis, stratified flows may be subdivided into two major categories: Two-layer systems and multilayer or continuous density gradient systems. The two-layer systems are by far the easier to treat analytically and within the realm of sediment transportation are the most common type encountered. The terms density current, gravity current, turbidity current, and underflow have been used interchangeably throughout literature describing the two-layer systems in which an interfacial region of rapid or discontinuous density change occurs. This type of motion is to be considered in the following section; a more general summary of stratified flow has been given by Harleman (1961).

A density current may be defined as the movement under gravity of a stream of fluid under, through, or over another fluid, the density of which differs by a small amount from that of the primary current. Further qualifications require that the two fluids be miscible, and that the density difference be a function of differences in temperature, salt content or sediment content, or both, of the two fluids. Density current manifestations in nature include stratified underflows in reservoirs and salt water intrusions in fresh water estuaries in addition to meteorological phenomena, e.g., cold fronts. The basic principles presented herein are mainly those concerned with the mechanics and sedimentation characteristics of sediment-laden underflows as found in reservoirs. This phenomenon is graphically described by Grover and Howard (1938) who reported several instances of the passage of a density current through Lake Mead due to a high suspended load concentration in the Colorado River as it entered the reservoir. A comprehensive introduction to this subject is also given in papers by Knapp (1942) and Bell (1942).

Fig. 2.141 shows the formation of a density current at the upstream end of a reservoir. The underflow is caused by the sediment-laden river water entering the reservoir with a higher specific weight than the clear water which has been impounded. The conditions at the so-called plunge point, where the river inflow disappears beneath the surface of the reservoir, are especially difficult to define. Intense local mixing at this point is due to the losses involved in the change of momentum as the velocities of the river are reduced to the much lower velocities of the underflow. Usually, the amount of energy available is not sufficient to produce complete mixing, however, the density current becomes diluted and the volume of the underflow per unit time may be larger than the river inflow. The rapid reduction of velocities near the entrance is also responsible for the deposition of the larger sediment particles and the formation of delta deposits. Over a period of years, the size of the delta increases in the downstream direction. In Lake Mead, e.g., during the period from 1935–1948 the delta was extended a distance of 42 miles. The shear stress at the interface of the moving density current generates a circulation pattern in the reservoir that results in small upstream surface currents and the collection of floating debris near the plunge point.

In general, the engineering problems associated with density currents are concerned with the passage of sediment through the reservoir and with the reduction of storage capacity by sediment accumulation. Ultimately, it should be possible to predict the occurrence and hydrodynamic characteristics of these currents. A complete solution would therefore require the following: (1) Determination of initial velocity, depth and sediment transport rate of density currents from known characteristics of the reservoir and the river inflow; (2) an estimate of the amount of interfacial mixing resulting in the decreasing sediment carrying capacity; (3) a determination of whether the density current will maintain its identity throughout the length of the reservoir (with the objective of providing sluicing devices to discharge sediment downstream of the dam.) Analytical and experimental results and quantitative field data are limited, and not all of the preceding questions can be answered. However, certain basic relationships among the major variables, e.g., velocity, depth, slope, density difference, and the boundary conditions of the underflow can be determined.

Experiments and analytical studies on two and three-dimensional density currents have been reported by Wood (1967) and Fietz and Wood (1967). In the latter study the density current was formed by emission from a point source, at the top of a downward sloping floor, of a fluid slightly denser than the surrounding fluid.

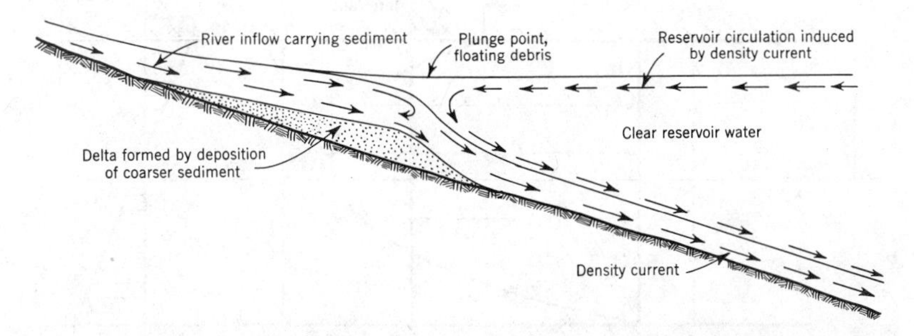

FIG. 2.141.—Upstream End of Reservoir Showing Formation of Density Current

81. Resistance Laws for Uniform Underflows.—A steady, uniform density current will occur along an incline when the driving gravity force per unit area due to the small density difference between the subsurface flow and the lighter liquid above is in equilibrium with the shear stresses exerted by the stationary boundary and by the moving interfacial boundary. The depth of the density current is assumed to be small in comparison with the depth of the lighter liquid above, and therefore the circulation pattern in the upper liquid may be neglected. For two-dimensional flow, the equilibrium equation is

$$\tau_o + \tau_i = \Delta\rho g d S \qquad (2.333)$$

in which $\Delta\rho$ = the density difference between the flowing and stationary liquids; d refers to the depth; S represents the slope of the channel; and τ_o and τ_i denote the bottom and interfacial shear stresses, respectively. The interface between the two liquids is smooth and distinct up to the point of mixing. In addition, the shear stress varies linearly from τ_o at the bottom (passing through zero at the point of maximum velocity) to τ_i at the interface. Letting $\tau_i = \alpha\tau_o$, the interfacial shear stress is a constant proportion of the bottom shear and depends only on the vertical location of the maximum velocity between the bottom and the interface. In the notation of the shear distribution shown in Fig. 2.142:

$$\alpha = \frac{1 - 2\dfrac{y_m}{d}}{1 + 2\dfrac{y_m}{d}} \qquad (2.334)$$

The limiting values of α are represented for open channel flow by $\alpha = 0$ and for flow between parallel stationary plates by $\alpha = 1$. Eliminating τ_i, Eq. 2.333 may therefore be written as

$$\tau_o = \Delta\rho g \frac{d}{1 + \alpha} S \qquad (2.335)$$

The shear stress, τ_o, may also be expressed in terms of the usual frictional resistance equation that is useful for both laminar and turbulent flow conditions:

$$\tau_o = \frac{f}{4}\rho\frac{V^2}{2} \qquad (2.336)$$

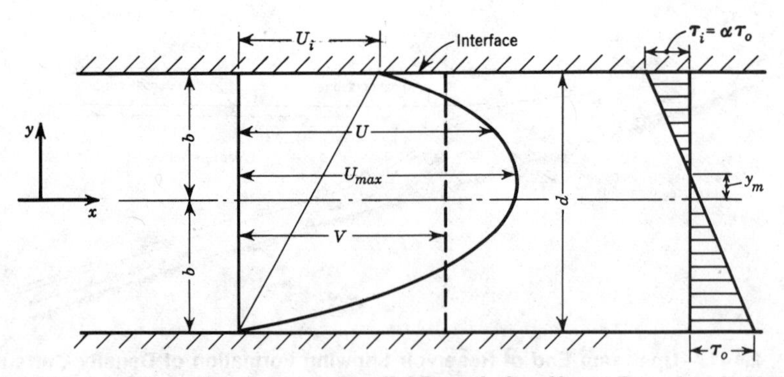

FIG. 2.142.—Laminar Flow between Parallel Boundaries: Upper Boundary in Motion

in which f is the familiar Darcy-Weisbach friction factor. Solving for V, the average velocity of the underflow, from Eqs. 2.335 and 2.336 is

$$V = \sqrt{\frac{8\frac{\Delta\rho}{\rho}g}{f}}\sqrt{\frac{d}{1+\alpha}S} \qquad (2.337)$$

A generalized form of the Chezy equation $V = C\sqrt{rS}$ is obtained in which the term $d/(1+\alpha)$ is the effective two-dimensional hydraulic radius of the density current.

Laminar Flow.—For the case of laminar flow, the variation of the resistance coefficient, f, with the Reynolds number can be determined analytically as shown by Ippen and Harleman (1952). The general form of the laminar velocity distribution can be expressed in terms of a single parameter, J, a dimensionless number indicating a ratio of viscous and gravity forces, by definition

$$J = \frac{\mathsf{F}^2}{\mathsf{R}S} \qquad (2.338)$$

in which F, the densimetric Froude number, is defined by

$$\mathsf{F} = \frac{V}{\sqrt{\frac{\Delta\rho}{\rho}gd}} \qquad (2.339)$$

and R, the Reynolds number is

$$\mathsf{R} = \frac{Vd}{\nu} \qquad (2.340)$$

The laminar velocity distribution, with the notation as shown in Fig. 2.142, is

$$\frac{U}{V} = 1 + 2\frac{z}{d} - \frac{1}{2J}\left[\left(\frac{z}{d}\right)^2 + \frac{1}{3}\frac{z}{d} - \frac{1}{12}\right] \qquad (2.341)$$

and the relation between U_i and the interface velocity and U_{max} found by differentiating the velocity distribution is

$$\frac{U_i}{U_{max}} = \frac{12J - 1}{12J^2 + 4J + \frac{1}{3}} \qquad (2.342)$$

For the two limiting conditions previously described, the associated values of J can be obtained. For flow between parallel stationary plates, $U_i = 0$, and from Eq. 2.342, $J = 0.083$. For flow with the upper surface free, $U_i = U_{max}$ and from Eq. 2.342, $J = 0.333$. Therefore, the value of J for uniform two-dimensional underflows must lie between these limits. However, it can be shown that the ratio U_i/U_{max} must be a constant depending on the fluid properties if the region near the interface is treated as a problem of laminar boundary layer development between parallel streams. This problem has been investigated analytically by Keulegan (1944) and for the usual range of densities and viscosities encountered, the ratio of interface to maximum velocity for a laminar density current may be taken as $U_i/U_{max} = 0.59$. Therefore, from Eq. 2.542, $J = 0.138$, and because J determines the velocity distribution, the corresponding values of $\alpha = 0.64$ (Eq.

2.334). By means of Eqs. 2.337 and 2.338, the friction factor, f, may be expressed as a function of the Reynolds number:

$$f(1 + \alpha) = \frac{8}{J\mathsf{R}} \quad (2.343)$$

In summary, therefore: (1) Flow between parallel stationary plates ($J = 1/12$)

$$f(1 + \alpha) = 2f = \frac{96}{\mathsf{R}} \quad (2.344a)$$

(2) laminar free surface flow ($J = 1/3$)

$$f(1 + \alpha) = f = \frac{24}{\mathsf{R}} \quad (2.344b)$$

(3) laminar density current ($J = 0.138$)

$$f(1 + \alpha) = 1.64f = \frac{58}{\mathsf{R}} \quad (2.344c)$$

Rewritting Eq. 2.337 for the laminar density current (case 3) yields

$$V = \sqrt{\frac{8\frac{\Delta\rho}{\rho}g}{f(1 + \alpha)}}\sqrt{dS} = \frac{\mathsf{R}^{1/2}}{2.7}\sqrt{\frac{\Delta\rho}{\rho}gdS} \quad (2.345)$$

Eq. 2.345 has been verified experimentally by Ippen and Harleman (1952) for the range of laminar flows from $\mathsf{R} = 15$ to 1,000. Additional experiments by Bata and Knezevich (1953), Raynaud (1951), and Bonnefille and Goddet (1959) have also verified Eq. 2.345, and the lower critical Reynolds number $\mathsf{R} = 1{,}000$.

Turbulent Flow.—The preceding analysis provides a method of approach for the problems of turbulent density currents because, in common with most turbulent flows, an analytic solution is not possible. A consideration of Eq. 2.343 leads to the conclusion that for two-dimensional flow, $(1 + \alpha)$ represents the amount by which the friction factor, f, is increased by the presence of the interfacial shear stress. Therefore, f, may be obtained from the resistance diagram for flow in conduits (Moody diagram) in which the hydraulic radius appearing in the conduit Reynolds number is the same as for free surface flow. The value of f thus obtained is to be increased by the factor $(1 + \alpha)$. Experimental information sufficient to determine the dependence of α on the Reynolds number is not available for turbulent density currents. However, experimental measurements (Bata and Kenezevich, 1953) of turbulent density current velocity distributions in the range of Reynolds numbers Vd/ν from 1,000 to 25,000 indicate that the maximum velocity occurs on the average at $y = 0.7d$. No systematic variation with Reynolds number was obtained. The corresponding value of α from Eq. 2.334 is $\alpha = 0.43$.

The experiments also indicated an average value of friction factor of the interface $f_i = 0.01$ when the fixed boundaries were hydraulically smooth. Keulegan (1949) has obtained a value of $f_i = 0.007$ based on approximate calculations of the resistance due to wave motion and mixing at the interface. Comparing the preceding values with the method previously proposed in which

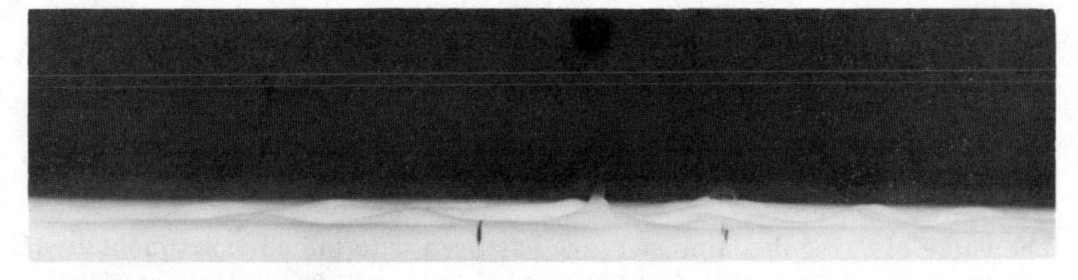

(a) Before mixing

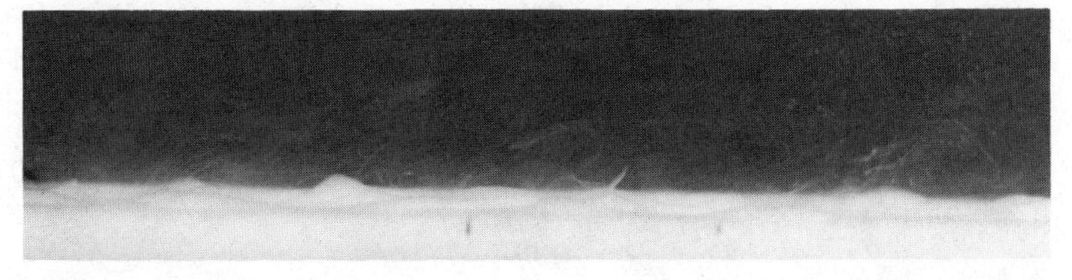

(b) Mixing with upper layer

FIG. 2.143.—Waves at Interface of Underflow: Flow is from Right to Left

$f + f_i = f(1 + \alpha)$ it is seen that $f_i = \alpha\ f = 0.43f$. If $f = 0.020$, e.g., then $f_i = 0.0086$, which is in agreement with the foregoing values. The conditions for uniform flow may then be obtained from Eq. 2.337 using the values of f and α appropriate to turbulent flows.

Analysis of experimental data for interfacial friction factors obtained from lock exchange flows, by Abraham and Eysink (1971), indicated values of α ranging from 0.4–0.5 for Reynolds numbers up to approx 10^5.

Any attempt to refine the analysis of turbulent flows is hindered by the fact that as the degree of turbulence increases, the interface becomes increasingly difficult to define, due to mixing and resulting vertical density variations.

Stability and Mixing in Stratified Flows.—The interface of a density current at very low velocities is smooth and distinct, and consists of a sharp discontinuity of density across which the velocity variation is continuous. As the relative velocity between the two layers is increased, waves are formed at the interface, and at a certain critical velocity the mixing process begins by the periodic breaking of the interfacial waves. Initially the amount of mixing is slight and does not appreciably affect the depths and velocities corresponding to the condition of uniform flow. Further increases in velocity result in sharp crested waves of greater height, and an increase in the frequency of eddies breaking from the crests. Fig. 2.143 shows photographs of a density underflow in a laboratory flume; the moving current is white and opaque due to the use of a bentonite solution. Fig. 2.143 shows the short crested waves at the interface near the point of instability. Fig. 2.143(b) shows the breaking of these waves and the resulting mixing with the clear water above, as shown by the increased cloudiness of the overlying layer.

Various criteria for determining the flow conditions at which mixing begins have been proposed. Keulegan (1949) has derived a stability parameter or criterion for mixing on the basis of viscous and gravity forces as

$$\theta = \frac{\left(\nu g \dfrac{\Delta\rho}{\rho}\right)^{1/3}}{V} \qquad (2.346)$$

in which V = the relative velocity of the underflow relative to the overlying fluid. This parameter can be transformed with the notation previously used into the form

$$\theta = \frac{1}{(\mathsf{F}^2\mathsf{R})^{1/3}} \qquad (2.347)$$

The value of θ_c for the laminar range has been determined experimentally (Ippen and Harleman, 1952) as $\theta_c = 1/\mathsf{R}^{1/3}$ which indicates that in laminar flow, mixing occurs when the flow is at critical depth, i.e., the Froude number equals unity.

The average experimental value of θ_c for the turbulent range (Keuligan, 1949) is $\theta_c = 0.18$. The maximum Reynolds number (Vd/ν) for these tests was approximately 10^4. In both cases, it is presumed that if the value of θ as computed from Eq. 2.347 exceeds the value of θ_c, no mixing should occur.

Example: Given $V = 0.1$ fps, $d = 20$ ft, $\Delta\rho/\rho = 0.010$ and $\nu = 1 \times 10^{-5}$ sq ft/sec, then

$$\mathsf{F} = \frac{V}{\sqrt{g\dfrac{\Delta\rho}{\rho}d}} = 0.04; \; \mathsf{R} = \frac{Vd}{\nu} = 2 \times 10^5$$

and $$\theta = \frac{1}{(\mathsf{F}^2\mathsf{R})^{1/3}} = 0.15$$

Therefore the value of θ is slightly below $\theta_c = 0.18$ for turbulent flow, and a small amount of mixing would be expected to occur. The application of such investigations on mixing to large bodies of water can only indicate the probability and degree of mixing to be expected. A fundamental experimental study of interfacial mixing has also been reported by Macagno and Rouse (1962). Their critieria for stability and amount of mixing are given in terms of Reynolds and densimetric Froude numbers, using an interlayer thickness and velocity. Thus, comparisons with the earlier results are difficult.

82. Transportation of Sediment by Density Currents.—A density current of the underflow type exists because its specific weight is greater than that of the surrounding fluid. In general, the increase is due to sediment carried in suspension, dissolved solids and temperature differences. Density currents due primarily to temperature have been observed in the system of the Tennessee Valley Authority (Fry, et al. 1953) in which velocities of the underflow ranging from 0.15 fps–0.35 fps have been observed. Of primary importance in the field of sediment transportation are those currents caused by suspended sediment. The relation between the sediment concentration and the difference in specific gravity is

$$\frac{\Delta\gamma}{\gamma} = \frac{C}{100}\left(1 - \frac{s_f}{s_s}\right) \qquad (2.348)$$

in which C = the percentage of concentration of sediment by weight; s_f denotes the specific gravity of water at given temperature; and s_s = the specific gravity of sediment. Studies of density currents in Lake Mead on the Colorado River (Howard, 1953) have shown that the influence of temperature and dissolved solids is small and that the density difference is primarily due to suspended sediment. These investigations have also shown that when appreciable underflows occur, they consist chiefly of particles in suspension of less than 20μ in diameter. Stokes' law gives a settling rate for particles of this diameter of approx 0.001 fps. Thus, transverse turbulent fluctuations of the order of 1% in a current having a mean velocity of only 0.1 fps would be sufficient to keep such particles in suspension. Sedimentary material of the size encountered in most density currents is commonly referred to as "wash load," that is, suspended material of very small size originating from erosion on the land slopes of the drainage area rather than from the bed of the stream itself. The concentration of such material is practically constant from bed to surface and is relatively independent of major changes in flow conditions such as occur at the plunge point of the reservoir.

The Lake Mead surveys have indicated the existence of underflows when the density difference between the current and the surrounding water is of the order of $\Delta\rho/\rho = 0.0005$. The average slope for Lake Mead is taken as 5 ft/mile, then $S = 0.0009$. If measurements indicate a depth of approx 15 ft for the moving current, an indication of the order of magnitude of the velocity can be obtained from Eq. 2.337.

Example: Assume $f = 0.010$ and $\alpha = 0.43$ for turbulent flow. From Eq. 2.337

$$V = \sqrt{\frac{8(0.0005)32.2}{0.010(1.43)}} \cdot \sqrt{15(0.0009)} = 0.35 \text{ fps}$$

The Reynolds number, assuming two-dimension flow, is therefore

$$\frac{Vd}{\nu} = 5 \times 10^5$$

The friction factor, f, for this Reynolds number is within 10% of the assumed value (for smooth boundaries) and a refinement of the calculation is not warranted. The computed velocity of 0.35 fps (approx 6 miles/day) is consistent with measurements that have been made in Lake Mead. The rate of sediment transport by the density current can also be estimated. From Eq. 2.348, $C = 0.08\%$, and if the average width of the current is taken as 500 ft, then the discharge represented by the density current is 2,600 cfs and that of the sediment 130 lb/sec, which is approximately equivalent to a sediment transport rate of 6,000 tons/day.

If the sediment concentration or volume of the inflow is increased suddenly, the front portion of the resulting underflow has been observed to have a characteristic "head" of greater thickness than the uniform current following. Because the stream moving along the bottom displaces the lighter fluid upward, a force must be provided to overcome boundary resistance. This initial force is larger than the gravity forces maintaining the subsequent uniform motion and requires a initial depth that may be almost twice as great as the uniform depth.

Density currents generated by saline and fresh water interactions are also important in the transport of sediment in estuaries. This subject is beyond the scope of this section inasmuch as it depends to a large extent on the dynamics of salinity intrusion. Reviews of the subject of estuary shoaling, together with bibliographies of related papers, have been given by Nichols and Poor (1967) and Harleman and Ippen (1969).

L. Genetic Classification of Valley Sediment Deposits

83. General.—Many problems in sediment engineering involve dealing with river valley deposits as a whole.

The genesis of valley deposits is intimately associated with the history of the stream by which they were formed. Therefore, a discussion of this subject is included in this section in the hopes that it will be of use to engineers concerned with such sediment problems. Like the stream history, the deposits themselves are very complex and it is not possible to describe them completely by any existing analytic means. However, a classification of the various kinds of deposits, their genesis and the relations between them will aid in understanding their structure. Such knowledge is helpful in planning investigations, evaluating the results thereof or in estimating probable behavior of a stream that is being modified by man-made works, e.g., dams, diversions, and other regulating structures.

84. General Characteristics.—Characteristics of the principal genetic types of fluvial sediments shown in Fig. 2.144 are summarized in Table 2.25, but such a tabulation can not represent adequately the complex gradational relationships between the types, none of which are entirely mutually exclusive.

Deposition of fluvial valley sediments is accompanied by sorting into coarser and finer parts, which correspond generally to bed load and wash or suspended load (Einstein, et al., 1940). The coarser bed load materials, mostly sand and gravel, accumulate in stream channels except in unusual or occasional local circumstances. Finer wash-load materials, mostly silt and clay, accumulate chiefly on flood plains, from overflows outside the channels. Meandering or other shifting of channels allows coarse channel deposits to be covered by finer overbank deposits, normally producing a relatively impervious surface blanket over more pervious subsurface materials. Such a vertical sequence in the lower Mississippi valley has been described as topstratum and substratum (Fisk, 1944, 1947, 1952).

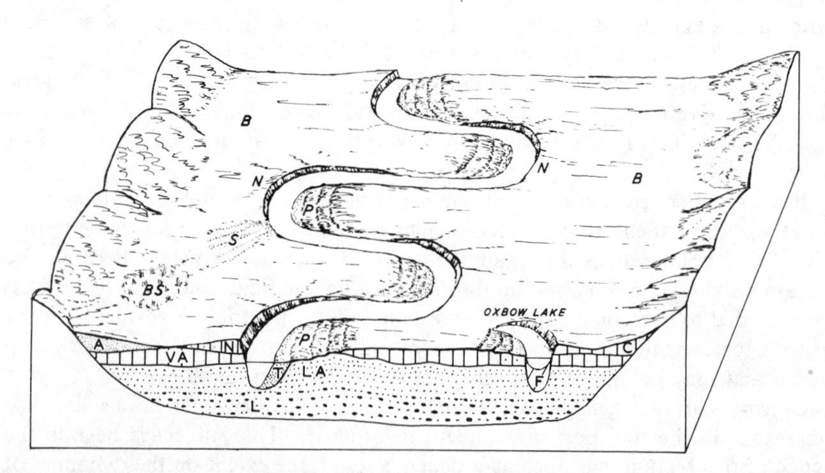

FIG. 2.144.—Typical Associations of Valley Sediment Deposits: VA, **Vertical Accretion;** N, **Natural Levee;** B, **Backland;** BS, **Backswamp;** LA, **Lateral Accretion;** P, **Point Bar;** S, **Splay;** A, **Alluvial Fan;** T, **Transitory Bar;** L, **Lag Deposit;** F, **Channel Fill;** C, **Colluvium**

TABLE 2.25.—Classification of Valley Sediments

Place of deposition (1)	Name (2)	Characteristics (3)
Channel	Transitory channel deposits	Primarily bed load temporarily at rest; part may be preserved in more durable channel fills or lateral accretions.
	Lag deposits	Segregations of larger or heavier particles, more persistent than transitory channel deposits, and including heavy mineral placers.
	Channel fills	Accumulations in abandoned or aggrading channel segments; ranging from relatively coarse bed load to fine-grained oxbow lake deposits.
Channel margin	Lateral accretion deposits	Point and marginal bars that may be preserved by channel shifting and added to overbank flood plain by vertical accretion deposits at top.
Overbank flood plain	Vertical accretion deposits	Fine-grained sediment deposited from suspended load of overbank flood water; including natural levee and backland (backswamp) deposits.
	Splays	Local accumulations of bed load materials, spread from channels on to adjacent flood plains.
Valley margin	Colluvium	Deposits derived chiefly from unconcentrated slope wash and soil creep on adjacent valley sides.
	Mass movement deposits	Earthflow, debris avalanche, and landslide deposits commonly intermix with marginal colluvium; mudflows usually follow channels but also spill overbank.

Similar sequences are characteristics of fluvial sediments in most other valleys, although finer topstratum and coarser substratum may vary greatly in thickness and texture, and either may be almost entirely lacking in places.

85. Transitory Channel Deposits.—Stream channel deposits are mostly transitory accumulations of bed load materials. Bed load particles move spasmodically, perhaps at long intervals, and are at rest on the stream bed most of the time. Individual particles may remain at rest until gradual shifting of other particles reduces frictional resistance, or water velocity increases because of rising discharge. All valley deposits are subject to repeated erosion, transportation and redeposition, and thus are transitory in a geologic sense, but those within existing stream channels are much less stable than overbank deposits.

Transitory bed load deposits may be spread quite uniformly along the channel, but their surface forms include ripples, waves, small dunes and larger bars. Transverse bars may migrate downstream relatively rapidly, and be essentially ephemeral features (Sundborg, 1956). Longitudinal bars extend in the direction of flow under various circumstances. Crossing bars occur between successive bends and are relatively stable in position although subject to alternating aggradation and degradation by rising and falling water stages. Some bars may persist for many years although subject to almost continual changes as individual particles are removed by currents at some places and replaced elsewhere by other particles, but varying flows may cause considerable changes in form, and even complete destruction of a bar during extreme stages.

86. Lag Deposits.—Lag deposits are formed by sorting out and leaving behind larger and heavier particles as smaller and lighter ones are moved farther and more rapidly downstream. Most lag deposits are included among transitory channel sediments, and are removed by shifting currents. Others are relatively stable and durable. Gold and other heavy mineral placers are notable examples of lag deposits, but gravel and boulder bars are more common. Relatively coarse gravels and boulders at the bottom of many valley deposits, immediately above the bedrock floor, are apparently lag deposits concentrated by repeated and progressive removal of smaller particles. Individual particles or boulders could be moved by the stream, at least during floods, yet the deposits as a whole are quite stable. There is a need for criteria to distinguish between such lag deposits of present streams, and coarse deposits accumulated in some earlier period of possibly greater stream competence or heavier sediment charge.

87. Lateral Flood-Plain Accretion and Point Bars.—Bed load deposits commonly accumulate on the convex inner sides of channel bends. Such lateral bars may be destroyed by later fluctuating flows, yet some persist until shifting of the channel diverts the main currents away from them. Continued growth of a bar reduces the depth and retards the velocity of high stage flows across it, and thus induces deposition of finer materials. The finer materials and less frequent overflows also favor establishment of vegetation, which helps stabilize and preserve it against erosion. The upper part of a bar is thus composed of finer sediment, with smaller proportion of bed load, and in time may be built up to the height of the adjoining older part of the flood plain (Wolman and Leopold, 1957). In this way bars become part of the overbank plain, which grows by lateral accretion (Fenneman, 1906).

Lateral bars formed inside shifting meanders of the lower Mississippi River are called point bars. The term was publicized by H. N. Fisk, although he acknowledged ambiguity between the local usage for individual bar ridges and his preferred application to accretionary meander points comprising many arcuate low ridges with swales between (Fisk, 1947, p. 34), which may cover several square miles. The latter meaning has come into wide usage (Leopold, et al., 1964; Allen, 1965) although the term also has been used by Sundborg for individual bar ridges (Sundborg, 1956).

The point forms may be inconspicuous on streams which do not meander strongly, or may be obscured among complex deposits of irregular "advanced-cut" meandering (Melton, 1936) on streams, e.g., the lower Missouri prior to artificial regulation (Happ, 1950; Schmudde, 1963). Also contrasting accretion opposite point bars, upstream from abrupt bends where the Mississippi River impinges against its bluffs, has been described under the name eddy accretion (Carey, 1969). Thus, lateral accretion appears to be the most appropriate term for the general process, although point bars of characteristic shape (by either definition of that term) are the conspicuous forms.

88. Vertical Accretion Deposits.—Vertical accretion deposits are formed by deposition of suspended load from overbank flood waters (Fenneman, 1906). All fluvial deposits might be considered vertical accretions in the sense that they are formed upon some preexisting surface, but the terminology is based on relation to growth of flood plains. In this respect overbank deposits build up the flood plain vertically, in contrast to lateral growth by channel-margin deposits. Flood plains may also be build up vertically by deposition of bed load materials spread locally from the channel, or by wind-blown deposits, mudflows, landslides, colluvial wash or other contributions from adjacent valley sides, but these other deposits are each sufficiently distinctive to justify more specialized names. The general term vertical accretion is applied to the common, widespread process of flood-plain aggradation by deposition of relatively fine suspended load from overbank flood waters. Vertical accretion deposits are commonly 5 ft–20 ft thick in many valleys, but may be less than 1 ft and are reported to be more than 100 ft thick in places along the lower Mississippi (Fisk, 1952).

89. Natural Levees and Backlands.—Velocity of flood water is checked abruptly where it leaves a stream channel. Thus the thickest and coarsest vertical accretion deposits form low ridges or natural levees immediately bordering the channel, although natural levees also include some bed load materials carried overbank either because of some local circumstances or because of overloading and general filling of the channel. Natural levees are reported to stand as high as 25 ft above the general flood plain along the lower Mississippi (Fisk, 1944, 1947) and to be as much as 5 miles wide (Kolb and Van Lopik, 1958). In small valleys they may be only a few feet wide and less than 1 ft high.

The most extensive vertical accretion deposits occur on relatively low and flat parts of broad flood plains, behind the natural levees. There is normally a gentle slope without sharp demarcation at the back of the natural levee, but the two environments are distinctly different although gradational from one to the other. The lowest parts of the flood plain may form "back-swamps", and deposits formed in that environment have been called back-swamp deposits. The broader term "backlands" is more widely applicable, however, and hence preferable for general usage. Backland deposits are of finer texture, and usually more uniform in composition and thickness than those of the natural levees. Ponding of flood waters behind natural levees favors deposition of even the finest sediment.

90. Flood-Plain Splays and Valley Plugs.—Splays are localized deposits of bed load materials, typically sand or gravel, spread out over the normally finer flood plain sediments. The name (Happ, et al., 1940) is taken from a common surface form which is fan-like but elongated along diverging axial lines. In other places splays of bed load materials may be left in isolated bars of various linear or arcuate forms, rather than spread fan-like from the channel bank.

Splay deposits are common on alluvial fans, but may form along any stream where currents spread bed load materials outward from the channel. They are formed chiefly during severe floods, or along streams which are aggrading their channels because of excessive bed load or obstructed flow. They were formed abundantly during severe floods of the Ohio River in 1937 and Kansas River in 1951, during floods of 1937 and 1941 in the Middle Rio Grande valley, New Mexico, and have been observed along many streams. Medium and coarse gravel splays formed both fan and bar patterns during a rare flood in the Graburn watershed, Alberta (McPherson and Rannie, 1969).

Splays are conspicuous where channels are filled with sediment, and all the additional bed load brought downstream is then spread out over the flood plain until a new channel has been formed. This extreme condition may result from some chance local obstruction, e.g., a log jam, or from tributary contribution of bed load which the mainstream cannot carry away, or from inadequate outlet for an artificially improved channel. The term "valley plug" has been used for such areas of local channel filling, with numerous bordering splay deposits, in small valleys affected by excessive channel filling from gullying of sandy upland subsoils (Happ, et al., 1940). Valley plugs look essentially like low-gradient alluvial fans, of which they might be considered a variety. They have been given a different name, however, to emphasize the difference in their topographic occurrence.

Splays are commonly formed by currents from crevasses in levees, but comprise only part of what should properly be included under the general term "crevasse deposits" (Johnson, 1891; Trowbridge, 1930; Welder, 1959).

91. Channel Fill Deposits.—Channel fill deposits result from channel avulsion, and always occur in linear forms although varying greatly in composition and manner of accumulation.

Aggrading streams heavily charged with bed load may fill their channels nearly to the top of the banks, and flow on a ridge higher than parts of the bordering flood plains. Avulsion to a new channel will then leave a low ridge of relatively coarse bed load materials, flanked by natural levee deposits. Lengthy avulsions on other types of streams result in channel fills including a lesser proportion of bed load materials, covered by finer deposits. Abandoned river courses in the Yazoo delta are reported to be filled partly by a wedge of sand, thinning downstream, overlain by finer materials largely deposited and reworked by smaller streams occupying the old course (Kolb, et al., 1958).

Cutoff of an individual meander bend usually involves plugging of the head of the abandoned section by bed load accumulation (Fisk, 1944, 1952) forming an oxbow loop, or lake, which fills with fine sediment. "Clay plugs" formed in abandoned bends along the lower Mississippi River are reported to reach a thickness of about 140 ft, but parts of bends abandoned by chute cutoffs, which result from gradual enlargement of high-water channels across point bars, are filled mostly by sand at the upper end although usually with a thick but narrow filling of fine material along the abandoned thalweg at the lower end (Fisk, 1947).

92. Mudflow and Related Mass Movement Deposits.—Mass movements of soil, rock, or older sediments are essentially hillside processes rather than valley processes, but the resulting deposits are formed largely in valleys.

Mudflows are the more fluid forms of mass movements, gradational from less fluid, generally drier processes of earthflows, debris avalanches, soil creep and landslides (Sharpe, 1938). Debris flow is sometimes used as a more general term (Leopold, et al. 1964; Sharp and Nobles, 1953), with mudflow then restricted to fine grained materials; but debris flow is also sometimes restricted to the coarser flows (Varnes, 1958; Beaty, 1963). Mudflow is the more common term and seems appropriate for the general process. Wetting of the finer materials, forming mud, appears to be the essential factor, although many flows consist chiefly of coarse materials.

Mudflow deposits occur chiefly along stream channels, most extensively on alluvial fans in semi-arid and mountainous regions, and also in arctic climates and volcanic ash areas. They are typically heterogeneous in composition, with little sorting, internal stratification or orientation of particles. They do, however, grade into sorted and stratified sediments where higher water content has produced conditions transitional to turbid stream flow (Fryxell and Horberg, 1943; Blissenbach, 1954; Bull, 1964). They range from fine silts and clays to coarse bouldery deposits with only a minor proportion of fines. Clay content has been reported to range from less than 5% up to 76% (Bull, 1964). Boulders may be of many tons, randomly distributed and oriented. Surfaces may be fairly smooth or irregular, hummocky or ridged, and may have a relief of 10 ft–20 ft (Leopold, et al., 1964). Boulders are commonly concentrated at the front and sides, and bouldery ridges formed along the sides have been called mudflow levees (Sharp, 1942). They characteristically form lobate tongues, and margins may be sharp or abrupt with distal slopes as steep as 40° (Wooley, 1946) and as high as 12 ft (Fryxell and Horberg, 1943).

Similar deposits are formed by debris avalanches and earth flows, also largely in valleys but extending up the bordering hillsides relatively more than do mudflow deposits. Both commonly occur under the same general conditions as mudflows, but are less known in semi-arid regions. Debris avalanches generally follow previous channels (Sharpe, 1938; Hack and Goodlett, 1960) down steeper slopes and in mountainous regions. Earth flows are common on soil slopes outside channels, and also are prominent where extensive alluvial terrace sediments are susceptible to erosion by mass movements, as in the St. Lawrence valley of Quebec and the Columbia valley in Washington.

93. Colluvium.—Colluvium or colluvial deposits are convenient terms for accumulations, along valley margins, of deposits derived from slope processes and extending up the hillsides above the valley bottoms. They are derived principally from unconcentrated slope wash and downhill soil creep (and solifluction where effective). Deposits from earth flows, land slides, avalanches and rock falls are commonly included, where they cannot be readily differentiated because of small size or other reasons. Small alluvial fans at mouths of wet weather drainage ways also are commonly intermixed indistinguishably. Such colluvial accumulations may comprise an important part of the sediment in some small valleys, as reported for a case in Pennsylvania (Lattman, 1960).

94. Alluvial Fans.—Alluvial fans or alluvial cones (cones are sometimes considered smaller than fans, but usage varies) are not a separate genetic type of deposit, but a distinctive association of genetic types. They are concentrations of sediment in fan-shaped patterns, formed where a stream enters an area of flatter slope, or emerges from a confined channel and is able to spread its deposits laterally. They commonly occur where a tributary enters a larger valley or lowland. Alluvial fans are especially large, steep, and prominent in dry and mountainous regions, but are also common at the mouths of hillside drainage ways and tributary valleys in more humid and flatter regions.

Splay deposits are abundant on alluvial fans, associated with channel filling and avulsions. Lateral and vertical accretion deposits are commonly developed but in greatly varying proportions under different circumstances; they probably are only minor components of some fans in semi-arid regions and smaller fans elsewhere, but may be more important in larger fans of more humid regions.

Mudflows form a major part of many alluvial fans in semi-arid regions (Blissenbach, 1954; Beaty, 1963; Bull, 1964; Blackwelder, 1928; Park, 1923; Hooke, 1967), and may be important contributors also in mountainous areas, arctic climates, and volcanic ash areas. Debris avalanches and earthflows also form fan-shaped deposits, but are less important than mudflows in fans built by larger streams. Colluvium from bordering hillsides is deposited on many alluvial fans, and may be a major component of small fans.

95. Delta Deposits.—Deltas are characteristic associations of types of sediments at places where stream velocity is checked by entry into a larger body of water.

They are usually built into the sea, or into estuaries or lakes, and hence are not considered to be essentially valley deposits although occurring in or at the mouths of valleys. Small deltas built within stream channels during floods (Jahns, 1947) must be included among valley deposits, but in such places they are subject to removal by later flows.

The upstream parts of deltas are extensions of valley flood plains, and may include any or all of the genetic types of valley sediments. Filled channels are apt to be unusually prominent, but mass movement deposits and colluvium are rare. Downstream the valley types of sediment are commonly mixed with shoreline, swamp or marsh, and shallow-water marine, estuarine or lacustrine deposits of various types.

In deeper water all bed load is deposited, together with coarser suspended load, tending to form foreset beds dipping downstream. In an idealized delta the foreset beds are built forward over finer, flatter bottomset beds deposited farther from the stream mouth, and the foresets in turn are covered by progradation of topset beds including typical valley deposits. In complex deltas, however, the idealized bottomset-foreset-topset relationship may be obscured by other areal variations (Kolb and Van Lopik, 1958; Russell and Russell, 1939; Mathews and Shepard, 1962).

96. Conclusions.—Most valley sediment deposits can be classified genetically as either transitory channel deposits, lag deposits, lateral accretion, vertical accretion, splays, channel fills, mudflows and related mass movement deposits, or as colluvium. Other terms are suitable and convenient for distinguishing various subdivisions, surface forms, or groupings of the genetic types.

M. References

Abdel-Rahmann, N. M., "The Effect of Flowing Water on Cohesive Beds," *Contribution No. 56,* Versuchsanstalt fur Wasserbau und Erdbau an der Eidgenossischen Technischen Hochschule, Zurich, Switzerland, 1964, pp. 1–114.

Abraham, G., and Eysink, W. D., "Magnitude of Interfacial Shear in Exchange Flow," *Journal of Hydraulic Research,* International Association for Hydraulic Research, Vol. 9, No. 2, 1971.

Acaroglu, E. R., "Sediment Transport in Conveyance Systems," thesis presented to Cornell University, at Ithaca, N.Y., in 1968, in partial fulfillment of the requirements for the degree of Doctor of Philosophy.

Ackerman, N. L., and Niyomthai, C., "Development of Solid-Liquid Flow Meter," *Journal of the Hydraulics Division,* ASCE, Vol. 90, No. HY2, Proc. Paper 3829, Mar., 1964, pp. 131-139.

Ahmad, M., "Experiments on Design and Behavior of Spur Dikes," *Proceedings,* Minnesota International Hydraulics Convention, Minneapolis, Minn., September 1–4, 1953a, pp. 145–159.

Ahmad, N., "Mechanism of Erosion Below Hydraulic Works," *Proceedings,* Minnesota International Hydraulics Convention, Minneapolis, Minn., September 1–4, 1953b, pp. 133–143.

Alam, A. M. Z., discussion of "Resistance Relationships for Alluvial Channel Flow," by R. J. Garde and K. T. Ranga Raju, *Journal of the Hydraulics Division,* ASCE, Vol. 93, No. HY2, Proc. Paper 5129, Mar., 1967, pp. 91-96.

Alam, A. M. Z., Cheyer, T. F., and Kennedy, J. F., "Friction Factors for Flow in Sand Bed Channels," *Hydrodynamics Laboratory Report No. 78,* Massachusetts Institute of Technology, Cambridge, Mass., June, 1966.

Alam, A. M. Z., and Kennedy, J. F., "Friction Factors for Flow in Sand Bed Channels," *Journal of the Hydraulics Division,* ASCE, Vol. 95, No. HY6, Proc. Paper 6400, Nov., 1969, pp. 1973–1992.

Albertson, M. L., "Effect of Shape on the Fall Velocity of Gravel Particles," *Proceedings,* 5th Hydraulics Conference, University of Iowa Studies in Engineering, Iowa City, Iowa, 1953.

Allen, J. R. L., "A Review of the Origin and Characteristics of Recent Alluvial Sediments, *Sedimentology,* Vol. 5, No. 2, 1965, pp. 89–191.

Altschaeffl, A. G., discussion of "Erosion and Deposition of Cohesive Soils," by E. Partheniades, *Journal of the Hydraulics Division,* ASCE, Vol. 91, No. HY5, Proc. Paper 4464, Sept. 1965, p. 301.

Ambrose, H. H., "The Transportation of Sand in Pipes. Part II Free-Surface Flow," *Proceedings of the Fifth Hydraulics Conference,* Engineering Bulletin No. 34, State University of Iowa, Iowa City, Iowa, 1953, pp. 77–88.

American Society of Mechanical Engineers, "Multi-Phase Flow Symposium," Philadelphia, Pa., 1963, Nov., pp. 17–22.

Anderson, A. G., "Distribution of Sediment in a Natural Stream," *Transactions,* American Geophysical Union, Part II, Washington, D.C., 1942, pp. 678–683.

Anderson, A. G., "The Characteristics of Sediment Waves Formed by Flow in Open Channels," *Proceedings,* Third Midwestern Conference on Fluid Mechanics, University of Minnesota, Minneapolis, Minn., June, 1953, pp. 379–395.

Anderson, A. G., "On the Development of Stream Meanders," *Proceedings,* Twelfth Congress of the International Association for Hydraulic Research, *Paper No. A46,* Vol. 1, Sept., 1967, pp. 370–378.

Anderson, A. G., "The Use of Submerged Groins for the Regulation of Alluvial Streams," *Proceedings,* Symposium on Current Problems in River Training and Sediment Movement, Academy of Sciences, Budapest, Hungary, 1968.

Ansley, R. W., "The Effect of Slimes on Open-Channel Transport of Fluidized Solids," *Transactions,* Engineering Institute of Canada, No. A–7, Vol. 6, 1962.

Ansley, R. W., "Open Channel Transport of Fluidized Solids," thesis presented to the University of Alberta, at Edmonton, Alberta, Canada, in 1963, in partial fulfillment of the requirements for the degree of Doctor of Philosophy.

Apperly, L. W., "Effect of Turbulence on Sediment Entrainment," thesis presented to the University of Auckland, Auckland, New Zealand, in 1968, in partial fulfillment of the requirements for the degree of Doctor of Philosophy.

ASCE, "Friction Factors in Open Channels," Progress Report of the Task Force on Friction Factors in Open Channels of the Committee on Hydromechanics of the Hydraulics Division, E. Silberman, Chmn., *Journal of the Hydraulics Division,* ASCE, Vol. 89, No. HY2, Proc. Paper 3464, Mar., 1963, pp. 97-143.

ASCE, "Nomenclature for Bed Forms in Alluvial Channels," by the Task Force on Bed Forms in Alluvial Channels, John F. Kennedy, Chmn., *Journal of the Hydraulics Division,* ASCE, Vol. 92, No. HY3, Proc. Paper 4823, May, 1966, pp. 51-64.

ASCE, closure to "Sediment Transportation Mechanics: Initiation of Motion," by the Task Committee on Preparation of Sedimentation Manual, Vito A. Vanoni, Chmn., *Journal of the Hydraulics Division,* ASCE, Vol. 93, No. HY5, Proc. Paper 5408, Sept., 1967a, pp. 297-302.

ASCE, "Erosion of Cohesive Sediments," by the Task Committee on Erosion of Cohesive Materials, F. D. Masch, Chmn., *Journal of the Hydraulics Division,* ASCE, Vol. 94, No. HY4, Proc. Paper 6044, July, 1967b, pp. 1017–1049.

Babcock, H. A., "The State of the Art of Transporting Solids in Pipelines," 48th National Meeting of the American Institute of Chemical Engineers, Denver, Colo., Aug., 1962.

Babcock, H. A., "The Sliding Bed Flow Règime," *Proceedings,* 1st International Conference on Hydraulic Transportation of Solids in Pipes, Paper H1, (Hydrotransport 1), British Hydromechanics Research Association, 1970, pp. 1–16.

Babcock, H. A., "Heterogeneous Flow of Heterogeneous Solids," *Advances in Solid-Liquid Flow in Pipes and its Application,* I. Zandi, ed., Pergamon Press, Inc., New York, N.Y., 1971, pp. 125–148.

Babcock, H. A., and Shaw, S., discussion of "Heterogeneous Flow of Solids in Pipelines," by Iraj Zandi and George Govatos, *Journal of the Hydraulics Division,* ASCE, Vol. 93, No HY6, Proc. Paper 5543, Nov., 1967, pp. 442–445.

Bagnold, R. A., "The Measurement of sand Storms," *Proceedings,* Royal Society of London, Vol. 167A, London, England, 1938, pp. 282–291.

Bagnold, R. A., *The Physics of Blown Sand and Desert Dunes,* William Morrow and Co., New York, N.Y., 1943, p. 265.

Bagnold, R. A., "The Movement of a Cohesionless Granular Bed by Fluid Flow Over It," *British Journal of Applied Physics,* Vol. 2, 1951, pp. 29–34.

Bagnold, R. A., "Experiments on a Gravity Free Dispersion of Large Solid Spheres in a Newtonian Fluid Under Shear," *Proceedings,* Royal Society of London, Series A, Vol. 225, London, England, 1954, pp. 49–70.

Bagnold, R. A., "Some Flume Experiments on Large Grains but Little Denser than the Transporting Fluid, and Their Implications," *Proceedings,* Institute of Civil Engineers, Part III, Vol. 4, Apr., 1955, pp. 174–205.

Bagnold, R. A., "Sediment Discharge and Stream Power," *United States Geological Survey Circular 421,* Washington, D.C., 1960.

Bain, A. G., and Bonnington, S. T., *The Hydraulic Transport of Solids by Pipeline,* Pergamon Press, Inc., New York, N.Y., 1970.

Barr, D. I. H., and Herbertson, J. G., discussion of "Sediment Transportation Mechanics: Initiation of Motion," by the Task Committee on Preparation of Sedimentation Manual, Committee on Sedimentation of the Hydraulics Division, Vito A. Vanoni, Chmn., *Journal of the Hydraulics Division,* ASCE, Vol. 92, No. HY6, Proc. Paper 4959, Nov., 1966, pp. 248-251.

Barr, D. I. H., and Herbertson, J. G., "Similitude Theory Applied to Correlation of Flume Sediment Transport Data," *Water Resources Research,* Vol. 4, No. 2, Apr., 1968, pp. 307—316.

Barton, J. K., and Lin, Pin-Nam, "A Study of the Sediment Transport in Alluvial Channels," *Report No. 55JRB2,* Colorado Agricultural and Mechanical College, Civil Engineering Department, Mar., 1955.

Bata, G., and Knezevich, B., "Some Observations on Density Currents in the Laboratory and in the Field," *Proceedings,* Minnesota International Hydraulics Convention, 1953, pp. 387–400.

Beaty, C. B., "Origin of Alluvial Fans, White Mountains, California and Nevada," *Annals,* Association of American Geographers, Vol. 53, Washington, D.C., 1963, pp. 516–535.

Bell, H. S., "Stratified Flow in Reservoirs and Its Use in Prevention of Silting," *Miscellaneous Publication No. 491,* United States Department of Agriculture, Washington, D.C., 1942.

Bender, D. L., "Suspended Sediment Transport in Alluvial Irrigation Channels," *Report CER No. 56DLB7,* Colorado Agricultural and Mechanical College, Civil Engineering Department, June, 1956.

Bennett, H. H., *Soil Conservation,* 1st ed., McGraw-Hill Book Co., Inc., New York, N.Y., 1939.

Benson, M. A., "Spurious Correlation in Hydraulics and Hydrology," *Journal of the Hydraulics Division,* ASCE, Vol. 91, No. HY4, Proc. Paper 4393, July, 1965, pp. 35–42.

Beverage, J. P., and Culbertson, J. K., "Hyperconcentrations of Suspended Sediment," *Journal of the Hydraulics Division,* ASCE, Vol. 90, No. HY6, Proc. Paper 4136, Nov., 1964, pp. 117–128.

Blackwelder, E., "Mudflow as a Geologic Agent in Semiarid Mountains," *Bulletin,* Geological Society of America, Vol. 39, Boulder, Colo., 1928, pp. 465–483.

Blaisdell, F. W., "Use of Sand Beds for Comparing Relative Stilling Basin Performance," *Transactions,* American Geophysical Union, Part II, Washington, D.C., 1942, pp. 633–639.

Blatch, N. S., discussion of "Works for the Purification of the Water Supply of Washington, D.C.," by A. Hazen and E. D. Hardy, *Transactions,* ASCE, Paper No. 1036, Dec., 1906, pp. 400–408.

Blench, T., "Mobile-Bed Fluviology," Department of Technical Services, University of Alberta, Edmonton, Alberta, Canada, 1966a.

Blench, T., discussion of "Sediment Transportation Mechanics: Initiation of Motion," by the Task Committee on Preparation of Sedimentation Manual, Committee on Sedimentation of the Hydraulics Division, Vito A. Vanoni, Chmn., *Journal of the*

Hydraulics Division, ASCE, Vol. 92, No. HY2, Proc. Paper 4891, Sept., 1966b, pp. 287–288.

Blench, T., "Coordination in Mobile Bed Hydraulics," *Journal of the Hydraulics Division,* ASCE, Vol. 95, No. HY6, Proc. Paper 6884, Nov., 1969, pp. 1871–1898.

Blench, T., "Regime Theory Design of Canals with Sand Beds," *Journal of the Irrigation and Drainage Division,* ASCE, Vol. 96, No. IR2, Proc. Paper 7381, June, 1970, pp. 205–213.

Blinco, P. H., and Partheniades, E., "Turbulence Characteristics in Free Surface Flows Over Smooth and Rough Boundaries," *Journal of Hydraulic Research*, Vol. 9, No. 1, 1971, pp. 43-69.

Blissenbach, E., "Geology of Alluvial Fans in Semiarid Regions," *Bulletin,* Geological Society of America, Vol. 65, Boulder, Colo., 1954, pp. 175–190.

Bohlen, W. F., "Hotwire Anemometer Study of Turbulence in Open-Channel Flows Transporting Neutrally Buoyant Particles," *Report No. 69–1,* Experimental Sedimentology Laboratory, Department of Earth and Planetary Sciences, Massachusetts Institute of Technology, Cambridge, Mass., 1969.

Bonnefille, R., and Goddet, J., "Study of Density Currents in a Canal," *Proceedings,* Eighth Congress International Association for Hydraulic Research, Vol. 2, Montreal, Canada, Aug., 1959.

Breusers, H. N. C., discussion of "Sediment Transportation Mechanics: Erosion of Sediment," Progress Report by the Task Committee on Preparation of Sedimentation Manual of the Committee on Sedimentation of the Hydraulics Division, Vito A. Vanoni, Chmn., *Journal of the Hydraulics Division,* ASCE, Vol. 89, No. HY1, Proc. Paper 3405, Jan., 1963, pp. 277–281.

British Hydromechanics Research Association, Hydrotransport 1, *Proceedings,* 1st International Conference on Hydraulic Transportation of Solids in Pipes, 1970a.

British Hydromechanics Research Association, *The Hydraulic Transport of Solids in Pipes—A Bibliography,* 1970b.

Brooks, N. H., "Laboratory Studies of the Mechanics of Streams Flowing Over a Movable Bed of Fine Sand," thesis presented to California Institute of Technology, at Pasadena, Calif., in 1954, in partial fulfillment of the requirements for the degree of Doctor of Philosophy.

Brooks, N. H., "Mechanics of Streams with Movable Beds of Fine Sand," *Transactions,* ASCE, Vol. 123, Paper No. 2931, 1958, pp. 526–594.

Brooks, N., "Flow Measurements of Solid-Liquid Mixtures Using Venturi and Other Meters," *Proceedings of the Institution of Mechanical Engineers,* Vol. 176, London, England, No. 6, 1962, pp. 123–139.

Brooks, N. H., discussion of "Boundary Shear Stresses in Curved Trapezoidal Channels," by A. T. Ippen and P. A. Drinker, *Journal of the Hydraulics Division,* ASCE, Vol. 89, No. HY3, Proc. Paper 3529, May, 1963a, pp. 327–333.

Brooks, N. H., "Calculation of Suspended Load Discharge from Velocity, and Concentration Parameters," *Proceedings of the Federal Interagency Sedimentation Conference,* 1965, *Miscellaneous Publication No. 970,* Agricultural Research Service, United States Department of Agriculture, Washington, D.C., 1963b, pp. 229–237.

Brown, C. B., "Sediment Transportation," Chap. XII, *Engineering Hydraulics,* H. Rouse, ed., John Wiley and Sons, Inc., New York, N.Y., 1950.

Buckley, A. B., "The Influence of Silt on the Velocity of Water Flowing in Open Channels," *Proceedings,* Institute of Civil Engineers, Vol. 226, Pt. II, London, England 1922-1923, pp. 183–197.

Bull, W. B., "Alluvial Fans and Near-Surface subsidence in Western Fresno County, California," *Professional Paper 437-A,* United States Geological Survey, Washington, D.C., 1964.

Burke, P. P., "Effect of Water Temperature on Discharge and Bed Configuration, Mississippi River at Red River Landing, Louisiana," United States Army Engineer District, New Orleans, La., 1965.

Callander, R. A., "Instability and River Channels," *Journal of Fluid Mechanics,* Vol. 36, Part 3, 1969, pp. 465–480.

Camp, T. R., "Sedimentation and the Design of Settling Tanks," *Transactions,* ASCE, Vol. 111, Paper No. 2285, 1946, pp. 895–958.

Carey, W. C., "Effect of Temperature on Riverbed Configuration: Its Possible Stage-Discharge Implications," *Proceedings of the Federal Interagency Sedimentation Conference, Miscellaneous Publication No. 970,* Agricultural Research Service, United States Department of Agriculture, Washington, D.C., 1963, pp. 237–272.

Carey, W. C., "Formation of Flood Plain Lands," *Journal of the Hydraulics Division,* ASCE, Vol. 95, No. HY3, Proc. Paper 6574, May, 1969, pp. 981–994.

Carey, W. C., and Keller, M. D., "Systematic Changes in the Beds of Alluvial Rivers," *Journal of the Hydraulics Division,* ASCE, Vol. 83, No. HY4, Proc. Paper 1331, Aug., 1957.

Carstens, M. R., "A Theory for Heterogeneous Flow of Solids in Pipes," *Journal of the Hydraulics Division,* ASCE, Vol. 95, No. HY1, Proc. Paper 6354, Jan., 1969, pp. 275–286.

Chamberlain, A. R., "Fine Sand Transport in Twelve-Inch Pipes," thesis presented to Colorado State University, at Fort Collins, Colo., in 1955, in partial fulfillment of the requirements for the degree of Doctor of Philosophy.

Chepil, W. S., "Relation of Wind Erosion to the Dry Aggregate Structure of a Soil," *Scientific Agriculture,* Vol. 21, Agricultural Institute of Canada, Ottawa, Canada, 1941, pp. 488–507.

Chepil, W. S., "Dynamics of Wind Erosion: I; The Nature of Movement of Soil by Wind," *Soil Science,* Vol. 60, 1945a, pp. 305–320.

Chepil, W. S., "Dynamics of Wind Erosion: II; Initiation of Soil Movement," *Soil Science,* Vol. 60, 1945b, pp. 397–411.

Chepil, W. S., "Dynamics of Wind Erosion: III; Transport Capacity of the Wind," *Soil Science,* Vol. 60, 1945c, pp. 475–480.

Chepil, W. S., "Dynamics of Wind Erosion: IV; The Translocating and Abrasive Action of the Wind," *Soil Science,* Vol. 61, 1946, pp. 167–177.

Chepil, W. S., "Properties of Soil which Influence Wind Erosion: II; Dry Aggregate Structure as an Index of Erodibility," *Soil Science,* Vol. 69, 1950, pp. 403–414.

Chepil, W. S., "Factors that Influence Cold Structure and Erodibility of Soil by Wind: IV; Sand, Silt, and Clay, *Soil Science,* Vol. 80, 1955, pp. 155–160.

Chepil, W. S., "Influence of Moisture on Erodibility of Soil by Wind," *Proceedings,* Soil Science Society of America, Vol. 20, Madison, Wisc., 1956, pp. 288–292.

Chepil, W. S., "Sedimentary Characteristics of Dust Storms: I; Sorting of Wind Eroded Soil Material," *American Journal of Science,* Vol. 255, 1957a, pp. 12–22.

Chepil, W. S., "Width of Field Strips to Control Wind Erosion," *Technical Bulletin 92,* Kansas Agricultural Experiment Station, Manhattan, Kans., 1957b.

Chepil, W. S., "Soil Conditions that Influence Wind Erosion," *Technical Bulletin No. 1185,* United States Department of Agriculture, Washington, D.C., 1958.

Chepil, W. S., "Equilibrium of Soil Grains at the Threshold of Movement by Wind," *Proceedings,* Soil Science Society of America, Vol. 23, Madison, Wisc., 1959a, pp. 422–428.

Chepil, W. S., "Wind Erodibility of Farm Fields," *Journal of Soil and Water Conservation,* Vol. 14, No. 5, 1959b, pp. 214–219.

Chepil, W. S., "Conversion of Relative Field Erodibility to Annual Soil Loss by Wind," *Proceedings,* Soil Science Society of America, Vol. 24, Madison, Wisc., 1960, pp. 143–145.

Chepil, W. S. "The Use of Spheres to Measure Lift and Drag on Wind-eroded Soil Grains," *Proceedings,* Soil Science Society of America, Vol. 25, Madison, Wisc., 1961, pp. 243–245.

Chepil, W. S., and Milne, R. A., "Wind Erosion in Relation to Roughness of Surface," *Soil Science,* Vol. 52, 1941, pp. 417–433.

Chepil, W. S., and Siddoway, F. H., "Strain-gage Anemometer for Analyzing Various Characteristics of Wind Turbulence," *Journal of Meteorology,* Vol. 16, 1959, pp. 411–418.

Chepil, W. S., Siddoway, F. H., and Armbrust, D. V., "Climatic Factor for Estimating Wind Erodibility of Farm Fields," *Journal of Soil and Water Conservation,* Vol. 17, 1962, pp. 162–165.

Chepil, W. S., and Woodruff, N. P., "Sedimentary Characteristics of Dust Storms: II; Visibility and Dust Concentration," *American Journal of Science,* Vol. 255, 1957, pp. 104–114.

Chepil, W. S., and Woodruff, N. P., "Estimations of Wind Erodibility of Farm Fields," *Production Research Report No. 25,* Agricultural Research Service, United States Department of Agriculture, Washington, D. C., 1959.

Chepil, W. S. and Woodruff, N. P., "The Physics of Wind Erosion and its Control," *Advances in Agronomy,* American Society of Agronomy, Vol. 15, Madison, Wisc., 1963, pp. 211–302.

Chepil, W. S., Woodruff, N. P., and Zingg, A. W., "Field Study of Wind Erosion in Western Texas," *SCS-TP-125,* United States Department of Agriculture, Washington, D.C., 1955.

Chien, N., "The Present Status of Research on Sediment Transport," *Transactions,* ASCE, Vol. 121, Paper No. 2824, 1956, pp. 833–868.

Chien, N., "A Concept of the Regime Theory," *Transactions,* ASCE, Vol. 122, Paper No. 2884, 1957, pp. 785–793.

Chiu, C. L., and Seman, J. J., "Head Loss in Spiral-Liquid Flow in Pipes," *Advances in Solid-liquid Flow in Pipes and its Application.* I. Zandi, ed., Pergamon Press, Inc., New York, N.Y., 1971.

Chow, V. T., *Open-Channel Hydraulics,* McGraw-Hill Book Co., Inc., New York, N.Y., 1959.

Colby, B. R., "Relationship of Unmeasured Sediment Discharge to Mean Velocity," *Transactions,* American Geophysical Union, Vol. 38, No. 5, Washington, D.C., Oct., 1957, pp. 707–717.

Colby, B. R., "Discontinuous Rating Curves for Pigeon Roost and Cuffawa Creeks in Northern Mississippi," *Report ARS41–36,* Agricultural Research Service, Apr., 1960.

Colby, B. R., discussion of "Sediment Transportation Mechanics: Introduction and Properties of Sediment," Progress Report by the Task Committee on Preparation of Sedimentation Manual of the Committee on Sedimentation of the Hydraulics Division, Vito A. Vanoni, Chmn., *Journal of the Hydraulics Division,* ASCE, Vol. 89, No. HY1, Proc. Paper 3405, Jan., 1963, pp. 266–268.

Colby, B. R. "Practical Computation of Bed-Material Discharge," *Journal of the Hydraulics Division,* ASCE, Vol. 90, No. HY2, Proc. Paper 3843, Mar., 1964a, pp. 217–246.

Colby, B. R., "Discharge of Sands and Mean Velocity Relationships in Sand-Bed Streams," *Professional Paper 462–A,* United States Geological Survey, Washington, D.C., 1964b.

Colby, B. R., and Hembree, C. H., "Computations of Total Sediment Discharge Niobrara River Near Cody, Nebraska," *Water-Supply Paper 1357,* United States Geological Survey, Washington, D.C., 1955.

Colby, B. R., Hembree, C. H., and Rainwater, F. H., "Sedimentation and Chemical Quality of Surface Waters in the Wind River Basin, Wyoming," *Water-Supply Paper 1373,* United States Geological Survey, Washington, D.C., 1956.

Colby, B. R., and Hubbell, D. W., "Simplified Methods for Computing Total Sediment Discharge with Modified Einstein Procedure," *Water-Supply Paper No. 1593,* United States Geological Survey, Washington, D.C., 1961.

Colby, B. R., and Scott, C. H., "Effects of Water Temperature on the Discharge of Bed Material," *Professional Paper 462–G,* United States Geological Survey, Washington, D.C., 1965.

Coleman, N. L., "Observations of Resistance Coefficients in a Natural Channel," *Publication No. 59,* International Association of Scientific Hydrology, Commission on Land Erosion, 1962, pp. 336–352.

Colorado School of Mines, "Theory of Solids Transportation in Pipelines," Golden, Colo., 1963.

Condolios, E., and Chapus, E. E., "Transporting Solid Materials in Pipelines," *Chemical Engineering,* June 24, 1963a, pp. 93–98.

Condolios, E., and Chapus, E. E., "Designing Solids-Handling Pipelines–Part II," *Chemical Engineering,* July 8, 1963b, pp. 131–138.

Cooper, R. H., and Peterson, A. W., "Analysis of Comprehensive Bed Load Transport Data from Flumes," presented at the August 21–23, 1968, ASCE Hydraulics Specialty Conference, held at Cambridge, Mass.

Craven, J. P., "The Transportation of Sand in Pipes," *Proceedings of the 5th Hydraulic Conference,* Engineering Bulletin No. 34, State University of Iowa, Iowa City, Iowa, 1953, pp. 67–76.

Cunha, Veiga da L., "About the Roughness in Alluvial Channels with Comparatively Coarse Bed Material," *Proceedings,* Twelfth Congress of the International Association for Hydraulic Research, Vol. 1, Paper No. A10, Sept., 1967, pp. 76–84.

Daily, J. W., and Bugliarello, G., "The Effects of Fibers on Velocity Distribution, Turbulence and Flow Resistance of Dilute Suspension," *Technical Report No. 30,* Hydrodynamics Laboratory, Massachusetts Institute of Technology, Cambridge, Mass., 1958.

Daily, J. W., and Chu, T. K., "Rigid Particle Suspensions in Turbulent Shear Flow: Some Concentration Effects," *Report No. 48,* Hydrodynamics Laboratory, Massachusetts Institute of Technology, Cambridge, Mass., 1961.

Daily, J. W., and Hardison, R. L., "Rigid Particle Suspensions in Turbulent Shear Flow: Measurement of Total–Head, Velocity and Turbulence with Impact Tubes," *Report No. 67,* Hydrodynamics Laboratory, Massachusetts Institute of Technology, Cambridge, Mass., 1964.

Daily, J. W., and Harleman, D. R. F., *Fluid Dynamics,* Addison–Wesley Publishing Co., Inc., New York, N.Y., 1966.

Danel, P., "Charriage en Suspension," International Association of Hydraulic Structures Research, Report of Second Meeting, Appendix 7, Stockholm, Sweden, 1948, pp. 113–144.

Dawdy, D. R., "Depth–Discharge Relations of Alluvial Streams–Discontinuous Rating Curves," *Water-Supply Paper 1948–C,* United States Geological Survey, Washington, D.C., 1961.

Dobbins, W. E., "Effect of Turbulence on Sedimentation," *Transactions,* ASCE, Vol. 109, Paper No. 2218, 1944, pp. 629–678.

Doddiah, D., Albertson, M., and Thomas, R., "Scour from Jets," *Proceedings,* Minesota International Hydraulics Convention, Minneapolis, Minn., September 1–4, 1953, pp. 161–169.

DuBoys, P., "Le Rohne et les Rivieres a Lit Affouillable," Annales des Ponts et Chaussees, Series 5, Vol. 18, 1879, pp. 141–195.

Dunn, I. S., "Tractive Resistance of Cohesive Channels," *Journal of the Soil Mechanics and Foundations Division,* ASCE, No. SM3, Proc. Paper 2062, June, 1959, pp. 1–24.

Du Plessis, M. P., and Ansley, R. W., "Settling Parameter in Solids Pipeline," *Journal of the Pipeline Division,* ASCE, Vol. 93, No. PL2, Proc. Paper 5340, July, 1967, pp. 1–17.

Durand, R., "Basic Relationship of the Transportation of Solids in Pipes–Experimental Research," *Proceedings,* Minnesota International Hydraulics Conference, Minneapolis, Minn., Sept., 1953, pp. 89–103.

Durand, R., and Condolios, E., "Experimental Investigation on the Transport of Solids in Pipes," *Le Journéls d'Hydraulique,* Société Hydrotechnique de France, Grenoble, France, June, 1952.

Durepaire, M. P., "Contribution a L'Etude du Dragage et du Refoulement des Deblais a L'Efat de Mixture," *Extrast des Annales des Ponts et Chaussees,* Fase II, Feurier, 1939, pp. 165–254.

Egiazaroff, I. V., "Coefficient f de la force d'entrainment critique des materiaux par charriage," *Proceedings of Academy of Sciences,* Armenian Soviet Socialist Republic, *Translation No. 499,* Electricité de France Service des Etudes et Récherches Hydrauliques, Paris, France, 1950.

Egiazaroff, I. V., "L'Equation generale du transport des alluvions non cohesives par un courent fluide," *Paper D43, Proceedings of the 7th Congress,* International Association for Hydraulic Research, Lisbon, Portugal, 1957, pp. 1–10.

Egiazaroff, I. V., "Calculation of Nonuniform Sediment Concentrations," *Journal of the Hydraulics Division,* ASCE, Vol. 91, No. HY4, Proc. Paper 4417, July, 1965, pp. 225–247.

Einstein, H. A., "Formulas for the Transportation of Bed Load," *Transactions,* ASCE, Vol. 107, Paper No. 2140, 1942, pp. 561–573.

Einstein, H. A., "Bed Load Transportation in Mountain Creek," *SCS–TP–55* United States Department of Agriculture, Soil Conservation Service, Washington, D.C., 1944.

Einstein, H. A., "The Bed Load Function for Sediment Transportation in Open Channels, *Technical Bulletin 1026,* United States Department of Agriculture, Soil Conservation Service, Washington, D.C., 1950.

Einstein, H. A., discussion of "Calculation of Nonuniform Sediment Concentrations," by I. V. Egiazaroff, *Journal of the Hydraulics Division,* ASCE, Vol. 92, No. HY2, Proc. Paper 4708, 1966, pp. 439–440.

Einstein, H. A., and Barbarossa, N., "River Channel Roughness," *Transactions,* ASCE, Vol. 117, Paper No. 2528, 1952, pp. 1121–1146.

Einstein, H. A., and Chien, N., "Second Approximation to the Solution of the Suspended Load Theory," Series No. 47, Issue No. 2, Institute of Engineering Research, University of California, Berkeley, Calif., January 31, 1952.

Einstein, H. A., and Chien, N., "Can the Rate of Wash Load be Predicted from the Bed Load Function," *Transactions,* American Geophysical Union, Vol. 34, No. 6, Washington, D.C., Dec., 1953a, pp. 876–882.

Einstein, H. A., and Chien, N., "Transport of Sediment Mixtures with Large Range of Grain Size," *MRD Sediment Series No. 2,* United States Army Engineer Division, Missouri River, Corps of Engineers, Omaha, Neb., 1953b.

Einstein, H. A. and Chien, N., "Second Approximation to the Solution of the Suspended Load Theory," *MRD Series Report No. 2,* University of California and Missouri River Division, United States Corps of Engineers, June, 1954.

Einstein, H. A. and Chien, N. "Effects of Heavy Sediment Concentration near the Bed on Velocity and Sediment Distribution," *MRD Series No. 8,* University of California, Institute of Engineering Research and United States Army Engineering Division, Missouri River, Corps of Engineers, Omaha, Nebr., Aug., 1955.

Einstein, H. A., and El-Samni, E. S. A., "Hydrodynamic Forces on a Rough Wall," *Review of Modern Physics,* Vol. 21, 1949, pp. 520–524.

Einstein, H. A., and Li, H., "The Viscous Sublayer Along a Smooth Boundary," *Transactions,* ASCE, Vol. 123, Paper No. 2992, 1958, pp. 293–313.

Einstein, H. A., Anderson, A. G., and Johnson, J. W., "A Distinction between Bed Load and Suspended Load in Natural Streams," *Transactions,* American Geophysical Union, Annual Meeting 21 (Part 2), Washington, D.C., 1940, pp. 628–633.

Elata, C., and Ippen, A. T., "The Dynamics of Open Channel Flow with Suspensions of Neutrally Buoyant Particles," *Technical Report No. 45,* Massachusetts Institute of Technology, Hydrodynamics Laboratory, Cambridge, Mass., Jan., 1961.

Ellison, T. H., "Turbulent Transport of Heat and Momentum from an Infinite Rough Plane," *Journal of Fluid Mechanics,* Vol. 2, Part 5, 1957, pp. 456–466.

Engelund, F., "Hydraulic Resistance of Alluvial Streams," *Journal of the Hydraulics Division,* ASCE, Vol. 92, No. HY2, Proc. Paper 4739, Mar., 1966, pp. 315–327.

Engelund, F., closure to "Hydraulic Resistance of Alluvial Streams," *Journal of the Hydraulics Division,* ASCE, Vol. 93, No. HY4, Proc. Paper 5304, July, 1967, pp. 287–296.

Engelund, F., and Hansen, E., "A Monograph on Sediment Transport in Alluvial Streams, *Teknisk Vorlag,* Copenhagen, Denmark, 1967.

Exner, F., "Uber die Wechselwirkung swischen Wasser und Geschiebe in Flussen, *Proceedings,* Vienna Academy of Sciences, Section IIA, Vol. 134, 1925, p. 199.

Fenneman, N. M., "Floodplains Produced without Floods," *Bulletin,* American Geographical Society, Vol. 38, New York, N.Y., 1906, pp. 89–91.

Field, W. G., "Effects of Density Ratio on Sedimentary Similitude," *Journal of the Hydraulics Division,* ASCE, Vol. 94, No. HY3, Proc. Paper 5948, May, 1968, pp. 705–719.

Fietz, T. R., and Wood, I. R., "Three-Dimensional Density Current," *Journal of the Hydraulics Division,* ASCE, Vol. 93, No. HY6, Proc. Paper 5549, Nov., 1967, pp. 1–23.

Fisk, H. N., "Geological Investigation of the Alluvial Valley of the Lower Mississippi River," United States Army Corps of Engineers, Mississippi River Commission, Vicksburg, Miss., 1944.

Fisk, H. N., "Fine-grained Alluvial Deposits and Their Effects on Mississippi River Activity," United States Army Corps of Engineers, Waterways Experiment Station, Vicksburg, Miss., 1947.

Fisk, H. N., "Mississippi River Valley Geology Relation to Regime," *Transactions,* ASCE, Vol. 117, Paper No. 2511, 1952, pp. 667–689.

Flaxman, E. M., "Channel Stability in Undisturbed Cohesive Soils," *Journal of the Hydraulics Division,* ASCE, Vol. 89, No. HY2, Proc. Paper 3462, Mar., 1963, pp. 87–96.

Flaxman, E. M., discussion of "Sediment Transportation Mechanics: Initiation of Motion," by the Task Committee on Preparation of Sedimentation Manual, Committee on Sedimentation of the Hydraulics Division, Vito A. Vanoni, Chmn., *Journal of the Hydraulics Division,* ASCE, Vol. 92, No. HY6, Proc. Paper 4959, No.v, 1966, pp. 245–248.

Fleming, G., and Poodle, T., "Particle Size of River Sediments," *Journal of the Hydraulics Division,* ASCE, Vol. 96, No. HY2, Proc. Paper 7068, Feb., 1969, pp. 431–439.

Fontein, F. J., "Research and Development Work on Vertical Hydraulic Transport," *Colliery Guardian,* Vol. 197, Holland, September 11, 1958.

Fortier, S., and Scobey, F. C., "Permissible Canal Velocities," *Transactions,* ASCE, Vol. 89, Paper No. 1588, 1926, pp. 940–984.

Fortino, E. P., "Flow Measurement Techniques for Hydraulic Dredges," *Journal of the Waterways and Harbors Division,* ASCE, Vol. 92, No. WW1, Proc. Paper 4683, Feb., 1966, pp. 109–125.

Fowler, L., "Notes on the Turbulence Function k," Missouri River Division, United States Army, Corps of Engineers, Nov., 1953.

Franco, J. J., "Effects of Water Temperature on Bed-Load Movement," *Journal of the Waterways and Harbors Division,* ASCE, Vol. 94, No. WW3, Proc. Paper 6083, Aug., 1968, pp. 343–352.

Free, E. E., "The Movement of Soil Material by the Wind," *Bulletin No. 68,* Bureau of Soils, United States Department of Agriculture, Washington, D.C., 1911.

Fry, A. S., Churchill, M. A., and Elder, R. A., "Significant Effects of Density Currents in TVA's Integrated Reservoir and River System," *Proceedings,* Minnesota International Hydraulic Convention, 1953, pp. 335–354.

Fryxell, F. M., and Horberg, L., "Alpine Mudflows in Grand Teton National Park, Wyoming," *Bulletin,* Geological Society of America, Vol. 54, Boulder, Colo., 1943, pp. 457–472.

Faddick, R. R., and Babcock, H. A., discussion of "Sediment Transportation Mechanics: J. Transportation of Sediment in Pipes," by the Task Committee on Preparation of Sedimentation Manual, Committee on Sedimentation of the Hydraulics Division, Vito A. Vanoni, Chmn., *Journal of the Hydraulics Division,* ASCE, Vol. 97, HY5, Proc. Paper 8081, May, 1971, PP. 745–748.

Garde, R. J., "Sediment Transport Through Pipes," thesis presented to Colorado A & M University, at Fort Collins, Colo., in 1956, in partial fulfillment of the requirements for the degree of Master of Science.

Garde, R. J., and Raju, K. G. R., "Resistance Relationships for Alluvial Channel Flow," *Journal of the Hydraulics Division,* ASCE, Vol. 92, No. HY4, Proc. Paper 4869, July, 1966, pp. 77–100.

Geiger, R., *The Climate Near the Ground,* Harvard University Press, Cambridge, Mass., 1957, P. 494.

Gessler, J., "The Beginning of Bed Load Movement of Mixtures Investigated as Natural Armoring in Channels," *Report No. 69,* Laboratory of Hydraulic Research and Soil Mechanics, Swiss Federal Institute of Technology, Zurich, Switzerland (in German), 1965.

Gessler, J., "Self-Stabilizing Tendencies of Alluvial Channels," *Journal of the Waterways and Harbors Division,* ASCE, Vol. 96, No. WW2, Proc. Paper 7263, May, 1970, pp. 235–249.

Gibert, R., "Transport Hydraulique et Refoulement des Mixtures en Conduit," *Anals des Pontes et Chaussees,* Vol. 130, No. 12, 1960.

Gilbert, G. K., "Transportation of Debris by Running Water," *Professional Paper No. 86,* United States Geological Survey, 1914.

Gill, M. A., "Rationalization of Lacey's Regime Flow Equation," *Journal of the Hydraulics Division,* ASCE, Vol. 94, No. HY4, Proc. Paper 6039, July, 1968, pp. 983–995.

Goldstein, S., *Modern Developments in Fluid Dynamics,* Vols. I and II, Oxford University Press, Oxford, England, 1950, pp. 256–265, 561–600.

Gradowczyk, M. H., "Wave Propagation on the Erodible Bed of an Open Channel," *Journal of Fluid Mechanics,* Vol. 33, 1968, pp. 93–112.

Graf, W. H., "A Modified Venturi Meter for Measuring Two-Phase Flow," *Journal of Hydraulic Research,* International Association for Hydraulic Research, Vol. 5, No. 3, 1967.

Graf, W. H., *Hydraulics of Sediment Transport,* McGraw-Hill Book Co., Inc., New York, N.Y., 1971.

Graf, W. H., and Acaroglu, E. R., "Homogeneous Suspensions in Circular Conduits," *Journal of the Pipeline Division,* ASCE, Vol. 93, No. PL2, Proc. Paper 5352, July, 1967, pp. 63–69.

Graf, W. H. and Acaroglu, E. R., "Sediment Transport in Conveyance Systems," *Bulletin of the International Association of Scientific Hydrology,* Vol. XIII, No. 2, 1968.

Gregory, W. B., *Mechanical Engineering,* Vol. 49, 1927, p. 609; see also S. S. Zabrodsky, *Hydrodynamics and Heat Transfer in Fluidized Beds,* MIT Press, Cambridge, Mass., 1966.

Griffith, W. M., "A Theory of Silt Transportation," *Transactions,* ASCE, Vol. 104, Paper No. 2052, 1939, pp. 1733–1748.

Grissinger, E. H., and Asmussen, L. E., discussion of "Channel Stability in Undisturbed Cohesive Soils," by E. M. Flaxman, *Journal of the Hydraulics Division,* ASCE, Vol. 89, No. HY6, Proc. Paper 3708, Nov., 1963, pp. 259–264.

Grover, N. C., and Howard, C. S., "The Passage of Turbid Water through Lake Mead," *Transactions,* ASCE, Vol. 103, Paper No. 1994, 1938, pp. 720–790.

Guy, H. P., Rathbun, R. E., and Richardson, E. V., "Recirculation and Sand-Feed Type Flume Experiments, *Journal of the Hydraulics Division,* ASCE, Vol. 93, No. HY5, Proc. Paper 5428, Sept., 1967, pp. 97–114.

Guy, H. P., Simons, D. B., and Richardson, E. V., "Summary of Alluvial Channel Data from Flume Experiments, 1956–61," *Professional Paper 462–I,* United States Geological Survey, 1966.

Hack, J. T., and Goodlett, J. C., "Geomorphology and Forest Ecology of a Mountain Region in the Central Appalachians," *Professional Paper 347,* United States Geological Survey, 1960.

Halbronn, G., "Remarque sur la théorie de l'Austausch appliqueé au transport des materiaux en suspension," *Proceedings,* International Association of Hydraulic Research, 3rd Meeting, 11–9, Grenoble, France, 1949, pp. 1–6.

Hanna, R. M., and Bidwell, O. W., "The Relation of Some Loessal Soils of Northeastern Kansas to the Texture of the Underlying Loess," *Proceedings,* Soil Science Society of America, Vol. 19, Madison, Wisc., 1955, pp. 354–359.

Hansen, E., "The Formation of Meanders as a Stability Problem," *Basic Research Report No. 13,* Technical University of Denmark, Hydraulic Laboratory, Jan., 1967, pp. 9–13.

Happ, S. C., "Significance of Texture and Density of Alluvial Deposits in the Middle Rio Grande Valley," *Journal of Sedimentary Petrology,* Vol. 14, Apr., 1944, pp. 3–19.

Happ, S. C., *Stream-Channel Control, Applied Sedimentation,* Parker D. Trask, ed., John Wiley and Sons, Inc., New York, N.Y., and Chapman & Hall, London, England, 1950, pp. 319–335.

Happ, S. C., Rittenhouse, G., and Dobson, G. C., "Some Principles of Accelerated Stream and Valley Sedimentation," *Technical Bulletin 695,* United States Department of Agriculture, Washington, D.C., 1940.

Harleman, D. R. F., "Stratified Flow," *Handbook of Fluid Dynamics,* V. Streeter, ed., McGraw-Hill Book Co., Inc., New York, N.Y., 1961.

Harleman, D. R. F., and Ippen, A. T., "Salinity Intrusion Effects in Estuary Shoaling," *Journal of the Hydraulics Division,* ASCE, Vol. 95, No. HY1, Proc. Paper 6340, Jan., 1969, pp. 9–27.

Harrison, A. S., "Report on Special Investigations of Bed Sediment Segregation in a Degrading Bed," Report Series No. 33, University of California, Institute of Engineering Research, Issue No. 1, Berkeley, Calif., Sept., 1950.

Hayashi, T., "Formation of Dunes and Antidunes in Open Channels," *Journal of the Hydraulics Division,* ASCE, Vol. 96, No. HY2, Proc. Paper 7056, Feb., 1970, pp. 357–366.

Haynie, R. B., and Simons, D. B., "Design of Stable Channels in Alluvial Materials," *Journal of the Hydraulics Division,* ASCE, Vol. 94, No. HY6, Proc. Paper 6217, Nov., 1968, pp. 1399–1420.

Hembree, C. H., Colby, B. R., Swenson, H. A. and Davis, J. H., "Sedimentation and Chemical Quality of Water in the Powder River Drainage Basin, Wyoming and Montana," *Circular 170,* United States Geological Survey, Washington, D.C., 1952.

Henderson, F. M., "Stability of Alluvial Channels," *Journal of the Hydraulics Division,* ASCE, Vol. 87, No. HY6, Proc. Paper 2984, Nov., 1961, pp. 109–138.

Henderson, F. M., *Open Channel Flow,* The Macmillan Co., New York, N.Y., 1966.

Highway Research Board, "Scour at Bridge Waterways," *Report No. 5,* National Cooperative Highway Research Program, Synthesis of Highway Practice, 1970.

Hino, M. "Turbulent Flow With Suspended Particles," *Journal of the Hydraulics Division,* ASCE, Vol. 89, No. HY4, Proc. Paper 3579, July, 1963, pp. 161–185.

Hjulström, F., "Studies of the Morphological Activity of Rivers as Illustrated by the River Fyris," *Bulletin,* Geological Institute of Upsala, Vol. XXV, Upsala, Sweden, 1935.

Homayounfar, F., "Flow of Multicomponent Slurries," thesis presented to the University of Delaware, at Newark, Del., in 1965, in partial fulfillment of the requirements for the degree of Master of Science.

Hooke, R. L., "Processes on Arid-Region Alluvial Fans," *Journal of Geology,* Vol. 75, 1967, pp. 438–460.

Houghton, G., "Particle Retardation in Vertically Oscillating Fluids," *The Canadian Journal of Chemical Engineering,* Vol. 46, No. 2, 1968, pp. 79–81.

Howard, G. W., "Transportation of Sand and Gravel in a Four-Inch Pipe," *Transactions,* ASCE, Vol. 104, Paper No. 2039, 1939, pp. 1334–1348.

Howard, G. W., "Effects of Rifling on Four-Inch Pipe Transporting Solids," *Transactions,* ASCE, Vol. 106, Paper No. 2101, 1941, pp. 135–137.

Howard, C. D. D., discussion of "Sediment Transportation Mechanics: Introduction and Properties of Sediment," Progress Report of the Task Committee on Preparation of Sedimentation Manual of the Committee on Sedimentation of the Hydraulics Division, Vito A. Vanoni, Chmn., *Journal of the Hydraulics Division,* ASCE, Vol. 88, No. HY6, Proc. Paper 3338, Nov., 1962, pp. 235–237.

Howard, C. D. D., "The Effect of Fines on the Pipeline Flow of Sand Water Mixtures," thesis presented to the University of Alberta, at Edmonton, Alberta, in 1962b, in partial fulfillment of the requirements for the degree of Master of Science.

Howard, C. S., "Density Currents in Lake Mead," *Proceedings,* Minnesota International Hydraulics Convention, 1953, pp. 355–368.

Hubbell, D. W., and Al-Shaikh A. K. S., "Qualitative Effects of Temperature on Flow Phenomena in Alluvial Channels," *Professional Paper 424–D,* United States Geological Survey, Washington, D.C., 1961, pp. D–21–D–23.

Hubbell, D. W., and Matejka, D. Q., "Investigations of Sediment Transportation, Middle Loup River at Dunning, Nebraska," *Water-Supply Paper No. 1476,* United States Geological Survey, Washington, D.C., 1959.

Hughmark, G. A., "Aqueous Transport of Settling Slurries," *Industrial and Engineering Chemistry,* Vol. 55, No. 5, May, 1961, pp. 389–390.

Hunt, J. N., "The Turbulent Transport of Suspended Sediment in Open Channels," *Proceedings,* Royal Society of London, Series A, Vol. 224, No. 1158, 1954, pp. 322–335.

Hunt, W. A., and Hoffman, I. C., "Optimization of Pipelines Transporting Solids," *Journal of the Pipeline Division,* ASCE, Vol. 94, No. PL1, Proc. Paper 6179, Oct., 1968, pp. 89–106.

Hurst, H. E., "The Suspension of Sand in Water," *Proceedings,* Royal Society of London, Series A, Vol. 24, 1929, pp. 196–201.

Hwang, Li-San, "Flow Resistance of Dunes in Alluvial Streams," thesis presented to California Institute of Technology, at Pasadena, Calif., in 1965, in partial fulfillment of the requirements for the degree of Doctor of Philosophy (also available from University Microfilms, Inc., Ann Arbor, Mich.).

Inglis, C. C., "Meanders and Their Bearing on River Training," The Institution of Civil Engineers, Maritime & Waterways Engineers Division, London, England, 1947, pp. 3–54.

Inglis, C. C., discussion of "Systematic Evaluation of River Regime," by Charles R. Neill and Victor J. Galey, *Journal of the Waterways and Harbors Division*, ASCE, Vol. 94, No. WW1, Proc. Paper 5774, Feb., 1968, pp. 109–114.

Inman, D. L., "Measures for Describing the Size Distribution of Sediments," *Journal of Sedimentary Petrology,* Vol. 22, No. 3, 1952, pp. 125–145.

Interagency Committee, "Some Fundamentals of Particle Size Analysis, A Study of Methods Used in Measurement and Analysis of Sediment Loads in Streams," *Report No. 12,* Subcommittee on Sedimentation, Interagency Committee on Water Resources, St. Anthony Falls Hydraulic Laboratory, Minneapolis, Minn., 1957.

Ippen, A. T., "A New Look at Sedimentation in Turbulent Streams," *Journal of the Boston Society of Civil Engineers,* Vol. 58, No. 3, July, 1971, pp. 131–163.

Ippen, A. T., and Harleman, D. R. F., "Steady-State Characteristics of Subsurface Flow," *Circular No. 521,* Gravity Waves Symposium, National Bureau of Standards, Washington, D.C., 1952, p. 79.

Ippen, A. T., and Verma, R. P., "The Motion of Discrete Particles along the Bed of a Turbulent Stream," *Proceedings*, Minnesota International Hydraulics Conference, Minneapolis, Minn., 1953, pp. 7–20.

Ismail, H. A., "Turbulent Transfer Mechanism and Suspended Sediment in Closed Conduits," *Transactions,* ASCE, Vol. 117, Paper No. 2500, 1952, pp. 409–454.

Iwanami, S., and Suu, T., "Pressure Losses in Right-Angled Pipe Fittings," *Proceedings,* 1st International Conference on Hydraulic Transportation of Solids in Pipes, (Hydrotransport 1), British Hydromechanics Research Association, 1970.

Jahns, R. H., "Geologic Features of the Connecticut Valley, Massachusetts, as Related to Recent Floods," *Water-Supply Paper 996,* United States Geological Survey, Washington, D.C., 1947.

Jeffries, H., "Transport of Sediments by Streams," *Proceedings,* Cambridge Philosophical Society, Vol. 25, Cambridge, England, 1929, p. 72.

Jobson, H. E., "Vertical Mass Transfer in Open Channel Flow," thesis presented to Colorado State University, at Fort Collins, Colo., in 1968, in partial fulfillment of the requirements for the degree of Doctor of Philosophy.

Johnson, C. O., "Similarity in Scour Below a Spillway," thesis presented to the University of Minnesota, at Minneapolis, Minn., in 1950, in partial fulfillment of the requirements for the degree of Master of Science.

Johnson, J. W., "The Importance of Considering Side-Wall Friction in Bed-Load Investigations," *Civil Engineering,* ASCE, Vol. 12, No. 6, June, 1942, pp. 329–331.

Johnson, J. W., "Laboratory Investigations on Bed-Load Transportation and Bed Roughness, A Compilation," *Mimeograph Publication SCS-TP-50,* United States Department of Agriculture, Soil Conservation Service, Washington, D.C., Mar., 1943.

Johnson, L. C., "The Nita Crevasse," *Bulletin,* Geological Society of America, Vol. 2, Boulder, Colo., pp. 20–25.

Jordan, P. R., "Fluvial Sediment of the Mississippi River at St. Louis, Missouri," *Water-Supply Paper 1802,* United States Geological Survey, Washington, D.C., 1965.

Kada, H., "Effect of Solid Particles on Turbulence," thesis presented to the University of Illinois, at Urbana, Ill., in 1959, in partial fulfillment of the requirements for the degree of Doctor of Philosophy.

Kalinske, A. A., "Criteria for Determining Sand Transportation by Surface Creep and Saltation," *Transactions,* American Geophysical Union, Part II, Washington, D.C., 1942, pp. 639–643.

Kalinske, A. A., "The Role of Turbulence in River Hydraulics," *Bulletin 27, Proceedings of the 2nd Hydraulics Conference,* University of Iowa Studies in Engineering, Iowa City, Iowa, 1943a, pp. 266–279.

Kalinske, A. A., "Turbulence and the Transport of Sand and Silt by Wind," *Annals of the New York Academy of Science,* Vol. XLIV, Article 1, 1943b, pp. 41–54.

Kalinske, A. A., "Movement of Sediment as Bed Load in Rivers," *Transactions,* American Geophysical Union, Vol. 28, No. 4, Washington, D.C., Aug., 1947, pp. 615–620.

Kalinske, A. A., and Pien, C. L., "Experiments on Eddy-Diffusion and Suspended-Material Transportation in Open Channels," *Transactions,* American Geophysical Union, Part II, Washington, D.C., 1943, pp. 530–536.

Kawashima, T. and Noda, K., "Hydraulic Transport of Solids in an Inclined Pipe: Theoretical and Experimental Studies in Pressure Loss," *Proceedings,* 1st International Conference on Hydraulic Transportation of Solids in Pipes, (Hydrotransport 1), British Hydromechanics Research Association, 1970.

Kennedy, R. G., "The Prevention of Silting in Irrigation Canals," *Proceedings,* Institute of Civil Engineers, Vol. 119, 1895, pp. 281–290.

Kennedy, J. F., "Stationary Waves and Antidunes in Alluvial Channels," thesis presented to California Institute of Technology, at Pasadena, Calif., in 1960, in partial fulfillment of the requirements for the degree of Doctor of Philosophy (also available from University Microfilms, Inc., Ann Arbor, Mich.).

Kennedy, J. F., "Stationary Waves and Antidunes in Alluvial Channels," *Report KH-R-2,* W. M. Keck Laboratory of Hydraulics and Water Resources, California Institute of Technology, Pasadena, Calif., 1961.

Kennedy, J. F., "The Mechanics of Dunes and Antidunes in Erodible-Bed Channels," *Journal of Fluid Mechanics,* Vol. 16, Part 4, 1963, pp. 521–544.

Kennedy, J. F., "The Formation of Sediment Ripples, Dunes, and Antidunes," *Annual Review of Fluid Mechanics,* W. R. Sears, ed., Vol. 1, Annual Reviews, Inc., Palo Alto, Calif., 1969, pp. 147–168.

Kennedy, J. F., and Koh, R. C. Y., "The Relation Between the Frequency Distribution of Sieve Diameters and Fall Velocities of Sediment Particles," *Journal of Geophysical Research,* Vol. 66, 1961, pp. 4233–4246.

Kennedy, J. F., and Koh, R. C. Y., discussion of "Sediment Transportation Mechanics: —Introduction and Properties of Sediment," by the Task Committee on Preparation of Sedimentation Manual of the Committee on Sedimentation of the Hydraulics Division, Vito A. Vanoni, Chmn., *Journal of the Hydraulics Division,* ASCE, Vol. 89, No. HY2, Proc. Paper 3468, Mar., 1963, pp. 171–174.

Kennedy, J. F., and Brooks, N. H., "Laboratory Study of an Alluvial Stream at Constant Discharge," *Proceedings of the Federal Interagency Sedimentation Conference,* 1963, *Miscellaneous Publication No. 970,* Agriculture Research Service, United States Department of Agriculture, Washington, D.C., 1965, pp. 320–330.

Kepner, R. A., Boelter, L. M. K., and Brooks, F. A., "Nocturnal Wind Velocity, Eddy-stability, and Eddy-diffusivity above a Citrus Orchard," *Transactions,* American Geophysical Union, Vol. 23, Part II, Washington, D.C., 1942, p. 240.

Keulegan, G. H., "Laminar Flow at the Interface of Two Liquids," *Journal of Research,* National Bureau of Standards, Vol. 32, No. 303, RP 1591, 1944.

Keulegan, G. H., "Interfacial Instability and Mixing in Stratified Flows," *Journal of Research,* National Bureau of Standards, Vol. 43, No. 487, RP 2040, 1949.

Knapp, R. T., "Density Currents: Their Mixing Characteristics and Their Effect on the Turbulence Structure of the Associated Flow," *Proceedings,* Second Hydraulics Conference, State University of Iowa, Iowa City, Iowa, 1942, pp. 289–306.

Koelzer, V. A., and Lara, J. M., "Densities and Compaction Rates of Deposited Sediment," *Journal of the Hydraulics Division,* ASCE, Vol. 84, No. HY2, Proc. Paper No. 1603, Apr., 1958, p. 1603.

Kolb, C. R., and Van Lopik, J. R., "Geology of the Mississippi River Deltaic Plain, Southeastern Louisiana," *Technical Report No. 3–483,* United States Army Corps of Engineers, Waterways Experiment Station, Vicksburg, Miss., 1958.

Kolb, C. R., Mabrey, P. R., and Steinriede, W. B., Jr., "Geological Investigation of the Yazoo Basin," *Technical Report No. 3–480,* United States Army Corps of Engineers, Waterways Experiment Station, Vicksburg, Miss., 1958.

Koloseus, H. J., and Davidian, J., "Flow in an Artificially Roughened Channel," *Professional Paper 424–B,* United States Geological Survey, Washington, D.C., 1961.

Komura, S., discussion of "Sediment Transportation Mechanics:—Introduction and Properties of Sediment," Progress Report by the Task Committee on Preparation of Sedimentation Manual of the Committee on Sedimentation of the Hydraulics Division, Vito A. Vanoni, Chmn., *Journal of the Hydraulics Division,* ASCE, Vol. 89, No. HY1, Proc. Paper 3405, Jan., 1963a, pp. 263–266.

Komura, S., discussion of "Sediment Transportation Mechanics: Erosion of Sediment," Progress Report by the Task Committee on Preparation of Sedimentation Manual of the Committee on Sedimentation of the Hydraulics Division, Vito A. Vanoni, Chmn., *Journal of the Hydraulics Division,* ASCE, Vol. 89, HY1, Proc. Paper 3405, Jan., 1963b, pp. 269–276.

Kramer, H., "Sand Mixtures and Sand Movement in Fluvial Models," *Transactions,* ASCE, Vol. 100, Paper No. 1909, 1935, pp. 798–878.

Krey, H., "Die Quergeschwindigkeitskurve bei Turbulenter Stroemung," Zeitschrift fur Angewandte Mathematik und Mechanik, Vol. 7, No. 2, 1927, pp. 107–113.

Krumbein, W. C., and Pettijohn, F. J., *Manual of Sedimentary Petrography,* D. Appleton-Century Co., New York, N.Y., 1938.

Lacey, G., "Stable Channels in Alluvium," *Proceedings of the Institution of Civil Engineers,* Vol. 229, Part I, London, England, 1929, pp. 259–292.

Lane, E. W., "Notes on Limit of Sediment Concentration," *Journal of Sedimentary Petrology,* Vol. 10, No. 2, Aug., 1940, pp. 95–96.

Lane, E. W., "Report of the Subcommittee on Sediment Terminology," *Transactions,* American Geophysical Union, Vol. 28, No. 6, Washington, D.C., Dec., 1947, pp. 936–938.

Lane, E. W., "Design of Stable Channels," *Transactions,* ASCE, Vol. 120, Paper No. 2776, 1955, pp. 1234–1279.

Lane, E. W., "A Study of the Shape of Channels Formed by Natural Streams Flowing in Erodible Material," *MRD Sediment Series Report No. 9,* United States Army Engineer Division, Corps of Engineers, Omaha, Neb., 1957.

Lane, E. W., and Eden, E. W., "Sand Waves in the Lower Mississippi River," *Proceedings of Western Society of Professional Engineers,* Vol. 45, No. 6, Dec., 1940, pp. 281–291.

Lane, E. W., and Kalinske, A. A., "The Relation of Suspended to Bed Material in Rivers," *Transactions,* American Geophysical Union, Part IV, Washington, D.C., 1939, pp. 637–641.

Lane, E. W., and Koelzer, V. A., "Density of Sediments Deposited in Reservoirs," *Report No. 9* of *A Study of Methods Used in Measurement and Analysis of Sediment Loads in Streams,* St. Paul United States Engineering District, St. Paul, Minn., 1953.

Lane, E. W., Carlson, E. J., and Hanson, O. S., "Low Temperature Increases Sediment Transportation in Colorado River," *Civil Engineering,* ASCE, Vol. 19, No. 9, Sept., 1949, pp. 45–46.

Langbein, W. B., "Geometry of River Channels," *Journal of the Hydraulics Division,* ASCE, Vol. 90, No. HY2, Proc. Paper 3846, Mar., 1964, pp. 301–312.

Langbein, W. B., and Leopold, L. B., "River Meanders—Theory of Minimum Variance," *Professional Paper 422–H,* United States Geological Survey, Washington, D.C., 1966.

Langham, W. H., Foster, R. L., and Daniel, H. A., "The Amount of Dust in the Air at Plant Height During Wind Storms at Woodwell, Oklahoma, in 1936–37," *American Society of Agronomy Journal,* Vol. 30, Madison, Wisc., 1938, pp. 139–144.

Larsen, I., discussion of "Heterogeneous Flow of Solids in Pipelines," by I. Zandi and G. Govatos, *Journal of the Hydraulics Division,* ASCE, Vol. 94, No. HY1, Proc. Paper 5703, Jan., 1968, pp. 332–333.

Lattman, L. H., "Cross Section of a Floodplain in a Moist Region of Moderate Relief," *Journal of Sedimentary Petrology,* Vol. 30, 1960, pp. 275–282.

Laufer, J. "Recent Measurement in a Two-Dimensional Turbulent Channel," *Journal of Aeronautical Science,* Vol. 17, No. 5, May, 1950, pp. 277–287.

Laursen, E. M., "Observations on the Nature of Scour," *Proceedings of Fifth Hydraulic Conference, Bulletin 34,* University of Iowa, Iowa City, Iowa, June 9–11, 1952, pp. 179–197.

Laursen, E. M., discussion of "Some Effects of Suspended Sediment on Flow Characteristics," *Proceedings, Fifth Hydraulic Conference, Bulletin 34,* University of Iowa Studies in Engineering, Iowa City, Iowa, 1953, pp. 155–158.

Laursen, E. M., "The Hydraulics of a Storm-Drain System for Sediment-Transporting Flow," Iowa Highway Research Board, 1956.

Laursen, E. M., discussion of "Application of the Modified Einstein Procedure for Computation of Total Sediment Load," *Transactions,* American Geophysical Union, Vol. 38, No. 5, Washington, D.C., Oct., 1957, pp. 768–773.

Laursen, E. M., "The Total Sediment Load of Streams," *Journal of the Hydraulics Division,* ASCE, Vol. 54, No. HY1, Proc. Paper 1530, Feb., 1958, pp. 1–36.

Laursen, E. M., "Scour at Bridge Crossings," *Transactions*, ASCE, Vol. 127, Pt. I, Paper No. 3294, 1962, pp. 166–209.

Laursen, E. M., and Toch, A., "A Generalized Model Study of Scour Around Bridge Piers and Abutments," *Proceedings*, Minnesota International Hydraulic Convention, Minneapolis, Minn., September 1–4, 1953, pp. 123–131.

Lazarus, J. H., and Kilner, F. A., "Incipient Motion of Solid Capsules in Pipelines," *Proceedings,* 1st International Conference on Hydraulic Transportation of Solids in Pipes (Hydrotransport 1), British Hydromechanic Research Association, 1970.

Lee, J. J., discussion of "Some Turbulence Measurements in Water," by F. Raichlen, *Journal of the Engineering Mechanics Division*, ASCE, Vol. 93, No. EM6, Proc. Paper 5620, Dec., 1967, pp. 287–293.

Lelliavsky, S., *An Introduction to Fluvial Hydraulics,* Constable and Co., Ltd., London, England, 1955.

Leopold, L. B., and Maddock, T., Jr., "The Hydraulic Geometry of Stream Channels and Some Physiographic Implications," *Professional Paper 252,* United States Geological Survey, Washington, D.C., 1953.

Leopold, L. B., and Wolman, M. G., "River Channel Patterns: Braided, Meandering and Straight," *Professional Paper 282–B,* United States Geological Survey, Washington, D.C., 1957.

Leopold, L. B., and Wolman, M. G., "River Meanders," *Bulletin of Geological Society of America,* Vol. 71, Boulder, Colo., 1960, pp. 769–796.

Leopold, L. B., Wolman, M. G., and Miller, J. P., *Fluvial Processes in Geomorphology,* W. H. Freeman & Co., San Francisco, Calif., 1964.

Liu, H. K., and Hwang, S. Y., "Discharge Formula for Straight Alluvial Channels," *Transactions,* ASCE, Vol. 126, Part I, Paper No. 3276, 1961, pp. 1787–1822.

Livesey, R. H., "Channel Armoring below Fort Randall Dam," *Proceedings of the Federal Interagency Sedimentation Conference,* 1965, *Miscellaneous Publication No. 970,* Agricultural Research Service, United States Department of Agriculture, Washington, D. C., 1963, pp. 461–470.

Long, D. V., "Mechanics of Consolidation with Reference to Experimentally Sedimented Clays," thesis presented to the California Institute of Technology, at Pasadena, Calif., in 1961, in partial fulfillment of the requirements for the degree of Doctor of Philosophy.

Lovera, F., and Kennedy, J. F., "Friction-Factors for Flat-Bed Flows in Sand Channels," *Journal of the Hydraulics Division,* ASCE, Vol. 95, No. HY4, Proc. Paper 6678, July, 1969, pp. 1227–1234.

Macagno, E. O., and Rouse, H., "Interfacial Mixing in Stratified Flow," *Transactions,* ASCE, Vol. 127, Part I, Paper No. 3289, 1962, pp. 102–128.

Mackin, J. H., "Concept of the Graded River," *Bulletin of the Geological Society of America,* Vol. 59, Boulder, Colo., 1948, pp. 463–512.

Maddock, T., Jr., discussion of "Recirculation and Sand-Feed Type Flume Experiments," by H. P. Guy, R. E. Rathbun, and E. V. Richardson, *Journal of the Hydraulics Division,* ASCE, Vol. 94, No. HY4, Proc. Paper 6015, July, 1968, pp. 1139–1147.

Maddock, T., Jr., "The Behavior of Straight Open Channels with Movable Beds," *Professional Paper 622–A,* United States Geological Survey, Washington, D.C., 1969.

Maddock, T., Jr., discussion of "Instability of Flat Bed in Alluvial Channels," by H. M. Hill, V. S. Srinivasan, and T. E. Unny, Jr., *Journal of the Hydraulics Division,* ASCE, Vol. 96, No. HY4, Proc. Paper 7189, Apr., 1970, pp. 1080–1081.

Maddock, T., Jr., discussion of "Sediment Transportation Mechanics: F. Hydraulic Relations for Alluvial Streams," by the Task Committee for Preparation of Sedimentation Manual, Committee on Sedimentation of the Hydraulics Division, Vito A. Vanoni, Chmn., *Journal of the Hydraulics Division,* ASCE, Vol. 97, No. HY11, Proc. Paper 8483, Nov., 1971, p. 1904.

Malina, F. J., "Recent Developments in the Dynamics of Wind Erosion," *Transactions,* American Geophysical Union, Washington, D.C., 1941, pp. 262–284.

Martin, T. R., discussion of "Experiments on the Scour Resistance of Cohesive Sediments," *Journal of Geophysical Research,* Vol. 67, No. 4, Washington, D.C., Apr., 1962, pp. 1447–1449.

Mathews, W. H., and Shepard, F. P., "Sedimentation of Fraser River Delta, British Columbia," *Bulletin,* American Association of Petroleum Geologists, Vol. 46, Tulsa, Okla., 1962, pp. 1416–43.

Matsunashi, J., "Researches on Bed-Load Transportation Under the Tractive Force Near the Critical Limit," *Memoirs of the Faculty of Engineering,* No. 4, Kobe University, Kobe, Japan, Mar., 1957, pp. 24–42.

Mavis, F. T., and Laushey, L. M., "Formula for Velocity at Beginning of Bed-Load Movement is Reappraised," *Civil Engineering,* ASCE, Vol. 19, No. 1, Jan., 1949, pp. 38–39 and p. 72.

Mavis, F. T., and Laushey, L. M., discussion of "Sediment Transportation Mechanics:Initiation of Motion," by the Task Committee for Preparation of Sedimentation Manual,Committee on Sedimentation of the Hydraulics Division, Vito A. Vanoni, Chmn., *Journalof the Hydraulics Division,* ASCE, Vol. 92, No. HY5, Proc. Paper 4891, Sept., 1966, pp. 288–291.

McLaughlin, R. T., Jr., "The Settling Properties of Suspensions," *Journal of the Hydraulics Division,* ASCE, Vol. 85, No. HY12, Proc. Paper 2311, Dec., 1959, pp. 9–41.

McNown, J. S., Lee, H. M., McPherson, M. B., and Engez, S. M., "Influence of Boundary Proximity on the Drag of Spheres," *Proceedings,* 7th International Congress for Applied Mechanics, Vol. 2, Part I, London, England, 1948, pp. 17–29.

McNown, J. S., and Lin, P. N., "Sediment Concentration and Fall Velocity," *Proceedings,* 2nd Midwestern Conference on Fluid Mechanics, Ohio State University, Columbus, Ohio, 1952, pp. 401–411.

McNown, J. S., and Malaika, J., "Effect of Particle Shape on Settling Velocity at Low Reynolds Number," *Transactions,* American Geophysical Union, Vol. 31, Washington, D.C., 1950, pp. 74–82.

McNown, J. S., and Newlin, J. T., "Drag of Spheres within Cylindrical Boundaries," *Proceedings,* 1st Congress of Applied Mechanics, American Society of Mechanical Engineers, 1951, pp. 801–806.

McNown, J. S., Malaika, J. and Pramanik, R., "Particle Shape and Settling Velocity," *Transactions,* 4th Meeting of the International Association for Hydraulic Research, Bombay, India, 1951, pp. 511–522.

McPherson, H. J., and Rannie, W. F., "Geomorphic Effects of the May 1967 Flood in Graburn Watershed, Cypress Hills, Alberta, Canada," *Journal of Hydrology,* Vol. 9, 1969, pp. 307–321.

Melton, F. A., "An Empirical Classification of Flood-Plain Streams," *Geographical Review,* Vol. 26, 1936, pp. 593–609.

Menard, H. W., "Sediment Movement in Relation to Current Velocity," *Journal of Sedimentary Petrology,* Vol. 20, No. 3, Sept., 1950, pp. 148–160.

Metzner, A. B., "Flow of Non-Newtonian Fluids," *Handbook of Fluid Dynamics,* Section 7, V. Streeter, ed., McGraw-Hill Book Co., Inc., New York, N.Y., 1961.

Meyer-Peter, E., and Muller, R., "Formulas for Bed-Load Transport," Report on Second Meeting of International Association for Hydraulic Research, Stockholm, Sweden, 1948, pp. 39–64.

Mih, W. C., Chen, C. K. and Osborn, J. F., *Bibliography Solid-Liquid Transport in Pipelines,* Albrook Hydraulic Laboratory, Washington State University, Pullman, Wash., 1971.

Miller, C. R., "Determination of Unit Weight of Sediment for Use in Sediment Volume Computations," *Memorandum,* Bureau of Reclamation, United States Department of the Interior, Denver, Colo., February 17, 1953.

Mining Magazine, "Hydraulic Transport of Minerals," Vol. 126, No. 4, Apr., 1972.

Monin, A. S., and Obukhov, A. M., *Trudy, Geofizicheskii No. 24(151),* Leningrad, USSR, 1954, p. 163.

Moody, L. F., "Friction Factors for Pipe Flow," *Transactions,* American Society of Mechanical Engineers, Vol. 66, No. 8, New York, N.Y., Nov., 1944, pp. 671–684.

Moore, W. L., and Masch, F. D., "Experiments on the Scour Resistance of Cohesive Sediments," *Journal of Geophysical Research,* Vol. 67, No. 4, Washington, D.C., Apr., 1962, pp. 1437–1449.

Mostafa, M. G., and McDermid, R. M., discussion of "Sediment Transportation Mechanics: F. Hydraulic Relations for Alluvial Streams," by the Task Committee for

Preparation of Sedimentation Manual, Committee on Sedimentation of the Hydraulics Division, Vito A. Vanoni, Chmn., *Journal of the Hydraulics Divison,* ASCE, Vol. 97, No. HY10, Proc. Paper 8407, Oct., 1971, pp. 1777–1780.

Nemenyi, P., "The Different Approaches to the Study of Propulsion of Granular Materials and the Value of Their Coordination," *Transactions,* American Geophysical Union, Part II, Washington, D.C., 1940, pp. 633–647.

Newitt, D. M., Richardson, J. F., Abbott, M., and Turtle, R. B., "Hydraulic Conveying of Solids in Horizontal Pipes," *Transactions,* Institution of Chemical Engineers, Vol. 33, 1955, pp. 93–113.

Newitt, D. M., Richardson, J. F., and Gliddon, B. J., "Hydraulic Conveying of Solids in Vertical Pipes," *Transactions,* Institution of Chemical Engineers, Vol. 39, 1961, pp. 93–100.

Nichols, M., and Poor, G., "Sediment Transport in a Coastal Plain Estuary," *Journal of the Waterways and Harbors Division,* ASCE, Vol. 93, No. WW4, Proc. Paper 5571, Nov., 1967, pp. 83–95.

Nordin, C. F. Jr., "Aspects of Flow Resistance and Sediment Transport: Rio Grande near Bernalillo, New Mexico," *Water-Supply Paper 1498–H,* United States Geological Survey, Washington, D.C., 1964.

Nordin, C. F. Jr., and Algert, J. H., "Spectral Analysis of Sand Waves," *Journal of the Hydraulics Division,* ASCE, Vol. 92, HY5, Proc. Paper 4910, Sept., 1966, pp. 95–114.

Novak, P., "Experimental and Theoretical Investigations of the Stability of Prisms on the Bottom of a Flume," *Proceedings,* Second Conference of International Association for Hydraulic Research, Stockholm, Sweden, 1948, pp. 77–91.

O'Brien, M. P., "Review of the Theory of Turbulent Flow and Its Relation to Sediment Transportation," *Transactions,* American Geophysical Union, Washington, D.C., April 27–29, 1933, pp. 487–491.

O'Brien, M. P., and Rindlaub, B. D., "The Transport of Sand by Wind," *Civil Engineering,* ASCE, Vol. 6, No. 5, 1936, pp. 325.

O'Brien, M. P., and Folsom, R. G., "The Transportation of Sand in Pipelines," publication in Engineering, Vol 3, University of California, Berkeley, Calif., 1937, pp. 343–384.

O'Loughlin, E. M., and MacDonald, E. G., "Some Roughness-Concentration Effects on Boundary Resistance," *La Houille Blanche,* No. 7, 1964, pp. 773–783.

Onishi, Y., Jain, S. C., and Kennedy, J. F., "Effects of Meandering on Sediment Discharges and Friction Factors of Alluvial Streams," *IIHR Report No. 141,* Iowa Institute of Hydraulic Research, University of Iowa, Iowa City, Iowa, Sept., 1972.

Otto, G. H., "A Modified Logarithmic Probability Graph for the Interpretation of Mechanical Analyses of Sediments," *Journal of Sedimentary Petrology,* Vol. 9, No. 2, Aug., 1939, pp. 62–76.

Paintal, A. S., and Garde, R. J., discussion of "Sediment Transportation Mechanics: Suspension of Sediment," by the Task Committee on Preparation of Sedimentation Manual, Committee on Sedimentation of the Hydraulics Division, Vito A. Vanoni, Chmn., *Journal of the Hydraulics Division,* ASCE, Vol. 90, No. HY4, Proc. Paper 3980, July, 1964, pp. 257–265.

Panofsky, H. A., Blackadar, A. K., and McVehil, G. E., "The Diabatic Wind Profile," *Quarterly Journal of the Royal Meteorological Society,* Vol. 86, No. 369, 1960, pp. 390–398.

Park, F. J., "Torrential Potential of Desert Waters," *Pan-American Geologist,* Vol. 40, 1923, pp. 349–356.

Partheniades, E., "Erosion and Deposition of Cohesive Soils," *Journal of the Hydraulics Division,* ASCE, Vol. 91, No. HY1, Proc. Paper 4204, Jan., 1965, pp. 105–139.

Partheniades, E., and Paaswell, R. E., "Erodibility of Channels with Cohesive Boundary," *Journal of the Hydraulics Division,* ASCE, Vol. 96, No. HY3, Proc. Paper 7156, Mar., 1970, pp. 755–771.

Pettijohn, F. J., and Potter, P. E., *Atlas of Primary Sedimentary Structures,* Springer-Verlag, Berlin, Germany, 1964.

Péwé, T. L., "An Observation of Wind-blown Silt," *Journal of Geology,* Vol. 59, 1951, pp. 399–401.

Pierce, R. C., "The Measurement of Silt Laden Streams," *Water-Supply Paper 400,* United States Geological Survey, Washington, D.C., 1917.

Prandtl, L., *Essentials of Fluid Dynamics,* Hafner Publishing Co., New York, N.Y., 1952.

Raichlen, F., "Some Turbulence Measurements in Water," *Journal of the Engineering Mechanics Division,* ASCE, Vol. 93, No. EM2, Proc. Paper 5195, Apr., 1967, pp. 73–97.

Rand, W., discussion of "Some Effects of Suspended Sediment on Flow Characteristics," *Proceedings,* Fifth Hydraulic Conference, *Bulletin 34,* State University of Iowa, Studies in Engineering, Iowa City, Iowa, 1953, pp. 156–158.

Rathbun, R. E., and Goswami, A., discussion of "Sediment Transportation Mechanics: Initiation of Motion," by the Task Committee on Preparation of Sedimentation Manual, Committee on Sedimentation of the Hydraulics Division, Vito A. Vanoni, Chmn., *Journal of the Hydraulics Division,* ASCE, Vol. 92, No. HY6, Proc. Paper 4959, Nov., 1966, pp. 251–253.

Rathbun, R. E., Guy, H. P., and Richardson, E. V., "Response of a Laboratory Alluvial Channel to Changes of Hydraulics and Sediment-Transport Variables, *Professional Paper 562–D,* United States Geological Survey, 1969.

Raudkivi, A. J., "Study of Sediment Ripple Formation," *Journal of the Hydraulics Division,* ASCE, Vol. 89, No. HY6, Proc. Paper 3692, Nov., 1963, pp. 15–33.

Raudkivi, A. J., discussion of "Sediment Transportation Mechanics: Initiation of Motion," by the Task Committee on Preparation of Sedimentation Manual, Committee on Sedimentation of the Hydraulics Division, Vito A. Vanoni, Chmn., Journal of the Hydraulics Division, ASCE, Vol. 92, No. HY6, Proc. Paper 4959, Nov., 1966, pp. 253–255.

Raudkivi, A. J., "Analyses of Resistance in Fluvial Channels," *Journal of the Hydraulics Division,* ASCE, Vol. 93, No. HY5, Proc. Paper 5426, Sept., 1967, pp. 73–84.

Raudkivi, A. J., discussion of "Sediment Transportation Mechanics: F. Hydraulic Relations for Alluvial Streams," by the Task Committee on Preparation of Sedimentation Manual, Committee on Sedimentation of the Hydraulics Division, Vito A. Vanoni, Chmn., *Journal of the Hydraulics Division,* ASCE, Vol. 97, No. HY12, Proc. Paper 8552, Dec., 1971, pp. 2089–2093.

Raynaud, J. P., "Study of Currents of Muddy Water through Reservoirs," Fourth Congress on Large Dams, New Delhi, India, Vol. 4, 1951, p. 137 (in French).

Reynolds, A. J., "Waves on the Erodible Bed of an Open Channel," *Journal of Fluid Mechanics,* Vol. 22, 1965, pp. 113–133.

Rossby, C. G., and Montgomery, R. B., "The Layer of Frictional Influence in Wind and Ocean Currents," *Papers in Physical Oceanography and Meteorology,* Massachusetts Institute of Technology and Woods Hole Oceanographic Institute, Vol. 3, No. 3, 1935.

Rouse, H., "Modern Conceptions of the Mechanics of Fluid Turbulence," *Transactions,* ASCE, Vol. 102, Paper No. 1965, 1937a, pp. 463–543.

Rouse, H., "Nomogram for the Settling Velocity of Spheres," Division of Geology and Geography Exhibit D of the Report of the Commission on Sedimentation, 1936–37, National Research Council, Washington, D.C., Oct., 1937b., pp. 57–64.

Rouse, H., "Experiments on the Mechanics of Sediment Suspension," *Proceedings, Fifth International Congress for Applied Mechanics,* Vol. 55, John Wiley and Sons, Inc., New York, N.Y., 1938.

Rouse, H., "An Analysis of Sediment Transportation in the Light of Fluid Turbulence," *Soil Conservation Service Report No. SCS–TP–25,* United States Department of Agriculture, Washington, D.C., 1939.

Rouse, H., "Criteria for Similarity in the Transportation of Sediment," *Proceedings of 1st Hydraulic Conference, Bulletin 20,* State University of Iowa, Iowa City, Iowa, Mar., 1940, pp. 33–49.

Rouse, H., "Critical Analysis of Open-Channel Resistance," *Journal of the Hydraulics Division,* ASCE, Vol. 91, No. HY4, Proc. Paper 4387, July, 1965, pp. 1–25.

Rouse, H., ed., *Engineering Hydraulics,* John Wiley and Sons, Inc., New York, N.Y., 1950, pp. 769–857.

Rubey, W. W., "Settling Velocities of Gravel, Sand and Silt Particles," *American Journal of Science,* 5th series, Vol. 25, No. 148, 1933, pp. 325–338.

Rubey, W. W., "The Forces Required to Move Particles on a Stream Bed," *Professional Paper 189B,* United States Geological Survey, Washington, D.C., 1948.

Russell, R. J., and Russell, R. D., "Mississippi River Delta Sedimentation, Recent Marine Sediments," American Association of Petroleum Geologists, Tulsa, Okla., 1939, pp. 153–177.

Sayre, W. W., and Albertson, M. L., "Roughness Spacing in Rigid Open Channels," *Transactions,* ASCE, Vol. 128, Paper No. 3417, 1963, pp. 343–372.

Schmidt, W., "Der Massenaustausch in freier Luft und verwandte Erscheinungen," *Probleme der kosmischen Physik,* Band 7, Hamburg, Germany, 1925.

Schmudde, T. H., "Some Aspects of Land Forms of the Lower Missouri River Flood-Plain," *Annals,* Association of American Geographers, Vol. 53, Washington, D.C., 1963, pp. 60–73.

Schumm, S. A., "The Shape of Alluvial Channels in Relation to Sediment Type," *Professional Paper 352–B,* United States Geological Survey, Washington, D.C., 1960.

Schumm, S. A., "Sinuosity of Alluvial Rivers in the Great Plains," *Bulletin of the Geological Society of America,* Vol. 74, Sept., 1963, pp. 1089–1100.

Scimemi, E., discussion of "Model Study of Brown Canyon Debris Barrier," *Transactions,* ASCE, Vol. 112, Paper No. 2319, 1947, pp. 1016–1019.

Settle, J. J., and Parkins, R., ICI Ltd., General Chemicals Division, Chief Engineers Department, 1951; see also "Correlation for Use in Transport of Aqueous Suspensions of Fine Solids Through Pipes," by K. E. Spells, *Transactions,* Institution of Chemical Engineers, Vol. 33, 1955.

Sharpe, C. F. S., *Landslides and Related Phenomena,* Columbia University Press, New York, N.Y., 1938.

Sharp, R. P., "Mudflow Levees," *Journal of Geomorphology,* Vol. 5, 1942, pp. 222–227.

Sharp, R. P., and Nobles, L. H., "Mudflow of 1941 at Wrightwood, Southern California," *Bulletin,* Geological Society of America, Vol. 64, Boulder, Colo., 1953, pp. 547–560.

Shen, H. W., "Development of Bed Roughness in Alluvial Channels," *Journal of the Hydraulics Division,* ASCE, Vol. 88, No. HY3, Proc. Paper 3113, May, 1962, pp. 45–58.

Shen, H. W., and Wang, J. S., "Incipient Motion and Limiting Deposit Conditions of Solid-liquid Pipe Flow," *Proceedings,* 1st International Conference on Hydraulic Transportation of Solids In Pipes (Hydrotransport 1), British Hydromechanic Research Association, 1970.

Sheppard, P. A., "The Aerodynamic Drag of the Earth's Surface and the Value of von Kármán's Constant in the Lower Atmosphere," *Proceedings,* Royal Society of London, Vol. 188A, No. 1013, 1947, pp. 208–222.

Shields, A., "Anwendung der Aenlichkeitsmechanik und der Turbulenzforschung auf die Geschiebebewegung," *Mitteilungen der Preussischen Versuchsanstalt fur Wasserbau und Schiffbau,* Berlin, Germany, translated to English by W. P. Ott and J. C. van Uchelen, California Institute of Technology, Pasadena, Calif., 1936.

Shih, C. S., "Hydraulic Transport of Solids in a Sloped Pipe," *Journal of the Pipeline Division,* ASCE, Vol. 90, No. PL2, Proc. Paper 4125, Nov., 1964, pp. 1–14.

Shih, C. S., closure to "Hydraulic Transport of Solids in a Sloped Pipe," *Journal of the Pipeline Division,* ASCE, Vol. 93, No. PL1, Proc. Paper 5125, Mar., 1967, pp. 49–51.

Shulits, S., "The Schoklitsch Bed-load Formula," *Engineering,* London, England, June 21, 1935, pp. 644–646; and June 28, 1935, p. 687.

Shulits, S., and Hill, R. D., Jr., "Bedload Formulas," *Bulletin,* Department of Civil Engineering, Hydraulics Laboratory, Pennsylvania State University, University Park, Pa., Dec., 1968.

Simons, D. B., and Albertson, M. L., "Uniform Water Conveyance Channels in Alluvial Material," *Transactions,* ASCE, Vol. 128, Part 1, Paper No. 3399, 1963, pp. 65–107.

Simons, D. B., and Richardson, E. V., "Resistance to Flow in Alluvial Channels," *Professional Paper 422J,* United States Geological Survey, Washington, D.C., 1966.

Simons, D. B., Richardson, E. V., and Haushild, W. L., "Some Effects of Fine Sediments on Flow Phenomena," *Water-Supply Paper 1498G,* United States Geological Survey, Washington, D.C., 1963.

Sinclair, C. G., "The Limit Deposit-Velocity of Heterogeneous Suspensions," *Proceedings of the Symposium on Interaction Between Fluids and Particles,* European Federation of Chemical Engineers, London, England, June, 1962, pp. 78–86.

Smerdon, E. T., and Beasley, R. P., "Critical Tractive Forces in Cohesive Soils," *Agricultural Engineering,* St. Joseph, Mich., Jan., 1961, pp. 26–29.

Smith, R. A., *Transactions,* Institution of Chemical Engineers, Vol. 33, 1955, p. 85; see also Sinclair, G. C., 1962.

Smith, R. A., and Carruthers, G. A., ICI Ltd., Billingham Division, Engineering Research Department, 1951; see also Spells, K. E., 1955.

Soo, S. L., *Fluid Dynamics of Multiphase System,* Blaisdell Publishing Co., Waltham, Mass., 1969.

Spells, K. E., "Correlations for Use in Transport of Aqueous Suspensions of Fine Solids Through Pipes," *Transactions,* Institution of Chemical Engineers, Vol. 33, 1955, pp. 79–84.

Stepanoff, A. J., "Pumping Solid Liquid Mixture," *Mechanical Engineering,* Vol. 86, No. 9, Sept., 1964.

Stepanoff, A. J., *Pumps and Blowers: Selected Advance Topics Two-Phase Flow; Flow and Pumping of Solids in Suspension and Fluid Mixture,* John Wiley and Sons, Inc., New York, N.Y., 1965.

Streeter, V. L., *Fluid Mechanics,* McGraw-Hill Book Co., Inc., New York, N.Y., 1962.

Straub, L. G., "Missouri River Report," *House Document 238,* Appendix XV, Corps of Engineers, United States Department of the Army to 73rd United States Congress, 2nd Session, 1935, p. 1156.

Sundborg, A., "The River Klaralven, a Study of Fluvial Processes," *Geografiska Annaler,* 1956, pp. 127–316.

Sutherland, A. J., discussion of "Sediment Transportation Mechanics: Initiation of Motion," by the Task Committee on Preparation of Sedimentation Manual, Committee on Sedimentation of the Hydraulics Division, Vito A. Vanoni, Chmn., *Journal of the Hydraulics Division,* ASCE, Vol. 92, No. HY6, Proc. Paper 4959, Nov., 1966, pp. 255–257.

Sutherland, A. J., "Proposed Mechanism for Sediment Entrainment by Turbulent Flows," *Journal of Geophysical Research,* Vol. 72, No. 24, 1967, pp. 6183–6194.

Sutton, O. G., "The Application to Micrometeorology of the Theory of Turbulent Flow over Rough Surfaces," *Quarterly Journal of the Royal Meteorological Society,* Vol. 75, No. 326, 1949, p. 342.

Sutton, O. G., *Atmospheric Turbulence,* John Wiley and Sons, Inc., New York, N.Y., 2nd ed., 1955, p. 111.

Sverdrup, H. U., "The Eddy Conductivity of the Air over a Smooth Snow Field," *Geofysiske Publikationer,* Vol. VII, No. 7, Met. Zeit. 53, Oslo, Norway, 1936.

Swaminathan, K. R., and Dakshinamurti, C., discussion of "Sediment Transportation Mechanics: H. Sediment Discharge Formulas," by the Task Committee for Preparation of Sedimentation Manual, Committee on Sedimentation of the Hydraulics Division, Vito A. Vanoni, Chmn., *Journal of the Hydraulics Division,* ASCE, Vol. No. 98, No. HY1, Proc. Paper 8620, Jan., 1972, p. 289.

Swineford, A., and Frye, J. C., "A Mechanical Analysis of Wind-Blown Dust Compared with Analysis of Loess," *American Journal of Science,* Vol. 243, 1945, pp. 249–255.

Tanaka, S., and Fugimoto, S., "On the Distribution of Suspended Sediment in Experimental Flume Flow," *Memoirs of the Faculty of Engineering,* No. 5, Kobe University, Kobe, Japan, 1958.

Tarapore, Z. S., "Scour Below a Submerged Sluice Gate," thesis presented to the University of Minnesota, at Minneapolis, Minn., in 1956, in partial fulfillment of the requirements for the degree of Master of Science.

Tarapore, Z. S., and Dixit, J. G., discussion of "Sediment Transportation Mechanics: H. Sediment Discharge Formula," by the Task Committee for Preparation of Sedimentation Manual, Committee on Sedimentation of the Hydraulics Division, Vito A. Vanoni, Chmn., *Journal of the Hydraulics Division,* ASCE, Vol. 98, No. HY2, Proc. Paper 8681, Feb., 1972, pp. 388–390.

Taylor, G. I., "Diffusion by Continuous Movements," *Proceedings,* London Mathematical Society, Vol. 20, 1921, pp. 196–211.

Taylor, G. I., *Some Recent Developments in the Study of Turbulence,* Fifth International Congress for Applied Mechanics, Cambridge, Mass., John Wiley and Sons, Inc., New York, N.Y., 1938.

Taylor, B. D., "Temperature Effects in Alluvial Streams," *Report No. KH–R–27*, W. M. Keck Laboratory of Hydraulics and Water Resources, California Institute of Technology, Pasadena, Calif., 1971.

Taylor, B. D., and Vanoni, V. A., discussion of "Effects of Water Temperature on Bed-Load Movement," *Journal of the Waterways and Harbors Division*, ASCE, Vol. 95, No. WW2, Proc. Paper 6528, May, 1969, pp. 247–255.

Taylor, B. D., and Vanoni, V. A., "Temperature Effects in Low-Transport, Flat-Bed Flows," *Journal of the Hydraulics Division*, ASCE, Vol. 97, No. HY8, Proc. Paper 9105, Aug., 1972a, pp. 1427–1445.

Taylor, B. D., and Vanoni, V. A., "Temperature Effects in High-Transport, Flat-Bed Flows," *Journal of the Hydraulics Division*, ASCE, Vol. 98, No. HY12, Proc. Paper 9456, Dec., 1972b, pp. 2191–2206.

Taylor, R. H., Jr., and Brooks, N. H., discussion of "Resistance to Flow in Alluvial Channels," *Transactions*, ASCE, Vol. 127, Part I, Paper No. 3360, 1962, pp. 982–992.

Tek, M. R., "Two-Phase Flow," *Handbook of Fluid Dynamics*, V. L. Streeter, ed., Section 17, McGraw-Hill Book Co., Inc., New York, N.Y., 1961.

Terzaghi, K., *Theoretical Soil Mechanics*, John Wiley and Sons, Inc., New York, N.Y., 1943.

Thomas, D. G., "Transport Characteristics of Suspensions: Part VI, Minimum Transport Velocity in Large Particle Size Suspensions in Round Horizontal Pipes," *American Institute of Chemical Engineers Journal*, July, 1962.

Thornthwaite, C. W., and Koser, P., "Wind Gradient Observations," *Transactions*, American Geophysical Union, Vol. 24, Part I, Washington, D.C., 1943, pp. 166–181.

Tinney, E. R., "A Study of the Mechanics of Degradation of a Bed of Uniform Sediment in an Open Channel," thesis presented to the University of Minnesota, at Minneapolis, Minn., in 1955, in partial fulfillment of the requirements for the degree of Doctor of Philosophy.

Tison, L. J., "Studies of the Critical Tractive Force for the Entrainment of Bed Materials," *Proceedings*, Minnesota International Hydraulics Conference, Minneapolis, Minn., 1953, pp. 21–35.

Tison, L. J., "Local Scour in Rivers," *Journal of Geophysical Research*, Vol. 66, No. 12, Dec., 1961, pp. 4227–4232.

Toffaleti, F. B., "A Procedure for Computation of the Total River Sand Discharge and Detailed Distribution, Bed to Surface," *Technical Report No. 5*, Committee on Channel Stabilization, Corps of Engineers, United States Army, Vicksburg, Miss., Nov., 1968.

Toffaleti, F. B., "Definitive Computations of Sand Discharge in Rivers," *Journal of the Hydraulics Division*, ASCE, Vol. 95, No. HY1, Proc. Paper 6350, Jan., 1969, pp. 225–248.

Tracy, H. J., and Lester, C. M., "Resistance Coefficients and Velocity Distribution in a Smooth Rectangular Channel," *Water-Supply Paper 1592–A*, United States Geological Survey, Washington, D.C., 1961.

Trask, P., "Compaction of Sediments," *Bulletin*, American Association of Petroleum Geologists, Vol. 15, Tulsa, Okla., 1931, pp. 271–276.

Trowbridge, A. C., "Building of Mississippi Delta," *Bulletin*, American Association of Petroleum Geologists, Vol. 14, Tulsa, Okla., 1930, pp. 867–901.

Turtle, R., "The Hydraulic Conveying of Granular Material," thesis presented to London University, at London, England, in 1952, in partial fulfillment of the requirements for the degree of Doctor of Philosophy.

Udden, J. A., "Erosion, Transportation, and Sedimentation Performed by the Atmosphere," *Journal of Geology*, Vol. 2, 1894, pp. 318–331.

United States Army Corps of Engineers, "An Analytical Study of the Degradation and Aggradation Problem at the Fort Randall and Gavins Point Projects, Omaha District," Omaha, Neb., May, 1949.

United States Army Corps of Engineers, "Sediment Transportation Characteristics—Study of Missouri River at Omaha, Nebr.," United States Engineer Office, Omaha, Neb., 1951.

United States Army Corps of Engineers, "Missouri River Channel Regime Studies," *MRD Sediment Series No. 13A*, United States Army Engineer District, Omaha, Neb., Nov., 1968.

United States Army Corps of Engineers, "Missouri River Channel Regime Studies, Omaha District," *MRD Sediment Series No. 13B,* Nov., 1969.

United States Bureau of Reclamation, "Total Sediment Transport Program, Lower Colorado River Basin, Denver, Colorado," Interim Report, Jan., 1958.

United States Department of Agriculture, "Soils and Men," *Yearbook of Agriculture,* United States Government Printing Office, Washington, D.C., 1938.

Van Driest, E. R., "Experimental Investigation of Turbulent Diffusion—A Factor in Transportation of Sediment in Open-Channel Flow," *Journal of Applied Mechanics,* June, 1945, pp. A-91–A-100.

van Olpen, H., *An Introduction to Clay Colloid Chemistry,* Interscience Publishers, New York, N.Y., 1963.

Vanoni, V. A., "Transportation of Suspended Sediment by Water," *Transactions,*ASCE, Vol. 111, Paper No. 2267, 1946, pp. 67–133.

Vanoni, V. A., "A Summary of Sediment Transportation Mechanics," *Proceedings,* Third Midwestern Conference on Fluid Mechanics, University of Minnesota, Minneapolis, Minn., 1953a, pp. 129–160.

Vanoni, V. A., "Some Effects of Suspended Sediment on Flow Characteristics," *Proceedings,* Fifth Hydraulic Conference, *Bulletin 34,* State University of Iowa, Studies in Engineering, Iowa City, Iowa, 1953b, pp. 137–158.

Vanoni, V. A., "Sedimentacion y Erosion en Tranques," *Revista Chilena de Ingenieria y Anales del Instituto de Ingenieros,* Vol. 75, No. 2, Santiago, Chile, March, April, May, 1962, pp. 17–23.

Vanoni, V. A., and Brooks, N. H., "Laboratory Studies of the Roughness and Suspended Load of Alluvial Streams," *Sedimentation Laboratory Report No. E68,* California Institute of Technology, Pasadena, Calif., Dec., 1957.

Vanoni, V. A., and Hwang, Li San, "Relation Between Bed Forms and Friction in Streams," *Journal of the Hydraulics Division,* ASCE, Vol. 93, No. HY3, Proc. Paper 5242, May, 1967, pp. 121–144.

Vanoni, V. A., and Nomicos, G. N., "Resistance Properties of Sediment Laden Streams," *Transactions,* ASCE, Vol. 125, Paper No. 3055, 1960, pp. 1140–1175.

Vanoni, V. A., Brooks, N. H., and Kennedy, J. F., lecture Notes on "Sediment Transportation and Channel Stability," Report No. KH-R-1, W. M. Keck Laboratory of Hydraulics and Water Resources, California Institute of Technology, Pasadena, Calif., 1960.

Varnes, D. J., "Landslide Types and Processes," *Special Report 29,* Highway Research Board, National Academy of Sciences—*National Research Council Publication 544,* 1958, pp. 20–47.

Vogt, E. G., and White, R. R., "Friction in the Flow of Suspensions (Granular Solids in Gases Through Pipe)," *Industrial and Engineering Chemistry,* Vol. 40, No. 9, Sept., 1948.

von Kármán, T., "Turbulence and Skin Friction," *Journal of the Aeronautical Sciences,* Vol. 1, No. 1, Jan., 1934, pp. 1–20.

von Kármán, T., "Sand Ripples in the Desert," *Technion Yearbook,* Vol. 6, 1947, pp. 52–54. 52–54.

Wadell, H. "Volume, Shape and Roundness of Rock Particles," *Journal of Geology,* Vol. 40, 1932, pp. 443–451.

Wadell, H., "Sphericity and Roundness of Rock Particles," *Journal of Geology,* Vol 41, 1933, pp. 310–331.

Wadell, H., "Volume, Shape and Roughness of Quartz Particles," *Journal of Geology,* Vol. 43, 1935, pp. 250–280.

Waggoner, P. E., and Bingham, C., "Depth of Loess and Distance from Source," *Soil Science,* Vol. 92, 1961, pp. 396–401.

Warn, F. G., and Cox, W. H., "A Sedimentary Study of Dust Storms in the Vicinity of Lubbock, Texas," *American Journal of Science,* Vol. 249, 1951, pp. 553–568.

Wasp, E. J., Ande, T. C., Seiter, R. H., and Thompson, T. L., "Hetero-Homogeneous Solids-liquid Flow in the Turbulent Regime," *Advances in Solid-liquid Flow in Pipes and its Application,* I. Zandi, ed., Pergamon Press, Inc., New York, N.Y., 1971.

Weisman, J., "Minimum Power Requirements for Slurry Transport," American Institute of Chemical Engineers Journal, Jan., 1963.

Welder, F. A., "Processes of Deltaic Sedimentation in the Lower Mississippi River," *Technical Report No. 12,* Coastal Studies Institute, Louisiana State University, Baton Rouge, La., 1959.

West Pakistan, "Canal and Headworks Data Observation Programme 1962–63 Data Tabulation," Part I, West Pakistan Water and Power Development Authority, Lahore, W. Pakistan, 1962.

White, C. M., "The Equilibrium of Grains on the Bed of a Stream," *Proceedings,* Royal Society of London, Series A, No. 958, Vol. 174, Feb., 1940, pp. 322–338.

Wicks, M., "Transport of Solids at Low Concentration in Horizontal Pipes," *Advances in Solid-liquid Flow in Pipes and its Application,* I. Zandi, ed., Pergamon Press, Inc., New York, N.Y., 1971.

Willis, J. C., discussion of "Sediment Transportation Mechanics: Initiation of Motion," by the Task Committee on Preparation of Sedimentation Manual, Committee on Sedimentation of the Hydraulics Division, Vito A. Vanoni, Chmn., *Journal of the Hydraulics Division,* ASCE, Vol. 93, No. HY1, Proc. Paper 5059, Jan., 1967, pp. 101–107.

Willis, J. C., and Coleman, N. L., "Coordination of Data on Sediment Transport in Flumes by Similitude Principles," *Water Resources Research,* Vol. 5, No. 6, Dec., 1969, pp. 1330–1336.

Wilson, K. C., "Bed Load Transport at High Shear Stress," *Journal of the Hydraulics Division,* ASCE, Vol. 92, No. HY6, Proc. Paper 4968, Nov., 1966, pp. 49–59.

Wilson, K. C., and Brebner, A., "On Two-phase Pressurized and Unpressurized Flow Behaviour Near Deposition Points," *Advances in Solid-liquid Flow in Pipes and its Application,* I. Zandi, ed., Pergamon Press, Inc., New York, N.Y., 1971.

Wilson, W. E., "Mechanics of Flow, with Non-Colloidal Inert Solids," *Transactions,* ASCE, Vol. 107, Paper No. 2167, 1942, pp. 1576–1594.

Wolman, M. G., and Leopold, L. B., "River Flood Plains: Some Observations on their Formation," *Professional Paper 282–C,* United States Geological Survey, Washington, D.C., 1957.

Wood, I. R., "Horizontal Two-Dimensional Density Current," *Journal of the Hydraulics Division,* ASCE, Vol. 93, No. HY2, Proc. Paper 5139, Mar., 1967, pp. 35–42.

Woodruff, N. P., "A Study of the Atmospheric Wind Gradient in the Layer Near the Ground," thesis presented to Kansas State University, at Manhattan, Kans., in 1952, in partial fulfillment of the requirements for the degree of Master of Science.

Woolley, R. R., "Cloudburst Floods in Utah 1850–1938," *Water Supply Paper 994,* United States Geological Survey, Washington, D.C., 1946.

Worster, R. C., "The Hydraulic Transport of Solids," *Proceedings of a Colloquium on the Hydraulic Transport of Coal,* National Coal Board, London, England, November 5, 1952.

Yotsukura, N., "Sediment Transport in a Pipe," thesis presented to Colorado State University, at Fort Collins, Colo., in 1961, in partial fulfillment of the requirements for the degree of Doctor of Philosophy.

Yubin, A. P., *Izyestiya U-Ser. Tekhn. Nauk.,* No. 8, Aug., 1949, p. 1146.

Zabrodsky, S. S., "Hydrodynamics and Heat Transfer in Fluidized Beds," ("Gidrodinamika i Teploperenos V Kipyaschem Sloye" by Fizmatgiz, Moscow-Leningrad, 1963) MIT Press, Cambridge, Mass., 1966.

Zandi, I., "Decreased Head Losses in Raw-Water Conduits," *Journal of the American Water Works Association,* Vol. 59, No. 2, Feb., 1967.

Zandi, I. *Advances in Solid-liquid Flow in Pipes and its Application,* Pergamon Press, Inc., New York, N.Y., 1971.

Zandi, I., and Govatos, G., "Heterogeneous Flow of Solids in Pipelines," *Journal of the Hydraulics Division,* ASCE, Vol. 93, No. HY3, Proc. Paper 5244, May, 1967, pp. 145–159.

Zandi, I., and Haydon, J. A., "A Pneumo-Slurry System of Collecting and Removing Solid Wastes," *Advances in Solid-liquid Flow in Pipes and its Application,* I. Zandi, ed., Pergamon Press, Inc., New York, N.Y., 1971.

Zenz, F. A., "Fluid Catalyst Design Data," *Petroleum Refiner,* Vol 36, Nos. 4–11, 1957; Russian Translation "Catalysts for Cracking in Fluidized Beds," (Katalizatory Krekinga v Kipyashchem Sloe), Izd. Gosinti, Moscow, USSR, 1958.

Zenz, F. A., and Othmer, D. F., *Fluidization and Fluid-Particle Systems,* Reinhold Publishing Corp., New York, N.Y., 1960.

Zernial, G. A., and Laursen, E. M., "Sediment Transporting Characteristics of Streams," *Journal of the Hydraulics Division,* ASCE, Vol. 89, No. HY1, Proc. Paper 3396, Jan., 1963, pp. 117–137.

Zingg, A. W., "The Intensity-frequency of Kansas Winds," *SCS–TP–88,* United States Department of Agriculture, Soil Conservation Service, Washington, D.C., 1950.

Zingg, A. W., "Evaluation of the Erodibility of Field Surfaces with a Portable Wind Tunnel," *Proceedings,* Soil Scinece Society of America, Vol. 15, Madison, Wisc., 1951, pp. 11–17.

Zingg, A. W., "Some Characteristics of Aeolian Sand Movement by Saltation Process," Editions du Centre National de la Recherche Scientifique 13, Quai Anatole France, Paris (7e), France, 1953a, pp. 197–208.

Zingg, A. W., "Wind Tunnel Studies of the Movement of Sedimentary Materials," *Bulletin 34, Proceedings of the 5th Hydraulics Conference,* Iowa Institute of Hydraulics, Iowa City, Iowa, 1953b, pp. 111–135.

Zingg, A. W., and Chepil, W. S., "Aerodynamics of Wind Erosion," *Agricultural Engineering,* Vol. 31, 1950, pp. 279–282, 284.

Znamenskaya, N. S., "The Analysis of Estimating of Energy Losses by Instantaneous Velocity Distribution of Streams with Movable Bed," *Proceedings,* Twelfth Congress of the International Association for Hydraulic Research, *Paper No. A4,* Vol. 1, Sept., 1967, pp. 27–30.

Chapter III.—Sediment Measurement Techniques

A. Fluvial Sediment

1. General.—Sediment movement in a natural stream, an extremely complex phenomenon, is treated in detail in Chapter II. It must be reemphasized that the development of adequate measurement equipment and techniques is dependent on a thorough understanding of the erosion, transportation, and deposition phenomena. The accuracy of sediment discharge determinations is dependent not only upon the field methods and equipment utilized in the collection of data, but upon knowledge of the distribution of the sediment in the flow. Particularly valuable is an understanding of the vertical and horizontal distribution of the sediment in a stream cross section, together with information on the size of the bed material and on the bed form.

In alluvial streams, the concentration of coarse material transported in suspension depends mainly upon velocity, concentration of fine sediment, bed configuration, and the shape of the measuring section. Therefore, the concentration of the coarse material in suspension may increase or decrease from section to section, even though the water and total sediment discharge remain uniform.

2. Sediment Transportation Characteristics of Streams

Vertical Distribution of Suspended Sediment.—The suspended-sediment concentration generally is at a minimum at the water surface, and at a maximum near the stream bed. The coarsest fractions of the suspended sediment, which are usually sand, exhibit the greatest variation in concentration from the stream bed to the water surface. The fine fractions, i.e., the silt and clay, are usually uniformly distributed over the depth of the stream, or nearly so. The bed material of a stream with a suspended load of sand, silt, and clay particles tends to be predominantly sand and will contain only traces of silt and clay material. For conditions of steady uniform flow, the concentration, C, of a given size-fraction of suspended sediment in a natural sand bed stream varies with depth in such a manner that a plot of $(d-y)/y$ against C is a straight line on log-log paper, in which d = stream depth; y = distance above bed to a point; and C = the concentration of the size-fraction at the point. (See Eq. 2.77 and Fig. 2.32, Chap. II, Section D).

The vertical distribution of velocity and total suspended sediment for a range of selected stream verticals for representative sand bed streams is shown in Fig. 3.1. Similar data, including concentrations for suspended sand, silt, and clay fractions, are shown in Figs. 3.2 and 3.3 for representative stream verticals for the Mississippi River at St. Louis, Mo., and the Rio Puerco near Bernardo, N.M., respectively. The Mississippi River is typical of a large sand bed stream and the Rio Puerco is an example of a steep, rapid, and heavily sediment-laden stream in the arid western United States.

The presence of dunes or antidunes in streams alternately decreases and increases the local depth. Changes in local depth are associated with large-scale eddies that bring about maximum percentage fluctuation in momentary concentration of sediment in suspension. In sand bed channels, the distribution and concentration of the coarse part of the suspended sediment (bed-material load) are related to the composition of the bed material. In channels with semirigid boundaries consisting of cobblestones or cobblestones and boulders, the distribution and concentration of the coarse part of the suspended sediment are not similarly related to the composition of the bed material. In such streams, the concentration of the coarse part of the suspended sediment generally is less than that for streams flowing in sand bed channels for comparable depths and velocities.

Momentary or instantaneous fluctuations in suspended-sediment concentrations are related to scale and intensity of stream turbulence and particle size of sediment in the bed of the stream. Maximum deviations from mean concentration values are to be found in streams where the suspended sediment consists largely of sand. In contrast, minimum deviations from mean concentration values occur in

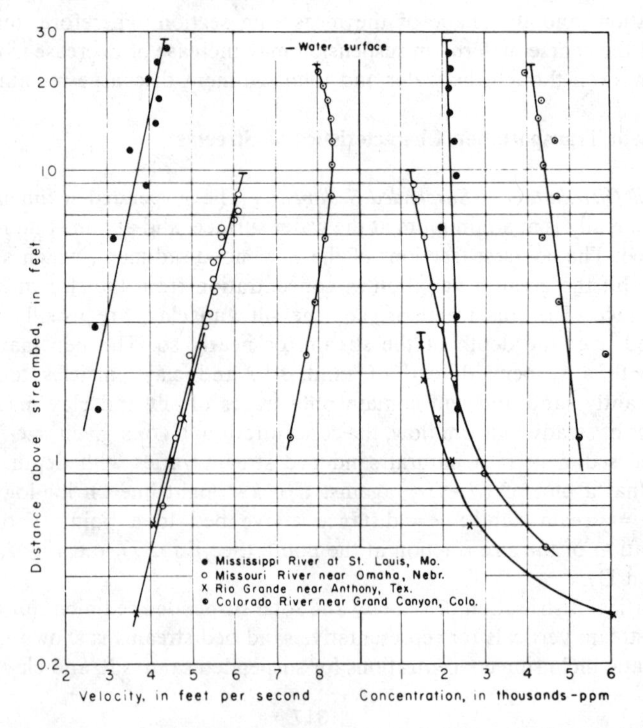

FIG. 3.1.—Distribution of Velocity and Suspended Sediment in Stream Verticals for Selected Rivers

streams where the suspended sediment consists largely of silt and clay. Thus, the amount of the momentary fluctuation in concentration of the suspended sediment at any depth varies with the ratio of the concentration of the sand fraction to the total concentration. The amount of fluctuation also is affected by bed configuration.

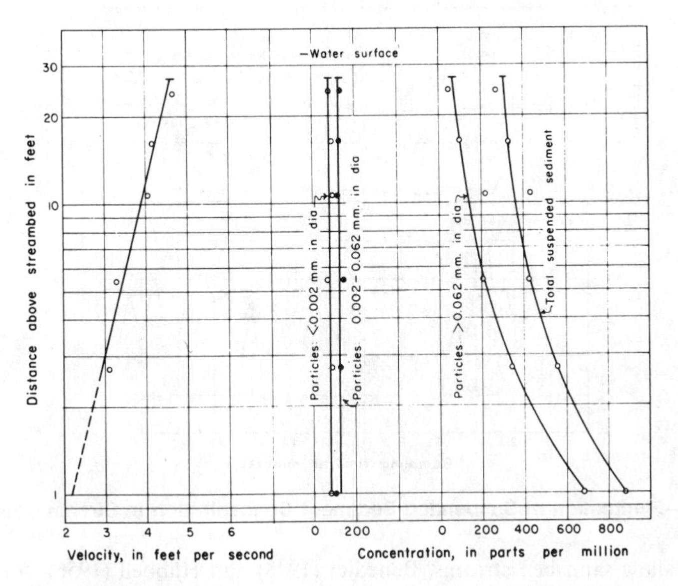

FIG. 3.2.—Distribution of Velocity and Suspended Sediment in Stream Vertical, Mississippi River, at St. Louis, Mo., April 24, 1956

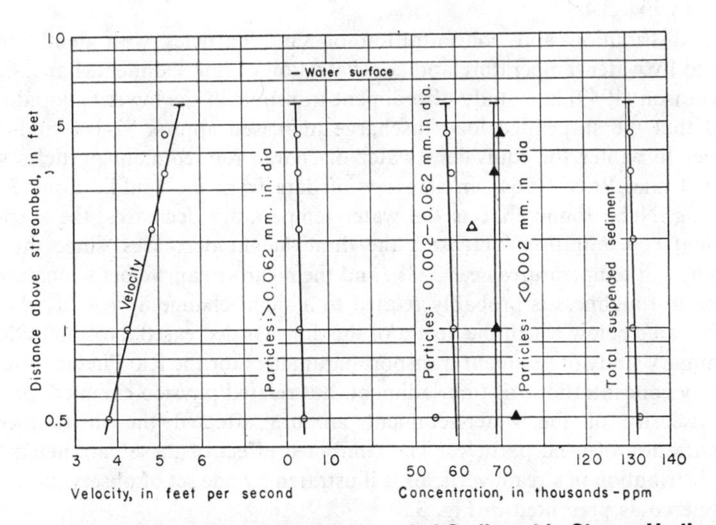

FIG. 3.3.—Distribution of Velocity and Suspended Sediment in Stream Vertical, Rio Puerco near Bernardo, N.M., August 19, 1961

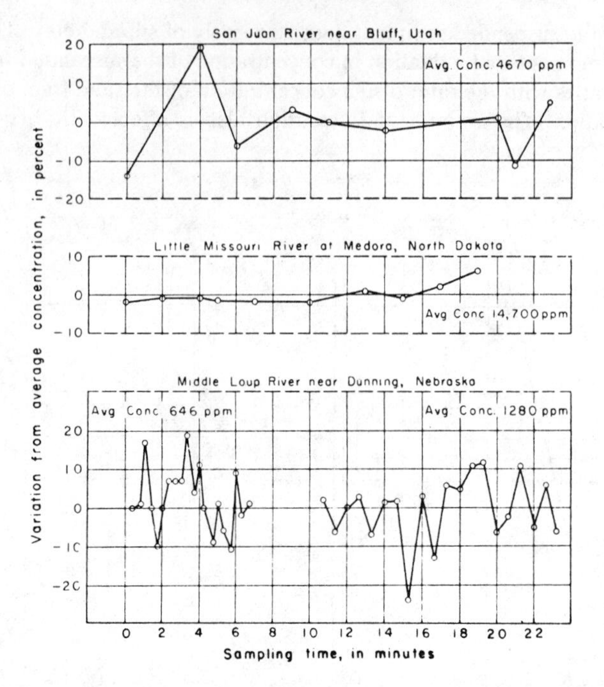

FIG. 3.4.—Fluctuation of Suspended-Sediment Concentration in Stream Vertical

In shallow sand bed streams, Benedict (1948) and Hubbell (1956)[1] found that the concentration for depth-integrated samples collected consecutively for selected stream verticals varied as much as 24% from the mean value. The fluctuation in the concentration at selected verticals for three sand bed streams is shown in Fig. 3.4.

The distribution and concentration of sand particles with depth are also affected by water temperature and concentration of fine sediment. Lane, Carlson, and Hanson (1949) in a study of sediment transport in the lower Colorado River, found that the suspended load discharge increased approx "2-1/2 times" from summer to winter for equivalent water discharge for sediment particles smaller than 0.3 mm. Brice (1964), in a review of data from the Middle Loup River at Dunning, Neb., found that as the water temperature decreases, the suspended-sediment concentration increases; the dune height decreases, since the bed is smoothest at temperatures near 32°F, and the mean stream velocity increases. The change in roughness is probably related to a slight change in velocity through a critical value below which the height of the dunes is decreased. Nordin (1963), in a preliminary study of sediment transport parameters for the Rio Puerco, found that not only concentrations of fine sediment, but related physico-chemical properties and viscosity of the water-sediment mixture affected the distribution and concentration of sand particles. The combined effect of these parameters on the sand distribution in stream verticals is illustrated by one set of observations for the Rio Puerco as presented in Fig. 3.3.

[1] References are given in Chapter III, Section F.

Lateral Distribution of Suspended Sediment.—The lateral distribution of suspended sediment at a stream cross section varies with velocity and depth, channel slope and alinement, bed form, particle size of sediment in transport, and inflow from immediate upstream tributaries. The variation in concentration in a cross section is given in terms of the average concentration for the cross section for several small rivers in Fig. 3.5. The concentration at each stream vertical is

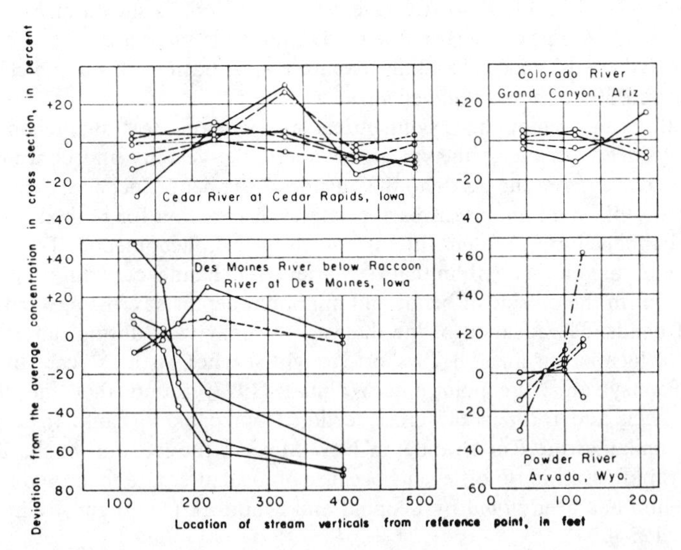

FIG. 3.5.—Variation of Suspended-Sediment Concentration in Cross Sections of Four Representative Small Rivers

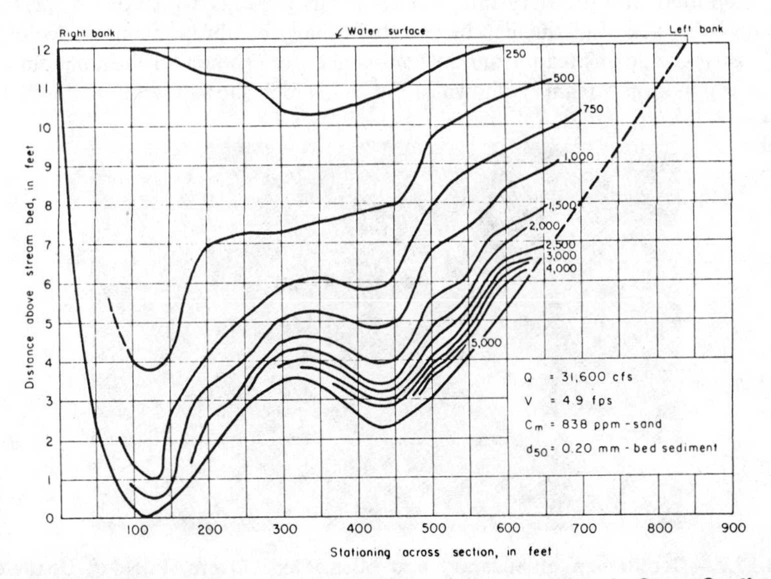

FIG. 3.6.—Distribution of Suspended Sand, in parts per million, in Cross Section of Missouri River near Omaha, Neb. (Bondurant, 1963)

determined from depth or point-integrated samples. Because the stream verticals were located at the centroids of sections of equal water discharge, the average of the concentrations for the stream verticals is assumed to be the discharge-weighted concentration, i.e., the sediment discharge concentration for the cross section. Each curve in each graph is based on data from concurrent observations and the curves cover a range of discharges. The distribution of suspended sand in a cross section in the Missouri River near Omaha, Neb., is shown in Fig. 3.6 for one sediment discharge measurement. During this measurement the water discharge averaged 31,600 cfs and the average concentrations of silt and clay were 611 ppm and 397 ppm, respectively.

Natural streams seldom have symmetrical cross sections or straight channels. In medium to deep streams with typical meanders, the velocity and concentration usually increases from the inside to the outside of the bend. Secondary circulation undoubtedly plays an important part in lateral distribution for both straight and curved channels, but at present little is known of this phenomenon. The location of the line of maximum depth in thalweg varies with channel curvature; maximum depths occur in the outside of bends and minimum depths at crossings and on the inside of bends. The location of the thalweg also may vary during runoff events because of local scour and fill. In short straight reaches, such as occur in Valley Creek, Pennsylvania, Leopold and Wolman (1957) found that the thalweg wanders back and forth in the cross section from bank to bank. In sand bed streams similar to the Rio Grande, in New Mexico, the location of the thalweg changes rapidly during runoff events because of local alternate scour and fill. This phenomenon has been noted by Leopold and Maddock (1953) and by Lane and Borland (1954).

Sediment contributed by inflow from tributary streams exerts a dominant influence on the distribution of sediment in the cross section. Numerous observers have reported that tributary inflow does not mix readily with flow in the main stream—the lack of mixing can be detected usually by abrupt change in color for some distance downstream from the confluence. An example of this phenomenon occurring in large streams is shown in Fig. 3.7, which shows the Mississippi River

FIG. 3.7.—Confluence of Missouri and Mississippi Rivers, Looking Upstream; Missouri Enters Mississippi from Left Foreground (Photograph by Massie—Missouri Resources Division)

below the mouth of the Missouri River. The latter stream, which transports large sediment loads, enters the Mississippi River from the right bank (looking downstream) at a point 15 miles above the MacArthur Bridge in St. Louis, Mo. Below the confluence, the Mississippi River flows in a sand bed channel with meanders typical of a flood-plain stream. At MacArthur Bridge, Jordan (1965) found that lateral concentration gradients have no apparent relation to the proportion of flow that is from the Missouri River or from the upper Mississippi River. Concentrations of suspended sediment in stream verticals near the right bank exceeded those near the left bank by as much as 2,400 ppm, or 0.24%, and at many times the difference has been greater than the mean concentration for the cross sections.

Lateral Distribution of Bed Load Discharge.—In sand bed channels under conditions of reasonably steady discharge and channel equilibrium, the percentage of total load moved as bed load varies as the cross section changes in the direction of flow. In narrow deep sections with high velocities, a larger percentage of the total load is moved as suspended load; in wide shallow sections, the converse generally is true. At a cross section, the bed load discharge per unit width is a function of the bed composition and of the local velocity; the maximum rate of transport generally is at the lateral location where the velocity is highest. Bed load discharge is also changed by viscosity of the water-sediment mixture, particularly if the flow-transport relationships are such that a slight change in velocity through a critical value results in a change in bed form (Brice, 1964).

In streams with lateral variation of bed roughness in the cross section, the rate of bed load movement may be expected to vary approximately as the concen-

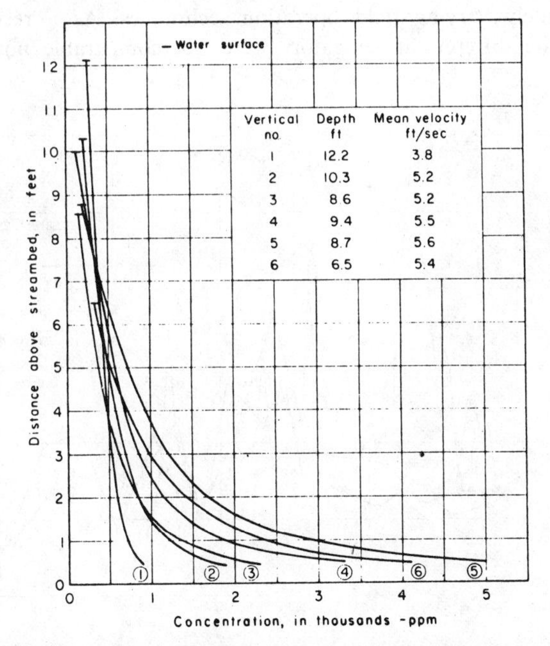

FIG. 3.8.—Distribution of Suspended Sand at Selected Stream Verticals in Cross Section of Missouri River near Omaha, Neb.

tration of suspended sand immediately above the bed. In the Missouri River at Omaha, Bondurant (1963) found that the concentration of the sand at a height of 0.5 ft above the stream bed varied about fivefold in the cross section. This variation in sand concentration in the cross section for six stream verticals is shown in Fig. 3.8. In shallow sand bed streams, Hubbell and Matejka (1959) found a similar variation in bed load discharge per unit width in observations of total transport at a turbulence flume in the Middle Loup River near Dunning, Neb. The lateral variation in total sediment concentration at this flume is shown in Fig. 3.9.

In alluvial channels composed mainly of coarse gravel, cobblestones, and small boulders, Benedict (1948) observed that the sand fraction of the bed load moves in streaks. Except in mountain streams that have very steep gradients, the movement of gravel, cobblestones, and small boulders is limited generally to that part of the cross section that has very high velocities.

For catastrophic floods such as occurred on Coffee Creek, Trinity County, California, Stewart, and LaMarche (1967) suggest that mean stream velocities may approach a value of 14.7 fps, which is adequate to transport boulders "as large as 6 ft in maximum diameter and more than 5 ft in intermediate diameter." Fahnestock (1963) in a study of the White River, Washington, a glacier stream, observed that a mean velocity of 7 fps was adequate to move boulders having an average diameter of 1.8 ft.

3. Equipment for Sediment Measurements.—Samplers used in fluvial sediment measurements during the 19th century were of the simplest type. After 1900, and particularly during the period 1925–1940, many investigators developed new equipment independently. Most of the samplers were used without precalibration and under widely varying field operation techniques. As a result, the data obtained by the different investigators were not comparable nor could their

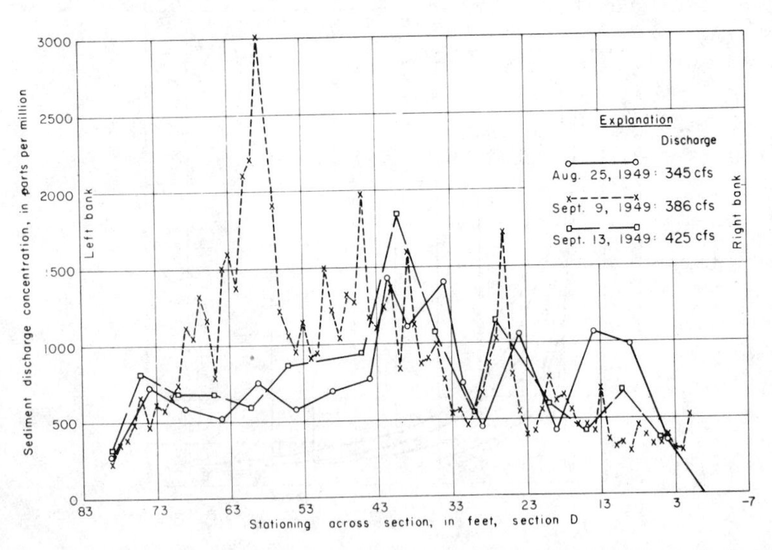

FIG. 3.9.—Lateral Distribution of Total Sediment Concentration in Turbulence Flume, Middle Loup River near Dunning, Neb.

accuracy be evaluated. This situation was recognized by several agencies of the United States Government and an Interagency program was organized in 1939 to study methods and equipment used in measuring the sediment discharge of streams and to improve and standardize equipment and methods where practicable (Interagency Committee, 1941b). The comprehensive study of sampling equipment included suspended sediment, bed load, and bed material samplers.

Instantaneous Suspended-Sediment Samplers.—In the initial stages of the development of samplers, an open container or bucket was used to dip samples from a stream. This rather crude instrument was followed by the bottle or closed container type, which could be opened and closed instantaneously at any selected depth by a drop-weight machanism. Later, instantaneous samplers of the horizontal trap type, that oriented themselves in the direction of flow, were developed and used by some investigators (Interagency Committee, 1940a).

A survey of sediment sampling equipment conducted under the Interagency program of the United States indicated that the 30 instantaneous samplers studied had very limited applicability either because of poor intake-velocity characteristics, or because of the short filament of water-sediment mixture sampled (Interagency Committee, 1940a, 1941b). Studies of their effectiveness indicated that instantaneous samplers were not suitable for general field use.

Time-Integrating Suspended-Sediment Samplers.—A time-integrating sampler is one that collects a sample during a finite time interval. Time-integrating samplers are of two types: depth-integrating and point-integrating. The depth-integrating sampler fills while it is moved at a uniform transit rate in a stream vertical. The point-integrating sampler fills while suspended at a selected point in a stream vertical.

The requirements of an ideal integrating suspended-sediment sampler, which have been summarized by Nelson and Benedict (1951) are as follows: (1) The velocity at the entrance of the intake tube should be equal to the local stream velocity; (2) the intake should be pointed into the approaching flow and should protrude upstream from the zone of disturbance caused by the presence of the sampler; and (3) the sample container should be removable and suitable for transportation to the laboratory without loss or spoilage of the contents. Furthermore, the sampler should; (4) fill smoothly without sudden inrush or gulping; (5) permit sampling close to the stream bed; (6) be streamlined and of sufficient weight to avoid excessive downstream drift; (7) be rugged and simply constructed to minimize the need for repairs in the field; and (8) be as inexpensive as possible, and consistent with good design and performance.

Some 35 time-integrating samplers were developed and used prior to 1940 on projects in the United States and elsewhere. The results of the studies conducted under the Interagency program indicated that none of these samplers met the criteria for an ideal sampler. In considering new designs, it was decided to develop both depth and point-integrating samplers (Interagency Committee, 1952, 1963).

United States Depth-Integrating Samplers.—Depth-integrating samplers are designed to accumulate a water-sediment sample as they are lowered to the stream bed and raised to the surface at a uniform rate. During transit, the velocity in the intake nozzle is nearly equal to the local stream velocity at all points in the vertical.

The air in the sample container is compressed by the inflowing liquid so that its pressure balances the external hydrostatic head. If the speed of lowering the sampler is such that the rate of air contraction exceeds the normal rate of liquid inflow, the actual rate of inflow will be higher than the local stream velocity and some inflow may occur through the air exhaust. If the speed of lowering the sampler is too slow, the sample container will be filled before the transit is completed, circulation will take place, and some outflow through the air exhaust line will occur. In practice, transit time should be adjusted so that the sample container is not completely filled; this will avoid error in sediment content of the sample due to outflow through the air exhaust. The transit velocities of the sampler for descending and ascending trips need not be equal, but the velocity for each trip must be constant to avoid error in sampling.

Theoretical maximum and minimum values of $R_T/\overline{U}$, the ratio of the sampler transit velocity, R_T, to the mean stream velocity at the sampling vertical, $\overline{U}$, are determined from Boyle's law and stream velocity. Values of $R_T/\overline{U}$ for a 1-pint container and typical vertical velocity distribution are shown in Fig. 3.10. The maximum raising or lowering rates should not exceed 0.4 of the mean vertical velocity to avoid excessive angles between the nozzle and the approaching flow. Laboratory tests indicate that the maximum transit velocity can be increased about 10% above those shown in Fig. 3.10 without introducing appreciable error in the sediment content of the sample. The practical maximum sampling depth for 1/8-in. and 3/16-in. diam nozzles is limited to about 15 ft when the depth-integrating sampler is lowered and raised at a uniform rate. For equipment used in

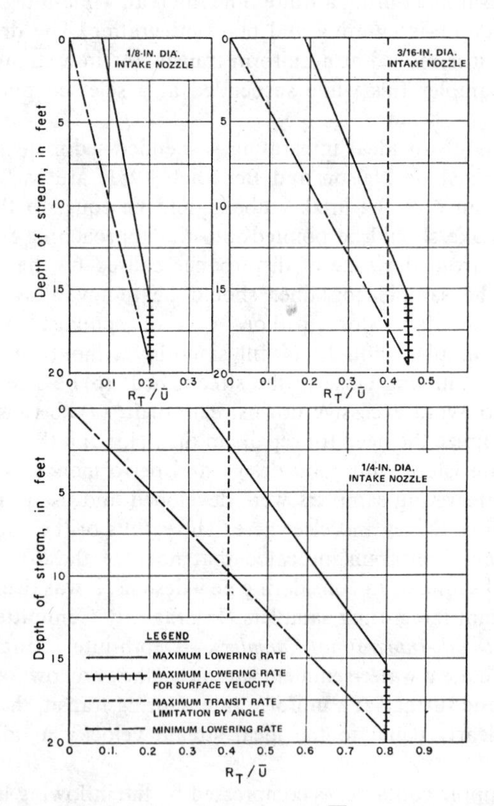

FIG. 3.10.—Maximum and Minimum Values of $R_T/\overline{U}$ for Depth-Integrating Sampler with 1-pint Container (0.473 ℓ)

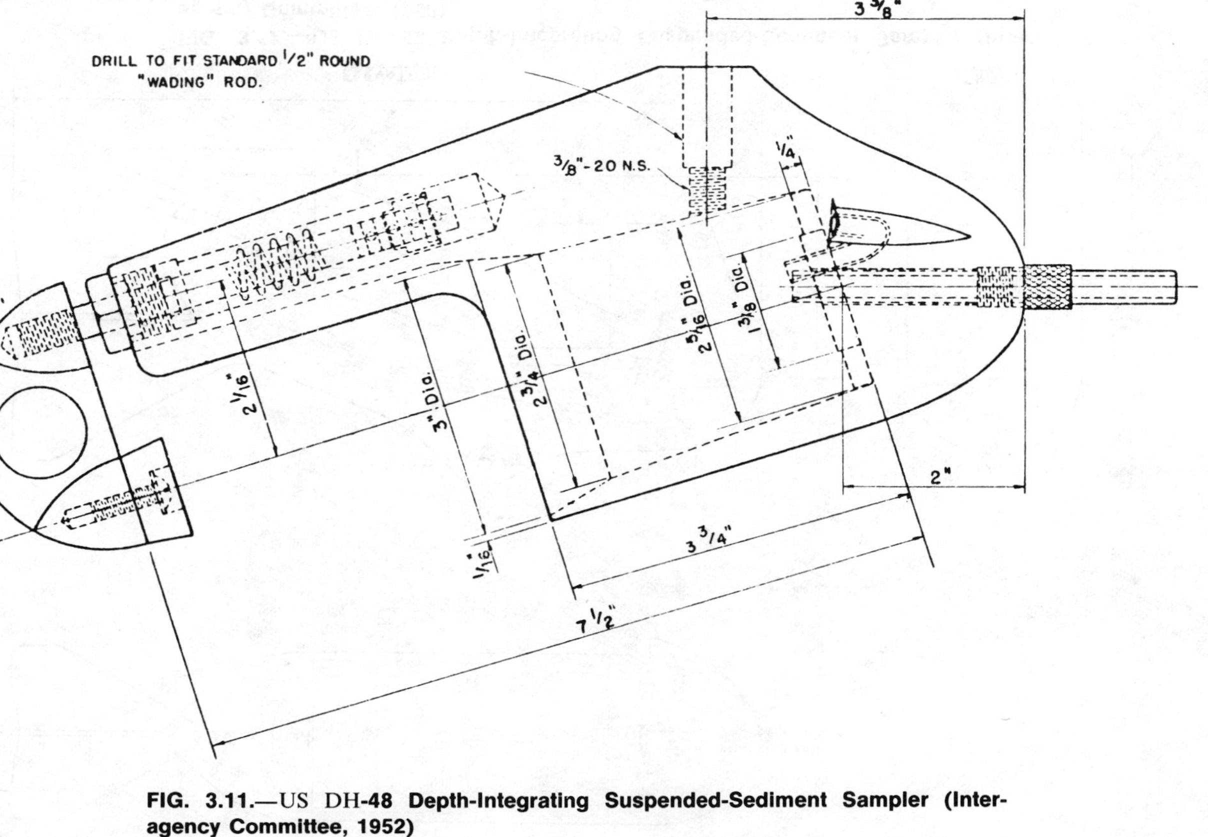

FIG. 3.11.—US DH-48 **Depth-Integrating Suspended-Sediment Sampler (Interagency Committee, 1952)**

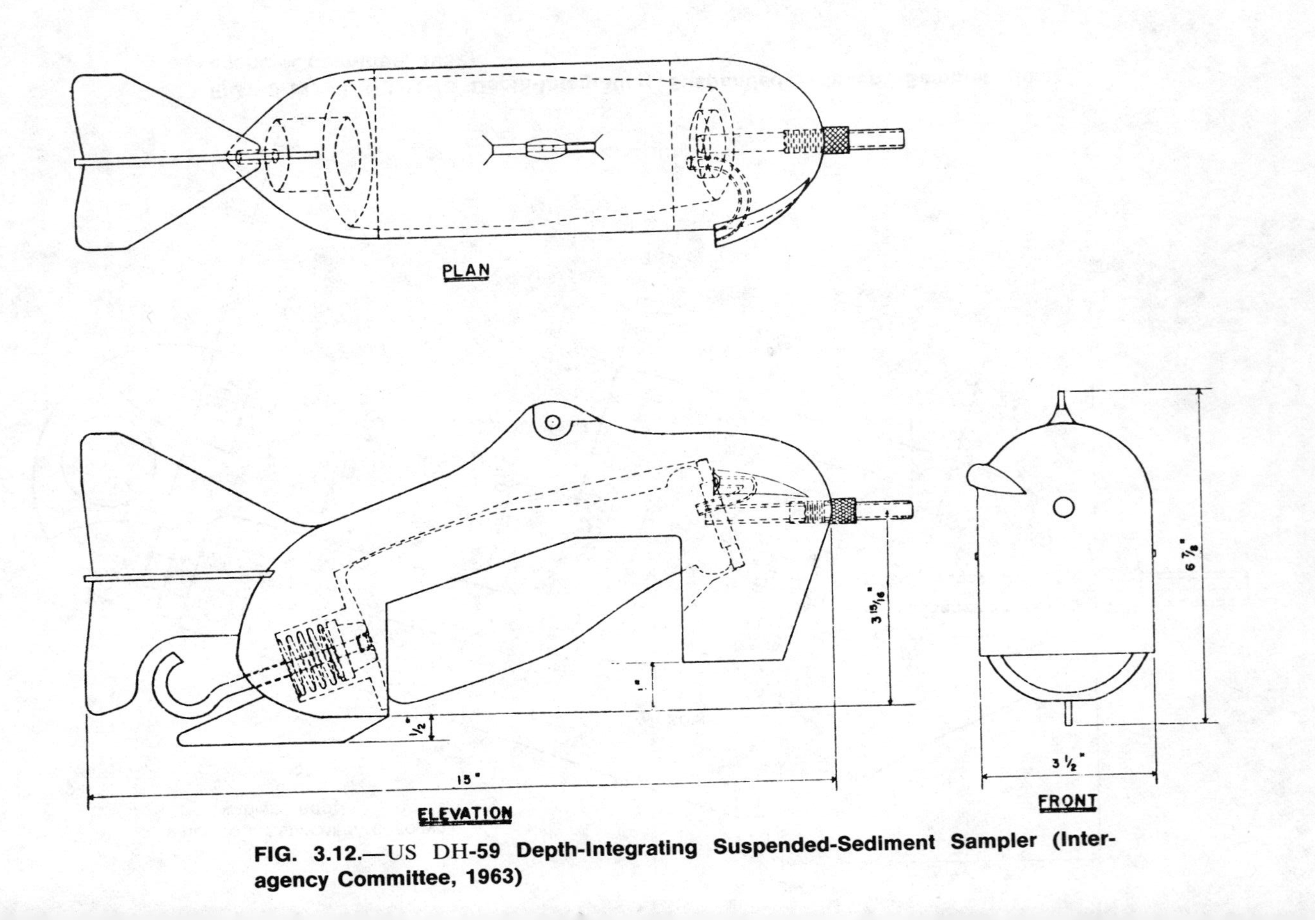

FIG. 3.12.—US DH-59 **Depth-Integrating Suspended-Sediment Sampler (Interagency Committee, 1963)**

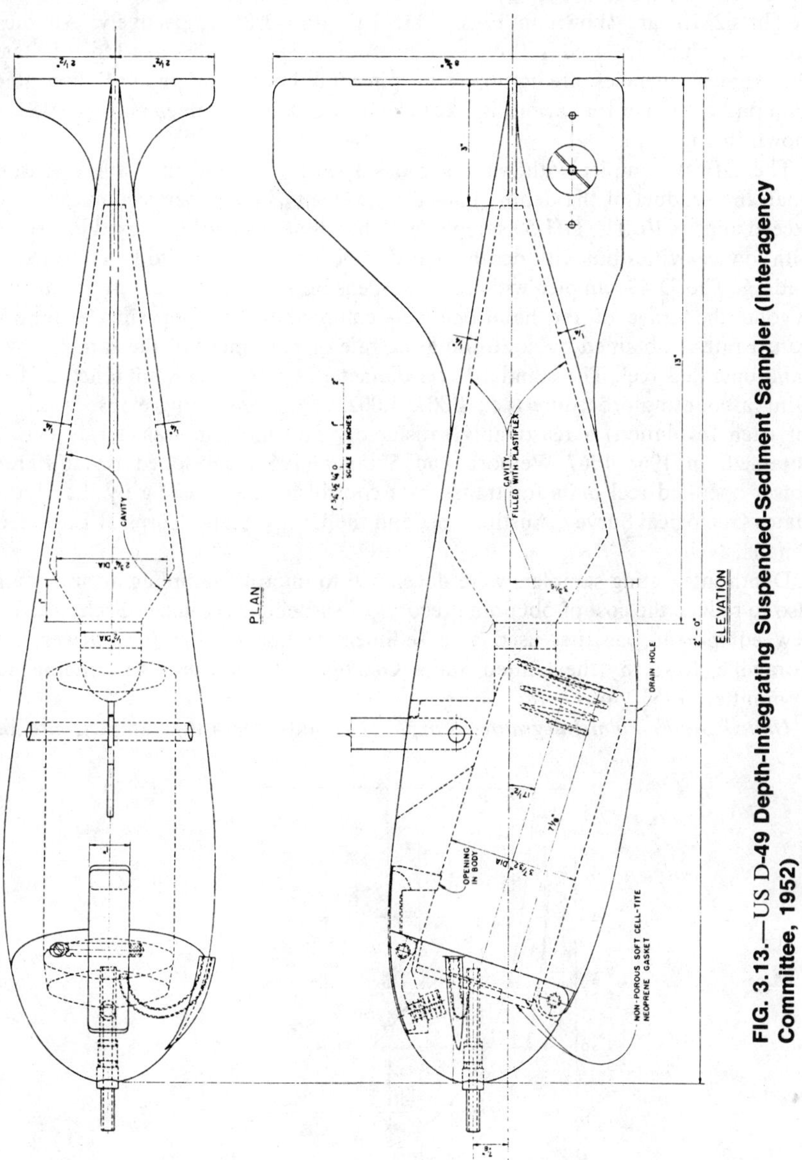

FIG. 3.13.—US D-49 Depth-Integrating Suspended-Sediment Sampler (Interagency Committee, 1952)

sampling streams deeper than about 15 ft, see the following section.

Depth-integrated samples may be collected by hand by wading in a stream or from a cable or bridge. Line drawings of the US DH-48 sampler, weight 4.5 lb, US DH-59 sampler, weight 24 lb, and the cable suspended US D-49 sampler, weight 62 lb, are shown in Figs. 3.11, 3.12, and 3.13, respectively. All these samplers, which are used in fluvial sediment investigations by most United States Government agencies, are equipped for pint milk bottle containers. Filling times for a pint container for various intake velocities and three diameters of nozzles are shown in Fig. 3.14.

The DH-48 sampler with wading rod suspension is used in shallow streams when the product of the depth, in feet, and velocity, in feet per second, does not exceed approx 10. The DH-59 sampler with handline suspension is used in streams with low velocities but with depths that do not permit samples to be collected by wading. The D-49 sampler with cable suspension is designed for use in streams beyond the range of the hand-operated equipment. An approximate uniform transit rate is obtained by controlling the rate of movement of the sampler by a hand-operated reel. The standard reel diameter is 1 ft in circumference and by using a counting procedure (i.e., 1,001, 1,002, etc., representing elapsed seconds for each revolution) a reasonably satisfactory and uniform transit rate can be obtained. In 1966–1967 Welborn and Skinner (1968) developed two different power-operated reel units for transit rate control for field testing by the United States Geological Survey, Austin, Tex., and the United States Corps of Engineers, Omaha, Neb.

Depth-integrating samplers were developed to improve sampling accuracy and also to reduce the cost of obtaining records of suspended sediment discharge. The new equipment was first used at a sediment station on the Iowa River near Coralville, Iowa, by the United States Geological Survey in 1943 (Interagency Committee, 1963, pp. 17–18).

United States Point-Integrating Samplers.—Point-integrating samplers are de-

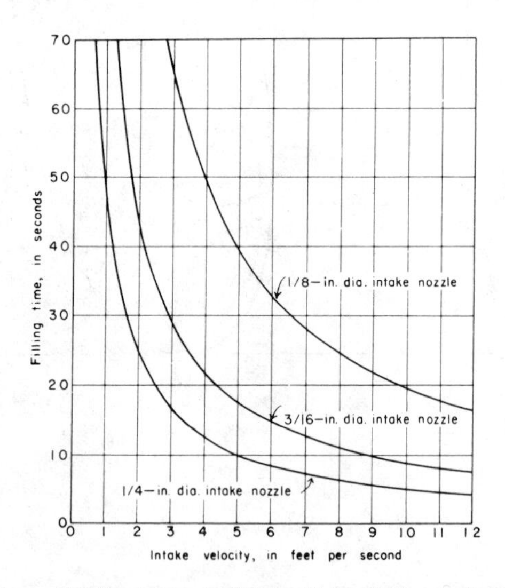

FIG. 3.14.—Relation of Filling Time to Intake Velocity for 1-pint Container (0.473 ℓ) (Interagency Committee, 1952)

signed to accumulate a water-sediment sample that is representative of the mean concentration at any selected point in a stream during a short interval of time. The air exhaust and intake characteristics are identical to those for the depth-integrating sampler. A rotary valve that opens and closes the sampler is operated by a solenoid energized by batteries at the surface. The current flows to the solenoid through a 1/8-in. two-conductor "reverse-lay current-meter cable," which suspends the sampler.

The body of the sampler contains an air chamber that is interconnected by tubing and a passage through the valve. The air chamber has an opening in the bottom of the sampler through which water can enter, thereby raising the air pressure in the sample container to the hydrostatic pressure at the sampling point. The passage to the air chamber is closed during the sampling periods and air from the container escapes through the air exhaust.

The point-integrating samplers can be used to collect depth-integrated samples by leaving the valve in the open position as the sampler is moved through the stream vertical. Depth-integrated samples can also be obtained with the samplers by opening the valve at the stream bed and integrating the vertical from the bed to the water surface in the usual manner. With this procedure, streams as deep as about 30 ft can be sampled. In deep streams, depth-integrated samples may be obtained with a point sampler by successively integrating parts of the vertical if a constant transit velocity is used throughout the depth.

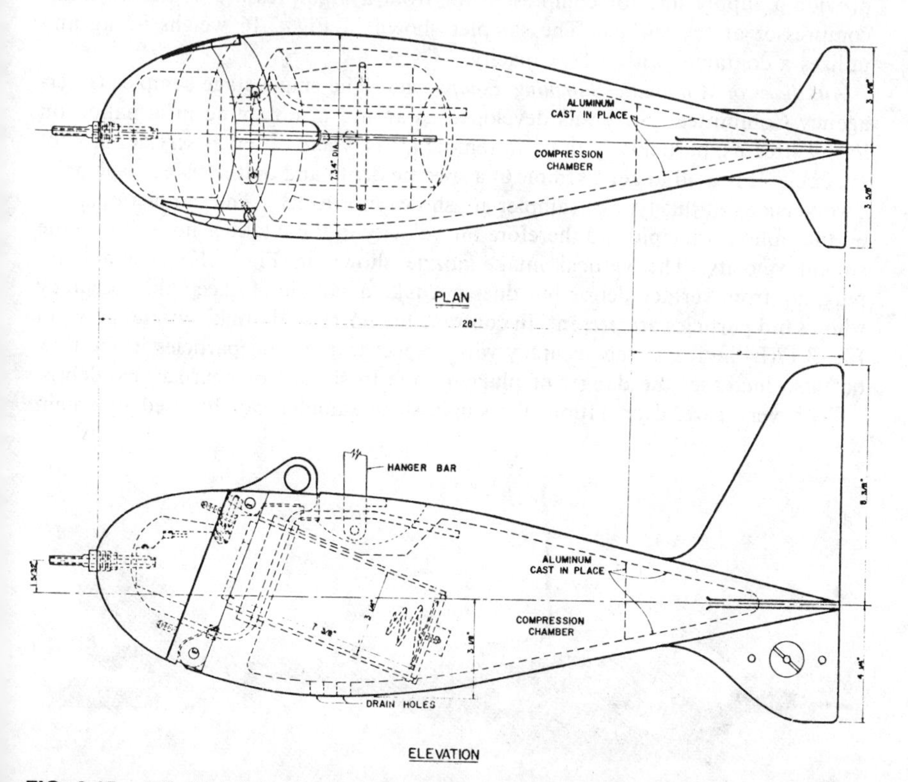

FIG. 3.15.—US P-61 Point-Integrating Suspended-Sediment Sampler (Interagency Committee, 1963)

The United States point-integrating samplers, P-46 and P-61 (100 lb) and P-63 (200 lb), have been adopted by most United States Government agencies and some private companies for investigations requiring information on sediment distribution in the stream vertical. The point-integrating sampler is also used to collect depth-integrated samples in streams that have depths and velocities beyond the range of the US D-49 sampler. The downstream drift of the P-61 sampler can be reduced by suspending a 100-lb stream-gaging weight below it. The P-46 and P-61 samplers utilize a pint milk bottle container. The P-63 sampler is designed to use either the pint or quart sample containers. A line drawing of the US P-61 sampler is shown in Fig. 3.15.

The US P-50 point-integrating sampler (300 lb) is similar to the P-63 sampler, except a sliding valve is used and the capacity of the container is a quart. This sampler was developed for use on major streams, e.g., the lower Mississippi River.

Neyrpic Sediment Sampler.—The Neyrpic sampler (1962) is designed to take an undisturbed water-sediment sample in flowing or quiescent water at selected points. A continuous flow of compressed air is used to prevent the water-sediment mixture from entering the container as the sampler is lowered or raised before and after sampling. During the sampling interval, the air is released from the container by two venturi tubes at a rate controlled by the stream velocity. This sampler can be used for the collection of either depth or point-integrated samples.

A special hollow cable 8.5 mm in diameter is used to support the sampler and to provide a supply line for compressed air from a high-pressure container or air compressor at the surface. The sampler shown in Fig. 3.16 weighs 93 kg and utilizes a container with a 1-ℓ capacity.

Auxiliary or Automatic Sampling Equipment.—The single stage sampler (Inter-agency Committee, 1961) was developed as an aid in obtaining information on flashy streams, particularly those in remote areas where observer services are not available. It is used to get a sample at a specific depth and on the rising stage only. Two versions of this type of sampler are shown in Fig. 3.17. The sampler operates on the siphon principle and therefore the velocity in the intake is not equal to the stream velocity. The vertical intake nozzle shown in Fig. 3.17(A) eliminates plugging from surface debris but does not take a sample of acceptable accuracy when sand particles are present. In contrast, the horizontal intake nozzle shown in Fig. 3.17(B) increases the accuracy with respect to the sand particles in the flow but also increases the danger of plugging due to surface or near-surface debris.

With very careful operation, the single stage sampler can be used to obtain

FIG. 3.16.—Neyrpic Sediment Sampler

supplemental information on sediment concentration at selected points in the stream vertical. Because of the limitations of the sampler, concentration data so obtained must be used with caution in the determination of the mean concentration in any stream cross section.

Research and development of pumping-type samplers and samplers utilizing radioisotopes, monochromatic light, electrical impedance and ultrasonic sound for sensing, is a part of the project activity being carried on at St. Anthony Falls Hydraulic Laboratory, University of Minnesota, under the sponsorship of the Committee on Sedimentation of the Federal Water Resources Council.

The pumping sampler under development and in field use by several agencies of the United States Government does not require an operator and is designed to obtain a continuous record of sediment concentration by sampling at a fixed point at specific time intervals. The velocity in the intake is not equal to the stream velocity, and the intake does not meet the requirements of an ideal sampler since it does not point into the flow. However, the pumping sampler can be rated or calibrated by comparing mean concentrations determined with it to mean values in the stream cross section determined from samples collected with standard depth or point-intergrating sampling equipment. By this means, an excellent continuous record can be obtained.

The pumping sampler permits collection of a sample of the pump discharge at each sampling interval. The bottle racks accommodate up to 145 pint milk bottles. The water-sediment mixture flows through a splitter mechanism that diverts an aliquot into each bottle. The pump is started by means of a float switch activated

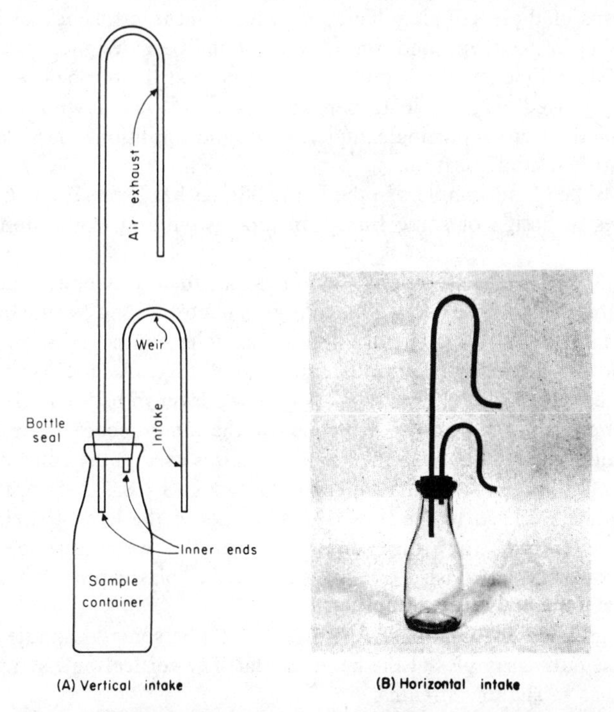

FIG. 3.17.—US Single Stage Suspended-Sediment Sampler (Interagency Committee, 1963)

by the water level in the stream. Samples are collected every hour during high stages and once every 12 hr during low stages.

A nuclear direct sensing device for suspended-sediment concentration has been developed by Parametrics, Inc. (1964) under contract with the United States Atomic Energy Commission. It utilizes a small amount of Cadmium-109 to determine the density of the water-sediment mixture. Correction for solute concentrations are determined from concurrent records of electrical conductivity.

Other equipment now under development for direct sensing of sediment concentration is not far enough advanced to determine practicability for either laboratory or field use.

Bed Load Samplers.—The development of bed load samplers has been closely associated with individual project studies. The samplers have been classified according to their type of construction and principle of operation—basket types, pan types, and pressure-difference types (Interagency Committee, 1940b). The basket and pan types cause an increased resistance to flow and resultant lowering of stream velocity at the sampler. This reduction in the stream velocity reduces the rate of bed load movement at the sampler, with the result that some particles accumulate at the entrance to the sampler and others are diverted away. The pressure-difference type is designed to eliminate the reduction in velocity, and thus any change in rate of bed load movement at the entrance to the sampler.

Bed load samplers have been developed and used in many countries in Europe including Switzerland, Germany, Hungary, Russia, and Poland; they have been used in India to determine rate of bed load movement of sediment particles varying in size from 1 mm–300 mm (Interagency Committee, 1940b) (Einstein, 1948). Calibrations of these samplers have indicated a mean efficiency of about 45% for the basket or pan type and about 70% for the best designed pressure-difference type. The efficiency of a sampler, i.e., the ratio of the trapped sediment to that actually moved as bed load per unit of time, varies with sampler characteristics, method of supporting sampler, hydraulic conditions, particle size, bed stability, and bed configuration.

Development of bed load samplers in the United States has been largely limited to minor changes in design of some European models without calibration tests (Hubbell, 1964).

Bed Material Samplers.—Bed material samplers, as first developed, may be divided into three types: the drag bucket, grab bucket, and vertical pipe (Interagency Committee, 1940b). The drag-bucket sampler consists of a weighted section of cylinder with an open mouth and cutting edge. As the sampler is dragged along the bed, it collects a sample from the top layer of bed material. The grab-bucket sampler is identical in principle to the drag-bucket sampler. It consists of a section of a cylinder attached to a rod and is used in the collection of samples from shallow streams. Both samplers are operated by dragging upstream so that the mouth is exposed to the flow, which results in the loss of some fine material while in transit from the stream bed to the water surface. The vertical-pipe or core sampler consists of a piece of metal or plastic pipe that can be forced into the stream bed by hand.

The drag and grab-bucket samplers are either too cumbersome to handle or do not obtain representative samples of the bed material. The vertical-pipe sampler is satisfactory for use in shallow streams.

United States Bed Material Samplers.—As a result of limited study under the Interagency program, three samplers have been adopted by United States

Government agencies for standard use where the bed material is predominantly sand or a mixture of sand and gravel (Interagency Committee, 1963).

The US BMH-53 sampler shown in Fig. 3.18 consists of a 9-in. length of a 2-in. brass or stainless steel pipe with cutting edge and suction piston attached to a control rod. The piston is retracted as the cutting cylinder is forced into the stream bed. The partial vacuum in the sampling chamber, which develops as the piston is withdrawn, is of assistance in collecting and holding the sample in the cylinder. This sampler can be used only in streams shallow enough to be waded.

The US BM-54 bed material sampler (weight 100 lb) is shown in Fig. 3.19. It is designed to be suspended from a cable and to scoop up a sample of the bed sediment that is 3 in. in width and about 2 in. maximum depth. When the sampler with the bucket completely retracted contacts the stream bed, the tension in the suspension cable is released and a heavy coil spring quickly rotates the bucket through 180°, scooping up the sample. At the close of the sampling operation, the cutting edge rests against a rubber stop, which prevents any sediment from being lost.

The US BMH-60 bed material sampler is similar to the US BM-54 and was developed for both handline and cable suspension. The sampler weighs 30 lb if made of aluminum and 40 lb if made of brass. It is used to collect samples in streams with low velocities but with depths beyond the range of the US BMH-53 sampler.

The bed material samplers previously described are used to obtain essentially "surface samples" of bed material. Samplers (Prych, Hubbell, and Glenn, 1965) have been developed to obtain core samples up to 6 ft in length in streams, particularly estuaries, where the mean velocity may be as much as 6 fps.

4. Sediment Discharge Measurements.—In the United States, the sediment discharge of streams is generally based on sediment and flow data obtained at or in the vicinity of existing gaging stations. Equipment developed as a part of the Interagency program makes it practicable to determine sediment concentration in a stream vertical except for a small segment just above the stream bed. With wading equipment, both velocity and concentration observations can generally be made to a point within 0.3 ft of the bed. For cable-supported equipment, the minimum distance above the bed for observations varies with the size of the equipment used. For velocity observations, it varies from 0.5 ft–1.0 ft depending upon the size of the stream-gaging weight; for concentration observations, it varies from 0.3 ft for the US D-49 sampler to 0.5 ft for the US P-61 sampler.

Suspended-Sediment Discharge Measurement.—Data normally obtained in the determination of suspended-sediment discharge consists of mean concentration, particle size, specific gravity of the suspended sediment, temperature of the water-sediment mixture, water discharge, and distribution of flow in the stream cross section.

The concentration data are usually obtained according to one of the following three schemes: (1) Depth-integrated samples collected at stream verticals representing areas of equal water discharge in the cross section; (2) depth-integrated samples collected at equally spaced stream verticals in the cross section; and (3) point-integrated samples collected at selected depths at stream verticals representing areas of equal water discharge. In the collection of samples [scheme (1)], the transit rate of the sampler must always be uniform but it need not be the same for all verticals; the concentration for the stream cross section is the mean of the concentrations of the stream verticals sampled. For scheme (2), the transit

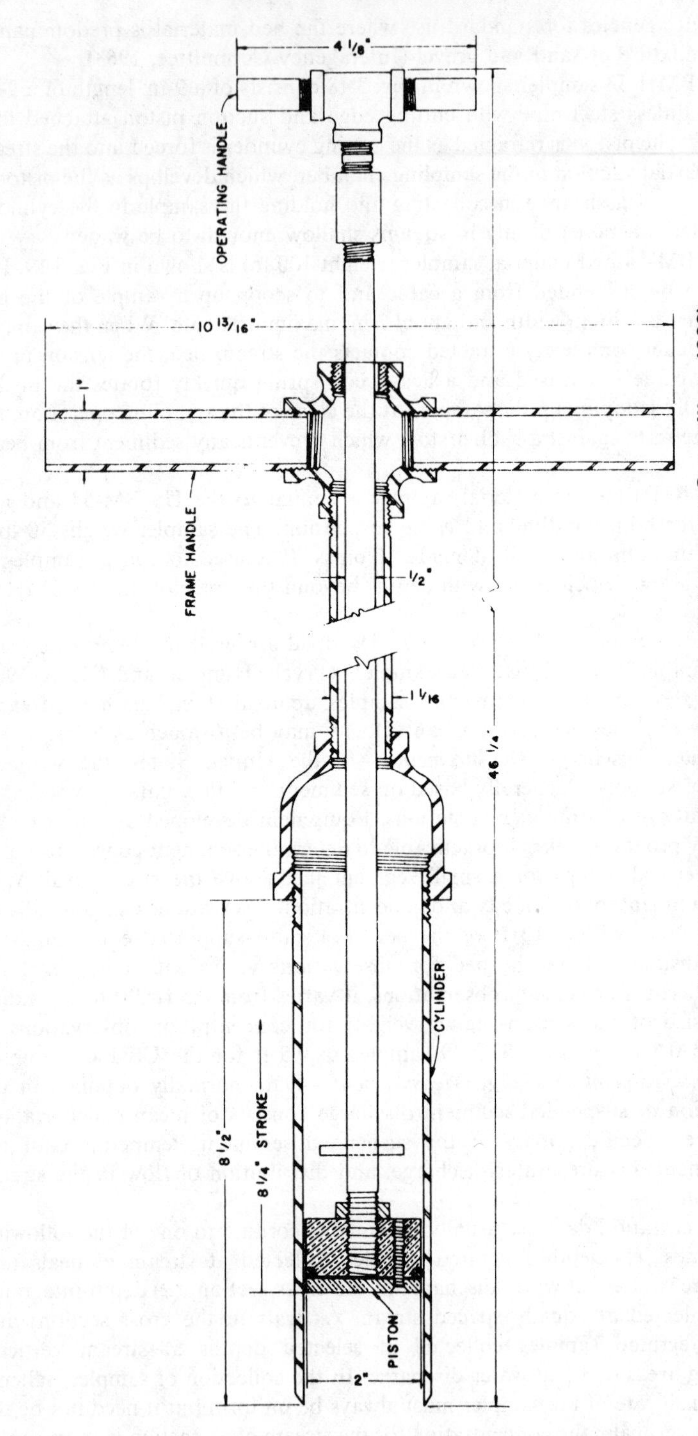

FIG. 3.18.—US BMH-53 Bed Material Sampler (Interagency Committee, 1963)

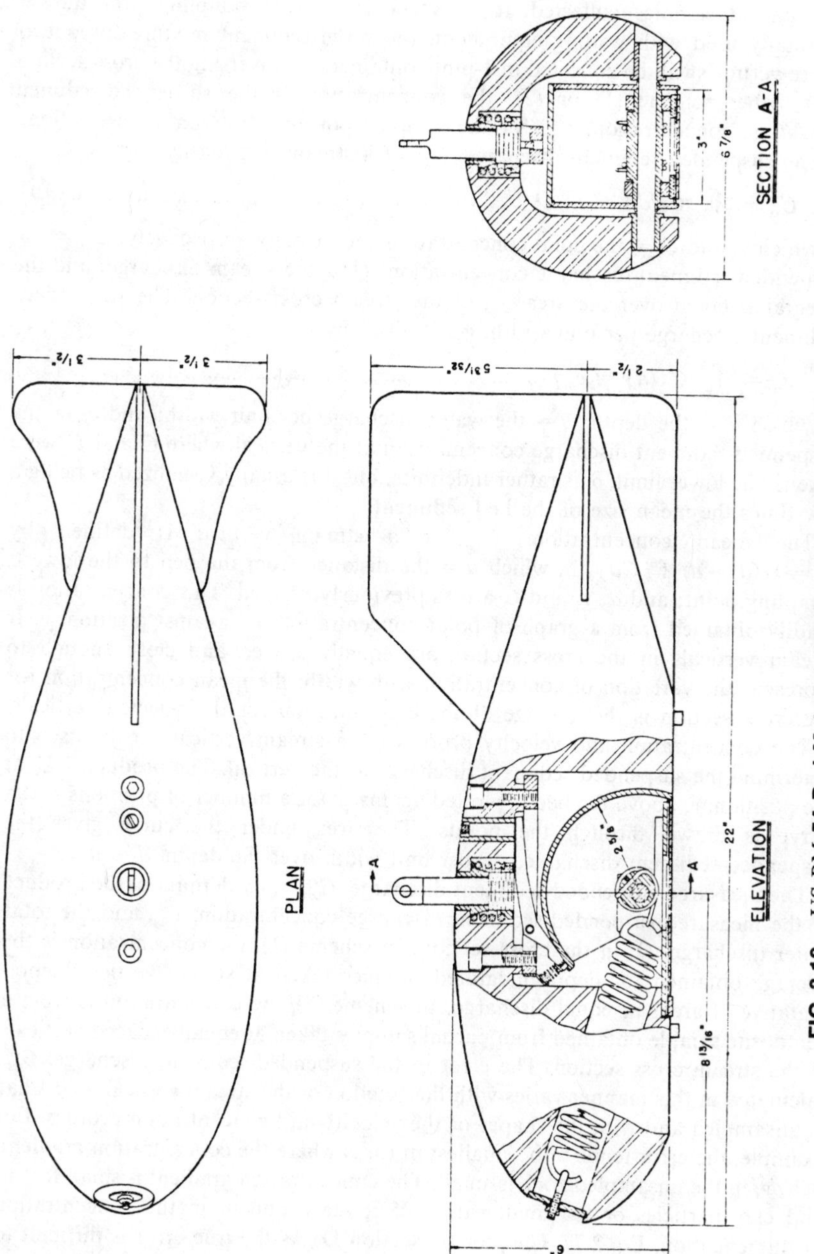

FIG. 3.19.—US BM-54 Bed Material Sampler (Interagency Committee, 1963)

velocity of the sampler must be uniform and the same for all verticals and the concentration for the stream cross section is a value obtained from a composite of the partial samples collected at all verticals. As the sampling procedure is normally used with wading equipment, the water-sediment mixture for one or more verticals will be collected in 1-pint containers in traversing the cross section. For either scheme (1) or (2), the concentration is the suspended-sediment discharge concentration, $\overline{C}'_m$, for the sampled portion of stream cross section.

The suspended-sediment discharge, G_{ss}, of a stream is given by

$$G_{ss} = \int_A CUdA = C_m Q \qquad (3.1)$$

in which C and U = the point concentration and velocity, respectively; C_m = the suspended-sediment discharge concentration; Q = the stream discharge; and the integral is taken over the area, A, of the stream cross section. The suspended-sediment discharge per unit width, g_{ss}, is given by

$$g_{ss} = \int_\delta^d CUdy = c_m q \qquad (3.2)$$

in which d = the depth; q = the water discharge per unit width; and c_m = the suspended-sediment discharge concentration at the vertical where C and U were taken. The lower limit, δ, is rather indefinite, but it is usually visualized as being a few times the mean size of the bed sediment.

The mean concentration, $\overline{C}$, for a stream vertical is defined by $\overline{C} = 1/(d - a) \int_a^d Cdy$, in which a = the distance from the bed to the lowest sampling point; and d, y, and C are as previously defined. This concentration is readily obtained from a graph of point concentration, C, against position, y. If stream verticals in the cross section are equally spaced and close enough to represent the variation of concentration with width, the mean concentration for the cross section is the average of the concentration for the selected verticals.

The concentration and velocity profiles for a stream vertical can be used to determine the suspended-sediment discharge at the vertical. The product, CU, at the position, y, above the bed is plotted against y for a number of positions and a curve is drawn through the points. The area under the curve gives the suspended-sediment discharge, g_{ss}, per unit width over the depth, $d - a$.

The measured suspended-sediment discharge, G'_{ss}, is, by definition, the product of the measured suspended-sediment discharge concentration, $\overline{C}'_m$, and the total water discharge, Q, in the cross section. In scheme (1), the concentration is the average obtained for depth-integrated samples taken at stream verticals representative of areas of equal discharge. In scheme (2), the concentration is from a composite sample obtained from partial samples taken at equally spaced verticals in the stream cross section. The error in the suspended-sediment discharge, G'_{ss}, calculated in this manner varies with the fraction of the stream vertical, a/d, that is unsampled and with the shapes of the velocity and concentration profiles. For example, the error tends to be smallest in cases where the concentration gradient, dC/dy, in the unsampled zone is small. The concentration gradient is small for silt and clay particles or for small values of z, the exponent in the concentration profile equation, Eq. 2.77, Chapter II, Section D. As the true error is difficult to estimate, corrections are seldom applied in the determination of the suspended-sediment discharge.

Water discharge and flow distribution in the cross section are based on velocity and depth observations at properly spaced stream verticals (Corbett, 1945). Centroids of areas of equal water discharge in the cross section are located from data from water discharge measurements. The water discharge in cubic feet per second for the individual areas, as represented by the selected stream verticals, is cumulated for the cross section. The cumulated discharge, expressed as a percentage of the total, is plotted against the lateral distance that each vertical is from a reference point.

The cumulative percentage of water discharge in the cross section for each of a selected number of verticals is given in Table 3.1. Given the number of sampling verticals, the location of each vertical is determined from the percentage values from Table 3.1 and the plot of cumulated water discharge versus location in the cross section. For streams with rapidly changing stage-discharge relationships, a water discharge measurement usually is made to determine the centroids of the areas of equal water discharge prior to collection of the sediment samples.

Data on particle-size distribution are obtained from samples selected to be representative of a range of sediment discharge and runoff conditions. Data on specific gravity normally are limited to a few samples representative of a range of sediment discharge. Data for both determinations can be obtained from samples collected for concentration analysis or from samples collected specifically for each determination. The latter procedure adds to the field work because of the time required to collect additional samples. However, the laboratory analyses are greatly simplified if representative individual samples are available for "size," "concentration," and "specific gravity determination," as current laboratory equipment for splitting samples does not provide an acceptable degree of accuracy if sand particles are present in the sample.

In actual practice, the frequency of sampling with respect to runoff and the number of sampling verticals in the cross section depends upon the scope of the study and the accuracy required. At measuring sections where there is a minimum amount of variation in the concentration in the cross section, adequate data can be obtained by the routine collection of depth-integrated samples at one vertical, supplemented by the periodic collection of samples at a number of equally spaced verticals or at several verticals located at centroids of equal areas of water discharge. Determination of the suspended-sediment discharge in the stream cross section is often called a sediment discharge measurement. If the concentration in the cross section fluctuates erratically, the accuracy can be improved only by increasing the number of sampling verticals and the frequency of sediment discharge measurements.

The accuracy of a sediment discharge record depends upon the accuracy of the determination of both the concentration and the related water discharge. During flood periods, a desirable degree of accuracy can be obtained only by making

TABLE 3.1.–Cumulative Percentage of Water Discharge in Cross Section

Number of sampling verticals (1)	Percentage of cumulative discharge at centroids (2)											
2	25	75										
4	12	38	62	88								
6	8	25	42	58	75	92						
8	6	19	31	44	56	69	81	94				
10	5	15	25	35	45	55	65	75	85	95		
12	4	12	21	29	38	46	54	62	71	79	88	96

concurrent water and sediment discharge measurements to define both the concentration and flow with respect to time. In addition, frequent reference samples at one stream vertical are required if rapid changes in concentration are to be recorded.

The type of field installations required for the collection of data to determine the suspended-sediment discharge depend primarily upon the size of the stream and accuracy of the record desired. For small streams, samples are collected by wading or from foot bridges with hand-operated samplers. For medium size rivers, samples are collected from cable or bridge measuring installations utilizing both depth or point-intergrating samplers, or both. At bridge measuring sections, a sampler and reel housed in a shelter attached to the side of a bridge is used if a record of daily concentration is to be obtained. Fig. 3.20(a) shows a typical bridge installation allowing a resident observer to collect frequent samples from the Bighorn River near Manderson, Wyo. A US D-49 sampler and four-wheel crane unit for the collection of samples at the sediment station on the Bighorn River near Kane, Wyo. is shown in Fig. 3.20(b). A 100-lb point-integrating sampler, as used in the Solomon River sediment investigations, is shown in Fig. 3.21. The valve in the sampler head is closed when the sampler touches the stream bed.

In large rivers, cable-suspended point-integrating samplers are used to collect both depth and point-integrated samples from bridges or boats. The equipment used at the sediment station on the Mississippi River at St. Louis, Mo., shown in Fig. 3.22, includes a power-operated car and crane assembly and US P-46 sampler. The 100-lb C-type stream gaging sounding weight, shown suspended below the sampler, was used to reduce the down-stream drift during flood runoff. The recent development of the US P-63 point-integrating sampler, weighing 200 lb, largely eliminates the need of using the C-type weight.

Total Sediment Discharge Measurement.—Direct observations of measurements of total sediment discharge, hereafter referred to as sediment discharge, cannot be made with existing sediment sampling equipment. Sediment discharge can be determined from direct observations of concentration, c_m, and particle size of suspended sediment, velocity distribution, stream depth, water temperature, and bed sediment size (Colby and Hembree, 1955) for selected segments of streams flowing in "sand bed channels."

In streams where the sediment load consists primarily of particles in the silt and

FIG. 3.20.—Equipment for Collection of Suspended-Sediment Samples: (A) US **D-49 Sampler Installation near Manderson, Wyo.; (B)** US **D-49 Sampler and Four-Wheel Crane near Kane, Wyo. (Photographs by United States Geological Survey)**

clay range and the concentration distribution in the vertical section is essentially uniform, the suspended-sediment discharge is very nearly equal to the total sediment discharge.

In some streams, the sediment discharge can be measured in a contracted section of a channel where the velocity and turbulence are sufficient to suspend all the particles in transport. Such sections are sometimes called natural turbulence flumes. The United States Geological Survey (Serr, 1950; Colby, et al., 1953; Benedict and Matejka, 1953; Colby and Hembree, 1955) utilized natural turbulence flumes in the Niobrara River, Nebraska, and Fivemile Creek, Wyoming, to determine the sediment discharge. In each instance, the suspended-

FIG. 3.21.—Equipment for Collection of Suspended-Sediment Samples, Solomon River at Beloit, Kansas: US P-46 Sampler and Four-Wheel Crane (Photograph by United States Geological Survey)

FIG. 3.22.—Equipment for Collection of Suspended-Sediment Samples, Mississippi River at St. Louis, Mo.; (A) US P-46 Sampler with 100-lb Sounding Weight Attached; (B) Power-Operated Car and Crane Assembly (Photographs by United States Geological Survey)

sediment discharge concentration, $\overline{C}'_m$, in the sampled depth was shown to be nearly representative of that for the total depth; thus, the observed concentration closely approximated the true discharge concentration.

In small streams, the sediment discharge can be measured if it is practicable to install an artificial turbulence flume. Fig. 3.23 is a photograph of this type of flume used on the Middle Loup River at Dunning, Neb., by the Geological Survey (Benedict, et al., 1955; Hubbell and Matejka, 1959). It consists of a series of baffles anchored to a concrete slab, the top of which is at average bed elevation for the stream at that cross section. The turbulence induced by the baffles is sufficient to transport in suspension almost the entire sediment load in the stream, thus enabling it to be sampled with a hand depth-integrating sampler.

The sediment discharge in small streams can also be measured at weirs or drops where it is practicable to collect samples by moving a depth-integrating sampler vertically through the nappe (Love and Benedict, 1948). A measuring station provided with a weir is shown in Fig. 3.24. It also may be measured on a continuous basis for individual storm periods by utilizing a modified weir and a splitting device similar or identical to that developed by the Tennessee Valley Authority (1961). In this type of installation, a continuous sample of the water-sediment mixture is taken by means of a series of splitting devices, in each of which a proportional part of the water and sediment passing that point is retained and the balance wasted. The accuracy of this type of sediment discharge measuring installation depends upon the accuracy with which the diverted flow represented the concentration of the total flow, the overall accuracy of the splitting device, and a related record of scour or fill in the weir pool, if any, for each runoff event.

A photograph of this type of installation on White Hollow Creek near Maynardville, Ten., is shown in Fig. 3.25. In this installation, the rate at which the actual sample was taken was shown to be 1/105,000 of the stream discharge.

The average annual sediment discharge of streams entering a reservoir can be obtained from reservoir sediment surveys if adequate information is available on sediment discharge from the reservoir. This method requires accurate topographic mapping of the deposits and measurements of the specific weight of the sediments deposited as discussed in Chapter III, Section B that follows.

Bed Load Discharge Measurement.—Direct field measurements of bed load have

FIG. 3.23.—Turbulence Flume, Middle Loup River near Dunning, Neb.: (A) View Looking Upstream towards Turbulence Flume, Water Discharge 350 cfs; **(B) Skeleton Model of Turbulence Flume (Photographs by United States Geological Survey)**

been made in Europe for only a few streams and for only short periods of time (Einstein, 1948). In the United States, several direct measurements of bed load movement under uniform flow conditions were made by the Soil Conservation

FIG. 3.24.—Streamflow and Sediment Measuring Station, Pine Creek above Barry Placer Division near Idaho City, Idaho; Broad-Crested Weir with Rectangular Flume for Low Stages; Overfall Structure for Sediment Sampling in Foreground (Photograph by United States Geological Survey)

FIG. 3.25.—Streamflow and Sediment Measuring Station, White Hollow Creek near Maynardville, Tenn.: (A) Flow Splitting Device with Water-Sediment Storage Tank; (B) Looking Upstream—Weir in Immediate Foreground, Flow Splitting Device at Left (Photographs by Tennessee Valley Authority)

Service in 1941. In these measurements, the sediment was accumulated in slot-traps placed across the channel and measured volumetrically or pumped into a weighing tank. Einstein (1944) points out that the volumetric method gives accurate results for small streams under uniform flow conditions.

United States Army Engineers have made preliminary studies of bed load movement with portable samplers similar in design to some of the European models. Since 1945, the experimental sampler shown in Fig. 3.26 has been used in the lower Arkansas River, where maximum depths of 60 ft and mean velocities of 14 fps are common occurrences. This sampler traps only that part of the sediment moving in a layer extending about 0.3 ft above the bed. The results from the sampler are uncertain when the bed is irregular, particularly when dunes or ripples are present. The sampler has been useful in obtaining qualitative information on sizes of sediment grains that may form a pavement and shield the bed from scour; it also has helped to obtain data for tractive force computations for deep flows, and information on when appreciable movement of the bed load begins. Information obtained also shows that "clay balls" that occasionally roll on the bed, clogged the entrance to the sampler. The samplers have not been used on a routine basis, as the investigations of bed load movement were entirely experimental in character (United States Corps of Engineers, 1958).

5. Records of Sediment Discharge.—Records of sediment discharge are generally computed on a daily or annual basis. The method used for computing daily or annual values depends upon the data collected, the frequency of the field observations, and the size of the bed material in transport.

Suspended-Sediment Discharge.—Suspended-sediment discharge can be computed for any selected time interval. Where a relatively high degree of accuracy is desired, daily intervals generally are used. The suspended-sediment discharge in tons per day is the product of the daily mean concentration, the daily mean water discharge, and a conversion factor. Concentration generally is expressed in grams of sediment per liter of the water-sediment mixture or in parts per million by weight. Concentration in parts per million by weight is obtained by dividing the weight of the dry sediment by the weight of the water-sediment mixture. In the United States, present practice for computing a daily record is to plot concen-

FIG. 3.26.—Experimental Bed Load Sampler: (A) View of Sampler with Baffles Removed; (B) View of Sampler with Sounding Weights Attached (Photographs by United States Army Engineer District, Little Rock, Ark.)

trations from depth-integrated samples in parts per million or milligrams per liter directly on a chart (or a print of the chart) of the gage height against time (Benedict, 1948). A smooth concentration curve is then drawn through the plotted points. Daily mean concentrations are then read from the concentration graph. During periods of rapidly changing concentration and water discharge, the concentration and gage height graphs are subdivided into smaller than daily increments of time. Fig. 3.27 shows a typical concentration and gage-height graph for the Bighorn River near Manderson, Wyo.

At sediment stations where depth-integrated sediment samples are collected frequently from a fixed sampling installation, the concentration at this reference vertical is related to the concentration in the cross section determined from concurrent water and suspended-sediment discharge measurements. The frequency of these concurrent measurements to evaluate the concentration in the cross section depends upon the accuracy desired, the number of runoff events, and the characteristics of the stream channel at and above the measuring section. Frequent measurements, particularly during periods of rapid increase in discharge, enhances the accuracy of the record.

In reconnaissance studies, a general relationship between suspended-sediment discharge or concentration and water discharge may be developed for a sediment station by plotting sediment discharge against water discharge for the period of record. The resulting curve is called the sediment transport curve. This approach is necessary when the interpolation of sediment concentration between infrequent measurements becomes impractical. If the sediment transport curve is a mean for a number of years of record, it can be used directly with a flow-duration curve (Searcy, 1959) to obtain an approximate amount of sediment passing the station

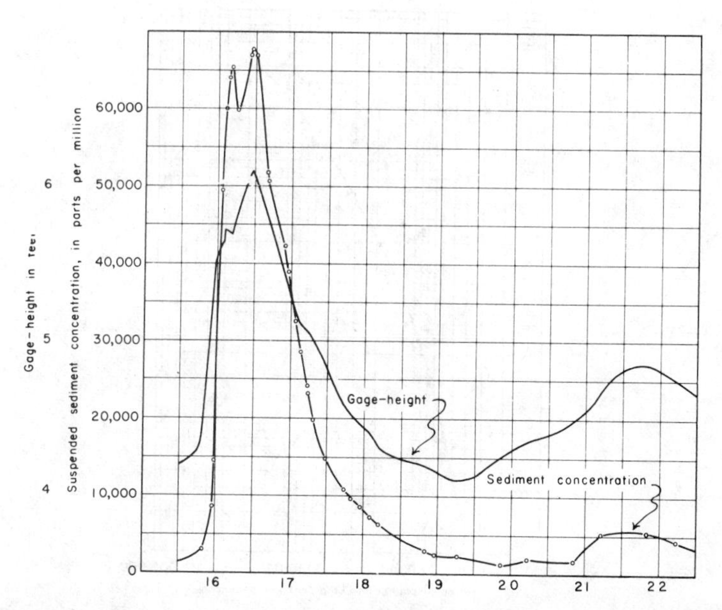

FIG. 3.27.—Graphs of Suspended-Sediment Concentration and Gage Height, Bighorn River near Manderson, Wyo., April 16-22, 1952

for the period of record for which observations are available (United States Bureau of Reclamation, 1951). This amount of sediment is called the sediment yield (see Chapter IV) for the period involved. Fig. 3.28 shows suspended sediment transport curve for the Colorado River near Grand Canyon, Arizona, for the water years 1948–1951. The scatter in plotted values is typical for streams transporting sediment consisting of clay, silt, and sand-size particles. Transport curves for suspended clay-silt and sand fractions are shown in Fig. 3.29 (Colby, 1956). Transport curves can also be developed for daily values of suspended load and water discharge or for selected runoff events from comparable data. In a study of "load-discharge relationships" for 20 stations, Leopold and Maddock (1953) found that the concentration of suspended sediment generally increases with discharge. For some flood runoff events, Heidel (1956) shows that the progressive lag in the peak concentrations behind the peak flow alters the normal relationship between sediment discharge and water discharge.

The transport curve method for computing suspended-sediment discharge has particular merit for streams flowing on alluvial beds with a high percentage of the sediment discharge in the sand-size range. The silt and clay fraction of the total load, sometimes referred to as fine material load or wash load, is not functionally related to stream discharge or competence. Therefore, adequately-defined relationships cannot be developed for a sediment station where an appreciable fraction of the sediment discharge is composed of fine material load (wash load) (Witzig, 1944; and Johnson, 1944). However, where sufficient particle size data are available, sand transport curves can be developed to give a greater degree of accuracy for that fraction of the total load. Details of this procedure will be further treated in the following section.

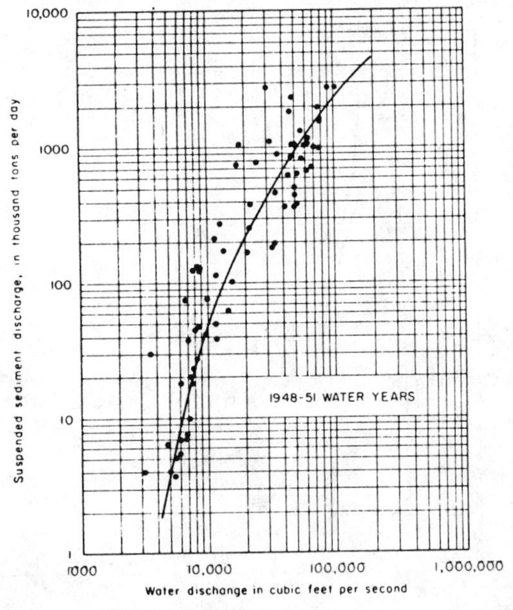

FIG. 3.28.—Suspended-Sediment Transport Curve, Colorado River near Grand Canyon, Arizona

Total Sediment Discharge.—In a study of the sediment discharge of the Niobrara River near Cody, Neb., Colby and Hembree (1955) developed a procedure for computing total sediment discharge based in part on Einstein's formulas and Geological Survey field data. This procedure that is discussed in Article 58 of Chapter II has been called the Modified Einstein procedure. It is based on field data including concentration, particle size of suspended sediment, stream velocity and depth, water temperature, and particle size of bed sediment for selected areas in a stream cross section.

Studies of sediment discharge of the Middle Loup River near Dunning, Neb., by Hubbell and Matejka (1959) confirm the results obtained in the Niobrara River study. Colby and Hubbell (1961) simplified the Modified Einstein computation procedure by the use of nomographs. Computations have been further simplified by the Bureau of Reclamation (1959) and Army Engineers by the use of a digital computer procedure.

Schroeder and Hembree (1956, 1957), Laursen (1957), Fowler (1957), and Jordan (1965) have concluded that the Modified Einstein procedure, which was based on studies of sediment transport in shallow streams, gives reasonable results in larger streams flowing in natural sand bed channels. This conclusion is further confirmed by studies of the United States Bureau of Reclamation in the Lower Colorado River (1958). Colby (1964), in a study of "dominant measures of bed-material discharge," concluded that either the relation between bed material discharge per foot of stream width and mean water velocity, or the relation between stream power and shear stress, can be used as a basis for practical computations of bed material discharge. He further concluded that sediment

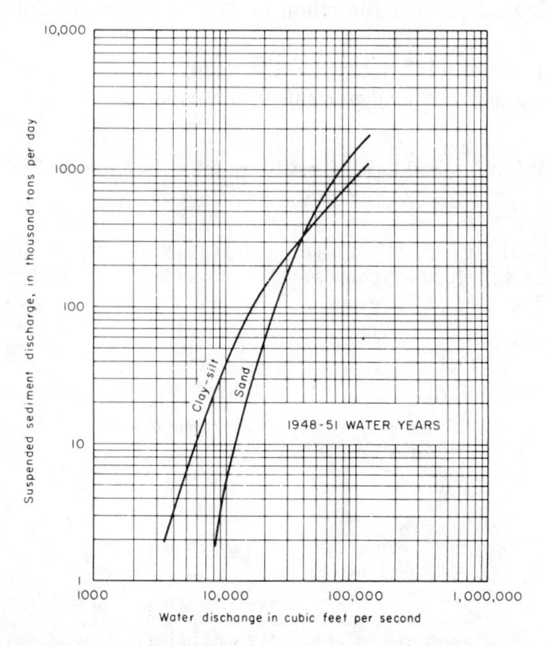

FIG. 3.29.—Suspended Clay-Silt and Sand Transport Curves, Colorado River near Grand Canyon, Arizona

transport curves based on the relationship between bed material discharge (0.1 mm–0.8 mm) per foot of width and mean water velocity are the most convenient to apply. Detailed discussion of this method for computing bed material discharge is presented in Chapter II.

Nordin and Beverage (1965), in a study of sediment transport in the Rio Grande, New Mexico, also concluded that a "sediment-transport curve" based on mean velocity and bed material discharge per unit of stream width provided a logical method for determining the total bed material discharge.

The Modified Einstein procedure or the sand-transport curve method provides a practical approach to determining the sediment discharge of streams flowing in channels consisting largely of sand. Frequency curves for water discharge or mean velocity per foot of width are used as appropriate. In streams where the coarse sediment load includes appreciable amounts of sand, gravel, and cobbles, the Meyer-Peter, Muller (Sheppard, 1960) formula has been used to compute that fraction of the load with a reasonable degree of success.

The average annual sediment discharge for long periods of time may be computed from reservoir surveys if adequate data on reservoir deposits and on sediment outflow are available. This procedure provides information on long-term sediment yields of drainage basins. The data, however, cannot usually be related to runoff events and therefore the procedure has limited use for determining sediment discharge.

Records of sediment discharge data published in the United States Geological Survey Water-Supply Papers have been limited largely to information on the suspended load. Practical methods for converting these data to total load vary in application. Stevens (1946) in a review of the data for the Colorado River basin, arbitrarily added a bed load correction of 15% to the suspended load to obtain total load.

Lane and Borland (1951) discuss the many factors to be considered in estimating the rate of bed load movement and conclude that it is not possible to

TABLE 3.2.—Maddock's Classification for Determining Bed Load

Concentration of suspended load, in parts per million (1)	Type of material forming channel of stream (2)	Texture of suspended material (3)	Bed load discharge, in terms of suspended sediment discharge, as a percentage (4)
Less than 1,000	Sand	Similar to bed material	25-150
Less than 1,000	Gravel, rock, or consolidated clay	Small amount of sand	5-12
1,000-7,500	Sand	Similar to bed material	10-35
1,000-7,500	Gravel, rock, or consolidated clay	25% sand or less	5-12
Over 7,500	Sand	Similar to bed material	5-15
Over 7,500	Gravel, rock, or consolidated clay	25% sand or less	2-8

develop a simple rule or formula that will give quantitative values for all streams. These conclusions have been summarized by Thomas Maddock, Jr. in Table 3.2.

The sand-transport curve method provides a procedure for converting daily values of suspended-sediment discharge to total sediment discharge for sand bed streams. Using a graphical procedure, correction coefficients can be computed from total sand transport determined from empirical graphs (Colby, 1964) and suspended sand transport based on field data, i.e., concentration and particle size of suspended sands, temperature, and mean velocity. For sand and gravel bedstreams, the Meyer-Peter, Muller bed load formulas (Sheppard, 1960) are often used to compute the bed load discharge.

B. Reservoir Deposits

6. General.—This section discusses the field measurement techniques to determine the volume occupied by sediments deposited in a reservoir. Also included are descriptions of some of the field equipment and office computational procedures used in the sediment volume determination. The process of applying the field measurement techniques, office procedures, and equipment results in a "reservoir sediment survey." The United States Army Corps of Engineers (1961) aptly defines a sediment survey as "a term applied to an individual reservoir sedimentation investigation, interpreted broadly to include office work, laboratory analyses of sediment samples, field measurements, and processing and analysis of data."

As a stream enters a reservoir, the flow depth increases and the velocity decreases. This causes a loss in the transporting capacity of the stream and the deposition of at least some of the waterborne sediments. Sediments carried into the reservoir may deposit throughout its full length. The pattern of deposition generally begins with the coarser sediments dropping in the reservoir headwaters area. This sedimentation process, described as aggradation, continues progressively until a delta is formed (see Chapter V, Section E). The finer sediment particles may be transported by density currents down to the dam, thus completing the depositional pattern. Fig. 3.30 shows a profile sketch of a typical reservoir delta (Bondurant, 1955). Aggradation in the stream channel can occur for long distances above the reservoir because of the reduction in velocity and sediment transporting capacity of the stream in this reach.

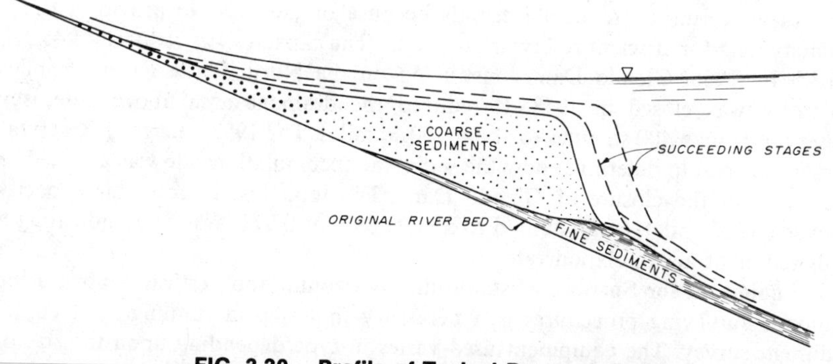

FIG. 3.30.—Profile of Typical Reservoir Delta

Sediment surveys determine not only the volumetric reduction in the reservoir but also furnish other valuable information and data. Such information includes how the sediment deposits are distributed in the reservoir and what change the stream channel has undergone owing to sediment transport and deposition. Field data collected during surveys are analyzed to include the specific weights of the deposits and their grain-size distribution, sediment yield rates of the drainage area, reservoir trap efficiencies, density currents, and other sedimentation features of lacustrine deposits.

7. Frequency of Surveys.—The frequency of surveys generally depends upon the estimated rate of sediment accumulation in the reservoir. It follows then that reservoirs with high depositional rates are surveyed more often than those with low rates. Financing the cost of the survey is often a controlling factor in how frequently surveys are made. Considering that the cost of the survey is justified by the need to continue updating the reservoir capacity, a suggested guide to the frequency is a 5-yr to 10-yr interval depending upon the runoff inflow volume and magnitude of flows. Also, the frequency might be fixed by the period when the total reservoir capacity is reduced by some given percentage based upon the estimated sediment accumulation rate. Survey intervals can be influenced by special circumstances such as: (1) Following the occurrence of a major flood; or (2) timing the survey to update the capacity of a reservoir that has had its drainage area reduced by the construction of a dam upstream.

Some of the following suggested methods can be used as aids in determining how frequently sediment surveys should be made to ascertain the reservoir storage depletion: (1) Records obtained at sediment stations above the dam; (2) field observations of the reservoir area when the reservoir is drawn down (Murthy, 1971); (3) a check on the accuracy of the reservoir capacity curve when computing inflow-outflow volumes during operation studies; (4) reconnaissance measurements on a few index reservoir sediment ranges; and (5) when special problems associated with reservoir sediment deposition are indicated.

Guernsey Reservoir operated by the United States Bureau of Reclamation on the North Platte River, Wyoming, is cited as an example of scheduling surveys. The original total reservoir capacity in 1927 was 73,810 acre-ft. Knowing the reservoir was located on a heavily sediment-laden river, it followed that the capacity should be depleted at a high rate. This would, in turn, affect the reservoir operational plan including both power generation and irrigation. Partial or full-scale surveys were run eight times between the 1927 and the 1966 survey. The surveys were run at frequent intervals because of the need to provide current capacity data for efficient reservoir operation. The capacity was reduced to 45,228 acre-ft by 1966. Glendo Dam constructed on the North Platte River 16 miles upstream was closed in 1958. It reduced the drainage area above Guernsey Reservoir from 5,400 sq miles to about 700 sq miles. The 1957 Guernsey Reservoir survey was run to determine how the sediment accumulation rate was affected as the result of the closure of Glendo Dam. The total loss in reservoir capacity between 1927 and 1966 amounted to almost 39% or 0.97% annually, indicating a high sediment accumulation rate.

8. Equipment for Surveys.—Establishing horizontal and vertical control using standard surveying procedures is a necessary first step in running an accurate sediment survey. The equipment used varies in type depending upon the size of the reservoir. Most reservoirs regardless of their size would require standard surveying equipment, e.g., transits, levels, plane tables and alidades, and stadia rods. Field books that contain data from prior surveys, a base map or aerial photograph, if available, are also necessary for making an adequate reservoir sediment survey. A dry reservoir could be surveyed with this equipment but other apparatus would be needed if sediment samples were to be obtained. For large reservoirs, it is usually more economical to delay the initial survey of the reservoir sediment ranges until the pool is at least partially filled so that much of the work can be done with sonic sounders. Where part or all of the reservoir basin is submerged, special equipment is needed including a boat or raft and auxiliary equipment, sonic sounding equipment, distance measuring instruments, and sampling equipment to determine the specific weight of deposited sediments.

With the advent of more sophisticated electronic instrumentation, survey equipment has been and is being developed to make field measurements and to prepare field data directly for computer processing. For example, direct readings of sounding depths and stationing location can now be stored on magnetic tape in the field for later processing by computer. Another example of such techniques is given in Engineering News Record (1966) which describes briefly an on-shore method of plotting the sounding boat positions using electronic equipment to measure distances and bearings. The author of this article subsequently developed a procedure that records the soundings of survey ranges to true scale on the sounder chart.

Boat or Raft and Auxiliary Equipment.—To run a survey of a partially or completely filled reservoir, a boat or raft is often necessary. The size of either vessel depends upon how large the reservoir is and other conditions, e.g., the physiographic setting of the lake. For small reservoirs, a shallow draft boat 14 ft long or longer would normally be adequate. Large reservoirs, correspondingly, would require larger boats and very often two boats are desirable, one to carry sounding equipment and the other to transport survey personnel. Selecting the size of a boat or raft, or both, is governed by the areal extent of the lake to be covered. For example, in very large reservoirs, such as Lake Mead, boats up to 45 ft long and barges up to 21 ft by 105 ft in plan dimension were used during the Lake Mead survey of 1963–1964 and the survey of 1948–1949 described by Smith, et al. (1960).

Rafts are generally more suitable than single boats for sediment sampling operations because of their stability. A suitable raft can be formed by two boats fastened together using planks or a platform as shown in Figs. 3.38 and 3.47. Launches 40 ft long or longer, equipped with booms for handling equipment, have proved adequate for surveys on very large lakes. In most cases the smaller lake vessels are propelled by an outboard motor. A set of oars is needed in case the motor fails or the water is too shallow to accommodate a motor.

Trailers, usually attached to a truck or other heavy vehicle, are required for transporting the large boats or rafts. Small lightweight boats can be transported satisfactorily on ordinary passenger cars by means of top racks. Two-way radios are generally required to maintain communication between shore parties and boat personnel.

Sounding Equipment.—Sonic sounders are currently (1973) used to run most sediment surveys of both small and large reservoirs. Manually operated sounding lines and sounding poles used in the past to survey small reservoirs are virtually outdated and replaced by modern sonic sounders.

A sonic sounder is very necessary where an extensive program is planned. It is a portable recording instrument designed to measure water depths by projecting a high energy acoustical signal downward, from just below the water surface, and receiving the signal reflected off the reservoir bottom. Measuring the depth of water depends on the speed of sound through the water. For known reservoir water temperature and salinity conditions, the speed of sound through water is governed so the time interval occurring between the projected and receiving signal relates directly to water depth. Through complex electronic and mechanical components, depths are continuously recorded on a chart as the boat transporting the sonic sounder traverses a predetermined course across the reservoir. Modern, although expensive, instrumentation also permits the recording of depths on magnetic tape for later direct processing into computer format.

The sonic sounding instrument consists of three units: (1) The recorder housed in a protective metal case includes the mechanical recording components and the electronic phasing, keying, and power circuitry; (2) the transmitting-receiving transducers can be mounted in the hull of the boat as shown in Fig. 3.31 or suspended over the side as shown in Fig. 3.32 (Rausch and Heinemann, 1968); and (3) a power source which may be either a wet cell battery or a generator-converter combination depending upon the size of vessel used. Transducers weighing less than 3 lb and much smaller in size than those shown in Figs. 3.31 and 3.32 are currently (1973) being manufactured. Necessary auxiliary components include a check bar to verify the sounder accuracy and a thermometer to measure the reservoir water temperatures. Usually the salinity content of fresh water reservoirs is negligible or constant for extended periods; however, if it is significant or variable, a conductance meter is needed to establish a correction factor for the speed of sound through water.

The depth of water is recorded continuously, at a rate of several hundred

FIG. 3.31.—Transmitting and Receiving Transducers Mounted in Boat Hull (Photograph by United States Bureau of Reclamation)

soundings per minute, on a chart that is moved at a uniform speed past the "printer" mechanism. Several chart paper speeds are available on some sounders to expand or contract the degree of individual sounding detail. The recorder paper is generally backed with a metallic coating. The printer mechanism causes an arc when the return echo is received and a print mark is burned into the paper representing the measured depth. Also recorded on this chart at selected intervals is the location of the sounding by a vertical fix line traced fully across the chart. The fix line actuated mechanically by either manual or automatic operation represents a known location along the sounding track. An example of a sonic sounding chart in Fig. 3.33 (Rausch and Heinemann, 1968) includes the appropriate notations of the water depths, fix distances, and reservoir bottom. The three lower recordings that decrease in intensity are echoes. These echoes or strays are generally the result of having the gain too high on the recorder.

Three additional comments on sonic sounder operation are considered pertinent. Firstly, the sonic wave projected from the transducer is emitted in the shape of a cone. The diameter of this cone enlarges with depth but is dependent upon the degree of power or "gain" setting of the recorder. At greater depths or lesser "gain" settings some of the sound wave is lost and does not return to the transducer. An important fact to acknowledge is that the highest reflecting point within the sound cone, or the first return echo of sufficient strength sensed by the receiving transducer, is the depth recorded for that individual sounding signal. Consequently, the depth recorded may not necessarily be for the vertical point

FIG. 3.32.—Transducer Attached to Side of Boat (Photograph by United States Bureau of Reclamation)

immediately beneath the transducer although it would represent some depth within the quadrant of the cone. For example, the distances recorded of water depths that change abruptly because of an extreme rise or fall in the reservoir bottom may not be true. In these situations the transmitted sound waves that bounce off the highest point located on the sloping ground face within the sound cone would travel a shorter distance and would return to the recorder sooner than those waves reflecting off the point directly below the sounder. If more accuracy is necessary at greater depths, runs should be made slowly from both directions along the sounding line with close horizontal control. On most reservoirs, however, sounding operations of such accuracy are unnecessary except where receding sheer vertical bluffs exist. Accuracy is important for some of these anomalies when using a system transducer designed with a smaller beam width.

Secondly, the sonic sounder is capable of sensing the differences in density of submerged deposits known as subbottom profiling in the seismic field. An example of this can be noted by the multiple traces in Fig. 3.34. This capability is attained if the deposits are of low enough density allowing sonic penetration. The penetration is dependent upon the low sonic frequency of the sounder, usually less than 10 kHz, and the composition of the deposited material. For example, a coarse sand layer will usually block further sonic penetration. The accuracy of the depths will, of course, exceed the 0.5%–1.0% error for water but should be within a reasonable range to verify the prevailing depths of deposit layers. Some sonic

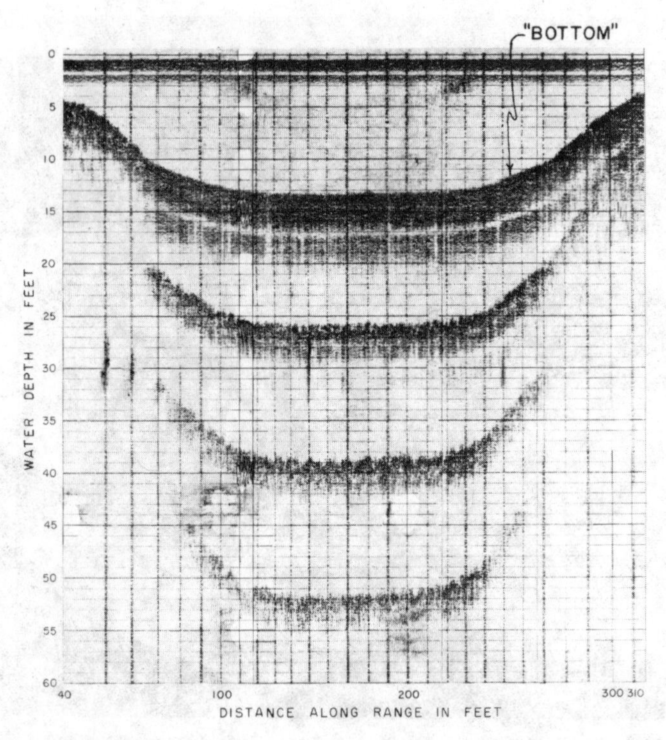

FIG. 3.33.—Sonic Sounder Chart Showing Profile of Reservoir Sediment Range along Bottom (Chart by United States Bureau of Reclamation)

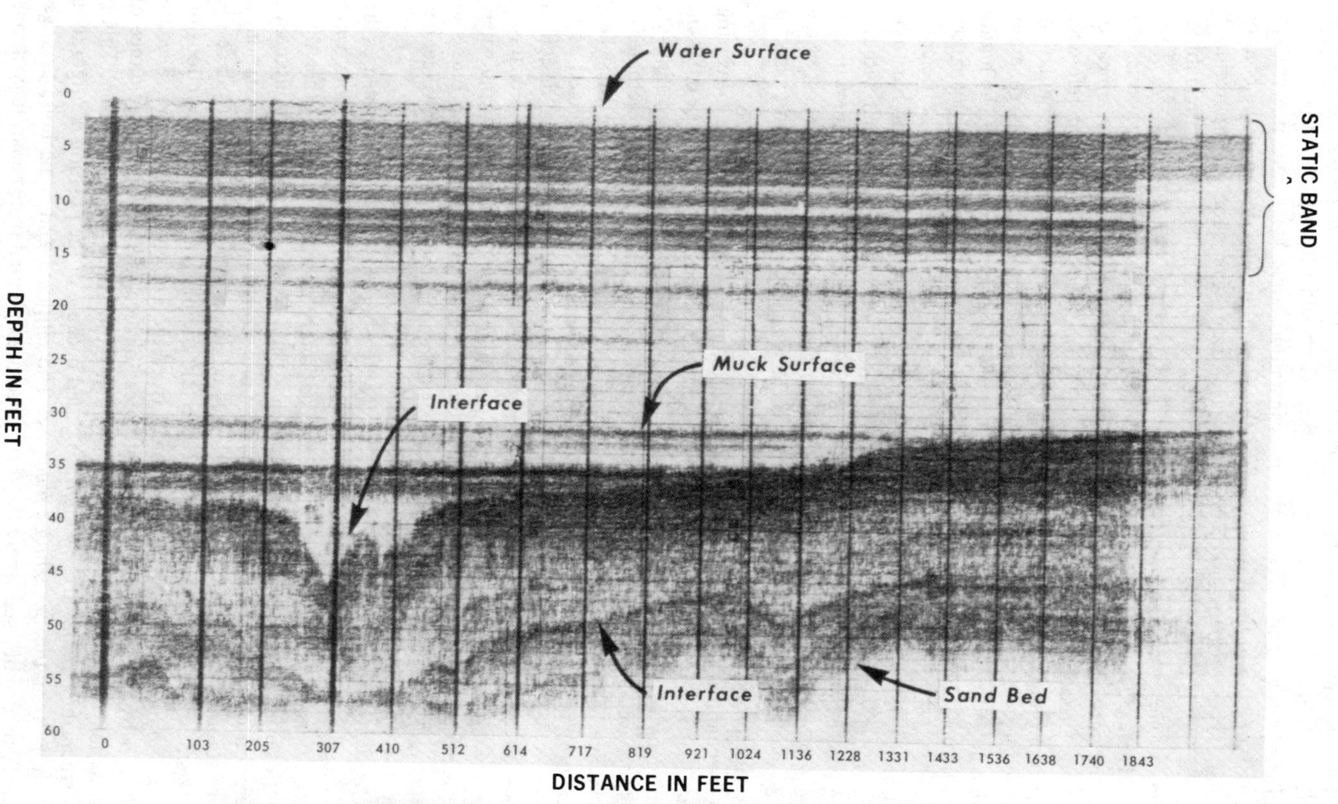

FIG. 3.34.—Differences in Sediment Densities Noted on Sonic Sounder Chart (Chart by United States Bureau of Reclamation)

sounding systems are designed to operate in one or simultaneously in two different frequency modes to obtain both bottom and subbottom profiles.

Thirdly, bubbles of air absorb the transmitted sonic wave and prevent the return of an echo to the transducer. For this reason the hull design of the boat or the method of suspension of the outboard transducer shoe is important. A "V"-type hull is the preferred design over the scow-type hulls because air bubbles are less prone to collect on it.

9. Distance Measuring Techniques.—The location of the boat as it traverses the sounding path can be determined by standard survey methods or mechanical or electronic distance measuring instruments. All methods require fixed reference points along the reservoir shoreline. The survey techniques include triangulation, intersection, sextant angles, or stadia distances. Mechanical methods generally involve measurement of a wire length or the towing of a current meter. The highly sophisticated electronic methods rely on either light or radio wave propagation. Regardless of the method used, the technique selected must be sufficiently accurate for maintaining the horizontal control. For instance, a decision has to be made in determining if it is necessary to locate the position of the boat within a given foot when the soundings being recorded represent the high point within a conical diameter of several feet on the reservoir bottom. Generally, the cost of the survey can influence the decision that must be made by the engineer-in-charge to select the proper technique for the job.

A mechanical device to maintain horizontal control is the distance measuring wheel that is 2 ft in circumference (Fig. 3.50). Encircling the wheel is a 0.029-in. diam piano wire wrapped to provide a maximum of friction and prevent slippage. As the wire is let out, it rotates the wheel whose shaft, in turn, actuates a mechanical counter indicating the distance. The wire is not stored on the wheel but simply passes around it as the sounding boat moves out from the fixed point on range located on shore. A special apparatus is available in the wheel design to allow fixes to be recorded at prescribed intervals on the sounding chart. For this, a microswitch is installed in the mechanical counter system of the measuring wheel that is actuated at any desired distance interval. By attaching floats or buoys to the piano wire at 500-ft to 1,000-ft intervals to keep it from sagging, the wheel has been used to measure distances to 40,000 ft.

The Raydist system, summarized briefly by Sheperdson (1965), measures the distance by the comparison of separate radio waves from two continuous wave transmitters. One must be located in a stationary position and the other on a mobile unit, e.g., the sounding launch. The system requires simultaneous operation of three transmitting-receiving units, consisting of a master station, a shore station, and a mobile transmitter. A common arrangement is to place the shore station at one end of the range line and the master and mobile stations on the sounding launch to record distances from the known shore station. Because only the distance from a known point is established, the center line guidance of the launch is maintained by transit through radio voice communication. This is known as the circular arrangement, so called because of the position configuration resulting from the operating procedure. The elliptical and hyperbolic arrangements can also be used where the three units are operated from separate positions.

The stadia method can be used for measuring distances by standard surveying techniques. However, when this technique is used, the boat must be stopped for the distance shots to be taken.

Distances can be measured by towing a current meter suspended from the boat. The number of revolutions of the current meter and the time can be noted on the depth recorder chart by using the "fix" button control and pencil notes.

Distance measurements by the intersection method are made by the use of the transit or plane table and alidade. Checkpoints along the continuous range profiles are made by fixes on the sounding charts that correspond with transit or alidade intersections of the range line.

Sextants can also be used as a means of measuring distances. Horizontal angles are measured by the sextants and the desired point to point distances are computed using trigonometric formulas.

Heinemann and Rausch (1971) describe a line measuring device, the productimeter, for use on small or medium-sized reservoirs of 1 mile or less in width. A small cable is stretched between the ends of a range and through the productimeter that is fastened to the boat. This keeps the boat on line between the range ends and measures the distance as the boat moves along the cable from one range end to the other. Rausch and Heinemann (1968), also installed a microswitch on the productimeter for automatically marking the sounding chart every 10 ft as the boat progresses along the range.

10. Communications.—The ability to communicate rapidly between members of a reservoir survey party is very important to a successful operation. Each shore and boat party requires a separate radio. Major use of the radios in the sounding operations is made when it is necessary to relay fix and position information for manual recording on the sounding chart. When considerable communication is necessary, e.g., between several separate parties during a major operation, the use of two or more frequency channels may be advisable. The radios provide another facility to maintain routine or emergency contacts with a base station.

Murthy (1971) mentions that a daylight signaling lamp is used in India to run reservoir surveys by the intersection method. The plane table man continuously sights through his alidade and intersects the points whenever a signal is flashed from the boat.

11. Sediment Sampling Equipment.—Sediment sampling equipment is required to obtain, whenever possible, undisturbed samples of reservoir deposits for determining the specific weight and grain-size distribution. Several special types of sediment samplers have been used with varying success, depending to a large extent on the grain size, the consolidation, and depth of the underwater sediment deposits. The depth of water over the sediment deposits and the thickness of the sediment beds are other factors to be considered in the selection of suitable samplers. Underwater sediments vary in texture and ordinarily they deposit in the pattern as indicated in Fig. 3.30.

Special equipment and samplers required to obtain undisturbed samples from deposits that are underwater include the gravity and piston core types. An example of a gravity core sampler is shown in Fig. 3.35. To operate, it is lowered from a boat or raft by cable off a power-operated reel. It is allowed to fall freely into the sediment deposits that are penetrated to maximum possible depth. Then the sampler is raised to the boat, the cutting shoe removed, a cap is placed over the exposed end of the plastic liner, and the plastic liner containing the sediment sample is withdrawn from the coring tube. The liner is cut off at the other end of the sample, capped, and identified for analyses. Experience has shown that the sampler can be operated without much difficulty in reservoir depths of 100 ft, and

10-ft penetration depths with full recovery have been attained in sediment deposits of very low specific weight often found near the dam. Components of the sampler include the top section made of a pipe on which lead has been molded to supply the driving force. At the base of the top section is a fitting that has an encapsulated ball valve to develop a vacuum necessary to retain the sample while the sampler is being raised. Coring tubes made of galvanized iron pipe in different lengths are connected to the top section base. The length of tube used depends upon the type of sediments at the reservoir bottom. Plastic liners of lengths equal to the coring tubes are inserted in the tubes. The liners are held in place by a steel cutting head at the bottom of the sampler as shown in Fig. 3.35.

A schematic drawing of a piston-type core sampler used to obtain samples of inundated sediment deposits is shown in Fig. 3.36. The sampler is operated from a large raft or from a barge. An example of a barge used in the 1963–1964 survey of Lake Mead (Lara and Sanders, 1970) is shown in Fig. 3.37. The sampler is operated by lowering it until the trigger weight touches the sediment surface. With

FIG. 3.35.—Gravity Core Sampler in Position for Lowering from Survey Boat (Photograph by United States Bureau of Reclamation)

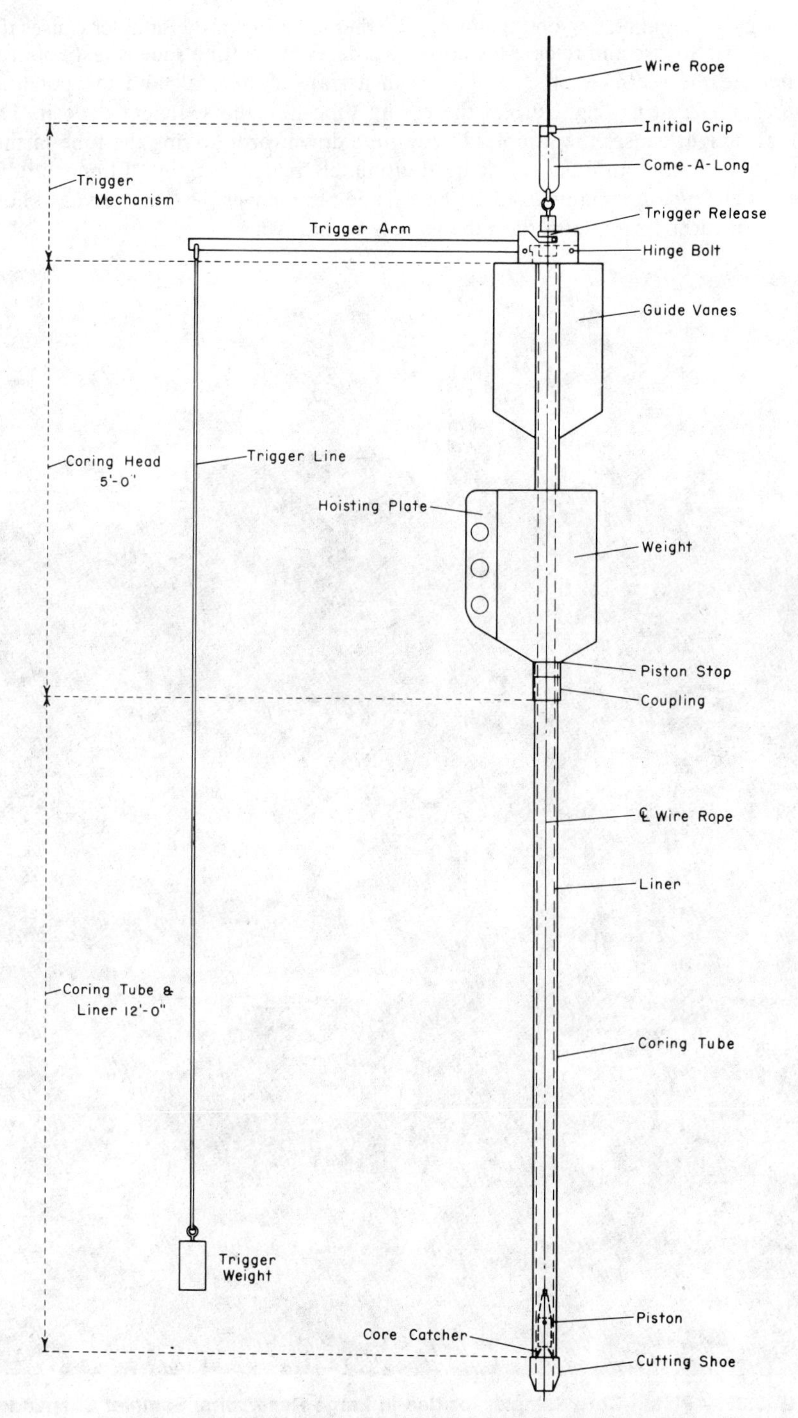

FIG. 3.36.—Schematic Drawing of Piston-Core Sampler

the trigger weight resting on the bottom, further lowering of the sampler causes the trigger arm to rise and release the coring head. As the cutting shoe is just about to penetrate the sediment, the sampler is in a state of free fall and the potential energy of its own weight drives the coring tube into the sediment deposit. The falling weight causes the sampler to continue downward, driving the tube farther into the sediment until the coring head ultimately reaches a point of unpenetrable depth. The piston remains fixed as the outside tube moves past and serves to hold the undisturbed sample in the tube as it is withdrawn.

FIG. 3.37.—Piston-Core Sampler for Use in Large Reservoirs; Sampler Suspended from A-Frame on Barge in Lowering Position (Photograph by United States Bureau of Reclamation)

An example of another type of sampler used to take underwater samples in small reservoirs is shown in Figs. 3.38 and 3.39. It is made of stainless steel tubes in lengths of 3 ft, 6 ft, or 9 ft. A piston and rod fit the inside of the tube to take the sample. The top of the tube is attached to an A-frame by means of tie ropes with snaps or hoist cables used to remove the sampler from the sediment deposits. This sampler is not commercially available but must be fabricated to order. To take a sample, the sampler is placed in a vertical position with the piston held at the sediment surface while the barrel is driven into sediment by repeatedly dropping the driving weight. When the sampler has penetrated the sediment or the barrel is full, it is withdrawn using a reel and A-frame on the boat, making sure the piston does not move in relation to the sampler. Relatively undisturbed samples can be obtained when the sediment is forced from the barrel by shoving on the piston rod. A 4-in. sample is taken from the core approximately every foot. Sample length and its depth in the deposit are measured on the piston rods. Samples are

FIG. 3.38.—Piston-Type Sampler for Use in Small Reservoirs with Boat Equipment: Man at Right Holds Sampler (Photograph by United States Bureau of Reclamation)

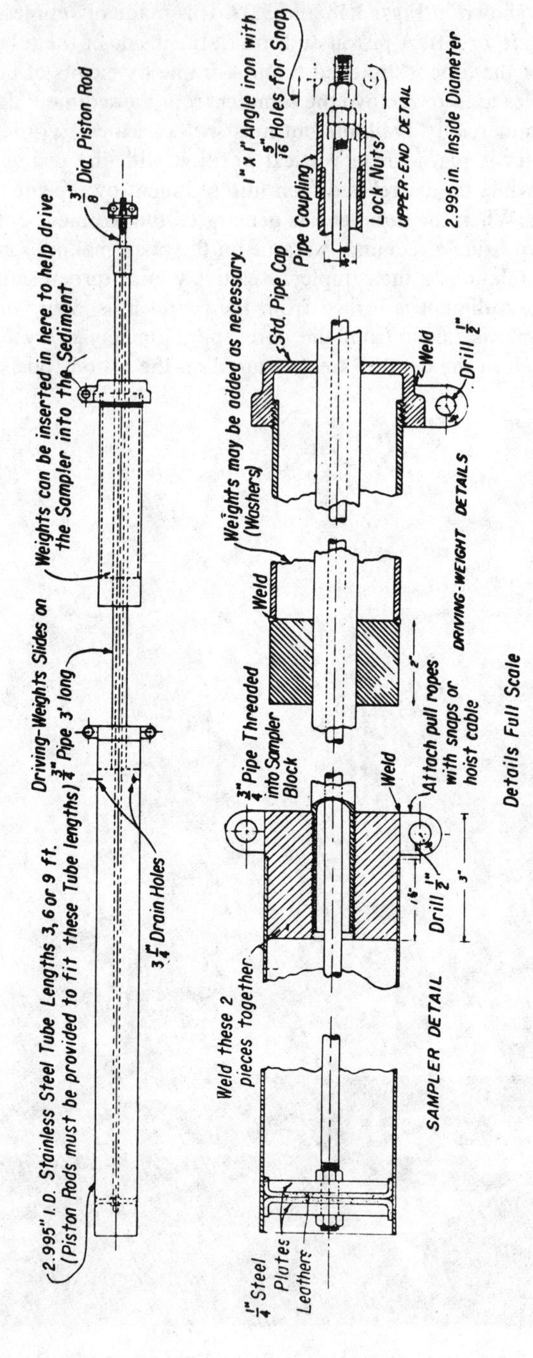

FIG. 3.39.—Piston-Type Sampler for Use in Small Reservoirs

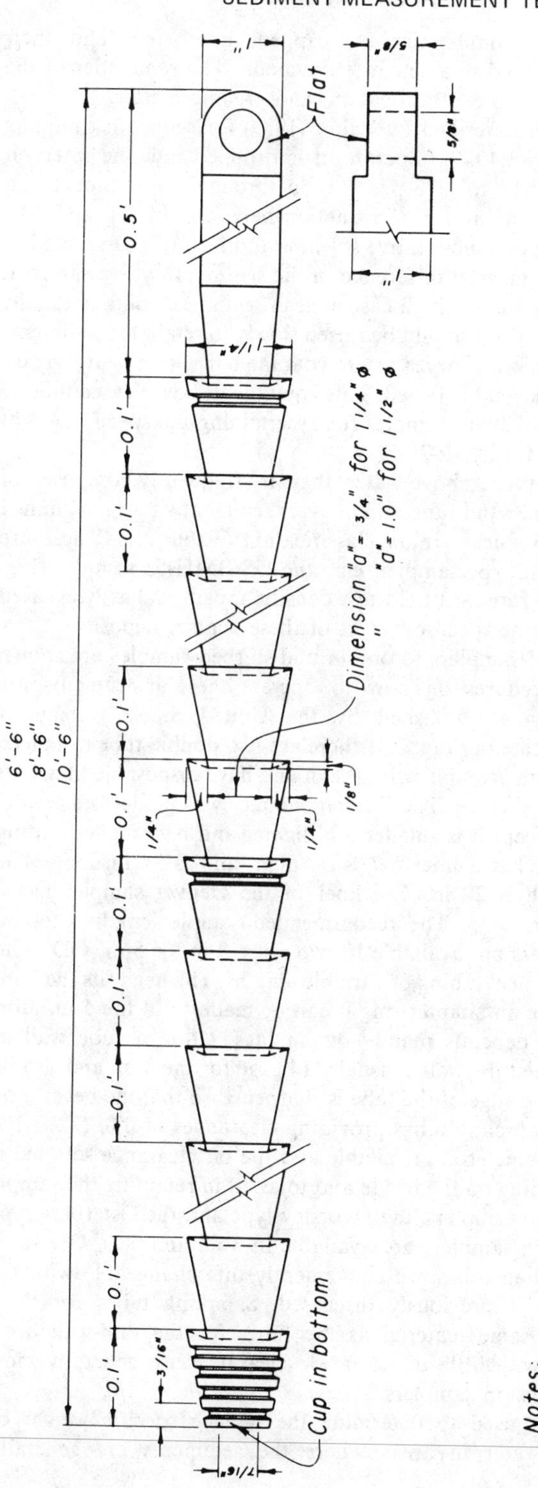

FIG. 3.40.—Spud Rod for Sediment Sampling

placed directly into plastic sample cartons, capped, numbered, and their description and location recorded in a separate notebook. The remainder of the core is observed closely for texture, stratification, and organic matter.

A drawing of the spud rod described by Eakin (1939) for sediment sampling underwater is shown in Fig. 3.40. To operate, it is dropped into the reservoir bottom from a sufficient height to give penetration through the sediment and slightly into the original underlying soil. Distinction between old soil and lake deposits generally depends upon comparative softness and lack of coarse sand in the deposits. Often there is a marked difference in the sediment; where sandy, it may fail to come up on the spud, in which case a clean section of spud marks the lower limit of the deposit. The spud should be raised slowly to retain the sediments in the notched cups. Sometimes it is necessary to coat the bar with a heavy grease to retain the sample when the spud is raised. This equipment, however, cannot be used to obtain undisturbed sediment samples for ascertaining the specific weight of the deposited sediment (Murthy, 1971).

When the sediment deposits are above water, they are frequently composed of coarse-grained materials in the sand range that covers sediment sizes 0.062 mm–2 mm by the American Geophysical Union classification (Table 2.1). These are usually sampled with a coring-type sampler, e.g., the US BMH-53 sampler (Fig. 3.18). In some situations the standard field test density apparatus has been used successfully for determining the specific weight of these coarser deposits.

Examples of some types of samplers to obtain undisturbed samples are shown in Fig. 3.41. Detailed procedures on how to operate these instruments are described in an earth manual published by the United States Bureau of Reclamation (1963). As indicated in Fig. 3.41 there are two double-tube samplers, the Denison and Denver. Both are similar in design and have disposable liners for handling and shipping the soil cores. The Denver sampler was modeled after the original Denison sampler except it is smaller and lighter and has smaller cutting teeth. The Denison sampler has a liner 5-7/8 in. diam and is 24 in. long. The recommended sample length is 20 in. The liner of the Denver sampler has a 5-5/8-in. diam and is 28 in. long. The recommended sample length is 24 in. Thin-wall open drive samplers are available in two sizes, 3-in. or 5-in. OD. The sampling tube is cold-drawn steel tubing of variable lengths. The head fits the 3-in. OD sampling tube and using an adapter ring it can be made to fit the 5-in. tube. The quality of the sample depends mainly on having as thin a tube wall as possible. The thickness of the tube wall is usually 14 gage for the 3-in. and 1/8 in. for the 5-in. tube. The cutting edge of the tube is sharpened so that the bevel is on the outside to form a bit. Sampling tubes providing clearances of 0%, 1/2%, 1%, and 1-1/2% of the inside diameter are available and the bit clearance selected is that necessary to minimize drag on the inside and to assist in retaining the sample in the tube. Two fixed-piston samplers, the Hvorslev type and the Osterberg type are shown in Fig. 3.41. Both samplers are available in 3-in. and 5-in. OD sizes. With these sizes, the sampling tubes are conveniently interchangeable with the thin-wall open drive sampler previously discussed. Sampling tubes for these samplers are made of the same material as the tubes for the thin-wall drive samplers. However, tubes with little or no inside clearance are generally most desirable for use with the piston samplers.

Another type of sampler used to determine the inplace specific weight of sediment deposits above water in areas where the sediments are generally composed of dry sandy soils is shown in Fig. 3.42. Procedures to run field tests with this apparatus are also fully described in the earth manual previously cited (United States Bureau of Reclamation, 1963). Briefly, the procedure consists of removing all loose soil from an area 18 in. sq–24 in. sq at the location to be tested. The exposed area is then leveled until a firm, smooth surface is obtained and the template, generally for a 6-in. diam hole, is selected and placed firmly on the test area. An excavation slightly smaller than the hole in the template is dug to about a 6-in. depth using hand tools or a soil auger. All material taken from the excavation is placed in an airtight container and subsequently weighed in the field or laboratory. The volume of the hole is determined by carefully filling the hole with a sand that will form a deposit of known specific weight using the sand cone device. The weight of the sand used to fill the hole is determined by weighing the sand and container before and after the sand is withdrawn from it. The volume of the hole can then be determined from the weight of the sand and its specific weight. The preceding steps complete the work specifically required at the test site to determine the inplace density.

Radioactive probes are used to determine specific weight of deposits without physically removing a sample as do some of the methods previously described. The measurement of the specific weights of submerged saturated sediment deposits using radioactive probes depends upon: (1) The homogeneous character of sediments having almost the same mass absorption coefficient; (2) the random

FIG. 3.41.—Examples of Different Types of Core Samplers (Photograph by United States Bureau of Reclamation)

emission, penetration, and scattering of gamma rays from a confined source; and (3) the statistical measurement of nonabsorbed or returning rays when the geometry between the source and the detection system remains constant. In brief

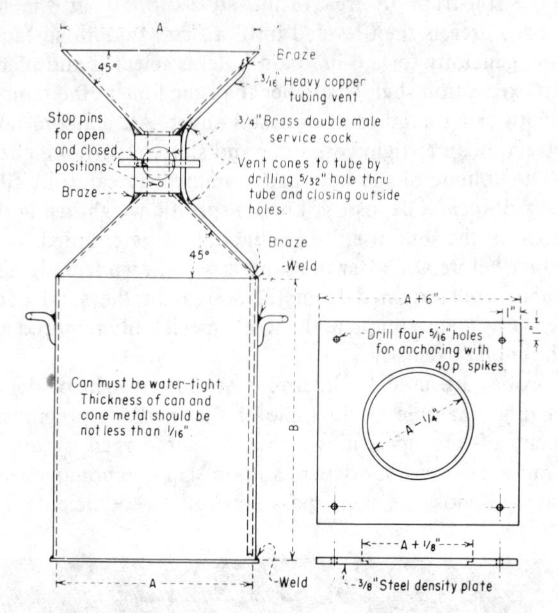

Dimensions of can for standard density holes are as follows.

DENSITY HOLE SIZE		DENSITY CAN SIZE		
DIAMETER INCHES	DEPTH INCHES	A INCHES	B INCHES	APPROX. VOL. CU. FT.
6	9	6	12	0.2
8	12 TO 14	8	16	0.5
10 TO 12	12 TO 14 *	12	16	1.0

* Conical shaped hole

(A) CONE DEVICE SHOWING SIZES FOR STANDARD DENSITY HOLES

(B) CONE DEVICE DESIGNED FOR USE ON SLOPES (REMOVABLE STANDARD 6-BY 12-INCH METAL CAN USED FOR SAND CONTAINER)

FIG. 3.42.—Apparatus for Determining Density of Exposed Sediment Deposits (Photograph by United States Bureau of Reclamation)

terms, when a radioactive source, e.g., Cobalt 60, is placed in a sediment deposit, the emitted ray or photon collides with an orbiting electron in an atom of the material, gives up some of its energy, and changes its direction of travel. Such a collision, known as the Comption effect, causes further random or secondary scattering of the photon until the energy of the photon is either reduced to a level where it is absorbed by the material or it encounters a detector mechanism such as a Geiger tube. The potential for a collision and further scattering increases proportionally with the density of the material; e.g., the number of electrons increases with the unit mass of the material. However, as the number of electrons increases, the probability also increases that the photon will be absorbed before it reaches a detector. Thus, as the density of the saturated sediment deposit becomes greater, a relatively fewer number of photons will be available for detection and

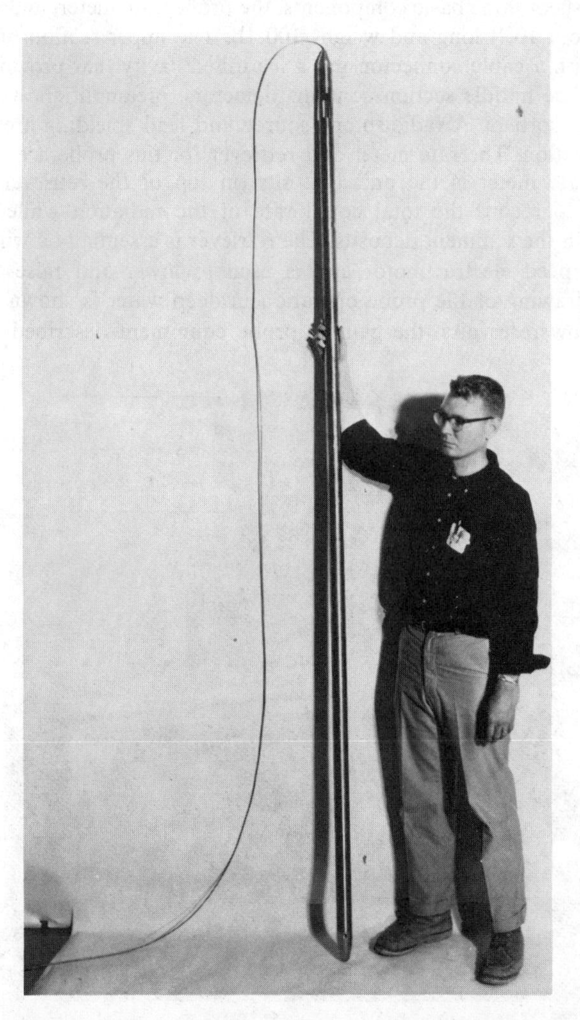

FIG. 3.43.—Gamma Probe for Measuring Density of Sediment Deposits Located under Deep Waters in Large Reservoirs (Photograph by United States Bureau of Reclamation)

the quantity of material representing the "sensitive volume" of measurement decreases. Because the density of most soils is proportional to the number of electrons present in a unit volume, this relative number of photons available for detection becomes through correlation a means for the direct measurement of the density and specific weight of the sediment deposit. It must be recognized here, however, that the physics of this phenomenon are much more complicated than described and that in reality the relationship between the counts measured by the detector and the density of the material is an empirical one. The preceding theoretical explanation was taken from a United States Army Corps of Engineers manual (1965).

An example of a radioactive probe for use in large reservoirs and described by Murphy (1964) is shown in Fig. 3.43. The instrument referred to as a gamma probe consists of three basic components, the probe, rate meter, and retriever. The probe is about 10 ft long and weighs 100 lb. The upper section of the probe is equipped with a cable connector and a leadfilled cavity that provides weight for the probe. The middle section contains detectors, preamplifiers, vertical sensor, and shock absorbers. A radioisotope source and lead shielding are contained in the lower section. The rate meter and retriever for this probe are shown in Fig. 3.44. The rate meter is the unit that sits on top of the retriever. Its primary function is to record the total count rate of the radiation while the probe is submerged in the sediment deposits. The retriever is essentially a winch driven by a variable speed electric motor and is used to lower and raise the probe. A schematic drawing of the probe operation in deep water is shown in Fig. 3.45.

For shallow reservoirs, the gamma probe equipment described by McHenry

FIG. 3.44.—Rate Meter and Retriever for Gamma Probe (Photograph by United States Bureau of Reclamation)

(1971) shown in Fig. 3.46 can be used. It is a dual probe system that consists of one probe containing the radioactive source and the other a detection system. The dual probe utilizes only attenuation as compared to a single source gamma probe system that relies on both attenuation and reflection or backscatter. Components of the dual probe systems are shown in Fig. 3.46. The radioisotope source is in the small probe and the large probe contains the scintillation crystal, photomultiplier tube, and preamplifier. At the lower left is the power supply for remote operation. On the right is the precision power supply (center), scalerratemeter (top) and pulse height analyzer (bottom). The system is shown being operated in Fig. 3.47.

12. Field Procedures.—Field procedures relating to methodology for determining the configuration of sediment deposits, running initial surveys and sediment resurveys, and sediment sampling techniques are discussed.

Methods Used to Locate Surface of Sediment Deposits.—Basically, there are two methods used to locate the surface of sediment deposits. They are the contour and range methods and in some situations they can be combined. The choice of methods depends upon the amount and distribution of sediment as indicated by reconnaissance, the availability of previous base maps, the purpose of the survey, and the degree of accuracy desired.

The contour method uses essentially topographic mapping procedures. Reservoir contours are mapped at given vertical intervals. The contour method is suitable for aerial surveys. Flights over the reservoir area can be scheduled for any known pool elevation; thus, the contour outlined by the water surface can serve as a vertical control base for progressively delineating the higher contours on maps using a stereoplotter. It is advisable to draw the reservoir down to its lowest level

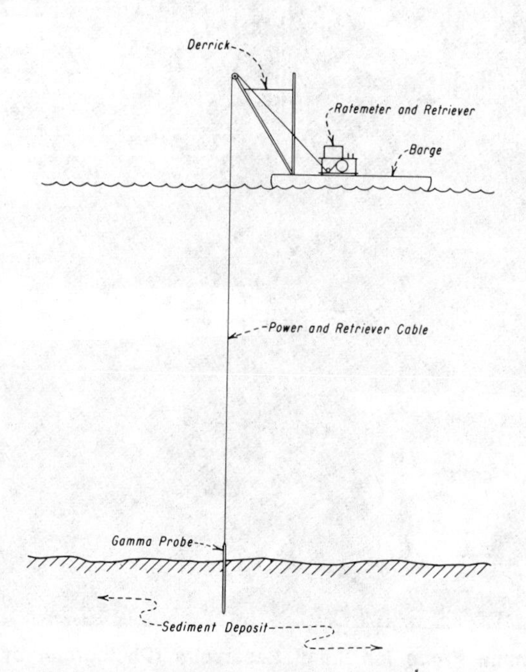

FIG. 3.45.—Schematic Drawing of Gamma Probe in Operation on Large Reservoir

before beginning the aerial survey to obtain the greatest map coverage of the reservoir area.

The range method consists of "profiling" across each range line. Sonic sounders are used to survey submerged portions of ranges. Standard land surveying techniques are used to profile the portions of ranges above waterlines.

The contour method has the advantage in cases where sediment accumulations are large in the area of special interest and are irregularly deposited. However, the method is too time consuming and usually too expensive for general application to large reservoirs, although it may be used to supplement range surveys where circumstances warrant the cost. On the other hand, the range method often requires less time and consequently is less costly. The range method permits frequent, precisely located, and often more representative measurements of the thickness and distribution of the sediment deposits. With permanent range end monuments, the same points of measurement can be resurveyed in the future to obtain profile data on the progressive development of the sediment deposits. These data are used to compute current reservoir sediment accumulation rates.

A combination of both the contour and range methods can be used where the reservoir topography varies significantly. The contour method would be used in the upper and flatter areas of the reservoir and the range method used in the lower and steeper areas. The 1963–1964 survey of Lake Mead described in a report by Lara and Sanders (1970) is an example using both contour and range methods. The contour method was used in the area of the major lake basins and the range

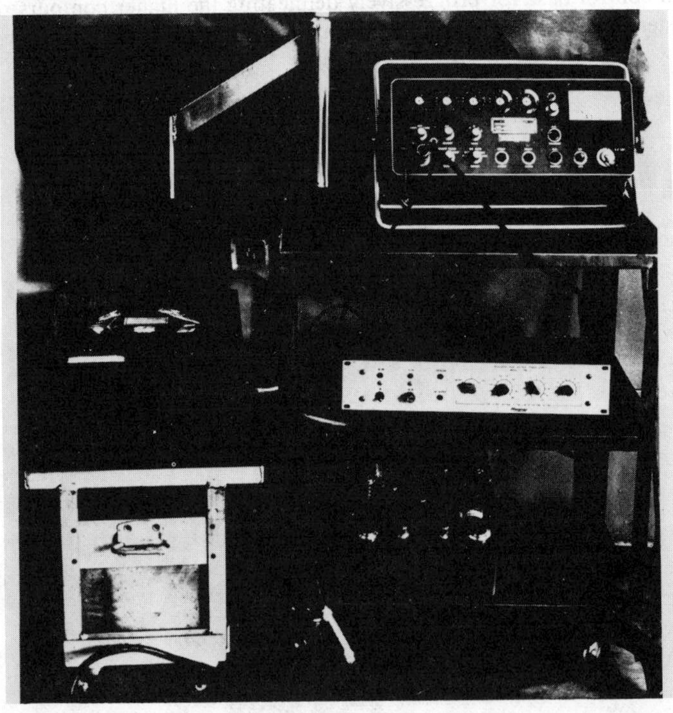

FIG. 3.46.—Gamma Probe for Small Reservoirs (Photograph by United States Agricultural Research Service)

method was used for Lower Granite Gorge, the part of Lake Mead located in the uppermost backwater area of the Colorado River.

Initial Surveys.—In running the initial or original survey of a reservoir by the contour method, standard surveying field procedures are followed in developing the topographic map. This survey would give an accurate reservoir contour map representing the conditions before closure of the dam. The map can be used to determine the reservoir capacity and may serve as the "base map" for planning sediment range networks.

If the range method is selected for the initial survey, a system of reservoir sediment ranges must be established. Laying out a system of ranges generally follows a set pattern. The location and number of ranges are determined for the particular reservoir. Where practicable, ranges are located normally to the streamflow and valley. They should be spaced so that the volume computed by the average end-area method using the cross-sectional measurements of each adjacent range would reasonably represent the volume of the valley between the ranges. How close they are spaced varies with the degree of accuracy required in volume estimates. Ranges also should be located across the mouths of all principal arms of the reservoir and the range network extended up the tributaries in a manner similar to that on the main stem. The range nearest the dam should be at a sufficient distance above the dam to represent the average stream and valley conditions and not be distorted by the toe of the dam. Closer spacing may be

FIG. 3.47.—Gamma Probe in Operation on Small Reservoir (Photograph by United States Bureau of Reclamation)

recommended in the drawdown zone of the reservoir than the spacings done near the dam (Murthy, 1971). This is necessary as greater movement of deposits can take place in this zone.

A suggested numbering system to identify the ranges in larger reservoirs is shown in Fig. 3.48. The range immediately above the dam is numbered 1 and the remaining ranges are numbered consecutively in upstream order on the main stream. The tributary ranges are also numbered in this manner from their mouth in the order that the streams empty into the reservoir above the dam. Their numbering, however, begins in the next unit of 10. For example, in Fig. 3.48, the last range on the main stem is numbered 13, then the range line at the mouth of the first tributary is numbered 20, and so forth. Each range end is marked "R" for right end and "L" for left end, the designation being made while facing in the downstream direction. If it is necessary to establish an additional range in a future survey, it may be identified by number and letter. Thus, in Fig. 3.48 Range 8A could be a range located between Ranges 8 and 9. Other numbering systems may be used. For example, the range numbers can represent the river or valley mileage. This is advantageous when determining distances between ranges or installing new ranges between the ranges already established.

The ends of each range should be permanently marked by a concrete monument and brass cap stamped to show the elevation and range number. They should be located above the maximum high waterline of the reservoir and away from any caving bank areas. The monuments should be accurately referenced to some permanent point, e.g., a section corner or an existing bench mark. If possible they should be tied to the existing State plane coordinate system. This will facilitate locating the range ends during later surveys and permit reestablishing them if accidentally disturbed or vandalized. A steel fencepost set above ground and adjacent to the monument provides a convenient witness reference point.

The initial profiling of reservoir ranges can be done in the field by running levels across the range lines. It may be advantageous, however, for large and shallow reservoirs to profile the range lines using a sonic sounder immediately after impoundment as suggested by Murthy (1971).

A range system suggested by Heinemann and Dvorak (1965) for small reservoirs is shown in Fig. 3.49. It is a combination contour-and-range system that consists of surveying the emergency spillway contour, a lower contour, and strategically placed ranges across the reservoir. The elevation of the lower contour is recommended to be about one-third the total reservoir depth below the emergency spillway elevation. It is advisable to establish a measured baseline along one side of the reservoir running generally parallel to the main valley. Ranges are struck off the baseline and identified by a number or letter system. A few permanent monuments are set at the point where the baseline intersects the range end. This facilitates the future reprofiling of range lines.

Sediment Resurveys.—The contour method of running a sediment resurvey can involve both surface and subsurface contour mapping. The surface contours are mapped by either a transit-stadia traverse or a plane-table survey. Photogrammetric methods by aerial surveys can also be used for this type of mapping. Subsurface contours are mapped usually by using a sonic sounder operated in a manner similar to the range method described in the following paragraph.

Running a sediment resurvey of a reservoir by the range method generally involves a land and hydrographic survey. The complete profile of a range extending between end monuments on opposite sides of the reservoir is required.

The range line on land from either end monument to the water's edge is profiled with a surveyor's level. Ordinarily the elevations need to be measured only to the nearest tenth of a foot.

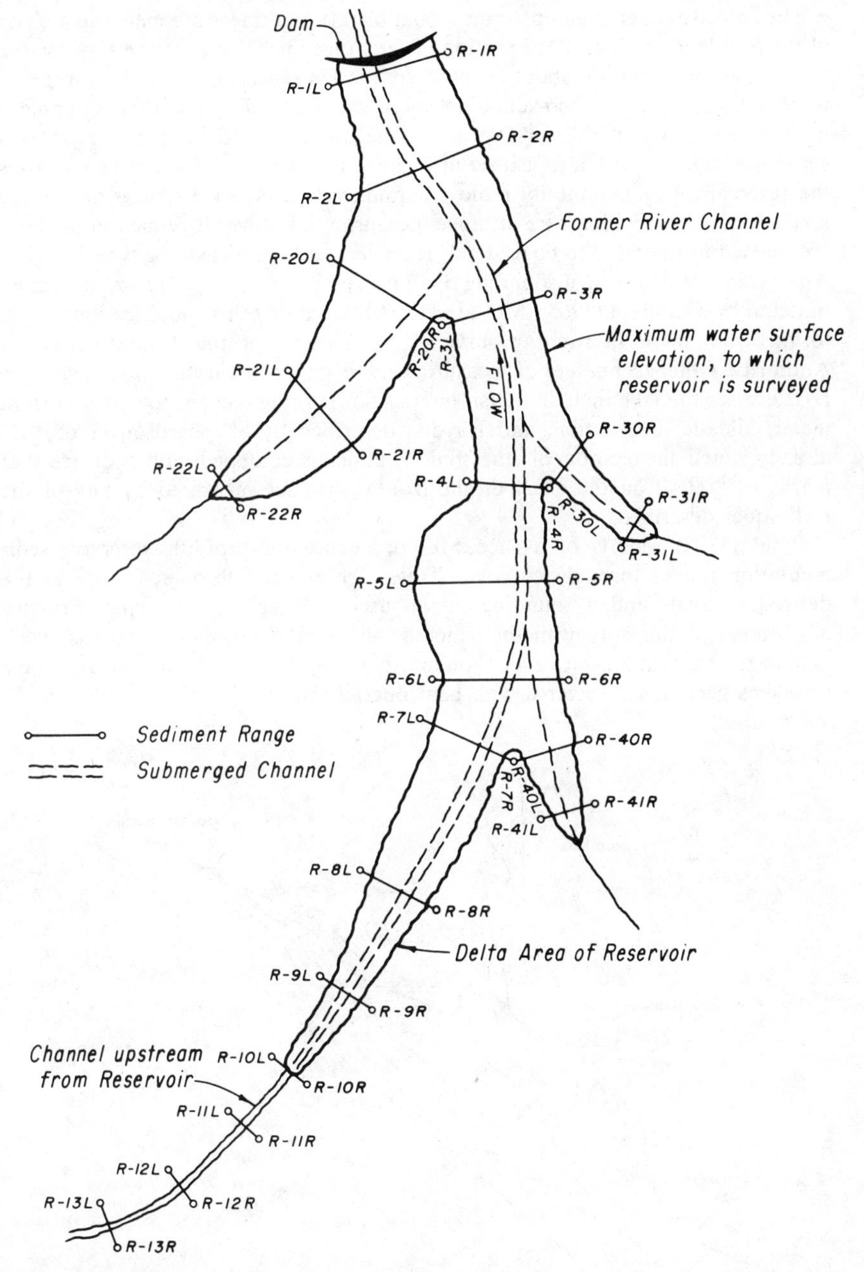

FIG. 3.48.—Suggested Numbering System and Layout of Sediment Ranges for Large Reservoirs

Distances to each level shot can be measured to the nearest foot by stadia or other instruments, e.g., the electrotape, tellurometer, or geodimeter. To complete the profile of the range that is submerged, a sonic sounder can generally be used on both large and small reservoirs.

The sonic sounder is mounted on a boat or raft and traces are made on a chart of the profile of the inundated portion of the range line. The boat can traverse the range line at speeds of about 3 fps–5 fps or faster depending on the type of transducer used in the echo-sounding operation. Depending upon the equipment used, a field party of four or five men is needed to sound the range. A man is stationed on shore with a transit set up to keep the boat on range line as it crosses the reservoir. This is done by radio communication. When distances across the lake are measured with a wire distance measuring machine, three men in the boat are needed to operate the boat, depth recorder, and the measuring wheel device. An example of this type of sounding operation is shown in Fig. 3.50 and described in detail by Shamblin (1965). McCain (1957) described other modifications made to this equipment in running surveys of reservoirs of the Tennessee Valley Authority. Other techniques of measuring the distance at varying intervals as the boat crosses the lake include transit intersection, productimeter, towing a current meter, alidade intersection, and Raydist described by Shephardson (1965). As already noted, the recorder of the sonic sounder is equipped with a device that marks or "fixes" on the chart the horizontal distance measured by any of the techniques described.

Jones (1973) reports of a successful procedure for profiling reservoir sedimentation ranges through ice cover. Holes are augered through the ice at the desired intervals and a sounding line is used to measure the depth. Primary advantages of this survey method include the accuracy obtained because each hole in the ice is at a positive location and not subject to wind, current, and wave problems generally associated with boat operations.

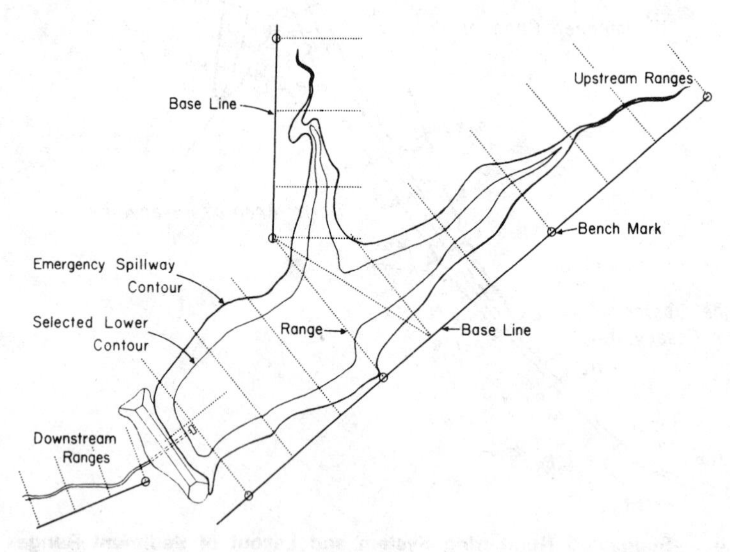

FIG. 3.49.—Suggested Layout of Sediment Ranges for Small Reservoirs

Sediment Sampling of Reservoir Deposits.—Determinations of specific weight and size distribution of reservoir sediments are generally made independently of other parts of the resurvey.

Data for this part of the survey are gathered by using the instruments described previously in the section on sediment sampling equipment. The physical samples removed by the coring-type samplers are taken to the laboratory to be analyzed for gradation and to determine the specific weight of the deposit. Data from the radioactive probes can determine only specific weights.

The specific weight data are used to compute the mean specific weight of the sediment accumulated in the reservoir. Mechanical analyses are made of the samples to give the size distribution of the material. The size-distribution data can be used to compute the specific weights of the sediment deposits using a procedure outlined by Lara and Pemberton (1965).

A sufficient number of samples should be taken to determine representative specific weights for various segments of the reservoir. The number of samples required is influenced by the reservoir size, the type and texture of the inflowing sediments, and the location and number of tributary streams.

13. Sediment Accumulation Computations.—The volume of the sediment deposits accumulated in a reservoir is generally obtained from the difference between the original reservoir capacity and the capacity computed from the resurvey. However, Heinemann and Rausch (1971) point out that compaction of the sediment deposits may occur to an unknown degree between individual surveys. The sediment accumulation rate between such surveys would not be truly representative when computed solely on the basis of differences in capacities. Under these circumstances the sediment volume at the time of the resurvey, computed by subtracting the present capacity from the original capacity, should be used for computing the sediment accumulation rate for the total period only.

Several methods used in computing the capacity of a reservoir are described. The studies of Heinemann and Dvorak (1965) provide an excellent source for some of these methods, most of which can be programmed for the electronic computer.

Methods of Reservoir Capacity Computations.—A basic procedure to compute the capacity of a reservoir is an adaptation of the average end-area method commonly used in computing earthwork quantities. When a topographic map is used, the procedure entails measuring, with a planimeter, the surface areas

FIG. 3.50.—Sonic Sounding of Reservoir Range Using Measuring Wheel to Determine Position (Photograph by United States Bureau of Reclamation)

enclosed by the contours of successive horizontal planes. An average is taken of these areas and this average is multiplied by the contour interval to compute the intermediate volume. When reservoir range data are used, the procedure involves measuring with a planimeter or computing the cross-sectional areas of adjacent ranges. These areas are averaged and the average multiplied by the distance between ranges to compute the intermediate volume. The use of cross sections (or ranges) has two defects. Firstly, the banks of reservoirs are frequently indented by embayments or inlets so that it is difficult to establish a range that is representative of any given reach. Secondly, the alinement of a reservoir is seldom straight, and it is frequently difficult to establish the effective distance between adjacent ranges. On the other hand, it is difficult to resurvey contours underwater.

The following are three of the more generally used methods of reservoir capacity calculation based upon contour area data (Heinemann and Dvorak, 1965). The fourth one listed is a mathematical method based on Simpson's rule. The accuracy of each method increases with the accuracy of the field measurements. For example, a topographic survey of a reservoir basin where the delineation of 2-ft contours is desired would need to be run by including more detailed information than one of 5-ft contours. Because of the greater detail required in field measurements, reservoir capacity computations by using the 2-ft contour interval data consequently would be more accurate.

The stage-area curve method requires establishing a well-defined stage-area or elevation-area curve. The area delineated by this curve between any two given elevations represents the reservoir capacity between the given elevations. For example, the shaded area in Fig. 3.51 would represent the reservoir capacity

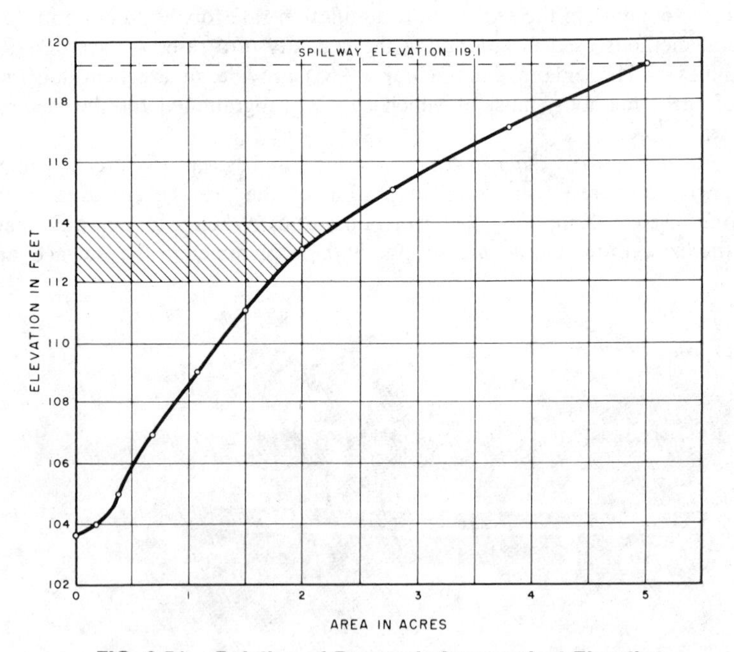

FIG. 3.51.—Relation of Reservoir Area against Elevation

between elevations of 112 ft and 114 ft. This area can be planimetered and converted directly to capacity.

An example of the modified prismoidal method is shown in Fig. 3.52. The

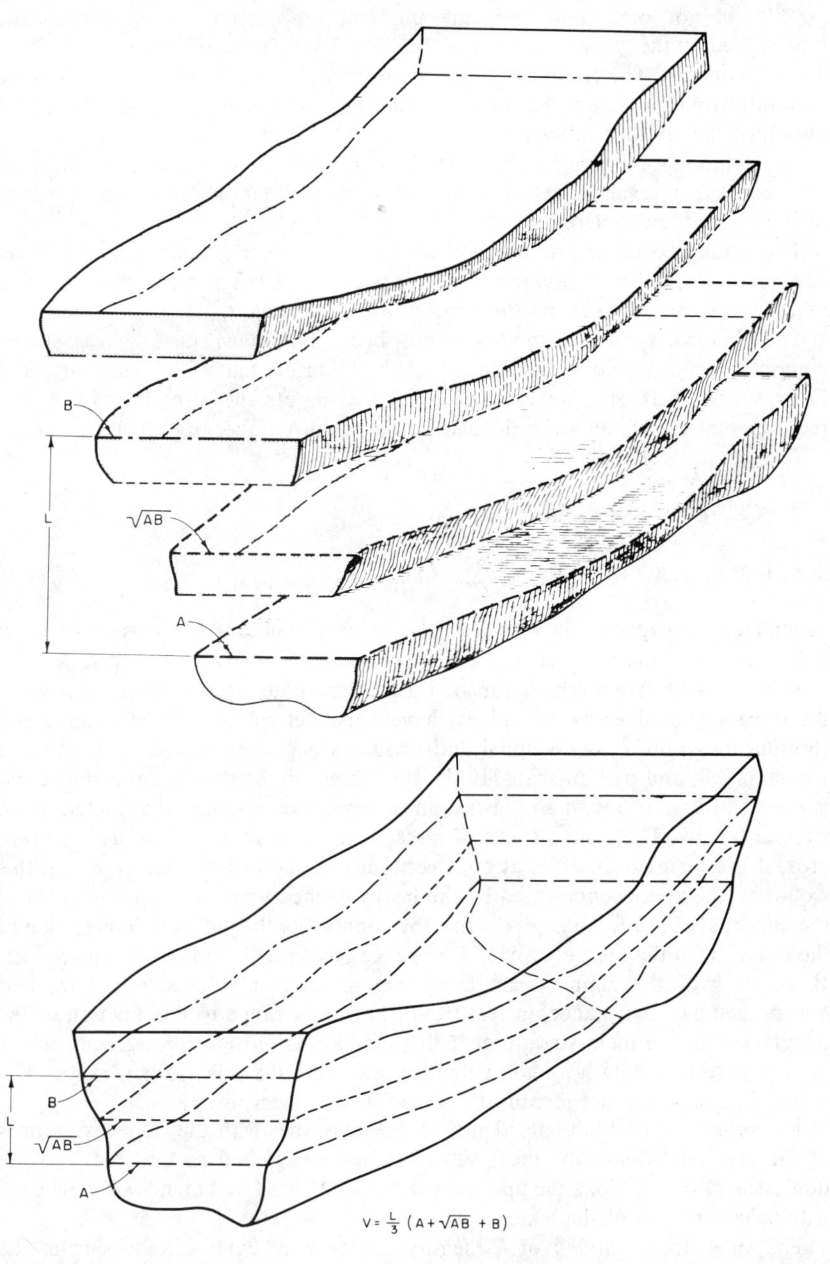

FIG. 3.52.—Terms of Modified Prismoidal Formula for Determining Capacity of Reservoir

equation appearing at the bottom of the sketch is for the volume of the segment between the horizontal surface areas generally one contour interval apart. The upper part of the sketch is a view of the segment in four separate tiered layers to illustrate analytically the principal elements of the equation. The middle area ($\sqrt{AB}$) is not measured from the physical topography; it is theoretically represented by the geometric mean of the two actually measured adjacent contour areas (A and B). The volume of the segment bounded by these areas is computed by multiplying the sum of the areas (A and B) and the geometric mean ($\sqrt{AB}$) by one-third the contour interval.

The symbols in Fig. 3.52 are defined as follows: V = capacity, in acre-feet; L = contour interval, in feet; A = area of lower contour, in acres; and B = area of upper contour, in acres.

The average contour area method consists of averaging the upper and lower contour areas and multiplying by the contour interval. The summation of such determinations approximates the capacity of the reservoir.

The Simpson's rule method has not been mentioned in reservoir survey computations as disclosed by a search of the literature made by Heinemann and Dvorak (1965). It can, however, be used to compute the reservoir capacity. It requires that the reservoir be divided into an even number of segments.

The general equation is

$$V = \frac{1}{3}h[A_o + A_n + 4(A_1 + A_3 + \ldots A_{n-1}) + 2(A_2 + A_4 + \ldots A_{n-2})] \qquad (3.3)$$

in which V = capacity, in acre-feet; A = area of contour or cross sections, in acres; and h = interval spacing between contour or cross sections, in feet.

The following five methods can be used to compute reservoir capacities when the cross-sectional areas (of ranges) have been determined. Eakin's range end formula uses both cross-sectional and surface areas as explained by Eakin and Brown (1939) and by Gottshalk (1951). It is based on the prismoidal formula and is shown in Fig. 3.53 with an illustration of a reservoir segment that includes two tributary arms. The symbols are: V = capacity, in acre-feet; A = total surface area of the segment, in acres at crest contour elevation (in the upper sketch this would be the area encompassed by the heavy dashed line); A' = surface area of the quadrilateral abcd, in acres (formed by connecting the points of intersection of the ranges with the crest contour); E = range cross-sectional area, in square feet; W = width of the main stream cross section, in feet at crest elevation; and h = perpendicular distance, in feet from a tributary range to the junction of the tributary with the main stream, or if this junction is outside the segment, to the point where the thalweg of the tributary intersects the downstream range. The letter, R, and a number identify the range in the reservoir system.

The subscripts of E and W identify these quantities with the respective ranges of the segment. Generally, these subscripts are designated so that No. 1 is the downstream range, No. 2 the upstream range, and No. 3 and higher are ranges on tributaries or arms of the lake.

The subscripts 3 and 4 of h identify the perpendicular distances from the tributary ranges to the respective points shown in Fig. 3.53. These are the values used in the basic equation quoted in the figure.

When the main stream ranges, R1 and R2, in Fig. 3.53 are not parallel, it becomes necessary to determine two additional distances, l_1 and l_2, to compute the area, A'. They are identified in the figure as follows: l_1 = the perpendicular distance, in feet, from the downstream range to the shoreline at the upstream range on the right side looking upstream; and l_2 = the perpendicular distance, in feet, from the upstream range to the shoreline at the downstream range on the left side looking upstream. Reservoirs with irregularly shaped segments may require as many as five different variations of this formula to compute the total capacity.

The cross-sectional area and distance from dam method is based on a plot of the cross-sectional area against distance from the dam. The accuracy of the method is improved if the distance from the dam is measured in a direction normal to the dam axis and all cross sections taken parallel to the dam axis. The

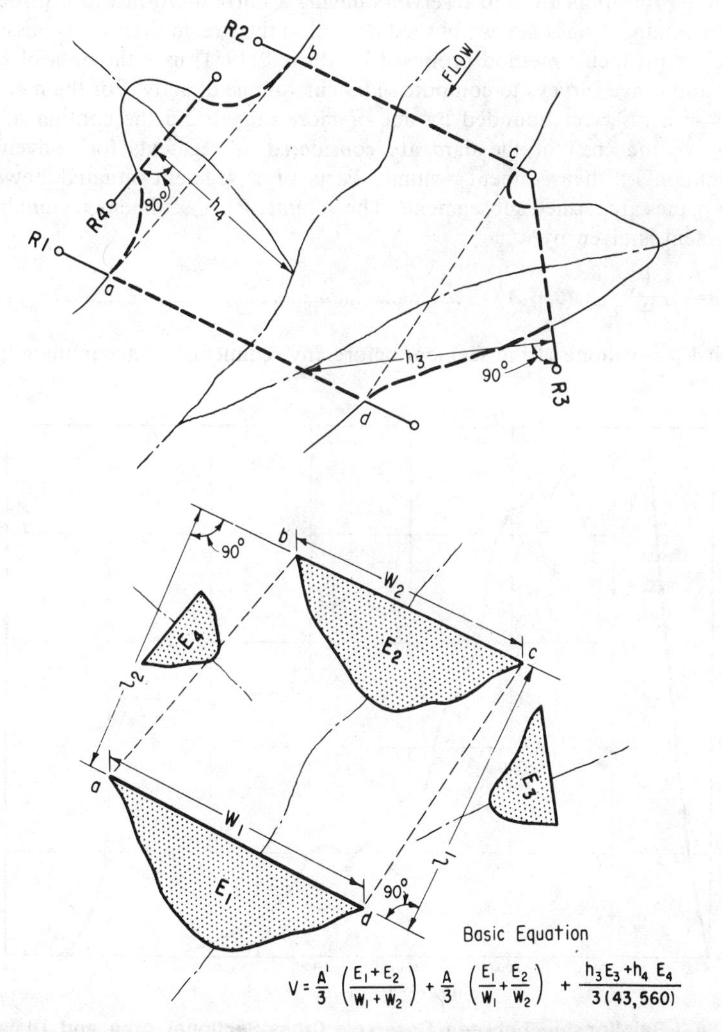

$$V = \frac{A'}{3}\left(\frac{E_1 + E_2}{W_1 + W_2}\right) + \frac{A}{3}\left(\frac{E_1}{W_1} + \frac{E_2}{W_2}\right) + \frac{h_3 E_3 + h_4 E_4}{3(43{,}560)}$$

FIG. 3.53.—Terms of Eakin's Range Formula for Determining Capacity of Reservoir

capacity is represented by the area between the curve and the "distance from the dam" axis. An example of one of these curves is shown in Fig. 3.54.

Simpson's rule, Eq. 3.3, can be used to calculate volumes by substituting the cross-sectional areas at ranges for $A_o, A_1 \ldots A_{n-1}$ and the normal distance between ranges for h. It is highly adaptable to small reservoirs having many evenly spaced parallel ranges in the range system. It may be necessary, however, to use another cross-sectional method to compute the volume of the final odd segment.

The average end-area method requires determining the average of adjacent range cross-sectional areas. This average is multiplied by the channel distance between ranges to compute the volume of the intermediate segment. The reservoir capacity is a summation of these segmented volumes computed to the elevation of the maximum designed water surface of the dam. The average end-area method is better suited for application to reservoirs having a fairly uniform width throughout its length and ranges are established normal to the stream thalweg as possible.

The constant factor method proposed by Burrell (1951) uses the data of both contour and range surveys to compute sediment volume directly. For the method, portions of a reservoir bounded by one or more ranges and the contour at the elevation of the crest of the dam are considered as segments for convenient determination of the sediment volume. Parts of a segment situated between contour planes are termed subsegments. The volume, V_s, of sediment accumulated in a segment is given by

$$V_s = \frac{V_o}{A'_o + A''_o}(A'_s + A''_s) \quad \text{.......... (3.4)}$$

in which V_o = volume of the segment before any sediment had accumulated; A'_o

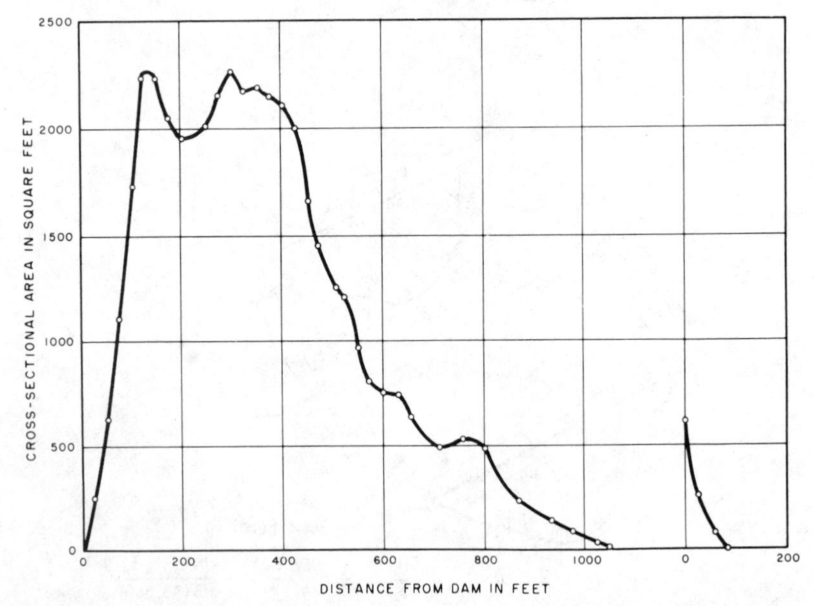

FIG. 3.54.—Relationship between Reservoir Cross-Sectional Area and Distance from Dam

and A''_o = area of downstream and upstream cross section, respectively; and A'_s and A''_s = the areas of the sediment deposit at the downstream and upstream ends of the segment, respectively. Burrell defines a factor, f, which is a constant for each segment

$$f = \frac{V_o}{A'_o + A''_o} \quad (3.5)$$

Substituting Eq. 3.5 into Eq. 3.4 gives

$$V_s = f(A'_s + A''_s) \quad (3.6)$$

Because f is a constant for each segment and subsegment this method is called the constant factor method. Once determined for each subsegment, the factor may be applied in computations for all sediment surveys. This method is the most practical for large reservoirs.

The studies of Heinemann and Dvorak (1965) showed the capacity computed by six different methods checked well within the desired practical accuracy for small reservoirs. A summary of the results of the computed capacities of one of these reservoirs, the Al Olsen Reservoir, is presented in Table 3.3. This reservoir was about 18.5 ft deep at the dam and about 1,400 ft long.

Based upon the field measurements, Heinemann and Dvorak concluded that

TABLE 3.3.—Summary of Capacity Computations: Al Olsen Reservoir

Methods of computation (1)	Contour interval, in feet (2)	Type of survey (3)	Remaining capacity, in acre-feet (4)
Stage-area	2	34 parallel ranges	68.06
	4	34 parallel ranges	68.01
	2	4 parallel ranges, 12-ft contour	68.56
	2	7 skewed ranges, 12-ft contour	69.16
Modified prismoidal	2	34 parallel ranges	67.69
	4	34 parallel ranges	67.36
	2	4 parallel ranges, 12-ft contour	68.74
	2	7 skewed ranges, 12-ft contour	69.33
Eakin's range formula	–	34 parallel ranges	68.23
	–	4 parallel ranges	72.43
	–	7 skewed ranges	64.07
Simpson's Rule (range cross-sectional area)	2	34 parallel ranges	67.81
	–	34 parallel ranges	67.69
	–	29 parallel ranges	67.54
Average contour area	2	34 parallel ranges	68.08
	4	34 parallel ranges	68.80
	2	4 parallel ranges, 12-ft contour	69.11
	4	4 parallel ranges, 12-ft contour	70.49
	2	7 skewed ranges, 12-ft contour	69.71
	4	7 skewed ranges, 12-ft contour	70.93
Cross-sectional area versus distance from the dam	–	34 parallel ranges	68.34

the stage-area method was considered the most direct, simple, accurate, and uniformly adaptable to determine the capacity of a small reservoir. Heinemann and Rausch (1971) also point out several methods for converting the volume of sediment to a weight basis using the data collected during the reservoir sediment survey. The methods involve combining sediment volumes with measured volume weights and calculating the average weight for each sampling location, range, and segment of reservoir. They outline a step procedure for making the calculations.

A word of caution is herewith given regarding the computation of reservoir capacities or sediment volumes by a method in which areas are averaged. Averaging of areas when there are changes in both dimensions that constitute the areas can result in erroneous values. The error, however, is not great unless a big difference exists in area sizes.

C. Accelerated Valley Deposits

14. General Remarks.—Culturally accelerated deposition of sediment in alluvial valleys and channels resulting from erosion of the land surface has been recognized in the United States for more than 100 yr. Decreased channel capacities and sand deposits on cultivated bottom land in the Eastern United States were reported in the beginning of the 19th century (Moore, 1801; Taylor, 1813). Modern accelerated valley deposition has been recognized by various American scientists in Mississippi, Tennessee, the southern Piedment, the driftless area of Wisconsin, and some valleys in California during the period from 1801–1931. All these early investigators noted the relationship between accelerated upland erosion and accelerated valley deposition.

With the expansion of national conservation and flood control programs in the United States, especially since 1930, it was recognized that more detailed knowledge of the problems connected with accelerated erosion and deposition was needed. The first comprehensive study of accelerated valley deposition in the United States was begun by Henry M. Eakin in 1935 for the Soil Conservation Service, United States Department of Agriculture. It included detailed studies of the character and volume of modern accelerated deposits and their relationships to extreme erosion in their drainage areas, as reported by Happ, et al. (1940). Techniques of study, measurement, and identification of accelerated alluvial deposits were developed during the investigation, and these techniques are standard for many similar investigations made subsequently in various parts of the United States by the Department of Agriculture. The techniques for measurement of accelerated valley deposits presented in this section are based mainly on those developed by Happ, et al. (1940). While these techniques have not been undertaken for more than 20 yr, the procedure is still appropriate. It is applicable to small and medium-sized valleys where accelerated deposition has aggraded flood plains that were stable long enough to develop a dark topsoil to serve as a reference plane.

As a result of studies of accelerated modern valley deposition, it was seen that careful measurement of a valley deposit provides a partial index to rates of upland erosion. These studies have shown certain criteria to be useful in recognizing and measuring modern sediment accumulation in valleys. Characteristics that are helpful in identifying modern deposits include color, compaction, distinctive minerals, buried artifacts and textural changes. Conditions that are correlated with accelerated deposition include drainage area, gully development, and texture of source material.

15. General Relationships.—Studies of many drainage basins throughout the United States have shown some typical relationships between upland environments and sites of modern valley deposition that are useful in measuring such depositions. In most instances, the valleys suffering the greatest bulk of accelerated alluvial deposition are headwaters valleys with tributary drainage areas not larger than a few tens or hundreds of square miles. Furthermore, conspicuous deposition has resulted where gully development has been an important aspect of accelerated erosion in the uplands and where the sedimentary debris derived from them has been relatively coarse-textured, containing at least a substantial proportion of sand.

16. Characteristics of Deposits.—Certain characteristics of deposits are useful in identifying them. For this reason descriptions of some of their characteristics follow.

Texture.—Distinctive differences in texture, i.e., the particle sizes and their proportionate distribution in the material, between a modern accelerated sedimentary deposit and the primeval soil in an alluvial valley, are commonly present. In most cases of accelerated erosion in sandy or gravelly deposits, the modern sediment accumulation is definitely coarser in texture than the underlying alluvial soil. In valleys developed by normal erosion and deposition in sandy belts of the Gulf Coastal Plain, the normal precultural soil is silt-loam or clay, even though the underlying formations of the upland are predominantly sandy. Thus, in boring test holes to penetrate the modern deposits, a typical section is seen to include sandy, sometimes gravelly, sometimes silty texture throughout the thickness of the modern deposits, beneath which lies the finer buried soil having a high organic content and often a cloddy soil structure, which in turn overlies a lighter-colored material grading downward to the normal valley alluvium. The underlying alluvium may show little or no difference in texture from that of the modern deposits. In valleys where coarse sand and gravel constitute an appreciable proportion of the modern deposit, textural differences between the accelerated deposits and underlying old soils are obvious.

In some areas such as the Redbeds Plains, valleys of the loess hills areas, and parts of the northern Great Plains of the United States, little difference in texture has been found between the modern deposits and the original soil. For example, in large parts of the Redbeds Plains section of Oklahoma, the upland soils and underlying country rock are predominantly silty. Likewise, the precultural valley soils are predominantly silt. In these cases identification of old soils beneath the accelerated deposits, which may have thicknesses of from 1 ft–12 ft in these areas, must be made on the basis of criteria other than texture. If a modern accelerated deposit has resulted from erosion of fine-textured silty clay and clay soils and subsoils, virtually no difference in texture between such deposits and the underlying valley soil can be detected even by laboratory analysis.

Color.—In most humid and subhumid areas a distinct color difference between the modern deposits and the normal buried valley soil is present. Generally, the buried soil is distinctly darker. This distinction may be a result of a white or light-colored sand over a very dark or nearly black soil of high organic content or it may be only a minor difference in shade between a reddish brown modern deposit and a slightly darker reddish brown original soil. Lack of color difference is typical of the Black Prairie belts of the Gulf Coastal Plain and the Northern Great Plains.

Compaction.—In most cases where substantial accelerated alluvial deposits have formed, a difference in compaction between the accelerated deposits and more compact older alluvium can be detected. For example, when samples are taken of the deposit by auger, the older, more-compacted alluvium is often visually recognizable. It is generally more dense, having less voids, and there may be a noticeable change in packing and fabric.

Distinctive Minerals.—A few preliminary borings in an accelerated valley deposit usually serve to indicate if there are distinctive minerals within the deposit compared to minerals in the buried soil profile. Some typical minerals common in greater quantities in the modern deposits are micas; small granules of unweathered country rock; and fragments of soft calcite, gypsum, or other slightly soluble minerals. Distinct differences in state of weathering usually are apparent. In some valleys, the modern deposit contains distinctive spots or stains of limonite, calcite, or manganiferous minerals. In the old soils and underlying alluvium, the same minerals may have consolidated concretionary form. In most valleys, firm concretions are not present in an appreciable quantity in the modern deposits. In most places, minerals that are readily subject to weathering are not found in large quantity in a true soil.

Buried Evidences of Cultural Activity.—If it is difficult to recognize the depth of modern deposition by any of the foregoing characteristics; fence posts, bridges, tools and implements, wooden boards, and other items found either partly or wholly buried in modern valley deposits may provide indicators of depth. In some valleys of the West Cross Timbers of Texas, as many as three or more "generations" of fence posts have been found, and in one instance three road bridges, the older two partly or completely buried by sediment, were found. Older tree stumps and trunks that have been engulfed by sediment are characteristic of accelerated deposits although they also may be found in older alluvium. In the valley of Salt Creek, a tributary of West Fork of Trinity River, Wise County, Texas, mature pecan trees that have their trunks buried by 6 ft–10 ft of a modern sand deposit have been able to maintain themselves by extending new roots into the new deposit several feet above the original root crown.

Stratification.—Generally the modern deposits show distinct stratification and irregularities, e.g., lenticular beds and cross-bedding. The buried soils and subsoils are usually more massive without conspicuous depositional stratification. In most valleys that contain an accelerated deposit, the buried soil can be traced from stream bank outcrop across the flood plain by borings to the edge of the modern deposit where it appears at the surface and continues to the lower part of the valley slope. This relationship should be certified in establishing the basis for a valley survey.

Chief Types of Accelerated Valley Deposits.—In studying a valley deposit, the various types of deposits, their respective areas, composition, texture, and other characteristics should be considered. In most valleys containing a substantial accelerated alluvial deposit, the most widespread type is the deposit of vertical accretion which includes the natural levees. Vertical accretion deposits are flood-plain deposits resulting from the settling of part of the suspended load of flood waters. Finer sediment is carried farther from the channel and is deposited as a thinner layer over the entire flood-plain surface. Normally, natural levees represent the thickest part of this type of deposit and they taper generally away from the channel to merge with the more widespread part of the vertical accretion deposit. Flood-plain splays and alluvial fans in many places contain relatively coarse sediment but their occurrence is irregular and usually limited in area. Soil borings usually do not measure lateral accretion deposits, but these are of small extent among modern deposits in most valleys where such surveys have been made. Colluvial deposits at the foot of valley slopes usually affect only narrow strips of valley bottom.

17. Survey of Accelerated Valley Deposits

Criteria.—In investigating accelerated valley deposits, the chief criteria for distinguishing between the modern deposit and underlying soils and alluvium must be established. They may include some or all the characteristics previously listed or others distinctive to a particular valley.

Base Map and Location of Ranges.—The best available base map should be used for a detailed valley survey. If available, aerial photographs are most suitable because of their accuracy and abundance of details. On the base map the drainage area boundary and the alluvial valley areas must be delineated. Field inspection of the alluvial valleys of the drainage area indicate the extent of the necessary range or cross section system. In general, the ranges should be placed essentially at right angles to the main axis of the valley and spaced at small multiples of the width of the flood plain. For example, in an alluvial valley in which the flood plain has an average width of 1/2 mile, a recommended interval between ranges is 1 mile–2 miles. Closer spacing may be necessary if identification of old soil profiles is difficult or if the modern deposit is highly irregular in thickness or extent.

Level Traverses and Borings.—The profile of the ground surface along each range line should be established by leveling of approximately third-order accuracy. Distances to points along the range are determined by steel tape or stadia measurement. Borings penetrating modern deposits on each range should be spaced at intervals ranging from 100 ft–300 ft, depending on irregularities. If channels are entrenched, a section of the modern deposit, underlying soil, deeper alluvium, and even bedrock may be revealed. The channel should be accurately measured and described at each range intersection. For use in further surveys all range ends should be monumented and marked with appropriate designation. Information on channel gradients and their relationships to channel conditions is obtained from a level traverse that is referred to the best local or sea level datum.

Mapping Flood Plains.—Concurrently with the range work or immediately thereafter, the margins of the modern deposits should be mapped. If the deposit is extensive the edge of sediment will conform roughly to the lower edge of a higher terrace or the foot of the valley slope. If the modern deposit is confined to local areas of natural levees, splays, alluvial fans, and channel avulsions these deposits must be mapped as separate features.

Mapping may include a comparison of textures of buried soils and modern sediment, approximate content of organic material, thicknesses of the modern deposit, disturbances of water table relationships including swamped areas, extent of channel filling and relationship to flooding, and undesirable changes of topography of the alluvial area.

Sampling and Laboratory Study.—Laboratory studies of the modern sediment deposits and underlying soils augment the information obtained from field descriptions and notes on logs of borings. Two types of laboratory studies are most useful: (1) Mechanical analysis to determine texture; and (2) mineralogical studies. Mechanical analyses are valuable in establishing average texture (grain size and grain size distribution) and differences in texture between modern deposits and underlying materials. Mineralogical analysis involves microscopic study to identify minerals, to determine the percentage of each mineral in the various grain sizes in the total sample, and to determine heavy mineral separates (Rittenhouse, 1944). Comparisons with similar studies of tributary or bedrock sources upstream may provide indications of major sediment sources. If the bedrock in the drainage area is fairly uniform in composition, mineralogical analysis will give no indication of sediment source and should not be undertaken.

Another type of laboratory analysis is determining available plant food constituents and organic matter in the modern sediment compared to those in the buried soil. The modern sediment may contain less available phosphorus, available potash, and organic matter than the underlying old soil. Differences in pH also may help distinguish modern sediments from old soils.

Such analyses have shown distinctly coarser textures in the modern sediments and less available plant nutrients than in the underlying finer buried soil.

Calculations and Report.—Office work and calculations involve: (1) Completing maps on a suitable scale showing primarily drainage area boundary, streams, and extent of the modern deposits; (2) plotting cross sections of ranges, showing range ends and present land surface including channels and the interface of the modern deposit and the old soil; (3) planimeter measurement of the drainage area, the segment areas (surface areas bounded by adjacent ranges) of the flood plain, and the cross-sectional areas of the sediment on the ranges; (4) determining the period of time of sediment deposition (may be based on local interviews); (5) calculating sediment volumes and rates of sediment deposition; and (6) completing summary tables and report.

The drainage area map must be on a scale large enough to permit accurate delineation of sedimentary deposits, e.g., alluvial fans, the larger splays, and the boundaries of the modern deposit and the flood plain. Range ends must be accurately plotted on the map and the surface areas of segments between ranges must be planimetered. The data on surface profile and depth of deposit must be plotted for each range (see Fig. 3.55), and the cross-sectional areas of the modern sediment deposit must be planimetered.

Calculations of sediment volumes are made using the surface areas of the segments and the cross-sectional areas of the sediment deposit at the ends of the segments according to the prismoidal formula (Soil Conservation Service, 1939):

$$V = \frac{A'}{3}\left(\frac{E_1 + E_2}{W_1 + W_2}\right) + \frac{A}{3}\left(\frac{E_1}{W_1} + \frac{E_2}{W_2}\right) \qquad (3.7)$$

in which V = sediment volume in acre-feet; A' = the quadrilateral area, i.e., the area in acres of the quadrilateral formed by connecting the points of range intersection with the edge of the flood plain between the upstream and downstream ranges; A = the measured surface area of the segment in acres, usually plainimetered from the base map; E = the cross-sectional area of the sediment deposit at the bounding range in square feet; and W = width of flood plain at bounding range, in feet. The subscripts 1 and 2 to E and W generally are used to indicate values at the downstream and the upstream range of each segment, respectively.

For calculating the volume of an upstream terminal segment, its area is regarded as a triangle with its base on the downstream range, and the upstream terminus is regarded as a point, i.e., E_2 and W_2 are equal to zero. The formula in this case becomes

$$V = \frac{A'}{3}\left(\frac{E_1}{W_1}\right) + \frac{A}{3}\left(\frac{E_1}{W_1}\right) = \frac{A' + A}{3}\left(\frac{E_1}{W_1}\right) \quad (3.8)$$

A somewhat simpler mean depth formula for valley surveys is

$$V = \frac{1}{3}\left[\frac{(E_1 + E_2)}{(W_1 + W_2)} + D_1 + D_2\right]A \quad (3.9)$$

in which V = volume of sediment, in acre feet; A = surface area of sediment between ranges, in acres; E_1 and E_2 = cross-sectional areas of sediment at ranges 1 and 2, in square feet; W_1 and W_2 = width of sediment on ranges 1 and 2, in feet; and D_1 and D_2 = average depth of sediment on ranges 1 and 2, in feet (Soil Conservation Service, 1940). If alluvial fans constitute an important part of a sediment deposit, the excess volume that they represent must be calculated separately and be included in the total volume. The volume of the fan is considered as a cone and is calculated using the formula

$$V = \frac{1}{3}Ad \quad (3.10)$$

in which V = volume, in acre feet; A = surface area, in acres; and d = the greatest depth of modern fan deposit, in feet. This may be estimated by measuring the total depth of modern deposit near the apex of the fan and subtracting from this the average depth of modern deposit in the valley or segment of the valley where the fan is located.

When calculations of volumes are completed, a summary table, e.g., Table 3.4, can be prepared showing the volumes and various rates of measured sediment

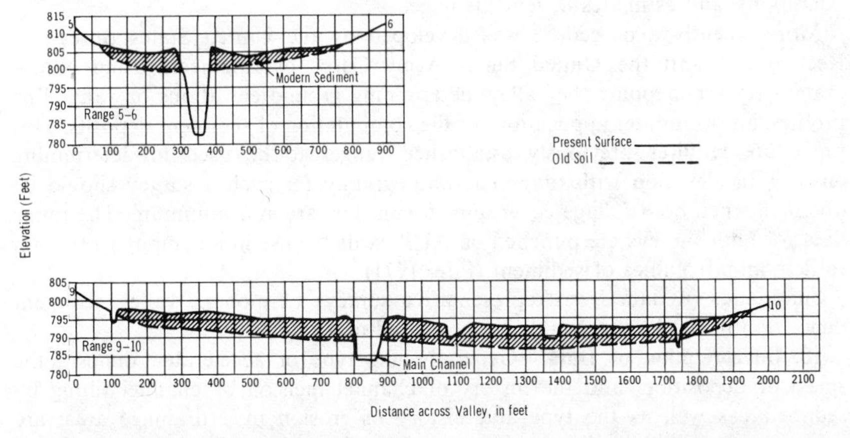

FIG. 3.55.—Representative Cross Sections of Modern Sediment Deposits in Valley

deposition. The indicated annual removal of soil from upland areas in the watershed, shown in this table, is only that accounted for in the measured sediment deposits. The actual removal of soil from the watershed is larger because an unmeasured amount of the material is carried farther downstream by runoff.

TABLE 3.4.—Summary of Data on Sediment Deposition Rates and Volumes in Briar Creek Basin, Wise County, Texas (Jones, 1948)

Factors measured (1)	Quantity (2)	Unit (3)
Age[a]	50.0	years
Drainage area	14.6	square miles
Sediment deposit		
Total sediment	3603.2	acre-feet
Average annual deposition		
From entire drainage	72.06	acre-feet
Per square mile of drainage area	4.93	acre-feet
Per acre of drainage area	335.0	cubic feet
Per acre of drainage area	14.2[b]	tons
Indicated average annual erosion	0.092	inches

[a]Period of accelerated erosion to date of survey (August, 1946) as determined from history of the region.
[b]At an estimated average volume weight of 85 pcf.

For determining approximate sediment volumes and preliminary estimates of land damages by a modern deposit, brief reconnaissance methods can be used. These include a rough determination of valley area and area of modern deposits and borings to determine thickness and character of a deposit on a cross-sectional basis without engineering control. Studies of channel conditions, chief sediment sources, and the nature of the sediment can be based on borings and pace and compass mapping. The cross sections can be more widely spaced than in detailed surveys. In most cases sediment volumes can be calculated by multiplying areas of segments or individual deposits by average depth. Reports can be prepared showing approximate volumes, chief sediment sources, history of channel conditions, and estimates of land damage.

More recently a procedure was developed by the United States Bureau of Reclamation and the United States Army Corps of Engineers using photogrammetry for mapping the valley, determining ground elevations on range line profiles, and computer application for the computation of sediment deposits. This procedure requires previously established ranges to be used in determining changes in elevation with time. The photography for such a survey should be obtained when both foliage cover and stream flow are at a minimum. The range data for both surveys are punched on ADP cards for use in a computer program to determine volumes of sediment (Fife, 1971).

Summaries of data from representative surveys of modern valley sediment deposition in the United States are presented in Table 3.5.

18. Interpretation of Data.—Generally, the type of accelerated deposit, the speed of deposition, and the history of channel incision or channel filling by sediment, as well as the type and history of erosion in a drainage area, are representative of a specific physiographic belt. Extensive gully development and deep sandy deposits of modern sediment are typical of a drainage area underlain by loosely consolidated sandy formations. In areas of deep loess, the growth of gullies and the volume of the accelerated valley deposits may be comparable to or even greater than those of the sandy areas. In areas of fine-textured soils, such as the Black Prairie of the Gulf Coastal Plain, sheet erosion usually predominates. The volumes of modern deposits may be at least as great as those in sandy areas, but damages to fertility and productivity of valley bottoms are less. Channel filling by sediment causes increases in flood heights and frequency of overflows. Similarly, the valley bottom topography may be adversely affected and swampy conditions may develop. In valleys where accelerated natural levee formation has been conspicuous, swampy conditions have developed at some distance from the channel. Furthermore, flood flows commonly develop secondary flood channels in these relatively low areas. Swampy conditions during a large part of the year have developed in such valleys in the Gulf Coastal Plain and other physiographic provinces. If channels have been completely filled they may become higher than the adjacent parts of the valley bottom. Since such conditions usually raise the water table in the whole valley bottom, the damage is a result of a combination of sediment deposition and disturbance of the normal ground-water relationships. In water-short areas, the high water table results in excessive water losses. Also the sediment deposits provide ideal seed beds to foster the spread of phreatophytes.

In calculating rates of erosion and deposition it is usually not possible to include the many small colluvial deposits that accumulate at the base of eroded slopes. Furthermore, sediment discharge records are not available for most small streams. Thus, the volume of the accelerated valley deposits is regarded as only an indication of minimum sediment yield. In all valley sediment studies, the rates of annual sediment deposition as calculated from survey data must be based on the best available records of the history of agriculture and accelerated deposition. Data from the survey can be used to indicate necessary channel improvements, requirements for drainage of swampy areas, necessary controls for erosion in active gully systems, and other necessary rehabilitation work.

D. Airborne Sediment

19. General.—Airborne sediment, commonly called dust and sometimes referred to as an aerosol contaminant or pollutant, is an important factor in all aspects of human life. Dust in excess of a critical limit is generally harmful to man. This critical limit varies with different kinds of dust. It is important not only to determine what the critical amounts are, but also to determine how to measure them.

Dust extends over the entire earth. It may be defined loosely as a powdered solid substance. It can originate from any solid material and, therefore, its nature is quite variable. It results from such multitudinous processes as weathering, erosion, volcanic action, crushing, drilling, or blasting. In nature, it is often unnoticeable except in dry regions affected by wind erosion. In industry, dust is a familiar substance and there is often a need to determine its amount and characteristics.

Dust ranges in size from colloidal to microscopic to macroscopic. Colloidal dust, which usually constitutes only a small proportion of the total, is kept suspended in air by the bombarding action of gas molecules; microscopic dust is commoner. It settles slowly in still air. The upward flow of eddies in turbulent air is sufficient to maintain microscopic particles suspended indefinitely. The maximum diameter of quartz particles maintained in atmospheric suspension usually does not exceed 0.1 mm (Udden, 1894).

Macroscopic dust is airborne only for short time periods. Wind-transplanted soil particles ranging from about 0.1 mm–0.5 mm or more in diameter, depending on wind velocity, are transported in a series of jumps known as saltation. Over 90% of the particles moved in saltation do so less than 1 ft high and relatively few jump several feet or more high. The saltating soil particles are most erodible by wind and cause duststorms by kicking up fine dust that is carried aloft by the

TABLE 3.5.—Representative Surveys of Modern

Valley surveyed (1)	Main drainage (2)	Physiographic area (3)	State (4)	Period of modern deposition, in years (5)	Drainage area, in square miles (6)
Kickapoo River	Wisconsin River	Driftless area	Wisconsin	90	768
Whitewater River	Mississippi River	Driftless area	Minnesota	80	250
Little Sioux River	Missouri River	Central Lowland Till Plain	Iowa	70	1,162[a]
Tobitubby Creek Hurricane Creek	Tallahatchie River	East Gulf Coastal Plain	Mississippi	100	78
Briar Creek	West Fork Trinity River	West Cross Timbers	Texas	50	14.6
Clear Creek	Elm Fork Trinity River	West Cross Timbers	Texas	62	347
Caney Creek	Cedar Creek Trinity River	Forested Coastal Plain	Texas	80	7
Sugar Creek	Washita River	Redbeds Plains	Oklahoma	40	326
West Barnitz Creek	Washita River	Redbeds Plains	Oklahoma	50	124
Battle Creek	Richland Creek Trinity River	Black Prairie (Sandy phase)	Texas	45	21

[a]Above West Fork and excludes 1,525 sq miles of glacial moraine noncontributing area.

turbulent wind (Chepil, 1945). They also bombard coarser grains on the surface and cause them to roll and slide along the surface, a movement known as surface creep.

Recently intense interest and concern about air pollution has resulted in an explosion of research on the problem and in developing and testing various devices to measure dusts or aerosol contaminants in the air. If a review under the headings: Aerosols, Dust Analysis, and Dust Measurement in referencing indexes is made for only the 5 yr ending in 1969, an extremely large number of references dealing with collecting and analyzing airborne contaminants (Engineering Index,

Valley Sediment Deposition in United States

Total sediment, in acre-feet (7)	Average Annual Accumulation: For entire area, in acre-feet (8)	Per square mile, in acre-feet (9)	Per acre, in tons (10)	Indicated annual erosion rate, in inches (11)	Source (12)
38,000	422	0.55	1.53	0.010	Happ, 1944
18,000	225	0.90	2.50	0.017	Happ, 1940
161,000	2,300	1.98	5.74	0.037	Happ, et al., 1941
20,881	209	2.7	7.7	0.05	Happ, et al., 1940
3,603	72	4.93	14.2	0.092	Jones, 1948
14,952	241	0.69	2.1	0.013	Jones, 1940–1950
497	6.2	0.89	2.3	0.017	Jones, 1940–1950
12,818	320	0.98	2.8	0.018	Jones, 1940–1950
7,924	158	1.27	3.3	0.024	Jones and Marshall, 1951
1,291	29	1.38	4.24	0.026	Jones, 1940–1950

Inc., 1964–1968) is found. This section on airborne sediment does not cover all those recent developments but presents some of the measuring sensors and collectors that have been used and some of the analysis methods that appear to be particularly applicable to measuring and analyzing aerial dusts.

20. Measuring Dustfall or Solid Matter Deposited from Atmosphere.—The amount of dust deposited from the atmosphere is important in studying air pollution and in soil genesis and renewal of agricultural soils (Abdel Salem and Sowelim, 1967b; Smith and Yates, 1968).

The sedimentation or settling technique is one that collects dust as it settles out of the air under the action of gravity. This technique is quite satisfactory for particles larger than about 5 microns (5μ) and is employed in most dustfall measurements. Collectors to measure dustfall may be any type in which dust will settle by gravity and become entrapped; however, in his investigations, Miller (1933) simply marked off a square yard on the ground or snow surface and brushed up or shoveled up all the dust-bearing layer. Battery jars and mason jars with plastic funnels having an area at the open end of about 930 cm^2 are frequently used (Hendrickson, 1968). Kampf and Schmidt (1967) used plastic trays, while Fairweather, et al. (1965) trapped dust for particle size analysis on a 12-sq ft plastic sheet. West (1968) indicated that a liquid or greasy coating is needed in dustfall collectors to prevent reentrainment of the collected dust by wind action. He suggested a collector similar to that shown in Fig. 3.56 in which perforations provide for rainwater drainage and minimize wind eddies. Abdel Salem and Sowelim (1967a) used a cylindrical glass beaker 17 cm high and 9.5 cm in diameter half filled with water and placed 50 cm above ground or rooftop in their Cairo, Egypt, studies. Smith and Twiss (1965) used the outer cylinder of standard Weather Bureau raingages in making extensive measurements of dust deposition rates. A 1-mm and a 6.35-mm standard 8-in. diam flat sieve was nested and placed in the top of the cylinder to prevent contamination by birds, insects, and plant fragments. Water or antifreeze was maintained in the cylinder and the collectors were placed in "nondusty" locations, protected for at least several hundred feet by grass or other vegetation from local soil blowing. Chatfield (1967) developed a dual-purpose sequential sampler to sample either dust concentration or deposition, depending on the kind of lid and sampling disk used. For deposition sampling, microscope slides were placed on a sampling disk

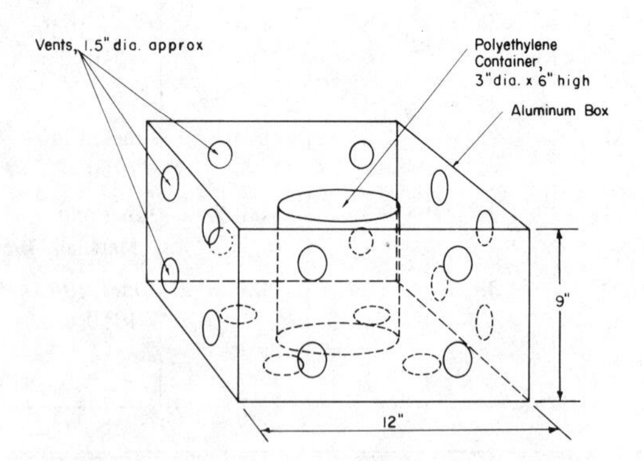

FIG. 3.56.—Dustfall Collector (West, 1968)

that was covered by a circular lid having one aperture. The sampling disk was rotated by a solenoid-operated ratchet mechanism so that each slide was exposed separately as it stopped beneath the aperture. A diagram of the sampler is shown in Fig. 3.57. Sanderson, et al. (1963) conducted extensive tests on five dustfall collectors including the Nalgene decanter, a 6-in. diam black plastic cylinder, the Nipher snow gage, the Toronto collector, and the Detroit collector. They found the Nipher gage, which was developed for the Meteorological Division, Department of Transport, Canada, gave the most consistently accurate measurement of dustfall. This gage consists of an inverted bell shield of aluminum attached to the top of a copper cylindrical container 5 in. in diameter and 20 in. high.

The principal measurements obtained from nearly all sedimentation type collectors are the quantity or amount of dustfall in a particular area over a period of time and the size distribution of the particles. Quantity is determined by properly exposing the devices, then periodically, usually once a month, collecting the container, rinsing the inside, and evaporating the liquid. The residue remaining is weighed and expressed in weight per unit of area and is considered the measure of the dustfall in a particular area over the period of measurement. Though such measurements are useful, they do not avoid numerous sources of inaccuracies. When results obtained from a small collecting area are expanded manyfold, inaccuracies are likewise expanded. Particles adhering to walls of collectors, deflection effects of collectors, wind eddies from nearby objects, and agglomeration of particles are sources of possible errors. Sizing is determined by sieving or microscopic techniques; however, particles that fall out of the atmosphere cannot usually be related to a given volume of air sampled and are not generally considered suitable for a count or determination of numbers of particulates analysis.

21. Measuring Aerosol Contaminants or Suspended Solid Material in Atmosphere.—Many types of apparatus have been used for this purpose. They can be

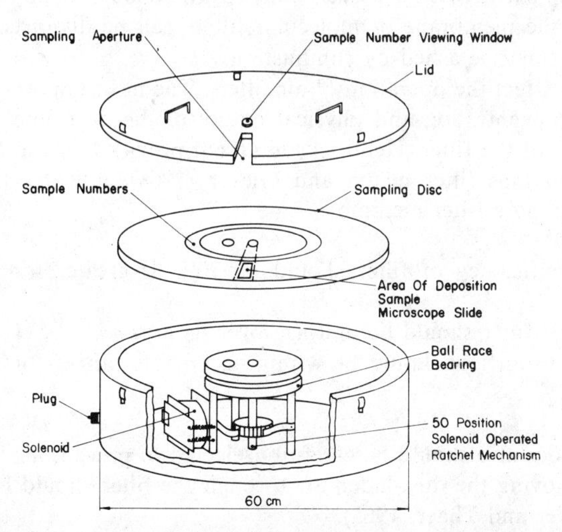

FIG. 3.57.—Sequential Deposition Sampler (Chatfield, 1967)

classified according to the principle of operation into five major groups: filtration, impingement, electrostatic precipitation, thermal precipitation, and centrifuging.

Filtration.—Filtration through a suitable material has been most widely used in collecting aerial dust. After collection, analyses are performed by weighing, determining chemical composition, or by particle sizing. The great variety of filters that have been used are of three general types: (1) Fiber filters including mineral wool, plastic, glass fibers, asbestos mats, wood fiber paper, and papers or mats made of other cellulose fibers; (2) granular filters such as porous ceramics, fritted glass or metal, sand, and where it is desirable to recover particulates, such soluble or volatile granular materials as sugar, salicylic acid, collodion, and resorcinol; and (3) controlled pore filters manufactured from cellulose esters, polyvinyl chloride, acrylic polyvinyl chloride copolymer, epoxy, fluorinated polyvinyl chloride, and silver (Hendrickson, 1968; Jacobs, 1960; Katz, et al. 1925).

Fiber filters are suitable for collecting dust for weighing and chemical composition analyses but not for collecting airborne particulate material for count or size determinations because the collected material is difficult to separate from the filtering medium. Generally, cellulose fibrous filters are not suitable for collecting particles less than 0.5μ–1.0μ in diameter, are difficult to stabilize for weighing in milligram quantities, and cannot be used under conditions of high moisture or temperature. Some of the newer glass or mineral fiber filters are effective for particles as small as 0.05μ and can be used at temperatures as high as 800°C.

Granular filters generally are not suitable for collecting particles less than about 1μ in diameter; however, since the granular materials may be obtained in a wide range of particle sizes, the effectiveness of collection covers a wide range above the 1μ size.

Controlled pore filters are commonly known as membrane or molecular filters. They are suitable for count and size determinations, are capable of uniformly high efficiencies, and are useful for collecting submicron particles (Giever, 1968). They have an advantage in particle sizing in that the filter can be made transparent with mineral oil so the particles may be seen. Like sugar, collodion, and other granular filters, some of the membrane or molecular filters can be dissolved with esters, ketones, or they may be ashed by combustion.

Many factors affect the operation of air filters. The most important are size of dust particles, concentration and physical nature of the dust, and porosity and loading capacity of the filter. Hendrickson (1968) provides an excellent table of efficiencies of various filter media and Giever (1968) indicates the following criteria for selecting a filter medium:

1. Collection efficiency of filter should be 90% or greater for smallest size ranges.
2. Efficiency of filter should be uniform over its area.
3. Resistance to airflow should be within acceptable ranges for the sampling procedure used.

The filter holder also should be carefully selected to prevent air leaks and the technique for moving the dust-laden air through the filter should be considered carefully (Mueller and Thaer, 1965).

Impingement.—When a dust-laden airstream is deflected around a body, the dust's greater mass and inertia resist the change in direction and dust collects on the body. This principle of inertial impingement has been used to develop several instruments to sample airborne sediment. The sediment may impinge either on a surface submerged in a liquid (wet impingers) or on a surface exposed to air (dry impingers). Greatest collection efficiency is obtained with larger particles because small ones tend to follow the streamlines, but good collection efficiency is obtained with dry impingers for particles down to 2μ with usual operating velocities. Wet impingers give good efficiencies for particles down to 1μ and, when operated at sonic velocities, have high collection efficiencies for particles as small as 0.1μ. The Greenburg-Smith (1922) and the midget impingers are examples of wet impingers (Fig. 3.58). The Greenburg-Smith impinger draws in 0.028 m^3/min of air through a 2.3-mm diam orifice and impinges against a flat surface 5 mm distant at a velocity of approx 113 m/s. The midget impinger draws in air at the rate of 2,800 cm^3/min through a 1-mm diam orifice and also impinges on a flat surface 5 mm distant but at a velocity of only 60 m/s. The dust collected in the water impinger is diluted as required and dust particles are counted in the microscopic field. Counting cells of convenient shape are used. A tyndallometer (apparatus used to measure the scattering or brightness of a beam of light when passed through a medium containing small suspended particles) can also be used

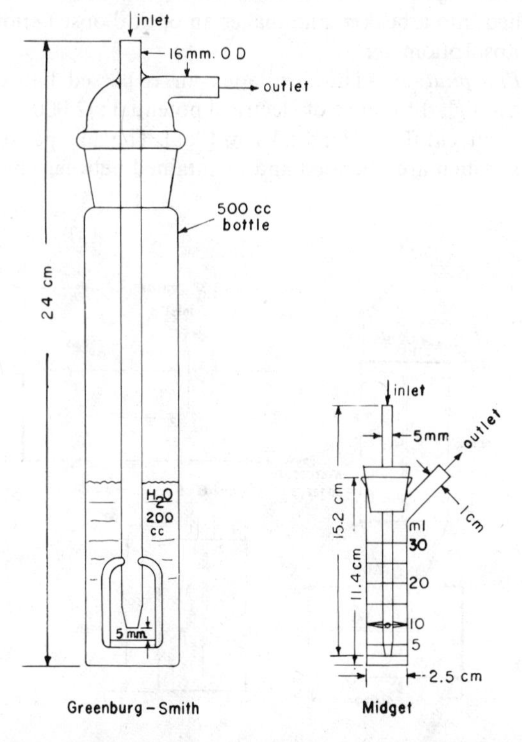

FIG. 3.58.—Examples of Wet Impingers (Hendrickson, 1968 and Katz, et al., 1925)

to measure the relative concentration of dust in suspension (Stuke and Rzeznik, 1964).

The Owens-jet dust counter, the Bausch and Lomb dust counter, the cascade impactor, and the Morrison dust meter are dry impingers (Morrison, 1959). Commercial versions of the dry impinger or impactor usually consist of a series of progressively smaller jets impinging on microscope slides (Hendrickson, 1968). An example of the cascade impactor is shown in Fig. 3.59. Progressively finer particles are collected by stages as the air velocity increases through the decreasingly smaller jet apertures.

Dust collectors that use a modified or partial impingement or impaction principle also have been devised. Examples include: (1) The Warn and Cox boxlike sediment trap in which dust particles in the airstream enter the open face of the trap, impact against the back side, and fall vertically through baffles to collection bottles (Warn and Cox, 1951); (2) the Any and Benarie collection device that employs acoustical detection by measuring the voltage impulses generated when dust particles strike a sensitive microphone (Any and Benarie, 1964); (3) the Baum, et al. wind-directional dust collector consisting of a 16-sided truncated pyramid with each side having a piece of sticky foil exposed to collect dust; a wind-vane-operated rotating cover alines a square opening with the wind direction (Baum, et al., 1964); (4) the Chatfield (1967) sequential sampler (Fig. 3.60); and (5) the Lucas and Moore (1964) CERL directional deposit gage (Fig. 3.61) in which dust impacts on a vertical surface, drops into a collection bottle, and then is washed into a beaker and makes an optical obstruction measurement with an E.E.L. absorptiometer.

Electrostatic Precipitation.—The dust-laden air is passed between two-spaced electrodes where a high difference of electrical potential (12,000 volts–30,000 volts d-c) establishes a current flow (Hendrickson, 1968). The dust particles collide with the charged ions, which are liberated and maintained between the electrodes and

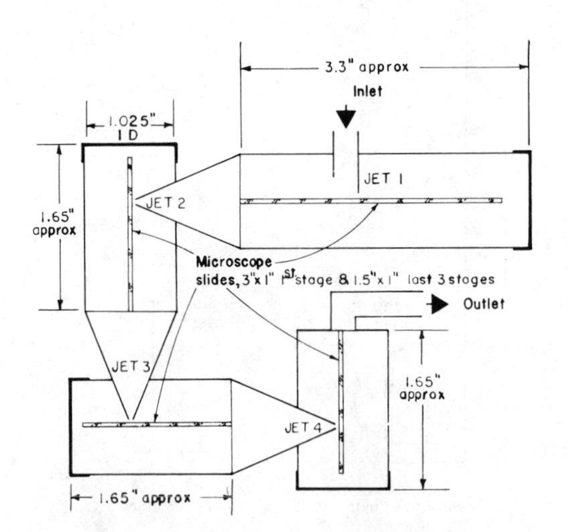

FIG. 3.59.—Four-Stage Cascade Impactor [Jet 1 Orifice Size is 19 mm × 6 mm; Jet 2, 14 mm × 2 mm; and Jets 3 and 4, 14 mm × 1 mm (Hendrickson, 1968)]

are thus made to assume a charge. The charged particles are transported to the collecting electrode by the electrical force field. The charge is then neutralized and the particles collected. The particle collection efficiency is a function of shape and size of particle, type of electrodes, electrode spacing, velocity and time of particle exposure, electrical energization, and electrical resistivity of the particle (Hendrickson, 1968). Efficiencies near 100% are attainable for particles in the 0.01μ–10μ range with electrostatic precipitators. The collected dust may be analyzed chemically, by weight, by counting or sizing, or by direct microscope examination. Several designs of electrostatic precipitators have been developed or are available commercially but all operate on the basic principle described herein (Drinker, 1934; Kobayashi, et al., 1963; Billings and Silverman, 1962).

Thermal Precipitation.—The thermal precipitator operates on the principle that a thermal force acts on a body suspended in a gas not in thermal equilibrium (Hendrickson, 1968). The force causes suspended particles to migrate from a zone of high temperature to one of low temperature. The thermal force is negligible if the gradient is less than about 750°C/cm but most commercial models operate with a thermal gradient above 3,000°C/cm. Commercial versions of the thermal precipitator generally consist of a hot wire suspended near a glass microscope slide with airflow directed between the wire and slide or an electrically heated plate is suspended above a water-cooled plate with the airflow entering the center of the heated plate and then flowing radially to its edges (Hendrickson, 1968). Thermal precipitators are well adapted for collecting particles for size evaluation and are said to be nearly 100% efficient for particles ranging from 0.001μ–100μ.

Centrifuging.—The dusty air is drawn and guided into a cylindrical tube in such a way that a whirling action is set up in the tube. The whirling action generates a

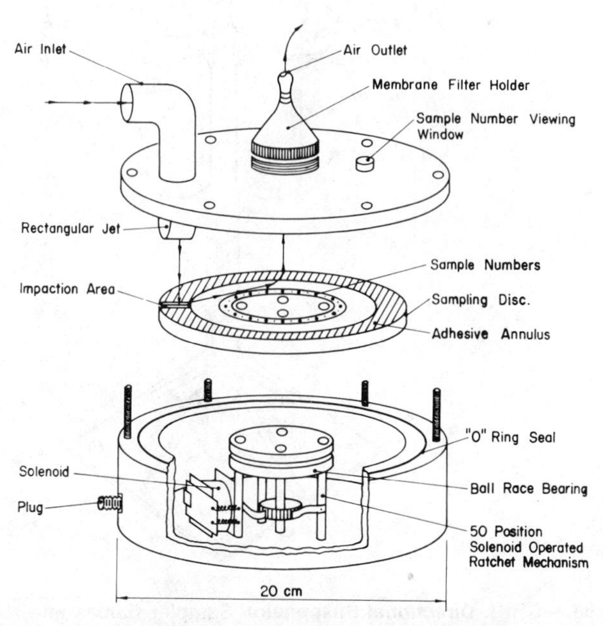

FIG. 3.60.—Sequential Suspension Sampler (Chatfield, 1967)

centrifugal force that tends to throw suspended particles toward the outside. Clean air is drawn from the center of the tube and dust settles down the circular wall of a tube onto a hopper underneath. Hendrickson (1968) describes the Goetz centrifugal sampler as consisting of a rapidly rotating helical channel with removable cone-shaped cover. About 20,000 g of centrifugal acceleration causes flow in the channel, and individual particles settle according to Stokes' law and are deposited on the inside of the cone. The device is effective for particle sizes down to 0.2μ. Many commercial air-cleaning plants are based on that principle.

22. Measurement of Aerial Dust Originating from Erosion of Soil by Wind.—Langham, et al. (1938) used a wet impinger method to measure dust at plant height during duststorms. Chepil (1945) measured total amount of soil eroded along the surface of the ground using the Bagnold catcher (Bagnold, 1943) to measure the quantity of coarse grains moved in saltation and the Langham, et al. method (1938) to measure amounts of dust carried in suspension.

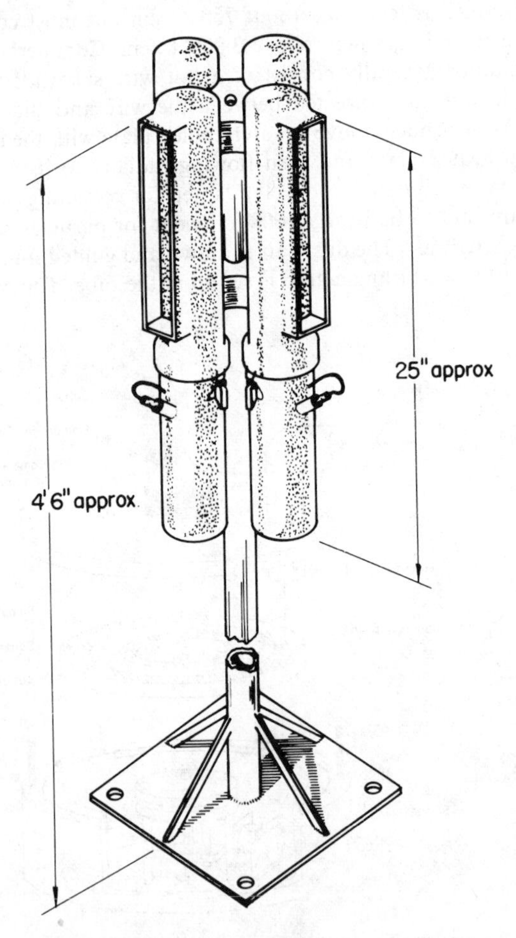

FIG. 3.61.—CERL **Directional Suspension Sampler (Lucas and Moore, 1964)**

Zingg (1951) devised a method based on vacuum cleaner cloth infiltration to measure total soil movement occurring near the ground surface for use in wind tunnel studies of soil erosion by wind. The technique involved the simultaneous sampling of the airstream at several heights above the soil surface. The velocity of air intake at elevations of sampling was made equal to the velocity of the airstream at the sampling heights. The total movement was calculated by integration of the product of concentration and velocity over the vertical distance sampled (Zingg, 1953).

The Zingg method, though valuable for wind tunnel work, was unsuitable for field work. The method was modified by Chepil and Woodruff (1957). One modification replaced the vacuum cleaner filters with fine glass wool filters packed into 1-in. diam round tubes. Hoses connecting the tubes to the suction units were lengthened so that dust could be collected at various heights up to 20 ft. Air meters and barometers were added for more accurate checks of the air volume intake. The glass wool filters collected 97.5% of the dust trapped by the wet impinger of Langham, et al. (1938) and had the advantage of allowing dry dust to be shaken out for determining its equivalent size distribution. (Size, shape, and bulk density of discrete soil grain or aggregate are expressed together as equivalent diameter which is the diameter of a standard particle that has an erodibility equal to that of a soil particle of any particular diameter, shape, and bulk density. It is approximately equal to σd per 2.65 in which $\sigma =$ bulk density of a discrete soil grain or aggregate of diameter d as determined by elutriation.)

The Chepil and Woodruff method was too complicated to facilitate measuring rate of soil movement at various locations across a field and crop strips. Chepil (1957) modified the Bagnold collector for this purpose. The modified apparatus is simple and convenient to handle but measures only rate of flow in saltation and surface creep.

23. Analysis of Airborne Sediment.—Quantity or concentration of aerial dust or airborne sediment is determined by weighing samples collected either alone after they are separated from the collecting medium or together with the medium, e.g., with a filter. Other analyses that are frequently performed include number count and size of particles and chemical and morphological analyses of particulates. A great amount of work has been done in those fields of analyses. Chapters 19, 21, and 22 prepared by West (1968), Giever (1968), and McCrone (1968) in the book *Air Pollution* give excellent descriptions of chemical analysis, analyses of number and size of particulates, and morphological analysis, respectively. Only a brief summary of those chapters and other pertinent literature is given herein.

Counting and Size Distribution of Particles.—Counting dust particles is usually done with a standard biological microscope or its equivalent. Sedgewick-Rafter or Dunn counting cells are most commonly used. Procedures and techniques for counting are fairly well standardized but vary somewhat, depending on collecting method used (Giever, 1968). Concentrations of dust determined from counting analyses are expressed in millions of particles per cubic foot of air, particles per unit surface area, or particles per cubic centimeter of air, depending again on the kind of collector used, i.e., liquid, dustfall, thermal precipitator, filter, or other.

Particle size distribution is determined by sieving, optical microscopy, sedimentation, elutriation, centrifugal classification, or by using an electron microscope or impactor.

Sieving may be done in air (dry sieving) or in liquid (wet sieving) by using one flat sieve at a time or a nest of sieves, each with progressively smaller openings, so that weighing the retained particles on each screen gives distribution (Cadle, 1955). Dry sieving may also be done using the rotary sieve described by Chepil (1952, 1962) which is essentially a rotating nest of concentric cylindrical sieves of progressively smaller openings operated on a sloping axis so particles larger than a given sieve size are moved to separate trays and may be weighed to obtain size distribution data. Wet sieving is useful for materials in the 37μ–74μ range and is particularly valuable for samples in which particles tend to agglomerate during dry sieving. Sieve size openings and wire diameters are standardized by the National Bureau of Standards and the American Society for Testing Materials (1964).

Particle size measurement by optical microscopy is done by first dispersing a bulk sample on a microscope slide and then making direct microscopic size measurements in terms of the "statistical" or "projected" size as defined by Feret's or Martin's diameters (United States Department of Health, Education, and Welfare, 1965). A filar micrometer, eyepiece reticle, or microprojector may be used as accessory equipment.

The sedimentation or settling method of particle size analysis is based on the terminal velocity of particles falling in a gas or liquid. The sedimentation diameter of a more or less gross grouping of particle sizes is computed using Stokes' law.

Elutriation is the inverse of settling. Air is introduced into a dust sample and particles smaller than the separation size for a given velocity pass through to a settling chamber where they are collected for weighing. Diameters are calculated using Stokes' law. Size classification is accomplished by changing the air velocity. Chepil (1951) found the elutriator an extremely useful extension to the rotary sieve when evaluating soil structure in connection with erodibility by wind.

Particle size analysis employing centrifuge techniques relies on centrifugal force to cause particles to deposit in the analyzer at specific locations peculiar to their settling velocity. The technique is useful for particles in submicron size ranges down to 0.025μ. Description of various centrifuge techniques and formulas for calculating settling velocities are given by Orr and Dalla Valle (1960).

The electron microscope allows direct observation of particulate matter down to 0.001μ in size. There are many techniques for sample collection and investigation by electron microscopy. Billings and Silverman (1962) have outlined many of the applicable methods.

Chemical Analysis of Inorganic Particulates.—It is often desirable to analyze dustfall and suspended-sediment samples to obtain specific information regarding the nature of the particulates. Determination of lead is one of the most important analyses made of airborne sediments in connection with air pollution studies. Sodium, potassium, calcium, nitrate, aluminum, copper, and many other metals often are also of special interest, so samples are analyzed for them. Methods used include titrimetric, colorimetric and spectrophotometric, ring oven, flame photometry, emission spectrometry, atomic absorption spectroscopy, and polarography. West (1968) gives a complete description of each and indicates that atomic absorption spectroscopy and ring oven methods are especially attractive for analysis of airborne particulates because of their specificity, convenience, sensitivity, and general reliability.

Morphological Analysis of Particulates.—Often it is important to identify the particles in airborne sediment to determine their composition because their composition may give information on the source of the sediment, and sediment is used as a tracer to study meteorological dispersion. Twiss, et al. (1969) used morphological analyses to identify grass phytoliths in dust deposition samples for possible use in tracing sources of atmospheric dust. Morphological analysis begins with a microscope to identify single small particles by such characteristics as shape, size, color, and transparency. The object as a whole or only significant components of it may be identified. The microscopist must be highly skilled and must have excellent reference slides to identify the thousands of different microscopic objects. In studying airborne sediment samples, it is most convenient to subdivide the possibilities into three classes: (1) Wind erosion products, including biological and mineral; (2) industrial dusts, including fibers, chemicals, foundry dusts, polymer, etc.; and (3) combustion products from incinerators, furnaces, boilers, etc. The experienced microscopist usually can recognize a given particle at sight or can identify it as one of a small group, e.g., cubic crystals, pollens, synthetic fiber, or polymer powders. The particle usually can then be identified by confirmatory tests based on refractive index, dispersion staining, birefringence, microchemical tests, melting points, etc. McCrone (1968) gives an excellent table of morphological characteristics and tests to identify particles. He included 32 color photomicrographs, which are excellent reference slides for identifying the wide variety of particulates found in airborne dusts.

E. Laboratory Procedures

24. General.—Sediment laboratories, i.e., laboratories equipped to measure the quantity and properties of sediment, are an essential part of measurement programs for investigations of fluvial or potentially fluvial sediments. Such programs may include research involving surface waters that are in contact with unconsolidated sediment, but probably would not involve sediment or porous media investigations of ground-water aquifers.

The two principal functions of a sediment laboratory are usually to determine the concentration of suspended-sediment samples and to determine the particle size distribution of suspended-sediment, stream bed material, and deposits associated with other surface waters. Other functions may include the determination of several physical aspects of stream bed material and soil samples. These physical aspects may include the particle size distribution, the roundness and shape of individual grains and their mineral composition, the amount of organic matter, the specific gravity of sediment particles, and the specific weight of deposits. As will be seen later and as indicated in Chapter II, Section B, the determination of particle size for fine sediments, often including the sand sizes, is accomplished on the basis of the fall velocity of grains rather than physical sizes.

The literature documents many methods for determining the concentration and physical and chemical properties of sediment. Such documents often reflect methods previously illustrated in the literature and usually contain some innovations made possible by advances in technology. The rather slow evolution of sediment laboratory methods by these innovations does not insure that standard methods and results will be obtained in every laboratory. The most comprehensive effort for improvement of methods in sediment laboratories was

"A Study of Methods Used in Measurement and Analysis of Sediment Loads in Streams," sponsored by the United States Interagency Committee on Sedimentation, Water Resources Council. Beginning with Report 4, "Methods of Analyzing Sediment Samples" (Interagency Committee, 1941a) several pertinent reports in this series are available. Much of the material on "Laboratory Procedures" to follow in this presentation is based on procedures developed and extensively used by personnel of the Water Resources Division of the United States Geological Survey (Guy, 1969).

This section on laboratory procedures includes a brief explanation of: (1) The reliability and accuracy requirements of fluvial sediment data; (2) the filtration and evaporation methods for determining the concentration of suspended-sediment samples; (3) the determination of particle size of sediment and its distribution by direct or indirect measurement, or some combination thereof, including the sieve, pipet, visual accumulation tube, bottom withdrawal tube, and hydrometer methods; and (4) other determinations related to fluvial sediment analyses and its characteristics, e.g., organic material, dissolved solids, related water quality analysis, specific gravity, specific weight, moisture content, and turbidity. The reader is referred to Section D of this chapter on airborne sediment for size analysis procedures of sediments transported and deposited by wind.

25. Reliability and Accuracy.—The reliability and accuracy of computations of suspended-sediment discharge, total-sediment discharge, and the probable shape characteristics, volume and specific weight of exposed and submerged reservoir deposits is directly related to the reliability and accuracy of the laboratory analysis. For example, in the determination of suspended-sediment concentration, the quantity and even the characteristics of the sediment, as well as the chemical quality of the water from the stream (native water), must be considered in processing the sample. Samples with too small a quantity of sediment tend to magnify errors caused by weighing or transfer from one container to another; samples with too large a quantity of sediment may cause problems with respect to splitting, drying, and weighing. Samples collected with certain types of mineralized water or containing colloidal clay, or both, result in difficult separation of sediment from the native water. Problems associated with these and other conditions will be treated in connection with the appropriate analysis as necessary.

The number of significant figures to be used in recording data from analysis in the sediment laboratory is the product of compromises between the need for some semblance of uniformity in the final tables of data, the precision of the measurement, and the importance of the precision for the use of the data.

It is practical and desirable to report sediment concentration to the nearest milligrams per liter up to 999 mg/l; for higher values, three significant figures should be used. Often, the third significant figure in concentrations ranging from 500–999, 5,010–9,990, etc. has very weak significance, but is justified on the basis of uniformity. Again, in some instances it may be desirable to report concentration to the nearest 0.1 mg/l up to 9.9 mg/l, and especially between 0.1 mg/l and 0.9 mg/l. These recommendations are based on the assumption that the net sediment can be weighed to the nearest 0.0001 g and the water-sediment mixture for a 300-ml or 400-ml sample can be weighed to the nearest 1 g.

With respect to size distribution, the goal should be to publish the percentage by weight of material in each fraction to the nearest whole percent. In some cases, where several critical fractions may contain 2% or less, it may be useful to report to the nearest 0.1%. Whenever practicable, the net quantity of sediment in a given fraction should be weighed to the nearest three significant figures. With balances weighing to the nearest 0.0001 g, the weight of quantities less than 0.0100 g can be determined only to two significant figures.

26. Concentration.—The common unit for expressing suspended-sediment concentration is milligrams per liter computed as 1,000,000 times the ratio of the dry weight of sediment in grams to the volume of water-sediment mixture in cubic centimeters. Other units, such as parts per million or percent by weight, have been used in the past to express suspended-sediment concentration. In the laboratory, it is more convenient to obtain the weight of the water-sediment mixture than to obtain its volume. When this is done, the following formula is used to calculate concentration:

Concentration, in milligrams per liter

$$= A \frac{\text{weight of sediment} \times 1{,}000{,}000}{\text{weight of water-sediment mixture}} \qquad (3.11)$$

in which the factor, A, is given by Table 3.6 and is based on specific weights of water and sediment of 1.000 g/cm³ and 2.65 g/cm³, respectively. Note that the ratio, weight of sediment × 1,000,000, to the weight of water-sediment mixture, defines parts per million of sediment in a sample as computed on the forms used in the laboratory (Fig. 3.62) and as reported by the United States Geological

TABLE 3.6.—Factors A in Eq. 3.11 for Computation of Sediment Concentration in milligrams per liter When Used with 10^6 Times Ratio of Weight of Sediment to Weight of Water-Sediment Mixture[a]

$\frac{\text{Weight of sediment}}{\text{Weight of sediment and water}} \times 10^6$ (1)	A (2)	$\frac{\text{Weight of sediment}}{\text{Weight of sediment and water}} \times 10^6$ (3)	A (4)
0- 15,900	1.00	322,000-341,000	1.26
16,000- 46,900	1.02	342,000-361,000	1.28
47,000- 76,900	1.04	362,000-380,000	1.30
77,000-105,000	1.06	381,000-398,000	1.32
106,000-132,000	1.08	399,000-416,000	1.34
133,000-159,000	1.10	417,000-434,000	1.36
160,000-184,000	1.12	435,000-451,000	1.38
185,000-209,000	1.14	452,000-467,000	1.40
210,000-233,000	1.16	468,000-483,000	1.42
234,000-256,000	1.18	484,000-498,000	1.44
257,000-279,000	1.20	499,000-513,000	1.46
280,000-300,000	1.22	514,000-528,000	1.48
301,000-321,000	1.24	529,000-542,000	1.50

[a]Based on density of water of 1.000 g/ml, plus or minus 0.005 in the range of temperature 0°C–29°C, dissolved solids concentration between 0 ppm and 10,000 ppm, and the specific gravity of sediment of 2.65.

Survey through the 1966 water year. Total sediment concentration has sometimes been expressed on a volume basis, especially for sands and other coarse material. Such volume measurements are generally not reliable or consistent because of variation in particle characteristics and compaction. Compaction is usually a function of pressure, vibration, and other sample acquisition and handling methods.

Each of several methods for separating sediment from the water in a sample has advantages and disadvantages. The two most commonly used (evaporation and filtration methods) are treated further herein. Filtration is faster than evaporation as long as the amount of sediment in a given sample is not too large. For samples having a low concentration of sediment, the evaporation method requires a correction if the dissolved solids content is high. Usually, the evaporation method is used where the sediment concentration of samples frequently exceeds 2,000 mg/l–10,000 mg/l. The lower limit applies when the sample consists mostly of fine material (silt and clay) and the upper limit when the sample is mostly sand. The filtration method is therefore best for lower concentrations.

Filtration Method.—The filtration method utilizes a Gooch crucible with one of several suitable types of filter material. The crucible is a small porcelain cup of

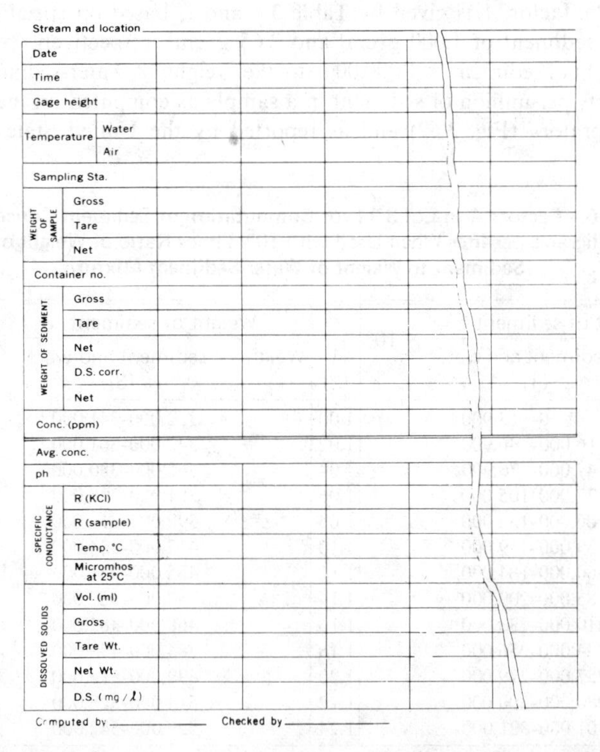

Stream and location ____________________

Date							
Time							
Gage height							
Temperature	Water						
	Air						
Sampling Sta.							
WEIGHT OF SAMPLE	Gross						
	Tare						
	Net						
Container no.							
WEIGHT OF SEDIMENT	Gross						
	Tare						
	Net						
	D.S. corr.						
	Net						
Conc. (ppm)							
Avg. conc.							
ph							
SPECIFIC CONDUCTANCE	R (KCl)						
	R (sample)						
	Temp. °C						
	Micromhos at 25°C						
DISSOLVED SOLIDS	Vol. (ml)						
	Gross						
	Tare Wt.						
	Net Wt.						
	D.S. (mg/l)						

Computed by ____________ Checked by ____________

FIG. 3.62.—Laboratory Form, Sediment Concentration Notes, Depth-Integrated Sample (Comprehensive Form) for Tabulation of Results of 10 Bottles, Adapted from Guy, 1969, p. 15

approx 25 ml capacity with a perforated bottom that is easily adapted or connected to various kinds of vacuum systems. The whole sample may be filtered while in a more or less dispersed state; but usually the sediment is allowed to settle to the bottom of the sample bottle, the supernatant liquid is decanted, the water is filtered out of the remainder as the sediment is washed onto the filter, and then the crucible is dried in an oven, cooled in a desiccator, and weighed.

The use of the crucible in the filtration method has several advantages over the use of the evaporating dish in the evaporation method, as follows: (1) It is lighter in weight and consumes less oven and desiccator space; (2) its tare weight is less likely to change during weighing due to sorption of moisture from the air; and (3) dissolved material passes through the crucible and thus eliminates the need for a dissolved-solids correction.

A commercial glass-fiber filter disk, e.g., Corning No. 934-AH, is satisfactory for most types of sediment. For some fine-grained sediments, a glass-fiber disk in conjunction with an asbestos mat is sometimes used. The crucible with this extra mat is prepared by placing the glass-fiber disk in the crucible while vacuum is applied and then pouring an asbestos slurry on top of the disk, also while vacuum is applied. The somewhat coarse asbestos mat retains much of the sediment that would ordinarily clog the glass-fiber disk.

If glass-fiber filters are not available, then it will be necessary to use the pure asbestos mat, even though some very fine sediment may be lost. The simplest way to prepare an asbestos mat is to prepare a slurry of shredded asbestos and distilled water and then to pour a small volume of this into the crucible while vacuum is applied. The resulting mat should be rinsed with distilled water while vacuum is still applied, after which the crucible is oven dried, cooled in a desiccator, and tare weighted. Because of the uniformity of the commercial mats, it is not necessary to obtain the tare weight each time a given crucible is used. Sometimes, especially for low concentrations of coarse sediment, it is possible to use the same mat two or more times, thus saving preparation time. Reused in this manner, the gross weight of the prior use becomes the tare weight for the next sample.

When the filtration method is used to determine suspended-sediment concentration, the laboratory should always be concerned over the relative amount of very fine sediment lost in the filtrate. Informal tests of filters previously described used in the United States Geological Survey laboratories seldom show losses of more than 5% for river samples, even those having mostly clay. The highest relative loss would logically occur for samples having low concentrations and a high percentage of well dispersed colloidal clay. With such low concentrations, the weighing error (±0.001 g) will be the dominant source of error.

Evaporation Method.—The evaporation method offers some advantages in simplicity of equipment and technique over that offered by the filtration method. In the evaporation method, the sediment is allowed to settle to the bottom of the sample bottle, the supernatant liquid is decanted, the sediment is washed into an evaporating dish, and then dried in an oven. The settling is accomplished by allowing the container to sit undisturbed for several hours or days, unless the sample contains naturally dispersed clay that requires special flocculating procedures. The supernatant liquid is decanted carefully to a point where a specific quantity of liquid (about 20 ml) is left with the sediment. Care must be taken so that none of the sediment in the bottom of the container will be removed. The sediment and remaining liquid are then washed into the evaporating dish with distilled water. The contents of the dish are then dried in an oven at a temperature of 5°C or 10°C lower than the boiling point. The dishes must be maintained in a dry state by use of a desiccator prior to and during weighing.

The natural dissolved solids in the supernatant liquid at the bottom of the sample container are retained with the dried sediment. Assuming, for example, that an aliquot of about 20 ml of native water is included and that the sample originally contained 400 g of water-sediment mixture, then a dissolved solids correction should be made to the net evaporated material when the dissolved solids content exceeds 200 mg/l and the sediment content is less than 200 mg/l. The dissolved solids content of a stream usually does not change appreciably from day to day for normal flow, but often diminishes rapidly with increasing runoff. Such increased runoff and low dissolved-solid periods can be determined by a study of the gage heights recorded on the sample containers and recorded on the forms discussed as follows. A single dissolved-solids concentration determination is usually adequate for several days of samples when the streamflow is uniform. This can be accomplished by compositing nearly equal small volumes of sediment-free water withdrawn from each sediment observation during the composite period. Note that each sediment observation is likely to consist of two or more sample bottles.

Errors in the evaporation method can usually be attributed to: (1) Loss of particles in the decanting of supernatant liquid; (2) weighing errors when concentrations are very low; or (3) needed corrections for dissolved solids in the supernatant liquid at the bottom of the sample container. These possibilities for error suggest that the evaporation method may require somewhat closer supervision than for the filtration method.

Displacement Method.—For special studies, where the amount of sediment is considerable and consists mostly of sand sizes, the displacement method may be used with the advantage of avoiding the problems of oven drying the sediment. The principle, as the name suggests, is similar to that used to determine specific gravity, in that the difference in weight between a volume of water and an equal volume of water plus solids is required. For the displacement method to determine the weight of solids in a sample requires that the specific gravity of the sediment be known. The formula for computation is

$$W_s = \frac{W}{1 - \dfrac{\gamma}{\gamma_s}} \qquad (3.12)$$

in which W_s = the weight of solids; W = the difference in weight of vessel with water only and water plus sediment; γ_s = the specific weight of solids; and γ = the specific weight of water. When $\gamma_s/\gamma = 2.65$, then $W_s = 1.606\ W$.

As mentioned in *Report 4* (Interagency Committe, 1941, p. 167), the accuracy of the displacement method depends on the sensitivity of the balance in relation to the total weight of solids in the sample, as well as the accuracy of the estimate for specific weight of the solids. The method is therefore limited mostly to samples from laboratory sediment experiments, especially where it is not desirable to oven dry the sediment.

Forms for Logging Sample and Laboratory Data.—Efficient and systematic operation for determining sediment concentrations requires the use of standard forms for logging samples into the laboratory, recording data during the laboratory operations, and for making the necessary computations. A form useful for achieving these objectives for the determination of depth-integrated suspended-sediment concentrations in the sediment laboratory is shown in Fig. 3.62 which has space for 10 bottles or samples. Each bottle should be recorded in the chronological order of sampling at the time of sample weighing. Appropriate notes can be made on the form if one or more samples are later withdrawn for particle size analysis or combined for the determination of a "composite" concentration. If more space is needed for remarks, reference by number should be made to the back of the sheet; expanded notes can be retained on the front by using space ordinarily used to record data for other bottles.

Some laboratories have found it desirable to redraft Fig. 3.62 where lines ph through D.S. (milligrams per liter) are not used, so that two sets of 10 bottles (lines Date through conc. ppm) can be accommodated on each page. Instead of a "comprehensive form," this is then called the "short form." For streams commonly having concentrations greater than 15,900 ppm, another line may be added to the form for space to convert parts per million to milligrams per liter (Eq. 3.11; Table 3.6).

With the foregoing in mind, specific laboratory procedures for determining sediment concentration may be developed depending on the number and kind of samples analyzed. Generally, though, samples should be inspected for loose caps on bottles, arranged in chronological order, and stored in a cool dark room to inhibit the growth of microorganisms. In the procedure where the supernatant liquid is withdrawn prior to washing the sediment into either the evaporating dish or the filtration crucible, it is desirable to store the weighed samples on a convenient rack or table for several hours to assure complete settlement of sediment prior to siphoning the supernatant liquid. At this point, samples to be analyzed for particle size distribution can easily be separated from those to be analyzed for concentration, because they are arranged so that the relative quantity of sediment can be seen and the relation to other size analyses during the record can be noted.

27. Particle Size Data and Methods.—Particle size information regarding sediment that may be eroded, is being transported, or has been deposited is useful in the application of predictive theory on erosion, transport, and deposition rates and quantities. Vice, Guy, and Ferguson (1968) computed the likely amount of erosion and deposition within a drainage area on the basis of the amount and size of sediment transported from the basin and the particle size of the eroding soils in the basin. In transport, the particle size determinations of both material on the bed and in suspension were used by Colby and Hubbell (1961) to compute the rate of transport in the unsampled zone. In deposition, particle size is one of the most important aspects in the evaluation of the amount of space a given weight or quantity of transported sediment will occupy. These examples illustrate some of the application of particle size data—others may involve sorption of dissolved constituents and radioactive materials, water treatment and other urban hydrology problems, stream morphology and hydraulic studies, and the distribution of suspended sediment in the cross section of the stream.

As indicated in Chapter II, Section B, the sizes of sediment particles vary over extremely wide ranges and therefore the size of an individual or some small number of particles has little or no meaning. Sediments are therefore measured in very large numbers and grouped into specific, but arbitrary, size classes or grades, as defined in Table 2.1 and as recommended by Lane (1947) of the subcommittee on Sediment Terminology of the American Geophysical Union. Sediment particles not only vary widely with respect to size, but also with respect to specific weight and shape; therefore, different particles of a given physical size will behave in the hydraulic environment as though they are larger or smaller, depending on how their shape and specific weight vary from the defined size class.

Because of the wide range of particle characteristics, particle size usually needs to be defined in terms of the method of analysis. Large sizes, including boulders, cobbles, and perhaps gravel, are measured directly by immersion, circumference, or diameter. Intermediate sizes, including gravel and sands, are measured semidirectly by sieves. Small sizes, including silts, clay, and perhaps sands, are measured hydraulically by sedimentation techniques, i.e., by observing their settling characteristics in water. There is a practical upper limit on the size of a particle that can be analyzed in a laboratory by settling, because the settling chamber must be at least 10 times the diameter of the particle (Fig. 2.11). The fall velocity of the particles in the chamber is converted to the standard fall diameter defined in Chap. II, Section B. The nominal diameter is only approximated when the particles are measured by hand or by sieve. As indicated by B. C. Colby (Interagency Committee, 1957a), the relationship of the nominal diameter to sedimentation diameter becomes a measure of the effect of shape, roughness, and specific gravity on the settling velocity of a particle. Note that the relative effect of shape varies with size and that the effect of shape is more significant with the sieve than with the sedimentation type of analysis, at least for the smaller sizes.

This leads to the fact that there are essentially only two kinds of particle size determinations: (1) Those using direct measurement; and (2) those using sedimentation methods. The direct methods would include immersion or displacement volume, some direct measure of circumference or diameter, or semidirect measure of diameter by sieves. Sedimentation methods most commonly employ the pipet, the bottom-withdrawal (BW) tube, the hydrometer, and the visual accumulation (VA) tube. The VA tube, used only for sands, operates as a stratified system where particles start from a common source and become stratified at the bottom of the tube according to settling velocities. The pipet and BW tube, used only for silts and clays, operate as dispersed systems where particles begin to settle from an initially uniform dispersion. The size distribution for the VA tube is determined from data on the amount of sediment accumulated at the bottom of the tube as a function of time. Size distribution by the pipet method is determined by measuring the concentration of sediment at specific intervals of depth and time, such that specific larger sizes will have settled past the pipet withdrawal zone in the settling medium. For the BW tube method, the distribution is obtained from the quantity of sediment remaining in suspension after various settling times when the coarser sizes and heavier concentrations are withdrawn at the bottom of the tube.

Sediment-Sample Characteristics.—Before proceeding into a discourse of the methods of particle size analysis, it is desirable to recognize some of the quantitative and qualitative aspects of sediment samples and to consider methods for their handling and analysis. The quantity of sediment in a sample may range from perhaps 100 representative boulders in a mountain stream measured in situ to 0.05 g of fine sand in a suspended-sediment sample suitable for measurement in a VA tube. Usually, only particles smaller than coarse gravel ($<$ 64 mm) are brought to the laboratory because of the bulk and weight of a suitable representative sample of the larger particles. Bed material samples consisting mostly of sand may range from a few grams to a kilogram and may be stored wet or air dried. Samples should be vapor proofed to prevent drying if they contain sufficient clay to cause the particles to bind together if dried.

Suspended-sediment samples are usually obtained with a sampler nozzle 3/16 in. or 1/8 in. in diameter, as explained in Section A of this chapter, and therefore seldom contain particles coarser than fine gravel (4.0 mm); usually, much of the sediment is in the silt and clay size range. Such suspended-sediment samples, as well as wet bed material samples, should be protected from light and preferably stored at a cool temperature to prevent organic growth that would interfere with analysis. As already indicated, clay particles in suspended-sediment samples are sometimes flocculated together by ions in the native water and require special treatment to insure that the particles fall singly during the analysis, as explained subsequently.

Also note that a sample containing silt, clay, and coarser material will often require analysis by two or more methods because of the limitation on the range of sizes that can be analyzed by a given method. A sample of coarse bed material may require hand measurement of a few large particles, the remainder of the sample sieved down to 1 mm or 2 mm, the sands finer than this analyzed by VA tube, and then the remaining silt and clay analyzed by pipet. Most suspended-sediment samples require that the sand (> 0.062 mm) be removed for analysis by sieve or the VA tube before the pipet or BW tube can be used for analysis of the silt and clay part.

Because suspended-sediment samples often contain a very small quantity of sediment, it is usually feasible to analyze for size only those samples obtained during periods of relatively high sediment discharge; even then, considerable judgment will be required by both the field and laboratory technicians to insure that sufficient sediment for a given observation is available for size analysis. On the other hand, some streams may at times contain too much sediment in even a single bottle for either the pipet or BW tube analysis, in which case it will be necessary to wet split the sample one or more times to obtain the desired quantity. Table 3.7 gives a guide to the range of sizes, analysis concentration, and weight of sediment for the sieves, VA tube, pipet, BW tube, and hydrometer methods and

TABLE 3.7.—Recommended Size Range, Analysis Concentration, and Quantity of Sediment for Commonly Used Methods of Particle Size Analysis

Method of analysis (1)	Recommended size range, in millimeters (2)	Desirable range in analysis concentration, in milligrams per liter (3)	Range, in optimum quantity of sediment, in grams (4)
Sieves	0.062-32	–	0.05[a]
VA Tube[b]	0.062- 2.0	–	0.05 -15.0
Pipet	0.002- 0.062	2,000-5,000	1.0 - 5.0
BW Tube[c]	0.002- 0.062	1,000-3,500	0.5 - 1.8

[a] Based on use of 3-in. diam sieves and a median size of 0.5 mm or less.
[b] See Table 3.8 for more detail.
[c] If necessary, may be expanded to include sands up to 0.35 mm, the accuracy decreasing with increasing size—the concentration and size increased accordingly.

should facilitate decisions on how many sample bottles are needed, how much splitting is required, or which methods are best to use for a given stream and sample.

It may be necessary to remove bothersome extraneous organic materials found in some samples. This is accomplished by adding about 5 ml of 6% solution of hydrogen peroxide for each gram of dry sample in 40 ml of water. It must be stirred thoroughly and covered for 5 min or 10 min. Large fragments of organic material may then be skimmed off, if it can be assumed that they are free of sediment particles. If oxidation is slow, or after it has slowed, the mixture is heated to 93°C, stirred occasionally, and more hydrogen peroxide solution added as needed. After the reaction has stopped, the sediment must be carefully washed two or three times with distilled water.

It is not expected that suspended-sediment samples will ordinarily contain too much sand for analysis (15 g for the VA tube). It is likely though that samples will frequently contain too much fines (silt and clay) for analysis, at least for the pipet and BW tubes. The best procedure then is to separate the sand from the fines by wet sieve (after the organic material is removed, if necessary) and then to wet split the fines as necessary with a Jones-Otto or BW splitter. There is no known method for obtaining a "true" split for relatively small quantities of sand-sized particles, but for the fines, either the Jones-Otto or the BW splitter is adequate. The Jones-Otto splitter consists of a hopper with an even number of uniformly spaced slots in the bottom. Alternate slots are open to opposite sides of the hopper so half the sample is discharged on each side. The BW splitter is a bottom-withdrawal tube (Fig. 3.66) to which an inverted Y is attached by a short length of rubber tube.

Direct Measurement.—Direct measurement techniques for sediment particle size are usually limited to large particles that cannot be readily brought to the laboratory. Direct methods for particles as small as sands are laborious and therefore suitable only for special situations. Measurements in the field for boulders and cobbles are usually made with a caliper type apparatus for diameter or a tape for measuring circumference. If the particles are not close to spherical, then it may be desirable to determine the shape factor, in which case the three mutually perpendicular diameters must be measured; the first or a must be the largest, and c must be the smallest. Shape factor is then defined as

$$\mathrm{SF} = \frac{c}{\sqrt{ab}} \qquad (3.13)$$

A thorough review of mathematical expressions for defining the sphericity of regular geometrical shapes of particles has been made by Rao (1970). Sphericity expressions require that the surface area of the particles be known, for which reliable procedures are apparently not available.

The three mutually perpendicular diameters used for Eq 3.13 are also used to obtain an indication of the nominal diameter, in which

$$d_n \sim \frac{a + b + c}{3} \qquad (3.14)$$

The true nominal diameter of a large particle can best be determined by the volume of fluid, V, that it displaces which in turn can readily be converted to an equivalent sphere by the equation

$$d_n = 1.24\ V^{1/3} \qquad (3.15)$$

The measurement can quickly be made with a water-filled cylinder having a volumetric scale on the side. For reasonable accuracy, the diameter of the immersion cylinder should not be more than about twice the nominal diameter of the particle; therefore, several diameters of cylinders will be required for use of the immersion technique in the field.

Ritter and Helley (1968) have described a less laborious method of inplace measurement of large particles that makes use of the Zeiss Particle-Size Analyzer. With this method, a photograph from 1 m or 2 m above the stream bed is made at low flow with a tripod supported 35 mm camera—the height depending on the size of the bed material and the lens system. A reference scale, e.g., a meter stick must appear in the photograph. The photographs are printed on the thinnest paper available. In the laboratory, an iris diaphragm illuminated from one side is imaged by a lens onto the plane of a plexiglass plate. The photograph is put on this plate. By adjusting the iris diaphragm, the diameter of the sharply-defined circular light spot appearing on the photograph can be changed and its area made equal to that of the individual particles. As the different diameters are registered, a puncher marks the counted particles on the photograph. An efficient operator can determine the size of particles at the rate of about 2,000/h. Diameters can be registered cumulatively or individually on exponential or linear scales of size ranges. After the data are tabulated, the sizes registered on the counter of the particle size analyzer must be multiplied by the reduction factor of the photograph, which is calculated from the reference scale in the photograph.

As already indicated, direct measurement of smaller sizes, e.g., sands and silts, is seldom attempted because semidirect and sedimentation methods are reasonably reliable and can be accomplished much more effectively. This is not to say that some photographic or microscopic inspections and records are not desirable for some special situations. An example of a comparison of the photographic with other kinds of analysis for sands used in flume experiments is given by Guy, et al. (1966). Microscopic methods may be the best where precise measurements are needed, especially in the development of other indirect methods (Interagency Committee, 1943, 1957b) but should be avoided for routine analysis because the methods still involve a certain amount of "art" in that results by different workers are not always the same.

Semidirect Measurement.—Sieving is considered a semidirect method of particle size measurement because it does not divide the particles precisely for at least four reasons: (1) Because of irregularities in shape, particles larger or smaller (larger for cylinders and smaller for disks) than the spherical equivalent diameter may pass through the sieve openings; (2) inaccuracies in size and shape of the sieve openings; (3) the sieving operation is for a finite time and therefore some particles may not have an opportunity to pass a given sieve opening; and (4) smaller particles may cling to large ones, thereby changing the percentage of material reaching the sieves with smaller openings.

As indicated in Table 3.7 a minimum of about 0.05 g of sand is needed for a sieve analysis—somewhat more if particles are large or if sieves larger than 3 in. in diameter must be used. For many suspended-sediment samples, the minimum quantity will not be available, in which case only the percentage of sediment coarser than 0.062 mm can be given. However, for bed material or other sediment deposits, the problem is usually one of obtaining a sample sufficiently small so as not to overload any particular sieve and yet be large enough to be representative of the stream bed or deposit. A fraction of a large moist sample may be obtained directly from its container with a thin-walled tube by inserting the tube to the bottom of the sample container at two or three locations. A large dry sample can be reduced by the well-known quartering method or by a sample splitter. If the large sample contains a considerable quantity of material coarser than the sand sizes, then these large sizes must usually be removed by hand or a coarse sieve, and their relative amount determined on the basis of the whole sample.

A wet-sieve method may be more desirable than a dry-sieve method. The recommended wet-sieve method uses a technique that keeps the sieve screen and sand completely submerged. The equipment may consist of six or more 10-cm ceramic dishes, a set of 3-in. (7.5 cm) sieves, and a thin glass tube. All sieves are washed with a wetting solution (detergent) and then rinsed gently with distilled water so that a membrane of water remains across all openings. The first or largest sieve is immersed in a ceramic dish with distilled water to a depth of about 1/4 in. (1/2 cm) above the screen. If the surface tension of the water across the openings is sufficient to trap a pocket of air beneath the screen, the thin tube is used to blow out a small group of the membranes near one edge of the screen. This will allow the air to escape if the open holes are kept above the water until the rest of the screen is immersed. The sediment is washed onto the wet sieve and agitated somewhat vigorously in several directions until all particles smaller than the sieve openings have a chance to fall through the sieve. Material passing the sieve with its wash water is then poured onto the next smaller size sieve in another crucible. Particles retained on each sieve and those passing the 0.062-mm sieve are transferred to containers that are suitable for drying the material and for obtaining the net weight of each fraction.

The dry-sieve method is less laborious than the wet-sieve method because a mechanical shaker can be used with a nest of sieves for simultaneous separation of all sizes of interest. It requires only that the dry sand be poured over the coarsest sieve and the nest of sieves shaken for 10 min on a shaker having both lateral and vertical movements.

For either the wet or dry-sieve methods, it is essential that order be established throughout the sieving operation and the results suitably recorded on a form that is either a part of the form used for the VA tube, the pipet, the BW tube, or the hydrometer methods, or a separate sheet.

The Coulter Counter has sometimes been used to obtain a semidirect measurement of relatively fine sediments in special investigations (Fleming, 1967). The Coulter Counter gives an indication of the physical size whereas all other techniques for measurement of the fine sediments indicate fall diameter. The equipment is relatively expensive, sample preparation may be more difficult, and the results may be more difficult to interpret than for the pipet or BW method. More specifically, the Interagency Committee (1964, p. 69) states "Electronic sensing is a relatively accurate method of determining the size distribution of sediments between 1 and 160 microns in diameter," and "... is a valuable laboratory instrument for determining size distribution as a standard to compare with other less accurate but faster methods, ... "

Indirect Measurement.—Methods of indirect measurement of sediment particle size will be limited herein to those involving liquid suspensions and fall velocity, e.g., the VA tube, pipet, BW tube, and hydrometer. A determination of fall velocity by one of these methods, from which a theoretical fall diameter may be obtained, is much easier than a physical determination of the particle size, shape, and weight needed to predict the behavior of the particles in a body of water. The fall velocity-fall diameter concept thus provides a desirable expression of the hydraulic characteristics of the very large number of particles in a water sample.

Most particle size analyses by sedimentation methods should be accomplished so that floccule formation of the clay and perhaps the silt-sized particles is minimized, and thus the results would be in terms of "standardized" ultimate sizes. To accomplish this, the native water is removed from the sample as completely as possible. Then the sample is washed into a soil dispersion cup, diluted to about 300 ml with distilled water, mixed for 5 min with a milkshake mixer, and then poured over a sieve with mesh openings of 0.062 mm and washed with distilled water to separate the sand from the silt and clay. To insure complete dispersion of the clay particles during the sedimentation analysis of the silt-clay part by the pipet, BW tube, or hydrometer methods, 1 ml of dispersing agent must be added for each 100 ml of the desired suspension; it must be mixed again for 5 min, and then transferred to the settling tube. The dispersing agent is usually made (Guy, 1969, p. 29) by dissolving 35.70 g of sodium hexametaphosphate and 7.94 g of sodium carbonate in distilled water and diluting to 1-ℓ volume. A determination of a dissolved solid's correction should be made each time a new solution of dispersing agent is prepared for use when pipet or BW tube withdrawals are evaporated.

Though most analyses should be made to yield the ultimate particle size, it is often desirable to make some comparative analyses in the native water so that "fall velocity" characteristics may better resemble those in the stream. Fluvial sediments are transported in natural waters having a wide variation in kind and concentration of chemical constituents that cause the discrete particles to form floccules to varying degrees. This will not be a problem in many streams that are low in dissolved solids or contain sodium as the dominant cation. Obviously, if a native-water analysis is to be made of the fine sediment, then the instructions given in the previous paragraph, concerning the use of a dispersing agent and mixing times, are not applicable because the sediment should be disturbed as little as possible. The main difficulty with trying to obtain particle size results for natural conditions is that both the concentration of sediment and the chemical character of the native water which affects flocculation, and thus settling velocity, changes with time and location in the drainage basin (Guy 1969, pp. 18–23).

VA Tube Method

The visual accumulation tube is a fast, economical, and reasonably accurate means of determining the size distribution of a sediment sample in terms of the fundamental hydraulic properties of the particles and the fall velocity or fall diameter. It is especially adapted for samples composed of sand, as shown in Table 3.7. The fine material less than 0.053 mm is removed by either wet sieving or by sedimentation methods, and analyzed by either the pipet or BW method. In a few instances, sieving must be employed to remove and measure a few particles that may be too large (greater than 2.0 mm) for the VA tube. Difficulty may arise in use of the VA tube on samples that contain considerable organic materials, e.g., root fibers, leaf fragments, and algae. Because of the fixed calibration of the VA tube, samples that contain large quantities of heavy or light minerals, e.g., taconite or coal, should be analyzed in both the VA tube and sieves so that their sieve and sedimentation diameters may be compared.

Equipment for the VA tube method of analysis consists primarily of the special settling tube and the recording mechanism, in addition to the usual laboratory equipment for sediment investigations. In somewhat more detail (Fig. 3.63), the device consists of: (1) A glass funnel with a stem about 25 cm long; (2) a rubber tube connecting the funnel and the main settling tube; (3) a manual pinch valve that serves as a "quick acting" valve; (4) removable glass settling tubes each having a different sized collector at the bottom; (5) a tapping mechanism that strikes against the glass tube to keep the accumulation of sediment uniformly packed; (6) a special recorder consisting of a chart cylinder that rotates at a constant rate; (7) a tracking carriage that can be moved vertically by hand on which is mounted the recording pen and an optical instrument for locating the surface of the accumulation; and (8) a recorder chart that is a printed form incorporating the fall-diameter calibration.

Sand particles should be in such a condition that the grains will fall as

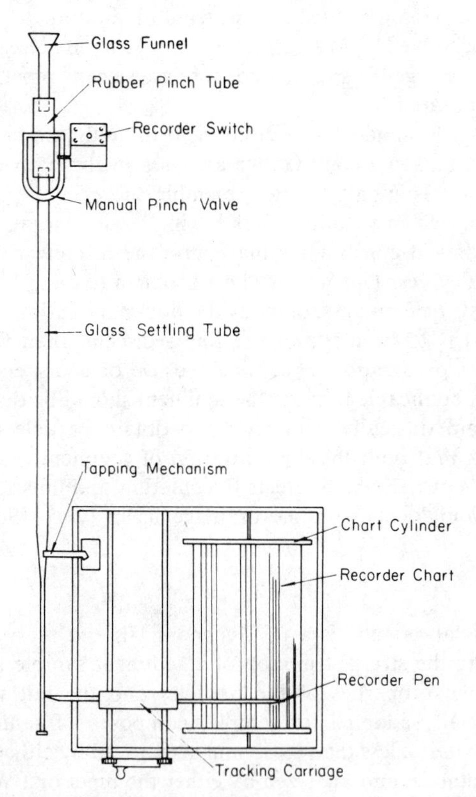

FIG. 3.63.—Sketch of Visual Accumulation Tube and Recording Mechanism

individual particles in the settling tube and therefore should be thoroughly wet and free of attached clay particles or air bubbles before analysis. They should be contained in 40 ml, or less, of water at a temperature within ±2°C of that of the water in the VA tube.

As already mentioned, the VA tube method employs the stratified system of settling as contrasted with the dispersed system of the pipet or the BW tube. The stratified system is one in which the particles start falling from a common source and become stratified according to settling velocities as they fall and are deposited. At a given instant, the particles coming to rest at the bottom of the tube are of one "sedimentation size" and are finer than particles that have previously settled out and are coarser than those remaining in suspension. Many details covering the equipment and methods of operation are not discussed herein because a clear and concise operator's manual in Interagency Committee (1958) is furnished with each of the apparatus. Likewise, the details of the development and calibration are contained in a report by the Interagency Committee (1957b).

The chart on which the trace of surface of the accumulated sediment is recorded is supplied by the manufacturer. A combined form has been developed by the United States Geological Survey that gives space on the back of the chart to record sieve analysis of sizes too large for the VA tube and pipet analysis for the fine material in the sample. The combined form is desirable because it keeps all data for a given sample together and facilitates computations when several kinds of analyses are made for a wide range of sizes in a given sample.

Results from the VA tube are best if the total height of accumulation in the bottom of the tube is between 4 cm and 12 cm. Table 3.8 gives a guide to the selection of an appropriate size of VA tube on the basis of sample weight, volume, and maximum particle diameter. The usable limits of the respective tubes overlap, so if a satisfactory size is not selected the first time, the sample can be rerun in another size of tube.

The VA tube analysis results in a continuous pen trace of the depth of sediment deposited as a function of time. The chart is calibrated so that, for a given temperature, the relative amount of each size in terms of fall diameter and percentage finer than a given size can be determined with a graduated scale laid on the chart so that 100 is on the base line and zero is on the total accumulation line. If some of the material finer than that analyzed in the VA tube was removed prior to the VA tube analysis, e.g., 30% of the original sample, then the 30 on the

TABLE 3.8.—Guide to Size Selection of VA Tube

Settling Tube		Sample Size		Approximate maximum particle size, in millimeters (5)
Length, in centimeters (1)	Diameter, in millimeters (2)	Dry weight, in grams (3)	Volume, in milliliters (4)	
120	2.1	0.05- 0.8	0.03-0.5	0.25
120	3.4	0.4 - 2.0	0.2 -1.2	0.40
120	5.0	0.8 - 4.0	0.5 -2.4	0.60
120	7.0	1.6 - 6.0	1.0 -4.0	1.00
180	10.0	5.0 -15.0	3.0 -9.0	2.00

scale is held on the total-accumulation line and direct readings of percentage finer made as aforementioned. Similarly, if coarse material has been removed, then the percentage coarser than the total is subtracted from the total of 100 and this number is held on the zero-accumulation line. The results of these readings in percentage finer than the given size are tabulated on the form for this purpose.

Pipet Method

The pipet method is considered the most reliable indirect method for routine use to determine the particle size gradation of fine sediments (less than 0.062 mm), especially for samples having small quantities of such fine sediment (see Table 3.7). In principle, the pipet method utilizes the Oden theory (Krumbein and Pettijohn, 1938, p. 115) and is organized to determine the concentration of a quiescent suspension at a predetermined depth as a function of settling time. Particles having a settling velocity greater than that of the size at which separation is desired will settle below the point of withdrawal after elapse of a certain time. The time and depth of withdrawal are predetermined on the basis of Stokes law

$$w = \frac{gd^2}{18\nu}\left(\frac{\gamma_s - \gamma}{\gamma}\right) \qquad (3.16)$$

in which w, g, d, ν, γ_s, and γ are, respectively, the settling velocity, acceleration of gravity, diameter of particle, kinematic viscosity of the liquid, specific weight of sediment, and specific weight of liquid.

Table 3.9 gives recommended times and depths of withdrawal to determine

TABLE 3.9.—Time of Pipet Withdrawal for Given Temperature, Depth of Withdrawal, and Diameter of Particles[a]

Temperature, in degrees Celsius	DIAMETER OF PARTICLE, IN MILLIMETERS					
	0.062	0.031	0.016	0.008	0.004	0.002
	Depth of withdrawal, in centimeters					
	15	15	10	10	5	5
	Time of withdrawal, in minutes and seconds[b]					
(1)	(2)	(3)	(4)	(5)	(6)	(7)
20	44	2 52	7 40	30 40	61 19	4 5
21	42	2 48	7 29	29 58	59 50	4 0
22	41	2 45	7 18	29 13	58 22	3 54
23	40	2 41	7 8	28 34	57 5	3 48
24	39	2 38	6 58	27 52	55 41	3 43
25	38	2 34	6 48	27 14	54 25	3 38
26	37	2 30	6 39	26 38	52 2	3 33
27	36	2 27	6 31	26 2	52 2	3 28
28	36	2 23	6 22	25 28	50 52	3 24
29	35	2 19	6 13	24 53	49 42	3 19
30	34	2 16	6 6	24 22	48 42	3 15

[a]The values in this table are based on particles of assumed spherical shape with an average specific gravity of 2.65, the constant of acceleration due to gravity = 980 cm/s^2, and viscosity varying from 0.010087 cm^2/s at 20°C to 0.008004 cm^2/s at 30° C.

[b]Data in Col. 2 are in seconds.

concentrations of sediment finer than each of six sizes for a range of water temperatures. Values given for the standard depths of withdrawal are: 15 cm for the 0.062-mm and 0.031-mm sizes; 10 cm for the 0.016-mm and 0.008-mm sizes, and 5 cm for the 0.004-mm and 0.002-mm sizes.

The pipet equipment described by Krumbein and Pettijohn (1938, pp. 165–167) is satisfactory when it is necessary to analyze only a small number of samples. However, to facilitate operations to analyze hundreds of samples each year, a more complicated apparatus shown in Fig. 3.64 is suggested. J. C. Mundorff was among the first to assemble and use such apparatus in a United States Geological Survey laboratory at Lincoln, Neb. The equipment shown in Fig. 3.64, except the vacuum bottle, but also including a mechanism for lowering and raising the pipet, is mounted on a movable carriage so that pipetting can be accomplished consecutively among several settling cylinders (500-ml or 1,000-ml cylinders) in accordance with the schedule determined from Table 3.9. Note that withdrawals can be made for all sizes given in Table 3.9 or that only withdrawals for fewer selected sizes need to be made when less detail on size distribution is needed. For many purposes, only the 0.031-mm, 0.008-mm, and 0.002-mm sizes are determined; this is true especially when a considerable percentage of the sample is sand analyzed by another method.

A 25-ml pipet is used that is equipped with a three-way stopcock and attached to sufficient tubing to allow the carriage to traverse the length of the rack. Pipetting is accomplished when the three-way stopcock is open through the tubing to the vacuum bottle maintained with a suction of about 1/5 of an atmosphere. A small adjustable screw clamp on a short length of tubing just above the pipet forms a constriction that helps to maintain a desired rate of withdrawal. Pipetting is stopped when the water level rises to the stopcock. An inverted Y-shaped glass tube is attached to the right stem of the three-way stopcock to join with a pressure bulb at one end and to a distilled water supply with a head of 1 m–1.5 m at the other end. The flow of distilled water is controlled by a water valve.

After a sample is drawn into the pipet, the pipet is removed from the cylinder,

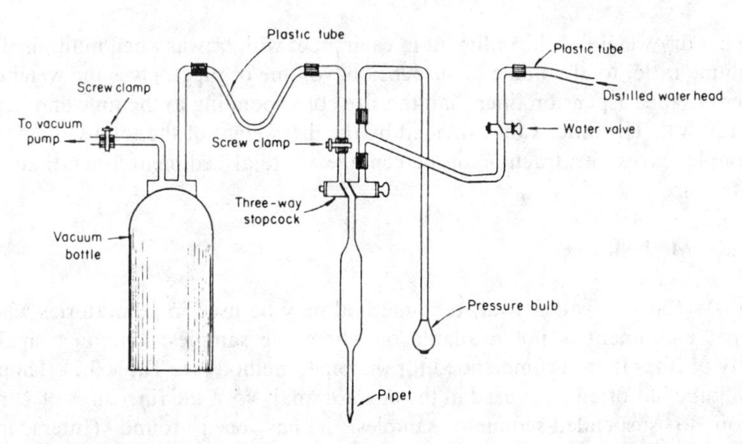

FIG. 3.64.—Relationship of Components for Pipeting among Several Settling Cylinders—Portion between Two Plastic Tubes is Mounted on Vertical Moving Rack and Horizontal Moving Carriage

the three-way stopcock plug is rotated 180°, and the sample is allowed to drain freely into an evaporation dish. The drainage can be accelerated by use of the pressure bulb. To insure complete removal of all sediment in the pipet, the distilled-water valve is then opened, and the pipet is washed out from the top. When the rinse is complete, it may be necessary to blow the remaining one or two drops of water from the tip of the pipet with the pressure bulb. At this time, the small quantity of the mixture that may have collected in the vacuum line above the three-way stopcock (resulting from overfilling the pipet) must be removed by allowing a small quantity of air to be sucked through the line from the open pipet, thus clearing the constriction that controls the vacuum on the pipet.

After the silt-clay fraction of a sample has been transferred to the settling cylinder, and before pipetting is begun, the temperature of the suspension, the depth of withdrawal, the settling time, and the weights of numbered containers for each withdrawal must be recorded on a form, e.g., the one shown in Fig. 3.65. The suspension is then stirred for 1 min with a hand stirrer of the plunger type illustrated in Krumbein and Pettijohn (1938, p. 167), and the stop watch is started when the stirrer is removed. The pipet for each withdrawal is filled in about 10 sec and then emptied into the respective evaporating dishes. One rinse of the pipet is added. The material in each evaporating dish is dried at 110°C, cooled in a desiccator, weighed, and the weight entered appropriately on the form.

The calculation of results from the pipet method requires the total weight of sediment in the sample. This can best be determined by two methods:

1. The weight of the silt and clay fractions can be determined from the mean concentration and volume of the pipet settling suspension. The average concentration of the suspension is determined by making a "concentration withdrawal" immediately after mixing. The weight of sediment in the suspension cylinder is then added to the weight of the sand, if any, which was determined separately.
2. Determine the dry weight of all sediment remaining in the suspension after all pipettings have been completed. To this add the dry weight of sediment in each pipet withdrawal and the dry weight of the sand fraction, if it was separated.

The net dry weight of the sediment in each pipet withdrawal when multiplied by the volume ratio, total volume of suspension/volume of pipet, gives the weight of sediment in the suspension finer than the size corresponding to the time and depth of withdrawal. The latter value, divided by the dry weight of the total sediment in the sample, gives the fraction or percentage of total sediment finer than the indicated size.

BW Tube Method

The BW (bottom withdrawal) tube method may be used in laboratories where the pipet equipment is not available or where the samples contain a smaller quantity of fines than recommended for the pipet method (see Table 3.7). Though the BW tube has often been used in the past for analysis of the finer sizes of sands common to suspended-sediment samples, it has been found (Interagency Committee, 1953) that such sand is usually a source of error to the whole analysis. Therefore, it is recommended that the sands be removed and analyzed by the VA tube, as described previously.

Special equipment beyond that ordinarily found in the sediment laboratory would consist of the BW tube or tubes with adequate provisions for mounting. The following specifications together with Fig. 3.66 should enable one to manufacture the tube.

The length of the BW tube is approx 122 cm, the inside diameter is 25 mm–26 mm, and the lower end of the tube is drawn down to 6.35 mm ± 0.25 mm ID with a wall thickness of 1.25 mm–1.75 mm and an angle of tapered portion of 60° ± 10° with the horizontal plane. The 2.0-cm diam straight nozzle at the lower end may be welded on instead of drawn from the tube if the weld is smooth on the inside.

The calibration of the tube is accomplished by marking off half centimeter

PARTICLE SIZE ANALYSIS, SIEVE-PIPET METHOD

File no. ____

ANALYSIS DATA	DISSOLVED SOLIDS	TOTAL SAMPLE DATA
Date ____ by ____	Volume ____ cc. dispersed / native	Stream ____
Portion used ____	Dish no. ____	Location ____
Disp. agent ____ cc.	Gross ____ gm.	Date ____ Time ____
Pipette suspen.: Sed. ____ gm.	Tare ____ gm.	G.H. ____ Sta. ____
Pipette suspen.: Vol. ____ cc.	Net ____ gm.	Composite: No. bottles ____
Pipette suspen.: Concen. ____ mg/ℓ	WEIGHT OF PORTION NOT ANALYZED	Composite: Wt. sample ____ gm.
Dry sand before dry sieve: Gross ____ gm.	Portion ____ Dish no. ____	Composite: Wt. sed. ____ gm.
Dry sand before dry sieve: Tare ____ gm.	Gross ____ gm.	Composite: Mean conc. ____ mg/ℓ
Dry sand before dry sieve: Net ____ gm.	Tare ____ gm.	Dis. solids ____ mg/ℓ
Weight: Sieve fract. ____ gm.	Net ____ gm.	Spec. conductivity ____
Weight: Sand fract. ____ gm.	Dis. solids ____ gm.	Other chem. qual. ____
Weight: Pipet fract. ____ gm.	Net ____ gm.	____
Weight: Silt-clay ____ gm.		
Weight: Total sed. ____ gm.		

Remarks:

SIEVE OR VA TUBE SIZES

Size, mm	4.0	2.0	1.0	0.50	0.25	0.125	0.062	Pan
Weight-gm.: Gross								
Weight-gm.: Tare								
Weight-gm.: Net								
% of total								
% finer than								

PIPET SIZES

Pipet no. Volume: Volume factor:

Size, mm	Conc.	0.062	0.031	0.016	0.008	0.004	0.002	Resid.
Clock time								
Temperature								
Fall distance								
Settling time								
Container no.								
Weight-gm.: Gross								
Weight-gm.: Tare								
Weight-gm.: Net								
Weight-gm.: D.S. corr.								
Weight-gm.: Net sediment								
Weight-gm.: Finer than								
% finer than								

FIG. 3.65.—Form Commonly Used for Recording Data in Pipet Method of Particle-Size Analysis

intervals from 5 cm near the bottom to 100 cm at the top. The exact location of the 10-cm line is determined as follows: On a completed tube, measure off 90 cm on the straight portion and ascertain the volume contained between these points. Add one-ninth of this volume to the tube. The bottom of the water meniscus will be the location of the 10-cm line. Any other 10-cm portion of the tube shall be equal to the volume below the 10-cm line ±2 ml. A pinch clamp is used on a short piece of rubber tubing at the lower end of the tube to provide a means of making quick withdrawals.

The recommended procedure for use of the BW tube is a combination of that reported by the Interagency Committee (1943, pp. 82–88) and more recent experience. Among other things, the more recent experience recognizes more specifically the limitations of the BW tube in sizing sand (Guy, 1969, p. 40, 47). Fig. 3.67 shows the form used to record data obtained in the BW tube analysis. After removal of the coarse fraction from the whole sample, the remaining fine fraction is split down, if necessary, to the desired 0.5 g–1.8 g of sediment, transferred to the BW tube, and diluted to about the 90-cm mark with distilled water. To insure complete dispersion of the sediment for the dispersed settling medium, add 5 ml of dispersing agent to the sample in the tube, transfer to the mixer, and mix for 5 min as for the pipet analysis. If the analysis requires the use of the "natural" settling medium, then only supernatant water from the sample can be used to handle the sediment and to dilute the mixture in the tube—the dispersing agent would not be used.

Before placing the tube in the rack to start the settling operation, mild mechanical mixing is accomplished by placing a cork in the upper end of the tube and tilting the bottom of the tube up about 10° from the horizontal. It must be

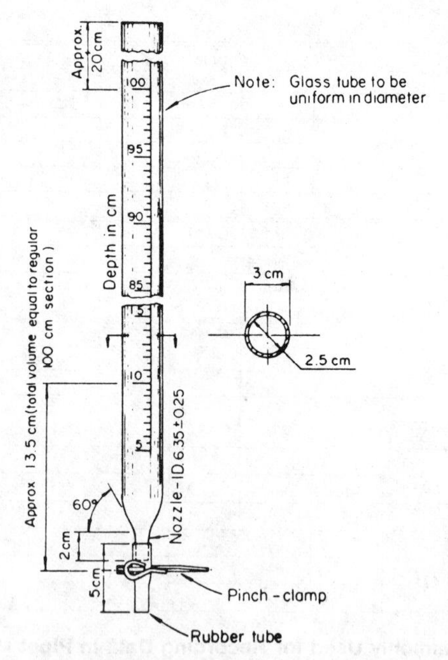

FIG. 3.66.—BW **(Bottom-Withdrawal) Tube (Interagency Committee, 1943)**

PARTICLE SIZE ANALYSIS, BOTTOM WITHDRAWAL TUBE METHOD

Tube no. ______ File no. ______

ANALYSIS DATA

Date ______ by ______ Portion used ______ Disp. agent ______ cc.

Dry sand before dry sieve:
- Gross ______ gm.
- Tare ______ gm.
- Net ______ gm.

Tube:
- Temperature ______ °C
- Sed. ______ gm.
- Vol. ______ cc.
- Concen. ______ mg/ℓ

Weight:
- Sand fract. ______ gm.
- BW fract. ______ gm.
- Total sed. ______ gm.

Composite:
- No. bottles ______
- Wt. sample ______ gm.
- Wt. sed. ______ gm.
- Mean conc. ______ ppm

DISSOLVED SOLIDS
- Volume ______ cc. dispersed / native
- Dish no. ______
- Gross ______ gm.
- Tare ______ gm.
- Net ______ gm.
- Concentration ______ mg/ℓ

TOTAL SAMPLE DATA
- Stream ______
- Location ______
- Date ______ Time ______ G.H. ______
- Station ______ Temperature ______ °F.
- Spec. cond. ______ pH cond. ______
- Remarks:

	Withdrawal no.	0	1	2	3	4	5	6	7	8	9	10
a	Clock time											—
b	Fall distance-cm	100	90	80	70	60	50	40	30	20	10	0
c	Settling time-min.											—
d	Volume-cc											
e	Container no.											
f	Weight-gm.: Gross											
g	Weight-gm.: Tare											
h	Weight-gm.: Net											
i	Weight-gm.: D.S. corr.											
j	Weight-gm.: Net sediment											
k	Weight-gm.: Cumulative											—
l	Depth factor 100/(b)	1.00	1.11	1.25	1.43	1.66					10.0	—
m	Sed. in susp. (k x l)											—
n	% sed. in suspension											—
o	Time for 100 cm (c x l)											—

FIG. 3.67.—Example of Laboratory Form Used to Record Data in Particle-Size Analysis by BW Tube Method

held in this position and shaken to wash all particles from the nozzle. The air bubble should then be at the nozzle end and all coarse particles should be distributed as uniformly as possible along the tube. The tube is then held in an upright position to allow the bubble to travel the full length of the tube (about 5 sec). The tube must be inverted from end to end in this manner for 1 min (3 min if the tube contains fine sand). At the end of this time, when the bubble is at the constricted end, the tube is turned immediately in an upright position and securely fastened to the stand. Time of settling is begun for the settling process when the bubble starts upward from the bottom. The cork should be removed after the bubble has reached the top.

Equal-volume fractions are usually withdrawn using time intervals chosen in such a way as to best define an Oden curve similar to Fig. 3.68. If tangents are drawn to the curve in Fig. 3.68 at any two points corresponding to times of withdrawal t_1 and t_2, and the tangents allowed to intersect the ordinate axis at W_1 and W_2, then the difference between the percentage, W_2 and W_1, will represent the percentage by weight of material in the size range determined by the settling times, t_1 and t_2. Obviously, the values, W_1 and W_2, are the percentages by weight of the sediment in the sample that are finer than sizes corresponding to t_1 and t_2, respectively.

Each withdrawal should represent a column height of 10 cm. However, the method can be varied considerably, whereby fractions of any desired depth and volume can be withdrawn as long as the particle size range is covered and enough points are obtained to define the Oden curve. If the preceding recommendations are followed concerning the use of the BW tube for the analysis of silt and clay only, then a suitable schedule would involve withdrawal times ranging from 3 min or 4 min to about 450 min. The schedule of the withdrawal times may be determined from Table 3.10 and the fall distance for each withdrawal. The last scheduled withdrawal time should be well past the settling time for definition by tangent of the 0.0195-mm size. At 20°C this should be about 520 min, and at 30°C about 420 min would be sufficient (for 10 cm).

The actual withdrawal is started 2 sec or 3 sec before the chosen withdrawal

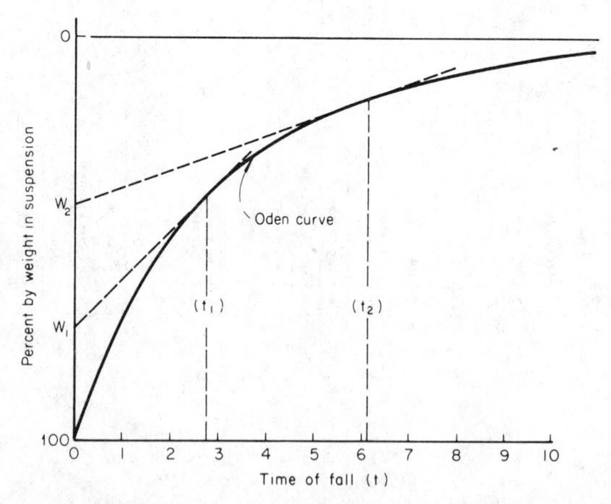

FIG. 3.68.—Oden Curve Showing Relative Amount of Sediment Remaining in Suspension with Time; Intersection of Tangent to Curve with Ordinate Represents Percentage by Weight of Sediment in Suspension in Time (t)

time. The pinch clamp is opened fully and then closed slowly as the last of the sample is being withdrawn. A full opening is required at the start so that a rush of water will clear any deposited sediment from the cone above the nozzle. Based on reasoning that material held on the meniscus does not fall in accordance with the Oden theory, the final withdrawal should be stopped while the meniscus remains in the nozzle of the tube. It must be remembered that the total settling time is not the time that the pinch clamp is opened, but the time it is closed.

Samples are withdrawn into a 100-ml graduate or flask to eliminate the possibility of losing any of the sample from splashing. The samples are then poured from the graduates or flasks into evaporating dishes and the sample containers washed with a stream of distilled water. The evaporating dishes are placed in the oven to dry at a temperature near the boiling point, but not hot enough to cause splattering by boiling. The small flask instead of the evaporating dish may be used for drying the sediment, if it is feasible to weigh and if cleaning is not too difficult. When the evaporating dishes or flasks are visibly dry, the temperature is raised to 110°C for 1 hr, after which the containers are transferred from the oven to a desiccator and allowed to cool. The weighing procedure is the same as that for sediment concentration determinations.

Since the temperature of the dispersion in the tube greatly affects the viscosity and settling velocity of the particles, the temperature reading of the suspension should be made between the 6th and 7th withdrawals. If the room temperature is not reasonably constant, more frequent readings will be necessary.

Computations on the form (Fig. 3.67) are described in reference to the lines lettered a–o. Entries on lines a–g, inclusive, are recorded for each withdrawal during the analysis. The net weight of sediment, line h, is obtained by subtracting the tare from the gross. The dissolved solids correction, line i, is determined on the basis of the withdrawal or evaporated volume and information as recorded in the dissolved solids block. The net sediment, line j, is then determined by subtracting i from h. The total sediment weight in suspension above each indicated depth is obtained on line k by adding the net weights cumulatively, starting with the last withdrawal. The depth factor, line l, has been obtained by dividing the fall heights, line b, into the standard or total depth of 100 cm. If the fall height is different from that shown in line b, then values different from those shown in line l must be used. The depth factor, l, is then multiplied by the cumulative weights, k, reducing them to the weight, m, that would be present in a 100-cm depth at the same average density. The percentage of sediment in suspension, n, is obtained as a ratio of sediment in suspension, m, to the total sediment weight of the sample including the fraction seived out as sand, if any. Line m can be omitted if n is computed directly by $k \times l \times 100$/total sediment weight. The time required for the average density above each observed height to be reached at the equivalent 100-cm fall, o, is the result of applying the depth factor, l, to the settling time, c. Thus, the computations reduce the observed times of settling and weights in suspension to a constant depth of 100 cm.

As indicated by Fig. 3.68 the Oden curve is plotted on a graph having rectangular coordinates. The complete plotting of the data from entries n and o to enlarged scale (0 min–7,000 min) results in a complete upper curve. Lower curves can be used to represent part of the withdrawals with scales such as 0 min–350 min and 0 min–70 min for better definition of the coarser fractions. If only silt and clay sizes are analyzed, then it may not be necessary to use a 0-min–70-min scale.

Other horizontal (time) scales may be used, providing it is convenient to draw smooth curves through the plotted points.

The intercept of the tangent from the point indicated by the given size to the ordinate (percentage in suspension) can be read as the percentage finer than the indicated size. Care needed in the construction of the Oden curve and the drawing of tangents is indicated by the fact that the shape of the curve will greatly affect the intercept of the tangent with the percentage scale. For most samples, the slope of the curve does not approach zero over the period of time covered by the analysis because many fine particles are still settling at this time of the last scheduled withdrawal. Obviously, the curve should never have a reverse slope. A tangent from a curve with too steep a slope or too sharp a curvature cannot be assumed to result in the desired accuracy. Proper use of the expanded time scales will alleviate some of this difficulty.

Hydrometer Method

Since Bouyoucos (1928) introduced the hydrometer in 1927, it has found considerable use for size analysis for samples having an abundant quantity of fine sediment (Table 3.7). Because of simplicity of operation and low cost of hardware, it has been extensively used in the study of soils. The principle of operation involves a correlation between the progressive decrease in density or buoyant effect that occurs at any given elevation in a disperse suspension and the sedimentation diameter of the particles. It is assumed that the difference in density between the suspension and pure fluid as measured with a hydrometer is proportional to the concentration.

Further detail on laboratory procedures with the hydrometer will not be

TABLE 3.10.–Bottom Withdrawal Tube

Temperature, in degrees Celsius (1)	Particle		
	0.25 (2)	0.125 (3)	0.0625 (4)
18	0.522	1.48	5.02
19	0.515	1.45	4.88
20	0.508	1.41	4.77
21	0.503	1.39	4.67
22	0.497	1.37	4.55
23	0.488	1.34	4.45
24	0.485	1.32	4.33
25	0.478	1.30	4.25
26	0.472	1.28	4.15
27	0.467	1.26	4.05
28	0.462	1.24	3.97
29	0.455	1.22	3.88
30	0.450	1.20	3.80
31	0.445	1.18	3.71
32	0.442	1.17	3.65
33	0.438	1.15	3.58
34	0.435	1.13	3.51

[a]Time in minutes required for spheres having a specific gravity of 2.65 to fall 100

presented herein because of its rather limited use in fluvial sediment work and because the procedures are well documented by the American Society for Testing and Materials (1966). However, to make a decision on whether to use the hydrometer for a given situation, it is well to keep the following list of precautions in mind (Means and Parcher, 1963):

1. The hydrometer should be removed from the suspension between readings and wiped off with a clean dry cloth. Grains of soil settle on the hydrometer bulb causing it to settle too low if left in the suspension between readings.
2. The hydrometer should be inserted into and removed from the suspension very slowly, to prevent excessive disturbance. It should require about 10 sec–15 sec to insert the hydrometer and the same time to withdraw it.
3. The cylinder containing the suspension should be fairly large in diameter compared to the diameter of the hydrometer bulb, to prevent excessive disturbance when the hydrometer is inserted and withdrawn.
4. Care should be taken to prevent convection currents in the suspension; they are produced by placing the cylinder in air currents or in the sunlight so that one side is heated more than the other.
5. The temperature should not change more than a few degrees during the progress of the test because such change will affect the viscosity of the water, thereby making the rate of settlement variable.
6. The hydrometer stem should be free of material that will prevent the development of a complete meniscus. Hydrometers are calibrated for correct reading with a fully developed meniscus.

As for the pipet and BW tube methods, the hydrometer results become more accurate as the particle size decreases. Therefore, the best results are obtained

Time of Settling to Be Used with Oden Curve[a]

Diameter, in millimeters				
0.0312 (5)	0.0156 (6)	0.0078 (7)	0.0039 (8)	0.00195 (9)
20.1	80.5	322	1,288	5,154
19.6	78.5	314	1,256	5,026
19.2	76.6	306	1,225	4,904
18.7	74.9	299	1,198	4,794
18.3	73.0	292	1,168	4,675
17.8	71.3	285	1,141	4,566
17.4	69.6	279	1,114	4,461
17.0	68.1	273	1,090	4,361
16.7	66.6	266	1,065	4,263
16.3	65.1	260	1,042	4,169
15.9	63.7	255	1,019	4,079
15.6	62.3	249	997	3,991
15.3	61.0	244	976	3,907
14.9	59.7	239	956	3,825
14.6	58.5	234	936	3,747
14.2	57.3	229	917	3,671
13.9	56.1	224	898	3,494

cm in water at varying temperatures.

when the sand fraction is removed and analyzed by another method. Also, because of the heavy sediment concentration of the suspension, it is necessary, even more so than for the pipet or the BW tube, to use a dispersing agent to prevent flocculation of the particles.

28. Other Determinations Related to Sediment Analysis

Organic Material.—Sediment samples may contain many organic materials ranging from macroscopic fibrous plant material and coal to microscopic colloidal humus. It is expected that neither the macroscopic nor the microscopic forms of organic matter will be an important factor in the yield of material from most drainage basins, because sediment concentration is defined in terms of the ratio of the weight of dry matter in the sample to the volume of the water-sediment mixture. It is expected, however, that exception to this may be found where streams are used for washing coal. Organic material does, however, affect average specific weight and greatly affects the particle size analysis if present in sufficient quantities.

Quantitative determination of organic material is usually recommended for about one-half of the samples analyzed for particle size and all that are analyzed by use of the native-water settling media if the material consists of 5% or more of organic matter. It must again be emphasized that the portion of the sample actually analyzed for particle size in a native-water settling medium should not be treated for removal of organic matter. The decomposition of the organic matter results not only in the formation of carbon dioxide and water, but also in the release of all ions incorporated in the organic material. Therefore, it is obvious that oxidation of organic material by chemical reaction could markedly affect the quality of the native water and the flocculating ability of the sediment particles.

In the process of analyzing sediment for particle size in a dispersed settling medium, it is usually desirable to remove even relatively small quantities of organic material if it is in the form of colloidal humus, which acts as a binding agent for aggregates or floccules. Robinson (1922) was the first to show that samples containing appreciable quantities of organic matter cannot be adequately dispersed unless they are removed. Fourfold increases in the percentage of clay were obtained for some samples by treatment with hydrogen peroxide. Other investigators (Baver, 1956) have also found that oxidation of organic matter with hydrogen peroxide is essential for the complete dispersion of soil particles.

For samples containing significant quantities of coal, it is essential that separation and quantitative determination be made on the basis of difference in specific gravity. This can be accomplished, as suggested by White and Lindholm (1950), with a mixture of bromoform and acetone adjusted to a specific gravity of 1.95. The sediment then either floats or settles into portions lighter or heavier, respectively, than a 1.95 specific gravity. In the programming for determination of particle size, attention should be given to the feasibility of analyzing both the mixture of all sediments and the part heavier than a specific gravity of 1.95 for some samples.

Dissolved Solids.—The term dissolved solids is theoretically intended to include the anhydrous residue of the dissolved substances in water not including gases or volatile liquids. In reality, the term is defined in a quantitative manner by the method used in its determination. For example, with the residue-on-evaporation method, both the drying temperature and the length of time of drying will affect the result. Rainwater and Thatcher (1959) suggest that the quantity of material in the evaporating dish is also a factor due to the fact that massive residues give up their waters of crystallization more slowly than their residue films and due to the possibility of entrapped pockets of water sealing over.

The dissolved solids determination in sediment laboratories should be made by the residue-on-evaporation method. A volume of sample that will yield less than 200 mg of residue is evaporated slowly just to dryness using a steam bath, if available. The residue is dried at 110°C for 1 hr, cooled in a desiccator, and immediately weighed. An efficient desiccant must be used because many of the salts in the residue are hygroscopic. An alumina desiccant with a moisture indicator is recommended. The dried residues should not be allowed to stand for long periods of time before weighing. Only a few dishes of residue should be included in one desiccator, due to the effect of contamination with outside air during the weighing. Under no circumstances should dissolved solids dishes be cooled in a desiccator containing sediment dishes, unless it is known that the sediment is mostly sand-sized particles.

The recommended calculation for concentration is

$$\text{dissolved solids, in milligrams per liter} = \frac{\text{grams of residue} \times 1{,}000{,}000}{\text{milliliter of sample}} \qquad (3.17)$$

Related Water Quality Analysis.—In connection with obtaining an understanding of the effects of environment on fluvial sediment, especially with respect to transportation and deposition, it is desirable to evaluate specific conductance, PH, the concentration of calcium, bicarbonate, sodium, potassium, and magnesium for all samples split for particle size and analyzed in both chemically dispersed and native water settling media. These determinations are most efficiently made in a chemical laboratory using standard methods and equipment (Rainwater and Thatcher, 1959). An aliquot of the native water consisting of at least 200 ml should be withdrawn just prior to splitting the sample and tightly stoppered for storage until analyzed by the chemical laboratory. The aliquot is withdrawn just prior to making the particle-size analysis because it is desirable to include the effects of storage. The results of these chemical analyses are then noted as constituents of the native water settling media for the size analysis and because of storage may or may not be representative of the stream at the time the sediment samples were collected.

Specific Gravity.—The measurement of the specific gravity of a sediment is accomplished by direct measurement of weight and volume. If the sample particles are large (20 mm or 30 mm), the volume is determined by noting the volume of a liquid before and after immersion of the particles. This direct method may result in considerable error due to air-filled pore space in or on the particle. B. C. Colby experienced considerable difficulty in making precise measurements of specific gravity on glass beads used in experimental work (Interagency Committee, 1953, pp. 26–28).

For fine sediment and where only small samples are available, the method employing the pycnometer is the most satisfactory. The pycnometer, which has a constant volume, is weighed; first, when it is filled with distilled water, and second, when sediment is added and the bottle refilled with distilled water. More specifically, the water for the initial weighing should be at 15°C and this weight labeled a_s. One milliliter or 2 ml of the water is removed and approx 1.0 g of the sample is inserted. Suction or boiling is used to remove air bubbles and the bottle is again filled with water for reweighing at the same temperature. The weight of the bottle with sediment and water is recorded as b_s. The specific gravity, SG, is then given by

$$SG = \frac{\gamma_s}{\gamma} = \frac{W_s}{a_s - b_s + W_s} \quad \text{(3.18)}$$

in which γ_s = specific weight of sediment; γ = specific weight of fluid; and W_s = weight of sediment. If some other liquid is substituted for water to avoid difficulty with air bubbles adhering to the sand or crushed material, the results should be converted into terms of water.

Specific Weight of Sediment Deposits.—Specific weight of a sediment deposit is weight of sediment per unit volume of deposit. In the metric system of grams per cubic centimeter, the specific weight would be equal numerically to specific gravity. The most common English system of dimensions used in connection with soils and sediment deposits or of water-sediment mixtures is that of pounds per cubic foot. The method of measurement is simple in that the dry weight of a known volume of the undisturbed material is necessary. The main problem is then one of sampling to obtain the correct amount of material for the given sample volume, the difficulty being that any sampling technique is likely to disturb the sample in some way.

Moisture Content.—The moisture content of a sediment sample is defined as

$$\text{Percentage of moisture} = \frac{\text{loss in weight}}{\text{weight of dried sample}} \times 100 \quad \text{(3.19)}$$

Samples for moisture content are usually collected and stored in metal cans with tight fitting lids. The moisture is evaporated from the sediment in an oven at 105°C–110°C for at least 24 hr.

Turbidity.—Turbidity measurements are of interest with respect to studies of sediment in natural waters because it has been obtained as a "record" of the stream condition by many water users where regular sediment data are not available. Turbidity, because of its adaptability to automation in present technology, has some promise as a tool for obtaining sediment information on surface waters where surveillance is needed for adherence to stream quality standards.

Turbidity is basically an expression of the amount of light that is either scattered or absorbed or both by the sediment-water mixture. Thus, different types of turbidimeters have different calibrations and therefore the results for different meters may not be comparable. Water color, though it does not register as turbidity, will reduce the sensitivity of most turbidimeters. Often, only the nephelometric effect (light reflection) is used with these instruments. Benedict (1946), in a study of the Iowa River at Iowa City, Iowa, found an approximate relationship between turbidity and suspended-sediment concentration. It must be noted, however, that if turbidity data are used, the analysis must include consideration of sampling conditions and the equipment used in the turbidity determinations.

F. References

Abdel Salem, M. S., and Sowelim, M. A., "Dust Deposits in the City of Cairo," *Journal of Atmospheric Environment,* Vol. 1, 1967a, pp. 211–220.

Abdel Salem, M. S., "Dustfall Caused by the Spring Khamsin Storms in Cairo: A Preliminary Report," *Journal of Atmospheric Enviroment,* Vol. 1, 1967b, pp. 221–226.

American Society for Testing and Materials, "E 11–61 Standard Specifications for Sieves for Testing Purposes, Wire Cloth Sievers, Round-Hole and Plate Screens or Sieves," *ASTM Standards Part 30,* 1964, pp. 161–162.

American Society for Testing and Materials, "Grain-Size Analysis of Soils," *D422–63, Book of ASTM Standards,* Part 11, 1966, pp. 193–201.

Any, A. P., and Benarie, M., "Akustischen Nachiveis von Staubteilchen," *Staub.,* Vol. 24, No. 9, 1966, pp. 343–344.

Bagnold, R. A., *Physics of Blown Sand and Desert Dunes,* William Morrow & Co., New York, N.Y., 1943.

Baum, F., Hermann, L., and Reichardt, I., "Praplesche Erfahrungen mit Windrichtungsabhangigen Staubnuderschlop-Messungen," *Gesundheits-Ingenieur,* Vol. 85, No. 3, 1964, pp. 80–83.

Baver, L. D., *Soil Physics,* 3rd ed., John Wiley and Sons, Inc., New York, N.Y., 1956.

Benedict, P. C., "A Study of the Error Resulting from the Use of Turbidity in Computing the Suspended-Sediment Discharge of Iowa Streams," Iowa Geological Survey, Iowa City, Iowa, 1946.

Benedict, P. C., "Determination of the Suspended Sediment Discharge of Streams," *Proceedings,* Federal Interagency Sedimentation Conference, Denver, Colo., United States Bureau of Reclamation, Washington, D.C., 1948, pp. 55–67.

Benedict, P. C., and Matejka, D. Q., "The Measurement of Total Sediment Load in Alluvial Streams," *Proceedings,* 5th Hydraulic Conference, *Bulletin 34,* State University of Iowa, Iowa City, Iowa, 1953.

Benedict, P. C., Albertson, M. L., and Matejka, D. Q., "Total Sediment Load Measured in Turbulence Flume," *Transactions,* ASCE, Vol. 120, Paper No. 2750, 1955, pp. 457–488.

Billings, C. E., and Silverman, L., "Aerosol Sampling for Electron Microscopy," *Journal of the Air Pollution Control Association,* Vol. 12, 1962, pp. 586–590.

Bondurant, D. C., "Report on Reservoir Delta Reconnaissance," *MRD Sediment Series No. 6,* Missouri River Division, Corps of Engineers, United States Army, Omaha, Neb., 1955.

Bondurant, D. C., personal communication, Missouri River Division, Corps of Engineers, United States Army, Omaha, Neb., 1963.

Bouyoucos, G. J., "The Hydrometer Method for Making a Very Detailed Mechanical Analysis for Soils," *Soil Science,* Vol. 26, 1928, p. 233.

Brice, J. C., "Channel Patterns and Terraces of the Loup Rivers in Nebraska," *Professional Paper 422–D,* United States Geological Survey, Washington, D.C., 1964.

Burrell, G. N., "Constant-factor Method Aids Computation of Reservoir Sediment," *Civil Engineering,* ASCE, Vol. 21, No. 7, July, 1951, pp. 51–52.

Cadle, R. D., *Particle Size Determination,* John Wiley and Sons, Inc., New York, N.Y., 1955.

Chatfield, E. J., "A Battery Operated Sequential Air Concentration and Deposition Sampler," *Journal of Atmospheric Environment,* Vol. 1, 1967, pp. 509–513.

Chepil, W. S., "Dynamics of Wind Erosion: I. Nature of Movement of Soil by Wind," *Soil Science,* Vol. 60, 1945, pp. 305–320.

Chepil, W. S., "An Air Elutriator for Determining the Dry Aggregate Soil Structure in Relation to Erodibility by Wind," *Soil Science,* Vol. 71, 1951, pp. 197–207.

Chepil, W. S., "Improved Rotary Sieve for Measuring State and Stability of Dry Soil Structure," *Proceedings,* Soil Science Society of America, Vol. 16, Madison, Wisc., 1952, pp. 113–117.

Chepil, W. S., "Width of Field Strips to Control Wind Erosion," *Technical Bulletin No. 92,* Kansas Agriculture Experiment Station, 1957.

Chepil, W. S., "A Compact Rotary Sieve and the Importance of Dry Sieving in Physical Soil Analysis," *Proceedings,* Soil Science Society of America, Vol. 26, No. 1, Madison, Wisc., 1962, pp. 4–6.

Chepil, W. S., and Woodruff, N. P., "Sedimentary Characteristics of Dust Storms: II. Visibility and Dust Concentration," *American Journal of Science,* Vol. 255, 1957, pp. 104–114.

Colby, B. R., "Relationship of Sediment Discharge to Streamflow," Open File Report, United States Geological Survey, Washington, D.C., 1956.

Colby, B. R., "Practical Computations of Bed-Material Discharge," *Journal of the Hydraulics Division,* ASCE, Vol. 90, No. HY2, Proc. Paper 3843, Mar., 1964, pp. 217–246.

Colby, B. R., "Discharge of Sands and Mean-Velocity Relationships in Sand-Bed Streams," *Professional Paper 462–A,* United States Geological Survey, Washington, D.C., 1964.

Colby, B. R., Matejka, D. Q., and Hubbell, D. W., "Investigations of Fluvial Sediments of the Niobrara River near Valentine, Nebraska," *Circular 205,* United States Geological Survey, Washington, D.C., 1953.

Colby, B. R., and Hembree, C. H., "Computations of Total Sediment Discharge, Niobrara River Near Cody, Nebr.," *Water-Supply Paper 1357,* United States Geological Survey, Washington, D.C., 1955.

Colby, B. R., and Hubbell, D. W., "Simplified Methods for Computing Total Sediment Discharge With The Modified Einstein Procedure," *Water-Supply Paper 1593,* United States Geological Survey, Washington, D.C., 1961.

Corbett, D. M., et al., "Stream-gaging Procedure," *Water-Supply Paper 888,* United States Geological Survey, Washington, D.C., 1945.

Drinker, P., "Alternating Current Precipitator for Sanitary Air Analysis," *Journal of Industrial Hydrology,* Vol. 14, 1934, p. 364.

Eakin, H. M. (rev. C. B. Brown), "Silting of Reservoirs," *Technical Bulletin 524,* United States Department of Agriculture, Washington, D.C., 1939.

Einstein, H. A., "Bed-Load Transportation in Mountain Creek," *SCS–TP–55,* United States Department of Agriculture, Soil Conservation Service, 1944.

Einstein, H. A., "Determination of Rates of Bed-Load Movement," *Proceedings,* Federal Interagency Sedimentation Conference, Denver, Colo., United States Bureau of Reclamation, Washington, D.C., 1948, pp. 75–90.

Engineering Index, Inc., See Aerosols, Dust Analysis, and Dust Measurement, *The Engineering Index,* New York, N.Y., 1964–1968.

Engineering News-Record, "Plotter Speeds Sounding Surveys," August 11, 1966, p. 106.

Fahnestock, R. K., "Morphology and Hydrology of a Glacial Stream—White River, Mount Rainier, Washington," *Professional Paper 422–A,* United States Geological Survey, Washington, D.C., 1963.

Fairweather, J. H., Sidlow, A. F., and Faith, W. L., "Particle Size Distribution of Settled Dust," *Journal of the Air Pollution Control Association,* Vol. 15, No. 8, 1965, pp. 345–347.

Fife, R. W., discussion of "Sediment Measurement Techniques: C. Accelerated Valley Deposits," by the Task Committee on Preparation of Sedimentation Manual, Committee on Sedimentation of the Hydraulics Division, Vito A. Vanoni, Chmn., *Journal of the Hydraulics Division,* ASCE, Vol. 97, No. HY4, Proc. Paper 8012, Apr., 1971, pp. 586–588.

Fleming, G., "The Computer as a Tool in Sediment Transport Research," *Bulletin,* International Association of Scientific Hydrology, Vol. XIIe, Annee No. 3, 1967, pp. 45–54.

Fowler, L. C., discussion of "Application of the Modified Einstein Procedure for Computation of Total Sediment Load," *Transactions,* American Geophysical Union, Vol. 38, No. 5, Washington, D.C., 1957, pp. 771–772.

Giever, P. M., "Number and Size of Pollutants," *Air Pollution,* Arthur C. Stern, ed., *Academic Press,* New York, N.Y., 1968, pp. 249–280.

Gottschalk, L. C., "Measurement of Sedimentation in Small Reservoirs," *Proceedings, Journal of the Hydraulics Division,* ASCE, Separate No. 55, Vol. 77, Jan., 1951, pp. 1–11.

Greenburg, L., and Smith, G. W., "A New Instrument for Sampling Aerial Dust," *Report of Investigations 2392,* United States Bureau of Mines, 1922.

Guy, H. P., Simons, D. B., and Richardson, E. V., "Summary of Alluvial Channel Data from Flume Experiments, 1956–61," *Professional Paper 462–I,* United States Geological Survey, Washington, D.C., 1966.

Guy, H. P., "Laboratory Theory and Methods for Sediment Analysis," *United States Geological Survey Techniques of Water Resources Investigations,* Book 5, Chapter C1, Washington, D.C., 1969.

Happ, S. C., Rittenhouse, G., and Dobson, G. C., "Some Principles of Accelerated Stream and Valley Sedimentation," *Technical Bulletin No. 695,* United States Department of Agriculture, Washington, D.C., May, 1940.

Happ, S. C., "Sedimentation in Whitewater River Valley, Minnesota," Soil Conservation Service, United States Department of Agriculture, Washington, D.C., 1940.

Happ, S. C., "Effect of Sedimentation on Floods in the Kickapoo Valley, Wisconsin," *Journal of Geology,* Vol. 52, No. 1, Jan., 1944, pp. 53–69.

Happ, S. C., Mortimore, M. E., Barnes, L. H., Brown, C. B., et al., "Sedimentation in Valleys of Little Sioux River and Tributaries, Iowa," Soil Conservation Service, United States Department of Agriculture, Washington, D.C., 19411

Heidel, S. G., "The Progressive Lag of Sediment Concentration With Flood Waves," *Transactions,* American Geophysical Union, Vol. 37, No. 1, Washington, D.C., 1956, pp. 56–66.

Heinemann, H. G., and Dvorak, V. I., "Improved Volumetric Survey and Computation Procedures for Small Reservoirs," *Proceedings of the Federal Interagency Sedimentation Conference 1963, Miscellaneous Publication No. 970,* United States Department of Agriculture, Washington, D.C., 1965, pp. 845–856.

Heinemann, H. G., and Rausch, D. L., discussion of "Sediment Measurement Techniques: B. Reservoir Deposits," by the Task Committee on Preparation of Sedimentation Manual, Committee on Sedimentation of the Hydraulics Division, Vito A. Vanoni, Chmn., *Journal of the Hydraulics Division,* ASCE, Vol. 97, No. HY9, Proc. Paper 8345, Sept., 1971, pp. 1555–5161.

Hendrickson, E. R., "Air Sampling and Quantity Measurement," *Air Pollution,* Arthur C. Stern, ed., Academic Press, New York, N.Y., 1968, pp. 3–52.

Hubbell, D. W., et al., "Investigations of Some Sedimentation Characteristics of a Sand-Bed Stream," United States Geological Survey, Open File Report, Washington, D.C., 1956, p. 12.

Hubbell, D. W., and Matejka, D. Q., "Investigations of Sediment Transportation, Middle Loup River at Dunning, Nebr., With Application of Data From Turbulence Flume," *Water-Supply Paper 1476,* United States Geological Survey, Washington, D.C., 1959.

Hubbell, D. W., "Apparatus and Techniques for Measuring Bedload," *Water-Supply Paper 1748,* United States Geological Survey, Washington, D.C., 1964.

Interagency Committee, "Field Practice and Equipment Used in Sampling Suspended Sediment," *Report No. 1,* Federal Interdepartmental Committee, Hydraulic Laboratory of the Iowa Institute of Hydraulic Research, Iowa City, Iowa, 1940a.

Interagency Committee, "Equipment Used for Sampling Bed Load and Bed Material," *Report No. 2,* Federal Interdepartmental Committee, Hydraulic Laboratory of the Iowa Institute of Hydraulic Research, Iowa City, Iowa, 1940b.

Interagency Committee, "Methods of Analyzing Sediment Samples," *Report No. 4,* Federal Interdepartmental Committee, Hydraulic Laboratory of the Iowa Institute of Hydraulic Research, Iowa City, Iowa, 1941a.

Interagency Committee, "Laboratory Investigations of Suspended Sediment Samplers," *Report No. 5,* Federal Interdepartmental Committee, Hydraulic Laboratory of the Iowa Institute of Hydraulic Research, Iowa City, Iowa, 1941b.

Interagency Committee, "A Study of New Methods for Size Analysis of Suspended-Sediment Samples," *Report No. 7,* Federal Interdepartmental Committee, Hydraulic Laboratory of the Iowa Institute of Hydraulic Research, Iowa City, Iowa, 1943.

Interagency Committee, "The Design of Improved Types of Suspended Sediment Samplers," *Report No. 6,* Subcommittee on Sedimentation, Federal Interagency River

Basin Committee, Hydraulic Laboratory of the Iowa Institute of Hydraulic Research, Iowa City, Iowa, 1952.

Interagency Committee, "Accuracy of Sediment Size Analyses Made by the Bottom Withdrawal Tube Method," *Report No. 10,* Subcommittee on Sedimentation, Federal Interagency River Basin Committee on Water Resources, St. Anthony Falls Hydraulic Laboratory, Minneapolis, Minn., 1953.

Interagency Committee, "Some Fundamentals of Particle Size Analysis," *Report No. 12,* Subcommittee on Sedimentation, Interagency Committee on Water Resources, St. Anthony Falls Hydraulic Laboratory, Minneapolis, Minn., 1957a.

Interagency Committee, "The Development and Calibration of the Visual Accumulation Tube," *Report No. 11,* Subcommittee on Sedimentation, Interagency Committee on Water Resources, St. Anthony Falls Hydraulic Laboratory, Minneapolis, Minn., 1957b.

Interagency Committee, "Operators Manual on the Visual-Accumulation Tube Method for Sedimentation Analysis of Sands," *Report No. K,* Subcommittee on Sedimentation, Interagency Committee on Water Resources, St. Anthony Falls Hydraulic Laboratory, Minneapolis, Minn., 1958.

Interagency Committee, "The Single-Stage Sampler for Suspended Sediment," *Report No. 13,* Subcommittee on Sedimentation, Interagency Committee on Water Resources, St. Anthony Falls Hydraulic Laboratory, Minneapolis, Minn., 1961.

Interagency Committee, "Determination of Fluvial Sediment Discharge," *Report No. 14,* Subcommittee on Sedimentation, Interagency Committee on Water Resources, St. Anthony Falls Hydraulic Laboratory, Minneapolis, Minn., 1963.

Interagency committee, "Electronic Sensing of Sediment," *Report R,* Subcommittee on Sedimentation, Interagency Committee on Water Resources, St. Anthony Falls Hydraulic Laboratory, Minneapolis, Minn., 1964.

[Note: Reports may be purchased by sending a request to the District Engineer, U.S. Army Engineer District-St. Paul, 1210 U.S. Post Office Bldg., St. Paul, Minnesota.]

Jacobs, M. B., *The Chemical Analysis of Air Pollutants,* Interscience, New York, N.Y., 1960.

Johnson, J. W., discussion of "Sedimentation in Reservoirs," by B. J. Witzig, *Transactions,* ASCE, Vol. 109, Paper No. 2227, 1944, pp. 1072–75.

Jones, V. H., "Sedimentation in Valleys of: Battle Creek of the Richland Creek Drainage, Trinity River, Texas; Caney Creek of the Cedar Creek Drainage, Trinity River, Texas; Clear Creek of the Elm Fork, Trinity River, Texas; Sugar Creek of the Washita River Basin, Oklahoma," Soil Conservation Service, United States Department of Agriculture, Ft. Worth, Tex., 1940–1950.

Jones, V. H., "Causes and Effects of Channel Floodway Aggradation," *Proceedings of the Federal Interagency Sedimentation Conference,* Denver, Colo., United States Bureau of Reclamation, Washington, D.C., 1948, pp. 168–178.

Jones, V. H., and Marshall, R. L., "Sedimentation Damages and Sediment Contribution Rates in Barnitz Creek Drainage Area, Washita River Basin, Oklahoma," Part of a Flood Control Work Plan, Soil Conservation Service, United States Department of Agriculture, Ft. Worth, Tex., 1951.

Jones, W. L., personal communication, "Conducting Reservoir Sediment Ranges Surveys Through the Ice," United States Bureau of Reclamation, McCook, Neb., 1973.

Jordan, P. R., "Fluvial Sediment of the Mississippi River at St. Louis, Missouri," *Water-Supply Paper 1802,* United States Geological Survey, Washington, D.C., 1965, pp. 13–14.

Kampf, W. D., and Schmidt, B., "Bestimmung des Staubniederschlages, Durch Transparente Haft Flaschen," *Staub-Reinhaltung der Luft,* Vol. 27, No. 9, 1967, pp. 395–399.

Katz, S. H., Smith, G. W., Myers, W. M., Trostel, L. J., Ingels, M., and Greenburg, L., "Comparative Tests of Instruments for Determining Atmospheric Dusts," *United States Public Health Bulletin 144,* 1925.

Kobayashi, S., Oba, T., and Arai, S., "Transistorized Dustmonitor," *Toshiba Review,* Vol. 14, 1963, pp. 40–46.

Krumbein, W. C., and Pettijohn, F. J., *Manual of Sedimentary Petrography,* D. Appleton-Century Co., New York, N.Y., 1938.

Lane, E. W., et al., "Report of the Subcommittee on Sediment Terminology," *Transactions,* American Geophysical Union, Vol. 28, No. 6, 1947, pp. 936–938.

Lane, E. W., Carlson, E. J., and Hanson, O. S., "Low Temperature Increases Sediment Transportation in Colorado River," *Civil Engineering,* ASCE, Vol. 19, No. 9, Sept., 1949, pp. 45–46.

Lane, E. W., and Borland, W. M., "Estimating Bedload," *Transactions,* American Geophysical Union, Vol. 32, No. 1, 1951, pp. 121–123.

Lane, E. W., and Borland, W. M., "River-Bed Scour During Floods," *Transactions,* ASCE, Vol. 119, Paper No. 2712, 1954, pp. 1069–1087.

Langham, W. H., Foster, R. L., and Daniel, H. A., "The Amount of Dust in the Air at Plant Height During Wind Storms at Goodwell, Oklahoma, in 1936–37," *Journal of the American Society of Agronomy,* Vol. 39, 1938, pp. 139–144.

Lara, J. M., and Pemberton, E. L., "Initial Unit Weight of Deposited Sediments," *Proceedings of the Federal Interagency Sedimentation Conference, 1963, Misellaneous Publication No. 970,* United States Department of Agriculture, 1965, pp. 818–845.

Lara, J. M., and Sanders, J. I., "The 1963–64 Lake Mead Survey," *Report REC–OCE–70–21,* United States Bureau of Reclamation, 1970.

Laursen, E. M., discussion of "Application of the Modified Einstein Procedure for Computation of Total Sediment Load," *Transactions,* American Geophysical Union, Vol. 38, No. 5, Washington, D.C., 1957, pp. 768–771.

Leopold, L. B., and Maddock, Jr., T., "The Hydraulic Geometry of Stream Channels and Some Physiographic Implications," *Professional Paper 252,* United States Geological Survey, Washington, D.C., 1953.

Leopold, L. B., and Wolman, M. G., "River Channel Patterns: Braided, Meandering and Straight," *Professional Paper 282–B,* United States Geological Survey, Washington, D.C., 1957.

Love, S. K., and Benedict, P. C., "Discharge and Sediment Loads in the Boise River Drainage Basin, Idaho, 1939–40," *Water-Supply Paper 1048,* United States Geological Survey, Washington, D.C., 1948, p. 23.

Lucas, D. H., and Moore, D. J., "The Measurement in the Field of Pollution by Dust, Air and Water Pollution," *International Journal,* Vol. 8, 1964, pp. 441–453.

McCain, E. H., "Measurement of Sedimentation in TVA Reservoirs," *Journal of the Hydraulics Division,* ASCE, Vol. 83, HY3, Proc. Paper 1277, June, 1957, pp. 1–12.

McCrone, W. C., "Morphological Analysis of Particulate Pollutants," *Air Pollution,* Arthur C. Stern, ed., Academic Press, New York, N.Y., 1968, pp. 281–301.

McHenry, J. R., discussion of "Sediment Measurement Techniques: B. Reservoir Deposits," by the Task Committee on Preparation of Sedimentation Manual, Committee on Sedimentation of the Hydraulics Division, Vito A. Vanoni, Chmn., *Journal of the Hydraulics Division,* ASCE, Vol. 97, No. HY8, Proc. Paper 8267, Aug., 1971, pp. 1253–1257.

Means, R. E., and Parcher, J. V., *Physical Properties of Soils,* Merrill Book, Inc., Columbus, Ohio, 1963, p. 58.

Miller, E. R., "The Dustfall of November 12–13, 1933," *Monthly Weather Review,* Vol. 62, 1933, pp. 14–15.

Moore, T., "The Great Error of American Agriculture Exposed: and Hints for Improvement Suggested," Baltimore, Md., 1801.

Morrison, C. A., "Dust Meter," *United States Patent No. 2,898,803,* 1959.

Mueller, E., and Thaer, A., "Ein Registrierendes Staubmessuerfohren auf Membranfilterbasis," *Staub.,* Vol. 25, No. 7, 1965, pp. 251–256.

Murphy, T. D., "Innovations in the Sediment Density Gamma Probe," presented at the August, 1964, ASCE Hydraulics Division Meeting, held at Vicksburg, Miss.

Murthy, B. N., discussion of "Sediment Measurement Techniques: B. Reservoir Deposits," by the Task Committee on Preparation of Sedimentation Manual, Committee on Sedimentation of the Hydraulics Division, Vito A. Vanoni, Chmn., *Journal of the Hydraulics Division,* ASCE, Vol. 97, No. HY6, Proc. Paper 8155, June, 1971, pp. 877–879.

Nelson, M. E., and Benedict, P. C., "Measurement and Analysis of Suspended Sediment Loads in Streams," *Transactions,* ASCE, Vol. 116, Paper No. 2450, 1951, pp. 891–918.

Neyrpic, *Hydraulic Measuring Instuments,* Neyrpic, Inc., Grenoble, France, 1962. Nordin,

Nordin, Jr., C. F., "A Preliminary Study of Sediment Transport Parameters, Rio Puerco Near Bernardo, New Mexico," *Professional Paper 462–C,* United States Geological Survey, Washington, D.C., 1963.

Nordin, Jr., C. F., and Beverage, J. P., "Sediment Transport in the Rio Grande, New Mexico," *Professional Paper 462–F,* United States Geological Survey, Washington, D.C., 1965, p. 16.

Orr, C., and DallaValle, J. M., *Fine Particle Measurement, Size, Surface, and Pore Volume,* The Macmillan Co., New York, N.Y., 1960, pp. 83–91.

Prych, E. A., Hubbell, D. W., and Glenn, J. L., "Measurement Equipment and Techniques Used in Studying Radionuclide Movement in the Columbia River Estuary," presented at the 1965 ASCE Conference on Coastal Engineering, held at Santa Barbara, Calif.

Radiological Mechanisms for Geophysical Research, Parametrics, Inc., Waltham, Mass., 1964.

Rainwater, F. H., and Thatcher, L. L., "Methods for Collection and Analysis of Water Samples," *Water-Supply Paper 1454,* United States Geological Survey, Washington, D.C., 1959.

Rao, Udayagiri, M., discussion of "Sediment Measurement Techniques: F. Laboratory Procedures," by the Task Committee on Preparation of Sedimentation Manual, Committee on Sedimentation of the Hydraulics Division, Vito A. Vanoni, Chmn., *Journal of the Hydraulics Division,* ASCE, Vol. 96, No. HY5, Proc. Paper 7248, May, 1970, pp. 1215–1220.

Rausch, D. L., and Heinemann, H. G., "Reservoir Sedimentation Survey Methods," *Research Bulletin 939,* University of Missouri, College of Agriculture, Columbia, Mo., 1968.

Rittenhouse, G., "Sources of Modern Sands in the Middle Rio Grande Valley, New Mexico," *Journal of Geology,* Vol. 52, 1944, pp. 145–183.

Ritter, J. R., and Helley, E. J., "Optical Method for Determining Particle Sizes of Coarse Sediment," *Techniques of Water-Resources Investigations,* United States Geological Survey, Book 5, 1969.

Robinson, G. W., "Notes on the Mechanical Analysis of Humus Soils," *Journal of Agricultural Science,* Vol. 12, 1922, pp. 287–291.

Sanderson, H. P., Bradt, P., and Katz, M., "A Study of Dustfall on the Basis of Replicated Latin Square Arrangements of Various Types of Collectors," *Journal of the Air Pollution Control Association,* Vol. 13, 1963, pp. 461–466.

Schroeder, K. B., and Hembree, C. H., "Application of the Modified Einstein Procedure for Computation of Total Sediment Load," *Transactions,* American Geophysical Union, Vol. 37, No. 2, 1956, pp. 197–212.

Schroeder, K. B., and Hembree, C. H., closure on discussions of "Application of the Modified Einstein Procedure for Computation of Total Sediment Load," *Transactions,* American Geophysical Union, Vol. 38, No. 5, 1957, pp. 772–773.

Searcy, J. K., "Flow-Duration Curves," *Water-Supply Paper 1542–A,* United States Geological Survey, Washington, D.C., 1959.

Serr, E. F., III, "Investigations of Fluvial Sediments of the Niobrara River Near Cody, Nebr.," *Circular 67,* United States Geological Survey, Washington, D.C., 1950.

Shamblin, O. H., "Reservoir Sedimentation Survey Methods in the United States Army Engineer District, Vicksburg, Mississippi," *Proceedings of the Federal Interagency Sedimentation Conference, Miscellaneous Publication No. 970,* United States Department of Agriculture, 1965, pp. 810–818.

Shepherdson, I., "Using Raydist for Sedimentation Surveys on Larger Reservoirs," *Proceedings of the Federal Interagency Sedimentation Conference, Miscellaneous Publication No. 970,* United States Department of Agriculture, 1965, pp. 869–879.

Sheppard, J. R., "Investigation of Meyer-Peter, Muller Bedload Formulas," United States Bureau of Reclamation, 1960.

Skinner, J. V., discussion of "Field Test of an X-Ray Sediment Concentration Gage," by C. E. Murphree, G. C. Bolton, J. R. McHenry, and D. A. Parsons, *Journal of the*

Hydraulics Division, ASCE, Vol. 95, No. HY1, Proc. Paper 6323, Jan., 1969, pp. 531–532.

Smith, R. M., and Twiss, P. C., "Extensive Gaging of Dust Deposition Rates," *Transactions,* Kansas Academy of Science, Vol. 68, No. 2, 1965, pp. 311–321.

Smith, R. M., and Yates, R., "Renewal, Erosion, and Net Change Functions in Soil Conservation Science," *Transactions,* 9th International Congress of Soil Science, Vol. 4, Adelaide, Australia, 1968, pp. 743–753.

Smith, W. O., Vetter, C. P., and Cummings, G. B., "Comprehensive Survey of Sedimentation in Lake Mead, 1948–49," *Professional Paper 295,* United States Geological Survey, 1960.

Soil Conservation Service, "Instructions for Stream and Valley Sedimentation Surveys," *Technical Letter SED–3,* United States Department of Agriculture, Washington, D.C., Feb. 21, 1939.

Soil Conservation Service, "Formula for Approximate Sediment—Volume," *Technical Letter SED–14,* United States Department of Agriculture, Washington, D.C., Oct., 1940.

Stevens, J. C., "Future of Lake Mead and Elephant Butte Reservoir," *Transactions,* ASCE, Vol. 111, Paper No. 2292, 1946, pp. 1231–1338.

Stewart, J. H., and LaMarche, V. C., "Erosion and Deposition Produced by the Flood of December, 1964, on Coffee Creek, Trinity County, California," *Professional Paper 422–K,* United States Geological Survey, Washington, D.C., 1967.

Stuke, J., and Rzeznik, J., "Der Tyndallograph ein Optisches Staubmessgeraet mit Elektrischer Anzeige," *Staub.,* Vol. 24, No. 9, 1964, pp. 366–368.

Taylor, J., "Being a Series of Agricultural Essays, Practical and Political: In Sixty One Numbers," by a citizen of Virginia, Washington, D.C., 1813.

Tennesee Valley Authority, "Forest Cover Improvement Influences Upon Hydrologic Characteristics of White Hollow Watershed 1935–58," *Report No. 0–5163A,* 1961, p. 87.

Twiss, P. C., Suess, E., and Smith, R. M., "Morphological Classification of Grass Phytoliths," *Proceedings,* Soil Science Society of America, Vol. 33, No. 1, Madison, Wisc., 1969, pp. 109–115.

Udden, J. A., "Erosion, Transportation, and Sedimentation Performed by the Atmosphere," *Journal of Geology,* Vol. 2, 1894, pp. 318–331.

United States Bureau of Reclamation, "Analysis of Flow-Duration, Sediment Rating Curve Method of Computing Sediment Yield," 1951.

United States Bureau of Reclamation, "Total Sediment Transport Program, Lower Colorado River Basin," Interim Report, 1958.

United States Bureau of Reclamation, "Determination of Total Suspended Load in a Stream by the Modified Einstein Procedure, *Electronic Computer Program Description No. HY–100,* 1959.

United States Bureau of Reclamation, *Earth Manual,* Denver, Colo., 1963.

United States Corps of Engineers, "Reservoir Sedimentation Investigations Program," *Manuals,* EM 1110–2–4000, United States Army, 1961.

United States Corps of Engineers, Official Communication from United States Army Engineer District, Little Rock, Ark., 1958.

United States Corps of Engineers, *Operation Manual for the Radioactive Sediment Density Probe,* United States Army Engineer District, Omaha, Neb., 1965.

United States Department of Health, Education, and Welfare, "The Industrial Environment, Its Evaluation and Control," *Public Health Service Pub. No. 614,* 1965.

Vice, R. B., Guy, H. P., and Ferguson, G. E., "Sediment Movement in an Area of Suburban Highway Construction—Scott Run Basin, Virginia—1961–64," *Water-Supply Paper 1591,* United States Geological Survey, Washington, D.C., 1968.

Warn, G. F., and Cox, W. H., "A Sedimentary Study of Dust Storms in the Vicinity of Lubbock, Texas," *American Journal of Science,* Vol. 249, 1951, pp. 553–568.

Welborn, C. T., and Skinner, J. V., "Variable-Speed Power Equipment for Depth-Integrating Sediment Sampling," *Water-Supply Paper 1892,* United States Geological Survey, Washington, D.C., 1968, pp. 126–137.

West, P. W., "Chemical Analysis of Inorganic Particulate Pollutants," *Air Pollution,* Arthur C. Stern, ed., Academic Press, New York, N.Y., 1968, pp. 147–185.

White, W. F., and Lindholm, C. F., "Water Resources Investigation Relating to the Schuylkill River Restoration Project," Pennsylvania Department of Forest and Waters, 1950.

Witzig, B. J., "Sedimentation in Reservoirs," *Transactions,* ASCE, Vol. 109, Paper No. 2227, 1944, pp. 1054–1057, 1099–1100.

Zingg, A. W., "A Portable Wind Tunnel and Dust Collector Developed to Evaluate the Erodibility of Field Surfaces," *Agronomy Journal,* Vol. 43, 1951, pp. 189–191.

Zingg, A. W., "Wind-Tunnel Studies of the Movement of Sedimentary Material," *Proceedings,* Fifth Hydraulics Conference, *Bulletin 34,* State University of Iowa, Iowa City, Iowa, 1953, pp. 111–135.

Chapter IV.—Sediment Sources and Sediment Yields

A. Introduction

1. Changing Landscape.—The continuing landform development occurring on the earth's surface necessarily implies the production and subsequent distribution of sediments. Earthy or rock particles are removed from one location and deposited at another, and there is a need to quantify these erosion and deposition rates. Because water is the prime entraining agent and mover of eroded materials, it is virtually impossible to plan, design, construct, or maintain river basin projects rationally without postulating the distribution of these materials to downslope and downstream locations. Typical erosion and deposition occurrences are indicated in Fig. 4.1.

Most individuals have observed soil erosion and deposition phenomena in nature and could ascribe general reasons for these occurrences with some degree of confidence. The interplay of the forces causing erosion is less obvious, however.

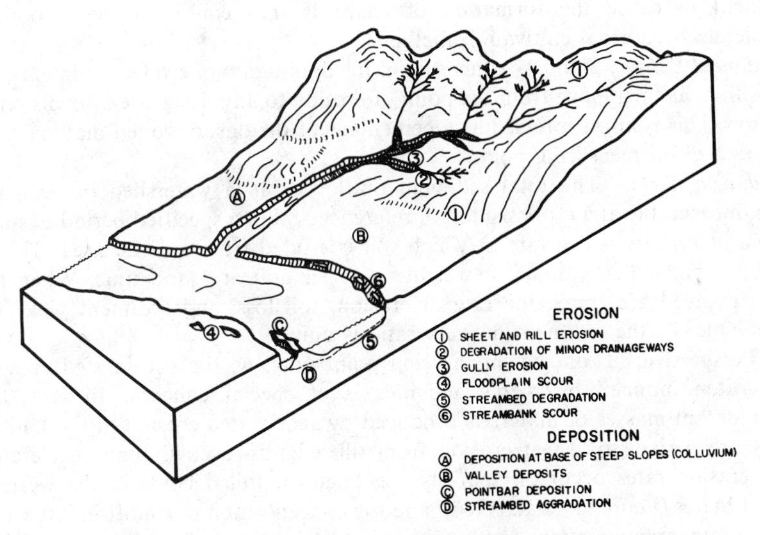

FIG. 4.1.—Typical Erosion and Deposition Occurrences

As with other natural occurrences, erosion and deposition rates could be accurately predicted if all causes were known and could be taken into account. In nature, and often in the laboratory, it is seldom possible, or practical, to measure all these variables in isolation so that a completely deterministic model can be developed. Therefore, probabilistic values based on measurements must be assigned to the major causative variables to obtain quantitative estimates of erosion and deposition for a given situation.

The adequacy of measurement records and the success in analyzing them determine the degree to which deterministic methods rather than probabilistic (stochastic) methods have been utilized for estimating rates of erosion, sediment transport, and deposition.

In this chapter, a few of the more commonly used methods for determining sediment production are examined, and references are made to publications that provide more detail on the subject.

2. Definitions.—Erosion, the loosening or dissolving and removal of earthy or rock materials from any part of the earth's surface, is often differentiated according to the eroding agent (wind, water, rain-splash) and the type or source (sheet, gully, rill, etc.). These terms are mostly self-explanatory, but some are included in the following definitions:

Sheet Erosion.—The wearing away of a thin layer of the land surface. This is interpreted to include rill erosion unless otherwise specified.

Rill Erosion.—The removal of soil by small concentrations of flowing water, with the formation of channels that are small enough to be smoothed completely by normal cultivation methods.

Soil Loss.—The quantity of soil actually removed by erosion from small test areas.

Gully Erosion.—The removal of soil by concentrations of flowing water sufficient to cause the formation of channels that could not be smoothed completely by normal cultivation methods.

Sediment Delivery Ratio.—A measure of the diminution of eroded sediments, by deposition, as they move from the point of erosion to any designated downstream location. This is also expressed as a percentage of the onsite eroded material that reaches a given measuring point.

Sediment Yield.—The total sediment outflow from a watershed or drainage basin, measurable at a cross section of reference and in a specified period of time.

Rate of Erosion.—The rate at which soil is eroded from a given area. This is usually expressed in volume or weight units per unit area and time. When the areas involved are small, the rate of erosion, soil loss, and sediment yield are equivalent, i.e., the sediment delivery ratio is unity.

3. Perspective.—From an engineering point of view, the accelerated erosion most often induced by man's activities is of special concern. In fact, the preponderant masses of materials produced by accelerated erosion in the United States are derived by sheet erosion from tilled lands, even though accelerated sheet erosion rates occur on sparsely vegetated, nontilled lands in the western United States (Leopold, et al., 1966)[1] and the concentration of runoff in large and small drainageways often causes extensive channel erosion damages. Other erosion, from such specialized activities as mining and the construction of homes,

[1] References are given in Chapter IV, Section D.

factories, highways, and utilities, is becoming increasingly important in some localities.

All of these erosion types combine to create perplexing problems in land management and in the design of sediment control measures. But the overall goal, as always, is the economic management of our land resource so that the rate of normal geologic erosion, plus an allowable rate of man-made erosion, will not exceed our ability to sustain the soil. Attainment of such goals would minimize the engineering and esthetic problems incurred by sediment movement downstream.

B. Erosion Rates from Sediment Sources

4. Sheet Erosion.—The need for factual, quantitative information to determine soil erosion rates under a variety of climatic, physiographic, land use, and soil management situations led to the establishment of small, fractional-acre test plots in the United States early in this century. The first erosion plots in this country were instrumented in Missouri in 1917 and were patterned after earlier German studies (Smith, 1966). The Public Works during the Great Depression, the droughts of the 1930's and the more intensive use of soil and water resources since that time gave further impetus to erosion studies of this type. Today, much factual information exists on the erosion rates and processes because instrumentation of these field plots usually included rain gages, precalibrated runoff-measuring flumes, and apparatus for sampling the outflowing sediments (Mutchler, 1963).

The field plots shown in Fig. 4.2(a) are 90 ft long and 10.5 ft wide, and are 7 ft apart, with a 12-in. corrugated metal cutoff wall driven to an 8-in. depth around the perimeter. The 90-ft plot length approximates the horizontal spacing of terraces on land slopes of 3%; plot width and spacing were based on even multiples of 42 in. between corn rows. Runoff from the plots is collected and measured in the sediment storage and outlet tanks shown in the foreground of Fig. 4.2(a). The outlet tank receives one-ninth of any overflow from the sediment storage tank. The soil loss from the plot is found by determining the soil concentration of the measured runoff, by weight.

For larger plots, as in Fig. 4.2(b), the runoff is measured by a water-level recorder housed at the outlet flume. The volume of runoff is calculated directly from the water-level recorder trace and the precalibrated stage-discharge relation of the flume. Soil loss is measured by trapping an aliquot portion of the runoff with a wheel-type sampler (Parsons, 1954), as shown at the lower right of Fig. 4.2(b). If the time distribution of soil loss is desired, it is necessary to sample the runoff throughout the duration of the storm.

In the United States, soil loss data are periodically summarized in special reports of the individual research stations. Records from many locations are available at the National Runoff and Soil Loss Laboratory of the Agricultural Research Service at Lafayette, Ind. These do not include experimental data from forests and western rangelands.

Findings from small plot studies at experiment stations across the United States conclusively show that a myriad of variables affect erosion. These findings have been the basis for recommendations leading to optimum agricultural use of our soil resources consistent with economic goals. One example of this type of information is tabulated in Chapter II C, which shows annual soil losses from five widely separated areas. For conditions of clean-tilled land planted to row crops, the soil losses average more than 200 times those for densely vegetated areas planted to good meadow.

Another study, intended to illustrate the effect upon erosion of such relatively subtle variables as fertility levels and cropping systems, is excerpted in Fig. 4.3. This 16-yr study of field-plot erosion on claypan soils with a slope of 3% at McCredie, Mo., shows that high fertility levels are an erosion deterrent for both the 4-yr crop rotation and continuous corn cropping. Soil losses for the unfertilized corn-oats rotation are of interest as one comparison standard because the worn-out croplands of this area were formerly subjected to this rotation.

The data from Fig. 4.3 are only a partial summary of a comprehensive report (Whitaker and Heinemann) that shows the effect of many other variables of the natural environment and land stewardship upon erosion. As with all natural phenomena, however, it is difficult to ascertain exact quantitative relationships between the many variables and to apply them to ungaged regions where different soil, topographic, and management conditions prevail. Reference will now be made to the types of data presently available and to procedures for predicting sheet erosion rates.

FIG. 4.2.—Fractional-Acre Erosion Plots at: (a) McCredie, Mo.; (b) Holly Springs, Miss.

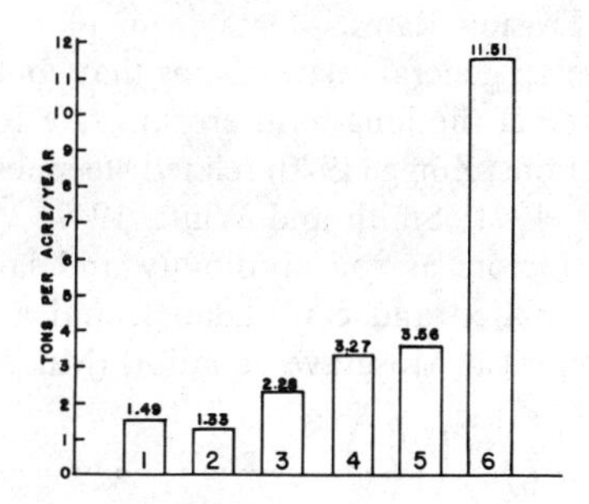

1. CORN-WHEAT-MEADOW-MEADOW ROTATION. FULL FERTILITY: PLOTS 4, 5, 8, 12, 14, 15, 16, and 21.
2. CORN-WHEAT-MEADOW-MEADOW ROTATION. STARTER FERTILITY ONLY: PLOTS 2, 7, 10, 11, 13, 17, 23, and 24.
3. CONTINUOUS CORN. FULL FERTILITY: PLOTS 6 and 22.
4. CONTINUOUS CORN. STARTER FERTILITY: PLOTS 9 and 20.
5. CORN-OAT ROTATION. NO FERTILITY: PLOTS 25 and 26.
6. A FALLOW CONDITION WAS MAINTAINED, 1959-1969, PLOTS 1 and 18.

FIG. 4.3.—Effect of Cropping System and Soil Fertility on Erosion Rates at McCredie, Mo., 1954–1969

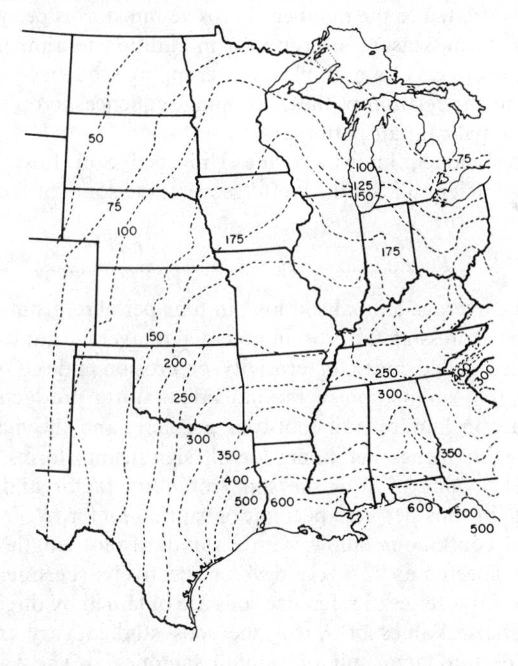

FIG. 4.4.—Average Annual Values of Rainfall Erosivity Factor *R* in Eq. 4.2 for Midwestern United States

5. Predicting Sheet Erosion Rates.—Data from plot studies of sheet erosion made it possible to develop general relationships that could be used by soil-water resource planners to predict the long-term erosion rate for a given field under a variety of land-use programs. Zingg (1940) related steepness and length of slope to soil loss. Others (Smith, 1941; Smith and Whitt, 1947; Van Doren and Bartelli, 1956) considered such factors as soil erodibility and land management. These factors were further evaluated and consolidated, and a rainfall parameter was added to obtain the empirical Musgrave equation (Musgrave, 1947) of the form

$$E = FR\left(\frac{S}{10}\right)^{1.35}\left(\frac{L}{72.6}\right)^{0.35}\left(\frac{P}{1.25}\right)^{1.75} \qquad (4.1)$$

in which E = the average topsoil loss, measured either in tons per acre per year or in inches per year; F is an erodibility factor for the specific soil considered, measured in the same units as E; R = a dimensionless cover factor that is the ratio of the erosion rate of the field under the existing cover to the erosion rate for a continuous fallow or row crop condition; S = the land slope, as a percentage; L = slope length, in feet; and P = the 30-min, 2-yr frequency rainfall, in inches, for the region being considered.

By 1960, the available soil loss records from test plots of cropland and pasture exceeded 10,000 plot-yr. These data verified the importance of the factors used in the foregoing equation (4.1), but the equation parameters were not readily adaptable to many of the land-use conditions that were encountered. A prediction model was needed that would: (1) Improve and extend the soil erodibility and cropping factors; (2) overcome the deficiency of a single rainfall-intensity factor that is not closely related to the number of erosive rainstorms per year; (3) predict erosion rates by storm, season, or crop year in addition to annual averages; and (4) account for the effects of a multiplicity of cropping sequences, crop yields, and crop residues, and the relation of these cropping sequences, crop yields, and crop residues to locational rainfall patterns.

The prediction model, known as the Universal Soil Loss Equation, was developed (Wischmeier and Smith, 1960) to overcome some of these deficiencies. It has the general form

$$E = R\;K\;L\;S\;C\;P \qquad (4.2)$$

in which E = the average annual soil loss, in tons per acre, from a specific field; and R = a factor expressing the erosion potential of average annual rainfall in the locality. It is also called index of erosivity or erosion index (Wischmeier and Smith, 1958). It is a summation of the individual storm products of the kinetic energy of rainfall, in hundreds of foot-tons per acre, and the maximum 30-min rainfall intensity, in inches per hour, for all significant storms on an average annual basis. The value of K = the soil erodibility factor and represents the average soil loss, in tons per acre per unit of rainfall factor, R, from a particular soil in cultivated continuous fallow, with a standard plot length and percentage slope arbitrarily selected as 72.6 ft and 9%, respectively. (Pertinent values of the erodibility factor for a series of reference soils are obtained by direct measurement of eroded materials. Values of K for the soils studied vary from 0.02 tons/acre/unit to 0.70 tons/acre/unit of rainfall factor, R.) The values of S and L = topographic factors for adjusting the estimate of soil loss for a specific land gradient (S) and length of slope (L). It should be emphasized that S and L in Eq

4.2 are factors representing functions of slope steepness and length. For most cases, the factor $S = (0.43 + 0.30S + 0.043S^2)/6.613$ and factor $L = (L_{actual}/L_{standard})^n = (L/72.6)^{1/2}$. The land gradient (slope) is measured as a percentage. Slope length is defined as the average distance, in feet, from the point of origin of overland flow to whichever of the following limiting conditions occurs first: (1) The point where slope decreases to the extent that deposition begins; or (2) the point where runoff enters well-defined channels. The value of C = the cropping management factor and represents the ratio of the soil quantities eroded from land that is cropped under specified conditions to that which is eroded from clean-tilled fallow under identical slope and rainfall conditions; and P = the supporting conservation practice factor (strip-cropping, contouring, etc.). For straight-row farming, $P = 1.0$.

A typical use for a sheet erosion equation, as taken from a handbook on the subject (Wischmeier and Smith, 1965), might be to calculate the expected average annual soil loss from a given cropping sequence on a particular field. Consider a field in west central Indiana, on Russell silt loam soil, having an 8% slope and a length of about 200 ft. Plan a 4-yr crop rotation of wheat, meadow, and two seasons of corn. Assume that all tillage operations are on contour and that prior crop residues are plowed down in the spring before row crops and are left on the surface when small grain is seeded.

The values of the variables of the equation are obtained as follows: The rainfall factor, R, derived from rainfall records for west central Indiana (see Fig. 4.4) is 185. The factor, K, is a measure of the erodibility of a given soil and is evaluated independently of the effects of topography (LS), cover and management (C), and supplementary practices (P). When these conditions of independence are met, K equals E/R and is represented by the slope of the trend line of Fig. 4.5, where soil loss is plotted versus R for a number of field plots. Factor K for Russell silt loam equals 0.38 tons/unit of erosion index. The soil erodibility factor for Russell silt loam was not evaluated directly from plot studies; instead, it was compared with other soils and judged equal in erodibility to the Fayette silt loam, which was directly evaluated at the LaCrosse, Wisconsin Research Station. Erodibility values for 23 "benchmark" soils are listed by Wischmeier and Smith (1965). Values for numerous other soils and subsoils have been approximated by comparison of soil characteristics with these 23 "calibrated" soils and by procedures enumerated in Section 8.

The topographic factors, L and S, are the outgrowth of regression analyses from many field plots throughout the country. From Fig. 4.6, the composite factor is found to be 1.41 for an 8%, 200-ft slope.

The cropping factor, C, is computed by crop stages for the entire 4-yr rotation period. Input for calculation of C includes average planting and harvesting dates, productivity, disposition of crop residues, tillage, and distribution curves of the erosion index (rainfall factor R) throughout the year. The ratio, C, of soil loss from cropland to correponding loss from continuous fallow for each period of crop growth or field condition is catalogued in tables too voluminous for inclusion herein (Wischmeier and Smith, 1965). The C value for each of these time periods is weighted according to the percentage of annual rainfall factor occurring in that period. The summation of these RC products for the entire 4-yr rotation is then converted to a mean annual, C. The erosion-index distribution curve for central Indiana is shown in Fig. 4.7. The value of C is computed to be 0.119.

The practice factor, $P = 0.6$, is based on the decision to contour and depends on land slope, as shown in Table 4.1. The expected average annual soil loss rate for this Indiana field is then $E = (185)(0.38)(1.41)(0.119)(0.6) = 7.1$ tons/acre. Other practices include contour strip-cropping, contour listing, and minimum tillage. Terracing is not always considered to affect the P factor, because soil loss reductions from terracing are reflected by changes in the LS factor. For erosion

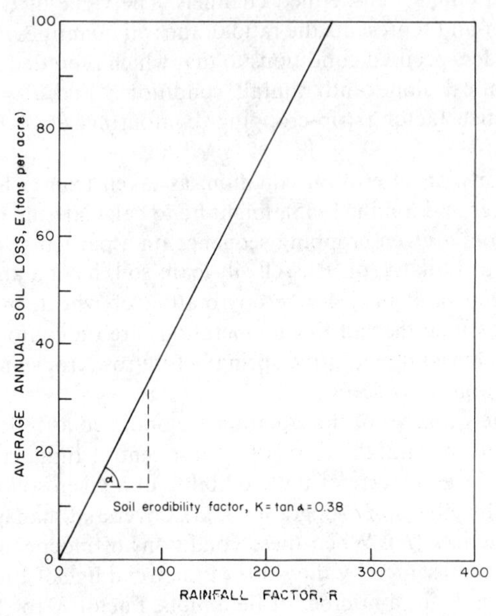

FIG. 4.5.—Erodibility Factor k in Eq. 4.2 for Fayette Silt Loam Soil, as Determined from Field Plot Studies

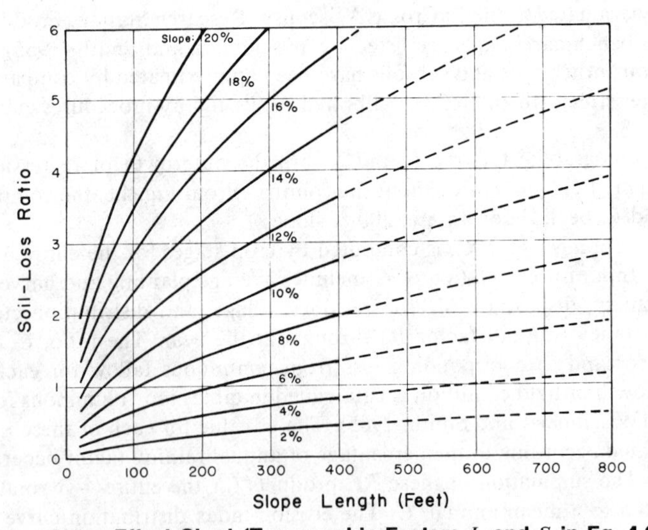

FIG. 4.6.—Slope Effect Chart (Topographic Factors L and S in Eq. 4.2)

control planning on farm fields, graded terraces are usually credited only for the reduction in slope length that they affect, because soil is moved toward the terrace channels. However, as much as 90% of the eroded soil can be deposited in the terrace channel and does not become a source of downstream pollution (Zingg, 1941).

It should be mentioned that if planting had been up and down slope, with all the factors evaluated, the sheet erosion rate for the field would be 11.8 tons/acre/yr. If minimum tillage of corn had been combined with contour planting, the cropping factor, C, would have been reduced to 0.075, and the annual soil loss by sheet erosion from the field would be 4.5 tons/acre.

The rainfall factor, R, represents the average annual erosivity of rainfall. If there is a need to design land management operations on the basis of the likelihood of occurrence of erosivities higher than the average annual value, this can be accomplished by analyses of Weather Bureau records. Data summarized for several Indiana locations, e.g., show that the value of R that will be exceeded 20% of the time is about 225, compared with 185, which is the average annual value for west central Indiana. From this consideration, it can be seen that 20% of the time the sheet erosion rates for the year being considered will exceed 225/185, or 1.22 times the average value. Table 4.2 is a summary of probability values of the annual erosion index for several locations.

The soil loss for any given rainstorm can be approximated in like manner if one determines the storm erosion index, R, and lets the cropping factor, C, in the equation equal the soil loss ratio predetermined for the specific crop-stage conditions on the field at the time of the rain. Table 4.3 shows exceedence values of erosion index R for individual storms at several locations.

6. Reliability of Prediction Equation.—One measure of the reliability of the Universal Sheet Erosion Equation would be to determine the fit of measured versus computed soil loss values for the 1,200 different field plots and many cropping patterns, located at 47 research stations in 24 states, that form the basis of the equation. This detailed information is not available, but Fig. 4.8 shows the measured versus the computed soil loss from field plots, for a wide variety of

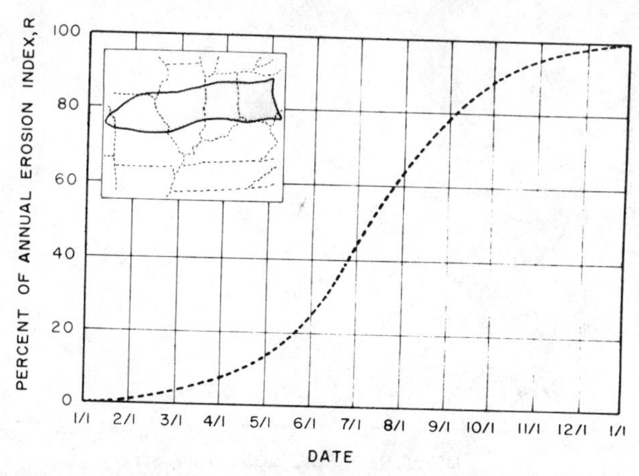

FIG. 4.7.—Seasonal Distribution of Rainfall Factor R for Parts of Missouri, Illinois, Indiana, and Ohio

environmental conditions, at seven geographic locations. These averages represent 1,082 plot-yr of data, a variety of crops, and many management systems.

Despite this large quantity of data, all combinations of conditions existing in the majority of design situations are not represented, and existing data must be extrapolated to other soil types and land-use situations. Aside from confidence in the equation by virtue of the sheer mass of data used, some measure of adequacy is obtained by evaluating the component variables (Wischmeier, 1973). Many other evaluations of the topographic, cropping, and management factors upon erosion rates have appeared in the literature (Wischmeier, 1960; Duley and Ackerman, 1934; Krusekopf, 1943) and the treatment of the rainfall factor, *R*, and soil factor *K* is analyzed as follows.

7. Evaluation of Rainfall Erosivity Factor, *R*.— In a study to determine the best rainfall "erosivity" index (Wischmeier and Smith, 1960), seven expressions, representing rainfall amount, intensity, and energy, were correlated with soil loss. The product of the kinetic energy of rainfall and the maximum 30-min rainfall intensity (the *R* factor) explained a larger percentage of soil loss variation than any other rainfall-derived parameter, with coefficients of determination from 0.71–0.89 for the five soils studied. A similar comparison (Dragoun, 1962), on a storm basis, for two field-size watersheds in Nebraska showed the *R* parameter to be better correlated with sediment yield than most variables based upon rainfall intensity and amount, but not so well correlated with sediment yield as several combinations of runoff volume, intensity, and antecedent conditions. It is reasonable to assume that the best erosivity parameter based on rainfall data would approach, but not equal, a factor that expresses runoff rates or volumes, or

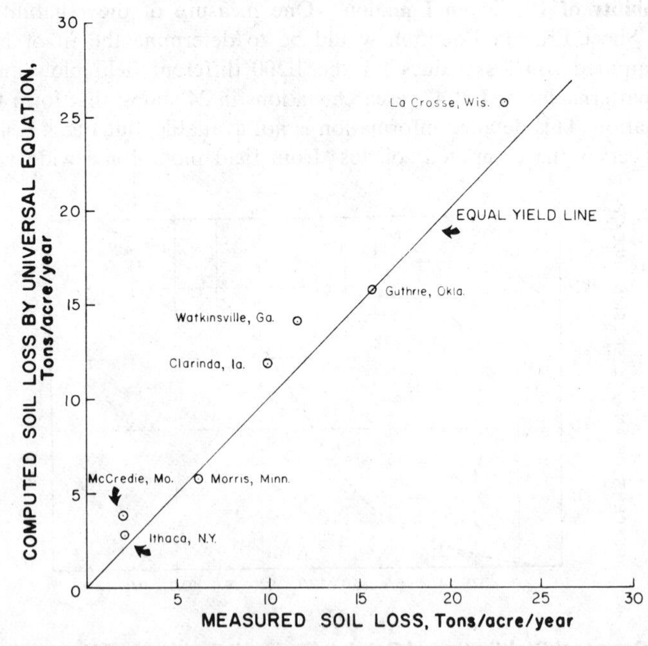

FIG. 4.8.—Comparison of Measured Soil Loss and Soil Loss Computed by Eq. 4.2 for 1,082 Plot-Years at Seven Locations

both, since runoff represents the integrated effect of all rainfall and antecedent moisture conditions.

8. Evaluation of Soil Erodibility Factor, *K*.—This soil erodibility factor is calculated after the other variables of the erosion equation have been evaluated. Basically, it is obtained by direct measurement of soil loss per unit of rainfall factor *R* on a standard test plot of 9% gradient and 72.6-ft length, tilled up and

TABLE 4.1.–Conservation Practice Factors[a] (*P* in Eq. 4.2)

Slope, as a percentage (1)	Contouring (2)	Strip-Cropping: Alternate meadows (3)	Strip-Cropping: Alternate close-grown crops (4)
1.1- 2.0	0.60	0.30	0.45
2.1- 7.0	0.50	0.25	0.40
7.1-12.0	0.60	0.30	0.45
12.1-18.0	0.80	0.40	0.60
18.1-24.0	0.90	0.45	0.70

[a]From Wischmeier and Smith, 1965.

TABLE 4.2.–Frequency Distribution of Values of Annual Erosion Index (*R* in Eq. 4.2) at Selected Locations in Indiana[a]

Location (1)	Values of Erosion Index (*R*): Observed 22-yr range (2)	Exceeded 50% of time (3)	Exceeded 20% of time (4)	Exceeded 5% of time (5)
Evansville	104-417	188	263	362
Fort Wayne	60-275	127	183	259
Indianapolis	60-349	166	225	302
South Bend	43-374	137	204	298
Terre Haute	81-413	190	273	389

[a]From Wischmeier and Smith, 1965.

TABLE 4.3.–Expected Magnitudes of Single-Storm Erosion Index, *R*, at Selected Locations in Indiana[a]

Location (1)	Index Values Normally Exceeded Once in: 1 yr (2)	2 yr (3)	3 yr (4)	10 yr (5)	20 yr (6)
Evansville	26	38	56	71	86
Fort Wayne	24	33	45	56	65
Indianapolis	29	41	60	75	90
South Bend	26	41	65	86	111
Terre Haute	42	57	78	96	113

[a]From Wischmeier and Smith, 1965.

down slope and continuously in fallow for at least 3 yr. Therefore, K is, in part, a function of residual errors from these variables in addition to being a soil erodibility function. Present attempts to refine the erodibility factor by simulated rainfall on small plots are intimately tied to more basic research to find specific soil properties that affect erodibility by water. Meyer has taken a basic step in defining the mechanics of soil erosion by rainfall splash and overland flow (Meyer and Monke, 1965), using simulated rainfall on simulated (noncohesive) soils. Grissinger has made extensive tests on the resistance to water erosion of a natural soil with various clay-mineral additives (Grissinger, 1966). Wischmeier, et al. (1971), proposed a short-cut method for determining the K factor on the basis of particle size analysis, organic matter content, soil structure, and permeability.

9. Concepts of Allowable Soil Loss, "Normal" Erosion, and Soil Renewal.—The Universal Soil Loss Equation is a versatile and useful tool that can be applied to many problems of soil and water management. Although sometimes used to delineate sediment problems far downstream from the source areas of erosion (as explained in subsequent sections on sediment delivery and yield), it is also an aid to land management planning in the agricultural source areas. In recognizing land of a given capability and wishing to develop an economic utilization of it, a soil loss tolerance, T, may be substituted for E in the equation; the solution would then be in terms of permissible values of C and P (cropping and practice factors) for any set of existing or assumed conditions.

Any program for long-term management of the land resource must, at once, consider its economic utilization and its permanent preservation or improvement. To accomplish these objectives, researchers must interpret erosion rates, for most contemplated land uses, in terms of soil renewal rates by normal geologic processes and by special fertilization and treatment programs. They must also determine the allowable soil loss that will sustain a given soil capability and learn something about the recuperative ability of soils that are subjected to varying degrees of abuse.

It is known, e.g., that an annual erosion rate greater than the estimated 0.1-ton/acre normal geologic rate on Marshall soil at Clarinda, Iowa, will not destroy essential soil properties because of the favorable depth and quality of subsoil; a 5-ton/acre allowable erosion rate has been proposed for this soil in state legislation. By contrast, a 3-ton/acre tolerance on the claypan soils of Missouri would aggravate the present deficiency of soil depth unless renewal practices (e.g., heavy fertilization and perhaps even uphill plowing) could restore the total soil depth.

Rates of normal geologic erosion have been approximated for numerous soils (Smith and Stamey, 1965) by measuring the soil loss on small plots planted to close-growing vegetation. Results of these investigations indicate that the normal geologic erosion rate from vegetated areas is commonly less, and usually much less, than 0.30 tons/acre annually. Knowledge of C factors for pasture, meadow, idle land, and woodland (Wischmeier, 1972) can provide closer approximations of normal geologic erosion rates for specific sites. Also, the erosion rate under conditions of natural vegetation varies with the mean annual rainfall of a region. It usually reaches a maximum when the mean annual rainfall is between 10 in. and 15 in., because improved vegetation inhibits erosion under wetter conditions, while sediment-entraining runoff becomes more rare when mean annual rainfall is less than 10 in. (Langbein and Schumm, 1958).

Other scientists have calculated geologic erosion rates for large regions on the basis of age, position, and thickness of ancient sedimentary rock (Krusekopf, 1958; Kulp, 1961) and unconsolidated Pleistocene glacial and loessial deposits (Ruhe, Meyer, and Scholtes, 1957). Menard (1961) has compared geologic erosion rates derived from the volume of deposits in "closed systems" with recent measurements of the sediment discharge of these river basins. He finds that geologic and present deposition rates of the Mississippi Basin are nearly equal, at 0.55 tons/acre/yr and 0.50 tons/acre/yr, respectively. Deposition rates of the Appalachian Region in geologic time were higher than present rates, 0.74 tons/acre/yr compared with 0.10 tons/acre/yr; geologic deposition rates of the Himalaya Region were lower than at present, 2.62 tons/acre/yr and 12.0 tons/acre/yr, respectively.

Soil renewal takes place through weathering of subsurface and surface rocks and soil and by surface deposition. Estimates of soil renewal have been based on broad geologic interpretations (Krusekopf, 1958) and on the determination of crop yield recoveries with good soil management (Free, 1960; Smith, et al., 1967). Smith, et al. (1967), reported a partial comeback of an Austin clay soil 30 yr after it had been desurfaced to a depth of 15 in. They also cited an Ohio experiment whereby the top 8 in. of Canfield silt loam was removed. In both cases, crop production improved rapidly when good fertility treatments were enacted during years of favorable rainfall; with less favorable seasons, or with no fertilization, little cropping improvement was shown.

Erosion and deposition rates during near-geologic, historic, and recent times have been determined for a number of locations. Table 4.4 lists rates for several areas. The Iowa valleys are relatively simple landscape units where erosion and deposition rates can be measured with some degree of confidence. The historic erosion rate at the Adair County site averaged 6 in./100 yr, or about 9

TABLE 4.4.—Erosion Rates during Near-Geologic Past and Historic Times, Determined for Selected Areas

Location (1)	Drainage Area: Total, in acres (2)	Drainage Area: Erosion, in acres (3)	Drainage Area: Deposition, in acres (4)	Time, in years (5)	Erosion rate—valley sides and head slopes, in inches per year (6)
Side Valley Adair County, Iowa	4.4	1.7	2.7	0-125[a]	0.060
		4.4	—	125-6,800[a]	0.006
Colo Bog Story County, Iowa	90	—	—	0-3,100[a]	0.001
				3,100-8,300[a]	0.004
Toby Tubby Watershed Lafayette County, Mississippi	36,100	31,300	4,800	Since settlement (about 1840-1937)	0.055[b]
Hurricane Watershed Lafayette County, Mississippi	20,500	18,400	2,100	Since settlement (about 1840-1937)	0.051[b]
Kickapoo Valley Wisconsin	492,000	437,000	—	1880-1940 (about 60 yr)	0.017[b]

[a] Before present.
[b] A minimum estimate.

tons/acre/yr; this was about 10 times the geologic erosion rate (Ruhe and Daniels, 1965). The Story County site illustrates the variation in erosion rate during near-geologic times as a result of changing vegetation, landform, and climate. At Colo Bog, the maximum rate of soil removal was about 0.5 in./100 yr. Walker (1966) assessed this to be less than the horizon development of Clarion

FIG. 4.9.-Gully Erosion on Cultivated Uplands in Western Iowa: (a) Aerial View Showing Most of 83-acre Watershed and Its Drainage Network; (b) Upstream Movement of Gully Head, 1964-1968 is 130 ft

and Webster soils on upper watershed slopes of the region during postglacial time, which was at the rate of 3 in./100 yr.

The drainage basin surveys in Mississippi show that recent erosion rates, 1937–1965, have been reduced by at least 50% from the agriculturally accelerated rates for the preceding 100 yr. Happ's review (1968) of the situation makes the point that severe soil erosion was noted in the region only 20 yr after white settlement and did not decline significantly until the 1940's.

10. Erosion in Gullies.—The concentration of runoff under some circumstances encourages the formation of gullies. Fig. 4.9 shows the development of a typical gullied drainage system. Overland runoff during periods of gully advance, which are shown in Fig. 4.9(b), is given in Table 4.5. The runoff over the headcut of Fig. 4.9(b) contributes to gully growth by: (1) Exerting forces on the channel boundary; (2) removing accumulated soil debris from the channel; and (3) eroding gully banks by undercutting them and by gravity (moisture) loading to a value above the critical shear strength (Piest, et al., 1975).

Gullies, or upland channels, are common to most regions, and their development is usually associated with severe climatic events, improper land use, or changes in stream base levels. Gully growth patterns can be cyclic, steady, or spasmodic and can result in the formation of continuous or discontinuous channels. Gullies also form on the perimeter of upland fields and actively advance into these fields. Most of the significant gully activity, in terms of quantities of sediment produced and delivered to downstream locations, is found in regions of moderate to steep topography having thick soil mantles. The total sediment outflow from eroding gullies, though large, is usually less than that produced by sheet erosion (Leopold, et al, 1966; Glymph, 1951), although the economic losses from dissection of upland fields, damage to roads and drainage structures, and deposition of relatively infertile overwash on flood plains are disproportionately large.

Gully advance rates typically have been obtained by periodic surveys, measurements to steel reference stakes or concrete-filled auger holes that are placed in the gully head and bank, or examination of gully changes from existing small-scale maps or aerial photographs, or combinations of these. The gully erosion process has been admirably described for several regions of the United States (Ireland, et al., 1939; Brice, 1966), but the cause-effect interrelationships of gully formation have never been put into proper perspective. Methods are,

TABLE 4.5.—Measured Gully Head Advance and Related Surface Runoff Volume for Intervals Shown in Fig. 4.9(b)

Period (1)	Surface runoff, in acre-feet (2)	Gully erosion, in tons (3)
December, 1964-April, 1965	25	130
May, 1965-September, 1966	45	1,120
October, 1966-December, 1968	79	1,560
January, 1969-December, 1973 (Not on photo)	85	1,160
Total	234	3,970

therefore, not available for any given locality and under any set of existing or assumed conditions, for accurately predicting rates of gully erosion or gully advance. However, studies are producing quantitative information and some empirical prediction procedures have been advanced. In Mississippi, the measured sediment production from active gullies with little drainage area other than the raw gully head ranges to 7 in. annually at the gully exit and is related to gully relief, exposed area, and the nature of exposed materials (Miller, et al, 1962).

For a severely gullied loessial area of western Iowa, the following relationship has been developed for predicting gully growth (Beer and Johnson, 1965):

$$X_1 = 0.01 X_4^{0.982} X_6^{-0.044} X_8^{0.7954} X_{14}^{-0.2473} e^{-0.036 X_3} \qquad (4.3)$$

in which X_1 = growth in gully surface area acres, for a given time period; X_3 = deviation of annual precipitation from normal, in inches; X_4 = index of surface runoff, in inches; X_6 = terraced area of watershed, in acres; X_8 = gully length at beginning of period, in feet; and X_{14} = length from end of gully to watershed divide, in feet. Some of the variables of Eq. 4.3, e.g., X_6, are not universally applicable, e.g., $X_1 = 0$ when $X_6 = 0$.

A field study of gully activity in several locations throughout the United States has resulted in the tentative relationship (Thompson, 1964)

$$R = 0.15 A^{0.49} S^{0.14} P^{0.74} E^{1.00} \qquad (4.4)$$

in which R = average annual gully head advance, in feet; A = drainage area, in acres; S = slope of approach channel, as a percentage; P = annual summation of rainfall, in inches, from rains equal to or greater than 0.5 in./24 hr; and E = clay content of eroding soil profile, as a percentage by weight.

The United States Soil Conservation Service (1966) basis for the solution of field design problems involving gullying is the equation

$$R = 1.5 A^{0.46} P^{0.20} \qquad (4.5)$$

in which R and A are defined as in Eq. 4.4 and P = the summation of 24-hr rainfall totals of 0.5 in. or more occurring during the time period, converted to an average annual basis, in inches.

The Agency recognizes that other inadequately defined factors influence the headward advance of gullies and prefers to account for these factors by utilizing past gullying rates calculated from maps or aerial photos. The actual prediction equation is then

$$R_2 = R_1 \left(\frac{A_2}{A_1}\right)^{0.46} \left(\frac{P_2}{P_1}\right)^{0.20} \qquad (4.5)$$

in which the subscripts 1 and 2 refer to past and future, respectively.

Seginer (1966), in a review of research finds that gully erosion has been most prominently related to size of drainage basin or some intercorrelated variable. He has suggested, therefore, that the gully erosion problems of a locality can be evaluated from an equation of the form

$$R = C A^b \qquad (4.6)$$

in which R = average annual lineal gully advance determined by historic or geologic study, or calculated; A = area of drainage basin; and C and b = constants. Further, since the headcuts of all continuous gullies in a drainage system reach their present position by migration from a common origin, C is the same for all gullies of the system. The relative gully advance rate then depends upon b. If $b = 0$, the average advance rate of all gullies would be the same; intermediate values of b would show the influence of watershed size which, in turn, represents the integrated effect of many primary hydrologic variables of the region. The b variation between regions and between watersheds of a given region could also reveal the effect of soils, topography, land use, and management. This procedure would allow conservation planners to place proper emphasis on problem areas.

11. Erosion in Channels.—Channel erosion, which includes stream bed and stream bank erosion, can be very significant under some circumstances. Accelerated stream bed erosion, e.g., can cause the lowering of ground-water levels, and in water-short areas, this can drastically reduce the yield of such bottomland crops as corn and alfalfa. It can also trigger downcutting cycles in tributary channels and gullies because of the lowering of the base level. Stream bank erosion, often occasioned by the clearing of protective cover from banks and from channel straightening and realinement measures, affects the flow through changes in slope and in stream competence.

For channels in noncohesive sediments, a tool for qualitative prediction of erosive channel conditions is Lane's relationship (Lane, 1955b)

$$QS \propto G_s d_s \qquad (4.7)$$

in which Q = stream discharge; S = longitudinal slope of stream channel; G_s = bed sediment discharge; and d_s = particle diameter of bed material. Alteration of one variable will indicate the effect upon the others. This proportion is particularly useful when two of the variables can be assumed to remain constant.

Quantitative estimates of channel erosion or deposition rates are obtained from time sequence comparisons of surveyed cross sections, from maps and aerial photographs, and from historical records. For example, the writer has channel cross sections taken by the United States Soil Conservation Service in Iowa that substantiate a degradation of more than 20 ft of the stream bed (and commensurate channel widening) of West Tarkio Creek above the Iowa-Missouri state boundary, 1920–1970. United States Geological Survey streamflow measurement summaries of the One Hundred and Two River near Bedford, Iowa, showed a 5-ft lowering of the average stream bed elevation.

Predictions of future channel changes are based on erosion or deposition rates as estimated by the preceding methods; when future changes in the flow regime are expected, rough estimates of scour or fill can be obtained from sediment discharge formulas (Lane and Borland, 1951; Einstein, 1950; Colby and Hembree, 1955), the application of principles of fluvial morphology (Leopold and Maddock, 1953), the use of Regime theory (Blench, 1957), or other methods that consider the forces exerted on the stream boundaries (Lane, 1955a).

12. Erosion on Western Rangelands.—Erosion rates on the western rangelands of the United States are highly variable because of extremes of rainfall, soils, geology, topography, and vegetation. Leopold, et al. (1966) and Lusby (1963) believe that sheet-rill erosion on these rangelands is generally more dominant than channelized erosion. In one study of paired watersheds (Lusby, 1963), with one set of watersheds located on steep shaly terrain and the other set having gentle slopes and sandy soil, runoff rates were similar because frost action at these high altitudes kept the shale-derived soils as permeable as the sandy surface. But erosion rates as evidenced by measured sediment yields from shale areas were much greater. A later report on these highly erodible Mancos shale watersheds (Lusby, 1970) showed the effect of normal grazing versus no grazing. Over a 12-1/2-yr period, grazed watershed sediment yields averaged 9.5 tons/acre/yr compared with 6.5 tons/acre/yr for ungrazed areas. No change was noted between grazed and ungrazed watersheds the first year; thereafter, the sediment yield from the grazed areas averaged 1.8 times those from ungrazed areas.

An 8-yr study of causes of reservoir sediment deposition in four rangeland watersheds also showed that the condition of range vegetal cover significantly affected an erosion parameter. Sediment yields for all watersheds were less than 0.1 tons/acre/yr.

13. Erosion along Rights-of-Way.— Since areas of exposed soil sustain disproportionately high erosion losses when compared with heavily vegetated areas, it is of some interest to examine the sediment production from roadways, banks, and road ditches. Erosion rates during right-of-way construction are even higher, and compare with those from urban construction and strip mining (Vice, et al., 1968). In the United States, there are more than 3,000,000 miles of rural public roads, and the interstate highway building program is developing another 41,000 miles of new or improved highways. With a minimum 200-ft right-of-way, at least 1,000,000 acres are being reshaped by man for the interstate system. Many state highway departments, in cooperation with the United States Bureau of Public Roads, have set standards for roadside erosion control (McCully and Bowmer, 1969).

Most erosion damages to rights-of-way occur on unprotected roadbanks and in roadside ditches that must convey large volumes of runoff. Studies of erosion rates from unprotected roadbanks that were cut into a clay subsoil in Georgia (Diseker and McGinnis, 1967) show good correlation with bank slope, slope aspect, annual average temperature below 32°F, and a soil moisture parameter. Fig. 4.10 shows that the largest soil losses over a 5-yr period occurred during the January-March period of low temperature and high soil moisture. Although the rainfall erosivity factor, R, explained some of the variability in soil loss, high R values were not necessary for high soil losses in the first quarter of the year.

Road ditch erosion is dependent upon the same hydraulic factors that pertain to stable channels. Roadway erosion is confined mainly to the secondary highway system in the United States, where some roads are not hard-surfaced or improved to the extent that they can resist erosion. Many secondary unpaved roads in the loessial areas of the country have eroded into gullies and waterways; Ziemnicki and Jozefaciuk (1965) have also observed this phenomenon in loessial areas of Poland. Forest logging roads and skid trails can be important sediment sources (Lull and Reinhart, 1963).

14. Urban Erosion Sources.—The task of evaluating urban erosion sources and controlling the damages resulting therefrom, both on-site and downstream, involves a consideration of similarities and differences between urban and rural erosion phenomena. Procedures and guidelines presently available for measuring, predicting, and controlling agricultural erosion rates are an invaluable starting point for dealing with urban erosion problems.

There are many sources of sediment in urban areas, but the recent burst of home building, highway construction, and other activities involving earthwork in metropolitan areas of the United States is causing special concern. Even the sediment from "stabilized" sections of a city, from streets and gutters, can be significant when it is delivered to stream channels incapable of transporting it, or to estuaries that must be maintained by dredging. This type of debris entering the Cuyahoga River at Cleveland, Ohio, e.g., adds to the sediment deposition in the

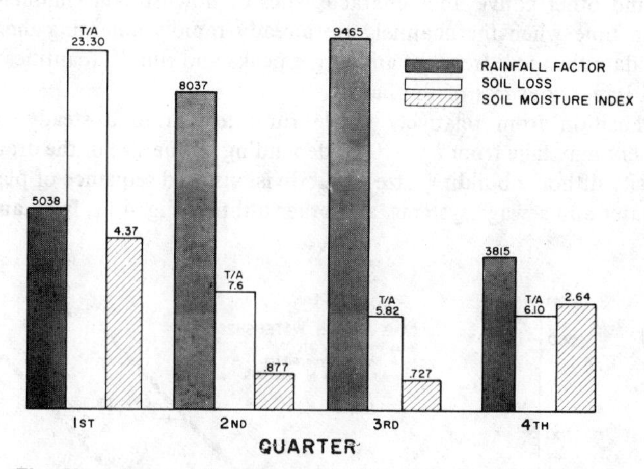

FIG. 4.10.—Five-Year Record (by Quarters) of Soil Loss, Rainfall Index, and Soil Moisture Index from Two Roadside Plots

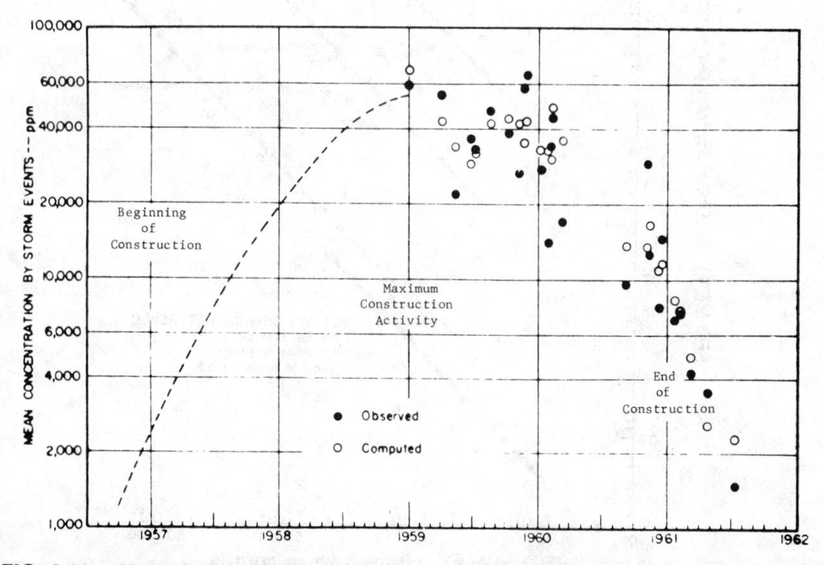

FIG. 4.11.—Mean Sediment Concentration of Storm Runoff from Area of Residential Construction at Kensington, Md., 1957-1962

river that must be removed, at a cost of about $1/cu yd, to maintain the navigation channel and harbor.

Erosion considerations in areas undergoing urban development are unique in some respects. The soil loss rates are many times the preconstruction rates. Damages are usually more serious to downstream landowners and to the affected municipality than to the developer of a new housing tract. Exposed subsoils often contain larger particles with less cohesive material and thus pose special problems because little is known of their permeability, erodibility, and related properties. The larger particle sizes of eroding subsoils can affect the width-depth relation and other conveyance characteristics of downstream channels (Dawdy, 1967) at a time when the channels are already rapidly enlarging themselves to accommodate the more frequent and larger peaks and runoff quantities due to the increased imperviousness of the basin.

The transition from relatively stable rural terrain to a steady-state urban environment may take from 2 yr–10 yr, depending on the size of the drainage area, the intensity of home building, size of subdivisions, and sequence of placement of streets, water and sewage systems, and other utilities. Fig. 4.11, from an article by

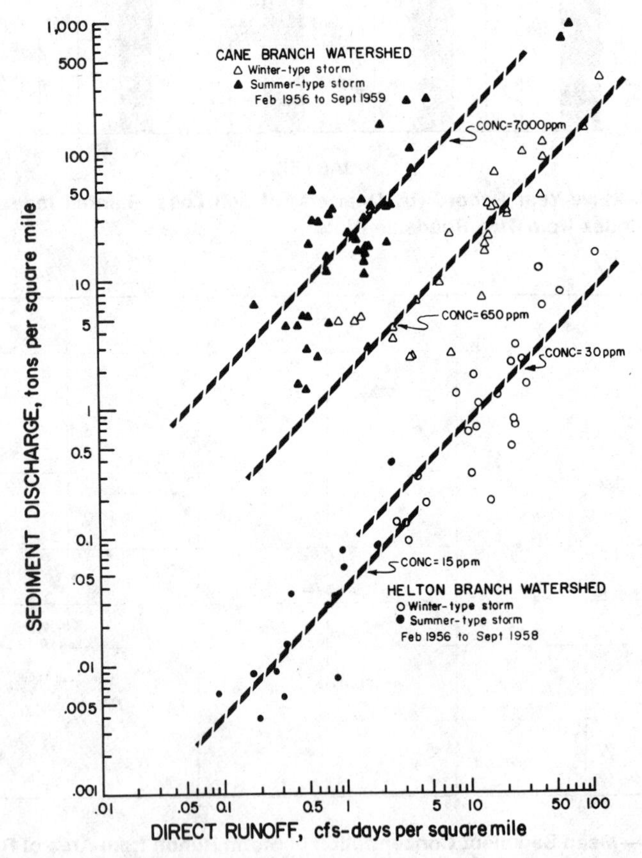

FIG. 4.12.—Relation of Sediment Discharge to Water Discharge, by Storm, for Cane, and Helton Creek Watersheds, Kentucky

Guy (1965), shows the mean sediment concentration of streamflow, as observed or estimated, for a 58-acre drainage area at Kensington, Md., during a 5-yr rural-to-urban cycle. The interpretation, based on 25 gaged and sampled storm runoff events during this period, is that the soil removed from 20-1/2 acres undergoing actual construction is 3,880 tons. About 70% of this quantity was discharged from July, 1958 to July, 1960, and corresponds to an annual soil loss rate of 66 tons/acre.

This rate is probably a realistic approximation of the norm during the peak of urban construction. Surveys of sediment deposits in Lake Barcroft, Virginia (Geiger and Holeman, 1959; Guy and Ferguson, 1962) when related to urban development in the upstream watershed, show an annual soil loss rate of 39 tons/acre. This lower figure has been attributed to such differences as a larger drainage basin, greater sediment storage in the upstream flood plain and stream channel above Lake Barcroft, the more methodical development of the latter subdivisions, and the building of a larger proportion of custom homes.

A direct relationship between urban and rural erosion rates cannot be cited, but a recent study of two 80-acre watersheds in western Iowa provides some basis for comparison. The watersheds, planted to corn in rows that are approximately on contour, have been intensively streamgaged and sampled since 1964. Interpretations (Piest and Spomer, 1968) that are based on a consideration of rainfall amounts, intensities, and kinetic energies show that a 20-ton/acre annual soil loss can be expected under average rainfall conditions. Further examination shows that the bulk of the sediments leaving these watersheds are transported during the early crop stage when the soil is freshly tilled and is particularly vulnerable to erosive action.

It can be concluded, therefore, that an average 40 tons–50 tons of sediment per acre per year can reasonably be expected during urban development when the soil surface is unvegetated and subject to constant reworking.

The implications of the foregoing comments are not lost on engineers concerned with urban erosion rates during this transient period of construction. It is difficult to design sediment retention works for urban watersheds with any degree of efficiency since, during the short period of instability, abnormal climatic conditions could cause large differences in soil loss. In the case of the Iowa watersheds, 92% of the total soil removed from the watershed, 1964–1973, by sheet and rill erosion took place during May and June. During the severe runoff season in June, 1967, the average soil loss from these two watersheds was 75 tons/acre (Spomer, et al., 1970).

15. Erosion and Sediment Movement from Strip Mining Operations.—Strip mining, particularly in coal fields of the eastern United States, is an important source of sediment. Spoil banks resulting from contour stripping are often steep-sided and consist of a loose, heterogeneous mixture of soil and unweathered rock materials that are devoid of vegetation. In forested areas of Appalachia where the natural sediment yields are low, erosion from extensive areas of unreclaimed spoil banks has caused serious sedimentation problems in the streams (Collier, 1969).

A study of the effects of strip mining on the hydrology of Cane Branch Watershed in southern Kentucky revealed that an average annual sediment yield of about 27,000 tons/sq mile was measured from that portion of the watershed with unreclaimed spoil banks (Collier, et al., 1964, 1971). By contrast, the annual sediment yield from unmined watersheds in the area was about 25 tons/sq mile.

Fig. 4.12 shows the sediment transport relation, for summer and winter storms, of Cane Branch Watershed (with about 10% of the area strip-mined) and the naturally vegetated Helton Branch watershed (Collier, 1969). The mean storm sediment discharge from the strip-mined watershed is approximately the same the year around, although the summer concentrations of storm flow are much higher. The slow, natural revegetation of the spoil banks has not been sufficient to appreciably reduce the rate of weathering and erosion of the spoil material. Whereas the sheet erosion rate has gradually decreased, the erosion rate from gullies cut into the steep slopes has increased with time.

16. Other Sediment Sources and Erosion Types.—Sediments also originate from construction activities, logging operations (Lull and Reinhart, 1963), excavation and dredging for sands and gravels, mining, and flood-plain scour. Regardless of the erosion source, the movement of sediment is maximized by factors that enhance the processes of erosion and overland flow. Factors that affect erosion do not necessarily affect overland flow in the same manner. High erosion rates are a function of soil erodibility, high rainfall energies and intensities, steep and long landslopes, sparse vegetal cover, and poor land treatment. Overland flow rates are related to most of these variables although such runoff-inducing soil properties as high clay content and low permeability are not well correlated with erodibility.

Often a careful choice of variables is possible that will lessen sediment movement. For example, a minimum or no-till cultural treatment will usually not reduce runoff from agricultural lands (and may increase it because of reduced evaporation and greater moisture retention in the soil surface profile) but will drastically reduce erosion rates and sediment movement.

Erosion by runoff has been treated in the foregoing sections. Wind is another important eroding agent in many localities, and in geologic time it played a major role in shaping the landscape.

Other erosion processes result in the gravity transport of materials along slopes, with or without the benefit of the dynamic forces of flowing water or wind. These mass movements can be subdivided into many types of flows, slides, and falls (Leopold, et al., 1964). Quantitative evaluation of these mass supply processes is usually lacking, especially for the more subtle debris flows. Studies (Rice and Foggin, 1971) in the San Dimas Experimental Forest in southern California showed that soil slippage occurred on one-sixth of the grass areas but on only one-twentieth of the brush-covered areas during the intense winter storms of 1969. Other factors affecting slippage were rainfall and aspect. The slip threshold was lowered from an 80% slope for the 1966 winter storm period to 60% for the more intense rainfall in 1969. North-facing slopes were less stable than southerly aspects. Plastic flows have been observed in South Dakota Badlands topography (Schumm, 1956) and many other locations. The mechanisms of mass movement are detailed in soil mechanics and rheology discussions. Scheidegger (1961) has reviewed theories of deformation and stability of earth masses that relate to geomorphological problems.

C. Yield from Sediment Sources

17. Sediment Routing and Delivery.—Not all of the eroded material is effectively sluiced through the river systems and delivered to the sea. The rate at which sediment is discharged to the oceans is usually less, and often much less, than one-fourth of the rate at which it is eroded from the land surface. The bulk of the sediment is deposited at intermediate locations wherever the entraining runoff waters are insufficient to sustain transport. It is scattered to adjacent down-slope positions; it is deposited at the base of eroding slopes and at the bluff line bordering major river valleys; it overlays the flood plains; and it clogs stream channels and reservoirs. Much of the deposited sediment is the result of geologic erosion; it is natural, inevitable, and sometimes is even beneficial. Even so, the amount and location of these distributed sediments, as they pause on their march to the sea, are of vital concern to man in the design of engineering works.

The percentage of sediment delivered from the erosion source to any specified downslope location is affected by such factors as size and texture of erodible material, climate, land use, local environment, and general physiographic position. In some cases, the collecting waterways immediately adjacent to the sediment source can retard 75% or more of the eroded soil (Forest Service, 1965; Williams and Berndt, 1972). It has been postulated that a particle in the stream bed of Seneca Creek, Maryland, may be in transport only about 0.3% of the time. This discontinuous movement would also apply to a particle settled on the flood plain by overbank flows or by lateral accretion of Seneca Creek, where it would take more than 1,000 yr, on the average, to be reexposed for further transport (Leopold, et al. 1964).

An extreme example of the downstream dimunition of sediment discharge because of channel aggradation is cited for a glacier-fed Alaskan stream (Borland, 1961). The average annual sediment discharge for an 868-sq mile drainage area was 9,120 acre-ft; 135 miles downstream, the annual sediment discharge decreased to 6,440 acre-ft, although the drainage area increased to 6,290 sq miles and the runoff was nearly trebled.

The change (per unit area) of this downstream sediment movement, from the source to any given measuring point, is expressed by the delivery ratio, D, defined by

$$D = \frac{Y}{T} \qquad (4.8)$$

in which Y = the sediment yield at the measuring point, in tons per acre per year; and T = the total material eroded from the watershed and drainage system upstream from the measuring point, in tons per acre per year. In common usage, the sediment delivery ratio is considered a percentage, or $100(Y/T)$.

A foreknowledge of the sediment delivery ratio is very useful in planning a wide variety of water utilization and control structures, e.g., dams, diversion channels, and debris basins. Under ideal circumstances where total upstream sheet, gully, channel, and other erosion can be calculated (by procedures already outlined), the design of a specific water-utilization project can be optimized by allowing for (or altering) the expected downstream sediment yield.

As with gully and channel erosion estimation procedures, there are no generalized delivery relationships available that can be applied to every situation. However, several researchers show trends in the sediment delivery ratio for specific areas.

Using data acquired from field investigations in the southeast Piedmont region of the United States, the following relationship for the delivery ratio was developed (Roehl, 1962):

$$\text{Log } D = 4.5 - 0.23 \log 10W - 0.51 \log \frac{L}{R} - 2.79 \log B \quad \text{......................} \quad (4.9)$$

in which D = the sediment delivery percentage; W = the drainage area, in square miles; L/R = the dimensionless basin length-relief ratio (watershed length, as measured essentially parallel to the main drainageway divided by elevation difference from drainage divide to outlet); and B = the weighted mean bifurcation ratio. (Bifurcation ratio is the ratio of the number of streams of any given order to the number in the next higher order.)

From the Piedmont, delivery ratios were found to vary from 3.7%–59.4% (see Table 4.6). A similar study for the Red Hills physiographic area of southern Kansas, western Oklahoma, and western Texas produced the relation (Maner, 1958)

$$\log D = 2.943 - 0.824 \log \frac{L}{R} \quad \text{...} \quad (4.10)$$

In Illinois (Ackerman and Corinth, 1962), studies of deposited reservoir sediment quantities have been related to drainage basin characteristics, i.e.

$$P = 0.43\, E\, T\, \frac{I^{1/3} S^{1/2}}{F^{3/4}} \quad \text{...} \quad (4.11)$$

in which P = sediment deposition in the reservoir, in tons per acre per year; E = total upstream erosion, in tons per acre per year; T = that portion of the total sediments in transport in a stream that are trapped by a given reservoir, expressed as a percentage; F = the dimensionless ratio of principal tributary length to equivalent watershed diameter (equivalent watershed diameter being the diameter of a circle that has an area equal to that of the watershed); I = the ratio of direct tributary annual inflow, in acre-feet, to the top foot of reservoir surface, in acre feet; and S = the mean slope of the first-order streams, minus the mean

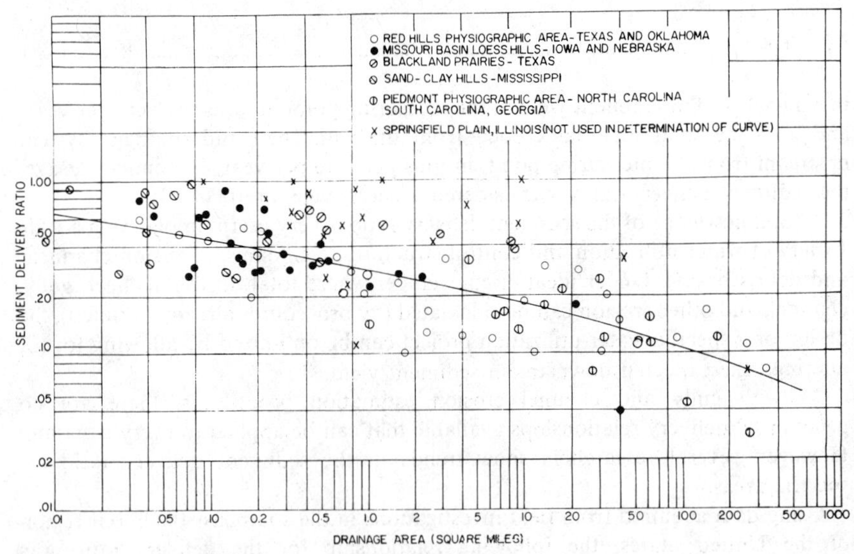

FIG. 4.13.—Relationship between Size of Drainage Basin and Sediment Delivery Ratio

slope of the second-order streams. The sediment delivery ratio for this area, P/ET, can then be expressed in terms of basin shape, slope, and runoff.

Relationships have been developed for the loess hills area of Nebraska and Iowa (Gottschalk and Brune, 1950) and Texas Blackland Prairies (Maner and Barnes, 1953). These are summarized in Fig. 4.13. This decrease in sediment delivery ratio with increasing drainage area is typical of most locations studied. The delivery ratio usually decreases with increasing drainage area in a basin that is relatively homogeneous with respect to soils, climate, and topography, but large downstream increases in erosion rates in a nonhomogeneous basin can increase the delivery ratio. In the future, it may be possible to accurately define curves like Fig. 4.13 for each "homogeneous" subregion of the United States and then to further explain the point scatter from these curves in terms of the major variables affecting erosion and soil movement across the land surface and through the channel systems. This space-time journey of average sediment particles from a point source (which can be considered synonymous with a fractional-acre "watershed") all the way to the ocean could then be portrayed in any desired manner.

The first step toward this goal could be the portrayal of the sediment yield (or sediment delivery ratio) versus drainage area relationship. Fig. 4.14 is based on work by Fleming (1969), Glymph (1951), and the Interagency Task Force (1967), and Fig. 4.15 is based on work by Renfro and Adair (1968), Roehl (1962), and Glymph (1954). The ordinates of both figures differ only by a constant if we assume that total erosion, T, is not a function of watershed size in a so-called homogeneous region. That is, the sediment delivery ratio, $D = kY$, where $k = 1/T$ in the relation $D = Y/T$ of Eq. 4.8.

TABLE 4.6.–Delivery Ratio and Related Variables for Reservoir Watersheds in Piedmont[a]

Name of reservoir (1)	Drainage area, W, in square miles (2)	Stream length, D, in miles (3)	Relief/length, R/L (4)	Weighted mean bifurcation ratio, B (5)	Sediment delivery, as a percentage, $100D$ (6)
Apex	2.2	1.99	0.00572	4.61	17.2
High Point	62.3	11.27	0.00417	4.13	11.9
University (N.C.)	30.3	4.18	0.01451	4.05	20.5
Roxboro	7.52	3.90	0.01116	4.51	12.7
Burlington	105.0	13.18	0.00302	4.76	3.7
Chester	15.92	5.83	0.00926	4.11	12.9
Lancaster	9.34	5.66	0.00669	4.46	10.2
Cannon	17.7	2.54	0.01568	4.60	17.9
Concord	4.54	1.39	0.02446	4.45	28.6
Lexington (N.C.)	6.66	7.09	0.00601	4.92	12.6
Issaqueena	13.86	6.35	0.01238	4.72	14.9
Michie	166.7	22.15	0.00440	4.00	9.2
Carroll Lake	6.9	2.42	0.01838	3.28	59.4
Temple Reservoir	0.61	1.01	0.03580	4.92	55.0
Lake Brandt	73.4	11.72	0.00396	3.98	8.5

[a]After Roehl, 1962.

Fig. 4.14 includes data taken from three sources. The outer envelope actually encompasses a series of 11 parallel lines, each of which best represents the observed sediment yields from one or more of the 17 land resource areas that comprise the 188,000-sq mile Upper Mississippi River Basin (Interagency Task Force, 1967). Fig. 4.16 (Interagency Task Force, 1967) summarizes sediment yield information for three land resource areas in the Rock River Subbasin; 19 suspended-sediment sampling locations and one surveyed reservoir were used in the study. Observed sediment yields for the entire basin consisted of streamflow sediment measurements at 121 sites and reservoir sediment deposition measurements at 132 locations. All data were adjusted (sometimes extrapolated) to a 20-yr base period, 1945–1964.

The inner envelope of Fig. 4.14 encloses similar data from 51 watersheds in 20 states (Glymph, 1951). The single line from Fleming (1969) represents an average trend based on 250 watersheds on four continents. All lines of Figs. 4.14 and 4.15 can be represented by an exponential relation such as $Y = aA^b$, in which Y is the ordinate. The ordinate is sediment yield in tons per square mile in Fig. 4.14 or (sediment yield ÷ total erosion rate) = sediment delivery ratio in Fig. 4.15; A is the drainage area in square miles. The slopes of all lines but Fleming's are about -0.12. Fleming's relationship has a slope (exponent) of only -0.04. His sediment yield records are from larger watersheds, and there is some evidence that the sediment delivery ratio versus drainage area relationship becomes asymptotic to some minimum sediment delivery ratio. This would cause a flatter slope.

With the foregoing assumptions that: (1) Both sediment yields and sediment delivery ratios vary approximately with the $-1/8$th power of drainage area; and (2) a fractional-acre field research plot can be considered the "point source" reference area that includes all soil eroded regardless of location or distance of

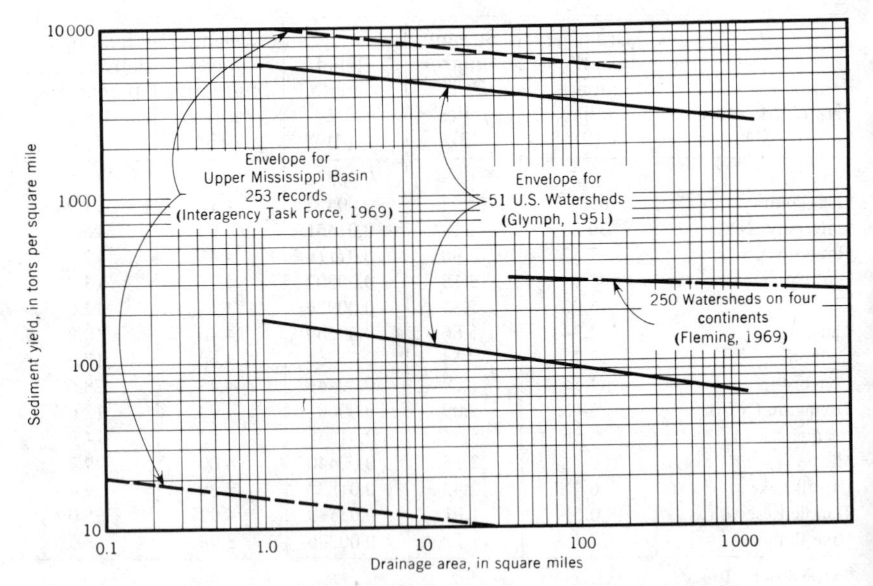

FIG. 4.14.—Effect of Watershed Size on Sediment Yield

movement—the rate of sediment movement for the average drainage area contributory to the sea is attenuated to about 10% of the erosion rate at the "point source." This is illustrated by computing the sediment delivery, $D = aA^b$, for a 10,000-sq mile area; D = delivery percentage; A = drainage area in square miles; b is assumed a constant, $-1/8$; and a = a constant to be evaluated by using a 0.001-sq mile "watershed" as a "point erosion source" with 100% delivery. Substitution of the constant into the equation for a 10,000-sq mile drainage area gives a delivery value of 13%. The selection of a 0.001-sq mile area as the "reference size" for point erosion is arbitrary, but the substitution of a smaller reference size will not greatly affect the computed sediment delivery ratio.

Sediment delivery ratios between any two points in a drainage system could be approximated by the type of information shown in Table G-7 of Interagency Task Force (1967). In that report, the annual sediment yields were tabulated according to land resource area and watershed size. The table shows, e.g., that the annual sediment yields for 10,000-sq mile subbasins of the Upper Mississippi River Basin are only about one-third of the sediment yields for 1-sq mile tributary basins. This reduction was approximated for all land resource areas in the Upper Mississippi Basin, even though sediment yields per unit area for some basins were several hundred times those for other basins.

Happ's early study (Happ, et al. 1940) of subwatersheds of the Tallahatchie River Basin, Mississippi, showed historic upland erosion rates varying from 0.051 in.–0.086 in. annually (8 tons/acre/yr–13 tons/acre/yr). But only about 0.28 in. of this soil in 100 yr (0.0028 in. annually) was estimated to reach the downstream sampling section at Greenwood, with a 7,700-sq mile drainage area. Sediment delivery from these upland erosion sites where soils are somewhat sandy was therefore about 6% or less.

Holeman (1971) has reviewed the background of efforts to determine sediment delivery rates. He concludes that the initial estimate by Brown (1950) of a 4-billion ton total erosion rate for the contiguous United States (of which perhaps 2 billion tons washes into streams and 1/2 billion tons reaches tidewater) has not been

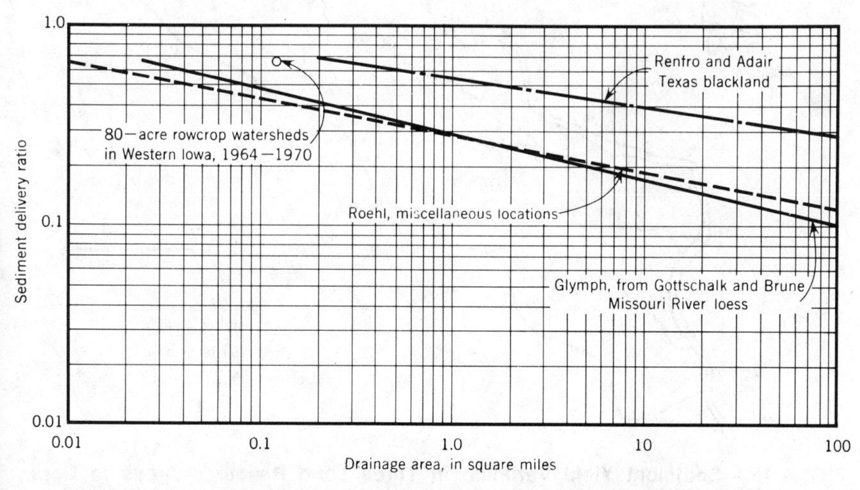

FIG. 4.15.—Effect of Watershed Size on Sediment Delivery Ratio

refuted but is probably conservative. He cites recent reservoir data to substantiate sediment yields, along with similar estimates of sheet erosion from cropland. About 2.4 billion tons of sediment is produced from sheet-rill erosion on the 683,000 sq miles of cropland in the contiguous United States. This is nearly 6 tons/acre/yr, which at first consideration seems high; but the actual soil loss rate from fields to water-courses would be only 3 tons/acre–4 tons/acre and, presumably, only about 1 ton/acre of cropland would reach the sea. Cropland is defined as cultivated land, orchards, and open land formerly cropped.

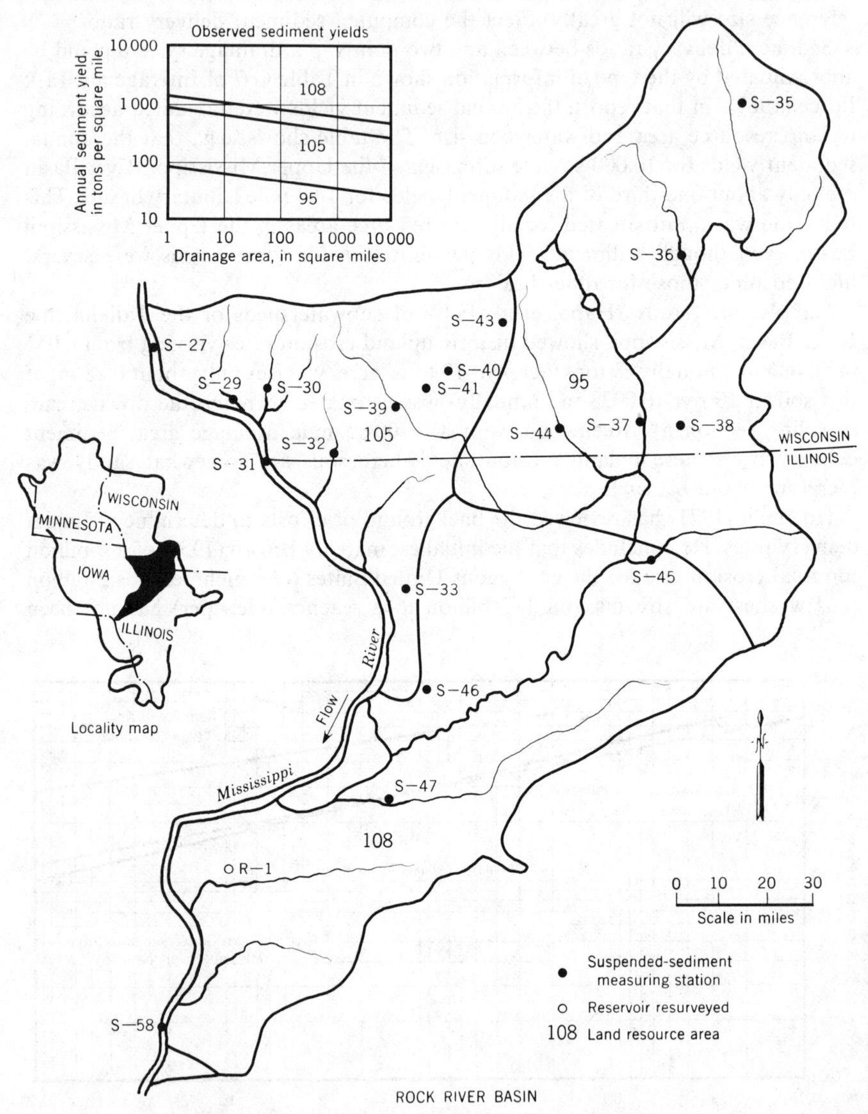

FIG. 4.16.—Sediment Yield Variation in Three Land Resource Areas in Upper Mississippi River Basin

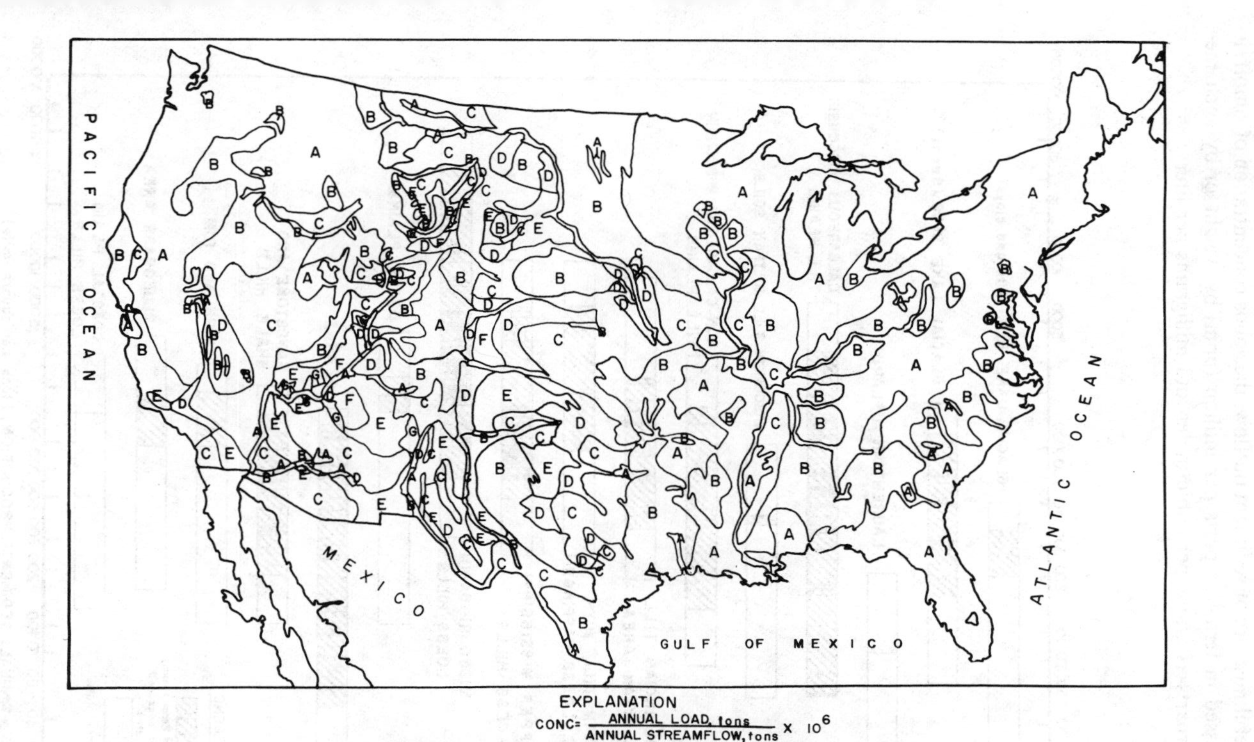

FIG. 4.17.—Map of Conterminous United States Showing Sediment Concentration of Rivers

18. Sediment Yields.—The sediment outflow, or yield, from a drainage area (see definitions) is commonly expressed in units of weight (tons), volume (acre-feet or cubic meters), or uniformly eroded depth of soil (inches or millimeters). Another useful measure of sediment outflow, the solids concentration of runoff, is often expressed in terms of parts per million (ppm) by weight or by such metric counterparts as grams per cubic meter and milligrams per liter.

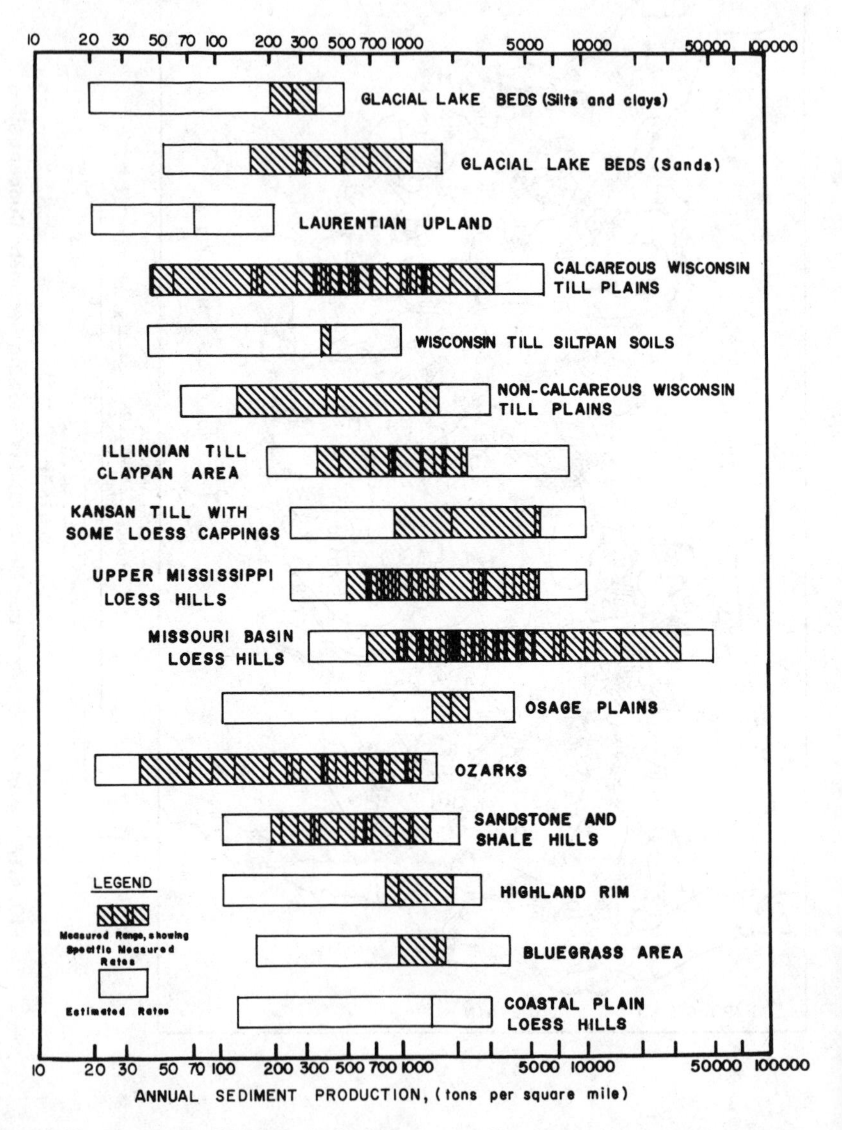

FIG. 4.18.—Sediment Yields in United States, by Physiographic Areas, in tons per square mile per year

Sediment yield is dependent upon the rate of total erosion in the contributing basin and the efficiency of transport of these eroded materials. Thus, solving Eq. 4.8 for Y, $Y = TD$.

It has been shown that these erosion and transport factors are so variable that any statements concerning sediment yield for a geographic region must be generalized into rather wide limits. Rainwater (1962) has indicated the variation in sediment concentration for streams throughout the conterminous United States (see Fig. 4.17). Drozd and Goretskaya (1966) have mapped the sediment concentrations of streamflow for the Ukraine, and other European researchers have compiled similar information for their areas. Brune (1953) portrays the sediment yield variation in selected physiographic areas of the United States in Fig. 4.18.

Although such compilations have limited usefulness for a specific problem, they allow a better understanding of general environmental factors that affect sediment yield. Langbein and Schumm (1958), for example, have utilized available sediment records to summarize the effect of such interrelated environmental factors as precipitation, runoff, temperature, and natural vegetation. Fig. 4.19 shows that the sediment concentration of streamflow decreases from arid to more humid regions. In dry regions, where the natural vegetative cover is sparse and the sediment supply limitless, the infrequent runoff events are capable of entraining very high concentrations of sediment. In the more humid regions, the vegetative cover limits

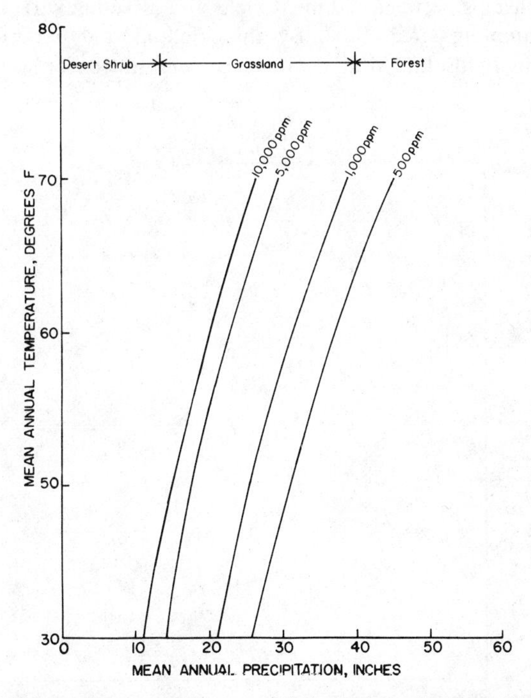

FIG. 4.19.—Sediment Concentration, as Affected by Temperature and Rainfall (Schumm, 1965)

the sediment supply. Also, the sediment concentration of runoff increases as the mean annual temperature increases—for any given rainfall.

Fig. 4.20 shows a more complex but logical trend of increasing sediment yield with decreasing aridity, up to the point where improved vegetative cover inhibits further increases. Sediment yield then decreases with increasing rainfall. Optimum conditions that allow maximum sediment entrainment and transport occur in regions where the "effective" precipitation is about 12 in. annually. (The actual precipitation amount is slightly greater than the temperature-adjusted "effective" value, or about 15 in.). Schumm (1965) also gives an excellent account of the effect of temperature on sediment yield.

The foregoing data were based on field determination of sediment yield by: (1) Measurement and sampling of streamflow; and (2) surveys of sediment deposits in reservoirs. These procedures for determining sediment yield are detailed in Chapter III, Sections A and B. Nearly all sediment yield investigations in the United States are performed by public agencies, which are coordinating their activities to simultaneously satisfy specific, present-day requirements, to meet future needs, and to provide for fundamental research. An idea of the quantity of data on hand can be obtained by reference to data collection programs in the United States. Suspended-sediment sampling was performed at 874 streamflow locations in 1971 (United States Geological Survey, 1971). Another publication (United States Department of Agriculture, 1973) lists reservoir sedimentation survey data from 1,518 locations.

19. Sediment Yields from Streamflow Sampling.—The best insights into the complex relationships between sediment yield and affecting variables are usually afforded by sampling streamflow. By this method, the pattern of sediment production from fields and drainageways is determined for individual runoff

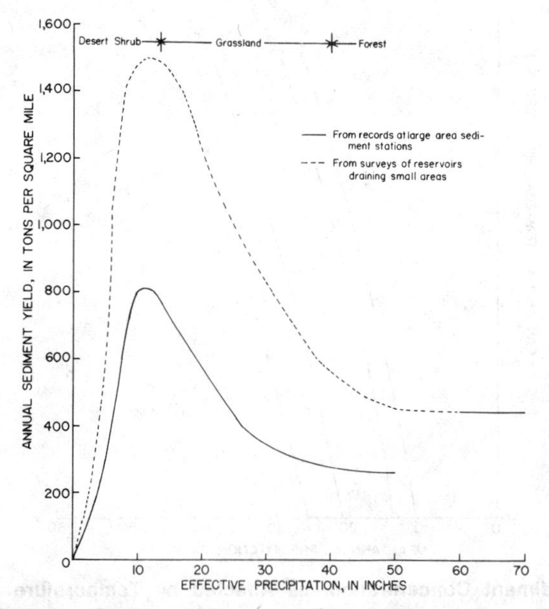

FIG. 4.20.—Sediment Yield, as Affected by Climate (Langbein and Schumm, 1958)

events and for any time during a specified event. However, present streamflow sampling methods are costly, and the suspended-sediment discharges thus obtained should often be adjusted upward by some estimation procedure to include sediment transported in close proximity to the stream bed (Chapter III, Section A).

20. Sediment Yields from Reservoir Surveys.—Sediment deposition rates for several reservoirs in the United States, along with accompanying information for the contributing watershed, are listed in Table 4.7. Sedimentation surveys of existing reservoirs provide a composite record of sediment yield from the upstream drainage area. This is a source of sediment yield information that is especially advantageous to persons engaged in reservoir design because the location of deposited sediments and the sediment quantities involved are equally important as determinants of design. The economic life of a reservoir is dependent upon both.

In its simplest form, a reservoir sedimentation survey entails only the measurement of the volume of accumulated sediment deposits. A more accurate determination of sediment yield would include such supplemental information as the distribution and volume-weight of deposits and the proportion of sediments trapped by the reservoir (trap efficiency). Additional data would include reservoir size and shape characteristics, density currents, tributary location and density, and other watershed and hydraulic characteristics that govern sediment movement or deposition.

21. Bases for Estimating Long-Term Sediment Yields.—The necessity for obtaining long-term sediment yields has been considered. Estimates in lieu of actual measurement are often necessary because financial and time considerations govern and because a given water resource project might involve basin or channel changes, or both, that would produce a sediment regimen different from the original. In addition to obtaining total sediment quantities, it is often necessary that the resource planner anticipate the physical characteristics of the sediment (particle size distribution, portion of wash load, bed load, organic matter, colloids, etc.) and the manner of delivery (storm, season, etc.). Such information requires a knowledge of the effect of hydrologic variables upon sediment yield. This knowledge is imperfect at present; many of the variables known to affect sediment yield are interrelated, and the quantitative influence of a specific variable in a given situation is difficult to ascertain.

Meteorologic factors affect sediment yield in several ways. Year-to-year or storm-to-storm precipitation differences, e.g., often account for larger variations in sediment yield from mixed-cover watersheds of a region than most other factors. When comparing sediment yields of the six continuously gaged and sampled watersheds in the 117-sq mile Pigeon Roost Creek Basin, Mississippi, for 11 yr, 1957–1967, the average annual sediment yield for the most erodible watershed was 6.5 tons/acre, or two and one-half times that of the least erodible watershed. However, when comparing sediment yields by year, the average for the six watersheds during 1957 was 7.0 tons/acre, or more than three and one-half times the average sediment yield for 1960. Rainfall between watersheds, as recorded by more than 30 rain gages, varied no more than 11% during any year, whereas the 1967 rainfall was 64% more than the 1965 amount.

The form and intensity of precipitation in different seasons of the year, along with antecedent soil conditions, are of primary importance in sediment production. For example, the sediment yield resulting from snowmelt runoff or from rainfall on frozen or snow-covered ground is usually much less than that resulting from a summer thunderstorm. Average sediment concentrations of direct runoff from winter storms in southern Wisconsin were one-third to one-half those resulting from direct runoff in the summer (Collier, 1963). During the record floods of January and February, 1959, in Ohio, the sediment concentrations were relatively low because the rains fell on frozen and snow-covered land (Archer, 1960). Similar seasonal differences were found in studies of erosion from barren strip mine spoil banks in Kentucky (Collier, 1969; Collier, et al., 1964). This is shown in the curves of Fig. 4.12. However, when large amounts of precipitation

TABLE 4.7.—Sediment Yields

Reservoir and location (1)	River basin (2)	Period of record, in years (3)	Reservoir volume, in acre-feet (4)	Reservoir volume, in acre-feet per square mile (5)
Sardis near Sardis, Miss.	Little Tallahatchie River	20.7	91,900	59.0
Bodeman near Brunswick, Ind.	Tributary of West Creek	13.0	52	19.0
Lake Tandy near Hopkinsville, Ky.	Little River	52.5	770	130.0
Kiser Lake near St. Paris, Ohio	Mosquito Creek	14.5	3,330	380.0
Upper Hocking #2 near Lancaster, Ohio	Hunter's Run	5.0	57	30.0
Lake Dante near Wagner, S. Dak.	Tributary to Choteau Creek	17.9	261	90.0
John Martin near Caddoa, Colo.	Arkansas River	15.2	423,000	22.0
Altus near Altus, Okla.	North Fork Red River	12.6	157,000	62.0
Bellerud Pond near Adams, N. Dak.	Dark River	13	17	15.0
Walnut Cove near Walnut Cove, N. C.	Dan River	9	970	2.5
Mission Lake near Horton, Kans.	Mission Creek	30.2	1,900	229.0
Guernsey near Guernsey, Wyo.	Platte River	30.3	74,000	14[d]

[a]Randomly selected from the Summary of Reservoir Sediment Deposition Surveys Miscellaneous Publication No. 964, Appendix A, Feb., 1965.

[b]Computed from Conservation Needs Inventories of Individual States—usually available ment of Agriculture.

[c]Estimated.

[d]Based on sediment-contributing area only.

and runoff occur during winter months in combination with minimal soil cover and freeze-thaw cycles that cause weathering of the soil surface, extremely high erosion rates have been recorded. This has been noted especially in the eastern United States at lower latitudes where winter and early spring moisture is plentiful and where the soil surface may be intermittently frozen and thawed throughout the winter. High erosion rates due to the foregoing conditions have been measured from small watersheds draining road ditches in Georgia (Diseker and McGinnis, 1967) and from the large watersheds drained by the Potomac and Monocacy Rivers in Virginia and Maryland (United States Geological Survey, 1959–1967). At higher latitudes of the United States, as in New England, land use is

Based on Reservoir Surveys[a]

Watershed Land Use, as a percentage[b]			Mean Annual Quantities		
Cropland (6)	Pasture-idle (7)	Forest (8)	Precipitation, in inches (9)	Runoff, in inches (10)	Sediment yield, in tons per square mile (11)
30	35	30	52	20	1,100
55	30	5	36	10	280
55	10	30	47	18	1,700
65	20	10	38	12	4,900
65	20	10	36	13	950
60	35	–	22	0.5	260
15	80	5	15	0.3	850
60	30	5	24	0.9	1,300
75	15	5	18	0.7	110
25	10	65	50	19	240
70	20	5	32	4.4[c]	740
15	80	–	8–24	1.3	210

made in the United States through 1960, United States Department of Agriculture,

by inquiry to state offices of Soil Conservation Service, United States Depart-

predominatly forest and meadow. Sediment yields are quite low but may be important to the design of water resource facilities. Essentially the only source of sediment in such areas is due to spring snowmelt carrying stream bank material that has been loosened by frost action. Summer storms yield comparatively small quantities of sediment. Measurements on the Sleepers River Basin, Vermont (Kunkle and Comer, 1972), show that sediment transport and dissolved solid transport are about equal—each about 1/4 ton/acre/yr.

Other within-year differences in sediment yield, on a storm or seasonal basis, have been attributed to rainfall intensity and energy, runoff amount and intensity, and moisture content of the soil prior to rainstorms (Guy, 1964). Comparisons of sediment yield between watersheds show the effect of such environmental factors as topography, soils and geology, and vegetative cover and land management, along with such transport efficiency factors as watershed size and shape and channel hydraulics (Lustig, 1965; Jones, 1966).

Circumstances would seem to indicate widespread use of a rainfall parameter to estimate sediment yields. Rainfall records are numerous, and many rain gages have been in operation for several decades. Although Dragoun (1962) has succeeded in obtaining a good correlation between a rainfall energy-intensity factor and sediment yield for several field-size watersheds, runoff (especially the overland flow portion) is the best single indicator of sediment yield characteristics. The graphical relation between runoff and sediment discharge (or concentration) is called the sediment transport curve and was used as early as 1940 for comparing sediment characteristics in the Southwest (Campbell and Bauder, 1940); it was used for extrapolation of records for the San Juan River, Utah, in 1951 (Miller, 1951). A comprehensive summary of the applications of the sediment transport curves was made by Colby (1956).

Advantages favoring the use of runoff rate for determining sediment yield are the availability of numerous short-term runoff records for drainage areas throughout the country and the presence of many long-term precipitation records that can be used—along with an assumed rainfall-runoff relationship—to extend these runoff records. After the runoff record has been adjusted to represent long-term conditions, some type of sediment transport relationship is utilized to compute sediment yields.

22. Sediment Transport Relationships.—The runoff entering stream channels promptly after rainfall or snowmelt is termed direct runoff. It forms the bulk of the "flood" or "storm" hydrograph and is generally responsible for most of the sediment in transport in all but the largest rivers. This direct runoff from a given area represents the integrated effect of most characteristics of the drainage basin and the superimposed environment as they relate to sediment production. The fine sediment fractions that form the wash load of a stream are readily entrained in runoff and are relatively insensitive to flow parameters, being mostly a matter of supply to the stream. The transport of the coarse sediment fraction, i.e., the bed sediment load, is dependent upon a balance between supply and flow parameters and may or may not be adequately sampled by suspended-sediment samplers. If a sediment transport curve can be deduced that correctly averages the effect of these differing sediment transport modes, this relation can be put to practical use. Such a relation may be obtained by simultaneously sampling and measuring the storm runoff rate from a watershed as outlined in Chapter III, Section A.

Characteristic curves for different types of storms can then be defined and a composite sediment transport curve evolved. For example, if the runoff event is of low intensity, the stream sediment concentration is comparatively low at the start of direct runoff and increases slowly with increasing runoff rate, to a point, and then slowly recedes, as shown in Fig. 4.21. If the storm is intense, the stream sediment concentration is comparatively high at the start of direct runoff and increases rapidly, as shown in Fig. 4.22. These conditions result in a "loop" effect on the sediment transport graph for a given storm, as shown in Figs. 4.23 and 4.24. (The suspended-sediment concentrations shown in Figs. 4.21 and 4.22 were measured with depth-integrating samplers and the sediment discharges in Figs 4.23 and 4.24 were obtained from the product of the measured concentration and the water discharge.) Under certain conditions, it is possible for high-intensity runoff events to attain a maximum, or near-maximum concentration at the beginning of runoff. This is true for watersheds where the weathering of soils or stream beds, or both, during long, dry periods has produced a large readily transportable load of fine material.

Consider the effect of these possibilities for variation of the sediment transport relation at low and medium rates of runoff. If random samples from storm runoff had been collected through many storms during the year, one would expect considerable scatter in the suspended-sediment concentrations derived from these samples. The sediment concentrations for a particular rate of storm runoff on Pigeon Roost Creek, e.g., could be 300 ppm on the rising side of a low-intensity storm hydrograph and about the same on the falling side. For a high-intensity event, the concentration on the rising side of the hydrograph could be 20,000 ppm

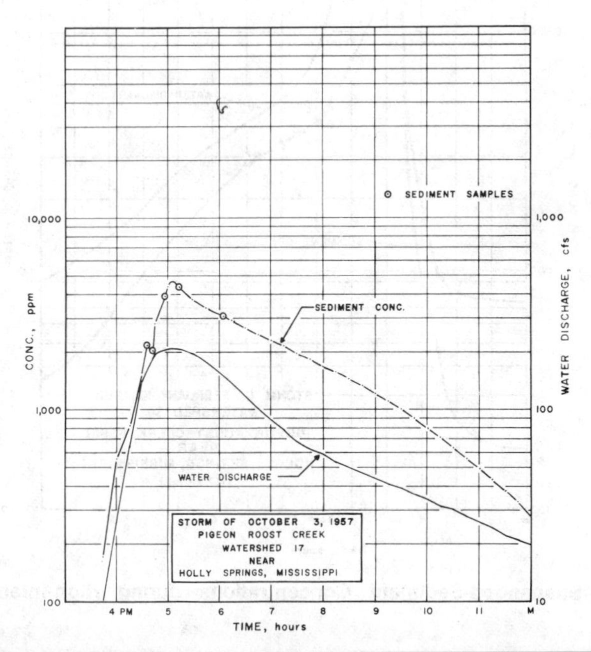

FIG. 4.21.—Suspended-Sediment Concentrations during Low-Intensity Storm Event

for this same runoff rate, while the comparable recession concentration could be as low as 1,000 ppm. The lower concentrations during falling stage are the result of a greatly reduced supply of fine sediments; the bed sediment is still in plentiful supply in these aggrading channels and comprises nearly one-third of the total sediment discharge. Fig. 4.25(a) shows the typical scatter that can occur from a single year of sampling. In the midportion of the figure, the seasonal effect upon point scatter is also discernible.

As shown in Fig. 4.25(a) the upper range of sediment discharge for a particular low or medium rate of direct runoff can be more than 50 times that of the lower range. This observed 50-times variation in the sediment discharge at Pigeon Roost Creek is typical of many locations. Inspection of streamflow sample records from a 75-acre watershed in cultivation near Treynor, Iowa, [Fig. 4.25(b)] shows a suspended-sediment concentration range from about 2,000 ppm to nearly 230,000 ppm for direct runoff rates of 5 cfs–10 cfs. It is probable that the concentration of some snowmelt runoff is much lower than 2,000 ppm, and that the concentration extremes for given low and moderate runoff rates can vary much more than 100:1. Despite this scatter, an average trend line through these points would accurately

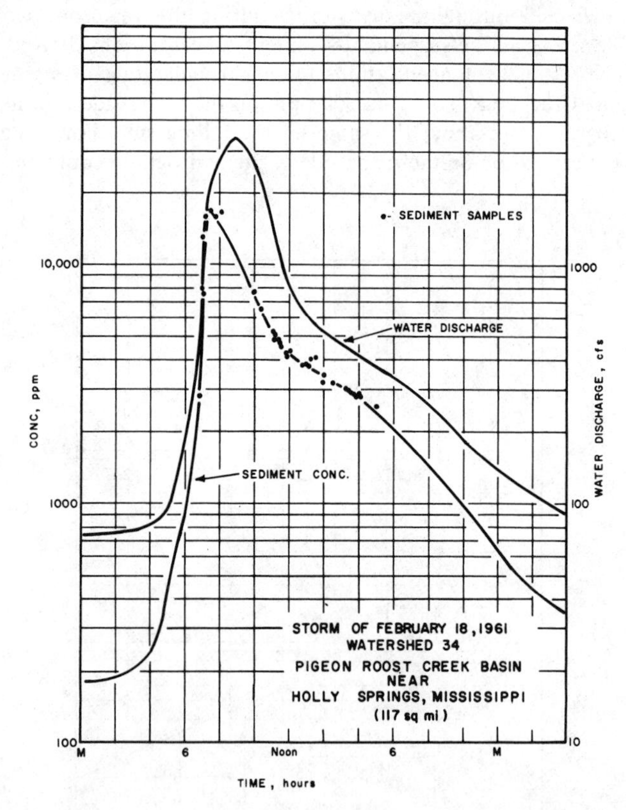

FIG. 4.22.—Suspended-Sediment Concentrations during High-Intensity Storm Event

characterize the overall water-sediment relation at these runoff rates if the sample points were representative.

Large relative variations in suspended-sediment discharge or concentration, for a given water discharge, usually occur for low to medium discharges. For extremely high discharges, observations indicate that this relative variation diminishes appreciably. This occurs because extremely high rates of runoff and sediment discharge can only be reached, for a given watershed, by a very limited combination of hydrologic circumstances. Any chance recurrence of these rates must be accompanied by much the same watershed and meteorologic conditions that previously occurred, and therefore must produce comparable sediment concentrations. The foregoing behavior is illustrated by streams in the central United States, where maximum rainfall amounts and intensities are mostly associated with thunderstorm activity in late spring and early summer and are less frequently the result of hurricane movements originating off the Atlantic and Gulf coasts in early autumn. The runoff response to these maximum rainfall occurrences varies greatly according to season and the condition of the land surface of this agricultural heartland. As a result, most large runoff events, and nearly all events with large sediment discharges, occur at crop planting time or in the early growing season when the land is most vulnerable to erosion. At other times of the year, the sediment supply is less and the runoff potential per unit of rainfall is low.

The suspended-sediment transport curve of Fig. 4.25(a) has its largest slope at

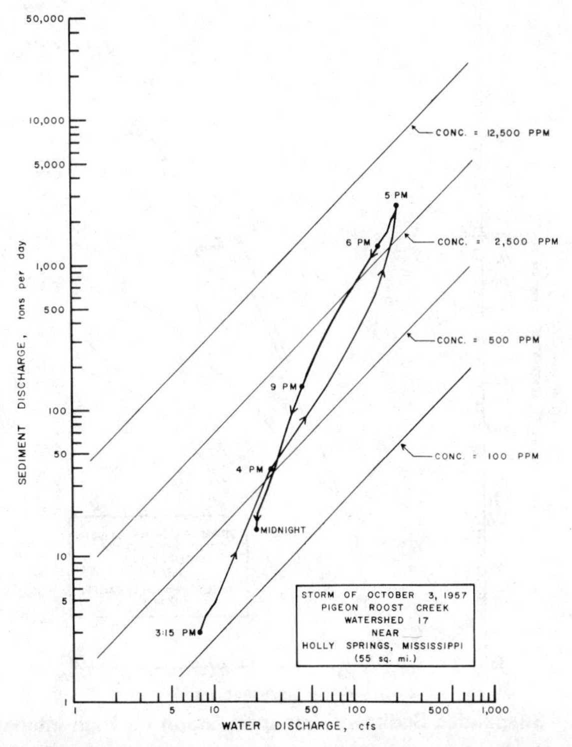

FIG. 4.23.—Suspended-Sediment Transport Graph for Low-Intensity Storm

low discharges and its smallest slope at the highest discharge so that the curve is concave downward. This shape is typical of suspended-sediment transport curves for many streams. Segments of a transport curve can be approximated by a power relation of the form

$$G_{ss} = LQ^n \qquad (4.12)$$

in which G_{ss} = suspended-sediment discharge; Q = water discharge; and L = a factor which can be taken as an index of relative erodibility. The exponent, n, is the slope of the curve on logarithmic paper measured in units of logarithmic cycles. If Eq. 4.12 is fitted to segments of a transport curve, e.g., the one in Fig. 4.25(a), the exponent, n, will diminish as the segments cover higher and higher ranges of water discharge.

In Fig. 4.25(a) the curve is drawn so that it approaches a slope of unity ($n \rightarrow 1$) at some high value of Q. For a slope of unity the sediment concentration $G_{ss}/Q = L$ is constant and independent of Q. The value of unity for n appears to be a common minimum value approached by many streams at very high flood discharge.

The values of sediment discharge for high flows are essential for estimating long-term sediment yield since an important fraction of the total sediment yield is produced by the high flows. Data for plotting the sediment transport curve in the high discharge range are sometimes scarce or lacking entirely and it is often

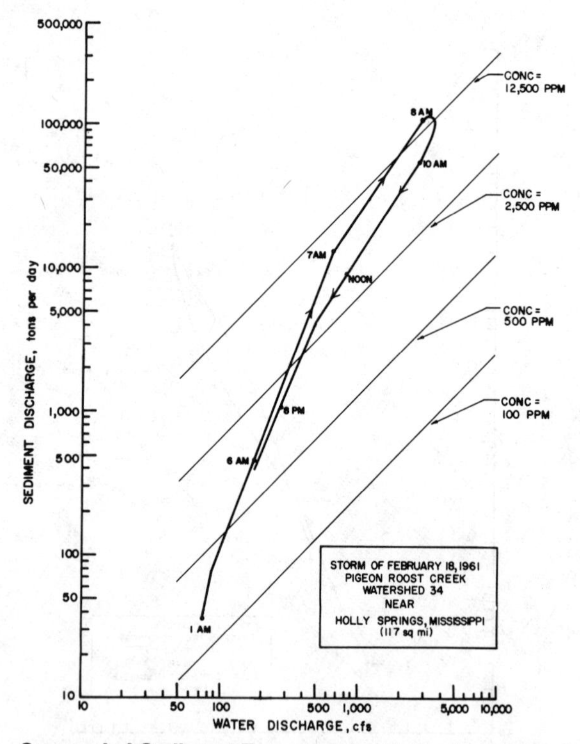

FIG. 4.24.—Suspended-Sediment Transport Graph for High-Intensity Storm

necessary to extrapolate the curve to cover this range. Information on the shape of the curve may be used as a guide in this extrapolation.

The wash load of a stream usually originates from sheet and rill erosion on the watershed and its concentration is determined by the amount brought to the stream and not by the capacity of the stream to transport it. On the other hand the concentration and discharge of bed sediment is determined by the hydraulic forces capable of moving it. For purposes of analyzing the relation between sediment discharge and water discharge for the high range of discharges it is convenient to treat the wash load and bed load separately. The wash load discharge, G_w, is assumed to be given by

$$G_w = BQ^m \quad \text{(4.13)}$$

and the concentration of wash load C_w is then

$$C_w = BQ^{m-1} \quad \text{(4.14)}$$

in which B = a basic erodibility constant for wash load, and for high discharges m has a value in the neighborhood of unity. Dawdy (1967) arrived at a relation between discharge of bed sediment, G_{bs}, and Q. To do so he assumed that $G_{bs} \propto V^3$, in which V = the mean velocity of the stream, and that the stream is wide and has a constant Manning roughness factor. The latter assumption gives $V \approx d^{2/3}$ or that $Q \sim d^{5/3}$. Eliminating V from the relation between G_{bs} and V gives

$$G_{bs} = KQ^{1.2} \quad \text{(4.15)}$$

in which K is a constant for any given channel. The total suspended-sediment discharge, G_{ss}, is then the sum of the wash load and bed sediment discharges or

$$G_{ss} = BQ^m + KQ^{1.2} \quad \text{(4.16)}$$

Since for high discharges m is probably not too different from 1.2 and G_w is usually several times G_{bs}, Eq. 4.16 can be approximated by Eq. 4.12 with n only slightly different than m.

This is an idealized construction of the sediment-transport relation. In practice, the sediment transport curve can have a slope exceeding the foregoing values because:

1. Some types of erosion can occur whereby sediment discharges and concentrations are increased during the course of a storm. Massive outwash from steep slopes, severe gully erosion, extensive slumping of streambanks, or a situation where most of the sediment load originates from the channel degradation of a truly alluvial stream are examples.
2. Many sediment transport curves are assumed to be linear on logarithmic paper even though the general shape of the curve is concave downward as in Fig. 4.25(a). The resulting slope value is therefore greater than the true slope of the upper portion of the curve, especially if base flows are considered.
3. Many curves are prepared from records that are of insufficient duration to experience representative storm events of high intensity.

To summarize, the relative variation of plotted points in the low runoff portion of a sediment transport curve based upon sediment samples can be one hundred fold (two log cycles), while the relative variation at the high runoff end is much less, the actual amount depending upon length of record and the occurrence and coverage of storm events during the period. A long-term sediment transport curve

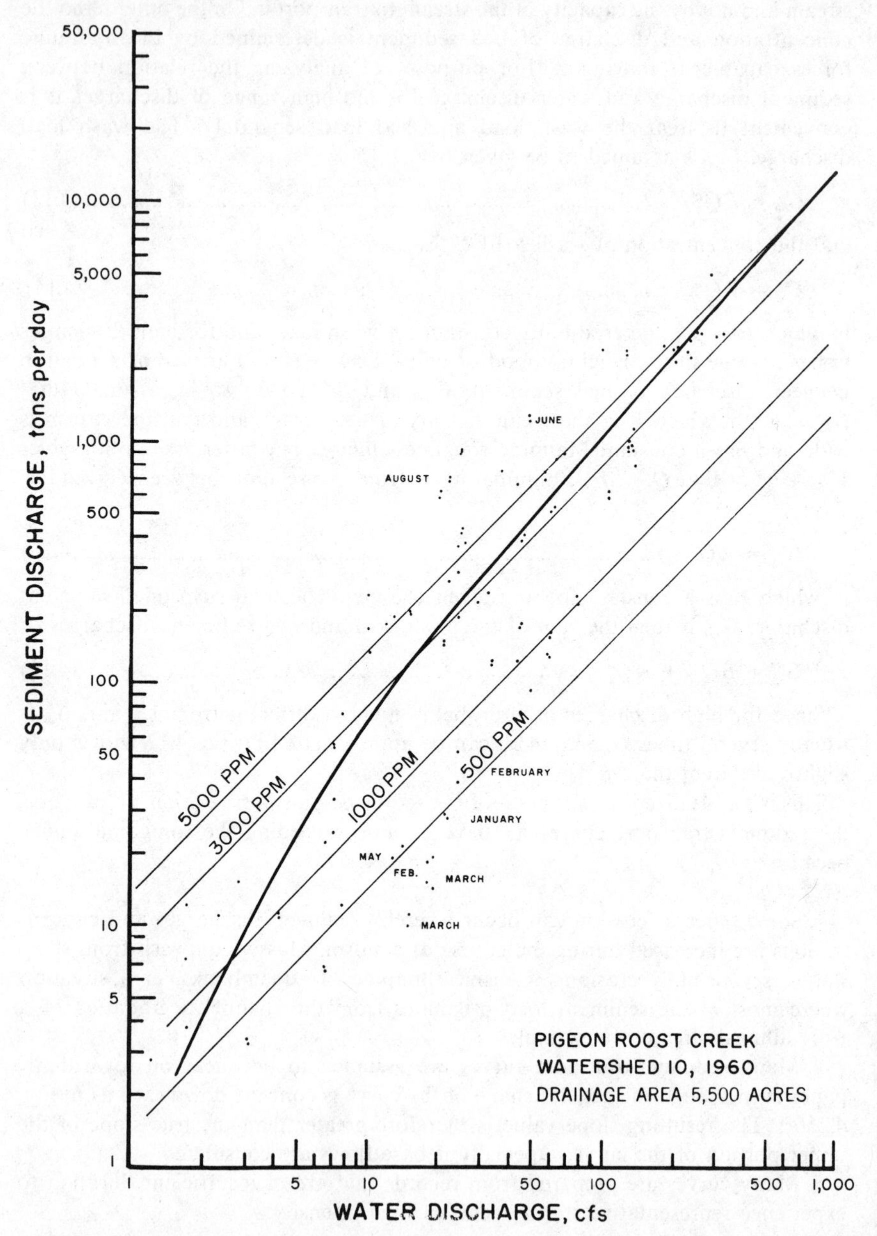

FIG. 4.25(a).—Suspended-Sediment Transport Relationships Showing Typical Scatter of Data from Random Samples of Storm Runoff Taken over 1 yr from Mixed Cover Watershed in Mississippi

can be constructed from streamflow measurements and sediment samples of a few storms by: (1) Considering the behavior of the runoff-sediment relation for the storms; and (2) a knowledge of how closely these storms represent the long-term experience. The slope of the upper portion of the sediment transport curve is further constrained within the theoretical limits cited. The Ecole Polytechnique (Bonnard and Bruschin, 1970) has a thorough discussion of seasonal and annual sediment transport curves for three Swiss streams.

Many of the attributes of an instantaneous sediment transport curve (i.e., a curve based upon sediment samples), apply to curves that are compiled from daily, monthly, annual, or storm discharges. Guy (1964) concluded that the storm is the most logical unit of study for many purposes and succeeded in relating sediment yields (and concentrations) to the various runoff and climatic variables that characterize storms. It is agreed that the storm is the best unit for study but the more prevalent daily records of runoff and sediment discharge can be similarly studied (Piest, 1965) if the uncorrelated subsurface and ground-water flows are

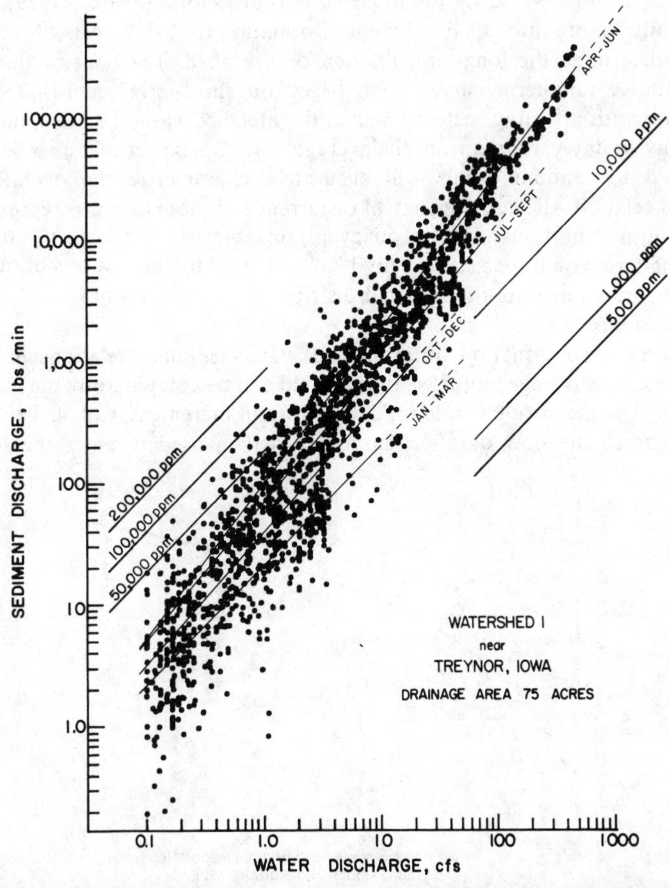

FIG. 4.25(b).—Suspended-Sediment Transport Relationships Showing Seasonal Scatter of 1,653 Samples, 1964-1970, from Row-Cropped, Iowa Watershed

first eliminated by inspection. Some procedures for estimating long-term sediment yields, based on hourly, daily, or annual runoff, are considered.

Many sediment prediction methods are based on the runoff-sediment relationship that exists in any given watershed (e.g., Fig. 4.26). In some cases, a reliable relationship exists and little extrapolation is needed. At other times, it may be necessary to synthesize the needed data. Once the estimated runoff-sediment relationship is obtained, it can be used, along with runoff-frequency data, to obtain a sediment yield prediction. This general procedure is now explained.

Runoff-frequency information is obtained in a variety of ways and depends on the base data available. Techniques include: (1) The extrapolation of annual-duration or partial-duration series of runoff; (2) the use of double-mass comparisons of runoff or precipitation records, or both, for the subject watershed and adjacent areas; and (3) an analysis of short-term rainfall-runoff relationships and a knowledge of the long-term rainfall pattern. Fig. 4.27 shows a long-term flow-duration curve that was synthesized, from a 3-1/2-yr runoff record at Pigeon Roost Creek Watershed 12 and a 32-yr record of the East Fork of the Tombigbee River near Fulton, Miss., by the index-station procedure (Searcy, 1959).

The daily runoff-suspended sediment information for 3-1/2 years of record was used to determine the long-term relation of Fig. 4.28. The construction of the representative long-term curve, when based on short-term records, requires a familiarity with affecting meteorologic and watershed variables. Despite considerable day-to-day variation from the average, one can expect that a smooth trend line fitted to random events will accurately characterize the overall water-sediment relation—if the frequency of occurrence of these events is representative of long-term conditions. The adequacy of this short-term record in portraying these long-term conditions can usually be determined by mass curves of rainfall or runoff, or by comparing the seasonal distribution of storm runoff with the long-time occurrence.

When the long-term flow duration and water-sediment relations have been synthesized, the average annual sediment yield can be computed by the method of Table 4.8. The mean runoff rate for each duration increment, Col. 4, Table 4.8, is obtained from the flow duration curve. It is then used to enter the sediment

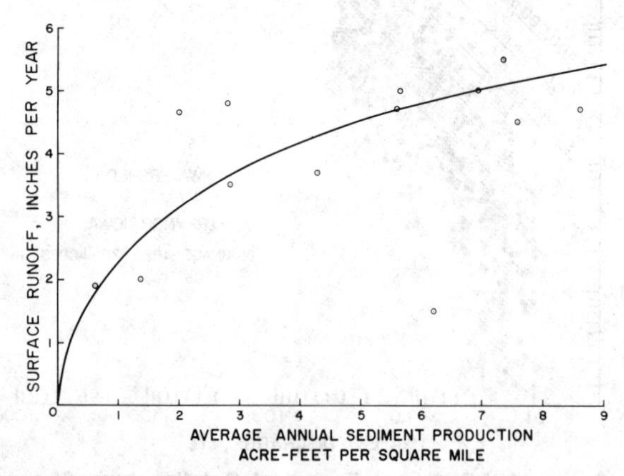

FIG. 4.26.—Runoff-Sediment Relation for 13 Watersheds, Based on Cultivated Acreages of Mixed Loess and Glacial Soils

transport curve and determine the sediment discharge rate for each duration increment (Col. 6, Table 4.8). Some long-term sediment yields estimated by this procedure are summarized in Table 4.9.

23. Sediment Yield Empiricisms.—Empirical equations for sediment yield, in terms of affecting variables, have been formulated for several areas. Anderson (1949) developed the following relation from forested drainage areas that vary from 4.5 sq miles–202 sq miles:

$$\log e = 1.041 + 0.866 \log q + 0.370 \log A - 1.236 \log C \quad (4.17)$$

in which e = annual sediment yield, in acre-feet per square mile; q = maximum yearly peak discharge, in cubic feet per second per square mile; A = area of main channel of watershed, in acres per square mile; and C = density of cover, as a percentage.

Most empirical equations are in terms of annual, monthly, or seasonal relationships between sediment yield and affecting variables. For two small cultivated watersheds in Texas (Baird, 1964), individual regression equations relating sediment concentration and runoff for a 4-yr period of record were expressed on an hourly basis as

$$\log C = 0.5340 \log H + 4.4238 \quad (4.18)$$

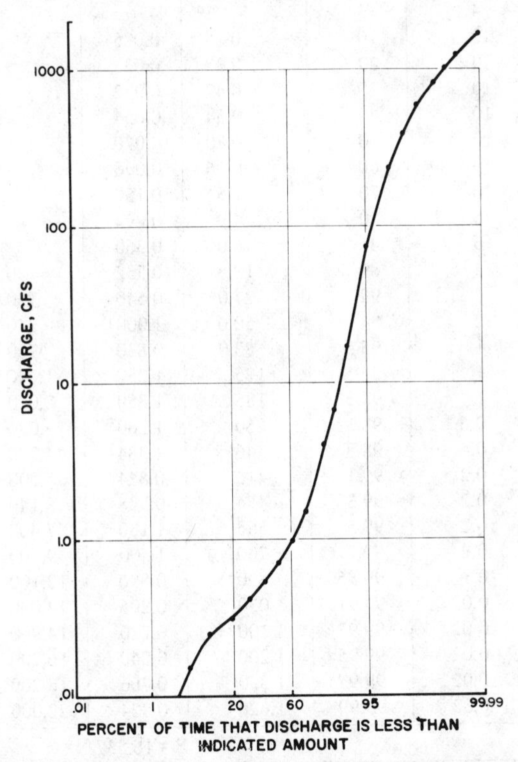

FIG. 4.27.—Long-Term Flow Duration Curve, Pigeon Roost Creek, Watershed 12 near Holly Springs, Miss.

and $\log C = 0.4269 \log H + 3.538$ .... (4.19)

in which C = the sediment concentration, in parts per million, and H = hourly runoff, in inches. Both relations were correlated at the 1% level, even though the variation between individual observations was large.

Empirical procedures were employed by the United States Army Corps of Engineers (Hill, 1965) to estimate rates of sediment yield for unsampled rivers in the Kansas City District. Short-term runoff-suspended sediment records were analyzed for 87 river basins, and the basins were divided according to soil type. An annual runoff-sediment discharge curve for each soil type was then con-

TABLE 4.8.–Long-Term Sediment Yield by Flow Duration-Sediment Rating Curve Method[a]

			Long-Term Yields			
Cumulative duration, as a percentage (1)	Duration, as a percentage (2)	Duration midpoint (3)	Flow at midpoint, in cubic feet per second (4)	Col. 2 X Col. 4 (5)	Sediment rate, in tons per day (6)	Col. 2 X Col. 6 (7)
1 - 0	1	0.5	–			
5 - 1	4	3	–			
15 - 5	10	10	0.25	0.025		
25 -15	10	20	0.32	0.032		
35 -25	10	30	0.42	0.042		
45 -35	10	40	0.54	0.054		
55 -45	10	50	0.70	0.070		
65 -55	10	60	0.96	0.096		
75 -65	10	70	1.5	0.150		
81 -75	6	80	4.9	0.294		
87 -81	6	84	6.0	0.360		
91 -87	4	89	13.8	0.552	40.0	1.60
93 -91	2	92	27.0	0.540	130	2.60
95 -93	2	94	50.0	1.0001	340	6.80
96 -95	1	95.5	83.0	0.830	700	7.00
97 -96	1	96.5	125	1.250	1,240	12.40
98 -97	1	97.5	185	1.850	2,020	20.20
98.6 -98	0.6	98.3	260	1.560	3,050	18.30
99.0 -98.6	0.4	98.8	346	1.384	4,290	17.16
99.2 -99.0	0.2	99.1	412	0.824	5,200	10.40
99.4 -99.2	0.2	99.3	474	0.948	6,140	12.28
99.6 -99.4	0.2	99.5	565	1.130	7,450	14.90
99.8 -99.6	0.2	99.7	700	1.400	9,300	18.60
99.9 -99.8	0.1	99.85	890	0.890	12,100	12.10
99.92-99.9	0.02	99.91	1,030	0.206	14,000	2.80
99.94-99.92	0.02	99.93	1,100	0.220	14,900	2.98
99.96-99.94	0.02	99.95	1,200	0.240	16,200	3.24
99.98-99.96	0.02	99.97	1.330	0.266	18,200	3.64
100 -99.98	0.02	99.99	1,620	0.324	22,000	4.40
				Σ = 16.537		Σ = 171.40

[a]Station 12, Pigeon Roost Creek Watershed, near Holly Springs, Miss.
Note: For Col. 5, average annual cumulation = 16.537 cfs X 365.25 days = 6,040 cfs-days.
For Col. 7, average annual cumulation = 171.40 tons/day X 365.25 days = 62,000 tons.

structed as shown in Fig. 4.26. The runoff represented an approximation of overland flow only, while the sediment yield was based solely on cultivated areas within the basin. This approach was justified because various researchers have reported that soil losses from most forests and rangelands are negligible compared

TABLE 4.9.–Long-Term Sediment Yields for Selected Watersheds in United States

Location (1)	Drainage area, in square miles (2)	Land Use, as a percentage: Crop-land (3)	Pasture–idle (4)	Forest (5)	Average Annual Quantities: Rainfall, in inches (6)	Runoff, in inches (7)	Sediment, in tons per square mile (8)
North Fork Clear Creek near Blackhawk, Colo.	55.8	–	10	90	24	2.8	340
Honey Creek near Russell, Iowa	13.2[a]	50	40	10	34	6.2	450
Mule Creek near Malvern, Iowa	10.6[a]	80	15	5	30	4.9	3,100
Plum Creek near Waterford, Ky.	31.9	50	25	25	50	15	1,100
Cuffawa Creek near Holly Springs, Miss.	31.3	25	60	15	52	10	4,200
Pigeon Roost Creek near Byhalia, Miss.	117.0	20	55	25	52	14	2,700
East Fork Big Creek near Bethany, Mo.	95.0	55	35	10	34	6.7	1,300
Mississippi River at St. Louis, Mo.	701,000	40	35	20	23 (15-40)	1.5	260
Dry Creek near Curtis, Nebr.	20.0	45	55	–	20	1.4	6,500
Stony Brook at Princeton, N.J.	44.5	35	15	25	45	12	140
East Fork Deep River near High Point, N.C.	13.9	30	15	45	44	12	1,000
Todd Fork Little Miami River near Roachester, Ohio	219.0	65	20	15	44	17	900
North Fork Little Miami River near Pitchin, Ohio	29.1	80	15	5	40	10	70
Bixler Run near Loysville, Pa.	15.0	25	10	60	42	16	93
Elm Fork Trinity River near Muenster, Tex.	46.0	19	81	–	35	6.4	43

[a]A dam-building program affected about 40% of the watershed area during the streamflow-sampling period.

with that from cultivated land. (Brune, 1948; Ursic and Dendy, 1965; Carter, et al., 1966). When the sediment yield is desired for an unsampled watershed, runoff is obtained by stream gaging or extrapolation routines, or both, and the cultivated area is tabulated from existing land use inventories, e.g., the one for Missouri (University of Missouri Agricultural Experiment Station, 1962). Then, with few adjustments, the required sediment yield information is obtained directly from the figure.

In some areas, there is such a paucity of sediment discharge and runoff data that no accurate quantitative estimates of sediment yield are possible from reservoir surveys or suspended-sediment measurements. For these situations, one can rely on calculations of sheet erosion by the various equations, with adjustment for added erosion sources (gully, channel, etc.). A delivery ratio is then applied to determine the sediment yield to any given downstream point. Such information serves to pinpoint source areas of erosion and is valuable for solutions of some design problems.

24. Mathematical Sediment Yield Models.—These are defined by Bennett (1974) as "hypothetical or stylized representation(s) of the physical processes occurring during the erosion and transport of sediment from a watershed." He includes "models as simple as a regression of measured suspended load on measured water discharge as well as models so complex as to contain components for computing overland flow and in-channel flow ... to predict sediment yield from the uplands and the stream channel." In these models, "an attempt is made to describe mathematically the pertinent hydrological, physiochemical, and biological processes occurring within the system so as to form a rational basis for

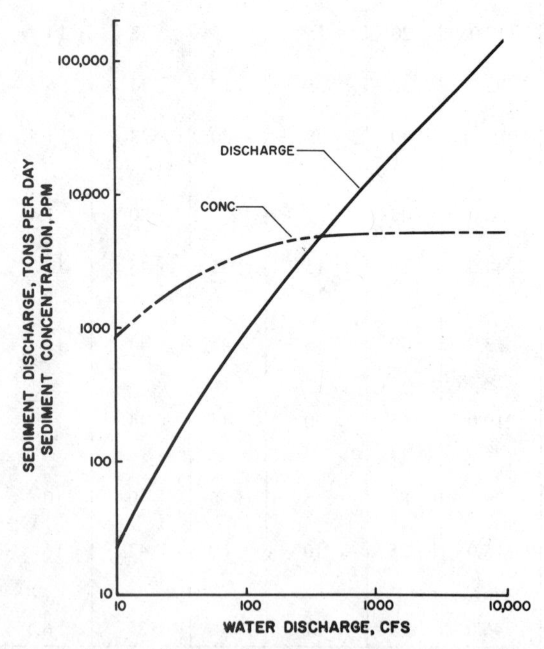

FIG. 4.28.—Daily Sediment Transport Relation, Pigeon Roost Creek Watershed 12

predicting, based on known or assumed inputs, the amount and composition of sediment output from the system in a given time period."

Most sediment yield modeling cannot be accomplished without use of a digital computer, and the recent impetus to develop workable models is the joint result of computer development and the mounting urgency to identify and evaluate sources of sediment and sediment-borne agricultural chemicals. Existing sediment yield models are varied in purpose; they have been designed for small acreages (Onstad and Foster, 1974) and for river basins (Negev, 1967). However, they all consider concepts whereby rainfall and runoff detach soil from field surfaces and transport it overland to drainageways (Rowlinson, 1971; Meyer, et al., 1975; Foster and Meyer, 1975).

Nearly all sediment yield models are second-stage components of runoff-generating models. Some runoff generation is by hydrologic budgeting methods using either empirical routing or linear reservoir routing techniques; other procedures involve kinematic routing by consideration of the hydraulics of overland flow. Refinements to these runoff-generation methods depend on the degree to which the natural processes are represented, and the reliance on parameter optimation and other statistical techniques.

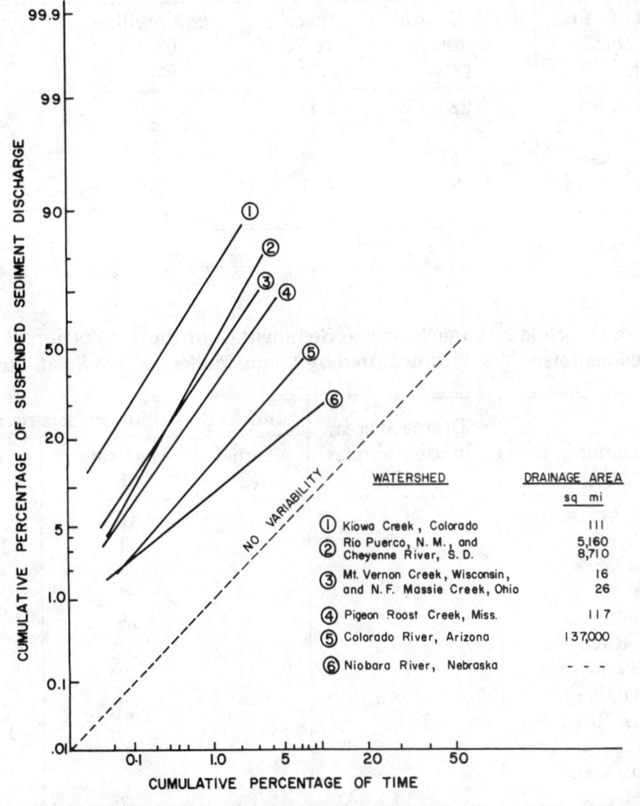

FIG. 4.29.—Plot of Cumulative Percentage Time Versus Cumulative Percentage of Total Suspended-Sediment Yield

The Universal Soil Loss Equation, usually with some modification, is the frequent basis for determining the quantities of soil that are detached from each small area of a watershed (Foster and Wischmeier, 1973). The downslope soil movement overland, and in rills and small channels, then becomes a function of sediment transport capacity. These aspects of sediment movement are not well documented, and some assumptions are necessary. For example, it is difficult to determine sediment transport capacity because the mode of hillslope transport —overland, in rills, and in channels—is usually unknown. It is also difficult to quantify the deposition that occurs in grassed waterways. Also, there are no present criteria for determining rill growth with time during a crop season. The role of rill and interrill processes is important, as they differently affect the movement of applied agricultural chemicals that may be surface applied or incorporated to some depth in the soil profile. The modeling experience is useful

TABLE 4.10.–Role of Large Storm as Sediment Contributor: Sediment Contribution by Storm Size[a]

Classification of storm size by amount of rain, in inches (1)	Percentage of Total Soil Loss Accounted for by Each Group of Storms			
	North Central States (2)	Ithaca, N.Y. (3)	Watkinsville, Ga. (4)	Guthrie, Okla. (5)
less than 1	22	43	13	7
1-2	47	33	22	29
2-3	21	23	38	26
3-4	9	1	7	7
greater than 4	1	0	21	30

[a]From Wischmeier, 1962.

TABLE 4.11.–Role of Large Storm as Sediment Contributor: Portion of Average Annual Sediment Yield Contributed by Large, Moderate, and Small Storms[a]

Location (1)	Drainage area, in square miles (2)	Storm Contribution, as a percentage		
		Large (3)	Moderate (4)	Small (5)
Kiowa Creek, Colorado	111	36	16	48
Honey Creek, Iowa	13.2	17	11	72
Pigeon Roost Creek, Mississippi	117	10	7	83
Dry Creek, Nebraska	20	46	15	39
Little Miami River, Ohio	50.6	11	7	82
Bixler Run, Pennsylvania	15	15	9	76
Big Elm Creek, Texas	68.6	28	14	58
Little Lacrosse River, Wisconsin	77.1	9	6	85
Muddy Creek, North Carolina	14.2	12	7	81

[a]From Piest, 1965.

for developing a better understanding of erosion and can delineate areas of needed research.

25. Magnitudes and Frequencies of Sediment-Producing Storms.—It is often necessary that the engineer determine the sediment yield of various rainstorms, or combinations of storms, and their probability of occurrence. This knowledge is needed to correctly assess and handle problems that occur in the operation of reservoirs and diversion dams; it is a necessary consideration in evaluating the effectiveness of terraces or an alternative conservation practice on the land; and it affects the degree of treatment of municipal and industrial water supplies and wastes.

The sediment contribution due to a given large rainstorm is highly variable, especially for smaller drainage basins. Love (1936) examined the sediment discharge records from 34 watersheds and found that as much as 53% of the sediment yield for a 15-month period occurred in a single day. The April 1, 1959, sediment yield from Mount Vernon Creek, Wisconsin, was more than 40% of that for the year and was greater than the annual sediment yield for each of the preceding three water years (Collier, 1963). In the case of the severe rainstorms on the 80-acre row-cropped watersheds in Western Iowa previously cited (Piest and Spomer, 1968), the June, 1967, sediment yield of 75 tons/acre was approximately equal to the yield for the remainder of the 4-yr period, 1964–1967. The June 20, 1967, sediment yield was half of the June total. In general, Neff (1967) found that only 40% of the long-term sediment load of arid regions was moved by runoff having a frequency of less than 10 yr; in humid regions, over 90% was moved.

These sediment discharges are not unusual, for Wolman and Miller record many occasions of this nature. Fig. 4.29 was modified from their report (Wolman and Miller, 1960) to show the trend of storm sediment yield for a number of large and small drainage basins throughout the United States. Daily sediment discharges were ranked, from highest to lowest, and accumulated in terms of total time and sediment quantity. Figures for the Rio Puerco, Cheyenne, Colorado, and Niobrara Rivers represent periods of record taken from United States Geological Survey Water Supply Papers; the remaining data were extrapolated from short-term records. It is noted, e.g., that in Kiowa Creek Basin, more than 20% of the entire suspended-sediment carried by the stream is discharged in 0.1% of the time. For the Niobrara River, only about 2% of the suspended sediment is discharged in 0.1% of the time.

Some sample storm statistics, for both fractional-acre plots and larger watersheds, are listed in Tables 4.10 and 4.11. For the small plots (Table 4.10), most of the sediment was contributed by rainstorms smaller than 3 in., although the Georgia and Oklahoma data showed that as much as one-third was from rainstorms of more than 3 in.

The data for watersheds listed in Table 4.11 tell much the same story. Most of the suspended sediment was produced by small storms, which are defined as those occurring at least once a year or more often, on the average. The large storms are defined as occurring less frequently than once in 2 yr, on the average.

D. References

Ackerman, W. C., and Corinth, R. L., "An Empirical Equation for Reservoir Sedimentation," *Publication 59,* International Association of Scientific Hydrology, Commission of Land Erosion, 1962, pp. 359–366.

Anderson, H. W., "Flood Frequencies and Sedimentation from Forest Watersheds," *Transactions,* American Geophysical Union, Vol. 30, No. 4, Washington, D.C., Aug., 1949, pp. 567–583.

Archer, R. J., "Sediment Discharge of Ohio Streams During Floods of January-February 1959," Ohio Department of Natural Resources, Division of Water, Columbus, Ohio, Aug., 1960.

Baird, R. W., "Sediment Yields from Blackland Watersheds," *Transactions,* American Society of Agricultural Engineers, Vol. 7, No. 4, St. Joseph, Mich., 1964, pp. 454–456.

Beer, C. E., and Johnson, H. P., "Factors Related to Gully Growth in the Deep Loess Area of Western Iowa," *Proceedings of 1963 Federal Interagency Conference on Sedimentation,* United States Department of Agriculture, *Miscellaneous Publication 970,* June, 1965, pp. 37–43.

Bennett, J. P., "Concepts of Mathematical Modeling of Sediment Yield," *Water Resources Research*, American Geophysical Union, Vol. 10, No. 3, June, 1974.

Blench, T., *Regime Behavior of Canals and Rivers,* Butterworth Scientific Publications, London, England, 1957.

Bonnard, D., and Bruschin, J., "Transports Solides en Suspension dans les Rivieres Suisses—Lonza, Borgne, Grand-Eau, 1966–1969," Ecole Polytechnique Federale De Lausanne, Lausanne, Switzerland, 1970.

Borland, W. M., "Sediment Transport of Glacier-Fed Streams in Alaska," *Journal of Geophysical Research,* Vol. 66, No. 10, Oct., 1961.

Brice, J. C., "Erosion and Deposition in the Loess-mantled Great Plains, Medicine Creek Drainage Basin, Nebraska," *Professional Paper 352–H,* United States Geological Survey, Washington, D.C., 1966.

Brown, C. B., *Sediment Transportation in Engineering Hydraulics,* H. Rouse, ed., John Wiley and Sons, Inc., New York, N.Y., 1950, p. 771.

Brune, G. B., *Proceedings,* Federal Interagency Conference on Sedimentation, Denver, Colo., 1948.

Brune, G. B., "Collection of Basin Data on Sedimentation," United States Department of Agriculture, Soil Conservation Service, Milwaukee, Wisc., 1953.

Campbell, F. B., and Bauder, H. A., "A Rating Curve Method for Determining Silt Discharge of Streams," *Transactions,* American Geophysical Union, Part 2, Washington, D.C., 1940, pp. 603–607.

Carter, C. E., Dendy, F. E., and Doty, C. W., "Runoff and Soil Loss from Pastured Loess Soils in North Mississippi," Mississippi Water Resources Conference, Jackson, Miss., Apr., 1966.

Colby, B. R., "Relationship of Sediment Discharge to Streamflow," Open File Report, United States Geological Survey, Washington, D.C., 1956.

Colby, B. R., and Hembree, C. H., "Computations of Total Sediment Discharge, Niobrara River near Cody, Nebraska," *Water-Supply Paper 1357,* United States Geological Survey, Washington, D.C., 1955.

Collier, C. R., "Sediment Characteristics of Small Streams in Southern Wisconsin, 1954–1959," *Water-Supply Paper 1669–B,* United States Geological Survey, Washington, D.C., 1963.

Collier, C. R., et al., "Influences of Strip Mining on the Hydrologic Environment of Beaver Creek Basin, Kentucky, 1955–1959," *Professional Paper 427–B,* United States Geological Survey, Washington, D.C., 1964.

Collier, C. R., personal communication, 1969.

Collier, C. R., et al., "Influence of Strip Mining on the Hydrologic Environment of Parts of Beaver Creek Basin, Kentucky, 1955–1966," *Professional Paper 427–C,* United States Geological Survey, Washington, D.C., 1971.

Dawdy, D. R., "Knowledge of Sedimentation in Urban Environments," *Journal of the Hydraulics Division,* ASCE, Vol. 93, No. HY6, Proc. Paper 5595, Nov., 1967, pp. 235–254.

Diseker, E. G., and McGinnis, J. T., "Evaluation of Climatic, Slope, and Site Factors on Erosion from Unprotected Roadbanks," *Transactions,* American Society of Agricultural Engineers, Vol. 10, 1967, pp. 9–14.

Dragoun, F. J., "Rainfall Energy as Related to Sediment Yield," *Journal of Geophysical Research,* Americal Geophysical Union, Vol. 67, No. 4, Washington, D.C., Apr., 1962.

Drozd, N. I., and Goretskaya, Z. A., "Map of Mean Sediment Concentration in the Waters of Ukrainian Rivers," Soviet Hydrology: Selected Papers, American Geophysical Union, No. 2, Washington, D.C., 1966.

Duley, F. L., and Ackerman, F. G., "Runoff and Erosion from Plots of Different Lengths," *Journal of Agricultural Research,* Vol. 48, No. 6, March 15, 1934.

Einstein, H. A., "The Bed Load Function for Sediment Transportation in Open Channel Flows," *Technical Bulletin 1026,* United States Department of Agriculture, Soil Conservation Service, Sept., 1950.

Fleming, G., "The Stanford Sediment Model I: Translation," *Bulletin* International Association of Scientific Hydrology, XII e, Annee 2, June, 1968, pp. 108–125.

Fleming, G., "Design Curves for Suspended Load Estimation," *Proceedings,* Institution of Civil Engineers, *Paper 7185,* No. 43, London, England, May, 1969, pp. 1–9.

Forest Service, Fort Collins, Colo. "Notes on Sedimentation Activities," United States Bureau of Reclamation, 1965.

Foster, G. R., and Meyer, L. D., "Mathematical Simulation of Upland Erosion Using Fundamental Erosion Mechanics," *Publication ARS–S–40,* United States Department of Agriculture, Agricultural Research Service, 1975.

Foster, G. R., and Wischmeier, W. H., "Evaluating Irregular Slopes for Soil Loss Prediction," *Paper 73–227,* presented at meeting of American Society of Agricultural Engineers, Lexington, Ky., June, 1973.

Free, G. R., "Research Tells How Much Soil We Can Afford to Lose," *Crops and Soils,* Vol. 12, 1960, p. 21.

Geiger, A. F., and Holeman, J. N., "Sedimentation of Lake Barcroft, Fairfax County, Va.," *SCS–TP–136,* United States Department of Agriculture, Soil Conservation Service, 1959.

Glymph, L. M., "Relation of Sedimentation to Accelerated Erosion in the Missouri River Basin," *Technical Paper 102,* United States Department of Agriculture, Soil Conservation Service, July, 1951.

Glymph, L. M., "Studies of Sediment Yields from Watersheds," *Publication No. 36,* de L'Association International D'Hydrologie, International Union of Geodesy and Geophysics, 1954, pp. 178–191.

Gottschalk, L. C., and Brune, G. M., "Sediment Design Criteria for the Missouri Basin Loess Hills," *SCS–TP–97,* United States Department of Agriculture, Soil Conservation Service, Oct., 1950.

Grissinger, E. H., "Resistance of Selected Clay Systems to Erosion by Water," *Water Resources Research,* Vol. 2, No. 1, 1966.

Guy, H. P., "An Analysis of Some Storm-period Variables Affecting Stream Sediment Transport," *Professional Paper 462–E,* United States Geological Survey, Washington, D.C., 1964.

Guy, H. P., "Residential Construction and Sedimentation at Kensington, Maryland," *Proceedings of 1963 Federal Interagency Conference on Sedimentation,* United States Department of Agriculture, *Miscellaneous Publication 970,* June, 1965.

Guy, H. P., and Ferguson, G. E., "Sediment in Small Reservoirs Due to Urbanization," *Journal of the Hydraulics Division,* ASCE, Vol. 88, No. HY2, Proc. Paper 3070, Mar., 1962, pp. 27–37.

Happ, S. C., "Effect of Sedimentation on Floods in the Kickapoo Valley, Wisconsin," *Journal of Geology,* Vol. LII, No. 1, Jan., 1944.

Happ, S. C., "Valley Sedimentation in North-central Mississippi," *Proceedings of Third Mississippi Water Resources Conference,* April 9–10, 1968.

Happ, S. C., Rittenhouse, G., and Dobson, G. C., "Some Principles of Accelerated Stream and Valley Sedimentation," *Technical Bulletin 695,* United States Department of Agriculture (original source, Rittenhouse, G., and Holt, R. D., unpublished; copies in SCS and ARS libraries), 1940.

Hill, A. L., "Rates of Sediment Production in the Kansas River District," report of Kansas City District, United States Corps of Engineers, 1965.

Holeman, J. N., personal communication, Apr., 1971.

Interagency Task Force, "Upper Mississippi River Comprehensive Basin Study," *Fluvial Sediment,* Appendix G, United States Corps of Engineers, Mar., 1967.

Ireland, H. A., Sharp, C. F., and Eargle, D. H., "Principles of Gully Erosion in the Piedmont of South Carolina," *Technical Bulletin 633,* United States Department of Agriculture, 1939.

Jones, B. L., "Effects of Agricultural Conservation Practices on the Hydrology of Corey Creek Basin, Pennsylvania, 1954–1960," *Water-Supply Paper 1532–C,* United States Geological Survey, Washington, D.C., 1966.

Krusekopf, H. H., "The Effect of Slope on Soil Erosion," *Missouri Agricultural Experiment Station Research Bulletin 363,* Apr., 1943.

Krusekopf, H. H., "Soils of Missouri—Genesis of Great Soil Groups," *Soil Science,* Vol. 84, 1958, pp. 19–27.

Kulp, J. L., "Geologic Time Scale," *Soil Science,* Vol. 133 (3459), 1961, pp. 1105–1114.

Kunkle, S. H., and Comer, G. H., "Suspended, Bed, and Dissolved Sediment Loads in the Sleepers River, Vermont," *ARS 41–188,* United States Department of Agriculture, Agricultural Research Service, June, 1972.

Lane, E. W., "Design of Stable Channels," *Transactions,* ASCE, Vol. 120, Paper No. 2776, 1955a, p. 1234–1260.

Lane, E. W., "The Importance of Fluvial Morphology in Hydraulic Engineering," Proceedings, ASCE, Vol. 81, Paper No. 745, July, 1955.

Lane, E. W., and Borland, W. M., "Estimating Bed Load," *Transactions,* American Geophysical Union, Washington, D.C., Feb., 1951.

Langbein, W. B., and Schumm, S. A., "Yield of Sediment in Relation to Mean Annual Precipitation," *Transactions,* American Geophysical Union, Vol. 39, Washington, D.C., 1958, pp. 1076–1084.

Leopold, L. B., Emmett, W. W., and Myrick, R. M., "Channel and Hillslope Processes in a Semiarid Area in New Mexico," *Professional Paper 352–G,* United States Geological Survey, Washington, D.C., 1966.

Leopold, L. B., and Maddock, T., "The Hydraulic Geometry of Stream Channels and Some Physiographic Implications," *Professional Paper 252,* United States Geological Survey, Washington, D.C., 1953.

Leopold, L. B., Wolman, M. G., and Miller, J. P., *Fluvial Processes in Geomorphology,* W. H. Freeman and Co., San Francisco, Calif., 1964, p. 328.

Love, S. K., "Suspended Matter in Several Small Streams," *Transactions,* 17th Annual Meeting, American Geophysical Union, Washington, D.C., 1936.

Lull, H. W., and Reinhart, K. G., "Logging and Erosion on Rough Terrain in the East," *Proceedings of 1963 Federal Interagency Conference on Sedimentation,* United States Department of Agriculture, *Miscellaneous Publication 970,* 1963, pp. 43–47.

Lusby, G. C., "Causes of Variation in Runoff and Sediment Yield from Small Drainage Basins in Western Colorado," *Miscellaneous Publication 970,* United States Department of Agriculture, 1963.

Lusby, G. C., "Hydrologic and Biotic Effects of Grazing Versus Non-grazing Near Grand Junction, Colorado," *Professional Paper 700–B,* United States Geological Survey, Washington, D.C., 1970, pp. 232–236.

Lustig, L. K., "Sediment Yield of the Castaic Watershed, Western Los Angeles County, California—A Quantitative Geomorphic Approach," *Professional Paper 422–F,* United States Geological Survey, Washington, D.C., 1965.

Maner, Sam B., "Factors Affecting Sediment Delivery Rates in the Red Hills Physiographic Area," *Transactions,* American Geophysical Union, Vol. 39, Washington, D.C., Aug., 1958, pp. 669–675.

Maner, S. B., and Barnes, L. H., "Suggested Criteria for Estimating Gross Sheet Erosion and Sediment Delivery Rates for the Blackland Prairies Problem Area in Soil Conservation," United States Department of Agriculture, Soil Conservation Service Publication, Feb., 1953.

McCully, W. G., and Bowmer, W. J., "Erosion Control on Roadsides in Texas," *Research Report 67–8F,* Texas Agricultural and Mechanical University, College Station, Tex., July, 1969.

Menard, H. W., "Some Rates of Regional Erosion," *Journal of Geology,* Vol. 69, 1961, pp. 154–161.

Meyer, L. D., Foster, G. R., and Romkens, M. J. M., "Sources of Soil Eroded by Water from Upland Slopes," *Publication ARS–S–40,* United States Department of Agriculture, Agricultural Research Service, 1975.

Meyer, L. D., and Monke, E. J., "Mechanics of Soil Erosion by Rainfall and Overland Flow," *Transactions,* American Society of Agricultural Engineers, Vol. 8, No. 4, St. Joseph, Mich., 1965.

Meyer, L. D., Monke, E. J., and Wischmeier, W. H., "Mathematical Simulation of the Process of Soil Erosion by Water," *Transactions,* American Society of Agricultural Engineers, Vol. 12, St. Joseph, Mich., 1969.

Miller, C. R., "Analysis of Flow-duration, Sediment-rating Curve Method of Computing Sediment Yield," United States Bureau of Reclamation Hydraulics Bureau, Apr., 1951.

Miller, C. R., Woodburn, R., and Turner, H. R., "Upland Gully Sediment Production," *International Association of Scientific Hydrology Publication No. 59,* Sept., 1962.

Musgrave, G. W., "The Quantitative Evaluation of Factors in Water Erosion, a First Approximation," *Journal of Soil and Water Conservation,* Vol. 2, 1947, pp. 133–138.

Mutchler, C. K., "Runoff Plot Design and Installation for Soil Erosion Studies," *Agricultural Research Service Report No. 41–79,* United States Department of Agriculture, Aug., 1963.

Neff, E. L., "Discharge Frequency Compared to Long-term Sediment Yields," Symposium on River Morphology, International Union of Geodesy and Geophysics, 1967, pp. 236–242.

Negev, M., "A Sediment Model on a Digital Computer," *Technical Report 76,* Stanford University, Stanford, Calif., Mar., 1967.

Onstad, C. A., and Foster, G. R., "Erosion and Deposition Modeling on a Watershed," *Scientific Journal Series 8537,* Minnesota Agricultural Experiment Station, 1974.

Parsons, D. A., "Coshocton-type Runoff Samplers, Laboratory Investigations," *SCS–TP–124,* Soil Conservation Service, United States Department of Agriculture, Washington, D.C., 1954.

Piest, R. F., "The Role of the Large Storm as a Sediment Contributor," *Proceedings of 1963 Federal Interagency Conference on Sedimentation, Miscellaneous Publication 970,* United States Department of Agriculture, June, 1965, pp. 97–108.

Piest, R. F., Bradford, J. M., and Spomer, R. G., "Mechanisms of Erosion and Sediment Movement from Gullies," *Publication ARS–S–40,* Agricultural Research Service, United States Department of Agriculture, 1975.

Piest, R. F., and Spomer, R. G., "Sheet and Gully Erosion in the Missouri Valley Loessial Region," *Transactions,* American Society of Agricultural Engineers, Vol. 11, St. Joseph, Mich., 1968, pp. 850–853.

Rainwater, F. H., United States Geological Survey Hydrologic Investigations Atlas, HA–61, 1962.

Renfro, G. W., and Adair, J. W., *Engineering and Watershed Planning Unit Technical Guide No. 12,* Soil Conservation Service, United States Department of Agriculture, Fort Worth, Tex., 1968.

Rice, R. M., and Foggin, G. T., "Effect of High Intensity Storms on Soil Slippage on Mountainous Watersheds in Southern California," *Water Resources Research,* Vol. 7, No. 6, Dec., 1971.

Roehl, J. W., "Sediment Source Areas, Delivery Ratios and Influencing Morphological Factors," *Publication 59,* International Association of Scientific Hydrology, Commission of Land Erosion, 1962, pp. 202–213.

Rowlison, D. L., and Martin, G. L., "Rational Model Describing Slope Erosion," *Journal of the Irrigation and Drainage Division,* ASCE, Vol. 97, No. IR1, Proc. Paper 7981, Mar., 1971, pp. 39–50.

Ruhe, R. V., and Daniels, R. B., "Landscape Erosion—Geologic and Historic," *Journal of Soil and Water Conservation,* Mar.–Apr., 1965.

Ruhe, R. V., Meyer, R., and Scholtes, W. H., "Late Pleistocene Radiocarbon Chronology in Iowa," *American Journal of Science,* Vol. 225, 1957, pp. 671–689.

Soil Conservation Service, United States Department of Agriculture, "Procedures for Determining Rates of Land Damage, Land Depreciation, and Volume of Sediment Produced by Gully Erosion," *Technical Release No. 32, Geology,* July, 1966.

Scheidegger, A. E., *Theoretical Geomorphology,* Prentice-Hall, Inc., Englewood Cliffs, N.J., 1961.

Schumm, S. A., "The Role of Creep and Rainwash on the Retreat of Badland Slopes," *American Journal of Science,* Vol. 254, Nov., 1956, pp. 693–706.

Schumm, S. A., "Quaternary Paleohydrology," The Quaternary of the United States, 1965.

Searcy, J. K., "Flow Duration Curves," *Water Supply Paper 1542–A,* United States Geological Survey, Washington, D.C., 1959.

Seginer, I., "Gully Development and Sediment Yield," *Research Report No. 13,* The Israel Ministry of Agriculture, Soil Conservation Division, Feb., 1966.

Smith, D. D., "Interpretation of Soil Conservation Data for Field Use," *Agricultural Engineering,* Vol. 22, 1941, pp. 173–175.

Smith, D. D., "Broad Aspects of Erosion Research," presented at the 1966 American Society of Agricultural Engineers Winter meeting.

Smith, D. D., and Whitt, D. M., "Estimating Soil Losses from Field Areas of Claypan Soils," *Proceedings,* Soil Science Society of America, Vol. 12, Madison, Wisc., 1947, pp. 485–490.

Smith, R. M., et al., "Renewal of Desurfaced Austin Clay," *Soil Science,* Vol. 103, 1967.

Smith, R. M., and Stamey, W. L., "Determining the Range of Tolerable Erosion," *Soil Science,* Vol. 100, No. 6, 1965.

Spomer, R. G., Heinemann, H. G., and Piest, R. F., "Consequences of Historic Rainfall on Western Iowa Farmland," *Water Resources Research,* Vol. 7, No. 3, June, 1971, pp. 524–535.

Thompson, J. R., "Quantitative Effect of Watershed Variables on the Rate of Gully Head Advancement," *Transactions,* American Society of Agricultural Engineers, Vol. 7, No. 1, St. Joseph, Mich., 1964, pp. 54–55.

United States Department of Agriculture, "Summary of Reservoir Sediment Deposition Surveys Made in the United States Through 1970," *Miscellaneous Publication 1226,* compiled under the auspices of the Sedimentation Committee, Water Resources Council, 1973.

United States Geological Survey, National Reference List of Water Quality Stations, Water Year 1971, Water Resources Division, Washington, D.C., 1971.

United States Geological Survey Water Supply Papers, "Quality of Surface Water of the United States," (published), and annual summaries (unpublished), Washington, D.C., 1959–1967.

University of Missouri, Agricultural Experiment Station, "Missouri Soil and Water Conservation Needs Inventory," Feb., 1962.

Ursic, S. J., and Dendy, F. E., "Sediment Yields from Small Watersheds Under Various Land Uses and Forest Covers," *Proceedings of 1963 Federal Interagency Conference on Sedimentation, Miscellaneous Publication 970,* United States Department of Agriculture, June, 1965, pp. 47–52.

Van Doren, C. A., and Bartelli, L. J., "A Method for Forecasting Soil Losses," *Agricultural Engineering,* Vol. 37, 1956, pp. 335–341.

Vice, R. B., Ferguson, G. E., and Guy, H. P., "Erosion from Suburban Highway Construction," *Journal of the Hydraulics Division,* ASCE, Vol. 94, No. HY 1, Proc. Paper 5742, Jan., 1968, pp. 347–348.

Walker, P. H., "Postglacial Environments in Relation to Landscape and Soils on the Cary Drift, Iowa," *Agriculture and Home Economics Experiment Station Bulletin 549,* Iowa State University, Nov., 1966.

Whitaker, F. D., and Heinemann, H. G., "The Effect of Soil Management Variables on Soil Losses," to be published in University of Missouri Experiment Station Bulletin.

Williams, J. R., and Berndt, H. D., "Sediment Yield Computed with Universal Equation," *Journal of the Hydraulics Division,* ASCE, Vol. 98, No. HY12, Proc. Paper 9426, Dec., 1972, pp. 2087–2098.

Williams, J. R., and Knisel, W. G., "Sediment Yield from Rangeland Watersheds in the Edwards Plateau of Texas," *ARS 41–185,* United States Department of Agriculture, 1971.

Wischmeier, W. H., "Cropping Management Factor Evaluations for a Universal Soil Loss Equation," *Proceedings,* Soil Science Society of America, Madison, Wisc., 1960.

Wischmeier, W. H., "Storms and Soil Conservation," *Journal of Soil and Water Conservation,* Vol. 17, Mar.–Apr., 1962.

Wischmeier, W. H., "Estimating the Cover and Management Factor for Undisturbed Areas," *Journal Paper 4916,* Purdue University, Lafayette, Ind., 1972.

Wischmeier, W. H., "Upslope Erosion Analysis," *River Mechanics III,* H. W. Shen, ed., 1973.

Wischmeier, W. H., Johnson, C. B., and Cross, B. V., "A Soil Erodibility Nomograph for Farmland and Construction Sites," *Journal of Soil and Water Conservation,* Sept.–Oct., 1971.

Wischmeier, W. H., and Smith, D. D., "Rainfall Energy and Its Relationship to Soil Loss," *Transactions,* American Geophysical Union, Vol. 39, Washington, D.C., 1958, pp. 285–291.

Wischmeier, W. H., and Smith, D. D., "A Universal Soil-Loss Equation to Guide Conservation Farm Planning," 7th International Congress of Soil Science, Madison, Wisc., 1960.

Wischmeier, W. H., and Smith, D. D., "Predicting Rainfall-erosion Losses from Cropland East of the Rocky Mountains," *Agriculture Handbook 282,* United States Department of Agriculture, Agricultural Research Service, May, 1965.

Wolman, M. G., and Miller, J. P., "Magnitude and Frequency of Forces in Geomorphic Processes," *Journal of Geology,* Vol. 68, Jan., 1960.

Ziemnicki, S., and Jozefaciuk, C., "Soil Erosion and Its Control," Translated from Erozja, Panstwowe Wydawnictwo Rolnicze i Lesne, Warszawa, 1965.

Zingg, A. W., "Degree and Length of Land Slope as it Affects Soil Loss in Runoff," *Agricultural Engineering,* Vol. 21, 1940, pp. 59–64.

Zingg, A. W., "Soil Movement Within the Surface Profile of Terraced Lands," unpublished report, United States Department of Agriculture, Soil Conservation Service, Dec., 1941.

Chapter V.—Sediment Control Methods

A. Introduction

1. General Remarks.—The control of sediments might present varying implications to individuals working in different fields. In the context of this publication it will be restricted to the prevention of the erosion of sediment particles from the land or from fluvial channels and to the controls applicable in fluvial channels or in the lakes, reservoirs, estuaries, or bays into which they flow. Items such as water supply or beach erosion, e.g., are not considered.

The subject is considered in two general areas, the land surface and the fluvial channels and associated water bodies. In order to conform with current practice, sediment control on land is discussed in terms of watershed areas, generally limited to 250,000 acres, and includes consideration of sheet erosion and the rills, guillies, or other fluvial channels included wholly therein. Larger channels are discussed separately in terms of natural and artificial channels, since control of the former is restricted generally by its existing form and the design of the latter is permitted many more degrees-of-freedom. Controls in reservoirs and harbors are also treated separately, since the concepts or purposes are apt to differ measurably.

B. Watershed Area

2. Concepts

Watershed Area.—A watershed area comprises all the land and water surface within the confines of a drainage divide. A watershed may be any size, ranging from a few acres above a farm pond, e.g., to the entire Mississippi River Basin. The United States Watershed Protection and Flood Prevention Act of 1954 (P.L. 566, 83d Cong., 68 Stat, 666, as amended) sets an upper limit of 250,000 acres as the maximum size area for Federally-assisted watershed project development in the United States. More and more, therefore, this has become recognized as the dividing line between so-called small watersheds and the larger areas that are appropriately termed river basins.

Watershed Protection Projects.—Watershed protection is an ancient concept. Roman and Greek engineers recognized the cause and effect relationship between deforestation and sediment deposition in harbors. Italian engineers in the late Middle Ages undertook primitive schemes of watershed protection to alleviate problems of flooding and sedimentation damages in the valleys of northern Italy. Washington, Jefferson, and others in early America wrote about the devastation of soil erosion on the rolling farmlands of Virginia. The concept was so well understood in scientific circles that creation of the National Forest reserves in 1892 was based on "Maintaining the flow of navigable streams." The material for this treatment of sediment control on watershed areas is based on practices developed by the United States Department of Agriculture.

Watershed projects in the United States are carried out on a national scale by the Federal Government in cooperation with local organizations and the States. About one-third of the land area of the Nation (including Alaska and Hawaii) is Federally-owned noncropland and is managed by Federal agencies. The principal agencies having custodial responsibility are the Forest Service of the United States Department of Agriculture, and the Bureau of Land Management, National Park Service, and the Fish and Wildlife Service of the United States Department of Interior. All of these land-managing agencies have soil and moisture conservation programs. The Forest Service and the Bureau of Land Management carry out much of this work in planned watershed management projects. Federal responsibility for the technical aspects of soil and water conservation in privately-owned lands, about 1-1/2 billion acres (or 75% of all United States land exclusive of Alaska and Hawaii), is vested in the Soil Conservation Service of the United States Department of Agriculture.

The present-day concept and the objective of watershed protection, management, and development are based on the following premises:

1. Using each acre of watershed land within its capability for sustained use without deterioration of the inherent soil resource.

2. Applying cultural, vegetative, and supporting mechanical practices on each acre of land as necessary to prevent soil deterioration and to obtain optimum soil, water, and vegetative management.

3. Protection from most floods and control of sediment production to the extent of: (a) Maximizing net economic benefits from floodwater and sediment damage reduction on flood-plain lands; (b) reducing reservoir and harbor deposition and other forms of sediment damages; and (c) improving water quality and related purposes.

4. Developing the water and related resources to the extent necessary to meet present and reasonably foreseeable future needs for municipal and industrial water supply, irrigation, drainage, fish and wildlife development, recreation, water quality management, and related purposes.

5. Improving and managing vegetative cover of trees and grass for sustained timber and forage yields as well as watershed protection where such cover represents the optimum land use.

6. Developing the most effective and economical combination of: (a) Vegetative and supporting mechanical practices on the surface of the land to influence the effect of precipitation where it strikes the earth; and (b) structural measures, e.g., reservoirs and channel improvement to control the movement of surface runoff.

7. Achieving a balance in water management between the needs of each acre, the watershed as a whole, and the river basin. The primary objectives of watershed protection programs, in broad terms, are: (a) Preservation and beneficial use of soil and water resources; and (b) reduction of erosion, floodwater, and sediment

damages. In some watersheds reduction of erosion and correction of existing and potential sediment problems are the primary purpose of a watershed protection project while in others floodwater damage reduction or water management may be of primary concern. In all watersheds, however, sedimentation processes and problems must receive consideration in keeping with their physical and economic significance wherever watershed protection projects are planned or installed.

Role of Planning Team.—Watershed project planning is a coordinated analysis by a team of technicians representing various disciplines (Tolley and Riggs, 1961).[1] The principal disciplines are economics, hydrology, geology, engineering, general soil science, and plant technology. Each is dependent on and interrelated with the others in developing and selecting an economically feasible system of improvements to meet the needs of a watershed.

In current United States Department of Agriculture practice, the planning party is headed by a staff leader who provides leadership and coordination for the activities of the group. The major technical disciplines represented by party members are economics, hydrology, geology, and engineering. Support in other disciplines previously enumerated is obtained from technicians representing their fields of endeavor in other programs of the Department.

The entire party collaborates in determining alternative preventive or corrective measures and selects and evaluates the most feasible protective or corrective measures to alleviate the problems encountered or anticipated in a watershed. The information gathered for a specific watershed is presented in a final watershed work plan that defines and describes the problems encountered; presents the recommended preventive or corrective measures for control of the problems; and evaluates the control measures for both physical and economic effectiveness as a basis for justification of a project.

Discussions of watershed surveys concerning the objectives, types of data collected, procedures or techniques involved, and types of control measures considered (all of the foregoing as it concerns sedimentation and erosion) are presented herein.

3. Watershed Survey

General.—The objective of a watershed survey, in the context of watershed protection, is to obtain the required information: (1) To quantitatively define the problems encountered; (2) to develop a program of works of improvement to reduce, alleviate, or eliminate the problems; and (3) to evaluate the effects of the selected works of improvement.

Sedimentation processes and consequences vary tremendously in different geologic, topographic, climatic, soil, and land-use environments. These processes and consequences also depend upon the intensity of economic developments. It is hazardous, therefore, to assume that conclusions arrived at through study of one watershed will apply equally well in another, even though the two areas may be adjacent. Each watershed is a case unto itself and must be so considered.

Despite the wide variation in the rates and significance of sedimentation processes among watersheds, there are six sequential steps that apply in dealing

[1] References are given in Chapter V, Section F.

with sedimentation problems by means of watershed protection projects. These are:

1. Identifying and determining sedimentation problems, existing and potential.
2. Determining the rates at which the problems occur or may be expected to occur.
3. Determining causal factors.
4. Determining alternative preventive or corrective measures.
5. Selecting and evaluating the most feasible preventive or corrective measures.
6. Installing and maintaining preventive or corrective measures.

Sedimentation Problems.—The unwanted deposition of eroded material is the most common expression of a sedimentation problem. Within the context of sediment control on watershed areas, however, sedimentation problems begin with the detachment of soil or rock materials by water or wind. They include also any associated adverse effects upon water quality and the stability and hydraulic efficiency of stream channel systems as well as problems resulting from the deposition of eroded material. Thus, geologists are faced with identifying sediment damages as well as sediment sources.

The many and varied problems that involve erosion, sediment in transport, and sediment deposition are well described in Chapter I and by Larson and Hall (1957). Not all of these exist in each small watershed but the possibility of their occurence must be investigated. Information must be obtained concerning the severity of the problems encountered. For instance, the extent, both as to area and depth, of damaging deposition on flood-plain lands; the extent of sediment deposition in channels and in irrigation and drainage ditches and its influence on watershed problems; deposition of sediment in existing reservoirs; the effect and extent of sediment in transport on water supplies; and the location of the sources of sediment in terms of erosion are among the types of data collected in formulating a program of sediment control in watershed areas.

Survey Procedures.—Deposition on flood-plain land is often obvious to the observer especially if overbank flow has recently occurred. In many instances, however, textural changes either to a composition coarser or finer than the original flood-plain soils are not obvious on the surface and must be determined by boring. The swamping of flood-plain land requires a knowledge of past use of the land. Diminished productivity, such as a regression to a less productive use brought about by deposition can be recognized by comparisons of aerial photographs taken in different years, or determined by interview or other historical information. Scouring of flood-plain land can be recognized by the existence of channels eroded in flood plains. Stream bank erosion, although obvious in some instances, may require comparisons of aerial photos. Other sedimentation problems encountered may include deposition in existing reservoirs, the severity of which must be determined by reservoir sedimentation surveys. Deposition in harbors and estuaries requires knowledge of previous capacities to compare with current soundings, but information on rates of filling in these facilities or areas may be obtained from records kept of recurring dredging operations. Detailed investigations to determine sedimentation damages may be determined by a valley range system as described by Larson and Hall (1957) or by aerial mapping of a flood plain. Sediment damages to facilities are determined as described previously.

As erosion is the source of sediment, the sedimentationist must be able to recognize the extent, location, and severity of the erosion. It is axiomatic that all land surfaces are subject to erosion—either geologic (normal) or accelerated. Accelerated erosion is of direct concern as most is man-induced. Controls can be applied to this type of erosion to reduce and minimize its effect. On sloping, cultivated land sheet erosion can be recognized by rills in fields, by deposition at the base of cultivated slopes, or by general appearance to the trained eye. On grassland and forested land sheet erosion may be inconspicuous although there is little doubt that soil is being moved by the erosion processes. Detailed procedures for quantitative determinations of accelerated soil movement by sheet erosion are contained in Chapter IV.

Active gullying is obvious to an observer, but rates of gullying to determine annual contributions to sediment yield must be developed either by comparison of aerial photographs taken in different years or by formulas described in Chapter IV. The volume of material produced by stream bank erosion, in general, must be determined by comparison of aerial photos taken at different times, although interviews with local landowners often provide excellent information on the rates of such erosion.

A stream bed that is degrading can be recognized by the lack of deposition in the channel or by obvious erosion of the bed. Comparisons of available differing-age cross sections afford a means of developing rates of erosion and volumes of material produced from this source. Where channel beds are composed of noncohesive materials, the application of bed load transport equations can give estimates of the volume of material moved each year.

Quantitative estimates of the production of eroded materials are required in order to develop, recommend, and evaluate watershed control measures that reduce generation of this downstream damaging material.

Reporting Erosion and Sediment Information.—The mere gathering of data concerning the erosion and sediment problems in any watershed is of little use unless the data are presented in a manner that allows an evaluation of the information. Generally, in United States Department of Agriculture practice, existing sedimentation problems are located and indicated on a map that is included in the final work plan. Critical sediment source areas also are shown. Damaging deposition is shown with a plus mark, swamping is shown by the conventional swamping symbol used by cartographers, and critical erosion sources generally are delineated by their areal limitations. Tabulations in work plans are summations of work sheets concerning these features developed during field and office work. Sediment yields as such are shown in the tabulations if they can be determined with reasonable accuracy. Both current sediment yields and expected yields after project completion are indicated in order that the effects of the program in a given watershed may be evaluated in terms of downstream interests if necessary.

4. Control Measures

General.—The reduction of erosion incident to cultivation, grazing, and timber production on watershed land is a first step to correcting most sediment problems. Sheet erosion (Fig. 5.1) and channel type erosion, including gullying (Fig. 5.2), on such lands are the principal sources of sediment in the more humid parts of the country where agriculture is the predominant land use. Generally, however, gullying and stream channel erosion are the greater sources of sediment in forest and range areas and in those parts of the country receiving less than 20 in. of precipitation annually (Brown, 1958). Erosion incident to cultural developments, including roads and super-highways, urbanization, mining, and industrial projects are also conspicuous sources of sediment in some watersheds and require increasing consideration in sediment control programs. Where upland sheet erosion problems prevail, control is usually by use of land treatment measures. Structural measures are normally needed for control of channel type erosion.

Land Treatment Measures for Watershed Protection.—Land treatment measures are the basic elements of watershed projects and are the first increment in project evaluation. They have a significant effect in reducing sedimentation damages, particularly in the humid agricultural areas of the country and in areas where damaging sediment is derived primarily from sheet erosion.

Land capability is a measure of the need and place for conservation practices and is an important consideration in the land use and choice of land treatment measures. There are eight major land capability classes (Soil Conservation Service, 1961). The first four (I–IV) are considered suitable for cultivation if specified management, cultural, and erosion control practices are observed although they are also suitable for grazing, woodland, or wildlife. As the class number becomes larger, increasing limitations are placed on the use of the land and more elaborate soil conserving practices are required. Land capability classes V, VI, and VII are generally considered suitable for grazing, woodland, or wildlife and class VIII is suitable for wildlife, recreation, or other watershed uses.

Thus it is obvious that care must be taken to plan use and treatment that is appropriate to the land's capabilities. In some instances, changes in land use must be considered. Upland areas presently being used beyond their capabilities usually can be converted to less intensive use in keeping with their capabilities. More intensive use can be shifted to flood plains when they are protected from frequent damaging floods by structural measures. Land in upland areas that cannot be cultivated safely or profitably can be shifted to permanent vegetative cover, e.g., grasses or trees. Such a land-use change both protects the land and provides more profit to an operator over the long-run.

1. Vegetative Treatment.—Vegetative land treatment measures are more effective in reducing erosion and resultant sediment yields than in reducing peak waterflow. Their effectiveness is accomplished, mainly, by improving the protective cover on the soil surface exposed to eroding forces and by increasing infiltration rates. Cover condititions may be improved by several standard agronomic and forestry practices (Soil Conservation Service, 1959) These include:

a. Conservation cropping systems that encompass the growing of crops in combination with needed cultural and management measures. Cropping systems involve the use of rotations that may include grasses and legumes grown in desirable sequence.

b. Cover cropping with close-growing grasses, legumes, or small grain in a cropping system primarily for summer or winter protection and for soil improvement.

c. Critical area planting that is achieved by establishing vegetative cover on excessive sediment-producing areas. Such stabilization may include planting

FIG. 5.1.—Sheet and Rill Erosion on Cultivated Land

FIG. 5.2.—Active Gully in Pasture Land

woody plants, or seeding or sodding adapted grasses or legumes to provide long-term ground cover.

d. Crop residue use that consists of incorporating into the soil plant remains left in cultivated fields or leaving plant remains on the surface during that part of the year when critical erosion periods usually occur.

e. Hayland planting that establishes long-term hay stands of grasses or legumes.

f. Mulching, in which plant or other suitable materials not produced on the site are applied to the soil surface.

g. Pasture planting in which adapted species of perennial, biennial, or reseeding forage plants are established on new pastureland converted from other uses.

h. Tree planting that includes planting tree seedlings or cuttings in open areas to establish a stand of forest trees.

i. Woodland interplanting that involves planting tree seedlings in sparsely or inadequately stocked stands.

Preservation and improvement of vegetative cover provides a fourfold means of reducing erosion and sediment yields in watersheds. The plant materials intercept rainfall and minimize the effect of raindrop impact, and increase infiltration and thereby reduce the rate of surface runoff; the roots and plant stems help to bind the soil into an erosion resistant mass; and vegetation increases the roughness of the ground surface reducing the velocity of overland flow and thereby its capacity to erode and transport sediment.

The expected ratio of soil loss under a specific crop and management practice to the corresponding loss from cultivated continuous fallow on the same soil and slope and under the same rainfall ranges from a ratio of 0.92 for continuous corn to a ratio of 0.006 for a well-established meadow (Agricultural Research Service, 1961). These ratios indicate the desirability of establishing agronomic and forestry practices as a means of reducing or maintaining minimum rates of erosion consonant with the agricultural needs of the area.

2. Protecting Existing Vegetative Cover on Forest and Grazing Land.—Fire protection as well as managed grazing are measures developed to protect existing vegetative cover and to control erosion and sediment yield on forest, brush, or grassland. In improving and maintaining each of these cover types, both offsite, or sediment control benefits, and onsite, or conservation benefits, may be derived. Frequently, protection of brush-covered watersheds is considered to have only offsite benefits because the brush, in itself, usually has little value.

a. Fire protection.—The type and intensity of fire protection varies widely depending on a number of factors. An intensive level of protection is used if the experienced average annual rate of burn is high, the associated erosion rates are also high, and there is a risk to life and property. Less intensive protection is used when studies reveal a decreasing hazard from fire or its influence.

Fire protection measures include installations, e.g., forest lookouts, guard stations, access roads and trails, fire-breaks water supplies, and heliports. Equipment includes fire fighting trucks and apparatus, aircraft such as helicopters, and suppression materials. Personnel must be available for supervising and coordinating fire fighting, for handling equipment and for special details such as parachuting to vantage points otherwise inaccessible.

The kinds of measures used depend on terrain, type and density of vegetation, and weather conditions that could exist during the fire season. Access roads are located for the purpose of reaching the proximity of fire breakout points with a minimum loss of time, providing the topography allows placement of such roads. In mountainous country, these roads usually must be supplemented by trails and heliports. Such measures are intended to bring fire under control as rapidly as possible. Measures such as firebreaks are for the purpose of isolating or limiting the spread of burns.

Considerable evidence is available on the effect of fires and following storms on some mountainous watersheds. Records of the Los Angeles County Flood Control District (Borke, 1938) and the United States Forest Service (Rowe, Countryman, and Storey, 1954) show that over 100,000 cu yd of sediment per sq mile have been produced under storm runoff on burned-over watersheds while the rate under favorable conditions is on the order of 5,000 cu yd/sq mile–7,000 cu yd/sq mile.

Measurements made after burns on three small chapparral brush-covered watersheds of Arizona and reported on by Glendening, Pase and Ingebo (1961) show that sediment yields of between 21,519 tons/sq mile and 64,446 tons/sq mile were produced the first 21 months after a fire. A 4-day storm on one of the watersheds before the fire produced an equivalent of only 50 tons/sq mile. A grass and brush fire occurred in 1951 in the Cottonwood Gulch Watershed near Boise, Idaho. Following two storms in 1959, which flooded portions of Boise, measurements were made of soil losses on the acres burned in 1951 plus those affected by two smaller fires in 1957 and 1958. An aggregate of 170,000 tons were lost from Cottonwood Gulch watershed slopes during the two storms, or a rate of about 11,000 tons/sq mile (Soil Conservation Service, 1959b). These gross erosion rates may be compared with sediment yield measurements made at the mouth of Cottonwood Gulch in 1939 and 1940 and reported on by Love and Benedict (1948). The average for two spring runoff seasons from the unburned watershed was 1,244 tons/sq mile.

b. Grazing management.—Managed grazing has the same objectives as fire protection, i.e., improving and maintaining vegetative cover. By implication, the need is to preserve plant density and cover adequate for protection against erosion and at the same time provide forage of economic value. The two objectives are compatible to the extent that conservatively grazed plants of any type provide materials that protect the soil. Grazing management directed toward these two objectives may involve several practices. They include grazing exclusion in critical runoff and sediment-producing areas, and temporary exclusion in less critical areas to firmly establish reseedings or to allow natural rehabilitation of overgrazed plants. A frequently employed practice is reduced stocking of range to permit plant recovery. Another is better distribution of stock over the range. Others include deferred, rotational, and seasonal grazing.

Managed grazing is facilitated by construction of fencing and stockwater developments, e.g., springs, wells, and ponds. Reseeding of range is sometimes done to relieve pressure on an adjacent area. Results of managed grazing on erosion and sediment yields are readily apparent, though not so spectacular as the effects of fire protection since the latter benefits may be derived from erosion control of much steeper slopes. Then, too, some plants and litter remain even after excessive grazing. Many range soils are very thin under undisturbed conditions, indicating the advantage of good cover density as protection against wind and water.

Several comparisons of sediment yield from unmanaged and managed watersheds are available. Rosa and Tigerman (1951) studied the results of sediment yield determinations in the Boise River Watershed and included a comparison of tons of sediment per unit discharge from two nearby creeks. They found that one, Grouse Creek, in a seriously depleted range condition produced more than three times as much as the other, Cottonwood Creek, described as in fairly good range condition. They found a similar difference between good and poor cover and sediment yield in the upper San Juan River Basin of New Mexico and Colorado.

3. Supporting Mechanical Field Practices.—Mechanical field practices are usually employed in connection with vegetative treatments. Some of these practices are more pronounced in their effect on erosion, particularly channel-type erosion, than are others. However, all of them afford added protection to the land. These practices include but are not limited to:

a. Contour farming.—Conducting farming operations on sloping, cultivated land in such a way that plowing, land preparation, planting, and cultivation are done on the contour. Depending on the steepness of slope, erosion from contoured fields may be reduced as much as 50% of that expected from up-and-down tillage (Agricultural Research Service, 1961). This practice alone, although applicable to all cultivated land, is most effective during storms of low rainfall intensity and on the flatter slopes encountered in land capability classes II and III. The slope length limits on which contouring alone is recommended (Agricultural Research Service, 1965) are shown in Table 5.1.

b. Contour furrowing on rangeland.—Plowing furrows on the contour at intervals varying with the slope and ground cover. This practice is applicable on: (1) Deteriorated ranges; (2) moderately fine, medium, and moderately coarse-textured soils; (3) slopes not exceeding 20%; and (4) where grazing management permits the vegetation to take advantage of the treatment.

c. Contour strip-cropping.—Growing of crops in a systematic arrangement of strips or bands on the contour (Fig. 5.3). The crops are arranged so that a strip of grass or close-growing crop is alternated with a strip of clean-tilled crop or fallow. This practice is adapted to well-drained cultivated soils where rainfall causes erosion. Steepness of slope, kind of soil, the usual amount and intensity of rainfall, and the size of farm equipment are factors to consider in determining the width of the strips. In many localities strips are made 100 ft wide on slopes of less than 7%, about 80 ft on slopes from 7%–12%, and 50 ft wide on slopes between 18% and 24% (Agricultural Research Service, 1965).

The effectiveness of this practice in reducing erosion and resultant sediment yields is indicated by the fact that erosion from contour stripcropped fields

TABLE 5.1.—Slope Length Limits for Contouring

Slope, as a percentage (1)	Maximum slope length, in feet (2)
2	400
4–6	300
8	200
10	100
12	80
14–24	60

averages from 45%–25% of that expected from up-and-down tillage, depending on the steepness of slope.

d. Gradient terraces.—Constructing earth embankments or a series of ridges and channels across a slope at suitable spacings and with accepted grades (Fig. 5.4). These are installed to reduce erosion damage and sediment yield by shortening slope lengths and by intercepting surface runoff and conducting it to a stable outlet at a nonerosive velocity. They are not constructed on deep sands or

FIG. 5.3.—Contour Strip-Cropping Showing Alternate Strips of Small Grain, Meadow, and Corn

FIG. 5.4.—Gradient Terraces Installed on Iowa Farm

on soils that are too stony, steep, or shallow to permit practical and economical installation and maintenance. Contour farming and strip-cropping usually are used in conjunction with gradient terraces. Gradient terraces may be used only if suitable outlets for runoff, either natural grassed waterways or vegetated areas, are or will be available.

The maximum vertical spacing of gradient terraces may be determined by the equation

$$VI = xS + y \qquad (5.1)$$

in which VI = vertical interval, in feet; x = a variable ranging in value from 0.4 in the southeast to 0.8 in a zone extending from northern Michigan into the areas north and west of central Colorado and New Mexico; S = land slope, in feet per 100 ft; and y = a variable with values from 1.0–2.0. The higher values apply to erosion-resistant soils and to cropping systems that provide significant protection to the soil during critical erosion periods. The vertical spacing may be increased by as much as 10% or 0.5 ft, whichever is greater, to provide better alinement or location or to reach a satisfactory outlet. The terrace should have sufficient capacity to handle the peak runoff expected from a 10-yr frequency storm without overtopping. The minimum cross-sectional area of a terrace channel is 8 sq ft for land slopes of 5% or less, 7 sq ft for slopes from 5%–8%, and 6 sq ft for slopes steeper than 8%. These minimums may be increased to accommodate the farm machinery being used. The channel grades may be either uniform or variable with a maximum grade of 0.6 ft/100 ft of length, and channel velocity should not exceed that which is nonerosive for the soil type.

e. Level terraces.—A series of ridges and channels that are constructed across a slope at suitable spacings, but have no grade. This practice is used to conserve moisture and to control erosion. Level terraces are constructed only on deep soils that are capable of absorbing and storing extra water without appreciable crop damage and in areas where the rainfall pattern is such that storage of rainfall in the soil is practicable and desirable.

The maximum spacing is determined by the equation

$$VI = 0.85\,S + y \qquad (5.2)$$

in which VI = vertical interval, in feet; S = land slope, in feet per 100 ft; and y = a variable with values from 1.0–2.0 as described for Eq. 5. 1. The vertical spacing may be increased as discussed under gradient terraces. The capacity of a level terrace must be adequate to handle runoff from a 10-yr frequency storm without overtopping. If a terrace is closed-end, the runoff volume of a 10-yr frequency, 24-hr storm is used in determining required storage capacity. The terrace cross section is proportioned to fit the land slope, the crops to be grown and the machinery to be used. The ridge height, as constructed, should include a reasonable settlement factor and have a minimum top width of 3.0 ft at the design height.

Level terraces may have open ends, partial end closures, or complete end closures. Partial and complete end closures are used only on soils and slopes where the stored water will be absorbed by the soil without crop damage. For level terraces of given dimensions, the volume of water stored above the terrace is proportional to the length. Therefore it is important that the length be controlled so that damages in the event of a break are minimized. The terrace length should not exceed about 3,500 ft unless the channel is blocked at intervals to provide segments not exceeding this length. If level terraces have open ends or partial end closures, adequate vegetated outlets must be provided to convey runoff from the terrace system to a point where the outflow will not cause damage.

f. Diversions.—Graded channels constructed across a slope with a supporting ridge on the lower side. They are used to divert water from areas of excess to sites where it can be used or disposed of safely. Diversions protecting agricultural land must have capacity to carry the peak runoff from a 10-yr frequency storm as a minimum, with a freeboard of not less than 0.3 ft. They are designed to develop velocities that are nonerosive for the soil materials through which they pass with consideration being given to the type and condition of the planned vegetative protection in the diversion. Table 5.2 indicates the range in velocities allowable under differing soil and vegetative conditions.

g. Grassed waterways.—Natural waterways or depressions that are reshaped or graded and on which suitable vegetation is established. They provide for disposal of excess surface water from terraces, diversions, contoured fields, or natural concentrations without damage by erosion or flooding. The minimum capacity should confine the peak runoff from a storm of a 10-yr frequency. The design velocities should not exceed those obtained by using procedures, friction coefficients, and recommendations contained in SCS-TP-61, Handbook of Channel Design for Soil and Water Conservation (Soil Conservation Service, 1947). Permissible velocities for waterways lined with vegetation, assuming average uniform stands of each type of cover, range from 8 fps on erosion-resistant soils with a channel slope less than 5% and Bermudagrass cover to 2.5 fps on mild (less than 5%) slopes in easily eroded soils supporting an annual type of vegetative cover.

h. Irrigation ditch and canal lining.—Installed to reduce water loss, to prevent waterlogging of land, or to prevent erosion. Nonreinforced concrete or asphalt lining may be cast in place or flexible membranes may be placed throughout the channel.

i. Grade stabilization structures.—Installed to stop the advance or prevent the formation of gullies in natural or constructed waterways, diversions and other points of concentrated waterflow. They also may be used to convey water safely from tributary watercourses or ditches into larger streams or canals. In this latter instance they provide the means of conducting the water through differentials in elevation without downcutting in the tributaries.

TABLE 5.2.—Allowable Velocities in Diversions

Soil texture (1)	VELOCITY, V, IN FEET IN SECONDS			
	Bare channel	Channel Vegetation		
		Poor	Fair	Good
	(2)	(3)	(4)	(5)
Sand, silt, sandy loam, and silty loam	1.5	1.5	2.0	3.0
Silty clay loam, sandy clay loam	2.0	3.0	4.0	5.0
Clay	2.5	3.0	5.0	6.0

There are so many varied conditions throughout the country that only general specifications for grade stabilization structures are presented herein. The structures may be drop inlets, chutes, or earth dams and each must be designed for its specific situation. They must be adequately supported and sufficient cut-off trenches or walls and toe drains provided to insure stability. Inlets must be adequately protected against velocities and trash. The design must provide adequate capacity to pass, without damage to the structure, the outflow commensurate with the purpose of the structure. Structures may be made of materials, e.g., concrete, rock, compacted earth fill, masonry, steel, or treated wood.

Several means of controlling gully erosion as mentioned here are contained in Farmer's Bulletin 2171, United States Department of Agriculture (Francis, 1961).

Structural Measures.—Watershed structural measures, the second element of a watershed project, are installed for land stabilization, waterflow control, or storage to provide user benefits. They may include: (1) Any form of earth work, either excavation or fill; (2) works of concrete, masonry, metal, or other materials; and (3) vegetative planting associated with such structural works.

Land stabilization measures primarily are used to prevent land destruction or to reduce production of damaging sediment. They are major structural measures that accomplish protection beyond that afforded by land treatment measures and produce community benefits. Waterflow control measures control damaging waterflows and water-borne sediment.

The major watershed structural measures commonly used include: (1) Reservoirs, both detention and multiple-purpose; (2) stream channel improvement and stabilization works; (3) debris and sedimentation basins; and (4) levees, dikes, floodways and floodwater diversions.

Reservoirs

Reservoirs are the basic structural measure in watershed projects in which flood prevention is a significant purpose. They reduce floodwater damages by reducing peak flows from storms of predetermined frequencies to allowable maximums. They also permit the use of more economical channel improvements or stabilizing structures in the channel downstream should such installations be necessary.

Reservoirs built in watershed protection projects are of two types: detention and multiple-purpose. Floodwater retarding structures are detention dams with a fixed-capacity principal spillway and an emergency spillway. The volume of storage between the inlet to the principal spillway and the emergency spillway is for detention of floodwaters that are released at a predetermined rate. Multiple-purpose reservoirs, in addition to floodwater detention, may embody additional capacity for such uses as municipal water supply, irrigation water, and recreation.

Both types of reservoirs are effective in two ways: (1) By the control of floodwater; and (2) by the impoundment of water-borne sediment. Analyses of some 22 runoff events in the North Fork Broad River in northeastern Georgia (Soil Conservation Service, 1962) indicate a reduction in the hydrograph peak at a given structure site to be as much as 95% due to the slow release of accumulated floodwater. The reduction in peak flow decreases in a downstream direction due to added inflow from areas not controlled by floodwater retarding reservoirs. Such reduction in peak discharges to downstream channels, except when the design storm is exceeded, prevents overflow of channels and destructive flood-plain damage by floodwater, sediment deposition, and scour.

In addition to their effect on waterflow, the structures provide barriers to the transport of sediment that originates in the areas draining into them. Although such reservoirs allow some of the sediment delivered to them to pass on through the outlet works, the portion carried on downstream is generally fine-grained and nondamaging. The coarse-grained material remains in the reservoir where storage space for it has been provided. The effectiveness of these reservoirs in trapping sediment has been shown in preliminary studies that indicate that the trap-efficiency may vary from 50% to nearly 100% depending on the grain-size of incoming sediment and other factors.

Detention structures used in watershed protection projects have little or no effect on the peak flows of extremely small flood events. Land treatment practices and measures have the opposite effect, i.e., their greatest influence is on smaller flood events. Thus, reservoirs, land treatment, and agricultural water management practices and measures complement each other to prevent flood and sediment damage in addition to conserving soil and water.

Structures are classified on the basis of the hazards of failure as follows:

1. Class a—Structures located in rural or agricultural areas where failure may damage farm buildings, agricultural land, or township or county roads.
2. Class b—Structures located in predominately rural or agricultural areas where failure may damage isolated homes, main highways or minor railroads, or cause interruption of use or service of relatively important public utilities.
3. Class c—Structures located where failure may cause loss of life, serious damage to homes, industrial and commercial buildings, important public utilities, main highways, or railroads.

Design criteria vary for each class. For example, the minimum storage and principal spillway capacity of a class a structure must be adequate to control a 25-yr frequency flood event without use of the emergency spillway. For class b and c structures, the minimum criteria require the control of 50-yr and 100-yr frequency flood events, respectively. The emergency spillway hydrograph used for designing the earth spillway and the freeboard hydrograph used to determine a minimum permissible height of dam will vary from the 100-yr event to the probable maximum event depending on the structure classification.

a. Detention Reservoirs.—A typical detention or floodwater retarding structure as shown in Fig. 5.5 consists of an earth fill dam designed to create a reservoir which will temporarily store flood flows. The stored water is released at a relatively uniform predetermined rate through a closed pressure conduit principal spillway. Rare flood events that may exceed the combined capacities of the reservoir and the principal spillway are passed through a vegetated emergency spillway. Minimum hydrologic design criteria for the emergency spillways for this type of dam have been outlined by Ogrosky (1964).

b. Multiple-Purpose Reservoirs.—If a detention reservoir is justified by reduction in downstream floodwater and sedimentation damages, additional storage may be added to it for other purposes. The most common multiple-purpose reservoirs to date are designed to provide water storage for irrigation, municipal, industrial, fish and wildlife, and recreational purposes. In watershed projects installed with the assistance of the United States Department of Agriculture, that portion of the cost of the structures that is allocated to municipal and industrial water supply or to water quality control is borne entirely by the local people. For other uses, e.g., irrigation, recreation, and fish and wildlife enhancement, the Federal Government cost-shares with the local people according to a fixed ratio.

c. Sedimentation Problems of Watershed Reservoirs.—Each reservoir included in watershed protection projects is designed to be fully effective over a selected period, usually 50 yr–100 yr. In order that the full design capacity is effective over the selected period, additional storage must be provided for the estimated accumulation of sediment. The amount of storage incorporated for sediment accumulation is determined on the basis of current and future sediment yields. The current sediment yield is that prevailing until land treatment measures and any structural measures have been installed in the drainage area of the structure and have become effective in reducing erosion and resultant sediment yield to the reservoir. After this period, which may range from 3 yr–15 yr, the reduced sediment yield is used in the sediment design for the remaining years of the structure's design life. It is essential, therefore, that the primary erosion and sediment source areas be delineated and the effectiveness of their potential treatment be evaluated to determine the degree of anticipated reduction in erosion and resulting sediment production.

General methods of estimating long-term sediment yields from watersheds have

FIG. 5.5.—Highland Creek Dam, California: Completed Floodwater Retarding Structure; Emergency Spillway is Located to Right of Dam

been outlined by Gottschalk (1958) and in Chapter IV. Briefly these methods include: (1) Transposition of available sediment yield data from comparable watersheds; (2) sediment-load sampling and development of sediment-rating and flow-duration curve; and (3) estimating gross onsite erosion in a watershed and delivery ratio of the sediment. Only the latter method provides an adequate basis for evaluating the influence of proposed watershed treatment measures on future sediment yields.

The sediment design of a detention or multiple-purpose reservoir requires consideration of other factors besides the sediment yield. Part of the sediment delivered to a structure may be vented through the outlet and it is unnecessary to provide capacity for it. From the standpoint of design, however, the portion of the total sediment yield that will be vented must be estimated. In general, the lower the capacity-inflow ratio of the structure and the finer the texture of the sediment, the greater the percentage of sediment which escapes downstream.

The location or distribution of deposits in a reservoir must be predicted. The concept of providing a "silt pocket" in the bottom contours of a reservoir basin to contain all the sediment is not applicable to small watershed reservoirs and can result in serious error in design. Studies of distribution of sediment in detention reservoirs indicate that, in extreme cases, as much as 50% of the total sediment yield may be deposited above the elevation of the principal spillway and that a deposition of 30%–40% above that elevation is commonplace. Proper design, therefore, dictates that adjustments be made in allocations for all storages commensurate with the amount of sediment expected to be deposited in each.

The total amount of sediment, in tons, that is estimated to be deposited in the reservoir during its design life is allocated by percent to the capacities that either will be permanently submerged or can be expected to deposit above a permanent pool elevation. These percentages vary according to the grain size of incoming sediment as well as to the mode of operation of the reservoir. After this allocation, the volume displaced by the sediment is estimated on the basis of the volume-weight of submerged and aerated sediment. The volume required to contain the submerged sediment establishes the elevation of the principal spillway. In detention reservoirs with the single purpose of floodwater retardation the capacity below this elevation is termed the sediment pool. In similar reservoirs with dry sediment storage provisions, the lowest port in the outlet is at or near stream bed elevation on the upstream toe of the dam.

If it were assumed that all of the sediment would deposit in a submerged condition, it could well be that the storage allocated for other uses would be depleted long before the sediment pool became filled. If it were not recognized that sediment would be deposited in the retarding pool, e.g., more frequent flows through the vegetated spillway would occur. This could cause higher maintenance costs and even endanger the structure as well as reduce its effectiveness as time progressed.

The construction of water regulation facilities inherently affects sediment movement by a stream. Reservoir backwater reduces inflow velocities and delta formation occurs just above and below the reservoir water surface elevation. Below the dam, the clear water releases may pick up materials from the stream bed and banks until a full sediment load compatible with the material available and the transporting capacity is attained. This latter action is commonly referred to as degradation.

The progression of delta formation and downstream channel degradation both begin with a flattening of stream bed gradient, adjustment in the bed material size, and change in channel cross section. In a strictly alluvial and homogenous situation the reaction would continue until the upstream and downstream gradients parallel the gradients existing at the time of structure installation (Lane, 1955a). Actually, the ideal situation seldom exists and bed armoring, base grade controls, varying transport capacity with channel width-depth change, and other factors control the final gradients (Hathaway, 1948) (Borland and Miller, 1960). Delta development and downstream degradation are deleterious or beneficial depending upon the project plan and the environment. The following indicates some of the factors to be considered:

In planning and designing structural measures on a stream course, it is necessary to estimate probable degradation. This controls, to some degree, the invert elevation of the outlet structure and serves as a guide to the need for downstream channel excavation. As has been indicated, degradation may be an important positive benefit in a project plan (Oliver, 1965; Bushy, 1961) or it may cause serious damage by bank scour, undermining bridge foundations, etc. If the primary purpose of the structure is for sediment detention or gully control, the development of a delta and creation of upstream channel filling is a desirable project feature (Woolhiser and Miller, 1963; Miller, et al., 1962; United States Bureau of Reclamation, 1959).

The guides and criteria available to realistically predict degradation and aggradation (delta development) are somewhat limited. However, a knowledge of the channel periphery and sediment transport material characteristics, flow conditions (inflow and regulated), and hydraulic factors can be combined with experience and engineering judgment to estimate the probable structure effects on the upstream and downstream channel situation.

Stream Channel Improvement and Stabilization

Stream channel improvement is an important structural measure used in watershed projects. It is used to supplement detention structures except in those watersheds in which the topography does not permit detention storage. (This subject is treated in more detail in Section C of this chapter.)

Stream channel improvement and stabilization may consist of excavation, which includes enlargement or straightening of existing channels, or removing brush and snags from existing channels without excavation. Enlargement by excavation increases the capacity for conveying water by increasing the cross-sectional area. Straightening a channel increases the gradient, giving rise to higher velocities thus allowing larger discharges of water in the same channel section. Removal of brush and snags gives rise to higher velocities by reducing the friction factor. It must be recognized that these higher velocities may cause channel scour and change in channel shape.

These kinds of stream channel improvement, although used to control waterflow, may affect flood peaks by increasing the velocity of flow for the same volume of water. They sometimes result in higher downstream flood peaks than would be experienced without them. Since they usually are used in combination with floodwater retarding structures, peak flows with the system of structural measures are significantly less than without the system. If such channel improvement is the only type of structural measure under consideration, it must provide sufficient capacity to prevent additional damage that could result from higher peak flows.

Channel stabilization is required if channels themselves are deteriorating by down cutting, bank erosion, or head cutting. Such deterioration results in production of damaging sediment and in loss by depreciation of productive land adjacent to the channel.

In general, channel improvement and stabilization methods fall into various categories. The method or combination of methods selected depends on the nature of the problem. In broad terms protective or control methods can be classed as permeable or impermeable, flexible or rigid, and permanent or temporary. Combinations of various methods are common (Silberberger, 1959; Miller and Borland, 1963; Stanton and McCarlie, 1962). The control may be for bank stability as shown in Fig. 5.6, stream gradient control, stream gradient reduction, or combinations of these purposes. The control can involve simple snagging or removing vegetation that impedes water conveyance or a complete new channel to replace a deteriorated natural drainageway. Control also can be accomplished by flow regulation. In general, the purposes of channel stabilization structures are: (1) To produce a protective blanket that resists forces of flowing water; (2) to create bank roughness and thereby reduce the velocity and erosive forces acting on a streambank; (3) to divert flow away from erodible banks; (4) to hold the stream thalweg gradient and prevent lowering with accompanying bank undercutting; (5) to convey the water and sediment load with maximum efficiency and minimum long-time maintenance requirements and; (6) to give water conveyance capacity and adjacent farmland drainage benefits. In all cases, continuity in design and treatment as opposed to spotty control is of prime importance.

FIG. 5.6.—Improved Reach of Channel, Mission Creek Protection Project, Cashmere, Wash.

In any channel rectification plans, full consideration is given to the stream hydraulics, probable flood frequencies, peaks and durations (with and without flood detention provisions), stream periphery materials, and the suspended and bed load characteristics. Hydraulic parameters e.g., velocity and tractive force are important limiting design factors (Fortier and Scobey, 1926; Lane, 1955b). Frequency of damaging floods and alteration of flow regime are items of considerable importance in selecting an improvement plan and establishing channel shape or capacity (Parsons, 1960). The properties of the channel's periphery materials are directly related to the extent of protection and limiting hydraulic force requirements. The existing guides and criteria for use in planning, designing, and maintaining channel control are inadequate in many respects, particularly in borderline cases. Installing stabilization works in anticipation that a problem will develop can result in needless costs and erroneous location and type of control. For this reason, delay of channel improvement or certain phases of improvement is sometimes desirable. The problem may be meander control, alteration in regime, reduction in sediment load, increase in conveyance capacity, rapid flood passage, or a combination of rectification needs. In any case, the methods used should be carefully selected to fit the conditions. They also must be incorporated in and be consistent with the other elements of the watershed project.

Debris Basins

A debris basin is a reservoir designed specifically to trap sediment and debris. Depending upon the volume of debris to be anticipated, the storage capacity available, and the economic magnitude of the project, the capacity to be provided may be equal to the volume of debris expected to be trapped at the site during the planned useful life of the structures of improvements it is designed to protect, or it may be equal to the volume of debris anticipated during only one or more major storms. In the latter case, it will be necessary to remove the accumulated material periodically. Debris basins are designed to reduce the amount of generally coarse-grained sediment and debris deposited in downstream channels and reservoirs or on highways, railroads, or urban and agricultural areas. Any reduction in peak flow that may take place as a result of temporary water storage in debris basins is incidental to their primary purpose.

In the design of a debris basin, if the structure is justified primarily by reducing downstream damages caused by coarse-grained sediment, it is desirable to construct a basin of low capacity-inflow ratio capable of trapping only that portion of the sediment yield causing the downstream damage. In these circumstances, the fine-grained sediment passes through and out of the structure since it is not creating downstream damages. This reduces the need for creating additional capacity to store the fine-grained sediment and decreases the cost of the structure without changing the benefits. (This subject is treated in more detail in Section E of this chapter.)

Other Structural Measures

In addition to reservoirs, channel improvement, and debris basins, other structural measures of various types are used in watershed projects. They may be designed to control erosion, provide proper drainage, distribute irrigation water, improve efficiency of water use, or for other purposes that provide desirable soil-water relations for agriculture. Among other supplemental structural measures that may be considered in a watershed protection project are levees and dikes, tide gates, floodways, pumping plants, and floodwater diversions. Some of these measures prevent overflow of flood-plain land thus eliminating the opportunity for deposition of damaging sediment on these lands. In other types the primary purpose is to prevent flooding and to remove excess surface water if sediment is of very minor concern either from the standpoint of damages or design of the structure.

Operation and Maintenance of Structural Works of Improvement

The ideal balance between initial cost and operation and maintenance costs for a structure are attained when the annual costs are a minimum for the planned life of the structure. This assumes that required maintenance is promptly done and that the function and purpose of the structure are not impaired during its economic life by malfunction of any part or element.

This ideal balance can be attained in the design of the structure only to the degree that accurate data are available on: (1) Durability of the materials involved; (2) cost of construction; and (3) operation and maintenance costs over the economic life of the structure. Seldom are these data adequate to permit a high degree of accuracy. However, the goal is worth the effort involved in making the best estimates possible with the information available.

Major river improvements are built by organizations that employ adequate maintenance forces and have budgeted money to perform timely maintenance. Such organizations have collected data on operation and maintenance costs for various types of structures and can approach the ideal situation in which there is a sound balance between initial investment and operation and maintenance expense.

In contrast, most watershed project structural measures are built by farmers or by local sponsoring organizations in cooperation with the Federal government. These individuals or organizations also operate and maintain the terraces, dams, channels, and other structures that constitute the sediment control part of a watershed project. They generally do not have money readily available for maintenance work. There are exceptions, but they are few.

As a result, cost data on maintenance of the various types of sediment control works are inadequate and difficult to obtain. Thus judgment based on general observations and experience must be used in computing anticipated maintenance costs. The accuracy of such judgment is limited by the relatively short history of use of sediment control structures in this country.

With the exceptions already cited, those generally responsible for maintaining sediment control structures are prone to put off required maintenance with the result that small maintenance jobs tend to grow into large ones. Some items of maintenance tolerate delay; others do not except at the risk of increased cost or impaired function of the structure.

If maintenance forces (men, materials, and equipment) are not readily available for timely work, the design should be such that maintenance requirements are reduced to a minimum if: (1) Continued operation of the structure element in question is critical to the proper performance of the whole structure; and (2) the anticipated cost of delayed maintenance, properly evaluated to current price, is greater than the cost of avoiding such maintenance. Elements of a structure that are essential to its operation but difficult to inspect or repair because of inaccessibility should use the most durable materials and foolproof machinery available.

For sediment control structures in relatively small watersheds, with short durations of discharge and low hazards of structure failure, savings can be made in original construction costs on certain elements of such structures by recognition of the fact that time will be available to repair damage should it occur.

For example, water tolerant vegetation has been used successfully in conjunction with an upstream berm on earth dam embankments in small watersheds to control wave erosion where the stage in the reservoir is fairly constant between periods of significant runoff. The use of such vegetation can be pushed to the limit of its capability because if it should fail there is ample time to correct the situation by use of rock riprap or by other measures.

Vegetated earth emergency spillways, properly designed to control velocity of flow, frequency of operation, duration of discharge, and layout have been used instead of reinforced concrete spillways with a great savings in cost. Experience to date indicates that the maintenance cost of such spillways is very low because of their infrequent use.

Watershed Plan.—Because land treatment measures are the basic element of any watershed project, it is essential that land treatment measures be considered as the initial increment in formulating any combined system of measures to meet watershed-wide objectives. The effects of land treatment measures in a watershed project are twofold—physical and economic. The physical effects result in a reduction of surface runoff and sediment yield from the treated areas. The economic effects are realized as onsite and offsite benefits. Onsite benefits are the increased net returns to the landowner or operator which result from application of appropriate land treatment measures in combination with other good management practices and maintenance of productivity of the land resource. The offsite benefits are the damage reduction benefits which accrue on downstream flood plains as a result of the reduction in surface runoff and sediment produced from the source areas. These benefits usually accrue to individuals other than the ones applying the land treatment measures.

Damaging sediment may be produced by either sheet or channel type erosion. The extent to which damage is caused, however, is contingent upon the nature and amount of sediment which reaches the damage area. Coarse-grained sediment is produced mainly by channel-type erosion such as gully erosion, roadside erosion, or stream bank and bed erosion. Sediment produced by sheet erosion is usually fine-grained. Land treatment measures provide the only effective means of reducing the rate of sheet erosion and resulting sediment production from this source in the watershed. Structural measures usually are required to reduce channel-type erosion and the sediment derived from it. For these reasons it is important to identify the nature of the sediment creating the damage and the source of that sediment in order to determine whether, broadly speaking, either land treatment or structural measures, or combinations of both, are necessary to reduce sediment damages.

In order that proposed structures function properly for the period of time for which they are designed and justified, consideration must also be given to sedimentation in their design. Thus it is the current policy of the United States Department of Agriculture to require that needed land treatment measures to reduce critical erosion and sediment production be installed in the watershed above the structure before or concurrently with its construction with Federal funds. This need also was recognized by the Congress in requiring that local organizations shall "obtain agreements to carry out recommended soil conservation measures and proper farm plans from owners of not less than 50% of the lands situated in the drainage area above each retarding reservoir" as a prerequisite to providing any Federal funds to be used in the construction of such a structure under Public Law 566.

After the physical and economic effects of land treatment measures have been evaluated, a system of interrelated structural measures is formulated to achieve project objectives. The kinds of structural measures vary depending on the nature and location of erosion, floodwater and sediment damages, and topographic and geologic conditions. The extent of each of these types of damages and, therefore, the types of structural measures needed to complement land treatment measures varies from one part of the country to another and from one watershed to another in the same part of the country.

If reducing floodwater damages is one of the purposes of a project, retarding flood runoff by temporary storage on the watershed is given first consideration in formulating a system of waterflow control measures. Other structural measures such as channels, floodways, or diversions usually are considered supplementary to floodwater retardation rather than as alternate means of accomplishing the same objective unless there is a significant difference in costs. If reducing sediment damages resulting from upstream channel erosion is of primary concern, grade stabilizing structures are given first consideration and such measures as sedimentation basins, periodic cleanout of reservoirs and channels, or incorporation of additional capacity for sediment accumulation are considered as alternate measures, depending upon results of comparative cost estimates.

In some land resource areas, upstream erosion damages such as land loss and land depreciation due to channel erosion represent the primary damages in a watershed. If these conditions exist, physical and economic evaluations of alternative structural measures may be made on selected representative areas. The total overall needs and benefit-cost determinations for the entire watershed are projected from these sample areas.

After the need for structural measures has been determined, potential sites are analyzed, with particular attention given to the location of each site with reference to the damage areas; the physical characteristics of each site with respect to storage efficiency and hydraulic conditions; availability of suitable construction materials; foundation conditions; and the presence of man-made improvements in the area, e.g., highways, railroads, pipelines, power lines, and buildings that would be influenced by the structure.

A physical and economic evaluation of the effects of the proposed structural system then is made. Land-use change and damage reduction benefits which will result from reducing erosion, reducing infertile deposition on flood plains, alleviating swamping conditions, and other sedimentation damages are determined. Monetary values of the improved quality of water for beneficial use are estimated if appropriate. The reduction of sediment yield at the mouth of a watershed is evaluated if benefits can be readily identified. Benefits of this type may accrue as a result of a measurable reduction in the rate of sediment deposition in reservoirs or harbors or in damage to oyster beds that are near to the watershed. However, no benefits for project justification are claimed for works of improvement remote from the watershed project. In addition, floodwater damage reduction benefits are determined as well as those benefits that result from land enhancement and more intensive use of flood-plain lands made possible by the reduction in magnitude and frequency of flooding.

All benefits are converted to an average annual equivalent basis and compared to annual cost. Annual costs are determined by adding the amortized installation cost to the annual operation and maintenance costs. If the benefits equal or exceed the costs, the project is feasible. If the benefits do not equal the costs, it may be necessary to revise the project objectives and modify the system of structures in order to develop a feasible project.

Often the economic evaluation analysis shows that the proposed system of structural measures is not only feasible but that higher levels of protection can be provided. When this condition exists, and the local people desire the higher level of protection, additional increments can be added so long as the incremental increase in benefits equals or exceeds the incremental increase in costs. This usually is accomplished by adding the most efficient of the alternate sites investigated to the basic system of structural measures. When all desired and feasible increments have been added, the formulation of the combined system of measures is complete.

This information, together with other information relative to the watershed problems and recommended solutions, is compiled into a work plan report that serves as the basis for project authorization.

C. Stream Channels

5. General.—Stream channels may be classified under various categories; however, only those formed in erodible materials will be considered herein. Such channels are complex features that cannot be described by analytical means except in general terms. Solutions of the problems encountered in the control of erodible channels usually depend more upon a thorough knowledge of channel characteristics than upon hydraulic theory.

For convenience in analysis, channels will be considered as either natural or artificial; a natural channel being defined for the purpose as one in which the form and dimensions have been established under natural conditions while an artificial channel will be similarly defined as one constructed to predetermined dimensions. The problems most frequently encountered in the first case deal with the maintenance or improvement of an existing channel. The basic problem in the second instance is the determination of stable dimensions.

6. Natural Channels.—The formation of a natural stream channel is influenced by numerous factors of climate, geology, and geography; the most important being the stream discharge, resistance of the land forms to erosion, geometric configuration, and the properties and amount of the sediment transported. The form and dimensions of such a channel, once established, are relatively invarient unless a major change occurs in one or more of the aforementioned factors. A channel may be eroded into a resistant formation, in which case its position, as well as its dimensions, tend to remain constant; or it may be formed in erodible materials, in which case there may be a continuing tendency toward a shifting of position by erosion and rebuilding of the banks. The latter type is most frequently found to be formed in materials previously transported and deposited by the stream, although they may also be formed in materials transported by other means, e.g., winds or glaciers.

Regimen.—Most natural streams are in regimen, i.e., the major dimensions of their channels remain essentially constant over an extended period of time. The condition of regimen does not preclude the shifting of channel alinement by erosion and rebuilding of the banks, but it requires a balance between these factors. It requires that the sediment discharged from any given reach be equal to that introduced into the reach; however, it is not necessary that there be an invarient relationship between sediment discharge and water discharge. For most mobile-bed streams, there will be a range of discharge values within which the stream can adjust to as much as a 10-fold variation in sediment discharge by variation in bed forms (ripples, dunes) and concurrent variation in flow depth and velocity without appreciably changing its slope, channel width, or average bed elevation. A stream may vary its channel dimensions locally, in time or space, without interfering with regimen as long as these variations fluctuate about a balanced average.

Aggrading and Degrading Streams.—When the amount of sediment introduced into a stream exceeds that which the stream can transport, the excess must be deposited, and the stream bed is thereby built up, or aggraded. Conversely, if the rate at which sediment is introduced into the stream is less than the transport capacity, and the channel bed and banks are erodible, the stream will erode or degrade the bed and banks to supply the deficiency.

The major dimensions of aggrading or degrading channels remain in a constant state of change until, by one means or another, equilibrium is established between the sediment inflow and discharge.

7. Channel Forms.—Natural channels occur in three general forms: straight, braided, and meandering. There are numerous factors that might influence a stream in assuming one form or the other, and their relationships are not completely known. There are, however, general concepts that should be considered in any works to rectify or control a channel.

Straight Channels.—Straight channels are those which follow essentially a straight alinement. They exist generally with either flat slopes, which are inadequate to provide erosive velocities, or steep slopes that produce relatively high velocities. In the latter case, it is possible that the straight alinement results primarily from momentum that discourages turning. It is also possible that the alinement might be influenced by a heavy concentration of sediment which required for its transport the full capacity provided by the steep slope; however, it is unlikely that this latter condition would be the controlling factor, for under those conditions a natural stream with its normally fluctuating discharge tends to be braided.

Braided Channels.—Braided channels [Fig. 5.7(a)] are those formed of random interconnected channels separated by bars and presenting the general appearance of a braid. Braided channels of streams in regimen are seldom found except on relatively steep slopes, indicated by Lane (1957) to be $S = 0.10 \sqrt[4]{Q}$ or greater and by Leopold and Wolman (1957) as $S = 0.06\, Q^{-0.44}$ or greater; Q in the first instance being the average discharge, in cubic feet per second and in the second, the bank-full discharge. Term S is the slope, in feet per thousand feet. In aggrading streams, braiding may occur on either flat, moderate, or steep slopes; but in either balanced or aggrading streams, it is believed to result primarily from random deposition of materials transported during high flows in quantities or sizes too great for continued transport during moderate or low flows. These deposits frequently form bars upon which vegetation flourishes to discourage further movement by succeeding high discharges, thus intensifying the braiding.

A further type of multiple channel stream is the distributary type found on delta formations or debris cones [Fig. 5.7(b)]. These are generally aggrading channels that divide to follow separate courses which finally either disappear into sheet flow or continue until again collected at the foot of the slope.

Meandering Channels.—Meandering channels [Fig. 5.8(a)] follow a winding or tortuous course. They are distinguished from a tortuous channel, the alinement of which is essentially fixed by geologic or geographic conditions, by the fact that the alinement of a meandering channel tends continuously to shift by local erosion and rebuilding of the banks. The majority of problems arising in channel control will be found to concern meandering streams, as these are the streams in which bank erosion most commonly occurs. As the solutions of these problems depend largely upon a knowledge of the channel characteristics, a somewhat detailed analysis will be presented.

Meandering streams are found to occupy valleys of moderate slope, expressed by Lane (1957) as between $S = 0.0017/\sqrt[4]{Q}$ and $S = 0.10/\sqrt[4]{Q}$, Q being the average discharge, in cubic feet per second. There is no general agreement as to the causes of meandering or of the relations between the factors believed to be involved. Claude Inglis (1949) represents, perhaps, the majority opinion in his statement that "meandering—is nature's way of damping out excess energy during a wide range of flow conditions; the pattern depending upon the grade of material, the relation between discharge and charge (sediment concentration), and the rate of charge and discharge." E. W. Lane (1957) is among those speaking for the opposition in citing the fact that his analysis of a large number of streams showed that "for the same average discharge and material size, streams in non-cohesive material tend to be more nearly straight when they are steeper than the meandering streams;" however, such straight channels would then be within his general classification of braided streams. Other observers are content to point out the fact that almost any obstruction in the bed can cause the current to be deflected into a bank where it might erode an irregularity that, in turn, can deflect the current to the opposite bank, thus establishing a chain reaction. Friedkin (1945) in reviewing laboratory flume tests states that the only requirement for meandering is bank erosion. More recently, Shen and Einstein (1964) and Shen and Komura (1968) have found that meandering tendencies develop with differences between the shear stresses at the two sides of the cross section. Other investigators have also mentioned the effects of secondary currents. Yang (1971) has analyzed stream meanders in terms of the least time rate of energy expenditure.

Insofar as practical channel control is concerned, agreement with one theory or the other as to the cause of meandering is much less important in the present state of knowledge than the realization that the meanders in any given stream were formed in accordance with implicit though imperfectly known laws of nature and that they should receive the proper consideration in any control works.

The basic stream meander is essentially a sinusoidal curve, as shown in Fig. 5.8(a) It is a dynamic form, tending constantly to shift its position by erosion of the concave banks and deposition along the convex banks of the bends. Under ideal conditions, a meander system will migrate downstream in an orderly

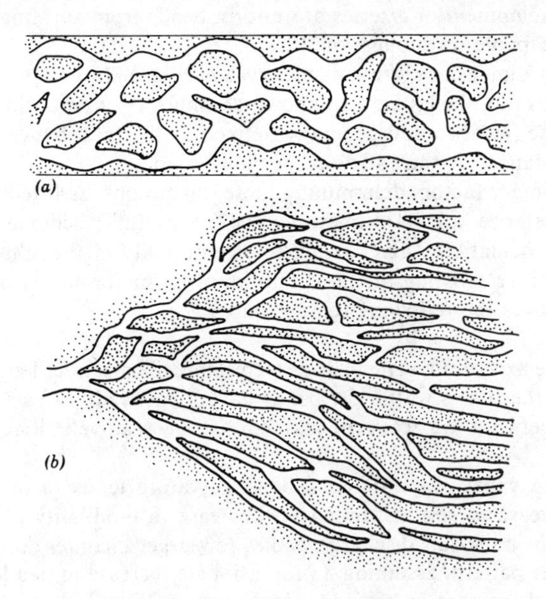

FIG. 5.7.—(a) Braided Channel; (b) Distributary Channel

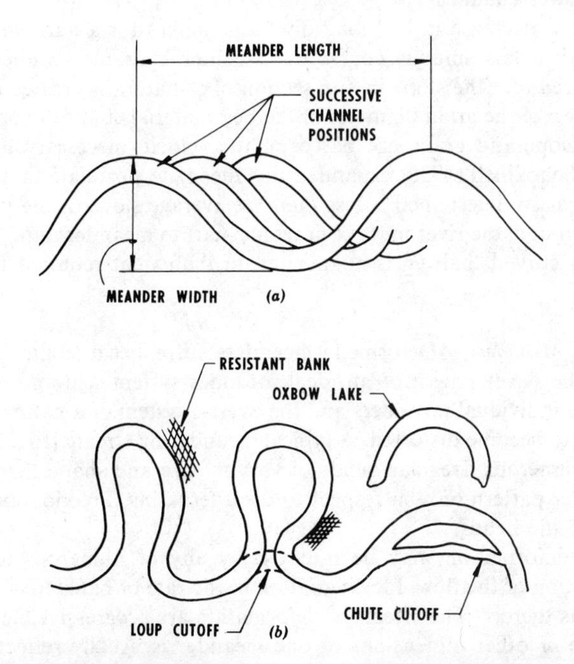

FIG. 5.8.—(a) Meander Channel; (B) Deformed Meanders

progression along a central axis. The basic relationships between meander dimensions and the various controlling factors are not known; however, Friedkin (1945) and others have demonstrated in laboratory flumes that "when all extraneous influences were eliminated and the elements reduced to their simplest forms, the development of a series of uniform bends from an initiating bend was positive and capable of duplication."

The primary dimensions of a meander system [Fig. 5.8(a)] are the length, width, and tortuosity ratio, the latter, also known as sinuosity or sinuosity ratio, being a resultant of the first two and generally defined as the ratio between the channel and valley distances or the valley and channel slopes.

The five primary factors determining these dimensions, as listed by Matthes in order of importance, are: (1) Valley slope; (2) bank-full discharge; (3) bed load; (4) transverse oscillations; and (5) degree of erodibility of the alluvium. Matthes cites the first three as being the most important in the formation of the meander pattern, and gives the following relationships:

1. Where the axis of the valley has a steep slope, the meander length is longer in proportion to the river's width, the bends have longer radii of curvature, and the cross-overs (between adjacent bends) consist of relatively long and shallow reaches.
2. Where the valley slope flattens, bend curvature tends to increase and the meanders increase in size, assuming no decrease in erodibility of the alluvium.
3. Increase in bank-full discharge makes for larger channel dimensions and a larger meander pattern, assuming a proportionate increase in bed load discharge.
4. A material increase in bed load discharge produces larger meanders, and a wider, shallower channel.
5. A marked decrease in bed load discharge makes for a narrower and deeper channel, with a less sinuous course on a lighter hydraulic gradient. Sternberg (1875) declared that the slope in any section of a naturally created river channel was a function of the grain diameter of the bed material, and developed a formula correlating slope and grain size based on the velocity necessary to initiate bed movement. Schoklitsh (1930) expanded this theory to postulate that "if the value of the slope as so determined is less than the average slope of the plain in which the river is eroded, the river must of necessity start to meander until the developed length of its curved thalweg is augmented to the extent required to make both slopes equal."

Deformed Meanders.—Deformed meanders form because the conditions required for the development of an ideal meander system seldom exist in nature, and both the individual meanders and the overall system of a natural meandering stream tend to become distorted. A typical meandering stream [Fig. 5.8(b)] will be formed of numerous irregular bends of varying size and shape that resemble an ideal meander pattern only in respect to the alternating direction and continuing migration of the bends.

Meander deformation may be initiated by any of numerous items; usually either deflection of the flow, local variation in the rate of bank erosion, or various combinations thereof. The effects of deformation are progressive, i.e., any change in the shape or other dimensions of one meander is usually reflected in one or more adjacent meanders, either as a local change in slope, as a variation in the angle of entry, or both. Either can cause a change that is further reflected to the next loop.

The most prominent deformations occur as cutoffs [Fig. 5.8(b)] either chute cutoffs, in which the flow cuts across the base of a meander loop; or as loop or neck cutoffs, in which a meander loop closes upon itself until the flow cuts across the narrow neck formed thereby.

A chute cutoff results generally when the slope becomes inadequate to support the flow and sediment discharge. Friedkin (1945) noted in his flume tests that there was a limiting size for each meander pattern and that, when for any reason this size was exceeded, chute cutoffs invariably occurred. In natural streams they have been observed to occur during flood flows that tended to follow a straighter path and during lesser flows when either the loop in question or an adjacent loop became excessively enlarged.

Loop cutoffs are the end result of a migrating meander impinging upon an erosion-resistant formation. Movement of the leading portion of the loop is restrained while the following position moves up until only a narrow neck separates the two. This neck may be breached either by continued erosion or by the erosive action of a flood flow cutting across.

The immediate influence of a loop cutoff will be noted in several adjacent meanders due to the local change in slope; however, there is apt to be a lasting influence that may affect the meander pattern permanently. When such a cutoff occurs, the ends of the old loop soon become filled with a deposit of sand, leaving the remainder as a crescent-shaped or horseshoe lake. Since sediment can then enter the lake only from local inflow or from over-bank flood flows, further sediment deposited consists of finer sediments, normally decreasing in size with distance from the new channel. Eventually, the bight of the loop is filled with a deposit of fine sediments. Unlike the usual accretion deposit that consists of a relatively thin layer of fine sediments over a mass of coarser material, the fine deposit in the bight of an abandoned loop extends to the full depth of the old channel, forming an erosion-resistant plug against which following loops may impinge. In one somewhat extreme case, an appreciable length of the lower Mississippi River is held in a straight reach by a succession of such plugs. The traces of old meander crescents, although they may be difficult to identify on the ground, will remain visible from the air or on aerial photographs for many years, perhaps even centuries. This results from a difference in the shading of the vegetative color due to difference in the soil type in the crescent and in adjacent land. Fig. 5.9, an aerial view of the Missouri River above Sioux City, Iowa, shows by these old crescents how the river has, at one time or another, occupied every part of the valley width.

Lesser, although sometimes equally important deformations occur with varying angles of entrance of the flow into the bend. Friedkin (1945) noted that meander width tended to increase as the angle of entry increased. In the development of an ideal meander, the entry angle (angle between the line of flow and the upstream tangent to the bend) would normally be a dependent variable rather than a causative factor; however, in a deformed system the entrance into any loop must necessarily be a function of the shape and position of the next upstream loop. In general, experience on natural streams shows that a small angle of entry generates a long flat meander, while a sharp entrance generates a short, wide meander. Within the multiple agglomeration of causes and effects on natural streams, other factors may, of course, overcome this generalized effect. One fairly common phenomenon is the case where the flow entering a bend at a sharp angle bounces or rebounds off a bluff or similar formation to create an additional minor meander.

A common cause of deformation is local bank erosion that creates a pocket or a false point (Fig. 5.10) which deflects the current into the opposing bank, thus initiating a completely new chain of effect. This is frequently followed by a lunated bend (Fig. 5.10). Here, the change in bank alinement permits the formation of a center bar with the result that the major flow tends to move from one side of the bar to the other with varying flow stages. Needless to say, this creates a very unstable condition downstream.

At times, a meander impinging against a bluff or other resistant formation may become partially perched on a shelf of this formation. Such a condition may create a section which is stable for a time; however, a flood flow may cut a deeper section off the edge of the shelf so that the channel falls off. This condition will affect downstream meanders seriously. If it tends to be intermittent, the downstream section becomes particularly unstable.

In any relatively straight reach of a stream with meander tendencies, sediment tends to move as a series of traveling bars (Fig. 5.11), alternating from one bank to the other. The location of the points of incipient bank attack will naturally vary continuously with the movement of the bars. These bars can be particularly effective in promoting erosion in streams carrying large amounts of drift or ice that might lodge and deflect the flow. A somewhat similar condition can develop in a wide, shallow reach. At times the bars will form in a completely random pattern. At other times the flow tends to collect along one bank until it is actually at a higher elevation than on the other side. The flow then bleeds off laterally in a series of random shallow subchannels to collect again on the other side and repeat the process. Flat or long radius bends tend to be unstable since the flow, instead of following the concave bank, may shift in a random manner or may spread with consequent formation of random bars.

FIG. 5.9.—Aerial View of Missouri River near Sioux City, Iowa, Showing Old Meander Crescents

8. Channel Control.—Channel control may be required for various purposes, e.g., the provision of greater capacity, maintenance of optimum depth, improvement of nonregimen channels, or simply the prevention of bed or bank erosion. The work may include stabilizing the alinement, the width, the bed, or any combination thereof. It may consist of the control of critical sections only, or of a complete program of alinement and bank stabilization.

The actual degree of control that may be applied to a given stream is frequently a matter of expediency or of economics. In many cases, the prevention of bank caving alone is adequate justification for a complete program, e.g., the Missouri River, in its uncontrolled state, eroded an average of 9,100 acres annually over a reach of 758 miles, the equivalent of the total destruction and rebuilding of the entire meander belt once every 70 yr. Annual bank erosion of the Arkansas River in the reach from Van Buren, Ark., to Little Rock, Ark., before control was initiated, was equivalent to a continuous strip 50 ft in width; while the average of the Mississippi River from the 770-mile reach from Cairo, Ill., to Angola, La., is equivalent to a continuous strip approx 65 ft in width. Locally, of course, the erosion of either of these streams may vary from a few feet per year to as much as 1,500 ft in 1 yr. Additional data on bank erosion have been tabulated by the Task Committee on Channel Stabilization Works of the American Society of Civil Engineers (1965). The term of reference is not given; however, it is probable that the values given are for local erosion reaches rather than a continuous strip as given previously.

On small streams the total incipient damage may be sufficient to justify the control of critical reaches only, while on the lower Mississippi or similar streams the very magnitude of the structures that would be required to move the channel to a predetermined alinement generally forces a procedure involving more or less temporary treatment of critical reaches together with stabilization of those bends, currently following a desirable alinement and subsequent stabilization of other reaches as natural erosion moves them into a proper position. Whether the control

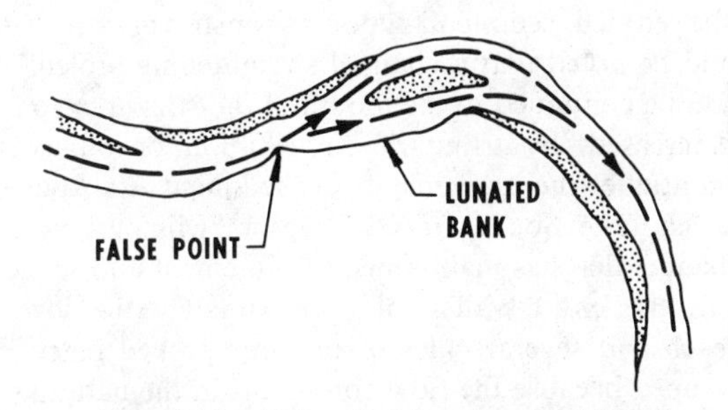

FIG. 5.10.—False Point or Lunated Bend

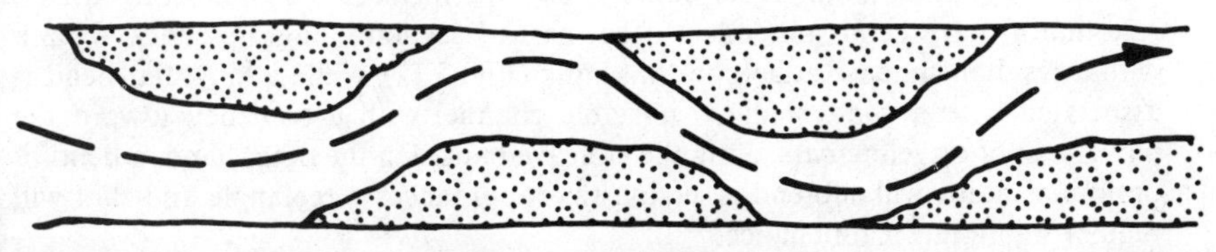

FIG. 5.11.—Traveling Bars

is partial, delayed, or complete, however, the general principles are the same and should be followed to the maximum practical extent.

The Missouri River presents probably the most extensive experience available in the control of meandering rivers, and this experience will be utilized in much of the following analysis. The initial works on this stream were isolated attempts by individual land owners or unrelated groups to halt bank erosion at specific locations. Some of these works may have been effective for a short time, but they were inevitably destroyed or bypassed within a few years, and their greatest value was in their demonstration of the need for a coordinated system. Subsequent experience has shown that a stable channel requiring minimum maintenance can be obtained in this stream only by establishing a controlled channel width, proper alinement, and optimum bend radii.

Alinement.—The primary requisite of alinement is that it be smooth and free from bends that are either too abrupt or too flat, lunate bends (Fig. 5.10), false points, or other irregularities in the bank line. The existing meander pattern, particularly the overall slope and channel length, should be generally maintained; however, within these limits channel changes or cutoffs may be made as desired to obtain an optimum alinement. It is sometimes desirable to make local changes to balance the slope within a reach in which an otherwise desirable alinement tends to be unstable.

Straight reaches should be avoided, although if a reach has been naturally fixed in a straight alinement by resistant formations and is otherwise satisfactory it should not be disturbed. The straightening of a truly meandering channel over long reaches should never be attempted, even though the anticipated rewards in terms of increased capacity are tempting, unless one fully understands and is prepared to accept the consequences. There are many examples of the successful straightening of tortuous channels in erosion resistant materials, but there are also many examples where the straightening of meandering channels in erodible materials has resulted in severe head-cutting in the channel and tributaries, excessive widening as the stream attempts to reassert its meanders, and the dumping of the eroded sediments upon downstream interests. In this latter respect, it should be noted that a channel straightening project, even in resistant materials, should be continued to the mouth of the stream or to a point where the flow will be increased, by tributary contribution or otherwise, to a degree adequate to continue the movement of sediment transported through the straightened reach. The South Grand River in Missouri is an example of a straightened channel that has maintained its alinement and section for more than 40 yr; however, the entire valley of approximately the lower 1-mile of the straightened reach and several miles of the unimproved reach downstream was completely swamped because the flow conditions in the natural channel were not adequate to continue the transportation of the sediment.

It will be found that meander dimensions are critical. It is an almost invariable rule that a sharp bend subtends a deep, narrow section along the concave bank with a resultant intense attack on that bank (Fig. 5.12), while a long, flat bend is associated with a wide, shallow, unstable channel with a tendency toward the movement of traveling bars along the concave bank. On the other hand, a bend of optimum radius will subtend a section that approaches a rectangle and that will require minimum maintenance.

In the present state of knowledge, it is usually necessary to determine, or at least check, optimum bend radii on the basis of experience, i.e., to observe the stream in question or a similar stream to locate bends that have been stable over an appreciable period of time and to govern the adopted dimensions thereby. It is particularly desirable to note the paths of the major currents of both high and low flows, for if the two coincide the bend is apt to represent optimum entrance and bend conditions, while if the two differ appreciably the bend is almost certain to contain some condition of instability. When an apparently stable bend suddenly begins to display a tendency toward lengthening by erosion or shortening by means of chute type cutoffs, it is well to observe downstream conditions for a distance of several complete meanders to see if a channel change there is disrupting the local slope.

There have been a number of attempts to derive formulas for meander dimensions, and, while none of them can be recommended for indiscriminate use, it is well to check them against available data and thus perhaps establish some range of dependability.

Leliavsky (1955) reviews a number of formulas that have been proposed to relate meanders to various channel functions. Inasmuch as a presentation of these would be lengthy, the reader is referred to the referenced publication. Inglis (1949) has analyzed data from a number of rivers, and has found an apparent correlation between meander length, M_L [Fig. 5.8(a)] and stream discharge, proposing the equation

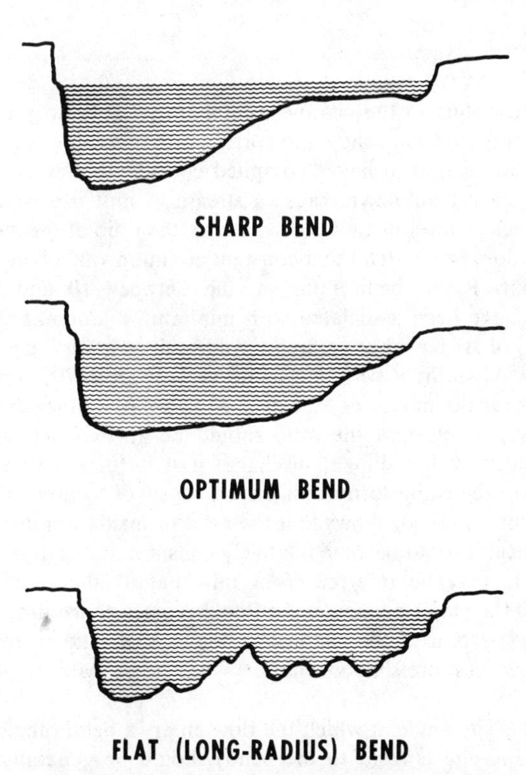

FIG. 5.12.—Typical Channel Cross Sections

$$M_L = 28\, Q^{1/2} \quad (5.3)$$

in which Q = maximum discharge, in cubic feet per second; and M_L is in feet. He also found some correlation between meander width and discharge, with a distinct separation between incised and spilling (flood-plain) streams, but apparently did not consider the relationship sufficiently firm to recommend a value.

Blench (1957) accepted Inglis' formula for meander length, but tentatively proposed an expanded equation including the effect of bed and bank materials as

$$M_L = m\left(\frac{F_b}{F_s}\right)^{1/2} Q_e^{1/2} \quad (5.4)$$

in which the factor, m, is to be determined by experience but may be assumed to fluctuate about a value of 12. Terms F_b and F_s = the bed factor and side factor, respectively, in Blench's regime formula (Chapter II, Section F), and Q_e = that portion of the peak flood that is effective within the meandering channel limits. For meander width M_w, Blench suggests that there is some merit in disposing of this function by noting that the meander ratio (M_w/M_L) is almost exactly 0.5 for the sinusoidal meandering of sand models, similar to flood-plain meanders, and 1.5 for incised meanders.

Bagnold (1960) has made a theoretical evaluation indicating that an optimum channel curvature should exist with a ratio of bend radius R to channel width b between two and three:

$$\frac{R}{b} = 2 \text{ to } 3 \quad (5.5)$$

in which R = the radius to the channel center line. Laboratory flume data by Leopold (1960a) and others are shown to corroborate this value, and Leopold and Wolman (1960b) are stated to have "compiled considerable evidence that when other conditions, as yet unknown, cause a stream of any size which flows in a deformable channel to develop a meander pattern, the ratio of the meander radius to channel width does in fact tend to a constant common value between 2 and 3."

On the Missouri River, bend radii varying between 10 and 20 times the controlled width have been associated with minimum maintenance while bends with an R/b ratio of six have been extremely difficult to hold. Since the optimum R/b ratios on the Missouri River, with width varying from 700 ft at Sioux City, Iowa to 1,000 ft near the mouth of St. Louis, Mo., are five times those developed by Bagnold, it is possible that the ratio should be an exponential function or should be associated with sediment discharge if it is to be extended to major streams. It is also interesting to note that an analysis of a number of European streams by Rzhanitsyn (1960) shows that the ratio of maximum depth to channel width in a bend decreases to become relatively constant at a radius to width ratio of from 10–14. It may be inferred from this that as the radius-width ratio decreases from 10 the major portion of the flow becomes concentrated in a deeper and more narrow portion of the channel; while as this ratio increases the flow tends to spread over a more shallow channel with the probability of random bar formation.

Angle of Entry.—The angle at which the flow enters a bend (angle between the major flow line and the tangent to the bend) should be generally limited to a maximum of 15°, although under extreme circumstances angles of up to 25° may be used. As the entrance angle increases above 25°, maintenance problems will increase disproportionately, and at angles approaching 45° the flow will tend to bounce and may create instability throughout several downstream meanders.

In a complete alinement project, the entrance angle normally results from the fact that the flow leaving the downstream limit of the concave bank of one bend must cross the channel to the upstream portion of the concave bank of the next adjacent bend. In this case, any necessary adjustment can usually be made by a spiral entrance and exit or by the use of a hyperbolic layout (Steed, 1956). In the case of partial controls, it is highly important to establish proper entrances by extending the controls upstream as far as may be necessary.

For either partial or complete alinement planning, it is preferable to work in reaches extending from one hard spot to another, i.e., to have each end fixed by a bluff contact, a clay plug (such as that found in the bights of previously abandoned meanders), or other natural controls. Due attention should be given to entrance and exit angles in any case, for if the head of the work is flanked or if a bounce is caused by an abrupt entry the work in the entire reach may be progressively destroyed.

One of the few exceptions to the entry rule is the case of a steep-slope stream in which a bend approaching a right angle may exist with the flow piling up in the bend and falling off to the side instead of flowing around the bend. In this case, the most successful type of control appears to be the provision of protection at the point of attack with the expectation that it will be necessary to extend this protection downstream periodically as the point bar moves forward.

Channel Width.—The optimum channel width is the maximum that will provide a cross section free from bar formations at normal low discharge. A channel that is too narrow will be subject to excessive bed scour and possible undermining of control works, while one that is wide enough to permit bar formations will be subject to random shifting of the flow lines. Bars are particularly dangerous in streams carrying heavy drift or ice loads that might lodge and obstruct a channel segment. It will also be desirable that the channel width of a mobile bed stream be designed so that the resultant of the sediment and water discharge conditions falls within the range in which dunes form on the bed. Within this range, the stream can readily adjust for variations in sediment inflow; while outside this range aggradation or degradation might occur. Reference may be made to Chapter II, Section G (Fig. 2.79) for an approximation of the dune range, or to Maddock (1970) for a somewhat different approach.

Note that the depth-width ratio of natural streams varies widely from stream to stream and may vary to a lesser extent from reach to reach in any given stream. It may also be noted that channels in erosion resistant materials are almost invariably narrow and deep while those in easily erodible materials tend equally to be wide and shallow. Schumm (1960) in an investigation of a number of natural streams, has found a reasonably close correlation between the width-depth ratio and the weighted mean percentage of silt and clay in the bed and banks. Since most natural streams are in regimen, it should follow that regimen can occur over a fairly wide range of width-depth ratios and that the width of a channel can be controlled within reasonable limits without disturbing the stream balance. This can be largely explained by the fact that a movable stream bed will assume variable bed roughness forms and thus permit the channel some latitude in adjusting to varied requirements for discharge and sediment transport. Brooks

(1958), Colby (1960), Kennedy (1961a, 1961b), Simons and Richardson (1961), Vanoni and Brooks (1957), Vanoni, Brooks, and Kennedy (1961), and Maddock (1970) have presented data showing variations in bed form roughness and its effect on the flow. Taylor (1961) has prepared a report in which he gives a preliminary review of the effect of laterally varying roughness. Bed forms are also reviewed in Chapter II, Section G.

Within the present state of knowledge, the determination of an optimum channel width is largely a matter of judgment that is most readily satisfied by observing a stable section of the stream in question or of a similar stream. Approximations of optimum width can be made, or judgment checked, by the use of transport formulas to compute sediment transport and bed erosion characteristics under various conditions of flow; however, as may be noted in the preceding references, this process of itself requires judgment in the selection of roughness and other coefficients.

Lane (1952) and Lane, et al. (1953) have presented an analysis of the shear on the banks by which one can compute the depth of flow at which unprotected banks become unstable, but this analysis was developed primarily for sloping banks of straight canal reaches in noncohesive materials. Reference may also be made to Chapter II, Section E. This procedure is not directly applicable in the case of channel controls that provide erosion resistance to the banks, but may be used to obtain a fair approximation of a desirable cross section if the banks are considered to be moderately resistant.

There have been a number of formulas developed, principally in Egypt and India, to correlate stable channel dimensions with various discharge and sediment functions. Leliavsky (1955) presents an excellent analysis of many of these, and Blench (1957) describes the Indian developments in his presentation of his own regime formulas. Leopold and Maddock (1953) proposed a similar set of formulas based on their investigation of streams in the United States. In as much as a complete description of these formulas would be too lengthy for the present paper, further details should be obtained from the references cited. Note, however, that the Blench regime formulas (Chapter II, Section F) are the only ones that include consideration of the bank materials, and, in view of the apparent correlation between the erosion resistance of the banks and the channel width, consideration of this factor appears to be justified.

These empirical or regime type formulas have not been widely accepted in the United States, partly because of lack of familiarity and partly because many engineers feel that because the Egyptian and Indian formulas were developed for canal conditions in those countries they are not necessarily adapted to the conditions found elsewhere. No attempt will be made herein to either encourage or discourage their use, but it is believed that any engineer working with stream channels should check these and other available formulas if for no other purpose than to become more familiar with the overall problem. In the case of controls provided in an attempt to develop regimen conditions in a nonregimen channel, transport formulas or regime type formulas provide the only available means of estimating a desirable channel section.

Note that in a meandering channel the critical width required applies to the crossing reach between bends, as the cross section in the bends will tend to be self adjusting in accordance with the meander dimensions. The width in the crossing should be measured normal to the flow line rather than normal to parallel tangents connecting the bends; otherwise, the actual flow width would be excessive, the flow would diverge going into the crossing, and deposition of sediment would occur in the crossing. As a general rule, the width measured normal to parallel tangents of a crossing should be not more than 85%–90% of the overall width in the bends. This can be accomplished by spiraling the entrance and exit of the bends or by laying out hyperbolic bends as described by Steed (1956).

The development of an optimum width is much more simple for a stream in which the discharge varies only moderately in contrast to a flashy type stream with a very wide range of discharge. The latter usually develops for high flows a wide, frequently incised, channel within which the low flows form a braided pattern or a random meander. The banks of such a stream are usually difficult to hold, as the low flow meander tends to change its position after each period of high flow, particularly upstream from constricted sections. The most successful attacks on this problem appear to be those which provided a dual channel, i.e., a controlled low flow channel within the major channel. This is essentially the equivalent of a flood-plain stream wherein normal flows are contained within the channel and flood flows are free to follow a more direct course over the flood plain.

Variations in channel width should be avoided. Where constrictions exist, flood flows may be partially ponded above the restriction with consequent deposition of sediment. As the flows decrease, an entirely new alinement may be eroded into the deposited material and severe bank caving may result.

Partial Controls.—As a practical matter, conditions making complete channel control either necessary or desirable are infrequently encountered. This may be either because the damage to be prevented will justify only local controls or because local variations in bank materials restrict the damage. In either case the aforementioned criteria will apply generally. In the event that control is to be restricted to individual bends, one need not be critical as to the width in the bend or the exit therefrom, but must be careful to obtain proper entrance conditions even though doing so requires an upstream extension of the controls. Likewise, in the case of isolated controls for the protection of bridge abutments or similar installations, the entrance into the control must be properly obtained to prevent possible flanking of the control works.

9. Channel Control Structures.—Control structures may be roughly divided into two groups: those designed primarily to prevent erosion of an existing bank, and those designed primarily to guide the flow or promote deposition of sediment in designated areas, or both. The two classifications may overlap or be combined in many cases, and either may utilize one or more of several types of solid or pervious construction.

Revetment.—Structures utilized primarily for bank protection are generally designated as revetment, and are most properly utilized on a bank which has been previously sloped and shaped to a desired alinement or which follows this alinement reasonably closely. There are four general types: blanket; pervious; solid fence; and groins. Blanket revetment is constructed of rock, concrete, asphalt, or other materials placed to form a protective cover with or without an accessory mattress extending to the thalweg of the stream. Pervious revetment consists of open fence, pile structures, cable connected jacks or baskets, and similar materials placed along the desired alinement both to prevent erosion of an existing bank and to build up the bank by deposition. Solid fence, usually one or more rows of fencing backed up by other materials, is useful on steep slope streams where the revetment must have strength to resist the forces of the flow rather than to guide it; and groins are short, usually solid, structures extending from the banks at approximately right angles to the flow.

Blanket Revetment.—The purpose of a blanket revetment is to prevent erosion of the bank upon which it is placed; and in order to accomplish this purpose it should ideally:

1. Be placed only on a bank that has been sloped and properly alined.
2. Possess adequate strength to resist movement by the flow or breakage due to the impact of debris, floating ice, etc.
3. Be sufficiently tight to prevent removal of the underlying soil materials through the interstices.
4. Be sufficiently flexible to conform to any irregularities in the bank.
5. Be sufficiently mobile to settle into any breaks that might develop, by soil settlement or otherwise, and thus be selfhealing.
6. For revetments that need not extend more than 15 ft vertically below the normal water surface, it is helpful if the revetment is sufficiently rough to create a zone of intensified turbulence and low velocity in the vicinity. This tends to hold the high velocity flow away from the revetment rather than in contact with it with the result that the revetment toe is less apt to be undercut by scour.

This final requirement is one that will not meet with universal agreement, for one result of the intensified turbulence along the rough face will be intensification of the tendency for leaching of the underlying materials through any interstices that might exist. Experience on the Missouri River with rough rock revetment, however, has been that the high velocity flow previously moving adjacent to the bank moves approx 50 ft riverward, thus holding the area of deepest scour away from the toe of the revetment and limiting the opportunity for undercutting. There are numerous examples on this stream where the deepest scour in the bend extends to as much as 30 ft below the toe of the revetment without any damage because the scour occurred away from the toe.

Contrary to some of the preceding criteria, Parsons and Apmann (1965) have developed a cellular concrete block revetment that has the advantages of deforming readily to accommodate subgrade elevation changes, a relatively rough surface, and adaptibility to the growth of vegetation in the cells to assist in stability. The dimensions of the cells are such that leaching of the underlying soil is unlikely, particularly if the cells are filled with gravel or crushed stone. It has a disadvantage in that severe deformation, or the removal of one block from the system, exposes it to complete deterioration. Also, it would be difficult to place underwater. It is not likely that this type revetment would be successful on large streams with readily erodible banks, particularly if underwater placement is required. It should be adequate, and might even be preferred to rock in smaller streams where placement would not be a problem or where rock was not readily available.

There have been many types of materials and placement methods utilized for blanket revetment. Solid pavement provides good resistance, but once a break occurs the entire pavement is usually destroyed. Loose rock, unless properly graded and placed, permits leaching and may be subject to movement by the flow. Riprap, hand placed and chinked, will still be subject to leaching and will tend to bridge until a serious break occurs rather than to exhibit any tendency to self-healing.

Experience with many types of blanket revetment on the Missouri River has led to the adoption of a rock fill, designated as toe trench (Fig. 5.13), characterized by a graded material so placed that all voids are filled and all rocks are well keyed into the mass. The median rock size is specified in accordance with maximum flow velocities and the maximum size is limited to about 1-1/2 times the median in order to eliminate undue protuberances. Minimum thickness should be 1-1/2 times the median rock size. Quarry run rock of a good limestone usually provides satisfactory gradation, but very hard or rough rock may require some selection. The material is placed by dragline and skip to avoid segregation. It is dressed to a reasonably regular surface by the dragline, but shifting or movement of the rock after placement is not otherwise permitted. The relatively large quantity of rock in the toe is provided so that it may drop and cover any scour that may occur. It is noteworthy that during a period of more than 25 yr under conditions of severe exposure there has not been a failure of consequence in the toe trench revetments on the Missouri River. Experience on the Arkansas River indicates that either

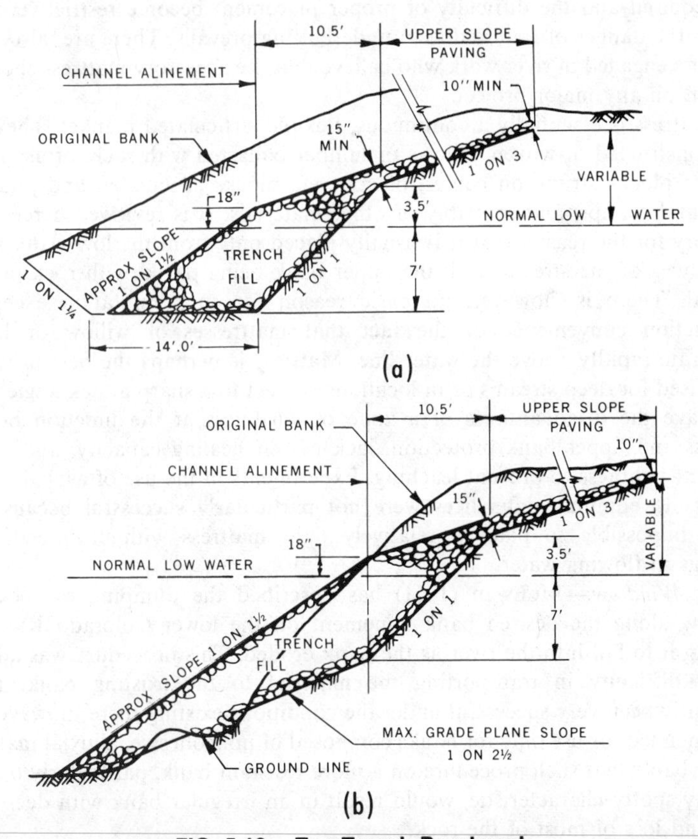

FIG. 5.13.—Toe Trench Revetment

additional toe rock is required in the initial placement for the stream or must be placed after the first flood flow. There are indications that this is due to the deep scour at the toe of bed forms moving along the revetment.

The size of rock required for a given revetment is a function of the flow velocity, usually taken as the velocity at a distance of 10 ft from the bank. Available formulas for computing the size of rock that will resist movement by the flow give results varying over a fairly wide range, and there appears to be no common agreement as to which might be best. The formula proposed by Isbash (1936) for the construction of dams by depositing rock in running water, modified to take into account the slope of the bank, gives results that are, in line with experience

$$W = \frac{4.1 \times 10^{-5}\, G_s V^6}{(G_s - 1)^3 \cos^3\phi_1} \qquad (5.6)$$

in which W = the weight of the stone in pounds; G_s = specific gravity of the stone; V = the velocity; and ϕ_1 = the angle of the pavement with the horizontal.

As a practical matter, the use of rock and similar materials for blanket revetment without an accessory mattress or other extension is restricted to depths of not more than 40 ft and preferably less than 30 ft and to locations where the angle of attack does not exceed about 30°. In the first instance, the quantity of rock required and the difficulty of proper placement become restrictive. In the second, the danger of failure due to undercutting prevails. There are, also, many engineers engaged in river work who believe that the accessory mattress should be required on any major project.

A mattress is essentially a continuous, flexible, articulated blanket. They have been constructed of woven willows or lumber ballasted with rock, of asphalt, of concrete blocks strung on cables, of concrete blocks articulated and joined by cables and clamps, and probably of other materials. It is reviewed herein as an accessory for the reason that it is usually placed only from the low water line to the thalweg of the stream, with the upper bank being paved with rock or other material. There is, however, no basic reason for such limitation except for construction convenience or the fact that mattresses of willow or lumber deteriorate rapidly above the water line. Mattress is perhaps the best protection yet devised for deep streams or in locations subject to a sharp attack angle, but it does have the disadvantages of a zone of weakness at the junction between mattress and upper bank protection, lack of self healing capacity, and lack of sufficient tightness to prevent leaching. Experiments in the use of asphaltic sheet mattress to eliminate interstices were not particularly successful because it is almost impossible to place a relatively light mattress without interstices or openings in flowing water.

Rock Windrow.—McEwan (1961) has described the dumping of rock in a windrow along the desired bank alinement on the lower Colorado River and allowing it to fall into the river as the bank eroded. This procedure was adopted due to difficulty in transporting the material to the existing bank. It has apparently been very successful under the conditions existing there; however, it is recommended for use only for banks composed of noncohesive, alluvial materials. It is probable that such procedure on a more resistant bank, particularly one of a typically spotty characteristic, would result in an irregular bank with deep scour holes and loss of most of the rock.

There have been many examples of revetment constructed on ungraded banks of old concrete, bed springs, and other junk of all types placed in a more or less haphazard manner. Frequently these installations provide adequate protection, particularly if located so that the angle of attack is not excessive, the bank materials are fairly resistant, and the stream is flashy so that exposure is neither frequent nor prolonged. Grout or soil-cement mix in various types of bags, gabions, and many other materials may be used successfully where they can be properly placed. Rock filled wire gabions are particularly useful where cobbles or stone unsuitable for rock revetment are readily available.

Pervious Revetment.—The primary function of a pervious revetment is to establish a zone of resistance that aids in guiding the flow while permitting deposition of sediment behind the revetment. It is particularly adapted to the formation of a smoothly alined bank, although often utilizied for protection where materials for blanket revetment are not readily available or where the cost of a blanket revetment cannot be justified. They are constructed of single or multiple row piling, jacks, old car bodies, fencing backed up with brush, or other material that can be driven or cabled to form a continuous unit. In locations where undercutting might occur they should be protected by a mattress placed prior to construction of the revetment or by rock dumped at their base. After a good fill has accumulated behind a pervious revetment, it is frequently wise to protect it with rock.

Being of open work construction, pervious revetment is relatively useless against direct attack, and is very apt to be destroyed if exposed to high velocity flow at attack angles greater than about 30°. Its efficiency is maximum when parallel to the line of flow, decreasing as the angle of attack increases. As it is primarily a resistance element, its efficiency can be increased by adding additional elements, either by closer spacing or by multiple rows. This latter expedient is frequently used when conditions require the use of revetment in an overly sharp bend. It is particularly important that pervious revetment be initiated sufficiently far upstream to obtain a proper angle of entry. In reaches where the revetment is located more than a few feet distant from the existing bank, cross connections are required to prevent accumulation of flow behind the structure.

L-Head Revetment.—In many cases, it is not necessary that a fence type revetment be continuous. On the Missouri and Arkansas Rivers, e.g., where it is desired to establish a new concave bank alinement riverward of the existing bank, either pervious or solid structures are constructed approximately normal to the existing bank and extending to the desired alinement. They are then continued downstream along the projected alinement for about two-thirds the distance to the next normal structure (Fig. 5.14). In streams carrying at least a moderate amount of sand-size sediment, these structures have been successful in engendering a fill within the structure field.

Jacks.—Jacks are the most common type of pervious revetment in relatively shallow streams. They are usually constructed of three equal lengths of steel angles, wood, concrete, or other material fastened together at the center points with their longitudinal axes at right angles, one to the other (Fig. 5.15). The individual units are cabled together and to anchors in the bank to form a continuous string. Several strings may be used, one above the other or one behind the other. There are no definitive specifications for their placement; however, several existing installations have been described in the literature (United States

Army Corps of Engineers, 1953; Woodson, 1961; Borland and Miller, 1962; Carlson and Dodge, 1962; Steinberg, 1960; Byers, 1962).

Pervious Fence.—Pipe and wire or timber and wire fence has been used with varying degrees of success. Generally, a single row fence is placed along relatively straight banks, while a double row with rock or brush filler is placed along the

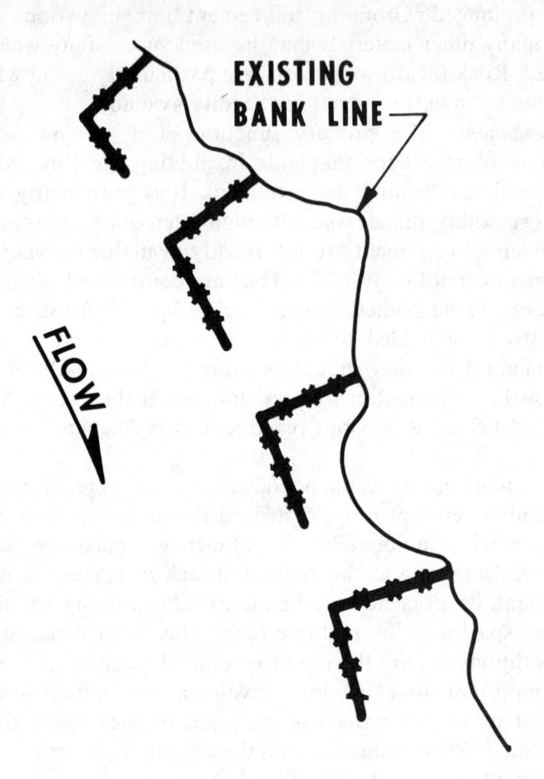

FIG. 5.14.—L-Head Dikes

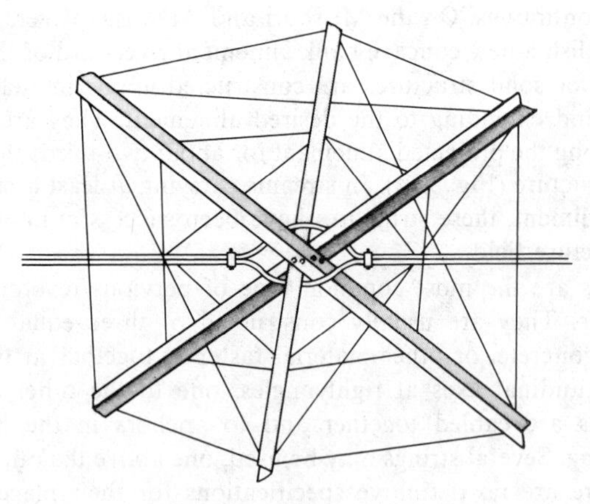

FIG. 5.15.—Steel Jacks

concave banks of bends (O'Brien, 1951). These installations are much more apt to be successful if a good stand of vegetation can be obtained behind them or if impervious bank connections are provided at frequent intervals to prevent the flow collecting behind the fence and continuing the erosion. One major problem, observed by the writer in Alkali Creek, Wyoming, was the formation of random gravel bars by high flows so that intermediate and low flows are diverted through the fence to attack the bank. Where double row fence is used, rock filler between the rows would be superior to brush as the rock could settle in the event of undercutting. Otherwise, some rock should be placed along the ground surface immediately behind the fence for the same purpose.

Miscellaneous.—Steinberg (1960) has described the use, in the Russian River, California, of gravel placed on a sloped bank and covered with wire to hold it in place. He also describes the use of a planting of willow trees or sprigs in a trench at the toe of the slope. Parsons (1965) has reviewed the use of vegetation for stream bank erosion control. All of these procedures have merit if utilized under the proper conditions.

Groins.—Groins are short, usually solid, structures placed at approximately right angles to the bank. They are not generally recommended, as they are apt to generate more damage than they prevent. It is a characteristic of a groin that an eddy will form immediately downstream of its outer end, and that this eddy may generate a scour pattern extending into the bank below the groin. The danger of bank scour from this source can be reduced by angling the groin upstream approx 10°–15°; however, in the event of overtopping by flood flows a groin angling upstream is more susceptible to damage at the root, or landward end. Groins should have a length of not less than three times the normal water depth at their outer end. In order to prevent an accumulation of flow at the root in the event of overflow, it is suggested that the crest should slope downward toward the stream at the rate of about 1 ft in 10 ft. There is no general agreement as to the optimum angle of a groin.

10. Training Structures.—The purpose of a training structure is to guide the flow in a manner such that an effective channel will be scoured and maintained along a desired alinement. As a rule, this purpose will be attained more readily with a pervious structure than with a solid structure, since the action of the pervious unit includes the generation of a turbulence zone that encourages, rather than forces, the flow to turn while permitting a portion of the flow and its sediment load to pass through to build a deposit behind the structure. The fill behind a solid structure is usually attained from flow passing over the unit during construction, from flow entering the area around the end of the unit, and from flow overtopping the unit during high discharges. The ultimate choice between a pervious and solid structure is apt to be based on available materials, depth of water in which the unit must be constructed, and the probability of heavy ice or drift runs. Heavy ice runs are particularly damaging to timber structures.

Training structures may be timber pile dikes, jacks, single or double row fence, dumped rock, old auto bodies or frames cabled together, broken concrete, or other available materials, depending largely upon the magnitude and economics of the project. In water depths of more than 10 ft, some form of pile dike is usually indicated as the more economical. This will be true even if rock is later dumped into the structure, for the piles will hold the rock so that the quantity of rock required will be substantially reduced.

Pile Dike.—Pile dikes may be either single or multiple rows of piling driven either singly or as clumps into the stream bed and connected by horizontal wales or stringers. Pile clumps are normally composed of three piles driven at the corners of a triangle and pulled together and cabled at the top (Fig. 5.16). At one time, it was common practice to sink a brush or lumber mattress along the dike alinement prior to driving in order to prevent scour. This practice is still followed in some instances, but, except in relatively deep and swift water, it is more economical and equally effective to dump rock along the dike line after driving.

Jacks.—The individual units are identical to those described for pervious revetment except that they are placed as a guide structure rather than as revetment. Their basic action is the same as that of a pile dike, and for flow depths up to about 6 ft they will probably be much more economical. For flow depths between 6 ft and 10 ft, their use will be dictated by convenience, and for depths in excess of about 10 ft they would not normally be used as the size of jack required would be excessive.

Rock Dikes.—Rock dikes may be placed by end dumping from trucks, dumping from a barge, or by clam shell or skip. The cross section is usually dictated by either the crown width necessary for a truck or by the section that will be attained by dumping in flowing water. A reasonably well-graded rock, usually quarry run, provides a keying action that helps stabilize the fill, but some leakage is acceptable except for some chute closure structures where the loss of water from the main channel might be objectionable.

Rock Filled Pile Dikes.—In depths of flow exceeding about 9 ft or 10 ft, and in areas where a solid rock dike is preferred, it is usually more economical to drive a

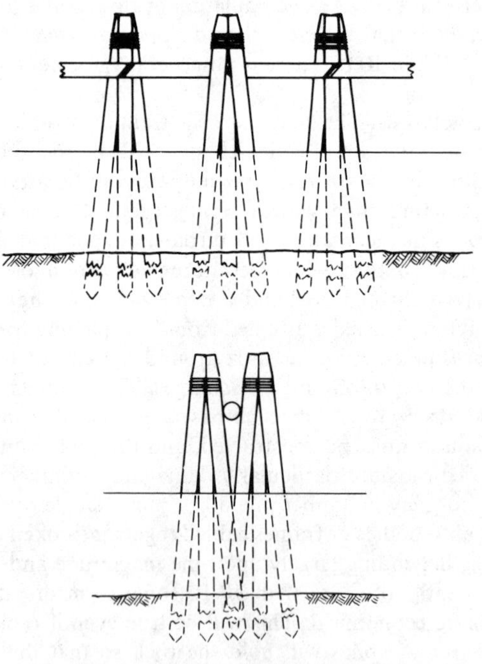

FIG. 5.16.—Pile Clump Dike

pile dike and fill it with rock rather than construct the initial structure of rock alone. The savings in rock quantities in this case will over balance the added cost of driving the piles.

Miscellaneous.—There may be cases where economics prevent the use of any of the previously described structures. In those cases, the use of any available materials will be dictated. They should be reasonably adequate if placed with due regard for alinement and the forces they must withstand.

11. Coordinated Project.—References to bank protection or stream training works are replete with instances where isolated structures, placed for the purpose of protecting some local area and without regard for a rational plan, were soon out-flanked or bypassed, and thus rendered useless. There have been many instances where an otherwise acceptable structure was destroyed because of too great an angle of entry. It is extremely important that even the smallest project be properly planned.

Anchor Points.—If at all possible, it is desirable that any control system be originated at an erosion resistant point. If a reasonably extensive system is involved, it should be laid out by reaches beginning and ending at such points. Lacking both of these conditions, the work should be originated well upstream of the point where the high velocity flow will contact it and should be well keyed into the bank to retard any possibility of flanking by an upstream channel change.

Angle of Entry.—If practical, the work should be originated at a location where it is parallel to the flow, being gradually turned thereafter into the planned alinement. In no case is it recommended that the angle of attack of the flow exceed 15° unless an auxiliary mattress extending to the thalweg is provided, in which case it is recommended that the angle of attack not exceed 25°.

Channel Contraction.—The flow in a wide, shallow channel tends frequently to meander within this channel, shifting its position at random intervals, particularly during or following a rise in stage. As a result, the location of bank erosion varies from time to time, and protection works placed at one time may be entirely bypassed at a later date while the attack is shifted to an unprotected location. For this reason, it is always recommended that the channel be contracted, or, at the least, adequate training works be provided to hold the flow in place. The contracted width should be the maximum for which the entire section is effectively occupied by the flow unless further contraction is required for navigation or other purposes. As previously mentioned, however, the contracted width should be such that the bed forms are in the dune range. If the channel is alined with alternating bends of proper radius, it is not necessary that both banks be continuously protected. It will be sufficient to revet the concave banks and to extend the revetment with training works adequate to direct the flow properly into the next bend. It may be necessary to also provide some training dikes at the head of a bend to insure a proper alinement, but, in any event, some overlap between the bends is desirable. The training dike downstream should never be terminated in a reverse hook or curve in an attempt to eliminate scour at the dike end as experience has shown that the flow will tend to follow the hook and produce a spread channel. The training works required may or may not be extensive, depending on the characteristics of the location.

There are instances where the protection of only a single, isolated location can be justified and where a shift in the erosion pattern would not be considered consequential. In such an instance, only a minimum protection might be recommended; however, it must be recognized that extensive maintenance may be required. Provision for a proper entrance and an adequate tie-in to the bank at the upper end should not be neglected in any case.

Layout of Training Structures.—Generally, the major training structure will be along, and will ultimately form, the planned alinement of the concave bank of a bend. It need not be continuous, but should have a total length equal to approximately two-thirds the overall length of the distance spanned. Adjacent downstream sections should be inset slightly, with an initial alinement such that extensions of the axes of the two adjacent sections form an angle of not more than 10°. Ties to the existing bank should be provided at spaces of from two to two and one-half times the length of the next upstream tie.

For the use of jacks as training structures it is usual practice to utilize one, or at most two, sizes without particular regard to the elevation of the top of the structure. With pile, rock, or similar structures, however, it is necessary to establish an elevation to which they will be constructed. In most instances of training works currently in place, the elevation has a constant relationship to either the low water profile or average top of bank. Recent experiments (Franco, 1967) have indicated that a more satisfactory fill may be obtained in a dike field if the elevation of each structure in the field is somewhat lower than that of the next

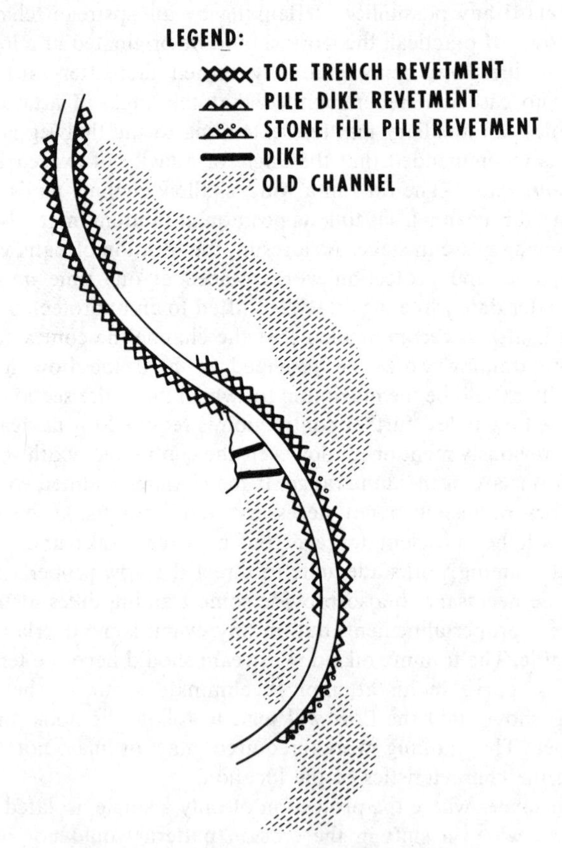

FIG. 5.17.—Section of Rectified Channel

upstream structure. The results of tests involving a sloping profile for the individual structures were indeterminate.

If, for one reason or another, it is necessary to accept an alinement with a long, flat (long radius) bend, it may be necessary to place structures normal to the flow line from the convex bank to prevent meandering or spread flow within the channel. Such structures are frequently placed to such elevation and length as required to hold average or low discharges within an effective channel width but permit high flows to occupy the entire channel.

In those cases where the desired alinement of a concave bank is landward of the existing bank, it is frequently more economical to excavate a trench and construct the revetment in the dry. The stream is then permitted, or encouraged by training works, to erode the existing bank back to the revetment. If a complete realinement is indicated, a pilot channel is excavated after the revetment is placed. Note, however, that clayey or other erosion resistant materials must be dredged because the pilot channel will enlarge slowly or not at all.

Fig. 5.17 shows a sketch of a complete realinement of an exceptionally wild, wide reach of the Missouri River. Fig. 5.18 is a sketch of the layout of pile dike training works utilized in another reach of that river. Note that ties to the bank are provided at intervals. These are important, because the flow passing through the main structure is permitted to accumulate and flow freely behind these structures and the deposition of sediment will be restricted or the sediment previously

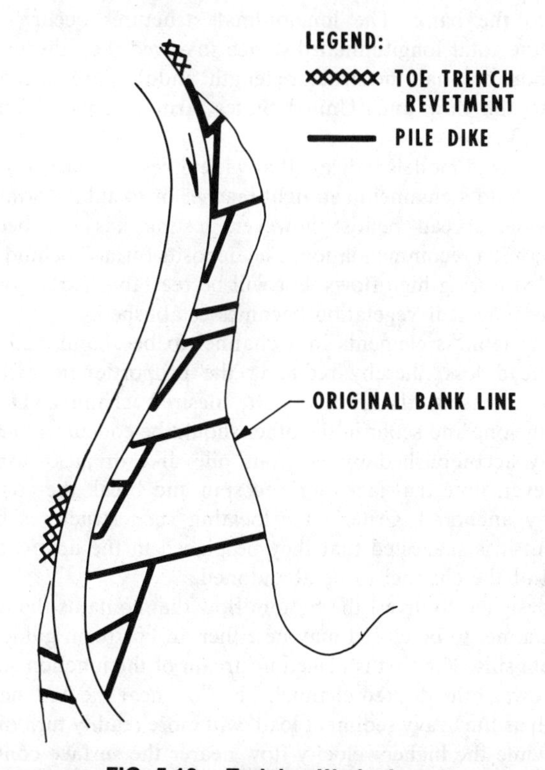

FIG. 5.18.—Training Works Layout

deposited will be eroded. There is no definite rule as to the spacing of the tie structures, but generally the spacing between any two adjacent ties should be approximately one and one-half to two times the length of the upstream tie. Inasmuch as the final layout of a training works field remains largely a matter of judgment and experience, the reader is urged to review published material such as the previously cited references for jacks and papers by the United States Army Corps of Engineers (1966a, 1966b, 1969); Bush (1962); Franco (1967); Parsons (1963); and Wall (1962).

12. Closure of Chutes or Secondary Channels.—It is frequently desired to close secondary channels, or perhaps, to divert the flow from an existing major channel to a minor channel or to a pilot cut. This has been accomplished by solid closure dikes, but the practice is not recommended except for final closure after the chute has filled with sediment since the total head loss through the channel to be closed is concentrated at one location and serious scour will occur if the structure is overtopped. A much more rational procedure is to either provide a complete system of training works, place roughness elements in the channel to be abandoned, or utilize structures designed to divert a high proportion of the sediment load into that channel.

The use of a system of training works would be as previously reviewed except that, if the channel to be closed is a chute or secondary channel, vane dikes (Fig. 5.19) across the chute may be adequate. As utilized in recent installations in the lower Mississippi River, these are essentially the frontal portion of the system without ties to the bank. The longitudinal structures occupy approximately two-thirds of the total longitudinal distance involved, i.e., the interval between structures is about one-half the structure length, and they are placed at an angle of about 10° into the flow line (United States Army Corps of Engineers, 1969, Chapter II, p. 23).

Preliminary tests in models indicate that vane dikes may be successfully utilized to contract and hold a channel in straight reaches or to aid in forming convex bar deposits in wide, spread bends; however, testing has not been sufficiently extensive to permit a recommendation. The deposits formed behind vane dikes are apt to be eroded during high flows, but will be reestablished at lower discharges and will be permanent if vegetation becomes established.

The use of roughness elements in a channel to be abandoned is designed to increase the head loss, thereby reducing the proportionate discharge in that channel and increasing the flow in the desired channel. Deterioration or deposition in the one and scour in the other should be concurrent and progressive. This is usually accomplished by pervious pile dike or jack strings across the channel; however, note that jacks tend to spin and break the connecting cables unless properly anchored. Criteria for locating such structures have not been established, but it is suggested that they be placed in the upstream one-third or near the head of the channel to be abandoned.

Structures designed to divert the bottom flow that contains the heavy sediment load into a channel to be closed may be either an upstream guide structure or a series of bottom sills. The first is placed upstream of the junction so as to turn the flow directly toward the desired channel. The flow near the bed, having the lower velocity as well as the heavy sediment load, will more readily turn into the channel to be closed while the higher velocity flow nearer the surface continues into the new channel. Bottom sills, or vanes, are placed on the bed, at or immediately above the head of the channel to be closed, at an angle such that the bottom flow and heavy sediment load will be guided into that channel. Installations of this type are reviewed and illustrated by Remillieux (1966) and Chabert, et al. (1961).

13. Emergency Bank Repairs.—Bank erosion during flood flows sometimes becomes critical and requires emergency measures, either to protect against major damage to an installation or to protect some adjacent structure. Such a condition might arise by direct attack on the bank, or it may result from an eddy formation generated by some upstream irregularity. In either instance the initial effort at control should be directed upstream from the actual area of erosion with a view toward deflecting the flow sufficiently to permit repair in the damage area. In the case of a direct attack, the effort to deflect the flow must be planned in accordance with the local geography of the channel and of the flow. In the case of local eddy scour, a roughness element should be placed immediately upstream. One of the most effective emergency controls in such a case, and one which is frequently available at the site, is a tree with the tip cabled to a deadman or other anchor upstream and the butt floating freely at or just below the upstream limit of the scour hole. Resistance units of this or similar types aid in guiding the flow, and they also appear to cause the development of a turbulence wake, or a trail of small eddy formation, which has the effect of breaking up the major eddy formation and permitting the scour hole to be filled. Attempts to fill the scour hole by dumping in material without prior placement of an upstream resistance unit will probably be useless.

14. Aggrading and Degrading Channels.—An aggrading channel is one that tends to fill with sediment, while a degrading channel is one in which a net loss of

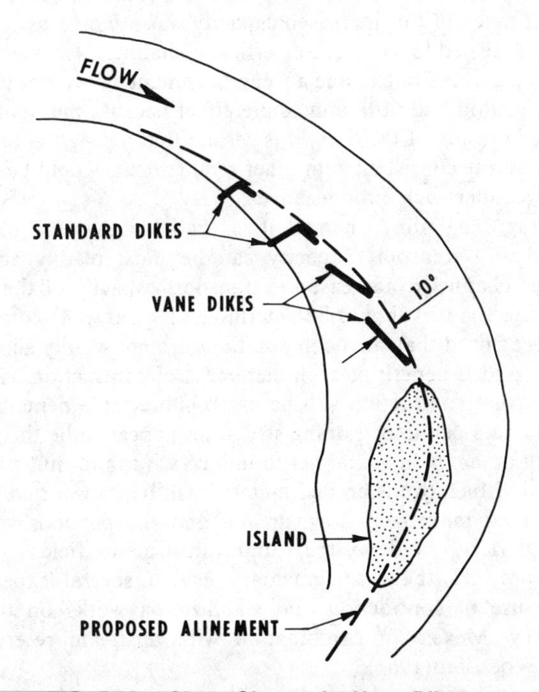

FIG. 5.19.—Chute Closure by Vane Dikes

bed or bank materials is progressively occurring. In either case, the problem is essentially one of an imbalance between the sediment load contributed to the channel and the ability of the flow to transport the load through the channel. The obvious solution is to restore the balance.

In an aggrading channel, restoration of the balance might be accomplished by either reducing the sediment inflow, increasing the transport capacity, or combining the two. In a relatively small drainage, an appreciable reduction in the sediment inflow might be accomplished by soil conservation measures, and in a stream served by an even larger area of severely gullied land, gully control would be a valuable adjunct to control of an aggrading downstream channel in addition to its value on the land. In a moderately large drainage, however, it must be recognized that the sediment that will be contributed to the stream in the reasonable future has already been eroded from the immediate land surface and is available in the tributary net. Measures to reduce sediment inflow in this case must include reservoirs or other means to trap the sediment as well as measures to control erosion that might be occurring in uncontrolled tributaries. In connection with the use of reservoirs as sediment traps, one must also consider the included effect of flow control, for the reduction of flood peaks will also reduce the transport capacity of the flow.

In larger basins where requirements for the use of water are not restrictive, it is sometimes possible to increase channel capacity by storing water and releasing it at high flow rates. On one test, e.g., in the Arkansas River in Kansas, which tends to fill with wind-blown sand between periods of high flow, the bank-full capacity was increased from 2,000 cfs–10,000 cfs over a 2-week period by releasing stored flood waters at the maximum rate that could be contained between the banks. It was noted that much of this increased capacity was attained by lateral shifting of the materials in the bed to form a more efficient channel. A similar test in a stream channel that had deteriorated due to encroachment of willow growth during a prolonged dry period had little immediate effect because the tightly laced willow roots prevented erosion of the bars; it is certain that a program of releasing flows at the maximum rate consistent with other requirements would be more beneficial than otherwise under such circumstances.

In the average aggrading channel it is probable that the balancing of the sediment load and transport capacity can be most readily accomplished by contracting the channel to increase the transport capacity of the available flow. The width of the channel should be determined by transport computations, even though it is recognized that such computations are not wholly accurate. It will be found that the added benefit of each increment of contraction will decrease to a point where further contraction will be of no appreciable benefit.

In most instances pervious training structures appear to be the most adaptable for contracting an aggrading channel to induce scour and thus provide the most efficient section. Much of the eroded material will be transported laterally, rather than in continued movement downstream, and the pervious structures better permit this material to be deposited within the structure field.

On major projects, the complementary use of several expedients may be justified. The use of contraction and stabilization works on the Middle Rio Grande in New Mexico in combination with upstream reservoirs has been described by Woodson (1961).

A degrading channel may be eroding either vertically or horizontally, depending upon the character of the materials in the bed and banks, and any remedial measures must be analyzed to insure that protection of one will not transfer the attack to the other. An interesting case in this respect was Alkali Creek in Wyoming, where degradation resulted from the introduction of return flow from irrigation. The channel degraded vertically until it reached a stratum of large gravel and cobbles; then started widening at a rapid rate. Over-contraction by groins in one reach reintroduced the vertical scour, which in this case was not important except that it required protection against headcutting in tributary drains; however, the use of groins in another reach to restrict the channel to an optimum width resulted in a good balance and reasonably stable channel.

In some instances where material is readily available, sediment can be introduced into the stream in sufficient quantities to satisfy the transport capacity. This could probably be most reasonably accomplished by periodically placing a volume of material in the channel at the head of the degrading reach. The added material would, of course, have to be of a size equivalent to the normal bed material. In the event that the channel banks are sufficiently resistant, material large enough to armor the bed could be utilized. The minimum quantity required would be an amount sufficient to form a single layer over about 60% of the bed. The channel characteristics should be such that the flow will be evenly distributed across the section; otherwise, bars might form and compound the problem.

The use of upstream storage to regulate the flow and thus reduce the total transport capacity could be helpful provided that the storage zones were located upstream from the primary source of the available sediment. For economic reasons, such a plan would usually require generous benefits from other uses of the storage structures.

Normally, the most expedient degradation controls will be sills or drop structures combined with such bank protection measures as may be required. The structures may be used singly, to protect only some critical location, or in series. In the latter case the spacing should be such that the slope from the tailwater below one structure to the head water over the crest of the next will be that required for the transport of the existing bed material supply.

Simple sills providing nothing more than a vertical control are not generally recommended, although it is recognized that there are circumstances where they will be adequate. There is always a dominant tendency for a scour hole to form immediately downstream and to cause failure of the structure or of the adjacent banks. When combined with downstream toe and bank protection, however, sills can be useful tools. Ferrell (1959) and Ferrell and Barr (1963) describe the design of check dams varying in height up to 40 ft above the stream bed, although Ferrell (1959) recommends a maximum height not exceeding 17 ft in the area in which he was working. Ferrell (1959) also reviews the design spacing, recommending a stabilized slope of 0.7 of the natural gradient for this area. Lombardi and Marquenet (1950) review the design of a stepped channel with loose bed stilling pools in a torrential stream, including criteria for spacing of sills. Linder (1963) gives a good account of the design of sheet pile and rock sills in the Floyd River, at Sioux City, Iowa and Stufft (1965) describes a series of stepped sills in the Gehring Drain (Nebraska). There are a number of other descriptions of lesser structures in the literature.

One type of rock sill, for which the writer has no reference, has been constructed on the Little Sioux River, Iowa. It consists of a rock weir with the downstream face on a 2:1 slope intersecting an apron having approximately a 10:1 slope. The effect of the rough rock surface on the downstream slope and the apron restricts the bottom velocities so that a reverse roller forms downstream and material scoured from the bed is moved toward, rather than away from the toe of the apron. Shortly after construction these structures were subjected to a flow equalling, or possibly somewhat exceeding, the design head of 5 ft over the crest. The depth of the scour hole downstream exceeded 30 ft, but no damage accrued to the structure from this section. Erosion of the bank upstream caused the flow to enter the upper structure at an angle, resulting in the loss of some rock from the crest and bank erosion downstream; however, this condition was corrected and the structures have since operated in a highly satisfactory manner. This experience emphasizes the requirement that the flow must be properly guided into this or any other similar type structure. It must also be emphasized that experience with this type structure is, to date, limited.

A further problem noted with rock weirs is that drift tends to land on the crest and to concentrate the flow into local zones so that the crest rock may be eroded. In areas where drift may be anticipated, the crest may be grouted. Gabions have been successfully utilized for the construction of weirs, particularly where available rock sizes are less than otherwise required for stability.

Drop structures are generally a conventional design with some type of stilling pool below the drop. Design criteria for concrete drop structures have been presented by Vanoni and Pollak (1959) and by Donnally and Blaisdell (1954). The former (California Institute of Technology Type) is suitable when the drop height is less than critical depth and the latter (Saint Anthony Falls Type) is used for drop heights exceeding critical depth.

D. Canals

15. General.—Sediment transported by natural streamflow has often presented the canal designer with major problems. In order to prevent clogging and costly maintenance operations, it must be removed from the water at the canal intake or transported through the canal system with a minimum of accumulation within the canal prism and structures. Complete elimination of the sediment at the diversion point is generally impractical and would be too costly in most cases. However, a combination of its control at the headworks and design of the canal hydraulics to minimize deposition through its length can be used to provide practical solutions to the problem. In general, three approaches can be taken in solving this problem:

1. Direct water only into the canal intake and return the sediment to the stream.
2. Design the canal system hydraulically so that the water with its sediment will be transported out onto the land with a minimum of sediment deposited in the canal.
3. Design the canal headworks to direct as little sediment as practicable into the system, and remove the sediment deposited by the most inexpensive method available.

Of these three approaches to the problem, No. 3 is the solution adopted in most designs. The ideal solution as stated in No. 1, would be to prevent all of the sediment from entering the canal. This can be partly accomplished by the use of special devices called sediment diverters placed at the canal intake. These devices exclude a large portion of the bed load sediment that would otherwise enter the canal. A great deal of research has been conducted in many countries in the development of such devices, and they have become very important in the control of sediment entering canals. India and Pakistan, because of their many large capacity canals, which take water from streams that are heavily laden with sediment, have contributed a great deal to our understanding of sediment diverters. These devices are described in detail subsequently in this section. However, it should be emphasized that the best of sediment diverters, while effective in removing part of the sediment, cannot completely remove all of it. Any operation or device capable of removing all sediment from water diverted into a canal or pipeline, e.g., filtering equipment, is very expensive. Thus, for irrigation water, this would not be operationally desirable or economically justified. In fact, fine-grained sediment carried in some irrigation water is beneficial in sealing the canal or lateral and if carried through the canal, may improve the texture and fertility of the croplands. Flooding rivers build rich soil areas. Examples of this may be found in the flood plains and delta areas of such rivers as the Mississippi in the United States, the Nile of Egypt, and many of the large rivers of India, Pakistan, and the Far East. In many of these river valleys the deposits from the river floods of rich organic matter have maintained the productive fertility of the land for centuries. Thus, transportation of sediment through canals and laterals and onto the land should not be overlooked as a possible solution to the sediment problem when factors e.g., soil, crops, land location, and canal conveyance, will permit designing the system for it.

Irrespective of the ultimate beneficial or detrimental effects of sediment on agricultural land, water directed into canals or pipelines from a natural watercourse always contains some sediment. The problem of controlling excessive sediment then must always be considered in canal and distribution system design. Selecting the proper approach is sometimes difficult. For instance, exclusion of sediment from the canal at the intake by removing as much of it as possible and returning it to the stream might at first glance appear to be the solution. However, returning the sediment to the stream increases the sediment concentration below the diversion point. This, in turn, could create additional problems in the channel downstream resulting in expensive maintenance or corrective action in the stream, or both. This corrective action may be so expensive that the approach of dumping the sediment back into the stream may have to be abandoned and another approach selected. To approach the problem in a more logical manner, it is necessary to make a study of the stream or lake, its sediment, changes in flow patterns that the diversion will make, and what means are available to control or limit its entry into the new water system.

The amount of sediment in the natural watercourse depends on the type of watershed above the point of diversion or, more specifically, the type of soil and its vegetative cover in the areas through which the stream flows, the regime condition of the stream (Chapter II, Sections F and H), and the stage of streamflow. At any given time at the point of diversion, the sediment discharge of the stream, including the size distribution, of the load can be measured or estimated from these factors. For example, in the Pacific Northwest region of the United States, the terrain is rocky, the soils largely clay, and heavy vegetative cover provides good erosion protection. The characteristic sediment discharge of the streams is small. On the other hand, in the Southwestern, Gulf Basin, and Great Plains states tributaries of three major rivers, the Colorado, Rio Grande, and the Missouri are heavily laden with sediment most of the time. On these streams the major problem is how to divert the sediment away from the canal entrance and not cause major problems in the stream below. For example, before the closure of Hoover and Glen Canyon Dams on the Lower Colorado River, the average sediment discharge was estimated to be about 100,000 acre-ft annually for a mean annual discharge of 11,610,000 acre-ft (United States Bureau of Reclamation, 1941).

The rivers of India and Pakistan carry very large quantities of sediment during the monsoon season. The sediment discharge in streams of the Himalaya Mountains is very heavy; often including boulders 24 in.–36 in. (61 cm–94 cm) in diameter as well as smaller cobbles, gravel, sand, and finer particles. The volume of sediment transported in a day by such streams may total several million cubic feet. Thus, waters diverted from these streams carry enormous quantities of sediment into the canals. At the Upper Bari Doab Canal headworks at Madhopur (in the years from 1939–1949), 168,000 cu ft–2,268,000 cu ft of sediment (particles larger than 0.075 mm in diameter) entered the canal in 1 day. Because of the high sediment content in the river at high stage, it was necessary to close the canal intake to eliminate some of the sediment (Uppal, 1951a).

Where no provision is made to divert sediment from canals, or where sediment is only partly diverted at the canal headworks, it is advisable to remove it from the canal by ejectors. Some examples of these structures will be described later in this section. Another way of removing sediment is by inducing it to deposit in specially constructed settling basins from which it can be removed by mechanical means. In the latter case much of the fine fraction of the sediment that might otherwise be transported through the canal, will be deposited in the settling basin. This will increase the volume of sediment to be removed and the spoil area required for disposal. From a practical standpoint, it is very difficult to transmit sediment-laden water through an irrigation canal system, even when the sediment is in suspension when the water enters the canal. To maintain it in suspension conditions must be such that the velocity in the various reaches of the canal or laterals, or both, are maintained at or above that required to maintain the condition of suspension. Operating conditions usually do not permit a consistent high velocity at all times. Changes in demand, or shortage of water may reduce the rate of flow in any reach far below that which will maintain the transporting velocity. Also, canal checking, at lower flow rates to meet required delivery elevations, will reduce velocities such that the sediment will be deposited in these settling basins between checks or barriers.

Canal design is usually based on an operation study to determine the pattern of the water demand over the project's full operating life. The sediment transportation characteristics are then determined from the operation pattern thus established. Sediment may have been deposited during one phase of operation and eroded during another with a balanced or poised resultant condition for the operating season. Reaching such a balance is complicated where the suspended load contains an appreciable percentage of cohesive clay particles. These particles, once they have settled, are difficult to reentrain. Some canals in fluvial material, even though their hydraulic properties are designed with great care, will scour their beds or banks, or both, in some reaches and deposit sediment in other reaches. Such canals require regular sediment removal and special erosion protection.

Numerous formulas for design of stable channels and canal sections have been developed in many countries based on research and experiments covering a variety of operating conditions. These formulas have been used with some success where soil types, sediment load, and conveyance conditions are constant and similar to those under which the formulas were developed.

16. Design of Canal Sections.—Experience shows that there is no set method of designing canals. Design of the slope and cross section for open channel flow in canals and laterals must to a large extent be based on experience and judgment. Stable channels can be developed where the discharge, sediment load, and earth materials through which the canal is to be constructed are uniform. However, these uniform conditions are seldom found—the regime formulas developed in India by Kennedy (1895) and Lindley (1919) and refined by Lacey, et al. (1929) have proved successful for the steady conditions for which they were developed.

Although the regime formulas have been widely used successfully in India and Pakistan, they have not been widely adopted in the United States. The primary reason for not fully adopting them is due to the great variation in soils usually found along the alinement of the canals. However, they do serve as a starting point for design, and are useful for that purpose. Even so, factors other than the transport characteristics of the canal enter into its design. Some of these factors are:

1. The terrain crossed by the canal alinement may extend for long distances through great variation in soils. The major types of soil can be considered separately in reaches, but within each soil class, e.g., clay, silts, sands, and gravel, there is a wide variation in size of particle and cohesion; and thus, in the erosion resistance of the soil.

2. Some soils in place can resist erosion but may erode easily when reworked in excavation and embankment. Locating a canal entirely in undisturbed materials would usually be uneconomical.

3. After construction, canal sections must hold water as well as be stable. To prevent excess leakage, compacted earth lining, hard surface lining, and membrane lining are often used. This materially affects the selection of the canal cross section. The cost of the lining selected has a tendency to reduce the area of the cross section for hard linings, and to increase the grain size of the cover material over membrane linings to resist erosion due to the higher water velocity. If the design of a canal is to include a compacted earth lining, the regime theory no longer applies and the plasticity of the lining material gives a more practical guide to the choice of the cross-sectional shape.

4. Land use and its state of development also affect the selection of the canal cross section. A new project area may require several years to develop a high water demand that approaches design capacity. Others may require the full design rate of flow at once. Thus a new canal can safely have a higher erosion potential at design capacity if serving an undeveloped area, than it would if serving a developed area, because of the longer time available to season the canal.

5. Methods of irrigation operation and maintenance also affect the canal design. Some canals remain in operation all the year, while others are operated only

during an irrigation season and are empty a period of months each winter. Maintenance and repairs can be made much easier and at less expense in the canal that operates only part of the year. Weed removal and removal of waterborne or windborne debris and sediment, or both, are maintenance items that may influence canal design. Turnout requirements that necessitate the use of checks to maintain a high water surface elevation at low flows is also a condition that affects design of the canal.

The considerations expressed previously all enter into the selection of canal cross section and slope. The type of canal referred to previously is the irrigation canal; the canal that conveys water from a natural stream source and distributes it to lateral and farm turnouts. Other canal systems, e.g., those obtaining their water from a large reservoir or natural lake source, divert mostly clear water and do not have the sediment problems expressed previously. Canals that have uniform flow rates with little or no fluctuation, e.g., power canals, main conveyance canals, and in some cases canals carrying municipal and industrial water, can be designed to maintain velocities that will carry through them any sediment diverted. Most of these canals are operated year round and have hard lining; thus, a minimum of sediment deposit is desirable to reduce maintenance costs. While irrigation canals may have hard linings, many have compacted earth lining or are unlined. Much of the sediment entering such canals must be removed during the off-irrigation season or, in some case, it may be sluiced out by design flow velocities with sediment passing out on the land through the turnouts. Hard lining will facilitate this type of operation. The membrane-lined canal with armored protection is also a type that may be used. Herein the design velocity may be kept high enough to carry the sediment on through the system, although the size of the armored protection particles must be increased. Cleaning this type of canal section is more costly than cleaning those with hard linings.

Design of Irrigation Canals.—Lined canals are canals lined with concrete and soil cement, and can be designed with high velocities at design rate of flow, such that any sediment taken into the system may largely pass through without excessive settling out. Large reductions in rate of flow or checking will cause slower velocities and depositing of sediment in the canal. Thus, only with flow close to the design rate can the sediment be transported through the canal system. At lower velocities provisions must be made for removal of the sediment before it accumulates to a volume that will reduce the canal's capacity. Hard lining provides a canal section that can usually be cleaned with heavy equipment.

Earth-lined or earth canals include many irrigation canals that are earth-lined or unlined. Many of these earth canals are in noncohesive materials. Much work has been done on the design of such canals to provide a stable section. A recent report of the United States Bureau of Reclamation (Scrivner, 1970) attempts to approach the design of such canal sections by incorporating such factors as water temperature, bed load discharge, and sediment particle properties and a critical sediment particle diameter to determine an effective transport rate. From the effective transport rate value computed, a probable bedform is determined. Then from curves and relations established in the report, section properties consisting of the hydraulic radius, velocity, depth of water, canal bottom width, critical slope, Reynolds number ratio, Manning's n, and sediment transport slope are determined. While the method is yet to be proven from practical results, it is a fresh approach and would seem to have promise. The section designed in this manner should be stable at the flow rate for which it is designed. With this method of obtaining a known rate of sediment transport through the section, such a section can be designed to carry, with a minimum of settling out, the sediment load present in a natural stream water course. It could also be designed for a part of the sediment, if a method was used to remove a known percentage of the sediment at the headworks of the canal.

As an example of present design procedures for earth-lined and unlined canals, the United States Bureau of Reclamation (USBR) selects preliminary canal sections from charts developed from relationships of variables empirically determined and experienced (United States Bureau of Reclamation, 1951, 1967). These sections are used in preliminary design studies, which include preliminary surveys of the proposed alinement and geological investigations. In preparing basic design data, the soils are classified and analyzed for particle size and plasticity. From these data, representative material size-distribution curves for the canal materials are constructed. Variations in soil textures with depth and the plasticity indexes, where applicable, are determined. Shear tests, and other tests as required, are made to determine stability of the canal section. The section is also checked for resistance to scour by applying the tractive force analysis if the section is to be earth-lined or unlined (United States Bureau of Reclamation, 1952; 1953; Chapter II, Section E). The section selected can be further evaluated from charts (United States Bureau of Reclamation, 1951) which are plotted to show the relationship of the canal base width, b, to water depth, d, for the various side slope values. Another test for earth section canals can be obtained from a report (United States Bureau of Reclamation, 1953) that develops a theoretically perfect shape and that evaluates the critical tractive force. Having satisfied the preceding criteria, the section is evaluated on the basis of experience for the probable effects of cohesion, which is still beyond exact analysis.

Other factors considered are: (1) Cost of original construction; (2) rights-of-way required; (3) risks involved if some scour is allowed; (4) the desirability of using scouring velocity to move wind-deposited sediment from sections of the canal, provided this condition can be predicted; (5) the number of section changes required to follow the theoretically different sections developed in a rapidly changing soil profile along the alinement; (6) cost and standardization of structures; and (7) need for lining.

Summary of Design Considerations.—The method of controlling the amount of sediment in the water diverted into an irrigation canal is a major problem to be considered in canal design. In this connection, some of the factors to be considered are the amount and type of sediment to be removed or carried into the canal system, type of earth materials through which canal is to be constructed, and type of lining, when used.

Most canal designs are presently based on judgment and experience. New approaches are being developed that will include sediment transportation factors and bedforms. However, this approach has yet to be proven in practical designs for canal construction. The Regime Theory has served as a beginning in some designs, although this theory is better adapted to conditions in India and Pakistan than elsewhere. It may be adapted to many conditions by alteration of coefficients and exponents as indicated in Chapter II, Section F.

Conditions at Point of Diversion.—The sediment transported by a stream is often referred to as suspended load and bed load. The suspended load usually consists of the finer particles, e.g., silt and clay. The bed load consists mainly of the coarser particles that move along the bed in dunes or by saltation. Actually, the sediment in suspension often contains particles ranging in size from fine clay to coarse sand. The sand is usually found in greater quantity and larger sizes near the bed of the stream with both the size and concentration decreasing from bed to water surface. The very fine sediment particles, e.g., silt and clay, which settle according to Stokes Law, are normally distributed uniformly throughout the depth of the flow. Fine sand may also be uniformly distributed through the flow prism. In highly turbulent flow, the coarser fraction that normally moves along the bed, is swept into suspension, creating a relatively dense suspension near the stream bed. This is the means by which much of the coarser sediment is carried from the stream into the canal. Fig. 5.20 is a graph prepared by Bondurant (1953) showing measured vertical distribution of sand size sediment and velocity in the Missouri River near Omaha, Neb., on October 18, 1951. It shows that the concentration of the larger sand sizes increase as the bed is approached, and this proportionate increase is

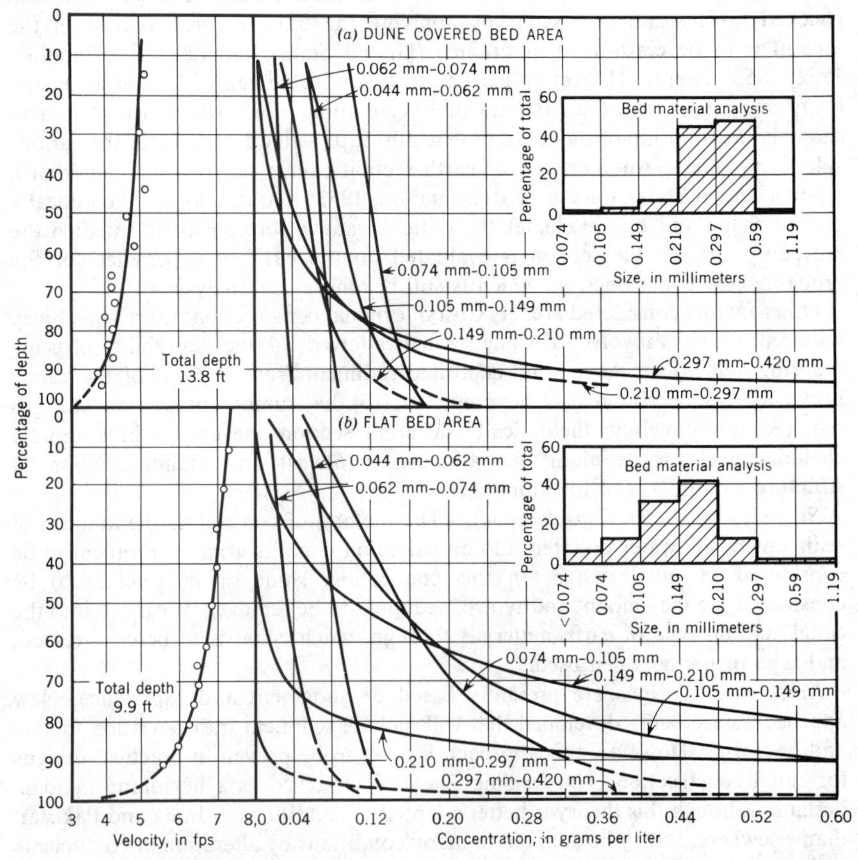

FIG. 5.20.—Distribution of Suspended Sediment and Velocity, Missouri River at Omaha, Neb.; Bondurant (1953)

greater as size of the particles increase. It also shows that there is a high correlation between the predominant sizes of the sands in the bed and in suspension.

The distribution and size of sediment particles in the vertical are important to the designer because most sediment diverters are designed to take advantage of the concentration of the heavier particles in the lower part of the water prism. These devices are designed to intercept and prevent water with this large concentration of sediment from entering the canal headworks.

Selection of Point of Diversion.—Careful selection of the point where the water is to be diverted from a stream is an important factor in the reduction of the quantity of sediment taken into a canal. In general, the outside or concave side of a curve has proven to be the best location (Joglekar, et al. 1951). This is true, because the heavy bed load is swept towards the inside of the curve and the sediment concentration at this point is lower than at other points in the stream. This effect is due to spiral flow, which was first explained by Thompson (1876) and is expressed by a number of authors, e.g., Ippen and Drinker (1962). This type of flow is shown in Fig. 5.21.

The helicoidal flow sweeps the bed load to the inside of the curve and forms point bars. This helicoidal flow is easily observed in the flow around a curve. Thus, it is this action that makes the outside of a stream curve the best place to divert water into a canal headworks. As will be shown later, this same flow condition can be included in the design of a canal diversion works as a part of the sediment exclusion device. It is a flow principle, sometimes referred to as principle of bed load sweep, that has been well known and widely used by irrigators of many countries for a long time.

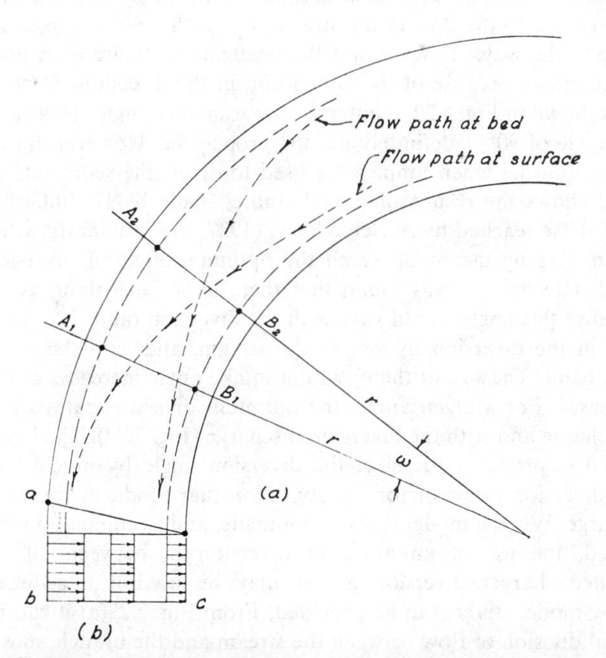

FIG. 5.21.—Schematic Diagram of Flow in Curved Channel

The principle of bed load sweep was very dramatically employed by Sir Claude Inglis (1949) in his recommendation of a curved approach to the Sukkur Diversion Dam on the Indus River in Sind, Pakistan. The result is that the bed sediment is deflected away from the canal headings and diverted over the dam. The principle is also used by the USBR at a number of its diversion sites, some of which are described in the following paragraphs.

Angle of Diversion.—The angle of deflection between the direction of flow in the parent channel and the direction of flow in the diversion channel is generally called the "Angle of Diversion." Egyptian engineers (Abdel, 1949) who studied the effect of the angle of diversion as a factor in the symmetry of flow at a diversion works, call it the "Angle of Twist," and attach considerable importance to its effect on the amount of sediment directed into the diversion channel. This conclusion is well justified and has been completely confirmed by a number of investigations.

From the preceding review relative to selection of point of diversion, it is easily seen that any diversion at an angle with the flow in the parent stream channel becomes in effect a curve with curvature opposite to that of the parent channel. The higher velocity surface water requires a greater force to turn it than the slower moving water near the bed. Consequently, the surface water tends, by its higher momentum, to continue with the parent stream; while the slower moving water near the bed, that carries the greater concentration of sediment, tends to flow into the diversion channel. Therefore, the diversion channel receives the sweep of the bed load, which flows from the outside to the inside of a curve, because for any angle of diversion the diversion takeoff is, in effect, on the inside of the curve created by the diversion.

Many canal diversions have been designed with the direction of flow into the headgates normal to the flow in the stream, i.e., with a 90° angle of diversion. In such a design the water flowing into the headgate structure does not follow the walls of the intake because of its momentum in the direction of the streamflow. Instead, as shown in Fig. 5.22, it enters at some smaller angle. This indicates that a diversion angle of 90° is definitely not the proper one. However, the use of a 90° angle is not unusual when tunnels are used to divert the sediment.

Fig. 5.22 shows the results of model studies made by H. Bulle (1926). These results and those reached by A. Schoklitsch (1937) independently, attempt to give some parameters by means of which the optimum angle of diversion could be determined. However, it was found that there is no such thing as an optimum angle, because this angle would vary with the diversion ratio; i.e., the ratio of the discharges in the diversion to that in the stream, and with the position of the intake in a bend. The size of the optimum intake angle increases as the diversion ratio decreases. For a given angle, the optimum diversion ratio varies with the stream discharge and with the diversion discharge (Fig. 5.22). The best solution to the problem is probably to select the diversion angle by model study for the dominant diversion ratio; or for the condition that produces the maximum bed load discharge. When a model study is not made, and a sediment diverter is not to be provided, the use of an angle of diversion of between 30° and 45° is recommended. Larger diversion angles may be used if a sediment diverter designed by model study is to be provided. From Fig. 5.22(a) it can be seen that for an equal division of flow between the stream and the branch, most of the bed load goes into the branch. Fig. 5.22(b) is a curve on which various divisions of flow between the stream and branch are plotted against the amount of bed load entering each for a constant angle of diversion of 30°. From the example plotted, it can be seen that for a diversion of only about 25% of the streamflow, as much as 50% of the bed load will enter the branch. Experiments conducted at the Poona Research Station, Bombay, India (Ingliss, 1949) confirm these findings.

17. Sediment Diverters

General.—A sediment diverter is a device or structure arrangement at a canal headworks that is designed to prevent the greater part of the stream sediment from entering the canal. The diversion and canal headworks structures for diverting water from a sediment-laden stream can be located and arranged so that only a

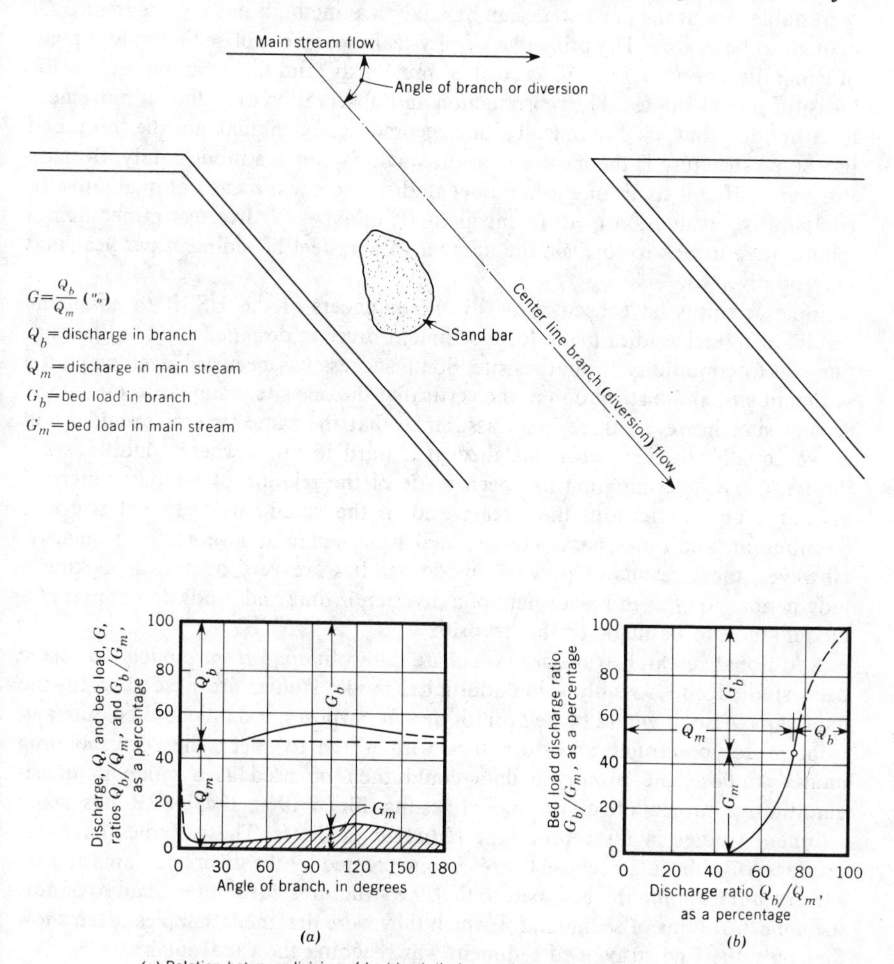

(a) Relation between division of bed load discharge and angle of branch. Subscripts *b* and *m* apply to branch and main channel, respectively (Schoklitsch, 1937).

(b) Relation between divisions of bed load and division of discharge. Plotted example for angle of branch of 30° as determined by H. Bulle (1926) and reported by Schoklitsch (1937).

FIG. 5.22.—Relation between Division of Bed Load Discharges, Division of Discharge, and Angle of Branch as Determined Experimentally by Bulle (1926)

small part of the bed load will be taken into the canal headgates. Historically, many ingenious arrangements and devices have been used to obtain this result. Some are based on theoretical approaches or model studies, or both; others are the result of many years of trial and error construction and operation of diversion structures for specific canal systems. Streams with most of the sediment load in suspension require settling basins in the system to remove it. These settling basins are designed to settle out and trap the sediment at predetermined locations along the canal where it can be removed expeditiously. Each type of device must be evaluated relative to the particular conditions of use.

Use of Hydraulic Models.—The most reliable way to design a diversion structure that includes a device to divert the sediment from the canal is by model study. It is regrettable, but at the present, design by analytical methods has not progressed far enough to be usable. The principles of physical movement of sediment at a point of canal diversion are known, as covered previously, and these can be used as the basis for model studies. The reproduction and observation of sediment movement in a model that is dynamically and geometrically similar to the proposed prototype structure is the most important single factor in a model study. Because it is very difficult to accomplish, model studies become a means of qualitative or comparative evaluation of different methods or types of diverters rather than a quantitative means to forecast the number of acre-feet of sediment per year that may be diverted.

Some attempts have been made by the engineers of the USBR to adapt the results of model studies made for a sediment diverter designed for conditions at one site to conditions at another site. Some success has been achieved when the sediment size and distribution in the vertical at the one site is similar to that at the second site; however, there is no assurance that the same type of excluder will prove equally efficient when installed at a third location where conditions are different. No determination has been made of the relation of the parameters of discharge and velocity in the stream and in the canal, the sediment size and distribution, and other parameters related to dynamic and geometric similarity. However, these results of previous model studies are very helpful in making a judgment as to type and placement of a diverter, if time and funds do not permit a model study to be made of the new site.

Additional research is needed to collate data obtained from models that have been studied and possibly some additional model studies are needed to fill the data gaps. Also, it would be helpful for design purposes, if data could be obtained from prototype structure performance with which to check the corresponding model studies. The prototype data could then be used as a guide in quasi-quantitative projection of the model results. Since 1960, the USBR has made sediment studies in three prototype sediment diverters. These studies have not produced usable data because the reservoirs created by the diversion dams at each site are still trapping the bed load so that the structures have not yet had to handle the concentrations of sediment for which they were designed. Samples taken show that only the fine suspended sediment was reaching the canal intake.

Training Walls and Banks.—As an example of the use of training walls to remove sediment, Fig. 5.23 shows the general plan and arrangement of the Woodston Diversion Dam on the South Fork, Solomon River, Pick-Sloan Missouri Basin Program, USBR. Note (Plan and Section B-B) that the inlet to the canal headworks consists of a 5-ft wide channel protected by a 10-ft high curved training wall. Section B-B shows the canal intake grade set well above the floor of curved intake channel. The curved walls of the intake channel create a curve in the flow artificially in which the helicoidal currents sweep the bed load to the

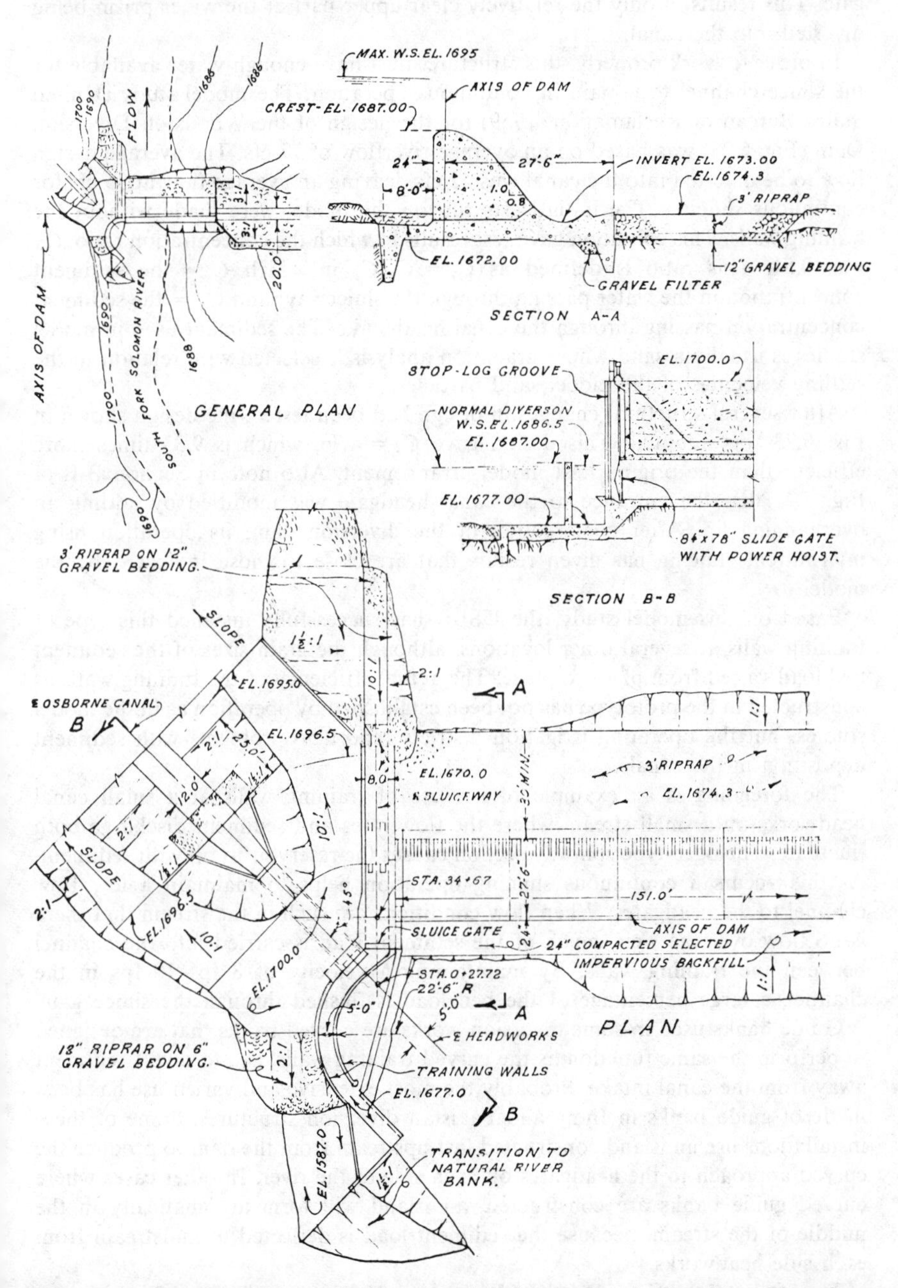

FIG. 5.23.—Traning Walls at Headworks of Osborne Canal, Woodston Diversion Dam, Pick-Sloan Missouri Basin Program, Kansas, USBR

inside of the curve and away from the headgates. Also, the velocity of flow in the channel between the two walls sweeps the bed load entering that channel by the canal headgate at an elevation below the intake sill and out through the sluice gate. This results in only the relatively clear upper part of the water prism being diverted into the canal.

In order to work properly, this structure must have enough water available for the sluice channel to remain in continuous operation. The model study (United States Bureau of Reclamation, 1959) for the design of the Woodston Diversion Dam (Fig. 5.23) was based on an average riverflow of 77 cfs. The average design flow to be diverted into the canal was 42 cfs, leaving an average flow at 35 cfs for continuous sluicing. The initial tests run on the model were made without the training walls. This condition gave test results in which the concentration ratio, C_r, was 0.51. This ratio is defined as $C_r = C_s/C_h$ in which C_s = the sediment concentration in the water passing through the sluiceway; and C_h = the sediment concentration passing through the canal headworks. The sediment used in model studies is usually a sand whose gradation analysis is selected with relation to the settling velocities of the model sand particles.

After several structural changes in design had been tested, the design shown in Fig. 5.23 was adopted. This design gave C_r = 4.76, which is 9.33 times more efficient than the original test model arrangement. Also note in Section B-B of Fig. 5.23 that the entrance to the canal headgate was modified by adding an overhanging lip. After construction of the diversion dam, its operation using intermittent sluicing has given results that are close to those indicated by the model.

Based on this model study, the USBR has successfully installed this type of training walls at several other locations, although the grain sizes of the sediment bed load varied from place to place. The actual efficiency of the training walls as constructed in the prototype has not been established by operation as in the model studies, but the operating irrigation districts report no problems with sediment deposition in the canals.

The foregoing is an example of the use of training walls at a small canal headworks on a small stream where the flow rates and sediment discharge both fluctuate. The ogee-type spillway section causes the reservoir to fill with sediment. As this occurs a continuous sluicing operation helps to maintain a low flow channel to the headgates. When flow conditions are such in the stream that there is no flow over the weir, much of the sediment load is carried into the channel between the training walls. By maintaining a velocity of 8 fps–10 fps in the channel, a large percentage of the bed load is flushed through the sluice gate.

Guide banks used in some diversion works are curved banks that are designed to perform the same function as the curved training walls, i.e., to divert sediment away from the canal intake. Probably the most extensive and varied use has been made of guide banks in India and Pakistan diversion structures. Some of these installations use an island constructed just upstream from the dam to produce the curved approach to the headgates on each side of the river. In other cases where curved guide banks are constructed, an island will form automatically in the middle of the stream, because the sediment load is deflected to midstream from each side headworks.

The model of the Kotri Dam (Joglekar, et al. 1951) shown in Fig. 5.24 shows the use of guide banks and a central island. The dam was designed in two parts, taking advantage of a natural island to provide a concave curvature on each side of the river. The arrangement was effective in excluding the bed load at the R. B. Feeder, and at the Pinyari Feeder Canal headings but was not satisfactory at the Fuleli Canal heading. This was because the latter heading was too far upstream to benefit from the favorable curvature induced by the guide banks. Operation of the dam gates to emphasize the curvature was tried in the model, but proved to be still ineffective. To improve the intake conditions to reduce sediment intake into the Fuleli Canal, the devices shown by Fig. 5.25 were installed on the model. These were: (1) Extension of the guide wall (divider wall); (2) variation of concave curvature of left guide bank; (3) alteration of the convex curvature of the island;

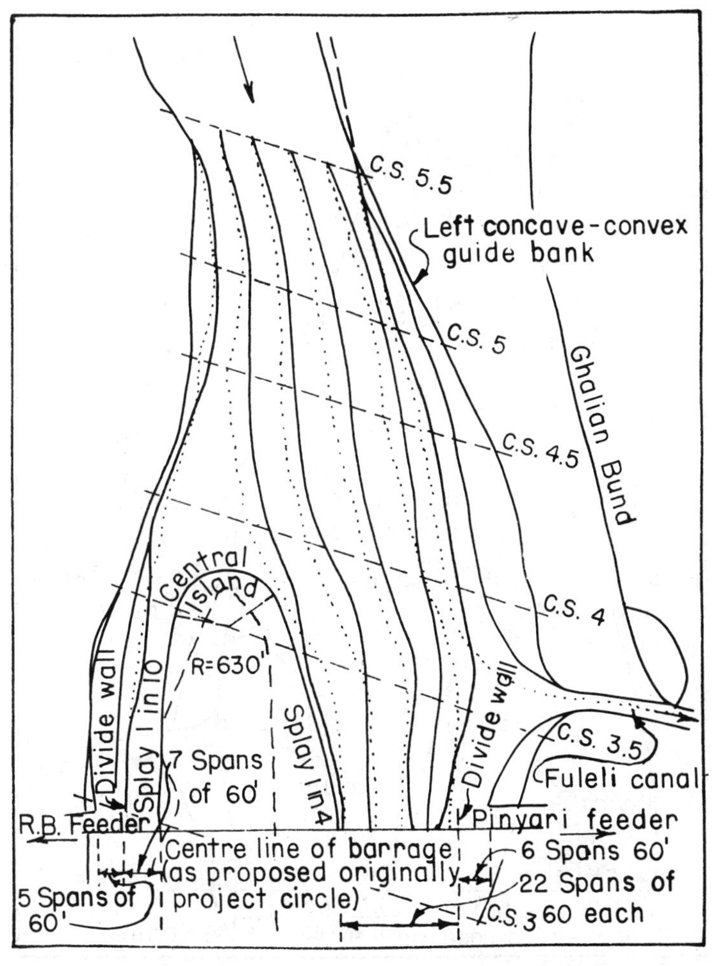

FIG. 5.24.—Guide Banks and Central Island in Model of Kotri Diversion Dam, Pakistan (Joglekar, et al., 1951)

and (4) submerged vanes of different length at various sites. However, none of these devices gave satisfactory results; thus, it was finally necessary to move the Fuleli Canal headworks near the downstream end of the curve, close to the headworks of the Pinyari Feeder.

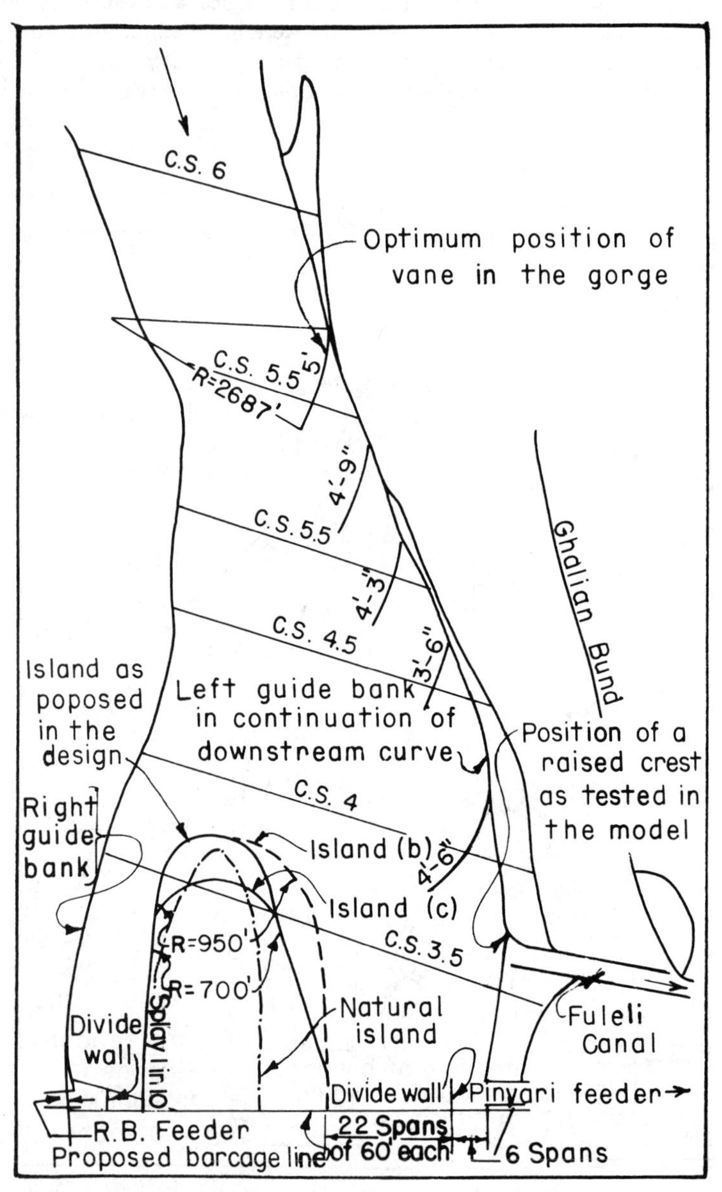

FIG. 5.25.—**Guide Banks and Central Island in Revised Model of Kotri Diversion Dam, Pakistan**

It is of interest to note that the results of the model study for Kotri Barrage (Ahmad, 1973) was not adopted completely for the design. Scouring turbulence required the construction of a protective groyne projecting upstream from the barrage. Other considerations dictated some structural variations from the model. The layout of the double diversion, as constructed, is shown in Fig. 5.26 (Ahmad, 1973). Additional data relative to the Pakistan experiments with headworks design for sediment exclusion from canals are presented by Ahmad (1951).

The model study for the Kotri Dam is referred to herein in detail because it shows that in some cases it is not enough to simply locate the point of diversion on the outside of a bend. It emphasizes that the point of diversion must be located downstream on the curve where the helicoidal pattern of flow, which produces the bed load sweep, has fully developed. This location will vary with the size and quantity of the bed load, the velocity of flow, and the geometry of the stream section. At the present it is not possible to predict what combinations of these parameters will produce satisfactory results. Consequently, the proper location of the point of diversion can best be determined by a model study. If it is not practical to construct a model, the point of diversion should be located at some

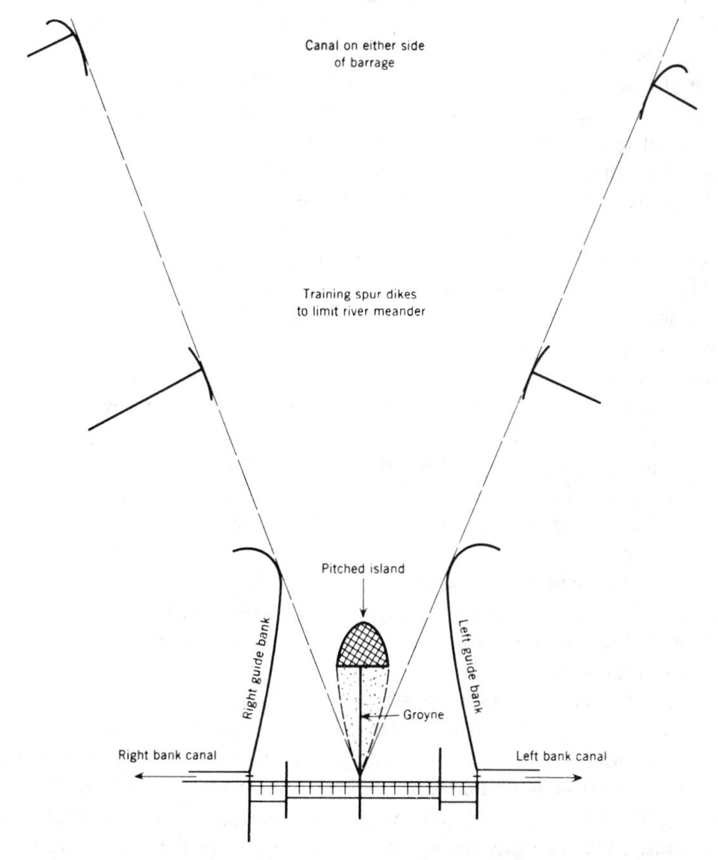

FIG. 5.26.—Layout of Kotri Diversion Dam (Pakistan) as Constructed

point located from two-thirds to three-fourths of the length of the curve from the beginning of curvature to be sure the secondary transverse currents will develop, and become fully effective upstream of the canal headworks structure.

Pocket and Divider Wall.—This structure arrangement (Uppal, 1951a) to reduce the amount of sediment taken into a canal headgate is produced by constructing a divider wall upstream from a dam so as to form a pocket in front of the canal intake (see Fig. 5.27). The wall divides the streamflow as it approaches the dam, so that part of the flow is directed to the sluiceway and part through the dam, where the dam is a gated structure. If the dam is an overflow weir, the main function of the divider wall is to form a sluicing pocket upstream from the sluice gate to produce a ponding area of low velocity in which the sediment will deposit rather than enter the canal headworks. The sediment thus deposited in the pocket, is subsequently scoured out by turbulent flow through the pocket when the sluice gates are opened.

The sluicing of sediment out of the pocket can be accomplished by the still pond method, which is intermittent sluicing, or by continuous sluicing if an adequate amount of streamflow is available. Where the still pond method is used, it is usually advantageous to close the canal headgates during the sluicing period. This is especially true where the canal is diverting a large percentage of the streamflow, because the turbulent flow conditions produced by sluicing puts most of the sediment into suspension.

Where the diversion dam is gated, the amount of opening of the dam gates can greatly affect the amount of sediment taken into the canal headworks. The dam gates most distant from the point of canal diversion should be opened the greatest amount; the gates nearest the pocket holding the sediment should be opened the least. Thus the increased discharge through the far gates creates a curvature effect that pulls the sediment away from the canal headgates. At structure installations where there is a canal headworks at each end of the dam, the gates in the middle of the dam should be opened the greatest amount.

The amount of sediment excluded from the canal headworks is increased to some extent by keeping a favorable ratio of velocities on each side of the divider wall that forms the sluicing pocket. This can be done by regulation of the openings of the sluicegates and the dam gates adjacent to the divider wall. An understanding of the velocity ratio, V_r/V_p, in which V_r = velocity of flow on river side of the wall; and V_p = velocity of flow on the pocket side of the dividing wall, is needed however. Where the ratio, V_r/V_p, is greater than unity, the bedflow tends to be bent around the nose of the divider wall and the bed load is directed to the river side of the wall. Where V_r/V_p is equal to unity, there is about an equal division of the bed load between the river side and the pocket side of the wall. However, where V_r/V_p is less than unity, most of the bed load is drawn into the pocket, and the amount of sediment that may be directed into the canal headworks is increased.

In general, the length of a divider wall is determined by its purpose and the ratio of flow in the stream to that being diverted into the canal or canals. If the purpose of the divider wall is to create a curved path to the canal headgate in a structure where the streamflow is limited, and most of the excess flow is used for sluicing (see Fig. 5.23), then the length, and degree of curvature, should be determined by a model study. On the other hand, where the purpose of the divider wall is to act as a splitter, and water in the stream greatly exceeds the requirement for the canal diversion, there the divider wall need not extend upstream farther than the upstream canal headgate. Generally, a wall extending from the dam to a point about two-thirds the distance to the upstream headgate will suffice.

The optimum width of the pocket and its depth below the canal headgate can best be determined by a model study. However, in general the cross section must be large enough to carry water for the canal diversion and for sluicing, and small enough to maintain sluicing velocities for dominant conditions or adopted design conditions of flow in the stream. A pocket that converges slightly towards the sluice gates is preferable to a straight pocket, as it helps the scouring action as water is taken off to the headgates. A convergence greater than one in 10 may hinder the proper flow in the stream to the adjacent dam gates. Also, a wide entrance decreases the effectiveness of sluicing at the upstream end of the pocket.

The still elevation of the canal headgates should be set at least one-third of the depth of flow in the pocket above the floor; more depth is desirable if obtainable. This raised position will help to exclude the bed load so that only the top water will be taken into the headgates. It will not, however, effectively exclude the bed load over a long period of time. This is true because the bed load will gradually deposit below the gate sill elevation and form a ramp up which the bed load will roll into the canal headworks. However, it does provide enough depth in the pocket to contain the bed load for a short period of time so that with either continuous sluicing or intermittent sluicing, it becomes a very effective bed load control.

Sand Screens.—Sand screens, which are used extensively in Egypt and India, are actually low barricade walls. They are called skimming weirs in the United States, because they do skim the clearer water off the top of the streamflow and direct the bed load away from the canal diversion. These devices are provided to direct the bed load away from the headgates of the canal and back into the flow of the stream. The relatively sediment-free water in the upper part of the prism flows over the wall into the canal diversion headworks. The tops of these walls are set at an elevation that will effectively divert the sediment for the dominant, or selected design, condition of flow. To adapt the walls to changes in stage of the streamflow, flash boards are often provided on top of the walls.

The sand screen has the greatest efficiency where there is a bed load, that is, when most of the sand and other coarse sediment is carried as a bed load. For these screen walls to be effective there must be sufficient flow past the point of diversion to carry away the bed load that tends to deposit at the upstream face of the sand screen or skimming weir. Otherwise ramps will be formed by the deposits of sediment, and the current will be deflected up and over the wall, carrying the sand into the canal headworks. This is particularly true, if the rate of flow over the skimming weir or sand screen is very high. The proper limiting concentration of such flow in any specific case should be determined by trial, because it would be a function of the size of sediment, hydraulic conditions, and the related, but unpredictable conditions. In some cases a thin horizontal foil or overhang at the top of the wall extending a few feet upstream has been used to increase the efficiency of this type of diverter. Herein the passing flow must be sufficiently fast to divert the bed load past the diversion inlet to the sluice gates.

Guide Vanes.—Diversion structures with guide vanes are shown diagrammatically by Figs. 5.28 and 5.29. The guide vanes produce localized helicoidal flow patterns that are similar to those generated naturally in flow around a curve as described previously. These guide vanes are of two types:

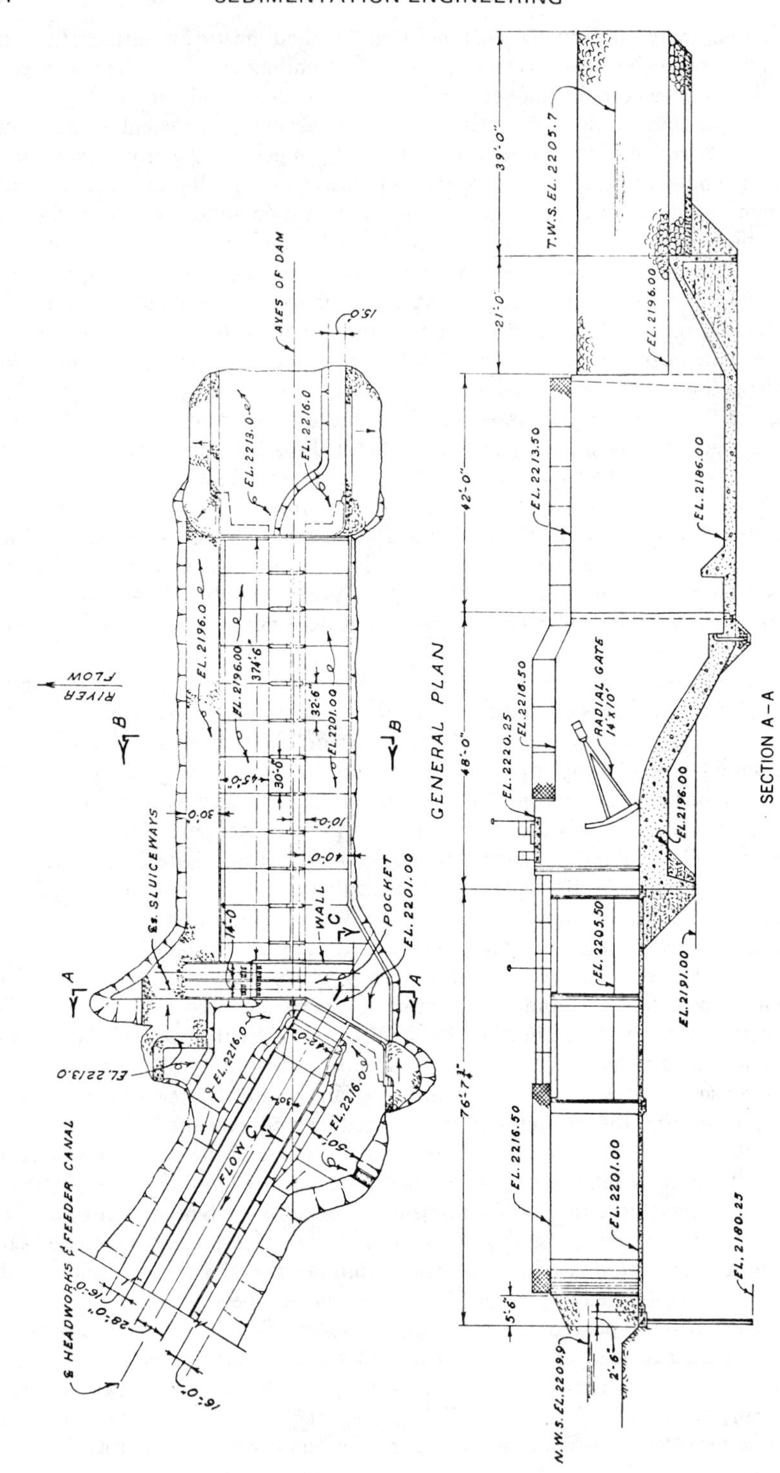

FIG. 5.27.—Schematic Diagrams of Pocket and Divider Wall at Canal Headworks

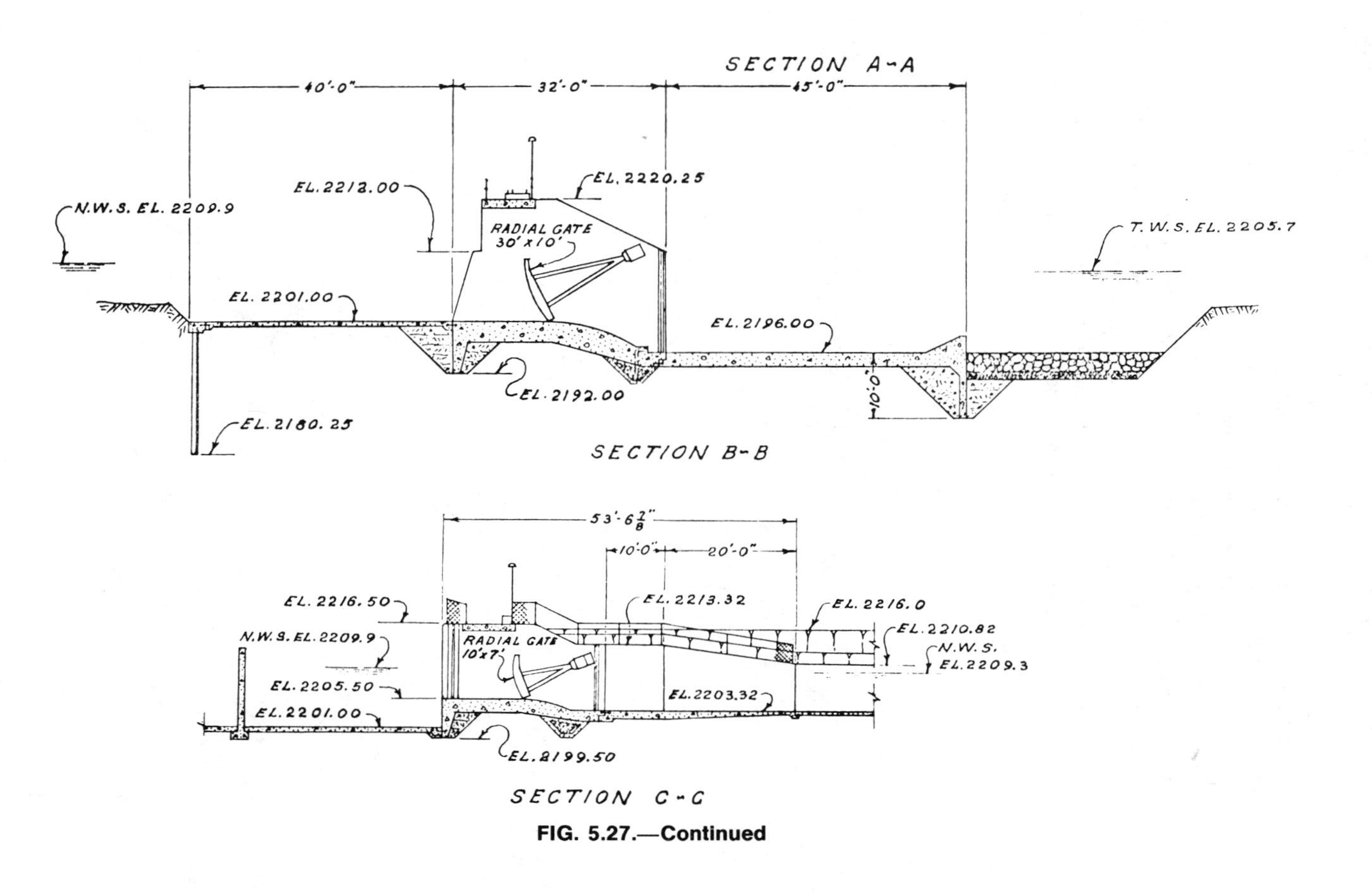

FIG. 5.27.—Continued

1. Bottom guide vanes placed so the bottom edge of the vanes are located at or near the stream bed (Fig. 5.28). The vanes then direct the direction of flow in the lower part of the stream prism.

2. Surface vanes that are generally supported by a raft arrangement from which the vanes project into the water from the surface far enough to influence the direction of flow of the surface water (Fig. 5.29).

As shown in Fig. 5.28 the submerged-type bottom vanes direct the water near the bottom away from the canal headgates. Because most of the heavy bed load is concentrated in this area of the flow, it is also deflected away from the canal intake. The surface vanes shown in Fig. 5.29 direct the surface water, which contains very little sediment toward the canal headgates. This induces a transverse flow of the bottom area water with its bed load away from the canal headgates.

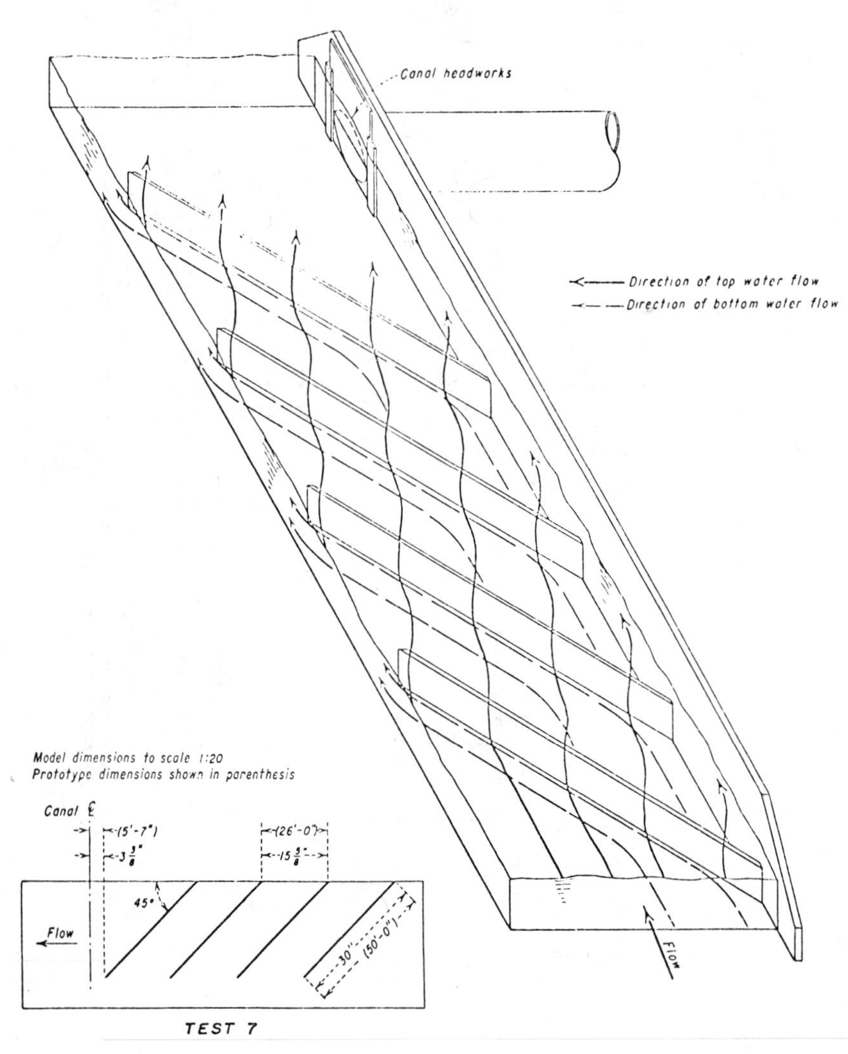

FIG. 5.28.—Bottom Guide Vanes—Hydraulic Model Study

Guide vanes have been studied in India (Central Board of Irrigation and Power, 1956), the United States (United States Bureau of Reclamation, 1962a; Carlson, 1963), and the USSR (Potapor and Pychkine, 1947).

Some early investigations of the use of vanes were made by King (1920, 1937) that are somewhat different from those discussed herein. His suggestions have been incorporated in structures in the canal system of the Indus Basin where they are reportedly working successfully.

A model study was made by the Bureau of Reclamation (1962a; Carlson, 1963) in 1962 of the probable effectiveness of the bottom and surface guide vanes shown in Fig. 5.28 and 5.29. This study is described because of the writers' familiarity with it, and also because it serves to illustrate the usefulness of guide vanes. Only

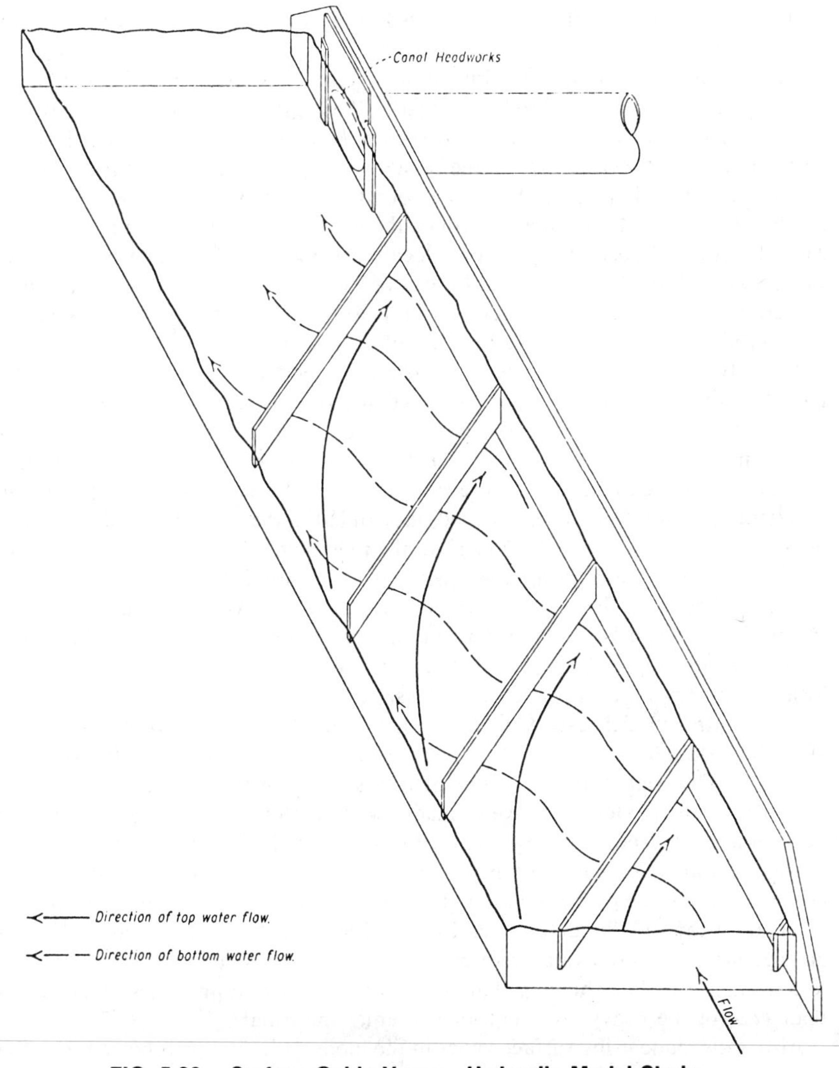

FIG. 5.29.—Surface Guide Vanes—Hydraulic Model Study

the essential findings are presented as developed for the San Acacia Diversion Dam, which is located on the Rio Grande, about 60 miles south of Albuquerque, N.M. These studies showed that although the guide vanes were not the best solution for this particular diversion dam, the potential usefulness of this model study at other locations has been demonstrated (Carlson, 1963).

The headworks of the Socorro Main Canal are located at the San Acacia Diversion Dam. The canal was originally constructed in 1936, with the headgate located in the position shown in Fig. 5.30(a). In 1957 and 1958, the inlet structure was moved about 255 ft upstream, as shown in Fig. 5.30(b) to provide space for the construction of the heading for a low flow river channel.

The low flow channel headgates were placed upstream from the old Socorro Main Canal headworks, as shown in Fig. 5.30(b) and downstream from the relocated headworks of that canal. After the canal headworks had been relocated, it was found that sediment accumulated in the canal prism, decreasing the capacity from 265 cfs to 35 cfs. This required a model study to determine what could be done to decrease the sediment being taken into the canal. (United States Bureau of Reclamation, 1962b; Carlson, 1963). It was recognized that the excessive conveyance of sediment into Socorro Main Canal was influenced by the close proximity of the relatively large capacity low flow channel headworks. Thus, it was desirable that some device be found that could be constructed at the headworks of the Main Canal to reduce the amount of sediment being diverted. The induction of secondary currents locally by the use of bottom and surface vanes was explored (United States Bureau of Reclamation, 1962b; Carlson, 1963).

Herein, use was made of the results of previous model tests which had been run to determine the most effective location of such vanes.

Actually these earlier Bureau of Reclamation model tests had been based on model studies made at the Moscow Academy of Science, USSR (Potapor and Pynchkine, 1947). The final arrangement used for the vane installation was as shown in Fig. 5.30(c) which was selected from the model study run for this location by the USBR. Tests were compared on the basis of the ratio of the sediment concentration in the diverted flow to that in the river. These showed that with a standard discharge of 8,760 cfs in the river, and 174 cfs being diverted into the Main Canal, the ratio obtained without vanes was 2.38. This was used as a datum to determine the improvement that resulted from various arrangements of the guide vanes (Carlson, 1963). These tests were made to determine the effects of vane spacing, angle, length, elevation or depth, cross section, number, and location in respect to the canal headgates.

Five control model tests were conducted without vanes, but with a 160-ft by 40-ft floor slab, 3 in. thick, with bottom elevation at approximately the riverbed elevation. This slab was later used as a base for the vanes [see Fig. 5.30(b)].

Thirty-seven model tests were made with bottom and surface vanes to determine a satisfactory arrangement with relation to the factors stated previously. It was demonstrated that within reasonable limits, these characteristics are not critical; but that the proper relation of these interrelated functions for any specific installation can only be determined by model tests. In this case the most efficient arrangement of bottom vanes reduced the sediment concentration ratio from 2.38 to less than 0.1, thus showing that the proper use of bottom vanes allowed less than 1/24 of the heavy sediment load to enter the canal.

Also, tests made using surface vanes in the model [Fig. 5.30(b)] showed they too are just about as effective as the bottom vanes in reducing the heavy bed load taken into the canal. However, a major drawback of surface vanes is their ability to trap and accumulate floating debris, which in turn can become a major maintenance problem, thus their use in this case would not be recommended. The results of these model studies show that either surface or bottom vanes are effective in reducing the heavy sediment taken into a canal headworks (Potapor

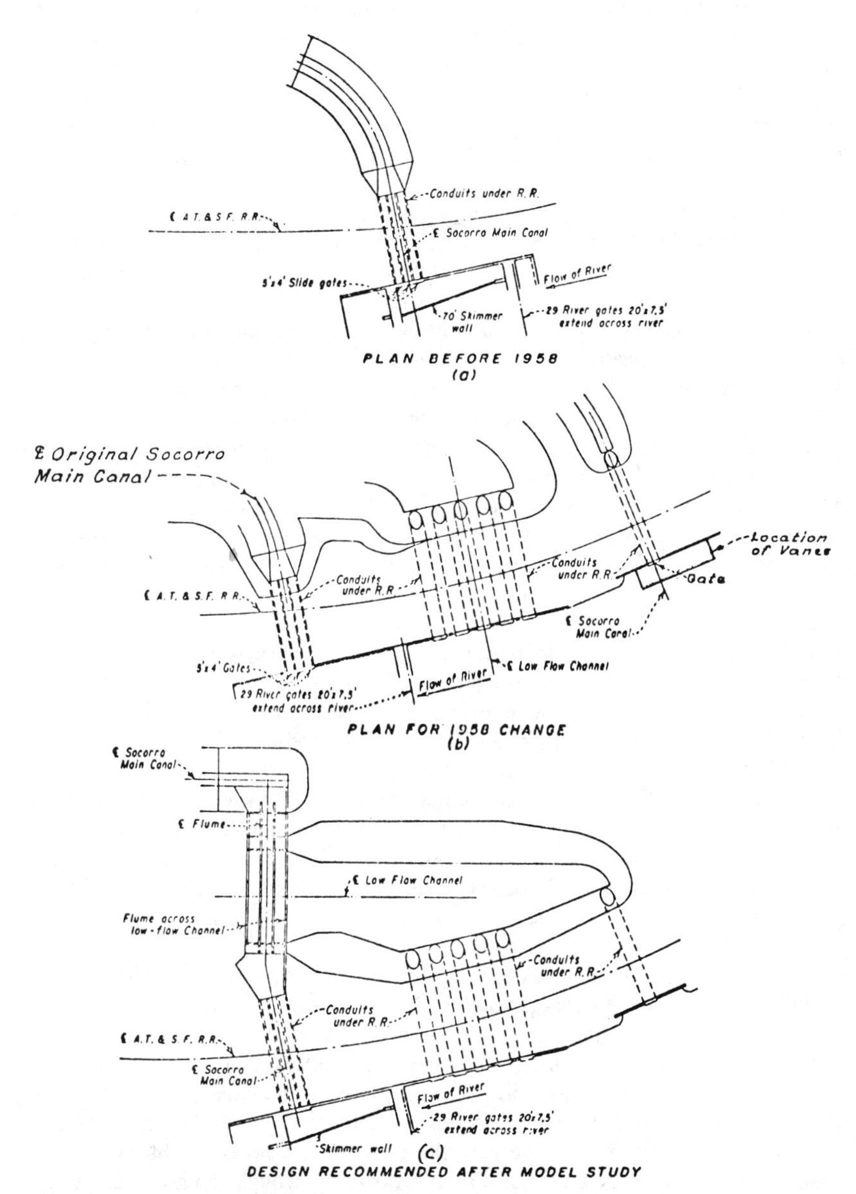

FIG. 5.30.—Use of Guide Vanes in Headworks Model of Main Socorro Canal—San Acacia Diversion Dam—Middle Rio Grande Project, New Mexico, USBR

and Pychkine, 1947; United States Bureau of Reclamation, 1962b). However, for other design considerations, the structure was modified as shown by Fig. 5.30(c).

Stream Inlets.—In general, stream inlets are drop inlet structures built in the bed of the stream. By nature of their construction they tend to trap sediment and cause serious maintenance problems unless some type of sluiceway is incorporated in the design. However, this type of structure is much less expensive to construct than diversion dams with specially arranged headgate structures. Consequently, stream inlet type structures are often built for small diversions in lieu of the higher priced diversion dams. Fig. 5.31 shows the layout for a stream inlet diversion structure that was designed with a sluice for removal of sediment. It will be noted from the figure that a slidegate controlled 12-in. diam pipe has been provided to remove accumulated sediment from the structure and to discharge it back into the stream below the diversion.

A steep stream gradient provides an opportunity to greatly improve the sluicing characteristics of a stream inlet because a short, high head sluice pipe can be used. Fig. 5.32 shows a stream inlet with a Dufour type sand sluice. The Dufour type sluice provides continuous sluicing and also increases the sluicing efficiency by the nature of the hoppers formed in the bottom of the sand catchment basin. During operation of the headworks the continuous sluicing makes removal of the sediment automatic. However, the structure should be inspected periodically.

Stream inlets are usually considered for construction of small canal diversion that are often on small mountain streams in relatively inaccessible places, or at least in places that are inconvenient to reach for manual operation or maintenance, or both. For these reasons care should be used to eliminate any features in the design that could cause brush and debris to be caught in the diversion structure. The trashrack is important in this respect; it should slope in the direction of streamflow at a grade of about 1 in./ft more than the drop of the stream bed.

The design of a stream inlet can be simple or very sophisticated. A good example of the latter was presented by Raynaud (Fig. 5.33) for a desilting structure, the Torrent Du Longon, located in the French Alps. Two sketches taken from Raynaud's paper (Figs. 5.33, 5.34) are reproduced herein to show the ingenious automatic features they contain.

Fig. 5.33 shows the general arrangement and Fig. 5.34 shows the details of the automatic desilting (sluicing) features. These figures are explained by Raynaud as follows: "The normal water level in the desilting basin is N. The desilting gate, C, is formed by two coaxial cylinders C_1 and C_2 of ϕ_1 and ϕ_2 diameters ..." (Fig. 5.34). The operation of this desilting gate is described as follows:

> The position of the ball in desilting gate C is adjusted so that the rate of outflow from the cylinders through orifice O is less than the inflow through the countersunk holes. Thus, when there is no sediment around the foot of the desilting gate, the holes are large enough that water flows *into* the gate faster than it flows out at orifice O; as a result, the water level inside and outside the gate remains the same and the gate stays closed. When sediment collects about the foot of the gate the countersunk holes are blocked and water flows *out* of the gate faster than it flows in through the holes, creating a buoyant force which lifts the gate, allowing the sediment to be sluiced out

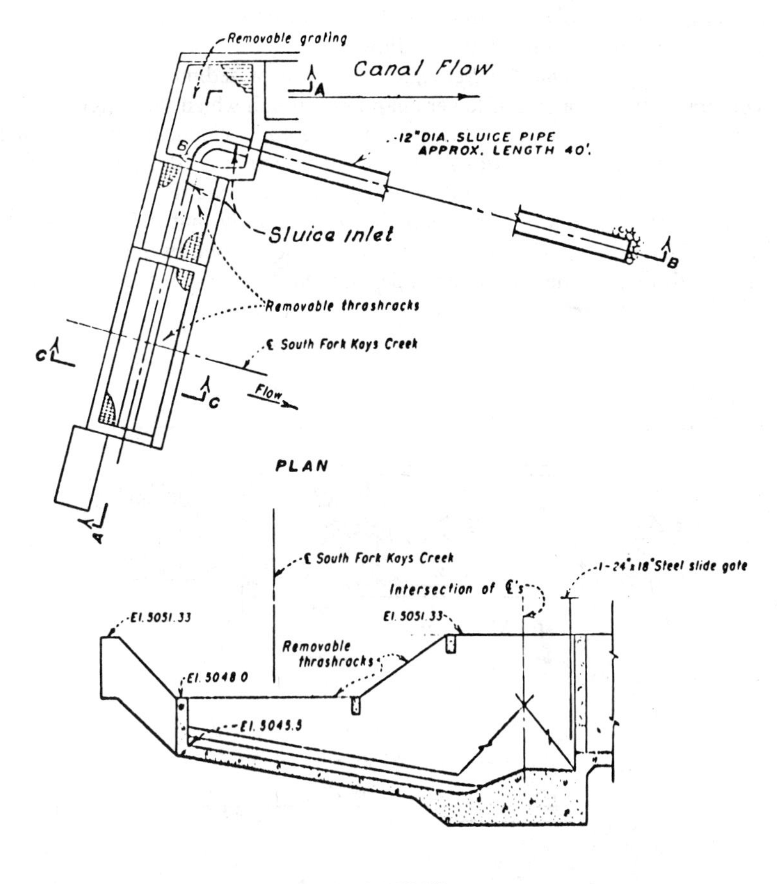

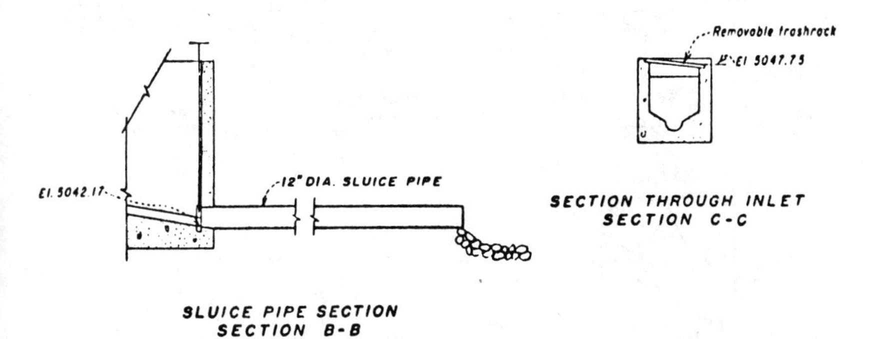

FIG. 5.31.—Stream Inlet Diversion Structure from South Fork Kays Creek Inlet—Weber Basin Project, Utah

through orifice O. When the countersunk holes are cleared by the sluicing action, the cylinder fills again and the gate settles its conical point into the orifice again to shut off the outflow.

Tunnel Type Sediment Diverters.—This type of sediment diverter is basically composed of an upper and lower chamber through which the water to be diverted is routed. The clearer water flows through the upper chamber into the canal, while the sediment laden water flows into the lower channel and is passed back into the stream. This type of sediment diverter has proven very efficient for both small and large canal diversions. It has been shown previously that the water in a flowing stream has more of the bed load sediment concentrated in the lower one-third of the depth than in the middle or upper one-third of the flowing prism. The coarse

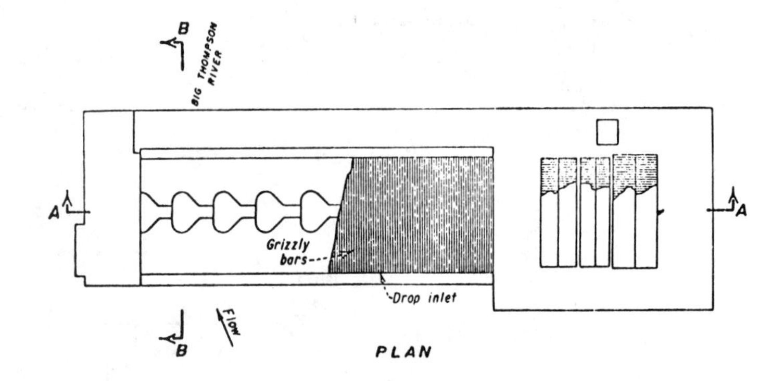

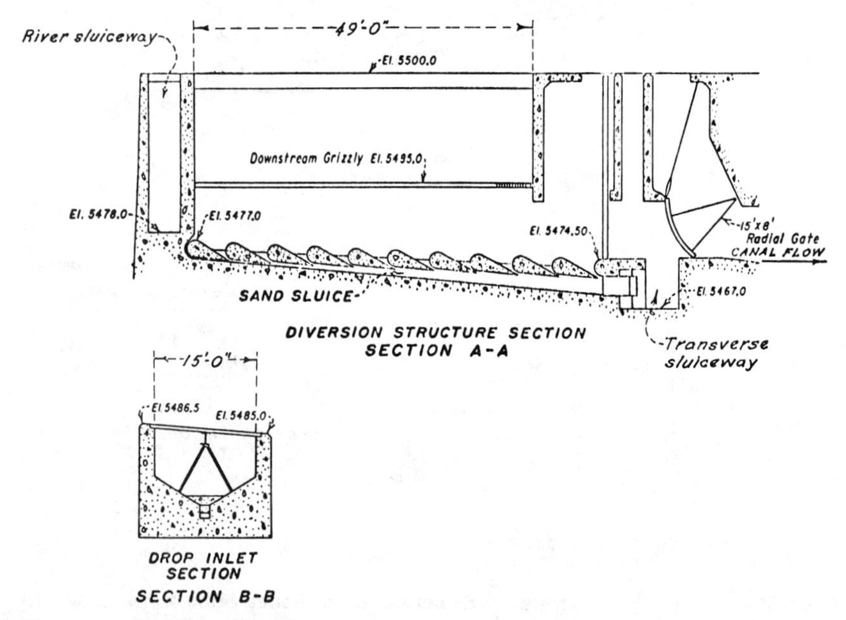

FIG. 5.32.—Stream Inlet with Dufour Type Sand Sluice—Horsetooth Supply Conduit—Colorado-Big Thompson Project, Colorado

sediment is almost entirely in the bed load. Thus, if the flow can be split horizontally so that the lower portion will pass through the sluiceway and the upper portion be diverted through the canal headgate into the canal, the sediment taken into the canal would be greatly decreased.

The idea of a tunnel-type sediment diverter was originally proposed by Elsden (1922). He proposed to use a closed-type chamber or conduit and divide the flow by using a horizontal diaphragm. The relatively clear water passed through the upper channel in the canal and the sediment-laden water flowed through the bottom channel, called the sediment diverter, back into the stream. However, it was not until 1934 that H. W. Nicholson built the first tunnel-type diverter at the headworks of the Lower Chenab Canal at Khanki, India (Joglekar, 1959). Many other tunnel-type diverters have since been built in India and Pakistan, all of them employing the same principles. In these structures, the lower layer of water must be separated from the rest of the flow without creating eddies that would throw the bed load into suspension. Research conducted at the Punjab Irrigation Research Institute (Joglekar, 1959) showed that for efficient operation, the discharge of the lower channel of the diverter should be at least 15%–20% of the canal diversion. If the canal diversion is small this percentage may need to be increased so that the velocity in the tunnels can be maintained at about 10 fps. Uppal's (1951) work with sediment diverters indicates that the efficiency may vary widely with discharge through the tunnels. It is advisable, if time and funds permit, to make a model study to determine the best operating conditions.

The openings or entrances for water into tunnel diverters should be limited to the upstream end of the tunnel. Some attempts have been made to provide entrances for water along the sides of the tunnels, but this type of intake has not proven to be successful. The tunnel section at the entrance should be bellmouthed and the hydraulic structure design of the tunnels should be such that at the design rate of flow, they run full. A sluicing velocity of about 10 fps should be maintained throughout the tunnel.

The tunnel section will not clog, because the high velocity maintained throughout its length will move the bed load particles through with no chance of settling out. As the sediment moves along the stream bed approaching the tunnel,

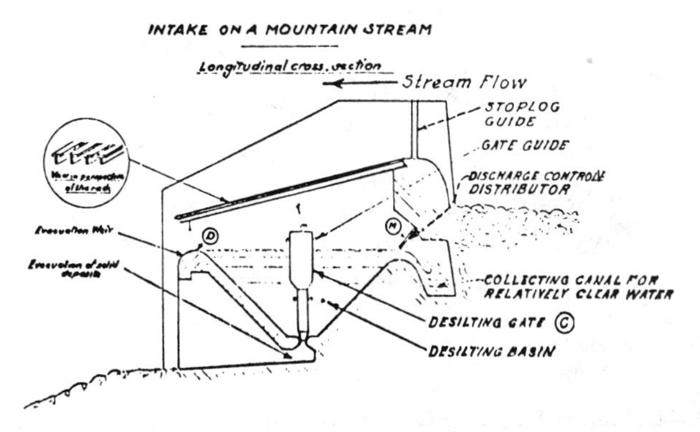

FIG. 5.33.—Stream Inlet—Intake on Mountain Stream—Torrent Du Longon, France (Raynaud, 1951)

it will form a ramp that slopes downward toward the tunnel which keeps the tunnel entrance open. The water velocity in the tunnel maintains the ramp slope above the tunnel inlet. This velocity effect extends only a short distance (a few feet) upstream from the tunnel entrance and is effective on either side of the

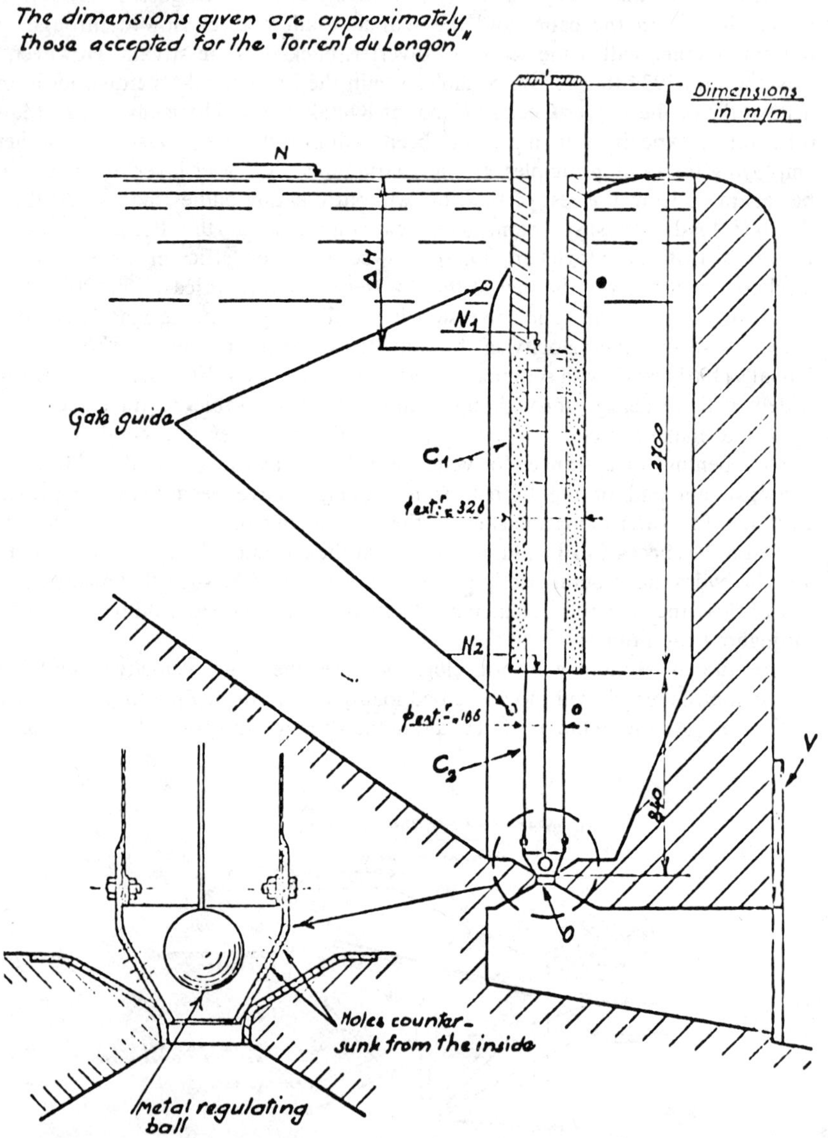

FIG. 5.34.—Details of Desilting Gate on Stream Inlet—Torrent Du Longon, France (Raynaud, 1951)

entrance for a distance about equal to the width of the tunnel. (The effect of a given velocity is dependent on size of the sediment particles. For the smaller size sediment particles the lateral effect of the velocity may be twice the width of the tunnel entrance or more.)

One of the most important items to remember about tunnel-type sediment diverters is that the outfall or discharge channel downstream from the outlet of the lower channel of the diverter must be kept clear so there is a free exit for the sediment-laden water. The addition of a specially constructed channel may be necessary that is separate from the stream channel, to provide a free exit for the sediment-laden water. Fig. 5.35 is a reproduction of pictures presented by Uppal (1951b) to illustrate the size and shape of a tunnel diverter for a large diversion such as many of those in India and Pakistan that have been the subject of much research by Uppal, Jogleka, and others.

All tunnel-type sediment diverters must be constructed in a very rugged manner because they are subjected to much buffeting by boulders, floating trees, logs, and other debris in the water. Often, because they are covered, the damage due to the impacting of these items against the structure is not evident until the diverter is unwatered.

An efficient tunnel-type diverter was developed by hydraulic model study by the USBR Hydraulic Laboratory (United States Bureau of Reclamation, 1954) for use at the Milburn Diversion Dam that is located on the Middle Loup River near Milburn, Neb. It was constructed in 1955. A similar design was made for the Arcadia Diversion Dam (USBR) that is also located at Middle Loup River and was constructed in 1961. The general plan and arrangement of the Milburn Diversion Dam is shown in Fig. 5.36.

The Milburn Diversion Dam is not a large structure, and the sediment diverter is a special adaptation of the tunnel-type diverter. As shown in Fig. 5.36 a horizontal shelf separates the bedflow and diverts it into the sluice gate back into the river. Note that the horizontal shelf does not form a closed conduit but that it is a very short, open-sided compartment that directs the sediment flow immediately into the sluiceway.

In the model study of this structure, it was found that some sediment accumulated on top of the shelf. For this reason the sluice gate was designed to open wide enough to sluice this sediment off the top of the horizontal shelf as well as removing the bed load passing through the compartment below the shelf. The operation of the Milburn Diversion Dam diverter has been very successful in bypassing the sediment and preventing its entry into the Sargent Canal. Also, the similar diverter at the Arcadia Diversion Dam has proven to be very efficient.

It would be very difficult to establish exact criteria for the design of these tunnel (shelf) type diverters. Hydraulic properties of the stream, type and amount of the bed load, physical conditions at the proposed structure site, and many other factors are different at each location. Thus a model study has proven to be the most dependable method of reaching a satisfactory solution to the problems presented by design of this type of diverter. The successful operation of the two-shelf-type diverters described previously show that this type of structure is practical and effective for small canal headwork diversions on the order of a flow range of 250 cfs to about 850 cfs, which are located on streams with a flow range of 3,300 cfs–20,000 cfs. The application of this type of structure to larger canal diversions on larger streams has not been proven. It would seem likely that for the larger canals with wide headgate structures, a closed tunnel-type of sediment diverter would probably work out to be the best.

18. Sediment Ejectors

General.—Sediment ejectors employ the same general principles of sediment removal as the diverters described previously except they are designed to remove sediment from the canal prism and are located downstream of the canal headgates. Regardless of the type of diverter employed in front of the canal headworks, some sediment will enter the canal prism. Diversion of a major part of the bed load back into the stream is all that can be expected of a diverter of any

FIG. 5.35.—Tunnel Type Sediment Diverter at Western Jumna Canal Headworks, India

type. Thus, some material, often that held in suspension due to eddies formed upstream from the headgate, will eventually be deposited in the canal prism. Some of this sediment may be gravel, some may be sand, and some may be fine clay. Very fine sediment particles (as clay) are detrimental to certain crops and should be removed from water that is used to irrigate such crops. This can be accomplished by a settling basin and some type of ejector. A number of different types of ejector devices have been developed, each designed for a particular combination of local conditions. Most of these structures have worked satisfactorily. Some of the more widely used types are described in the following paragraphs.

Tunnel Type Ejector.—The most commonly used type of ejector in the large canals of India and Pakistan is the tunnel type. This structure is very similar to the tunnel type diverter, previously described, except that it is located in the canal

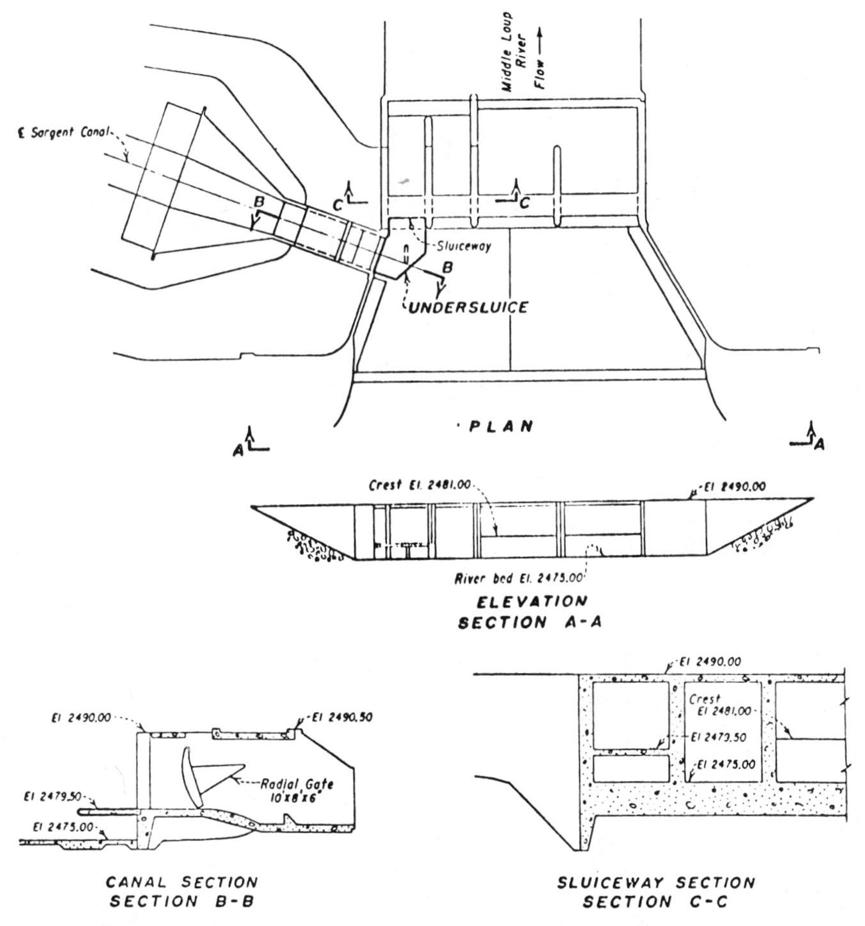

FIG. 5.36.—Undersluice Tunnel Type Sediment Diverter at Sargent Canal Headworks—Milburn Diversion Dam—Pick-Sloan Missouri Basin Program, Nebraska

prism to intercept and remove the residual canal bed load. Fig. 5.37 is a sketch of one tunnel type ejector that has been redrawn from one presented by Joglekar, Gholankar, and Kulkarnl (1951). As shown in the figure, the width of the canal is divided into a number of main tunnels. These tunnels curve to the right and converge to the point where they pass through the bank. At this point gates are provided to regulate the discharge. Each of the main tunnels may contain partitions that subdivide the cross section into smaller tunnels. These divider walls extend to the point where the structure passes through the bank. The height of the tunnel at its mouth is generally made 20%–25% of the canal water depth. The roof of the tunnel usually extends 1.5 ft–2 ft upstream of the entrance that helps to guide the sediment into the tunnels if there are local eddies present that cause the larger sized particles to lift off the bottom. Generally, a minimum head differential of at least 2.5 ft is required to operate the ejector. However, a greater head differential provides better efficiency but requires a higher flow rate. A velocity of

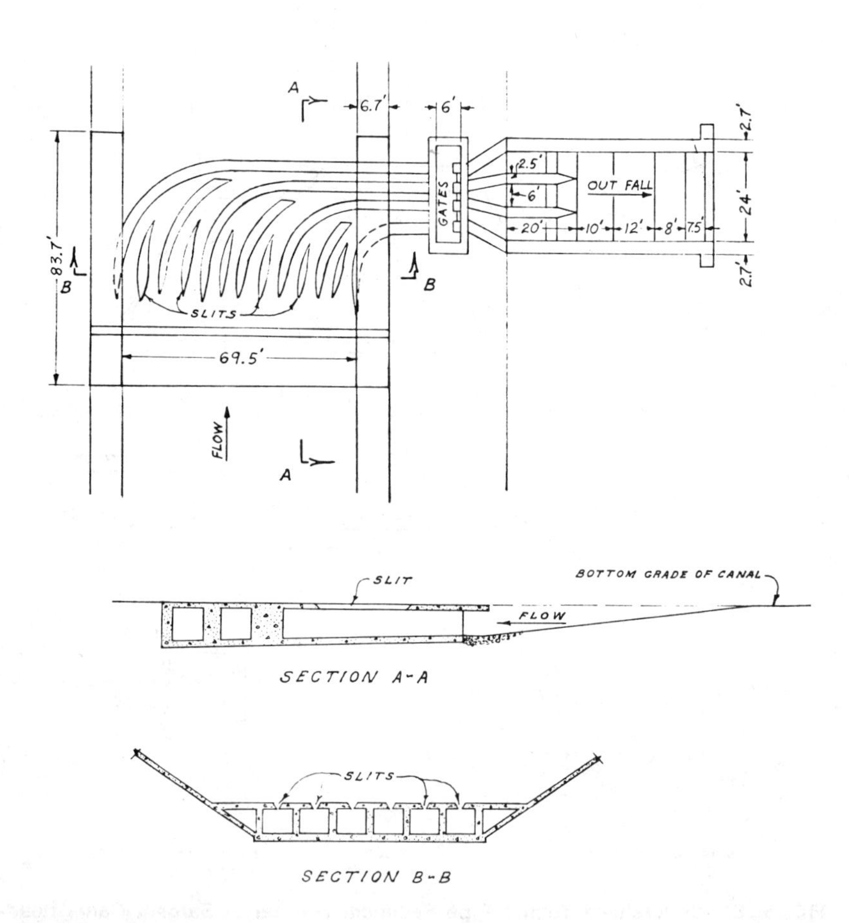

FIG. 5.37.—Tunnel Type Sediment Ejector at Salampur in Ubd Canal, India (Joglekar, et al., 1951)

8 fps–10 fps through the ejector will be adequate to move the sand size sediment through the tunnels. The amount of water diverted into the canal headworks must be increased 20% to 25% to operate an ejector of this type. In very large canals more than one ejector may be required, which will increase proportionally the amount of water needed to operate the additional ejector. The requirement to divert additional water from the river causes additional sediment to be drawn into the canal that must be removed by the ejector. This may cause plugging of the outfall channel from the ejector, unless there is enough sluicing water to handle the increased sediment. In some cases the ejector can be combined with a wasteway type of structure that regulates the flow rate into the canal and discharges the excess amount of water diverted with the sediment from the ejector back into the stream. Some cost savings may be realized from the combining of these two structures. In such a combined structure, the undershot-type wasteway gate has been found to be quite efficient for the removal of sediment from a canal headworks. Another type of wasteway uses a self-priming siphon to remove excess water back into the stream. This type of wasteway can be used in combination with a tunnel-type ejector. However, one of the major objections to this type of

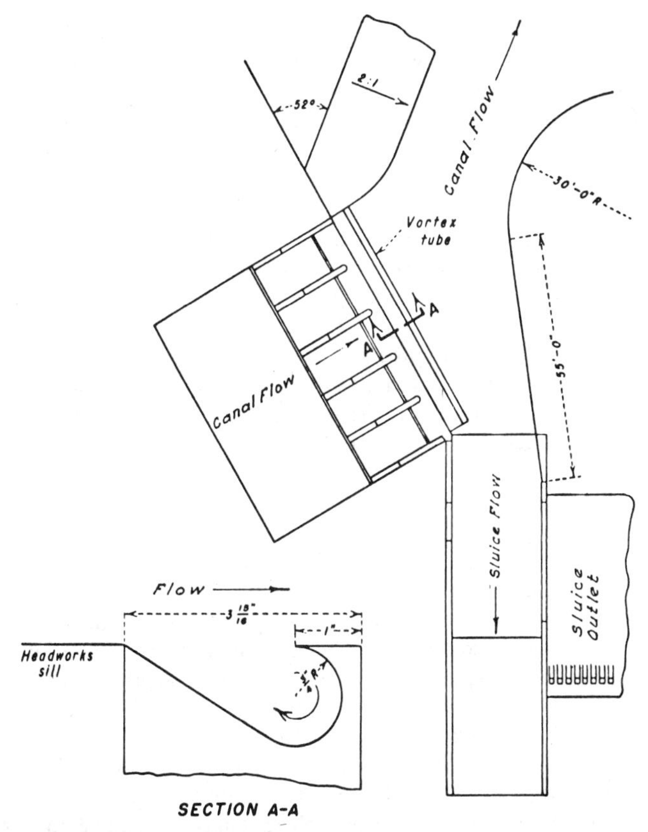

FIG. 5.38.—Vortex Tube Type Sediment Ejector—Hydraulic Model Study—Superior Courtland Diversion Dam—Pick-Sloan Missouri Basin Program, Kansas

structure is its periodic slugging type of discharge, so characteristic of a siphon. This type of discharge, because of its very large volume for a short period of time, may cause damage in the channel below which is out of proportion to the amount of water released.

Vortex Tube Ejectors.—This type of ejector consists of an open-top tube or channel with a cross section as shown by Section A-A of Fig. 5.38. This open-top tube is installed at an angle of 45° to axis of flow in the canal headworks channel. The edge, or lip, of the tube is set level with the bottom grade of the canal, whether it be lined or unlined. Water flowing over the opening induces a spiraling flow in the tube throughout its length. The spiral flow picks up the bed load and moves it along the tube to an outlet at the downstream end of the structure.

Experiments on vortex tube-type ejectors have been made by several recognized authorities in a number of countries. Possibly the best known of these in the United States were made by Parshall (1950) whose objective was to eliminate bed load sediment from channels, especially power channels. Fig. 5.38 shows the ideal vortex tube used in his studies. Various shapes, sizes, and length of tubes were studied by Parshall in the laboratory. From these studies he concluded:

1. For a 4-in. diam tube with a velocity of about 2–1/2 fps over the lip, 10%–15% of the total water diverted was wasted through the outlet.
2. The rate of spiral rotation was about 200 rpm, which was capable of moving gravel the size of hen's eggs.
3. Further studies of the 4-in. wide tube showed that the maximum rotation of the spiral flow to be about 300 rpm which moved cobblestones weighing 7–1/2 lb along the tube at a uniform velocity of about 1/2 fps. The axis of the tube had been set at an angle of 30° with the direction of flow and the tube was given a slope of 2 in. in 4 ft towards the outlet. Water discharged through this tube was only 3% of the total flow entering the channel. The efficiency of the tube would have been increased if its angle with the flow in the channel had been changed from 30° to 45°.
4. The action of flow in the tube can be improved by having the floor raised slightly just upstream of the lip and also to have the floor slope away from the lip. However, if such slope is too great the sediment may skip over the tube opening. Thus, any such special features should be tested in a model before it is constructed.

Fig. 5.39 shows a special adaptation of the vortex tube principle, but with an odd-shaped sluice pipe or box. This installation was reported to be very efficient, but no actual test figures of its efficiency are available.

Parshall's experiments were conducted using relatively small diameter vortex pipes with rather wide sediment slots as shown in Section A-A of Fig. 5.38. However, the canals were small with relatively shallow depths. Consequently, the resulting efficiency of sediment removal was good and the loss of water by sluicing through the vortex tube was only 10%–15%.

This seems to indicate that Parshall was interested in ejecting bed load material of large particle size such as may be found in mountain streams; thus, the wide sediment slot in the vortex tube. Research in Pakistan relative to vortex tube ejectors (subsequently referred to) has shown that for bed loads of small particle size the vortex tube slot should be narrower to prevent excessive waste of water.

It was found by Ahmad, et al. (1960) and Ahmad (1963), in their investigations of vortex tubes for use in canals in Pakistan, that a much narrower slot that was used by Parshall was desirable to decrease the amount of sluicing water taken from the canals. The most efficient width of slot would need to be established by model study in relation to other physical and hydraulic characteristics for a particular installation. However, it appears that the particle size of the bed load would have an appreciable influence on the slot width.

Ahmad (1973) also reports that vortex-type ejectors have been constructed in D. G. Kahn canal (9,000 cfs) and Hydel Project that have the vortex tube set in the crest of a flume with multiple withdrawal tubes that divide the vortex into two or more parts. These are reportedly working successfully (See Fig. 5.40).

Ali (1973), in discussion of the Pakistan investigations previously referred to (Ahmad, et al., 1960; Ahmad, 1963; Ahmad, 1973), reports that for very large canals it has been experienced that vortex tube-type silt ejectors with one outlet tend to choke. To counter this effect the vortex pipe was divided into equal parts

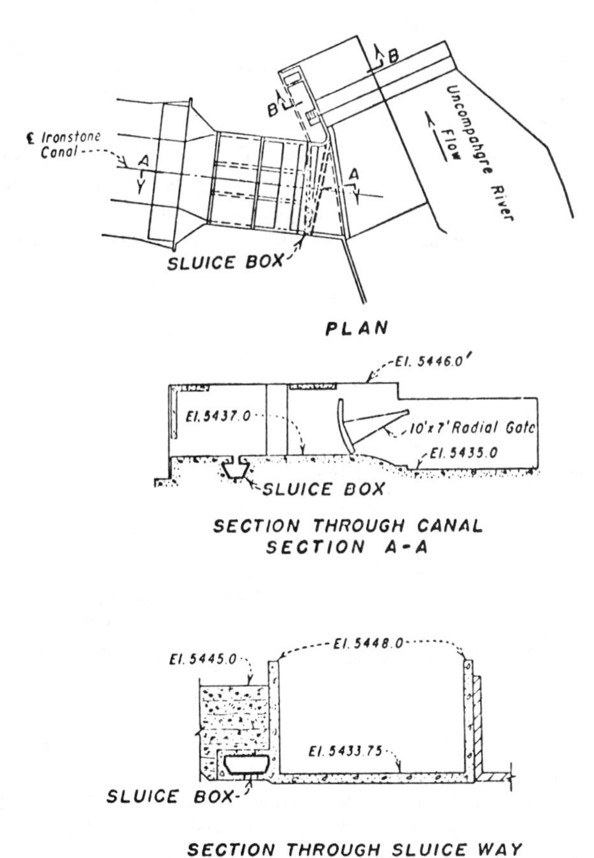

FIG. 5.39.—Vortex Tube Type Sediment Ejector in Ironstone Canal Headworks on Uncompahgre River—Uncompahgre Project, Colorado

by withdrawal tubes as previously indicated and in Fig. 5.40. Also, by placing the vortex at right angles to the flow it became easier to maintain a uniform flow in the tube.

It was also found in the Pakistan investigations that the maximum efficiency of the vortex ejector is at Froude number 0.8, but, in most cases in unlined canals, the Froude number is seldom more than 0.2 or 0.3. To obtain a Froude number of 0.8, the flow must be near the critical stage. To achieve this stage of flow the canal would have to be placed in a flume with a crest created by raising the bottom of the flume above the canal bottom grade, as previously suggested; or, the flow could be forced toward the critical stage by transition to a narrower section (Fig 5.40). In either case, there will be some change in the canal regime upstream that would most probably cause deposition of sediment.

It becomes quite clear that it is very impracticable to design a vortex tube-type of silt ejector without careful model study that takes cognizance of the many related physical and hydraulic factors involved.

19. Settling Basins

General.—The structures used to remove bed load and other sediment from the water at a canal headworks are of little value in the removal of suspended small particles. This protection is usually supplied by a settling basin of some type placed in the canal just downstream from the headworks. The settling basin consists of an oversized section of the canal or other arrangement in which the velocity is low enough to permit the suspended particles to settle out. Only clear water will then enter the canal and the sediment is removed from the settling basin by sluicing or mechanical means.

An inherent and desirable characteristic of all settling basins is that they can trap a large percentage of the fine fraction of the suspended sediment, along with the coarser fraction, which would have been carried through the canal system if a less efficient type of device had been used. Trapping of the very fine suspended sediment to keep it off the land can be of considerable value to the land owner, if the land being irrigated is a tight impervious-type soil. However, if the soil being irrigated is loose and sandy, the fine sediment fraction could be beneficial to the soil. The settling basin can be designed to control the amount of suspended sediment removed by varying the dimensions of the structure and thus the time that the water is retained in the basin.

Design Relation.—One of the basic relations for the design of a settling basin was given in the (USBR) Boulder Canyon Project Final Report (1949). This report establishes the following relation for the design of a settling basin (Vetter, 1940)

$$W = W_o e^{-wx/q} \qquad (5.7)$$

in which W = weight of sediment leaving the basin, in pounds; W_o = weight of sediment entering the basin, in pounds; e = natural logarithm base; w = settling velocity of a particle, in feet per second, considering water temperature; x = length of settling basin, in feet; and q = discharge per unit of width of settling basin, in cubic feet per second per foot of width.

Where the settling basin is a circular tank with radially outward flow, this relation becomes

$$W = W_o e^{-w(r^2 - r_o^2)/q} \qquad (5.8)$$

in which Q = the total discharge, in cubic feet per second; r = radius to the point considered, in feet; and r_o = radius at the influent circle, in feet.

From the preceding relation the rate of deposition of each grain size of sediment can be computed for basins in which there is a corresponding low velocity. Good results have been obtained in removing clay, sand, and silts with velocities up to 0.7 fps.

This formula was applied in designing the desilting devices for the headworks of the Gila Main and the All-American Canals that divert water from the Colorado River above Imperial Dam. Fig. 5.41 shows the layout of the Imperial desilting works for both canals.

Examples of Settling Basins.—The All-American desilting works consists of three basins, each containing twenty-four 125-ft diam rotary scrapers (Forester, 1938). The basins are arranged in lines inclined at a 60° angle from the alinement of the canal intake channel leading from the headworks to the basins. The intake channel terminates in a section that is V-shaped in plan. Each side of the divided basins is approx 270 ft wide by 770 ft long, as measured along the sides of the parallelogram. The levees forming the long sides of the basins are skimming weirs which discharge broadside into the adjacent effluent channels that discharge directly into the All-American Canal. In each half basin there are two rows of six scrapers each that sweep a 125-ft diam circular area. These sweeping mechanisms consist of two diametrically opposed arms revolving about a central pedestal on which the motor is mounted. Diagonal scraper blades move sediment toward a central pedestal to which the sludge pipes are connected. The flow out is controlled by small stand pipes inside the pedestal. The sludge pipes are connected to an interceptor line that extends lengthwise under the center of each

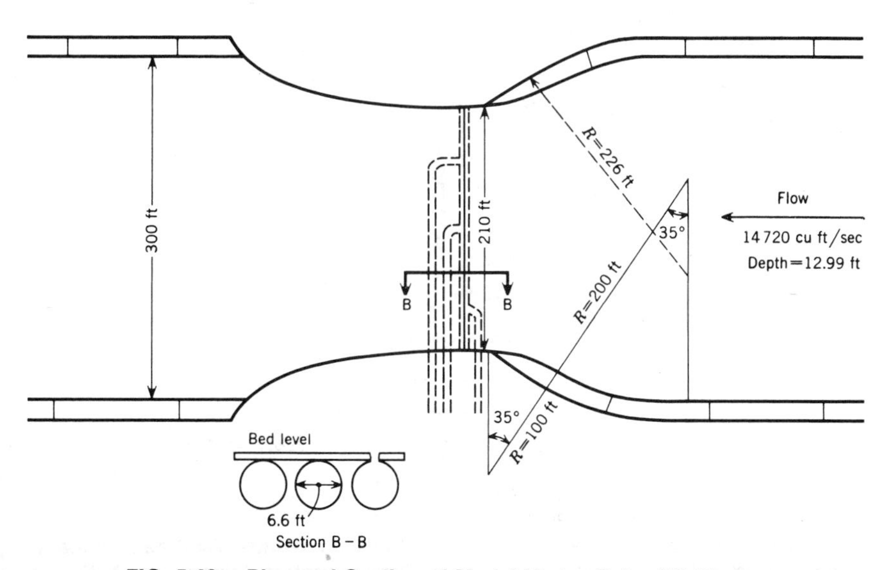

FIG. 5.40.—Plan and Section of Model Vortex Tube Silt Ejector

half basin and empties into the sluicing channel. Small triangular areas of the basin floor not covered by the scrapers are serviced by small sludge hoppers from which sediment accumulations are also flushed into the interceptor. The streamline entrance reduces turbulence, so that the effective velocity will closely approach the computed average of 0.25 fps.

The layout of the Gila Main Canal desilting basin is also shown in Fig. 5.41. The ultimate plan of development of this canal headworks included the construction of three concrete-lined basins with a total capacity of 6,000 cfs. To date (1972), only one basin has been constructed, which is designed for a capacity to remove the sediment from a flow of 2,000 cfs which was the original design

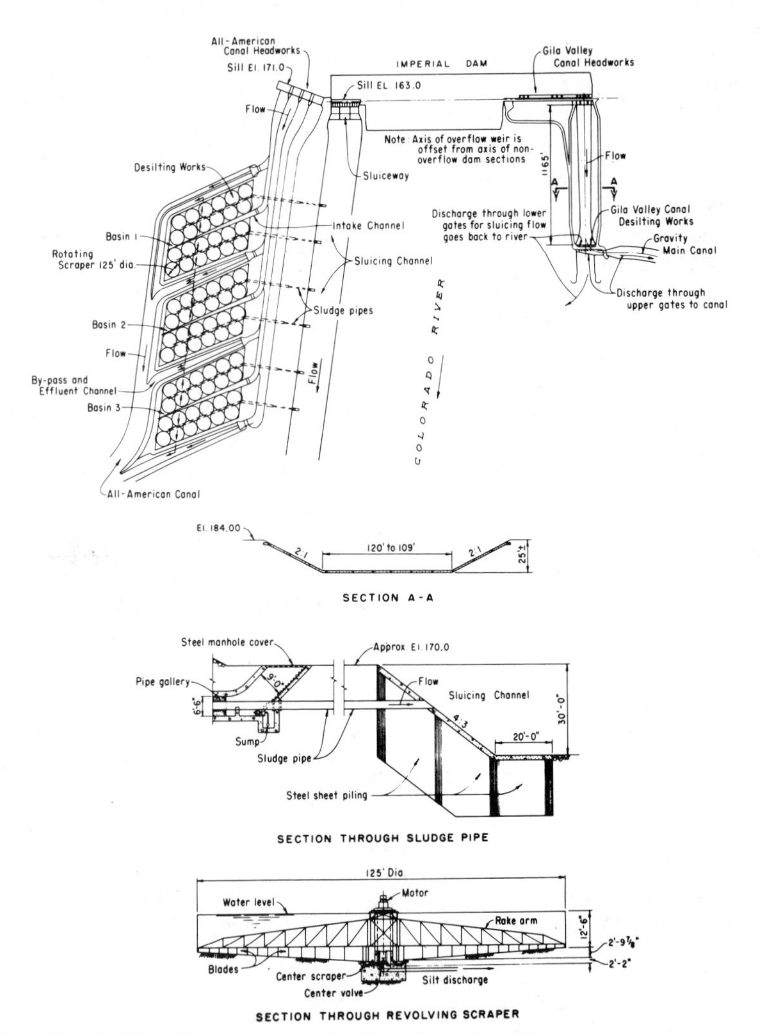

FIG. 5.41.—Settling Basins—Imperial Dam Desilting Works—All-American Canal and Gila Valley Canal Headworks—Boulder Canyon Project, Arizona and California

capacity of the Main Canal. Later operation has increased the capacity to 2,200 cfs, but this additional flow did not seriously decrease the efficiency of the basin. The one Gila Main Canal Desilting Works basin, which has been constructed, is a trapezoidal cross section structure some 1,165 ft long including transitions with a bottom width of about 115 ft. It is located roughly parallel to the Colorado River and perpendicular to the axis of the Imperial Dam. The bottom grade drops some 5 ft from the upstream inlet and to the outlet transition. The outlet control structure at the downstream end of the basin contains two sets of gates. These are located so that one set of gates is directly above the other. Flow from the upper gates discharges directly into the canal, while the lower gates are designed to pass water and sediment at sluicing velocities into a channel passing under the canal and back into the Colorado River. This type of desilting system was selected for this canal in preference to the type installed for the All-American Canal, because it can be readily enlarged in the future as operational development requires. It also needs less hydraulic head for operation, which in turn reduces the pumping costs in part of the distribution systems in which water is pumped.

When the Colorado River flow is 21,000 cfs or greater, it is estimated that the sediment discharged into the Gila Main Canal desilting works is 10,500 tons/day, while that for the All-American desilting works is 90,000 tons/day. The diversion into the Gila Main and All-American Canals is 2,000 cfs and 12,000 cfs, respectively, for this rate of sediment transport to occur. The design of the All-American Canal desilting works allowed a 50% sediment overload factor. It was considered unnecessary to remove small size particles of sediment from the canal because the land to be irrigated is sandy and the finer sediment particles passing through the canal in suspension would not be harmful to this type of soil. The desilting works were designed to remove only sediment particles that were larger than 0.05 mm.

Table 5.3 shows the predicted amount of sediment removed daily by the Gila Main Canal desilting works as about 8,000 tons; and that for the All-American desilting works to be near 70,000 tons/day. This is sediment material enough to

TABLE 5.3.–Maximum Design Performance of Settling Basins for Sediment Removal

Sieve mesh per inch (1)	Size of particles, in millimeters (2)	Percentage removal (3)	Tons of Sediment Removed per day	
			Gila Main Canal (4)	All-American Canal (5)
270	0.053	51	1,070	9,270
230	0.062	61	960	8,200
200	0.074	75	1,180	10,050
170	0.087	86	1,350	11,580
140	0.104	94	1,670	14,400
100	0.147	99	1,670	14,300
60	0.246	100	230	1,800
		Total	8,130	69,600

load fourteen 100-car freight trains daily from just the All-American Canal desilting works. Both of these basins have performed very satisfactorily since their completion in 1938; however, their actual efficiency has not been calculated. Due to its size, type of construction, and high costs it is unlikely that desilting works of the type constructed on the All-American Canal headworks will be constructed in the future for the removal of sediment from irrigation water. However, this type of costly desilting structure may well be justified for use for industrial or domestic water systems, or both.

Another technique that may be considered in designing settling basins is an application of a procedure developed by Einstein (1968). The procedure resulted from studies on the behavior of fine sediments carried by flows in suspension over a gravel bed. Einstein made these studies in connection with the design of artificial spawning grounds for salmon. Given a site where the water discharge, sediment discharge, average flow velocity, and water temperature are known, the procedure is applied to the design of a settling basin by first assuming the basin depth and length. The sediment deposited in the basin is then computed as a cumulative volume for a given set of individual sediment particle sizes. Upon computing the sediment volume deposited, the average width of the settling basin is determined next. The final design dimensions are based on the selected prismatic cross-sectional shape of the basin which usually is trapezoidal.

Generally, settling basins are constructed simply by widening and deepening the canal section to produce an enlarged cross section that will produce a lower velocity that permits the sediment particles to settle. The USBR-constructed Socorro Main Canal, located about 65 miles south of Albuquerque, N.M., has a settling basin of this type. The settling basin section is 2,000 ft long and 90 ft wide. The total bank height, including freeboard and water depth, is 13 ft. This compares with a canal section which is only 16 ft wide and has a water depth of 4 ft. The settling basin and canal sections have side slopes of 2:1. The design rate of flow is 275 cfs with a water depth of 10 ft in the settling basin. This produces a velocity in the settling basin of about 1/4 fps compared to about 2.86 fps in the canal.

A number of canal headworks at other locations have been constructed with similar settling basins. This type of settling basin requires removal of the sediment by mechanical means, usually a hydraulic dredge. This type of disposal is expensive in that it requires the purchase of a dredge, the expense of its operation, the purchase of a disposal area and additional right-of-way for equipment storage, maintenance, and access. At the Socorro settling basin, the USBR operates a 10-in. hydraulic dredge, which is capable of moving 110 cu yd of solid material per hour from 20 ft below the water surface to 10 ft above the water surface through a discharge line that is 2,000 ft long. In 1956, the USBR removed 170,000 cu yd of sediment from the Socorro settling basin using the 10-in. hydraulic dredge at a cost of $0.18/cu yd, exclusive of depreciation of the equipment and administrative costs. Due to increased labor costs it is estimated that the unit cost per cubic yard is now (1972) roughly twice the 1956 figure.

At Granite Reef Diversion Dam on the Salt River in Arizona, the Water Users' Association removes an average of 110,000 cu yd of sediment annually from the settling basin by dredging. The dredge used was a 10-in. suction type with cutterhead, with a rated discharge of 35,000 gpm at 81 ft of head. Computed for the years 1954–1963, the cost of removing sediment at this site was about $0.19/cu yd, exclusive of equipment depreciation. Current (1972) cost for removing sediment at this site is estimated to be about $0.35/cu yd.

The examples cited previously will serve to indicate the magnitude of the annual expense involved in removing sediment with a cutterhead type suction dredge, when the sediment was placed directly in the disposal area of the effluent. Where the sediment is placed in spoil piles and then rehandled, the cost is increased accordingly. As an example, the variations in costs of cleaning desilting basins on two canals in the Frenchman-Cambridge Division of the Pick-Sloan Missouri Basin Program, Neb., are cited. In 1963, on the Culbertson Canal about 21,000 cu yd of sediment were removed from the desilting basin by dragline at a unit cost of about $0.18/cu yd. The waste piles placed by the dragline were then pushed back by a bulldozer and a scraper, increasing the cost of removal to $0.39 per cu yd. In the earlier year of 1961, a similar operation of sediment removal from a settling basin was performed on the Cambridge Canal. Herein, however, a long haul by truck to the disposal area required double handling. The cost for removing some 16,000 cu yd of sediment increased to about $0.60/cu yd. These examples of cost figures are given only to demonstrate the additional expense involved in removing sediment from settling basins by mechanical means.

While removal of sediment from settling basins by mechanical means is costly, often there is no other feasible method. Where streamflow and channel conditions are favorable, sluicing is by far the cheaper method of sediment removal. However, at many canal diversion sites conditions do not permit its use. To use sluicing as a method of removal, there must be a large enough streamflow to permit the diversion of extra water to move the sediment back into the stream. Also, the amount of streamflow, its velocity, and channel conditions must be such that the sediment is moved downstream without settling out or clogging the stream channel in the downstream reach. Any sluicing of sediment that settles out in the stream bed or which causes it to aggrade downstream of the canal headworks must be abandoned for some other method of removal.

E. Reservoirs

20. Introduction.—When a dam is constructed across a stream to form a reservoir, the velocity of the flow entering the pool thus formed will be reduced, or essentially eliminated, and the major part, or all, of the sediment transported into the reach will be deposited in the reach of backwater influence and in the reservoir. Considering the limitations of current technology it must be recognized that, with few exceptions, ultimate filling of reservoirs is inevitable and must be taken into account in project planning and design. Most major reservoirs today are designed to provide sufficient storage to hold a 100-yr accumulation of sediment without encroaching on the storage provided for their functional operation.

In an age that has progressed from the first automobile to a landing on the moon in much less than a 100-yr span, it is possible that in time either the reservoirs of today will no longer be needed or that more effective methods of retaining their capacity will be developed; however, until then, it will be necessary to carefully analyze the sedimentation characteristics of reservoirs to ensure their continued functioning.

21. Sediment Sources.—Basically all of the sediment transported to a reservoir by a stream is derived from erosion of the land. It is not moved in one giant step, however. Much of the material eroded from the land surface at any given time will be moved only a few feet before lodging against a clump of vegetation or in a furrow. A portion will be moved somewhat farther, only to be trapped by fence rows or other obstructions. The remaining portion may be transported into rills or other channels; however, some of this material will be deposited in debris cones where each channel enters a larger channel. In some instances, the remainder of the eroded soils are finally moved into a stream network; but in other instances, material moved into (or eroded from) hillside channels debouch directly onto relatively flat valley floors where they remain indefinitely. As a result of these processes, the quantity of sediment transported to a given location in a stream channel during a given period averages from about 70% to less than 10% of that eroded from the land surface during that period. The higher value is for very small drainage areas, or, perhaps, from watersheds with high channel density or channels with extensive bank erosion or headcutting. The lower value is an average for drainage areas in excess of about 10 sq miles. In a drainage where hillside erosion is moved to a flat valley floor, the percentage entering a stream system may be much less. The ratio of sediment transported at a specific location in a stream to the total erosion from the drainage above that location is termed the sediment delivery ratio. The preceding phenomena have been discussed in much greater detail in Chapter IV, but have been repeated herein in view of their importance in estimating the probable sediment inflow into a reservoir.

As mentioned previously, in some instances the major immediate source of sediment may be erosion of the bed and banks or headcutting in a degrading channel. Conversely, the sediment discharge of a stream may be reduced by aggradation of the channel. Geologic investigations of some stream valleys indicate that, over the ages, the stream channel has been alternately aggraded and degraded, perhaps a number of times. These occurrences may well have been associated with periods of drought or of heavy rainfall. On the other hand, reconnaissance over several such basins indicates a close correlation of channel changes with headwater erosion. In one instance, in Oklahoma, a farmer stated that about 20 yr before, the very spot where he was plowing had been a gully big enough to put his house in. In this case the hills above the valley were severely incised with deep headcutting currently evident. In several other instances, degrading channels originated in well-rounded headwater areas with very little evidence of erosion.

Channel erosion, in recent time, may be associated with the works of man. A notable case is the 5-Mile Creek in Wyoming (Maddock, 1960) where the introduction of return flows from irrigation was concurrent with the beginning of severe channel degradation. A result of this condition was the rapid formation of an extensive delta in the Boysen Reservoir. Other incidents of upland channel degradation or headcutting result from poor land management practices or from downstream channel lowering.

22. Sediment Discharge Quantities

Sediment Discharge Measurement Stations.—For the determination of the sediment quantities to be anticipated at a reservoir site, it is most desirable to have records from a sediment discharge measuring station on the stream at, or near, the site in question. More often than not, however, such data are not available, and, if they are, the sediment record is too short to be used without extensive evaluation. Since both water and sediment discharge usually vary over a wide range from year to year, a period of record of from 10 yr or more is required to establish even a reasonably dependable average. In most areas, at least a 30-yr record is desirable. Sediment discharge measurement stations are discussed in more detail in Chapter III, Section A and Chapter IV.

Often there are adequate water discharge data available at the site in question, but the sediment record is of considerably shorter duration. The application of the flow-duration, sediment rating curve method (Chapter IV) can provide reasonable results in many instances, although in some areas there are streams in which the sediment discharge in a 1-day flood may exceed that for the rest of the year, and this method should be used with judgment. Sediment records of as short a period as 3 yr–5 yr may be used if the full range of anticipated discharges have been sampled. Daily sediment sampling is not required to define a sediment rating curve because all too often certain data points for base flows are repeated excessively, but samples at frequent intervals are desirable during flash flood periods. Even with short periods of water discharge records this method can be applied by extrapolation of the record by correlation with other available information.

In some cases, the planner may find sediment discharge records for stations controlling areas similar to that with which he is working. He will need to accumulate such data, and, by comparing these data with available streamflow and rainfall data, to develop as best he can a hypothetical record for his project. A sediment discharge measurement station should be established at the site as soon as the need for one becomes evident, as even a short record will provide a basis for comparison with other information.

Reservoir Sedimentation Surveys.—Since about 1940, the practice of surveying existing reservoirs for sediment deposition has become widespread, particularly in the Corps of Engineers, the Bureau of Reclamation, and the Soil Conservation Service of the United States Government. The primary data from these surveys are forwarded to the Federal Interagency Committee on Sedimentation, and are then distributed to all offices of the member agencies. These data are extremely valuable in delineating the sediment potential of various areas, and have been used, together with other information, to develop sediment production information to be included in a comprehensive investigation of the entire United States by a cooperative study by a number of Federal Agencies. The complete report has not been published as yet (1972), but the data therefrom should be available for inspection at offices of any of the aforementioned agencies.

Sedimentation Reconnaissance.—A skilled sediment specialist can usually derive a reasonably good estimate of sediment quantities to be anticipated from any given drainage on the basis of a careful reconnaissance of the entire drainage area. He will carefully note, on a map of the drainage areas, his estimate of sediment production from each of the various subareas. He will note erosion features, e.g., rills, gullies, and channels, and whether or not hillside rills or gullies enter directly into a stream channel or debouch onto a valley floor. He will also note headwater topography and land-use practices, as well as channel characteristics. From these data he may prepare an integrated estimate of the basin yield, which should be compared with any available sediment discharge data and adjusted as may be indicated. It should be noted, however, that attention must be given to the history of the area, including variations in cropping, conservation, and similar items such as weather variations.

The United States Soil Conservation Service uses formulas involving land slope, soil characteristics, rainfall, land use, and other factors in estimating sheet erosion from small watersheds, as outlined in Chapter IV. These are also used on large drainage areas on a sample basis. In view of their special training, the assistance of a trained soil conservation specialist in a sediment reconnaissance is invaluable regardless of the skill of the planner.

23. Trap Efficiency.—The actual accumulation of sediment in a reservoir will also depend on the proportion of the inflowing sediment that will be retained in the reservoir. In large reservoirs, possibly those with 10,000 acre-ft or more of storage capacity, it may be assumed that the trap efficiency will be 100%; i.e., all the sediment entering the reservoir will remain there. There may be some sediment moved through the pool in density flows or during periods of very high discharge, but, considering the basic approximations involved in most estimates of the sediment quantities to be anticipated, it is best to ignore this factor.

In small dry reservoirs, sometimes most of the inflowing sediment may be transported through the pool. This may also occur during high inflow periods when a reservoir discharges over the spillway and there is an appreciable velocity of flow through the reservoir. The proportion of sediment passing through the reservoir will depend primarily on two factors: the average velocity of flow through the pool and the character of the sediment. In respect to the latter, fine sediments (the silt and clay sizes) may remain in suspension long enough to pass through the reservoir. Sand sizes will not.

Two analytical methods are generally available for estimating the trap efficiency of reservoirs, both of which are based primarily on a function of the ratio of reservoir volume to inflow rates. Neither includes an analysis of sediment characteristics, and, possibly for this reason, some sedimentation specialists prefer simply to use a judgment factor.

Brune (1953) presented an empirical relationship based on the records of 44 normally ponded reservoirs. His curves, relating trap efficiency and the ratio

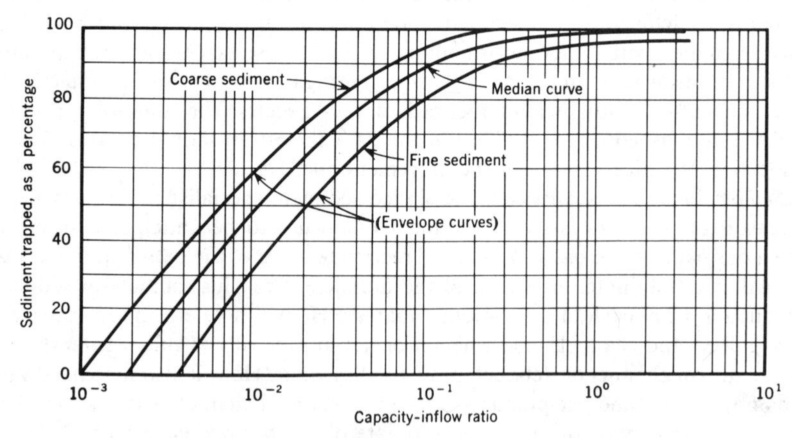

FIG. 5.42.—Trap Efficiency Curve; Brune (1953)

between reservoir capacity and mean annual water inflow, both in acre-feet, are shown in Fig. 5.42.

Churchill (1948) presented a relationship based on Tennessee Valley Authority reservoirs. His method relates the percentage of incoming sediment passing through a reservoir and the sediment index of the reservoir, i.e., the period of retention (capacity, in cubic feet, at mean operating pool level divided by average daily inflow rate, in cubic feet per second) divided by velocity (mean velocity, in feet per second, obtained by dividing average cross-sectional area, in square feet, into the inflow). The average cross-sectional area in this case is computed by dividing capacity, in cubic feet, by length, in feet. Churchill's curve is shown in Fig. 5.43.

Borland (1971) verified Churchill's method with reasonable accuracy by applying known data from a number of sources, including reservoirs with a capacity of several hundred thousand acre-feet, and concluded that Churchill's method gave better results than the Brune curves.

24. Location of Reservoir Deposits.—When the flow of a stream enters the head of the backwater reach above a reservoir, the flow velocity will immediately begin to decline, and the coarsest sediments in transport will begin to be deposited. This process will be continued, on a progressive scale, until, at some distance within the reservoir, the flow velocity has been sufficiently reduced so that all the sediments of sand size or larger have been deposited. The silt and clay-size sediments will be transported farther into the pool, and the location and manner of their deposition will depend on several factors, e.g., the longitudinal slope of the original stream bed, the shape of the reservoir, the mineral characteristics of the clay size sediments, and the chemistry of the water. Thus, the deposit in general will consist of a backwater deposit, a sand and gravel delta, and a bottom deposit of silt and clay materials. Many of the reports of reservoir sedimentation surveys, published

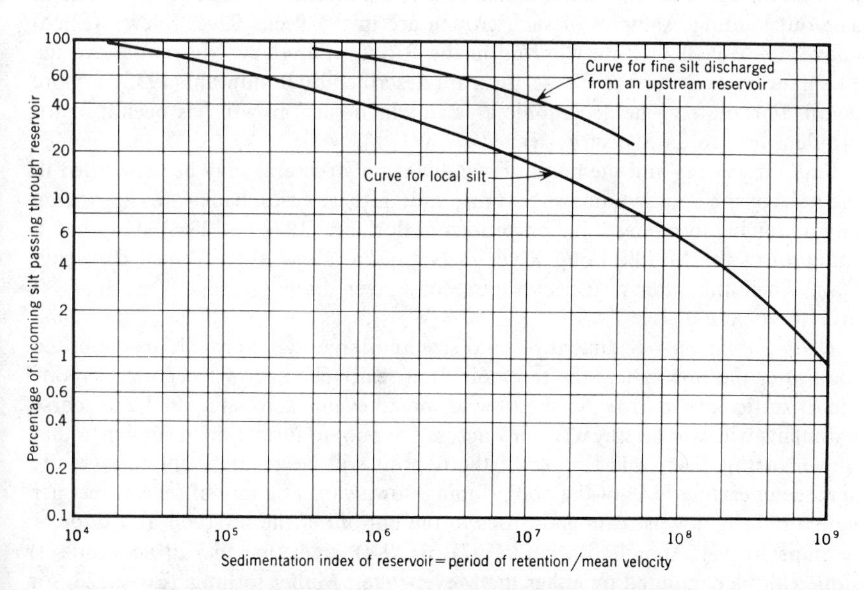

FIG. 5.43.—Churchill's (1947) Trap Efficiency Curve

or distributed by previously mentioned Federal Agencies, show the location and extent of sediment deposits in those reservoirs. Heinemann (1962) shows the clay content in deposited sediment and the effect of side tributaries on clay content versus distance from the dam. Stall (1964) shows the sediment movement and deposition patterns in four Illinois impounding reservoirs.

Backwater Deposit.—The backwater deposit is that material deposited in the backwater reach of the stream above the reservoir level. In theory, it should grow progressively, both into the reservoir and upstream, because, as the deposit grows, the backwater effect is extended; however, its growth will be limited as the stream adjusts its channel through the deposit by eliminating meanders, by forming a channel having an optimum width-depth ratio, and by varying the bed form roughness so that these factors in combination, enable the stream to transport its sediment load through the reach. Above small dams or diversion structures, and ultimately above almost all reservoirs, the backwater deposit will fill the channel and result in higher flow levels and possibly in flooding problems. In the case of large reservoirs operated at varying pool levels, the backwater deposit will probably not present a problem until the reservoir is substantially filled with sediment. If there are important facilities at the head of the reservoir they may, of course, be affected within a short time by a rising ground-water table or by increased flooding; however, material deposited in the backwater reach during high pool levels is normally eroded and moved into the pool during low pool levels. During at least the first half of the useful life of a large reservoir, the backwater deposit will seldom extend upstream above the maximum level attained by the pool.

Severe backwater deposition problems have resulted in some areas because of the growth of phreatophytes, e.g., salt cedars. These plants grow very thickly and impede the flow of the water and cause the sediment to be deposited. This, in turn, permits further growth, and, in a chain reaction, additional deposition upstream. The outstanding examples of such growth are in the Pecos River (New Mexico) above the McMillan Reservoir and in the Rio Grande (New Mexico) above the Elephant Butte Reservoir. These areas are described by Bondurant (1955) in more detail. Fortunately, the geographic area in which such growth has been a major problem has, to date, been restricted.

Small reservoirs and the pools above diversion structures may become filled to the level of the dam within a short time, and, in these cases, backwater deposition might quickly become a major problem. Borland (1971, p. 29-4) describes a situation of the Middle Loup River in Nebraska, above the Milburn Diversion Dam, where the water surface elevation, for a given discharge, rose as much as 5 ft in a 16-yr period.

Delta Formation.—Sediment of sand size and larger will normally be deposited soon after the flow enters the reservoir. In a relatively narrow reservoir; i.e., one that is sufficiently narrow for the flow to spread evenly across the pool, the coarse sediments will also be spread evenly across the pool to form a delta similar to that shown in Fig. 5.44. The surface of the deposit will be at, or slightly above, the average operating level of the pool, sloping downward at a rate of several feet per mile, until the downstream end drops to the bottom of the reservoir at a slope of perhaps 10 ft–15 ft/mile. Borland (1971, p. 29-4) states that the surface (top-set) slope can be computed by either the Meyer-Peter, Muller formula (Eq. 2.228) for beginning transport, or the Schoklitsch equation (Eq. 2.226) for zero bed load transport. He further states, however, that for most reservoirs the top-set slope closely approximates one-half the original channel slope, and he points out that Bureau of Reclamation surveys indicate that the downstream (fore-set) slopes average 6.5 times the top-set slope.

A rough estimate of the extent of a delta formation of this type at any future time can be made by determining the accumulated coarse sediment inflow to be anticipated; approximating the volume by assuming a dry weight, in place, of about 85 pcf; assuming top-set and fore-set slopes as described previously; and fitting the sediment volume into the limits determined by these slopes and the reservoir cross sections.

If, on the other hand, the stream enters a wide pool, the flow tends to enter the pool much as a jet, and a finite velocity of flow will continue along this line for an appreciable distance. Part of the sand in transport will be carried into the pool to be deposited along this flow line to form an underwater ridge subtending a submerged channel. As this procedure progresses, this channel will be built up to the surface, and will extend farther into the reservoir. Vegetative growth encourages additional deposition on the banks during high flows. Such a channel may extend for several miles into the reservoir, along the west or south banks in the northern hemisphere, until a bank is breached at some indeterminate location (avulsion) during a high inflow, and a new channel starts to form. During this process, a large portion of the fine sediment transported beyond the lower end of the channel will be carried by a reverse eddy back into the head of the pool and will be deposited there. Avulsions cause a new sand channel to be built up over the previously deposited fine sediment. Succeeding avulsions result in a delta consisting of a matrix of fine sediment and sand fingers, frequently with a number of entrapped pools. This formation is usually an unattractive swamp but offers an

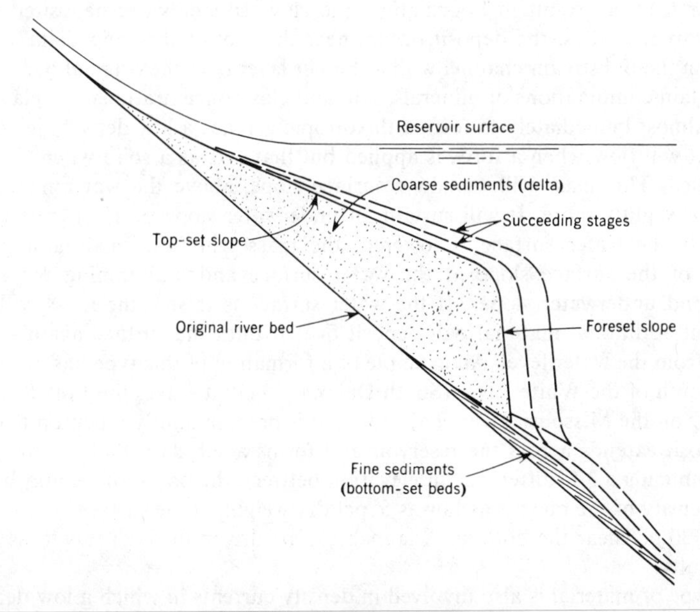

FIG. 5.44.—Reservoir Delta Form

excellent area for the development of a wild-fowl refuge. Since it would be extremely difficult to predict the proportion of fine sediment that might be involved, or the sequence of avulsions, prediction of the formation or extent of a delta of this type at any given future time would be hopeless.

Bottom Deposits.—The silt and clay-size sediments, except as noted in the foregoing paragraph on wide pools, will be transported into the pool beyond the delta and be deposited on the bottom of the reservoir. The form of the deposit will depend primarily on the mineral characteristics of the clay material and the chemistry of the water. Clays may be generally divided into three major mineralogical groups: the montmorillonites, including bentonites; the hydrous micas or illites; and the kaolinites.

The montmorillonites are the most active; i.e., they are the most susceptible to ionic exchange with a surrounding fluid medium that, in most natural waters, causes the particles to be mutually attractive so that they tend to coalesce or form floccules that settle out of the fluid at the same rate as coarse silt or fine sand particles. With some chemicals in the fluid, however, montmorillonites may have an ionic exchange so that the particles become mutually repulsive and remain in such a finely divided state that they are held in suspension indefinitely even in the quiescent waters of a reservoir. A recent investigation of a number of small reservoirs in Nebraska (unpublished data), some of which are clear and some of which are normally turbid, showed that in those that are the most turbid the inflowing sediments have the highest percentage of montmorillonite and the water contains the highest percentage of soluble phosphates. It must be emphasized, however, that this was a preliminary study and that no definitive proportions were developed. It is also noted that phosphates are the active ingredient in detergents. Montmorillonites also tend to be in a dispersed (repulsive) phase in seawater.

In those cases in which the combination of clay mineral and dissolved salts in the water tends to result in flocculation, the clay sediments are deposited with relative rapidity; thus, the deposit occurs near the foot of the sand delta and is thickest in the old stream channel with a thinner layer over the original overbank. With certain combinations of minerals, salt, and clay concentrations, the clay will deposit almost immediately to form a thixotropic gel, i.e., a low density, jelly-like mass that will flow when a force is applied but that acts as a solid when force is not applied. This mass will sustain a surface slope, above the water line, of 1 ft/mile or slightly more. It will sustain an underwater slope of 15 ft/mile to 18 ft/mile. If the water surface is lowered, the mass will flow, maintaining the junction of the surface slope at the water surface and maintaining the same surface and underwater slopes. If the water surface is raised, the mass will not move, but additional material will deposit over it until the surface again slopes upward from the water level. An example of a formation of this type has occurred at the mouth of the White River (South Dakota) where it enters the Fort Randall Reservoir on the Missouri River. The sediment is predominantly a bentonite clay. The deposit extends across the reservoir and forms a sub-dam that, at low pool stages, subtends a 3-ft difference in elevation between the pools above and below it. The density of the mass is as low as 5 pcf dry weight, at the surface, increasing to about 30 pcf near the bottom. A launch can be driven through this mass with no difficulty.

This type of material is also involved in density currents in which a low density mass of clay sediment will flow down the bed of a reservoir (Chapter II, Section

K) to the dam, where it may accumulate in a soft deposit (up to about 30 pcf dry weight) or be, in part, discharged through the outlets. The deposit thus formed will have a surface slope of about 1 ft/mile or more; thus, it is usually assumed that a density flow will not occur if the slope of the bed of the reservoir is less than 1 ft/mile.

The kaolinites are relatively inactive in terms of ionic exchange, and generally remain dispersed unless present in a sufficient concentration for mass attraction to cause flocculation. In a reservoir, they will be transported well into the pool and will be deposited over the bed of the reservoir at a uniform depth across a section. Kaolinites are frequently involved in a turbid reservoir.

Density of Deposits.—Sediment discharge data are normally reported in terms of weight, i.e., tons per day or per year. In order to utilize these data to estimate storage loss in reservoirs, it is necessary to convert the weight values into volumes. This is a very inexact procedure, because the volume of a given weight of sediment, when deposited in a reservoir, will vary with the proportions of sand, silt, and clay-size materials, the depth of the deposit, the mineralogical and chemical characteristics of the clay sediments and water, and variations of the pool level that might expose the deposits to alternate wetting and drying. The average overall weight of sediment deposits, based on samples and on reservoir sediment surveys will vary in different reservoirs from as low as 30 pcf (dry weight) to 100 pcf (dry weight). The dry weight of sand in place in several delta deposits was about 85 pcf; however, a more commonly used value is 95 lb. Heinemann (1962) shows how the weight of deposited sediment varies with depth at random locations in a reservoir, how it varies by reservoir segments, how it varies at constant depth versus distance from the dam, and how it varies in relation to percentage of clay. Several methods for converting the volume of deposited sediment in a reservoir to a weight basis are given by Heinemann and Rausch (1971) in a step-by-step procedure.

One method of computing the anticipated volume-weight ratio of a reservoir sediment deposit has been circulated as *Report No. 9* of the Federal Interagency Sedimentation Project, "Density of Sediment Deposited in Reservoirs" by Koelzer, dated November, 1943. The results thus obtained are reasonable, and it is well for even an experienced engineer to at least check his estimates by this or similar means. The procedure of Koelzer is summarized in Chapter II, Section B. This section also gives a relation developed by Miller (1953).

Borland (1971, p. 29–2) refers to Miller's procedure and also (pp. 29–34) points out that the Koelzer values shown in Table 2.6 of Chapter II were updated by Lara and Pemberton (1963). The reader is referred to several tables and charts in Borland's (1971) appendix.

McHenry (1971), McHenry and Dendy (1964), and McHenry, et al. (1969) have measured the density of continuously submerged sediments in several reservoirs at intervals over periods of years. The sediments involved were predominantly fines with clay (less than 2μ) contents upward of 30% or 40% and negligible sand content. Their data showed that as additional sediment accumulated with time, the density increased until reaching a maximum varying from about 46 pcf–76 pcf, depending on the characteristics and origins of the sediment. These maximums occurred at depths of from 2.5 ft–6 ft, and the density remained relatively constant below these depths. Similar measurements by the United States Army Corps of Engineers (unpublished data) of the previously mentioned thixotropic deposits in the Fort Randall Reservoir showed that the density increased from about 5 pcf (dry weight) at the surface, approached a maximum of about 30 pcf some 5 ft below the surface, and remained relatively constant below that depth.

It thus appears obvious that a knowledge of the characteristics and origins of the sediments for a given area should be very helpful in estimating ultimate sediment storage based on consolidated sediment densities. Unfortunately, definitive knowledge of the relationship between clay characteristics and deposit densities is not available, and estimates of this relationship must be based on experience. The data given by McHenry and his associates will be helpful in this respect. They are shown in Tables 5.4–5.7.

25. Prediction of Distribution of Reservoir Sediment Deposits.—As has been noted, the distribution of sediment in a reservoir depends on a number of factors,

TABLE 5.4.—Measurements of Sediment Density Made with Single Gamma Probe, Cow Bayou, No. 4, McLennan County, Texas, for Stations on Range R1-R2

Depth[a] below water surface, in meters	DEPTH[a] OF WATER, IN METERS			
	1.53	1.92	1.98	1.98
	Total wet density, in grams per cubic centimeter			
1.53	1.208			
1.68	1.568			
1.83	1.716	1.060		
1.98		1.257	1.281	1.166
2.14		1.493	1.517	1.464
2.29		1.596	1.531	1.494
2.44		1.609	1.569	1.512
2.59			1.602	1.559
2.75				1.613

[a]All reported depths are to the center of sensitive volume of the single gamma probe (McHenry, 1971.)

TABLE 5.5.—Measurements of Sediment Density Made with Single Gamma Probe, Dawson City Reservoir, Navarro County, Texas, for Stations on Range 1

Depth, below water surface, in meters	DEPTH OF WATER, IN METERS				
	1.13	1.16	1.13	1.10	1.07
	Total wet density, in grams per cubic centimeter				
1.07	1.137	1.089	1.088	1.151	1.428
1.37	1.348	1.354	1.335	1.338	1.340
1.68	1.389	1.373	1.386	1.370	1.368
1.98	1.418	1.395	1.416	1.401	1.409
2.29	1.433	1.425	1.423	1.420	1.434
2.59	1.450	1.410	1.435	1.437	1.426
2.90	1.462	1.437	1.424	1.459	1.437
3.20		1.446		1.445	1.437
3.51		1.458		1.447	1.437
3.81					1.456

and because the effect of the various factors is understood only qualitatively at best, the prediction of this distribution is an inexact procedure. However, Lara (1962) has presented a procedure, revised from that developed by Borland and Miller (1958), based primarily on the division of reservoirs into four different types according to the shape of the reservoir. For each shape, there appears to be a reasonable relationship between the percertage of total reservoir depth and total reservoir sediment volume.

In the Lara method (1962), reservoirs are classified by shape as follows:

Reservoir type	Classification	m
I	Lake	3.5–4.5
II	Flood-plain foothill	2.5–3.5
III	Hill	1.5–2.5
IV	Gorge	1.0–1.5

in which m = the reciprocal of the slope of the depth versus capacity plot on logarithmic paper. Fig. 5.45 presents the relative sediment area versus relative depth curves for each of the reservoir types (see also Fig. 5.46). The following example is an application of the Empirical Area-Reduction Method for a reservoir having a total capacity of 67,100 acre-ft and a predicted sediment inflow of 9,300 acre-ft. Table 5.9 is a summary of the computations used to arrive at the sediment distribution. The following steps were followed to complete the table:

1. Plot depth obtained from Col. 1 of Table 5.8 versus the original capacity in Col. 3 on logarithmic paper. Since the slope of this plot is approximately three, the reservoir is classified as Type II.
2. Compute Col. 4 by dividing the original depth represented by elevations in Col. 1 by 50.6, the original total depth.
3. Plot depth against surface area on logarithmic paper and complete Table 5.9 using data from Cols. 1, 2, and 3 of Table 5.8.

TABLE 5.6.—Changes in Total Wet Density, in grams per cubic centimeter, of Sediment Profiles Measured with Single Gamma Probe, Powerline Reservoir, Lafayette County, Mississippi

Depth below water surface, in meters (1)	RANGE M-N 15.25 m 1962 (2)	15.25 m 1967 (3)	25.93 m 1962 (4)	25.93 m 1967 (5)	30.5 m 1962 (6)	30.5 m 1967 (7)
2.29		1.000				
2.44		1.005				
2.59	1.000	1.336		1.000		1.000
2.75	1.383	1.660	1.000	1.146	1.000	1.155
2.90	1.688	1.722	1.305	1.554	1.290	1.621
3.05	1.696	1.717	1.620	1.693	1.663	1.684
3.20	1.736	1.712	1.679	1.723	1.703	1.711
3.36	1.738	1.767	1.718	1.716	1.716	1.724
3.51	1.759	1.774	1.729	1.737	1.715	1.725
3.66			1.736	1.753	1.724	1.752
3.81			1.767	1.777	1.767	1.763
3.97					1.765	

TABLE 5.7.—Measurements of Sediment Density Made with Single Gamma Probe in Lake Guayabal, Puerto Rico

Depth below water surface, in meters (1)	Total Wet Density, in grams per cubic centimeters	
	Site 15 (2)	Site 18 (3)
14.56	1.000	
14.87	1.063	
15.17	1.278	1.016
15.48	1.346	1.000
15.78	1.407	1.136
16.09	1.407	1.313
16.39	1.490	1.352
16.70	1.548	1.419
17.00	1.527	1.419
17.31	1.518	1.509
17.61		1.500
17.92		1.530
18.22		1.543

TABLE 5.8.—Sediment Disposition Computations

Reservoir________ Project________

Total Sediment Inflow 9,300 acre feet Computed by________ Date________

① Elevation (feet)	② Original Area (acres)	③ Original Capacity (acre-feet)	④ Relative Depth	⑤ A_p Type __	⑥ Sediment Area (acres)	⑦ Sediment Volume (acre-feet)	⑧ Accumulated Sediment Volume (acre-feet)	⑨ Revised Area (acres)	⑩ Revised Capacity (acre-feet)
1889.6	3445	67,100	1.00	0	0		9300	3445	57,800
						440			
1885.0	2960	52,380	.909	0.875	179		8860	2781	43,520
						1015			
1880	2487	38,790	.810	1.113	227		7845	2260	30,945
						1195			
1875	2060	27,420	.711	1.230	251		6650	1809	20,770
						1280			
1870	1678	18,100	.613	1.277	261		5370	1417	12,730
						1300			
1865	1270	10,700	.514	1.267	258		4070	1012	6,630
						1265			
1860	906	5,300	.415	1.210	247		2805	659	2,495
						1070			
1855.5	473	2,184	.326	1.117	228		1735	245	449
						956			
1851	197	779	.237	0.975	197		779	0	0
						699			
1845	48	80	.119		48		80	0	
						80			
1839	0	0	0		0		0	0	

SUPPLEMENT

1851	197		.237	.975	$K_1 = 197 \div 0.975 = 202$

$K_2 = 202\ (9300/9210)$

4. Plot data in Cols. 2 and 6 of Table 5.9 on Fig. 5.45 and draw a curve through the points. At the intersection of this curve with the curve for Type II reservoirs read $p_o = 0.237$, the relative elevation of the new reservoir bottom, and compute $p_oH = 12$ ft. The elevation of the new reservoir bottom at the dam is now 1,839 + 12 = 1,851 ft.

5. Read relative area values, A_p, from the Type II curve of Fig. 5.45 or calculate them from the formula given in the figure and place them in Col. 5 of Table 5.8.

6. Compute $K_1 = 197/0.975 = 202$ by dividing the original area at the new zero elevation by the relative area at that elevation.

7. Complete Col. 6 by multiplying each of the values in Col. 5 by the K_1 value.

8. Compute sediment volumes in Col. 7 with the average end area method using areas from Col. 6.

9. Accumulate values in Col. 7 to complete Col. 8. If the accumulated total sediment volume, S_1 does not agree with the observed value of S, compute a new K as follows:

$$K_2 = K_1\left(\frac{S}{S_1}\right) \qquad (5.9)$$

and repeat steps 6 and 7.

10. Compute Col. 9 as the difference between Cols. 2 and 6.

11. Complete Col. 10 with the differences between values in Cols. 3 and 8. New area and capacity curves can be drawn from the data in Cols. 9 and 10. Another empirical procedure for estimating the sediment distribution in small

TABLE 5.9.—Direct Determination of Elevation of Sediment Deposited at Dam

Reservoir ______ Project ______

S = 9,300 acre-feet H = 50.6 feet

① ELEV (ft.)	② p	③ V (pH)	④ S-V(pH)	⑤ HA(pH)	⑥ h'(p)
1848	0.178	331	8969	5971	1.502
1850	0.217	603	8697	7792	1.116
1852	0.257	996	8304	12093	0.687
1854	0.296	1575	7725	17204	0.449

p_o = 0.237
p_o H = 12.0
Bottom elevation = 1839.0
Elevation of sediment deposited at dam = 1851.0

NOTATION OF SYMBOLS

p = relative depth of reservoir.
V (pH) = reservoir capacity in acre-feet at a given elevation.
S = total sediment inflow in acre-feet.
H = height of dam in feet.
A (pH) = reservoir area in acres at a given elevation.
h'(p) = a function of the reservoir and its anticipated sediment storage expressed as follows:

$$h'(p) = \frac{S - V(pH)}{HA(pH)}$$

floodwater retarding reservoirs was developed by Heinemann (1961). Its use should be limited to small reservoirs in the Missouri Basin Loess Hills of Iowa and Missouri. A similar procedure can be developed from similar data on other reservoirs in other areas.

26. Reservoir Sedimentation Problems.—The major problem of reservoir sedimentation is, of course, the loss of storage capacity. In reservoir planning, it is necessary to predict the rate of storage loss and the ultimate useful life of the project. For multiple purpose reservoirs, where the capacity is allocated to each

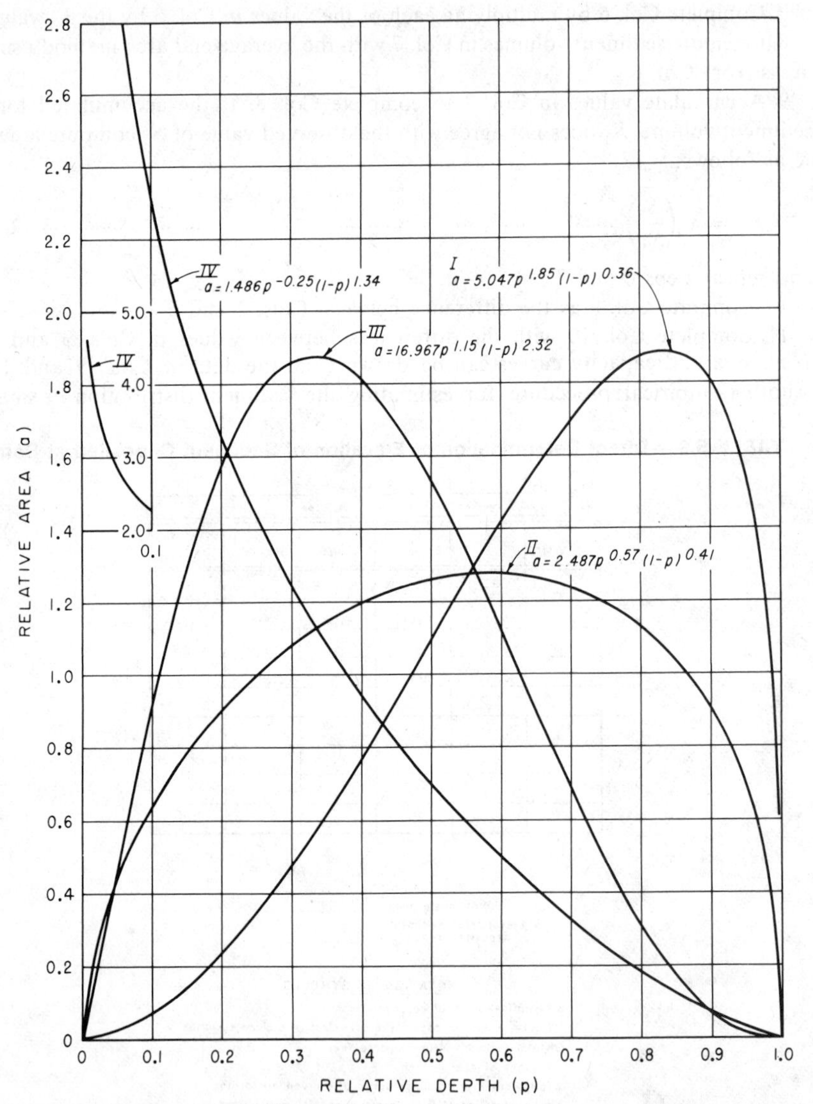

FIG. 5.45.—Relative Horizontal Area as Function of Relative Depth for Reservoir Types I, II, III, and IV; Lara (1962)

purpose by vertical zones, it is necessary either to predict the rate of storage loss in each zone, or to recognize that the proportionate allocations may have to be corrected from time to time.

With the current emphasis on recreation, it must be realized that in some areas urban development will occur adjacent to the reservoir, and, if the tributary drainage of the reservoir is relatively small, the proportionate storage loss from sediment deposition during the development period may be increased many times. Guy and Ferguson (1962) have published a report on "Sediment in Small Reservoirs Due to Urbanization" and Holeman and Geiger (1959) have documented the storage loss in Lake Barcroft, Fairfax County, Virginia due to urban development. After the development is completed, pavement placed, and lawns matured, however, the rate of sediment yield may be appreciably less than the original rate. Those individuals interested primarily in recreation facilities tend to consider their plans in terms of the surface area of the available pool. If a sedimentation analysis indicates that the pool available to them has a useful life of 100 yr, they assume that the full surface area should be available for 100 yr, failing to realize that the area will decrease as the delta deposit increases. A useful planning concept for discussing recreational facilities is a depth-area-time relationship.

A further problem, which became important with the development of recreation, is the shoaling of the inlets and coves that are so attractive for boat launching and docking. Such areas may become totally useless for these purposes in a short time as deposition occurs from sediment delivered by a tributary stream or as the inlet is blocked by a bar formed by lateral transport by wave action on material eroded from adjacent high banks. In the first instance, if a sedimentation

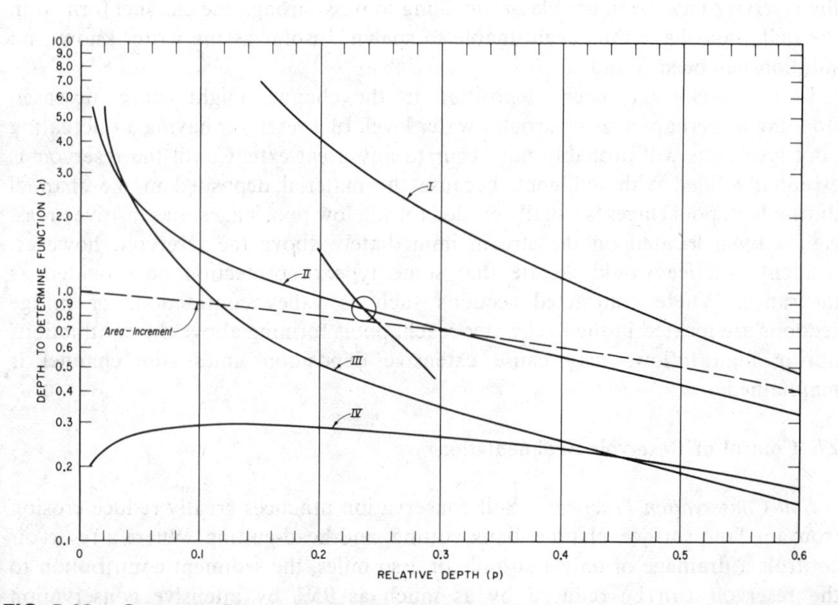

FIG. 5.46.—Curves to Determine Depth of Sediment in Reservoir at Dam for Type I, II, III, and IV Reservoirs; Lara (1962)

analysis indicates a high rate of productivity in the tributary drainage, development in the embayment should be prohibited. In the second case, development should be permitted only if relatively deep water is adjacent to the entrance.

Fine sediments settling on the bed of a reservoir may cover the biota which provide the basic fish food chain and prevent hatching. In shallow waters, wind and wave action may agitate the fine bottom deposits with sufficient frequency to prevent hatching and destroy the area for biological reproduction. There is currently no solution to this problem.

Turbidity resulting from very fine clay particles remaining in suspension may be a problem. In most instances turbidity will be undesirable because it tends to inhibit the growth of the microorganisms necessary in the fish food chain. In relatively shallow reservoirs where swimming and boating are the predominant activities, however, turbidity inhibits the growth of algae that sometimes seriously interfere with such activities. A turbid reservoir may be desirable in this case.

Growth of the delta deposit will form an area that is above the pool level when the reservoir is at or below normal maximum operating level. Depending on the characteristics of the sediments, of the reservoir, and of the climate, this area may develop into a sandy waste, an extensive swamp, or, in a few instances, a soft jelly-like (thixotropic) mass. In any event, a barren, unattractive area will be exposed during low pool levels. In some instances, this exposed area can be troublesome. In the Buffalo Bill Reservoir, in Wyoming, e.g., wind blowing down the canyon during low pool stages has carried large quantities of sand from the delta into the nearby town of Cody, Wyo. Various methods have been used in an attempt to stabilize this sand delta, with questionable results. Insofar as the writer knows, there have been no published reports on this phenomenon.

In several instances, fish that normally spawn in running water upstream from the reservoir have been unable or unwilling to pass through the channel formed in the delta, and have thus been unable to spawn. Insofar as the writer knows, no solution has been found.

In the backwater reach, deposition in the channel might cause increased flooding or perhaps a rising ground-water level. In a reservoir having a fluctuating pool level, this will probably not occur to any great extent until the reservoir is essentially filled with sediment, because the material deposited in the channel during high pool stages is usually eroded during low pool stages. In sensitive areas, e.g., a town located on the stream immediately above the reservoir, however, prudent practice would dictate that some type of protection be provided as insurance. Where contracted sections such as valley contractions or bridge sections are located in the backwater reach, pools forming above the contractions during high inflows may cause extensive deposition unless the channel is maintained.

27. Control of Reservoir Sedimentation

Soil Conservation Practices.—Soil conservation practices greatly reduce erosion from the land surface, channel bank cutting, and head-cutting. Where a reservoir controls a drainage of only 1 sq mile or 2 sq miles, the sediment contribution to the reservoir can be reduced by as much as 95% by intensive conservation measures. In some areas as large as 100 sq miles where the sediment contribution is derived primarily from channel erosion or where the eroded material is delivered almost immediately to a well developed stream net where the delivery ratio is unusually high, reductions of as much as 80% may be made by a complete conservation program, including channel stabilization, road ditch stabilization, and related measures.

Stall (1962) cites a number of cases (reservoir sedimentation and watershed suspended sediment) where conservation programs have reduced the sediment volume. The eight reservoirs cited have watersheds ranging from 1.4 sq miles–1,649 sq miles. Allen and Welch (1971) state that: "Sediment yield dropped sharply on each of four watersheds after flood retarding structures were installed. The reductions ranged from 48%–61%. These watersheds are tributaries to the Washita River basin and range in size up to 206 square miles."

For large reservoirs with large drainage areas, however, it should be recognized that it may not be economically feasible to control or greatly reduce the sediment inflow. Much of the sediment eroded from the land in previous years has been deposited in channels or on other areas where it is available for further movement. In addition, conservation must depend on continuous cooperation of the land owners, and conservation measures must be economically justified before they can be activated. The latter is an important consideration because large drainage basins usually have some areas of high sediment production that cannot be controlled within reasonable economic bounds.

There is a growing tendency toward cooperative projects, i.e., those on which the Soil Conservation Service plans and constructs structures in headwater areas and the Corps of Engineers plans and constructs the larger downstream flood control structures. In the planning of the latter, full advantage is taken of the effect of the upstream work in reducing the total area from which flood waters may be generated, but a similar sediment reduction is not assured as noted previously. Other groups may follow similar cooperative plans.

Control of Wind Erosion.—Wind erosion is a serious problem in areas of the United States with low precipitation, frequent drought, and where temperatures, evaporation, and windspeeds are high. The Great Plains are most susceptible to wind erosion, but it also occurs around the Great Lakes, the eastern coastal plains, and in the Northwest. Woodruff, et al. (1972, p. 4) recognize five basic principles of wind-erosion control:

1. Produce, or bring to the surface, aggregates or clods large enough to resist the force of the wind.
2. Roughen the land surface to reduce the wind velocity and to trap drifting soil.
3. Reduce field widths along the prevailing wind direction by establishing wind barriers or trap strips at intervals to reduce wind velocity and soil avalanching.
4. Establish and maintain vegetation or vegetative residues to protect soil.
5. Level or bench land where economically feasible to reduce effective field widths and erosion rates on slopes and hilltops where converging streamlines of windflow increase velocity and wind force.

Woodruff, et al. (1972) provide much helpful information on methods of controlling sediment moved by wind.

Control of Distribution of Deposits.—When a stream enters a wide reservoir, the location of the deposits can be controlled to some extent by judicious breaching of the self-formed channel previously described in the section on delta formation. In this manner, the major part of the sediment, including silts and clays, can be retained in the upstream portion of the pool. If such action is deemed desirable, the formation of this type delta can be managed within reasonable limits to ultimately provide an excellent refuge for water-fowl or other wildlife or to return the exposed delta to agriculture.

Other than the possible manipulation of a delta form in a wide basin, there is currently nothing that can be done to control the distribution of sediment deposition in a reservoir, except for miniscule and seldom profitable control that might be attained by varying the pool elevations.

Control of Backwater Deposition.—Backwater deposition is normally self-controlled, at least during an appreciable portion of the life of a reservoir, since the material deposited in the channel during high pool levels is scoured and moved further into the reservoir during low pool levels. Where extensive deposition occurs above a contracted section of the stream or valley, maintenance of a channel through the reach may be indicated.

Debris Basins.—The Corps of Engineers and the Soil Conservation Service have designed and installed debris basins in several mountainous areas of the United States. These are essentially reservoirs, generally located in canyons, designed to catch the coarse sediments and prevent their being carried downstream onto residential or other critical areas. In some cases where the topography permits adequate storage, they are designed to hold a 5-yr or more sediment inflow, but they are more often designed to hold the sediment from a 100-yr frequency storm plus space for 2 yr or 3 yr average annual sediment yield. The basins are maintained by periodically removing the accumulated sediment by mechanical means. Where these structures are above reservoirs they can be effective means of reducing the amount of sediment transported to the reservoirs.

Evacuation of Sediments.—Except in the case of debris basins that have sufficient economic value to justify the evacuation of sediments, the removal of the material from a reservoir is seldom, if ever, practical. In a few instances large holes have been opened in dams in the hope that the sediment would be flushed out of the reservoir, but in every case the only action was the scouring of a deep, narrow channel into the deposit. With current methods, the cost of mechanical excavation is prohibitive and a location to spoil the material is usually unavailable.

In reservoirs with a sufficiently steep longitudinal slope (greater than about 1 ft/mile), that proportion of the sediment susceptible to the formation of density flows may sometimes be discharged through outlets in the dam. Normally, this material is such a small part of the total that the effort is not worthwhile. In Algeria, however, Thevenin (1959) reports that such action is an integral part of project operation, and outlets are provided in the dam for this purpose. Here, the major part of the material is susceptible to density flows, additional reservoir sites are not available, and the water is urgently needed for the economy of the region. Duquennois (1959), also in Algeria, has suggested that a subsidiary dam be constructed at the head of a reservoir to store sediments, including coarse materials, with gates being opened periodically to flush the sediments through the main reservoir in a surge flow.

Control of Turbidity.—It is possible to control or reduce turbidity by flocculating fine sediment with chemicals or by placing organic matter (hay, etc.) in the lake. Normally, this would not be economical; however, in at least one instance the use of organic material has successfully controlled turbidity in a reservoir having a total capacity of 336,000 acre-ft and a permanent pool of 29,000 acre-ft. Mathis (1967) reports that the Nimrod Reservoir, Fourche La Fave River, Arkansas, became extremely turbid; bottom feeding fish (carp, buffalo, etc.) became predominant; and the inwashed sediment had so covered the bottom that the constant roiling of the sediment by wind and wave action and the feeding of the enormous numbers of bottom feeding fish kept the water muddy for 3 yr. With the concurrence of local residents, the lake was drawn down in the fall; the rough fish were removed; and the exposed mud flats were seeded to rye grass to produce organic matter to flocculate the suspended colloidal particles. The treatment gave excellent results. It was necessary to repeat it 4 yr later, and it is understood that the routine management of this project now includes such treatment every 3 yr or 4 yr. Sorghum is now planted instead of the rye grass originally used.

F. References

Abdel, A. I. B., "Treatment of Heavy Silt in Canals," Report to Second International Technical Congress, Cairo, Egypt, 1949.

Agricultural Research Service, "A Universal Equation for Predicting Rainfall-Erosion Losses," *Agricultural Research Service 22–06,* United States Department of Agriculture, Washington, D.C., Mar., 1961.

Agricultural Research Service, "Predicting Rainfall-Erosion Losses from Cropland East of the Rocky Mountains," *Agricultural Handbook No. 282,* Agricultural Research Service in Cooperation with Purdue Agricultural Experiment Station, United States Department of Agriculture, Washington, D.C., May, 1965.

Ahmad, M., "Some Aspects of Design Headworks to Exclude Sediment from Canals," *Proceedings,* International Association for Hydraulic Research, Fourth Meeting, R. 4, Question 2, Bombay, India, 1951.

Ahmad, M., "Design of Silt Excluders and Ejectors," West Pakistan Engineering Congress (Golden Jubilee Publication) Oct., 1963.

Ahmad, M., discussion of "Sediment Control Methods: Control of Sediment in Canals," by the Task Committee on Preparation of Sedimentation Manual, Committee on Sedimentation of the Hydraulics Division, Vito A. Vanoni, Chmn., *Journal of the Hydraulics Division,* ASCE, Vol. 99, No. HY7, Proc. Paper 9823, July, 1973, pp. 1176–1178.

Ahmad, M., Ali, M., and Kaliq, A., "Sediment Exclusion Methods and Devices at the Intake of Canals," *Proceedings of the West Pakistan Engineering Congress,* Paper No. 341, Vol. 44, Part 1, 1960.

Ali, I., discussion of "Sediment Control Methods: Control of Sediment in Canals," by the Task committee on Preparation of Sedimentation Manual, Committee on Sedimentation of the Hydraulics Division, Vito A. Vanoni, Chmn., *Journal of the Hydraulics Division,* ASCE, Vol. 99, No. HY 7, Proc. Paper 9823, July, 1973, pp. 1180–1182.

Allen, P. B. and Welch, N. H., "Sediment Yield Reductions on Watersheds Treated with Flood-Retarding Structures," *Transactions,* American Society of Agricultural Engineers, Vol. 14, No. 5, Madison, Wisc., Sept.–Oct., 1971.

American Society of Civil Engineers, "Channel Stabilization of Alluvial Rivers," Report of Task Committee on Channel Stabilization Works, Harvill E. Weller, Chmn., *Journal of the Waterways and Harbors Division,* ASCE, Vol. 91, No. WW1, Proc. Paper 4236,

Bagnold, R. A., "Some Aspects of the Shape of River Meanders," *Professional Paper 282–E,* United States Geological Survey, Washington, D.C., 1960.

Blench, T., *Regime Behavior of Canals and Rivers,* Butterworths Scientific Publications, London, England, 1957.

Bondurant, D. C., discussion of "Diversions from Alluvial Streams," by C. P. Lindner, *Transactions,* ASCE, Vol. 118, Paper No. 2546, 1953, pp. 278–282.

Bondurant, D. C., "Report on Reservoir Delta Reconnaissance," *MRD Sediment Series*

Report No. 6, Missouri River Division, United States Army Corps of Engineers, Omaha, Neb., 1955.

Borland, W. M., "River Mechanics," *Reservoir Sedimentation,* H. W. Shen, ed., Water Resources Publications, Fort Collins, Colo., 1971.

Borland, W. M., and Miller, C. R., "Distribution of Sediment in Large Reservoirs," *Journal of the Hydraulics Division,* ASCE, Vol. 84, No. HY 2, Proc. Paper 1587, Apr., 1958, pp. 1587–1 to 1587–18.

Borland, W. M., and Miller, C. R., "Sediment Problems of the Lower Colorado River," *Journal of the Hydraulics Division,* ASCE, Vol. 86, No. HY4, Proc. Paper 2452, Apr., 1960, pp. 61–87.

Borland, W. M., and Miller, C. R., "Stabilization of Five Mile and Muddy Creeks," presented at the May, 1962, ASCE Meeting, held at Omaha, Neb.

Brooks, N. H., "Mechanics of Streams with Movable Beds of Fine Sand," *Transactions,* ASCE, Vol. 123, Paper No. 2931, 1958, pp. 526–594.

Brown, C. B., *Engineering Hydraulics,* Chapter XII, Fourth Hydraulics Conference, Iowa Institute of Hydraulic Research, Hunter Rouse, ed., John Wiley and Sons, Inc., New York, N.Y., 1958, pp. 770–771.

Brune, G. M., "Trap Efficiency of Reservoirs," *Transactions of the American Geophysical Union,* Vol. 34, No. 3, Washington, D.C., 1953, pp. 407–418.

Bulle, H., "Untersuchungen Über die Geschiebeableitung bei der Spaltung von Wasserlaufen," *Forschungsarbeiten auf dem Gebiete der Ingenieiurwesens,* No. 283, Verein Deutscher Ingenieure, Publishers, Berlin, Germany, 1926.

Burke, M. F., "Flood of March 2, 1938," Los Angeles County Flood Control District, May, 1938.

Busby, C. E., "Some Legal Aspects of Sedimentation," *Journal of the Hydraulics Division,* ASCE, Vol. 87, No. HY4, Proc. Paper 2867, July, 1961, pp. 151–180.

Bush, J. L., "Channel Stabilization on the Arkansas River," *Journal of the Waterways and Harbors Division,* ASCE, Vol. 88, No. WW2, Proc. Paper 3126, May, 1962, pp. 51–67.

Byers, W. G., "Stabilization of Canadian River at Canadian, Tex.," *Journal of the Waterways and Harbors Division,* ASCE, Vol. 88, No. WW3, Proc. Paper 3214, Aug., 1962, pp. 13–26.

Carlson, E. J., "Sediment Control at a Headworks Using Guide Vanes," *Proceedings of the Interagency Sedimentation Conference, Miscellaneous Publication No. 970,* United States Department of Agriculture, Paper No. 33, Agricultural Research Service, Washington, D.C., 1963, pp. 287–304.

Carlson, E. J., and Dodge, R. A., Jr., "Control of Alluvial Rivers by Steel Jetties," *Journal of the Waterways and Harbors Division,* ASCE, Vol. 88, No. WW4, Proc. Paper 3332, Nov., 1962, pp. 53–81.

Central Board of Irrigation and Power, New Delhi, India, *Manual on River Behavior, Control and Training,* December 24, 1956.

Chabert, J., Remmillieux, M. and Spits, J., "Application de la Circulation Transversale a'la Correction des Riviers et la Protection des Prises D'Eau," International Association for Hydraulic Research, 9th General Assembly (Translation by Francis F. Escoffier), Dubrovnick, Yugoslavia, 1961, pp. 1216–1233.

Churchill, M. A., discussion of "Analysis and Use of Reservoir Sedimentation Data," by L. C. Gottschalk, *Proceedings of the Federal Interagency Sedimentation Conference,* Denver, Colo., 1947, pp. 139–140 (Published by United States Bureau of Reclamation, Denver, Colo., 1948).

Colby, B. R., "Discontinuous Rating Curves, Pigeon Roost and Cuffawa Creeks," *ARS41–36,* United States Department of Agriculture, Agricultural Research Service, Apr., 1960.

Donnally, C. A., and Blaisdell, F. W., "Straight Drop Stilling Basin," *Technical Paper 15,* Series B, St. Anthony Falls Hydraulic Laboratory, University of Minnesota, Minneapolis, Minn., Nov., 1954.

Duquennois, H., "Sedimentation in Reservoirs and Methods of Combat," *Proceedings of the International Conference on Dams and Reservoirs,* University of Liege, Liege, Belgium, May, 1959 (in French) pp. 197–226.

Einstein, H. A., "Deposition of Suspended Particles in a Gravel Bed," *Journal of the*

Hydraulics Division, ASCE, Vol. 94, No. HY5, Proc. Paper 6102, Sept., 1968, pp. 1197–1205.

Elsden, F. V., "Irrigation Canal Headworks," Punjab Irrigation Branch Papers, No. 25, Lahore, India, 1922.

Ferrell, W. R., "Mountain Channel Treatment in Los Angeles County," *Journal of the Hydraulics Division,* ASCE, Vol. 85, No. HY11, Proc. Paper 2242, Nov., 1959, pp. 11–20.

Ferrell, W. R., and Barr, W. R., "Criteria and Methods of Use of Check Dams in Stabilizing Channel Banks and Beds," *Proceedings of the Federal Interagency Sedimentation Conference, Miscellaneous Publication No. 970,* United States Department of Agriculture, Agricultural Research Service, Jan., 1963, pp. 376–386.

Forester, D. M., "Desilting Works for the All American Canal," *Civil Engineering,* Vol. 8, No. 10, Oct., 1938, pp. 649–652.

Fourtier, S., and Scobey, F. C., "Permissible Canal Velocities," *Transactions,* ASCE, Vol. 89, Paper No. 1588, May, 1926, pp. 940–984.

Francis, C. J., "How to Control a Gully," *Farmer's Bulletin 2171,* Soil Conservation Service, United States Department of Agriculture, Government Printing Office, Washington, D.C., Oct., 1961.

Franco, J. J., "Research for River Regulation Dike Design," *Journal of the Waterways and Harbors Division,* ASCE, Vol. 93, No. WW3, Proc. Paper 5392, Aug., 1967, pp. 71–87.

Friedkin, J. F., "A Laboratory Study of the Meandering of Alluvial Rivers," United States Waterways Experiment Station, Corps of Engineers, Vicksburg, Miss., May, 1945.

Frogge, R. R., "Stabilization of Frenchman River Using Steel Jacks," *Journal of the Waterways and Harbors Division,* ASCE, Vol. 93, No. WW3, Proc. Paper 5396, Aug., 1967, pp. 89–108.

Glendening, C. E., Pase, C. P., and Ingebo, P., "Preliminary Hydrologic Effects of Wildfire in Chaparral," *Proceedings,* Fifth Annual Arizona Watershed Symposium, Sept., 1961.

Gottschalk, L. C., "Predicting Erosion and Sediment Yields," International Union of Geodesy and Geophysics, XIth General Assembly, Toronto, Canada, Tome I, Vol. I, Gentbrugge, Belgium, 1958, pp. 146–153.

Guy, H. P., and Ferguson, G. E., "Sediment in Small Reservoirs Due to Urbanization," *Journal of the Hydraulics Division, ASCE, Vol 88, No. HY2, Proc. Paper 3070, Mar., 1962, pp. 27–37.*

Hathaway, G. A., "Observations of Channel Changes, Degradation and Scour Below Dams," Report of the Second Meeting of the International Association of Hydraulic Research, 1948, pp. 481–497.

Heinemann, H. G., "Sediment Distribution in Small Flood-Water Retarding Reservoirs in the Missouri Basin Loess Hills," Agricultural Research Service 41–44, Feb., 1961.

Heinemann, H. G., "Volume-Weight of Reservoir Sediment," *Journal of the Hydraulics Division,* ASCE, Vol. 88, No. HY5, Proc. Paper 3274, Sept., 1962, pp. 181–198.

Heinemann, H. G., and Rausch, D. L., discussion of "Sediment Measurement Techniques: Reservoir Deposits," by the Task Committee on Preparation of Sedimentation Manual, Committee on Sedimentation of the Hydraulics Division, Vito A. Vanoni, Chmn., *Journal of the Hydraulics Division,* ASCE, Vol. 97, No. HY9, Proc. Paper 8345, Sept., 1971, pp. 1555–1561.

Holeman, J. N., and Geiger, A. F., "Sedimentation of Lake Barcroft, Fairfax County, Virginia," *Technical Publication 136,* United States Soil Conservation Service, Mar., 1959

Inglis, Sir Claude, "The Behavior and Control of Rivers and Canals," *Research Publication No. 13,* Chapter 6, Central Waterpower, Irrigation and Navigation Research Station, Government of India, Poona, Bombay, India, 1949.

Ippen, A. T., and Drinker, P. A., "Boundary Shear Stresses in Curved Trapezoidal Channels," *Journal of the Hydraulics Division,* ASCE, Vol. 88, No. HY 5, Proc. Paper 3273, Sept., 1962, pp. 143–179.

Ishbash, S. V., "Construction of Dams by Depositing Rock in Running Water," *Transactions of the Second Congress on Large Dams,* Vol. V, Communication No. 3, United States Government Printing Office, Washington, D.C., 1936.

Joglekar, D.V., "Control of Sand Entering Canals," *Irrigation and Power,* Apr., 1959, pp. 177–190.

Joglekar, D. V., Ghotankar, S. T., and Kulkarni, P. K., "A Review of Some Aspects of the Design of Headworks to Exclude Coarse Bed Sand from Canals," Report on Fourth Meeting, International Association for Hydraulic Research, Bombay, India, 1951.

Kennedy, J. F., "Stationary Waves and Antidunes in Alluvial Channels," *Report No. KH–R–2,* W. M. Keck Laboratory of Hydraulics and Water Resources, California Institute of Technology, Pasadena, Calif., Jan., 1961a.

Kennedy, J. F., "Further Laboratory Studies of the Rougness and Suspended Load of Alluvial Streams," W. M. Keck Laboratory of Hydraulics and Water Resources, California Institute of Technology, Pasadena, Calif., Apr., 1961b.

Kennedy, R. G., "The Prevention of Silting in Irrigation Canals," *Paper No. 2820, Proceedings of the Institute of Civil Engineers,* London, England, Vol. CXIX, 1895.

King, H. W., "Silt Vanes," Punjab Engineering Congress, Vol. 8, 1920.

King, H. W., "Silt Exclusion from Distributories," Punjab Engineering Congress, Vol. 21, Paper No. 169, 1937.

Lacey, G., et al., "Stable Channels in Alluvium," *Paper No. 4736, Proceedings of the Institute of Civil Engineers,* Vol. 229, Part I, London, England, 1929–1930, pp. 259–292.

Lane, E. W., "Progress Report on Results of Studies on Design of Stable Channels," *Hydraulics Laboratory Report No. Hyd–352,* United States Bureau of Reclamation, Denver, Colo., 1952.

Lane, E. W., "The Importance of Fluvial Morphology in Hydraulic Engineering," *Proceedings Paper No. 745,* ASCE, June, 1955a.

Lane, E. W., "Design of Stable Channels," *Transactions,* ASCE, Vol. 120, Paper No. 2776, 1955b, pp. 1234–1260.

Lane, E. W., "A Study of the Shape of Channels Formed by Natural Streams Flowing in Erodible Material," *MRD Sediment Series Report No. 9,* Missouri River Division, Corps of Engineers, Omaha, Neb., 1957.

Lane, E. W., and Koelzer, V. A., "Density of Sediment Deposited in Reservoirs," *Report No. 9,* Federal Interagency Sedimentation Project, St. Anthony Falls Hydraulic Laboratory, Minneapolis, Minn., 1943.

Lane, E. W., Carter, A. C., and Carlson, E. J., "Critical Tractive Forces on Channel Side Slopes," *Hydraulic Laboratory Report No. Hyd–366,* United States Bureau of Reclamation, Denver, Colo., 1953.

Lara, J. M., "Revision of Procedures to Compute Sediment Distribution in Large Reservoirs," United States Bureau of Reclamation, Denver, Colo., May, 1962.

Lara, J. M., and Pemberton, E. L., "Initial Unit Weight of Deposited Sediments," *Proceedings of the Federal Interagency Sedimentation Conference,* Jackson, Miss., United States Department of Agriculture, *Miscellaneous Publication No. 970,* 1963, pp. 818–845.

Larson, F. H., and Hall, G. R., "The Role of Sedimentation in Watersheds," *Journal of the Hydraulics Division,* ASCE, Vol. 83, No. HY3, Proc. Paper 1263, June, 1957.

Leliavsky, S., *An Introduction to Fluvial Hydraulics,* Constable, London, England, 1955.

Leopold, L. B., and Maddock, T., Jr., "The Hydraulic Geometry of Stream Channels and Some Physiographic Implications," *Professional Paper No. 252,* United States Geological Survey, Washington, D.C., 1953.

Leopold, L. B., and Wolman, M. G., "River Channel Patterns, Braided, Meandering, and Straight," *Professional Paper 282–B,* United States Geological Survey, Washington, D.C., 1957.

Leopold, L. B., and Wolman, M. G., "River Meanders," *Bulletin of the Geological Society of America,* 1960.

Leopold, L. B., et al., "Flow Resistance in Sinuous and Irregular Channels," *Professional Paper 282–D,* United States Geological Survey, Washington, D.C., 1960.

Linder, W. M., "Stabilization of Stream Beds With Sheet Piling and Rock Sills," *Proceedings of the Federal Interagency Sedimentation Conference,* Jackson, Miss., United States Department of Agriculture, *Miscellaneous Publication No. 970,* United States Department of Agriculture, Washington, D.C., 1963, pp. 470–484.

Lindley, E. S., "Regime Channels," *Proceedings,* Punjab Engineering Congress, Vol. VII, Lahore, India, 1919.

Lindner, C. P., "Diversions from Alluvial Streams," *Transactions,* ASCE, Vol. 118, Paper No. 2546, 1953, pp. 245–288.

Lombardi, J., and Marquenet, G., "Rational Design of a Stepped Channel Consisting of a Series of Loose Bed Stilling Pools for Regularizing a Torrential Stream," *La Houle Blanche,* Nov., 1950 (in French).

Love, S. K., and Benedict, P. C., "Discharge and Sediment Loads in the Boise River Drainage Basin, Idaho, 1931–1940," *Water-Supply Paper 1048,* United States Geological Survey, Washington, D.C., 1948.

Maddock, T., Jr., "Erosion Control of Five Mile Creek, Wyoming," International Association for Scientific Hydrology, No. 53, 1960, pp. 170–181.

Maddock, T. Jr., "Indeterminate Hydraulics of Alluvial Channels," *Journal of the Hydraulics Division,* ASCE, Vol. 96, No. HY 11, Proc. Paper 7696, Nov., 1970, pp. 2309–2323.

Mathis, Wm. P., "Nimrod Lake, Fishery Management," United States Army Corps of Engineers, Little Rock, Ark., 1967.

Matthes, G. H., *American Civil Engineering Practice,* Abbett, ed., Vol. II, Sect. 15, pp. 15–56 to 15–60, John Wiley and Sons, Inc., New York, N.Y., 1956.

McEwan, J. S., "Bank and Levee Stabilization, Lower Colorado River," *Journal of the Waterways and Harbors Division,* ASCE, Vol. 87, No. WW4, Proc. Paper 2987, Nov., 1961, pp. 17–25.

McHenry, J. R., discussion of "Sediment Measurement Techniques: Reservoir Deposits," by the Task Committee on Preparation of Sedimentation Manual, Committee on Sedimentation of the Hydraulics Division, Vito A. Vanoni, Chmn., *Journal of the Hydraulics Division,* ASCE, Vol. 97, No. HY 8, Proc. Paper 8267, Aug., 1971, pp. 1253–1257.

McHenry, J. R., and Dendy, F. E., "Measurement of Sediment Density by Attenuation of Transmitted Gamma Rays," *Proceedings of the Soil Science Society of America,* Vol. 25, No. 6, Madison, Wisc., Nov.–Dec., 1964, pp. 817–822.

McHenry, J. R., Hawks, P. H., and Gill, A. C., "Consolidation of Sediments in a Small Reservoir in North Mississippi Measured in situ with a Gamma Probe," *Proceedings of the Mississippi Water Resources Conference,* 1969, pp. 101–112.

Miller, C. R., "Determination of the Unit Weight of Sediment for Use in Sediment Volume Computations," United States Bureau of Reclamation, 1953.

Miller, C. R., Woodburn, R. and Turner, H. R., "Upland Gully Sediment Production," *Publication No. 59,* International Association of Scientific Hydrology, 1962, pp. 83–104.

Miller, C. R., and Borland, W. M., "Stabilization of Five Mile and Muddy Creeks," *Journal of the Hydraulics Division,* ASCE, Vol. 89, No. HY 1, Proc. Paper 3392, Jan., 1963, pp. 67–98.

O'Brien, J. T., "Studies of the Use of Pervious Fence for Streambank Revetment," *TP-103,* United States Department of Agriculture, Soil Conservation Service, Feb., 1951.

Ogrosky, H. O., "Hydrology of Spillway Design: Small Structures—Limited Data," *Journal of the Hydraulics Division,* ASCE, Vol. 90, No. HY 3, Proc. Paper 3914, May, 1964, pp. 295–310.

Oliver, P. A., "Some Economic Considerations in River Control Work," *Proceedings of the Federal Interagency Sedimentation Conference,* Jackson, Miss., *Miscellaneous Publication No. 970,* United States Department of Agriculture, United States Government Printing Office, Washington, D.C., 1963, pp. 442–449.

Parshall, R. L., "Experiments in Cooperation with Colorado Agricultural Experiment Station," Fort Collins, Colo., July, 1950.

Parsons, D. A., "Effects of Flood Flow Effects on Channel Boundaries," *Journal of the Hydraulics Division,* ASCE, Vol. 86, No. HY 4, Proc. Paper 2443, Apr., 1960, pp. 21–34.

Parsons, D. A., "Vegetative Control of Streambank Erosion," *Proceedings of the Federal Interagency Sedimentation Conference,* Jackson, Miss., Miscellaneous Publication No. 970, United States Department of Agriculture, United States Government Printing Office, Washington, D.C., 1963, pp. 130–136.

Parsons, D. A., and Apmann, R. P., "Cellular Concrete Block Revetment," *Journal of the*

Waterways and Harbors Division, ASCE, Vol. 91, No. WW2, Proc. Paper 4311, May, 1965, pp. 27–37.

Potapor, M., and Pychkine, B., "Methods of Transverse Circulation and its Adaption to Hydrosediment," Moscow Academy of Science, USSR, *Translation No. 46 of Service Des Etude et Recherches Hydraulics,* Paris, France, 1947.

Raynaud, A., "Water Intakes on Mountain Streams, Example of Application to the Torrent Du Longon," International Association for Hydraulic Research, Fourth Meeting, Bombay, India, 1951, pp. 1–9.

Remillieux, M., "First Experimental Closure of a Secondary Channel Carried Out by Means of Bottom Panels," *Bulletin,* Permanent International Association of Navigation Congresses, Vol. 4, No. 2, 1966, pp. 43–73.

Rosa, S. M., and Tigerman, M. H., "Some Methods for Relating Sediment Production to Watershed Conditions," *Paper No. 26,* United States Forest Service Research Intermountain Forest and Range Experiment Station, May, 1951.

Rowe, P. B., Countryman, C. M., and Storey, H. C., "Hydrologic Analysis Used to Determine Effects of Fire on Peak Discharge and Erosion Rates in Southern California Watersheds," California Forest and Range Experiment Station, Forest Service, United States Department of Agriculture, Feb., 1954.

Rzhanitsyn, N. A., "Morphological and Hydrological Regularities of the Structure of the River Net, Gidrometoizdat, Leningrad," Translated by D. B. Krimgold, United States Department of Agriculture, Agricultural Research Service, 1960.

Schoklitsch, A., *Der Wasserbau,* Verlag Julius Springer, Vienna, Austria, 1930.

Schoklitsch, A., *Hydraulic Structures,* Vol. 2, Translated by S. Shulits, American Society of Mechanical Engineers, New York, N.Y., 1937, pp. 722–751.

Schumm, S. A., "The Shape of Alluvial Channels in Relation to Sediment Type," *Professional Paper 352–B,* United States Geological Survey, Washington, D.C., 1960.

Scrivner, M. W., "Hydraulic Performance of Noncohesive Channels," *Report No. DA–2,* Technical Engineering Analysis Branch, Division of Design, Office of Chief Engineer, United States Bureau of Reclamation, Denver, Colo., 1970.

Shen, H. W., and Einstein, H. A., "A Study of Meandering in Straight Alluvial Channels," *Journal of Geophysical Research,* December 15, 1964.

Shen, H. W., and Komura, S., "Meandering Tendencies in Straight Alluvial Channels," *Journal of the Hydraulics Division,* ASCE, Vol. 94, No. HY4, Proc. Paper 6042, July, 1968, pp. 997–1016.

Silberger, L. F., "Stream Bank Stabilization," *Agricultural Engineering,* Vol. 40, No. 4, Apr., 1959, pp. 214–217.

Simons, D. B., and Richardson, E. V., "Forms of Bed Roughness in Alluvial Channels," *Journal of the Hydraulics Division,* ASCE, Vol. 87, No. HY3, Proc. Paper 2816, May, 1961, pp. 87–105.

Soil Conservation Service, *Handbook of Channel Design for Soil and Water Conservation, SCS–TP–61,* United States Department of Agriculture, Washington, D.C., Mar., 1947 (Revised June, 1954).

Soil Conservation Service, *National Catalog of Practices and Measures Used in Soil and Water Conservation,* United States Department of Agriculture, Washington, D.C., July, 1959a.

Soil Conservation Service, Unpublished Measured Data, Files of the Soil Conservation Service, United States Department of Agriculture, Boise, Idaho, Aug., 1959b.

Soil Conservation Service, "Land-Capability Classification," *Agricultural Handbook No. 210,* United States Department of Agriculture, Washington, D.C., 1961.

Soil Conservation Service, Unpublished Measured Data, Files of the Soil Conservation Service, United States Department of Agriculture, Athens, Ga., 1962.

Stall, J. B., "Soil Conservation Can Reduce Reservoir Sedimentation," *Public Works Magazine,* Sept., 1962.

Stall, J. B., "Sediment Movement and Deposition Patterns in Illinois Impounding Reservoirs," *Journal of the American Waterworks Association,* Vol. 56, No. 6, June, 1964.

Stanton, C. R., and McCarlie, R. A., "Streambank Stabilization in Manitoba," *Journal of Soil and Water Conservation,* Vol. 17, No. 4, July–Aug., 1962.

Steed, C. V., "Hyperbolic Channel Bends," *Civil Works Engineer Bulletin 56–15,* United

States Army, Corps of Engineers (Based on Translation by O. Jergens of an Engineering Doctorate Thesis by Gunther Garbrecht, Technical University, Karlsrue, Germany), August 3, 1956.

Steinberg, I. H., "Russian River Channel Works," *Journal of the Waterways, Harbors and Coastal Engineering Division,* ASCE, Vol. 86, No. WW4, Proc. Paper 2647, Nov., 1960, pp. 17–32.

Sternberg, H., "Zietschrift fur Bauwasen," 1875, p. 483.

Stufft, W. E., "Erosion Control for Gehring Valley," presented at the Aug. 27, 1965, ASCE Hydraulics Division Conference, held at Tuscon, Ariz.

Taylor, R. H., Jr., "Exploratory Studies of Open Channel Flow Over Boundaries of Laterally Varying Roughness," *Report No. KH–R–4,* W. M. Keck Laboratory, California Institute of Technology, Pasadena, Calif., July, 1961.

Thevenin, J., "Study of Sedimentation in Reservoirs in Algeria and Methods for Preserving Their Capacities," *Proceedings,* Interanational Conference on Dams and Reservoirs, University of Liege, Liege, Belgium, May, 1959 (In French), pp. 227–264.

Thompson (Thomson), J., "On the Origin and Winding of Rivers in Alluvial Plains," *Proceedings,* Royal Society of London, London, England, May 4, 1876, p. 5.

Tolley, G. S., and Riggs, F. E., *Economics of Watershed Planning,* Chapter 7, The Iowa State University Press, Iowa State University, Ames, Iowa, pp. 87–95.

Uppal, H. L., "Sediment Excluders and Extractors," Report on Fourth Meeting, International Association for Hydraulic Research, Bombay, India, 1951a, pp. 261–315.

Uppal, H. L., "Sediment Excluders and Extractors," Fourth Meeting, International Association for Hydraulic Research, Question 2, Bombay, India, 1951b, pp. 22–41.

United States Army, Corps of Engineers, "Use of Kellner Jetties on Alluvial Streams," *Civil Works Investigation Report 509–a,* Albuquerque District, Albuquerque, N.M., June, 1953.

United States Army, Corps of Engineers, Symposium on "Channel Stabilization Problems," *Technical Report No. 1,* Committee on Channel Stabilization (Vol. 1, Sept., 1963; Vol. 2, May, 1964: Vol. 3, June, 1965; Vol. 4, Feb., 1966) (Published at the United States Waterways Experiment Station, Vicksburg, Miss., 1966a.

United States Army, Corps of Engineers, "Laboratory Investigation of Under Water Sills on the Convex Bank of Pomeroy Bend," *Report No. 2,* Mead Hydraulic Laboratory, Omaha District, Omaha, Neb., Nov., 1966b.

United States Army, Corps of Engineers, "State of Knowledge of Channel Stabilization in Major Alluvial Rivers," *Technical Report No. 7,* Committee on Channel Stabilization (Published at the Waterways Experiment Station, Vicksburg, Miss.), Oct., 1969.

United States Bureau of Reclamation, "Reclamation Instructions," *Design Standards No. 3, Canals and Related Structures,* Chapter 1, "Canals and Laterals," Paragraphs 1.6 to 1.13 and Fig. 3A, paragraph 1.7. Denver, Colo., Dec. 8, 1967.

United States Bureau of Reclamation, "Design and Construction of Imperial Dam and Desilting Works," *Boulder Canyon Final Reports,* Part IV, Denver, Colo., 1949, pp. 122–124.

United States Bureau of Reclamation, "Design and Construction," *Boulder Canyon Project Final Reports,* Part IV, Bulletin 1, General Features, 1941, pp. 238–241.

United States Bureau of Reclamation, "Stable Channel Profiles," *Hydraulic Laboratory Report No. HY 325,* Denver, Colo., Sept., 1951.

United States Bureau of Reclamation, "Progress Report on Results of Studies on Design of Stable Channels," *Hydraulic Laboratory Report No. HY 352,* Denver, Colo., June, 1952.

United States Bureau of Reclamation, "Critical Tractive Forces on Channel Side Slopes," *Hydraulic Laboratory Report No. HY 366,* Denver, Colo., Feb., 1953.

United States Bureau of Reclamation, "Milburn Diversion Dam Model Study, Missouri Basin Project, Nebraska," *Progress Report No. 4,* "General Studies of Headworks and Sluiceway Structures," *Hydraulic Laboratory Report No. 385,* Engineering Laboratories, Office of the Assistant Commissioner and Chief Engineer, Denver, Colo., Apr. 5, 1954.

United States Bureau of Reclamation, "Work Plan, Sheep Creek Cooperative Field

Demonstration Area, Garfield and Kane Counties, Utah," Interagency Report Prepared by the United States Bureau of Reclamation, Boulder City, Nev., Sept., 1957.

United States Bureau of Reclamation, "Hydraulic Model Studies of the Headworks and Sluiceways of the Woodston Diversion Dam—Missouri River Basin Project, Kansas," *Hydraulic Laboratory Report No. HY 451,* July 8, 1959.

United States Bureau of Reclamation, "Hydraulic Model Tests of Bottom and Surface Vanes to Control Sediment Inflow into Canal Headworks," *Hydraulic Laboratory Report No. 499,* Aug. 15, 1962a.

United States Bureau of Reclamation, "Hydraulic Model Study to Determine a Sediment Control Arrangement for Socorro Main Canal Headworks," *Hydraulic Branch Report No. HY 479,* San Acacia Diversion Dam, Middle Rio Grande Project, United States Bureau of Reclamation, Mar., 1962b.

Vanoni, V. A., and Brooks, N. H., "Laboratory Studies of the Roughness and Suspended Load of Alluvial Streams," *MRD Sediment Series Report No. 11,* Missouri River Division, United States Army Corps of Engineers, Omaha, Neb., 1957.

Vanoni, V. A., Brooks, N. H., and Kennedy, J. F., "Lecture Notes on Sediment Transportation and Channel Stability," *Report No. KH–R–1,* W. M. Keck Laboratory, California Institute of Technology, Pasadena, Calif., Jan., 1961.

Vanoni, V. A., and Pollak, R. E., "Experimental Design of Low Rectangular Drops for Alluvial Flood Channels," California Institute of Technology Sedimentation Laboratory, *Report No. E–83,* Pasadena, Calif., Sept., 1959.

Vetter, C. P., "Technical Aspects of the Silt Problem on the Colorado River," *Civil Engineering,* Vol. 10, No. 11, Nov., 1940, pp. 698–701.

Wall, W. J., "Stabilization Works on the Savannah River," *Journal of the Waterways, Harbors and Coastal Engineering Division,* ASCE, Vol. 88, No. WW 1, Proc. Paper 3063, Feb., 1962, pp. 101–116.

Woodruff, N. P., et al., "How to Control Wind Erosion," *Agricultural Information Bulleitn No. 354,* Agricultural Research Service, United States Department of Agriculture, June, 1972.

Woodson, R. C., "Stabilization of the Middle Rio Grande in New Mexico," *Journal of the Waterways, Harbors and Coastal Engineering Division,* ASCE, Vol. 87, No. WW 4, Proc. Paper 2980, Nov., 1961, pp. 1–15.

Woolhiser, D. A., and Miller, C. R., "Case Histories of Gully Control Structures in Southwestern Wisconsin," *Agricultural Research Service 41–60,* United States Department of Agriculture, Jan., 1963.

Yang, C. T., "On River Meanders," *Reprint Series No. 186,* Illinois State Water Survey, 1971.

Chapter VI.—Economic Aspects of Sedimentation

A. General Comment

1. Introduction.—Problems associated with the movement of sediment, either as erosion or deposition, were discussed in Chapter I. In Chapter VII legal problems related to erosion and sedimentation will be reviewed. Obviously, if such problems exist there are economic consequences.

In Chapter I, some data on cost of prevention of damage or cost of correcting an adverse condition were given. These were merely examples and do not provide sufficient information to evaluate the cost of erosion and sedimentation. Presentation of such information is difficult because the sediment problem is so ubiquitous that costs or benefits are hard to identify or classify.

For example, all land with an appreciable slope will experience geologic erosion though the rate may be small. When disturbed by agricultural practices the rate may increase or accelerate. Economic losses associated with accelerated erosion, or the cost of modification of agricultural practices to prevent accelerated erosion, must be considered in the overall economics of crop production. The natural Missouri River eroded its banks but at the same time built an equal amount of new land in areas formerly eroded. The economic response to erosion hazard was in the price of land, the greater the hazard, the lower the price.

Thus the economic effect of erosion, and deposition too, may be an increase in some annual cost or a loss of capital. Presumably these effects can be reduced to a single value by the use of some interest rate. Unfortunately, it must be realized that erosional and depositional processes are so general and apparently are so natural that they are rarely considered to be something that can be evaluated; therefore, from an economic standpoint, they are frequently ignored.

Nevertheless, some land areas in the world, e.g., parts of the near East and the limestone-dolomite region of Yugoslavia, have become a total loss, economically during historic times. Some areas of the southeast coastal plain of the United States have become practically useless for farming activities through active erosion.

2. Land Erosion.—A surprisingly small amount of information is available on the economics of land erosion. The information that is available expresses the economic cost of land erosion in terms of: (1) Loss of plant nutrients and the cost of replacing them by fertilizers; and (2) the increased cost of land tillage caused by erosion. Benefits from an erosion-control program usually are expressed, as far as the "on site" benefits are concerned, in terms of increased crop production.

A study of 24 flood-control reports by the United States Department of Agriculture covering an area of 185,200,000 acres gave an average of about $2.50/acre/yr as benefits from increased crop production to be derived from an erosion-control program (Leopold and Maddock, 1953).[1] Because this was a nationwide sample, there was a considerable economic variation among drainage basins—an average of about 3¢/acre/yr in the arid Virgin River basin in Utah, Arizona, and Nevada; about 23¢/acre/yr in the semi-arid Fountain Creek watershed in Colorado; and about $9.00/acre/yr in humid river basins, such as the Apalachicola in Alabama, Florida, and Georgia, the Green in Kentucky, and the White in Arkansas and Missouri.

It should not be construed that reduction in crop income or loss of land value should be the sole measure of the economic loss due to land erosion. There are also the economic losses resulting from the reduced economic activity associated with agricultural production-processing, marketing, and service activities of all kinds. Weinberger (1965), evaluated the economic cost of the destruction of agricultural land in the Deep Loess Hills area of Iowa and Missouri. He concluded that on about 125,000 acres of land subject to gully damage the capitalized value to society of all losses was about $603/acre. Of this amount only about $215 is the market value of an acre of land in good condition. A threefold multiplier of land values to express total social value is not unusual, in fact is conservative. However, the $603 by no means represents a sum that could be spent per acre to prevent erosion.

Beer (1962) also studied the cost of gully erosion in Iowa and concluded that the average annual damage from gullying was about $1.12/acre. This capitalized, would yield a figure a good deal less than Weinberger but the two are not really comparable.

Smith et al. (1967) studied the effect of sheet erosion on the Austin clay soil of Texas as represented by a continual decline in crop returns as erosion proceeded. After 18 yr the loss in income, as compared with uneroded areas in the same soil type, was $252/acre, just about equal to the land value.

B. Deposition of Sediment

3. General.—Deposition of eroded material is the counterpart of erosion. In the past the terms deposition and sedimentation have been used interchangeably by many writers. One of these was Brown (1948) who in 1948 reported on the annual economic cost of sedimentation as follows (Sedimentation as used in this quotation implies deposition of sediment: as used in the text it covers erosion, transportation, deposition, and compaction as defined in Chapter I.):

> The total average annual damage from all forms of sediment and sedimentation comes to approximately $174,000,000. The limit of error in this estimate probably is little if any greater than the error in estimates of the total annual flood damages in the United States. United States Weather Bureau data show an average annual flood damage of $100,657,000 for the 20 yr period, 1925–1944. The Forest Service has estimated the total annual flood and sedimentation damages at $250,000,000, based on data contained

[1] References are given in Chapter VI, Section F.

in 121 flood control survey reports, 6 by the Department of Agriculture and 115 by the Corps of Engineers, plus 99 preliminary examination reports by the Department of Agriculture. Altogether these reports cover 51% of the area of continental United States. The data were expanded on an areal basis. It is known, however, that several major classes of sedimentation damage, such as damage to reservoirs and agricultural land resources, were not evaluated in many of these reports.

The total sedimentation damage estimate includes these items:

1. Damage to agricultural land resources from overwash of infertile material, impairment of natural drainage, and swamping due to channel aggradation, associated flood-plain scour and bank erosion—$50,000,000. This estimate is based on data collected in field surveys of 34 watersheds on which flood control surveys have been completed, or nearly completed, by the Department of Agriculture. These watersheds contain about one-eighth of the land area of the United States, and being widely distributed should represent a fair average. The damage data were computed on a per-acre basis and expanded to the total land acreage of the country. If this figure seems large, it should be borne in mind that sedimentation damage to agricultural land is cumulative, whereas flood-water damage is recurrent. For example: on a given piece of valley land which is overflowed on an average of once in 5 years, sufficiently to produce a total crop loss, the loss is 1/5 of the potential net return. Take the same piece of land, however, and allow channel aggradation to raise the water table under the land so that crops cannot be grown at all, and the loss of return every year is 5 times as great.
2. Damage from sedimentation in storage reservoirs used for power, water supply, irrigation, flood control, navigation, recreation, and multiple purposes—$50,000,000. This estimate is based on the results from surveys of some 600 out of approximately 10,000 reservoirs now existing in this country. The national investment in reservoirs up to 1940 was at least 5 billion dollars. Thus, the damage figure is equivalent to about 1.0% annually of the investment. This should not be taken to imply, however, that the average rate of silting is 1.0%. The estimate is based essentially on the annual cost of fully maintaining the services now provided by these reservoirs through construction of additional storage or equivalent facilities as needed, plus the net losses if replacement is not possible. Additional storage must be provided in many drainage areas when 15 to 40% of the capacities of existing reservoirs are depleted. Often the supplementary storage will cost 2 to 10 times as much as the initial storage per acre-ft.
3. Cost of maintenance or progressive impairment of the capital value of drainage enterprises—$17,000,000. Studies of sedimentation in ditches of various drainage districts have led to the conclusion that an average value of 20 cents per acre annually for 86 million acres in drainage enterprises would give a fair measure of damages due to impairment of drainage canals or the cost of maintaining them.
4. Cost of maintenance of irrigation enterprises—$10,000,000. This is approximately 25% of the annual total operation and maintenance charge

for irrigation enterprises as reported in the 1940 census. This figure is believed to be on the conservative side.

5. Cost of maintenance of harbors and navigable channels as a result of sedimentation caused by erosion on watershed areas and stream-bank and channel erosion within rivers, but excluding sedimentation caused by tidal currents in harbors—$12,000,000. This figure is based on 1940 costs obtained from the U.S. Army Engineers Annual Report of that year. The costs chargeable to sediment transported from inland areas have been estimated on the basis of location of the project work.

6. Cost of water purification as a result of excess turbidity—$5,000,000. This estimate is based on studies of the costs of treatment in representative filter plants, and the amount of turbid water treated annually from surface sources.

7. Sedimentation damages, partly or wholly included in flood-damage estimates, including increased crop loss due to deposits of sediment on plants, increased damage to property, roads, streets, etc., by costs of cleaning off sediment, increased flood heights due to channel aggradation, at least 20% of the minimum estimate of total flood damage, or $20,000,000. This estimate is admittedly based on meager data. It is the estimated difference in damage from a few historic floods with and without the sediment load transported by the flood waters. Flood damages from the Ohio River flood of January 1937, and the southern California flood of 1938, and several others of record, show that a relatively high proportion of the flood damages in urban areas are due to the cost of cleaning sediment from streets, houses, furniture, etc.

8. Other damages not yet evaluated separately including the added cost of maintenance of highways, railroads, oil and gas pipe lines, communication lines, damages to power turbines, damages to the fish and oyster industry, damages to wildlife, damages to recreation, and impairment of public health from malaria due to clogging of channels—are believed to total at least $11,000,000, and may be much higher.

The benefits derived from the prevention of damages from sediment deposition by the United States Department of Agriculture erosion control programs in the 24 drainage basins reported on by Leopold and Maddock (1953) were estimated to be about 1% of the crop-production benefits, which is about $5,000,000 annually, or 2.7¢/acre. Again, extrapolating to the United States as a whole, this would be about $43,000,000/yr. It should be noted, however, that in 1953, Ford (1953) estimated the annual damage due to deposition of sediment to be about $160,000,000 in "upstream" areas of the United States. Much of the difference between the two estimates represents sediment damage that cannot be prevented. Too much reliance cannot be placed on these estimates because of inadequate data.

4. Deposition of Sediment in Reservoirs.—Special attention should be given to the economic problems associated with the deposition of sediment in reservoirs. Reservoirs, until they are filled, will trap most of the sediment moved by streams unless some specific provision is made to prevent it. Consequently the life of a reservoir is dependent upon the rate at which sediment is moved into it and upon the measures that are taken for sediment removal.

Reservoirs built in the United States from 1924–1941 having capacities of less than 200,000 acre-ft cost from about \$2.25/acre-ft–\$39.50/acre-ft—an average cost of about \$19.00/acre-ft. Reservoirs having storage capacities of more than 1,000,000 acre-ft built during the same period cost between \$1.50/acre-ft–\$14.50/acre-ft—an average cost of about \$5.00/acre-ft. At the present time these reservoirs all have large sediment accumulations, and replacement storage, if available at all, can no longer be provided at these costs. Present costs range from \$10/acre-ft–\$125/acre-ft. The increasing cost is caused not only by growing construction costs but more importantly by the fact that sites for low-cost dams are disappearing and will soon be gone.

Although it may have been advisable in the past to use additional storage as a means of solving sediment problems, this may no longer be possible. At \$10/cu yd an acre-ft is worth \$161.33; therefore, it is clear as we near prohibitive reservoir-construction costs, under favorable conditions, we are beginning to think about removal of sediment. Lest it be presumed that sediment removal can be adopted as a general practice, the experience of reservoir operation where removal of sediment has been undertaken should be considered. Ferrell and Barr (1965) reported on the operations of the Los Angeles County Flood Control District and stated:

> Trapping silt and debris in the basin is only part of the job. These erosion products must then be disposed of, and along with the excavation and hauling costs comes the problem and expense of locating and purchasing disposal sites. The latter must also be located near the canyon mouths to reduce haul distances.
>
> In the larger canyons, reservoirs built and maintained for the purpose of flood protection and water conservation become silted to the point where storage necessary for flood regulation or worthwhile conservation activities becomes inadequate. Desilting is an expensive and frustrating operation that involves sluicing and subsequent removal from downstream works or mechanical excavation and disposal similar to that of a debris basin. It can readily be seen that competition for available land in the foothill zone is entered into by both public agencies and private citizens.
>
> Primarily as a result of a medium-sized storm in the 1961–62 storm season, the Los Angeles County Flood Control District excavated 1,235,600 cu yds of debris from its debris basins at a cost of \$1,044,600 or about \$0.85 per yd, which does not include various extras such as disposal site costs. In actuality the total costs are concurrently running close of \$1.50 per yd. The great increase in cost is manifest when compared to the 1938 cost of \$0.33 per cu yd. . . . During the same storm the storage loss in 11 of the District's major flood-regulating and water-conservation reservoirs was computed to be 1,720 acre-ft (2,768,500 cu yds). These figures would be increased many fold for a capital storm. The need for stabilizing erosion in southern California, therefore, is based primarily on economic considerations.

Note that wherever the continuous operation of a facility depends on the removal of a sediment, whether in a reservoir or in a stream, the disposal problem is eventually a limiting factor. Insofar as costs are concerned the Los Angeles area is somewhat atypical. The large population density in an unusual topographic and geologic setting makes high costs feasible and increases the cost of disposal sites.

5. Deposition of Sediment in Estuaries, Harbors, and Coastal Areas.—Estuaries and bays are like reservoirs. They are being filled with sediment brought in by tributary streams. Deposition of sediment in estuaries, bays, and harbors, however, may be complicated by the movement of marine sediment by littoral currents, and all sediment problems in these areas are associated with tidal movements and density currents.

The Savannah River in Georgia is on the tidal estuary where the Savannah River empties into the Atlantic Ocean. Harris (1965, p. 669) reports that the annual shoaling rate for the inner harbor averages about 7,000,000 cu yd and by 1963 the maintenance program was costing more than $1,000,000/yr.

Many estimates have been made of the amount of sediment that moves into the San Francisco Bay system; the estimates range from about 8,000,000 cu yd/yr–4,000,000 cu yd/yr. Estimates by Smith (1965, p. 675) and Homan and Schultz (1965, p. 709) are complicated by the movement of debris from mining operations and by the construction of works on the Sacramento and San Joaquin River systems, which greatly modify the stream behavior. Maintenance dredging costs about $2,270,000/yr.

As in the removal of sediment from reservoirs, areas for the disposal of sediment dredged from channels become costly and difficult to obtain. In a discussion of sediment problems in Baltimore Harbor, Kolessar (1965) stated:

> The problem is one of the disposal of the dredged material. Waterfront property on the periphery of the harbor is highly industrialized and valued, in some cases, as high as $15,000 per acre. Large land areas for dumping are simply not available within economic distances from the channels to be dredged. Water areas adjacent to the shoreline are not often usable because depths of water and poor foundation conditions make the cost of necessary retaining dykes prohibitive. The dredged material composed of silt and clay with a dry weight of about 35 lb per cu ft is not ideal fill, and property owners are not eager to obtain such material for enhancement of their lands.

Not all costs of sediment control in estuaries are associated with dredging. Early efforts for the control of the bar at the mouth of the Columbia River were confined to jetty construction. This program was summarized by Lockett (1965). He reported that the initial work in 1885–1895 on the South Jetty resulted in the placement of 947,000 tons of stone at a cost of $1,969,000. The South Jetty was temporarily successful, but conditions deteriorated; a second project, consisting of an extension and enlargement of the South Jetty and the construction of the North Jetty, was completed in 1917. The project resulted in the placement of more than 4,800,000 tons of stone in the South Jetty at a cost of $5,657,000 and nearly 3,000,000 tons of stone in the North Jetty at a cost of $4,315,000. The South Jetty was rehabilitated between 1931 and 1936 and again in 1941, and the North Jetty was reworked in 1939. The cost of rehabilitation was more than $6,000,000.

Much of the growing cost of maintenance of navigation channels is caused by the increased size and draft of ships. In 1882, a 30-ft deep channel was adequate at the mouth of the Columbia River. By 1905, the channel depth had been increased to 40 ft and by 1954 to 48 ft.

6. Beach Maintenance.—Throughout the world and particularly the United States, beaches are rapidly increasing in economic value. This is because the number of fine recreational beaches is limited and the number of people using them continually increases. Thus beach erosion may have marked economic implications.

Most of the sands that make up the more desirable beaches have been brought to the sea by rivers. Whether a beach is stable depends on the movement of this material along the coast. This movement may be disturbed by storing the river sediment in inland reservoirs, interfering with the movement of sediment by the construction of harbors, breakwaters, etc., and finally by diverting the flow of sands into submarine canyons.

Watts (1965) discusses the problems resulting from the reduction in sediment outflow of the Santa Clara, Ventura, Los Angeles, San Gabriel, and Santa Ana Rivers of Southern California. Maintenance of beaches requires borrow of sand from various sources for replacement of that eroded. Watts gives no cost figures but indicates that the sediment contribution to the beaches of these streams has been reduced by over 1,000,000 cu yd/yr and states that stabilization of beaches will require an annual rate of borrow of about 300,000 cu yd.

The effect of harbor development was discussed by Herron and Harris (1966). They state:

> Sand bypassing, to alleviate the adverse effect of harbor structures, is becoming an increasingly important problem in southern California and in many other areas of the United States. Along the 300 miles of coastline between Point Conception, near Santa Barbara and the boundary between the United States and Mexico there presently exists 13 harbors and there are potential sites for 16 additional harbors. Nine of the existing harbors have interfered with natural littoral sand movement and 14 of the proposed harbors will require bypassing efforts. The population of this area is increasing at the rate of about 5% a year and the use of recreational boats is increasing even more rapidly. It is very likely that all of these 16 harbors will be developed within the next 35 yr. Thus, with presently available techniques and with an average annual littoral movement of 200,000 cu yd, southern California is facing an annual bypass cost of over $2,500,000. This situation will also be true of other coastlines where population and economic pressures will require full use of the coastline.

Saville (1961) reports that the first phase of a stabilization program on the Fire Island inlet to Jones Beach, Long Island, N.Y. cost about $2,700,000. No estimates of annual cost were given but the unit cost for sand movement was $0.66/cu yd.

7. Beneficial Aspects of Sediment Deposition.—Some phases of the sedimentation problem are beneficial to man, and the fact that "good" can result from the movement of sediment should not be overlooked. The rich bottom lands that border rivers and oceans are the result of sediment deposition in earlier times, and to some extent sediment is being deposited at the present time. The deltas of streams may be growing and slowly increasing the amount of useful land. Sediment-bearing water that overflows streambanks deposits silt and humus, which improve the top soil. Swamps, refractory clays, or saline soils may be buried by sediment, and the covered land is then available for agriculture.

Man has abetted nature in the building up and improvement of lowlands by spreading sediment-bearing water over the land, thereby building up soil or burying salt marshes with fertile sediment. In England this process is called "warping," and in France it is called "colmatage." In these countries, many thousands of acres of land subject to tidal or river overflow have been improved by the deposition of sediment. The same process has been used in parts of America; along the Nile, Tigris, and Euphrates; in China; and in other parts of the world. Salt marshes around San Francisco Bay and similar bodies of water in Oregon and Washington have been reclaimed and built up by applying sediment-laden water. Sandy soil along the Colorado River has been improved by clay deposited by irrigation water. Even the debris from the gold placer operations of the California 49ers resulted in an increase in land productivity in some places.

The most important effect of sediment deposition is the maintenance of fertility of land inundated annually by floods. The Nile River, mentioned in Chapter I is a case in point. In the United States, the elimination of annual flooding of the San Joaquin River in California was accompanied by a sharp drop in barley production from land that was formerly flooded (Waggenor, 1936). In this respect, a report prepared for the Kansas Industrial Development Commission is of interest (Wolman, et al., 1953):

> With respect to damages in rural or agricultural areas of the Kansas River basin, a soil survey has been made by the Department of Agriculture of the Kansas State College at Manhattan, to determine what damage and change in subsequent values may be due to flood inundation in such areas. Of the total area examined, 86.2% represents agricultural land to which no damage was done to the soil itself in the 1951 flood, even though included in all current official documents as showing material direct, or indirect, monetary damages. Damage to growing crops due to inundation, of course, was material.
>
> Observation of crop growth during the 1952 season, as well as chemical analyses of many soil samples confirm the conclusions deduced from the soil survey. All of these data with respect to soil damage, or primarily to the absence of such damage, are in agreement.
>
> The conclusions stated above have been confirmed by studies made elsewhere. For example, the Soil Conservation Service made some 200 cross-section surveys of deposition and soil removal following the 1937 flood in the Ohio Valley. Although their results have never been published, they are in the files of the Service. Of the Valley lands reviewed in that study, 22.8% suffered damage, of which 11% was due to deposits in urban areas. Only 11.8% of the rural area was adversely affected. The area benefited represented 37.4%, while the area unaffected accounted for 39.8%. In other words, 77.2% of the total area flooded was either unchanged or benefited. Gilbert F. White, who reviewed this situation in 1942, concluded: 'It seems clear, however, that the net effect in the valley as a whole was more beneficial than detrimental.'

C. Alluvial Channels

8. Channel Characteristics.—The characteristics of an alluvial channel are determined by the discharge of water, width, slope of the channel, and sediment concentration, grain size, and grain size distribution. A change in any one of these variables changes the characteristics of the channel.

One of the most important channel characteristics is stability, i.e., the degree to which bank positions are maintained and whether or not the bed of the channel degrades or aggrades. Consequently, changes in discharge, channel width and slope, and sediment concentration and size all affect channel stability; in some instances the effect is beneficial, but usually it is adverse. Almost all problems in the hydraulics of alluvial streams are intimately associated with the movement of too much or too little sediment.

9. Channels below Reservoirs.—The works of man that cause the greatest changes in the behavior of stream channels are dams. Reservoirs behind dams modify the discharge-duration relations of a stream, and they trap all of the coarser fraction of the sediment load and part of the fine fraction. The effects of reservoirs are seen most clearly on streams that normally move large sediment loads.

The Elephant Butte Dam was completed in 1915 on the Rio Grande in New Mexico. Stevens (1936) states that after closure, degradation began downstream from the dam and by 1925 extended for about 100 miles downstream; at El Paso about 127 highway miles below Elephant Butte Dam and farther downstream, the bed of the channel began to aggrade because floods originating above the reservoir were eliminated and no longer moved the deposits of coarse material brought in by tributaries (Stevens, 1938). The problems caused by channel instability of the Rio Grande resulted in a "canalization project" between Caballo Dam about 23 miles below Elephant Butte Dam and El Paso and the "rectification project" between El Paso and Fort Quitman, Texas originally about 150 river miles downstream. The projects, undertaken by the International Boundary and Water Commission, consisted of a leveed channel capable of carrying floods from local sources that cut the river distance between Caballo Dam and Fort Quitman about in half. Although this steepened channel is capable of moving much more sediment than the natural one, a large amount of sediment still is being deposited and must be removed (Bettle, 1965). In 1966, 800,000 cu yd of sediment was removed from the stream channel at a cost of about $210,000. Of this amount, about 40% had been deposited near the mouths of the arroyos that carry sediment to the river. The remaining 60% was incoming sediment that moved some distance downstream before being deposited.

As already observed, one of the problems associated with the removal of sediment from a stream channel is its disposal. Because sediment tends to concentrate in certain locations, areas for disposal become increasingly distant, which results in an increase in the annual maintenance cost. Through 1967, the International Boundary and Water Commission of the United States and Mexico had to purchase 35 acres of land for the sole purpose of providing places for the disposal of sediment.

The United States Section, International Boundary and Water Commission has made money available to the Soil Conservation Service for planning a program of dam construction to store sediment on some of the arroyos producing the largest sediment yields. At the present time (1967), a project for the control of three Rio Grande tributaries to the canalized section, Placitas Arroyo and Broad and Crow Canyons is authorized at a cost of more than $2,500,000.

The Colorado River has been controlled by Hoover Dam and by Davis, Parker, and Imperial Dams, 64, 156, and 303 miles, respectively, below it. The reservoirs above Davis and Parker Dams are known as Lake Mohave and Lake Havasu, respectively. The lake originally above Imperial Dam is largely filled with sediment.

This system of dams on the Colorado River has resulted in marked changes below Hoover Dam. When first seen by early explorers, the river was a slow aggrading stream in some of its reaches particularly in Arizona in the Yuma area below the present Imperial Dam and above Topock at the head of the present Lake Havasu. After dam construction, scour occurred below Hoover, Davis, Parker, and Imperial Dams, and aggradation took place above Lake Havasu and Imperial Dam. In 1965 Oliver reported that from the time of closure of Parker Dam in 1938 until October, 1944, 84,000,000 cu yd of sediment eroded below Hoover Dam. Oliver states:

> Surveys showed that 67% of the total originated in the reach between Hoover and the site of Davis Dam and that the yearly contribution of this reach averaged 500,000 cu yd by 1944. Although a large portion of the material eroded was deposited in Lake Havasu, some 48% of the total eroded up to December, 1944 was deposited in the flood plain between Needles and Topock.

Aggradation at Needles, Calif., 14 miles upstream of Topock and 60 miles from Parker Dam, was greatest in 1944 and decreased slowly until 1951, when the completion of a dredged channel dropped the stream bed to an elevation comparable to that existing in 1931. At Topock, gage heights continued to increase until about 1955 and have held steady since that time.

Oliver reported that control of erosion below Davis Dam and channel stabilization between the dam and Topock have been accomplished at a cost of $7,500,000. The principal benefit has been the reduction in evapotranspiration losses, which were estimated to be 65,000 acre-ft of water per year. This is in addition to the usual benefits of river bank stabilization. The estimated cost of channelization through Topock Gorge is $5,000,000, which probably will salvage an additional 60,000 acre-ft of water per year.

Degradation of the river channel below Parker Dam was of such magnitude that by 1944 it had become impossible to make gravity diversions to the Palo Verde Irrigation District intake 65 miles downstream from the dam. Between 1944 and 1957, diversions were maintained by a temporary rock weir. The weir was replaced in 1957 by a permanent structure costing about $5,000,000.

Although there has been a decrease in the magnitude of the aggradation and degradation problems in the 150 miles of river between Parker and Imperial Dams, Oliver (1965) stated:

> Degradation of the upper 90 to 100 miles has resulted in rather severe intrenchment, and, as the clear water is continually attacking the high banks, the sediment contribution is rather large. Water loss is also quite high in this 150 miles as a result of the excessive amount of braided, shallow, and excessively wide channel areas existing through the reach.

It is estimated that the cost of channel stabilization in this reach of river will be about $16,000,000. The principal benefit from this work will be the salvage of about 90,000 acre-ft of water per year.

Water is diverted at Imperial Dam into the All-American Canal of the Imperial Irrigation District on the west bank and into the Gila Gravity Main Canal of the Gila Project. Between the canal intakes at the Dam and the canals are desilting works to remove sediment that entered the canal system. Neither of these desilting works operated until the reservoir behind Imperial Dam became filled with sediment.

During this period the Colorado River channel below Imperial Dam degraded. Once the reservoir filled and the desilting works went into action the sediment removed from the canals was returned to the river and sluiced downstream. The river then stopped degrading and now has a slight trend toward aggradation. About 26 miles below Imperial Dam and a short distance below the California-Baja California border is the Morelos Dam, the Mexican point of diversion of gravity flow from the Colorado River. The sediments moving down the river from Imperial Dam were diverted here and, although considerably less in quantity than before dam construction, deposited in the Alamo Canal downstream from Morelos Dam.

To reduce the amount of movement of sediment into Mexico a dredge is now operating in a pool between Imperial Dam and Laguna Dam, about 5 miles downstream. Sediment is pumped from the river to fill old sloughs and ponds in the area between the two dams. The cost of this remedial program is not yet evaluated.

Thus the control of erosion through the stabilization of the Colorado River below Hoover Dam will cost about $30,000,000 with an annual maintenance cost not yet evaluated. The principal benefit will be the salvage of more than 150,000 acre-ft of water per year.

In many places regulated flows have been unable to cope with the encroachment of vegetation and the deposition of sediment from tributaries. An example is the Republican River below Harlan County and Trenton Dams for flood control in Nebraska. According to Northrup (1965):

> ... by analysis of flow records and field observations, it has been established that the usable channel capacity for non-damaging reservoir releases below Harlan County Dam, which had been assumed to be 15,000 c.f.s. in 1943 to 1945, was reduced to 4,000 c.f.s. in 1957. It is evident that about half the 11,000 c.f.s. loss in operational capacity is due to lowering the flood stage a little more than one foot since that time, and the remaining loss is an actual reduction in the carrying capacity of the channel.

The lowering of the flood stage is caused by encroachment on the river channel by riparian owners. The cost of a remedial program was not evaluated.

10. Channels above Reservoirs.—Aggradation above Imperial Reservoir and Lake Havasu has been discussed in connection with problems on the Colorado River. However, note that the behavior of streams above reservoirs and dams is somewhat uncertain (Woolhiser and Lenz, 1965; Maddock, 1965). An example is found in the Rio Grande above Elephant Butte Reservoir where aggradation above the level of the full reservoir has resulted in considerable difficulty and expense in maintaining the track of the Atchinson Topeka and Santa Fe Railroad

(1966) that was built in the valley of the Rio Grande. The suit of the railroad against the United States is discussed from a legal standpoint in Chapter VII. The railroad claimed that stream bed aggradation was associated with water levels in Elephant Butte Reservoir, that it had been necessary several times to raise its bridge on the Rio Grande at San Marcial and its track through the valley for several miles upstream from the town and that by 1954 it was considered advisable to move the railroad location out of the Rio Grande Valley. The Railroad asked for recovery of damages which in total amounted to $3,705,910.81.

In its defense the Government showed that the Rio Grande at San Marcial which, at the time of construction of Elephant Butte Reservoir was very near the full reservoir elevation, was aggrading before the construction of Elephant Butte Dam. Plaintiff and defendant agreed that this "natural" rate of aggradation averaged 0.27 ft/yr. In the period 1895–1954, accumulated aggradation at this rate was only 2-1/2 ft less than the actual aggradation. The court decided that for this reason, along with others presented during the trial, the construction of the dam was not essentially the cause of the Railroad's problems and damage recovery was denied.

A unique aspect of this court action was the presentation of a very large amount of basic data on the behavior of the stream. One source was the stream gaging station established in 1895 on the railroad bridge at San Marcial but even of more importance were data from a series of cross sections, established in 1914, located at intervals upstream from San Marcial. Note that this system of river ranges, first established by the Bureau of Reclamation has been, with time, extended for 150 miles upstream with the cooperation of the Corps of Engineers and the Soil Conservation Service. The Rio Grande in this reach is undoubtedly the most surveyed stream in the world.

11. Results of Other Types of Flow Modification.—One of the best-known sediment-oriented problem areas is the Middle Rio Grande Valley above and below Albuquerque, N. M. The consumptive use of water for irrigation in areas upstream from the Middle Rio Grande Valley and a period of unusual sediment contribution from tributaries has resulted in about 180 miles of aggradation of the Rio Grande channel. Woodson and Martin (1965) reported that from 1936–1953 about 17,500,000 cu yd of sediment was deposited on about 11,000 acres of channel and flood plain of the Rio Grande in the 154 miles between Cochiti and San Antonio, respectively, about 50 miles above and 100 miles below Albuquerque, N. M. The Corps of Engineers proposes to remedy the situation by storing sediment in four reservoirs all in New Mexico: Cochiti on the Rio Grande, Abiquiu on the Rio Chama, Jemez on the Rio Jemez, and Galisteo on the Rio Galisteo. It is estimated that the reservoirs will reduce the sediment inflow at Bernalillo about 20 miles upstream of Albuquerque, by 70% after about 20 yr.

The reservoir above Albiquiu Dam has a capacity of 1,211,000 acre-ft, and the one above Jemez Dam has a capacity of 120,000 acre-ft; the dams were built at costs of $21,297,000 and $4,177,000, respectively. The capacity allocated for sediment storage is 600,000 acre-ft at Abiquiu and 47,000 acre-ft at Jemez. Galisteo Dam cost about $15,800,000; the reservoir above the dam has a capacity of 90,000 acre-ft, of which 10,000 acre-ft is allocated for sediment storage. The construction of Cochiti Dam has been authorized and the cost is about $50,000,000; the reservoir has a capacity of about 602,000 acre-ft, of which 110,000 acre-ft is for sediment storage. Thus out of a total capacity of 2,023,000 acre-ft costing about $91,000,000, over one-third or 767,000 acre-ft are allocated to sediment control.

Another type of sediment problem occurs in Five Mile Creek, a tributary to the Wind River near Riverton, Wyo. The problem was created when canal wastes and drainage effluent from the Riverton Irrigation Project were disposed of in Five Mile Creek. The increase in runoff from about 4,000 acre-ft/yr under natural conditions to more than 80,000 acre-ft in 1950, resulted in the erosion of about 43,500,000 cu yd of alluvium between 1935 and 1950. By 1952, about 49% of the sediment being deposited in Boysen Reservoir by the Wind River was being contributed by Five Mile Creek. A corrective project for this problem was completed in 1953 by the Bureau of Reclamation at a cost of about $400,000. Upkeep, improvement, and maintenance costs have averaged about $4,000/yr. Maddock (1960) stated that by 1959, the sediment discharge of Five Mile Creek had been reduced by about 1,600,000 tons/yr. The cost of the reduction was about $13/acre-ft of sediment, which is cheaper than the per-acre-foot cost for Boysen Dam; in fact, few reservoirs can be constructed at this price.

12. Navigation on Rivers.—The navigability of rivers is associated closely with the way in which the stream moves its sediment load. In some places dredging of the channel is required to provide adequate depth. For example, the lower Columbia River navigation channel is 98.5 miles long from its mouth at the Pacific Ocean to the mouth of the Willamette River in the Portland-Vancouver area; dredging of about 9,800,000 cu yd of sediment is required annually through 26 river bars, a distance of about 50 miles. In addition to this, an average of 2,400,000 cu yd of sediment is dredged each year from the Columbia River entrance bar (Hyde and Beeman, 1965).

Considerable shoaling occurs in the lower Hudson River in the New York Harbor, and sediment dredged from only two shoal areas is 900,000 cu yd/yr. The problem is worse today than it was before World War II. In part, this may be due to the movement of material that was dredged from New York Harbor and dumped into deep water in the river during the War (Panuzio, 1965).

An example of the random behavior of sediment deposition by rivers and its economic effects was given by Smith (1965):

> During 27 to 30 May 1962, 28 motor vessels with 272 barges were delayed a combined total of 831 hr and 25 min due to the loss of project depth in Grand Lake Crossing (mile 504) on the Mississippi River. Until dredged, the controlling depth in this river crossing was 7 ft whereas the drafts of the loaded barges and towing vessels ranged from 8 to 10 ft. Again on 15, 21, and 22, August 1962 the navigation channel was blocked by sediment at three other separate shoaled crossings. These channel blocks occurred as the Mississippi River fell rapidly from middling stages and before the Corps of Engineers dredge plant could deepen a number of heavily shoaled crossings. These 1962 low-water navigation difficulties attest to the fact that sediment in the channel remains the major problem in providing a dependable navigation channel on the Lower Mississippi River.

Therefore, there are costs other than those directly attributed to dredging that must be charged to sedimentation.

According to Smith, in the Mississippi River between Cairo, Ill. and Baton Rouge, La., there are about 200 river crossings. From 40–85 of these do not scour at low flow and require annual dredging. The amount of sediment removed from the crossings each year ranges from 20,000,000 cu yd to 45,000,000 cu yd. For 1958–1962, inclusive, the amount of sediment removed annually averaged about 31,800,000 cu yd.

D. Municipalities

13. Planning Water-Course Modification.—The economic problems associated with the control of turbidity in municipal and industrial water supplies are well known. They were briefly discussed in Chapter I and will not be discussed herein. Equally or perhaps more important, however, every community has its water courses. As the community grows, it seems inevitable that there will be a decision of some kind that will modify the behavior of these streams. Discharges are diminished or increased, stream channels are straightened or confined, and sediment loads are modified. These modifications generally result in problems that are solved at relatively high expense. The expense for one modification is not very great, but there are so many modifications that the aggregate costs are large.

The writer has discussed this phase of the erosion and sediment problem with engineers whose practice is largely in the municipal engineering field. Almost without exception, they all say that the control of natural drainage is one of the most irritating and aggravating problems they have to deal with. Many high-cost drainage projects result from an inability to cope with what appear to be relatively simple problems. Thus an alluvial channel must be transformed into a pipe or a lined channel because its slope is too steep for the amount of water it is expected to carry. Straightening alluvial channels seems to be a minor adjustment but it inevitably leads to more serious problems. A realization that most natural channels respond to the movement of both water and sediment would do much to prevent obvious mistakes.

E. Sediment Removal

14. Cost of Works.—A recitation of the problems connected with the modification of the movement of water and sediment in stream systems leads to one conclusion: The problems arise as a natural consequence of the modification of a water-sediment regime for desirable and beneficial purpose.

If row-cropping is a necessary agricultural practice, some degree of increase in upland erosion must be expected. If roads and trails are to be built in the alluvial valleys in the Southwestern desert and if cattle are to be grazed in these areas, some degree of gullying must be expected. If water is to be diverted and consumptively used from major stream systems, aggradation may be expected. If permanent flows of water are introduced into normally dry waterways, erosion can be expected. Flooding and erosion are a normal expectancy on flood plains. Ill-considered occupancy inevitably results in loss. If natural channels are deepened, shoaling is sure to happen. A predictable reaction to a previous action may hardly be called "damage," although the reaction is, in its overall nature, injurious; rather, these reactions should be considered in the nature of costs for benefits brought about by the original action.

The responsibilities of sediment engineers are threefold:

1. To identify the problem and to forecast the nonbeneficial effects of the solution. Sometimes cause and effect relations are not known to those who make

decisions. A method of farming on flatland may be undesirable when applied to sloping ground. The diversion of stormflow from city streets to natural watercourses may be harmful. The breaching of a power dam filled with sediment will cause filling of channels downstream. All of these statements seem obvious, but such actions have been taken, frequently more than once, by people who did not know any better.

2. To design projects with desirable purposes to mitigate the undesirable effects, and when this is not possible.

3. To provide measures to control the undesirable reaction to an action for a desired purpose. Therefore, the engineer is confronted, as always, with the necessity of recognizing the problem, preparing a suitable plan to accomplish a specific purpose, and assessing all costs involved in the plan.

F. References

"The Atchinson, Topeka and Santa Fe Railway Company v. The United States," *United States Court of Claims No. 90–54,* June 8, 1960.

Beer, C. E., "Relationship of Factors Contributing to Gully Development in Loess Soils of Western Iowa," thesis presented to Iowa State University, at Iowa City, Iowa, in 1962, in partial fulfillment of the requirements for the degree of Doctor of Philosophy.

Bettle, A. F., personal communication, International Boundary and Water Commission.

Brown, C. B., "Perspective on Sedimentation—Purpose of Conference," *Proceedings,* Federal Interagency Sedimentation Conference, 1947; United States Department of Interior, Bureau of Reclamation, 1948, p. 307.

Ferrell, W. R., and Barr, W. R., "Criteria and Methods for Use of Check Dams in Stabilizing Channel Banks and Beds," *Proceedings,* Federal Interagency Sedimentation Conference 1963: *Miscellaneous Publication No. 970,* United States Department of Agriculture, Agricultural Research Service, 1965, p. 376–386.

Ford, E. C., "Upstream Floodwater Damages," *Journal of Soil and Water Conservation,* p. 240–246.

Harris, J. W., "Means and Methods of Inducing Sediment Deposition and Removal," *Proceedings,* Federal Interagency Sedimentation Conference, 1963, *Miscellaneous Publication No. 970,* United States Department of Agriculture, Agriculture Research Service, 1965, p. 669–674.

Herron, W. J., and Harris, R. L., "Littoral Bypassing and Beach Restoration in the Vicinity of Port Hueneme, California," *Proceedings,* Tenth Conference on Coastal Engineering, Vol. 1, 1966, p. 652.

Homan, W. J., and Schultz, E. A., "Model Tests of Shoaling and of Dredge Spoil Disposal in San Francisco Bay," *Proceedings,* Federal Interagency Sedimentation Conference 1963, *Miscellaneous Publication No. 970,* United States Department of Agriculture, Agricultural Research Service, 1965, p. 708–722.

Hyde, G. E., and Beeman, O., "Improvement of the Navagability of the Columbia River by Dredging and Construction Works," *Proceedings,* Federal Interagency Sedimentation Conference 1963, *Miscellaneous Publication No. 970,* United States Department of Agriculture, Agricultural Research Service, p. 454–461.

Kolessar, M. A., "Some Engineering Aspects of Disposal of Sediments Dredged from Baltimore Harbor," *Proceedings,* Federal Interagency Sedimentation Conference 1963, *Miscellaneous Publication No. 970,* United States Department of Agriculture, Agriculture Research Service, 1965, p. 613–618.

Leopold, L. B., and Maddock, T., Jr., *The Flood Control Controversy,* Ronald Press, New York, N.Y., 1953, p. 200–5.

Lockett, J. B., "Phenomena Affecting Improvement of the Lower Columbia Estuary and Entrance," *Proceedings,* Federal Interagency Sedimentation Conference 1963, *Miscellaneous Publication No. 970,* United States Department of Agriculture, Agricultural Research Service, 1965, p. 629–669.

Maddock, T., Jr., "Erosion Control of Five Mile Creek, Wyoming," International Association for Scientific Hydrology, No. 53, 1960, p. 170–181.

Maddock, T., Jr., "Behavior of Channels Upstream from Reservoirs," International Association for Scientific Hydrology, No. 71, 1966, p. 812–823.

Northrup, W. L., "Republican River Channel Deterioration," *Proceedings,* Federal Interagency Sedimentation Conference 1963, *Miscellaneous Publication No. 970,* United States Department of Agriculture, Agricultural Research Service, 1965, p. 409–424.

Oliver, P. A., "Some Economic Considerations in River Control Work," *Proceedings,* Federal Interagency Sedimentation Conference 1963, *Miscellaneous Publication No. 970,* United States Department of Agriculture, Agricultural Research Service, 1965, p. 442–449.

Panuzio, F. L., "Lower Hudson River Siltation," *Proceedings,* Federal Interagency Sedimentation Conference 1963, *Miscellaneous Publication No. 970,* United States Department of Agriculture, Agricultural Research Service, 1965, p. 512–550.

Saville, T., "Sand Transfer, Beach Control, and Inlet Improvements, Fire Island Inlet to Jones Beach, New York," *Proceedings,* Seventh Conference on Coastal Engineering, Vol. 2, 1961.

Smith, A. B., "Channel Sedimentation and Dredging Problems, Mississippi River and Louisiana Gulf Coast Access Channels," *Proceedings,* Federal Interagency Sedimentation Conference 1963, *Miscellaneous Publication No. 970,* United States Department of Agriculture, Agriculture Research Service, 1965, p. 618–626.

Smith, B. J., "Sedimentation in the San Francisco Bay System," *Proceedings,* Federal Interagency Sedimentation Conference 1963, *Miscellaneous Publication No. 970,* United States Department of Agriculture, Agricultural Research Service, 1965, p. 675–708.

Smith, R. M., et al., "Renewal of Desurfaced Austin Clay," *Soil Science,* Vol. 103, No. 3, 1967.

Stevens, J. C., "The Silt Problem," *Transactions,* ASCE, Vol. 101, Paper No. 1927, 1936, p. 207–250.

Stevens, J. C., "The Effect of Silt Removal and Flow-Regulation on the Regime of Rio Grande and Colorado River," National Research Council, *Transactions,* American Geophysical Union, Part 2, Washington, D.C., 1938, p. 653–659.

Waggoner, W. W., discussion of "The Silt Problem," by J. C. Stevens, *Transactions,* ASCE, Vol. 101, Paper No. 1927, 1936, pp. 268–271.

Watts, G. M., "Sediment Discharge to the Coast as Related to Shore Processes," *Proceedings,* Federal Interagency Sedimentation Conference 1963, *Miscellaneous Publication No. 970,* United States Department of Agriculture, Agricultural Research Service, 1965, p. 738–747.

Weinberger, M. L., "Loss of Income from Gullied Lands," presented at Annual Meeting of Soil Conservation Society, Philadelphia, Pa., August 25, 1965.

Wolman, A., Howson, L. R., and Veatch, K. T., "Report on Flood Protection Kansas Basin" Kansas Industrial Development Commission, Kansas City, Mo., 1953, p. 23.

Woodson, R. C., and Martin, J. T., "The Rio Grande Comprehensive Plan in New Mexico and Its Effects on the River Regime through the Middle Valley," *Proceedings,* Federal Interagency Sedimentation Conference 1963, *Miscellaneous Publication No. 970,* United States Department of Agriculture, Agricultural Research Service, 1965, p. 357–365.

Woolhiser, D. A., and Lenz, A. T., "Channel Gradients Above Gully-Control Structure," *Journal of the Hydraulics Division,* ASCE, Vol. 91, No. HY3, Proc. Paper 4333, May, 1965, p. 165–187.

Chapter VII.—American Sedimentation Law and Physical Processes

A. Introduction

1. Outline of Problem.—This chapter deals with some aspects of sedimentation law of interest primarily to the lawyer, engineer, geologist, ecologist, and soil scientist. It concerns rights in land, as deposited sediment, and rights to water, as containing or transporting sediment; and rights to be free from undue damage caused by artificial changes in the movement and effects of water, involving sedimentation as a process.

The process of sedimentation—weathering, erosion, transportation, deposition, and consolidation—will be followed in this statement, insofar as the positive law will permit. At each point the persuasive rules of local custom and practice will be anticipated or referred to when summarizing the coercive rules of law. However, it should be recognized that many of the rules of property and damage rights are really legal policy guides governing relationships between people, as individuals and groups, and their resources, and that the rules of custom and rules of law are interwoven in the fabric of social controls and agreements.[1]

Each such rule of policy has built into it features of: (1) Objective, purpose, and use that serve to channel human energy toward social goals; and (2) definitions, conditions, and limitations that serve to set depth, width, slope, and velocity guidelines for control of that human energy.

The implementation of these policy rules in terms of administrative organizations and programs, through grants of governmental powers, involving responsibilities in the field of sedimentation, is so large a subject that only certain aspects of regulation can be dealt with in this chapter. Again, however, these delegated powers are merely broad rules governing the relations among people and their resources in a governmental setting. They add up to the force that moves human energy through the policy channel.

The vast field of water law is treated herein only in skeleton form to bring out certain of its sedimentation aspects, recognizing, however, that every major quantity depletion, replenishment, and storage of water may have sedimentation and other consequences by altering the natural process at various points. Key questions are these: (1) What rights in and to sediments, as land (property), are recognized in law as arising out of natural and artificial changes in the movement of water; (2) What rights to legal damages are recognized in law as arising out of artificial changes in the movement or effects of water and wind, with special

[1] Numbers in parentheses refer to corresponding items in Chapter VII, Section E.

reference to sedimentation; and (3) What powers of government are recognized in law as necessary to regulate land and water use to prevent undue damage by sedimentation to resources and to the health, safety, and welfare of the community?

Answers to these questions usually revolve around concepts of public necessity and interest that change with the changing times. It is important, then, that scientists recognize the changing character of the physical and social processes (2).

2. Legal Concepts Applied to Water, Air, and Land.—It is commonly recognized that one must have possession (of property) or the right to possession as the basis for rights in land and water. The nature of what is meant by possession differs somewhat for the two types of resources, separately and together. In some cases physical possession is impossible or impractical, and so the legal idea of constructive possession comes into play.

When negligence, nuisance, and compensation are involved, it is also recognized that an injured party must be able to trace and prove as a matter of cause and effect that damage initiated and brought about by the action of another is the cause of material and substantial injury to others or their resources.

Furthermore, people who bring about such damage are liable for all the material consequences of their unlawful acts, save those outlawed by the statute of limitations or by some other legal defense, e.g., contributory negligence, laches, or estoppel. Such consequences may thus extend the effects of sedimentation damage to other types of damage, e.g., some forms of pollution, flooding, and waterlogging. Erosion damage is herein considered as a part of the sedimentation process.

In American jurisprudence, one cannot own the water as it runs in a stream or moves in the air, for one cannot legally possess it in these natural states. This has given rise to legal concepts as old as Roman Law; that these moving waters are the property of no one (res nullius) or of all people (res communes). All one can own of such moving resources is the right to take possession of and use them under certain conditions; the right to the fruits of such use; and the right to transfer these rights to others. There is also the legal concept of public property (res publici) that emphasizes public control of the resource within a given jurisdiction, such as a state (3).

Thus, most land and water rights are recognized as real property, and, in the case of water, until it is taken into possession, when it is considered personal property. But what constitutes personal property in water varies somewhat among the several jurisdictions; one can only own a "usufructory" right to water as it flows in a stream.

Presumably, these concepts apply to the sediment in suspension or otherwise, but this has not been well settled by the court decisions because definite values have not yet been attached to it. However, note that some streams carry certain types of sediments in suspension that are valuable as mineral and organic fertilizer. The Nile and Mississippi are good examples.

3. Erosion and Sedimentation Process Varies Geographically.—In considering these aspects of our natural resources, it may be well to keep in mind that the Midwest, Southeast, and Southwest vary greatly as to rock and soil material susceptible of erosion, transport, and deposition. And, also, that these geographic sections differ markedly as to the rate at which the sedimentation process is going on, being measured, and recorded.

The difference between the natural rate and any artificial rate produced by man is significant in fixing legal liability. But this may be a difficult matter of proof calling for the best scientific methods and techniques, both in engineering and in the law of evidence because the settlement of actions at law calls for reasonably ascertainable facts as to damage and some reasonable finality as to a settled accounting. Otherwise, fairness is not really possible and multiplicity of suits is to be anticipated.

The facts of sediment damage are presently more evident and measurable in larger stream valleys, e.g., the Mississippi, Sacramento-San Joaquin, Rio Grande, Colorado, and Red. Here the requirements of legal proof can more easily be satisfied than in small stream valleys owing to better and more reliable records. The public interest in these large valleys is greater, too, e.g., in terms of navigation, flood control, and drainage.

But the units of government and their agencies, owning and controlling land and water, cannot be sued for damages without their consent. Even if they do consent, their legal liabilities may be limited where the stream is a public or navigable one. Today the concept of what is a public stream is changing, as the needs of the public change. And so is the public interest in small streams and watersheds.

One of the most difficult problems in the field of sedimentation law is how to arrive at a final accounting of legal damage in the face of a physical process that changes over a span of years, during which natural processes of control become established.

4. Land Pattern Affects Process and Legal Consequences.—The land, from which sediments arise, occurs in natural patterns of rock and soil that reflect the geology, climate, vegetation, and topography of the earth's crust, now and in times past. But superimposed on this natural pattern is a cultural one that affects sediment production through land and water use. In like manner, the land that sediments produce also occurs in natural patterns and various cultural patterns are gradually superimposed on these.

In the civil law jurisdictions (Louisiana, Puerto Rico, parts of the Southwest and Michigan, and Quebec), the private land pattern often consists of long, narrow strips that extend back from the water's edge for various distances. Often these individual strips are fenced to permit exclusive use. This pattern reflects the riparian right of access to the water. But it tends to limit the opportunities for diversion and use of water (and included sediments) on the land owing to watershed and fence boundaries; it also tends to lead to subdivision into ever-narrower strips by reason of inheritance laws. At Taos, N.M., some strips are only a few feet wide and 1/2 mile long.

As a major example, the lower half of Louisiana is one vast sediment deposit (having its source in upstream states), subject to civil law rules of water use and management. Here is a vast area in which sediment deposits have created fertile lands, with the coarse red sands along the rivers and bayous, and the fine dark clays in low swales away from the channels. But along the south seacoast, the soils formed from such deposits are subsiding under salt water due to superimposed new sediments. Extensive efforts are being made there to prevent encroachment of contaminating salt water.

In the common law jurisdictions, the shape of land ownership tracts for settlements of the original states is often determined by metes and bounds (using stream thread or bank, as bounds, in some cases), or by sections and townships that ignore drainage lines for younger settlements of the newer admitted states. The Indian land pattern seems to have a rough resemblance to both, as at Taos. There are certain areas in the Mormon communities of the West that resemble the urban, wheel-like pattern of some European settlements.

In addition, there are public geographic boundaries of land areas, e.g., counties, municipalities, states, and the nation; and such as national forests, public parks, wildlife refuges, and so forth. These are either ownership boundaries or boundaries wherein governmental (police) powers may be exercised over natural resources and people. Any jurisdiction may declare public ownership of water within its boundaries and within its authority. But this has reference mainly to sovereign control of development and use, rather than to actual possession of the resource.

5. Water Pattern Affects Process and Legal Consequences.—The natural pattern of water occurrence over these natural and cultural land areas depends on slope, soil, bedrock, gullies, and stream channels. But superimposed on this is the invisible cultural pattern of water supply and rights of use, as defined and classified in law, e.g., diffused surface waters, vagrant flood waters, and defined stream and lake waters (watercourses). The latter are further defined and classified as either navigable or non-navigable, or just public for various reasons reflecting the current use needs of the people. The original common law concept of navigable waters was based on the ebb and flow of the tide, e.g., to float a sailing ship.

There is some difference as to how vagrant flood waters are presently defined from state to state, and the same is true for navigable waters. The federal test of navigability is especially important in determining title to stream beds and to sediments that have come to rest in these beds and on the flood plains.

Throughout the English-speaking world, there is usually a distinction made between the positive or existing law of record and the law as it ought to be in the minds of scholars. This division will be observed herein with special reference to the need to bring law and science closer together in terms of reality as process, e.g., the water cycle, the sedimentation process, the human rather than supernatural source of rights, and the interests they are designed to secure, and so forth. It is interesting to note in passing that much of the law of sedimentation is more nearly in line with the process character of reality than most other aspects of resource law. This is important to resource management.

These factors of natural and cultural patterns of resource ownership and use are outlined herein because they affect engineering surveys, appraisals, plans, and programs, as well as matters of law and administration. And because rights are often a matter of means of measurement and proof, as well as of definition and classification. The facts usually make the case.

6. Sources of Law.—There are two major sources of the positive law of sedimentation, civil and common, as modified by statute, compact, treaty, and constitution. The first, originating in Rome, was codified by Justinian. But the development of the law was somewhat different in respect to rights to water and sediments in France under the Code Napoleon and in Spain under Las Siete Partidas, etc., and their respective possessions (Louisiana, Puerto Rico, and Quebec).

In both areas local customs had a marked influence on the law as it evolved.

The French version tends to emphasize nonconsumptive uses of water, whereas the Spanish version tends to emphasize consumptive uses, though the latter was affected by the former after the Code Napoleon was promulgated. Differences in climate may have had an influence, too.

In these civil law areas, the legal method of defining water supplies or sources has been rather uniformly one of dividing up of the water cycle into many parts, and expressing each division in terms of land catchment areas; then the private and public rights and responsibilities for each such division and area were assigned, as well as for the cultural land pattern superimposed on it. This is not fully in line with scientific fact and, therefore, partially unsound because it tends to prevent unified administration of the law, conservation, and wise use of the supply in certain areas.

The process character of reality is the vital factor that is overlooked. The old legal definitions seem to reflect efforts to achieve greater certainty and stability but in the face of a process that is constantly in a state of change. However, these matters could be adjusted quite readily by applying more modern legislative methods of definition and by improving certain methods of judicial interpretation.

Louisiana, Puerto Rico, and Quebec emphasize water rights in watercourses mainly for those who own lands touching stream or lake banks. Nearly all of our eastern states follow these general rules, too. Some of these states permit diversion and limited uses on nonriparian lands, but do not usually permit seasonal storage of water by private persons, unless especially authorized by statute in the public interest. Specific authorization by new state statutes have been effective since 1954 in Mississippi, Virginia, Florida, Arkansas, and Iowa with some improvements in other eastern states.

In the western states, save for some remnants of the riparian rights of use, the prior appropriation law has been adopted as of necessity, to make possible needed depleting uses on suitable lands in a dry climate, together with the structural means of conserving water by storage. These uses and controls tend to deplete the natural flow to some extent in time and place, and, therefore, can appreciably affect sediment transport and deposition and also water quality.

Most of our states still follow some form of the English common law of diffused surface waters (precipitation runoff) which is not at any particular moment part of a stream. "Surface waters" is the term used in most official law reports. This is based on the theory that one who has title to the soil also owns all waters on and under it, as well as the space above and the minerals below.

This theory is not fully in line with scientific fact either, for many moving resources, e.g., water, oil and gas, air, and wildlife. For that reason it is partially unsound. However, the rates of ground-water movement are much slower than for streamflow, so the legal concept of possession should be adjusted accordingly. This old "cujus est solum" theory affects all water supplies because it affects every land ownership tract.

The various civil and common law concepts, including definitions of water supplies, have found their way into our State and United States Constitutions through various legal processes. These important provisions affect rights in water and land and rights to damage arising from sedimentation. The supremacy, due process, and commerce clauses are the most important. The due process clause is especially important as to sediment damage and its consequences. But its application is affected or limited by the commerce clause.

These constitutional provisions have been implemented by statutes concerning navigation, flood control, and drainage; hydroelectric power, recreation, and fishing; pollution, salt water contamination, and sediment control; and water allocation and use. They are too numerous to summarize herein, even though they could be included under the subject heading. Only special ones will be mentioned in this manual.

Suffice it to say, the broad trend of the law is away from older unscientific concepts and judicial administration toward more scientific concepts and executive administration. In this transition, the engineering and legal professions are playing a very important role by developing and applying basic scientific data within the broad framework of legal administrative processes and standards of fair play.

Much remains to be done, however, to bring definitions of water supplies, sedimentation and saline processes, and theories of property ownership and use in line with the process character of reality. One can see these trends and needs by examining, progressively, some of the broad rules of law and how they have changed over the years and over the nation.

In this summary statement it will be necessary to use much secondary material in presenting the subject and, at the same time, to make some assumptions regarding rights in and liabilities applied to sediments and sedimentation as a process, that seem to follow from legal rules applied to water and land. But it is hoped that eventually the subject can be presented in greater detail, so as to cull from the body of sedimentation law, as it is developing over the nation, the full story of this challenging field of social engineering.

B. Rights in Land and Diffused Surface Waters

7. Definitions of Supplies and Interests.—Precipitation accumulated on or running off the land is known as diffused surface waters in law, though the courts usually use the term "surface waters." Presumably, these have no defined channels in a legal sense, thus the term "diffused." Engineers know that water is usually in some sort of "channel" as soon as it starts to move over the land.

A channel ("watercourse") in law is one having bed and banks where the water flows when it normally does flow. Though this can be intermittent, there must be a "dependable" source of supply. Actually, the channel is merely a physical means of draining surface and ground waters. And the channel ought to be distinguished in law from the water supply source, as such, rather than confused with it. The channel affects the external properties of water but it is the internal properties of water that are most important (4).

Thus, the courts have divided the water cycle into diffused and defined surface waters. And then made an even more unscientific division, separating these waters from ground waters ("percolating" and "in defined subterranean watercourses and reasonably ascertained basins"). However, this was done when judges, as well as people generally, didn't understand the water cycle. And when there were few good precipitation and flow records. A distinguished judge has said that it is these men of authority who govern and not our legislatures; that they are the ones who formulate public policies and our basic law (5). If so, they should become better acquainted with applicable engineering principles.

Because land is commonly referred to in law as the absolute property of the owner, so, too, diffused surface waters are sometimes referred to as the absolute property of the landowner, in which they occur, for the title arises from ownership of the land (6).

Exceptions to that are the statutory, constitutional, treaty, and compact limitations in several western states, to the effect that all waters or all surface or running (stream) waters are the property of the state (or people of the state), or the respective states that are parties to the compact or treaty (7).

Relative rights to the use of these diffused surface waters have not been settled in the laws of our states, though the type and size of dams and purpose of reservoirs built to conserve these waters are regulated by statute in some western states from the standpoint of health and safety and to a limited extent from the standpoint of storage uses. More of this type of regulation may be anticipated in the more critical supply areas, especially in respect to livestock and recreation ponds. California adopted such a law in 1971.

There seems to be a tendency for some western courts to give the stream user a legal interest in diffused surface waters and in underground waters originating above and within the watershed to protect the established stream appropriator's interest. This is especially true in Colorado. But how to make this fully effective is a difficult administrative problem when the ground-water aquifer resembles a broad porous blanket.

In reviewing this aspect of the law, as applied to sediments carried and deposited by these waters, it may be well to think of diffused surface waters as: (1) Occurring in depressions; or (2) draining off the land through rills, rivulets, and small swales; and (3) combinations of these physical features. Rights to possession of the land also provide the primary legal basis for rights in and to these waters and sediments, as presently defined, except as indicated previously.

Thus, it would appear that sediments that have been deposited in these depressions and related physical features become the property of the landowner by reason of his possessory interest. By the same token, this could be true in large part for sediment deposits in downstream valleys below, and resulting from this type of runoff, even though these deposits may be in one's "possession" during one year and in transit and out of possession during another year. There seems to be little case law that specifically spells out these types of rules, as yet, for diffused surface waters and their "host" land. The law of streams is more definite in this respect.

In many densely populated areas of the world, fertile soil is so valuable that where accumulated as sediment on lower slopes, it is picked up in baskets and carts and carried upwards and then redeposited by hand on higher slopes to improve the eroded soil. In these circumstances, the "sediment-soil" has great value as property. Presumably, its possession affords a legal basis necessary to redistribute, conserve, and use it on the land. Of course, once it has left one's land and been deposited on that of another, the right of possession is probably lost. This aspect seems to have been recognized in Roman Law (8).

8. Common Enemy Rule.—From these "guideposts" of what the positive law is or might be in terms of definitions of supplies and interests under conditions of human occupation and need for land, one can now examine the existing rules on rights to use and change the movement of diffused surface waters and their included sediments over the land surface. In other words, these rules define the rights that apply to the interests that supplies can serve.

The so-called common enemy rule prevalent in Arizona, Arkansas, Connecticut, District of Columbia, Indiana, Maine, Massachusetts. Mississippi, Missouri, Montana, Nebraska, New Jersey, New York, New Mexico, North Dakota, Oklahoma, Rhode Island, South Carolina, Virginia, Washington, West Virginia, and Wisconsin (9) holds that the landowner (by reasons of his possessory interest) can treat these runoff waters as a common enemy and do with them as he pleases in order to protect his land and its use. Apparently, New Zealand and most of the Canadian provinces, except Quebec, have the common enemy rule, too. Iowa, Kansas, Pennsylvania, and Texas adopted this rule initially and then changed to the civil law rule. This was said to be the common law rule and was adopted by some states as rule of decision for the courts when the states were admitted to the Union or thereafter (10).

Such landowners can throw these waters back on higher or adjacent lands and should this result in (water and sediment) damage to these lands, no legal remedy is afforded the injured party. The courts refer to this as damage without legal injury (damnum absque injuria). But the public policy used to justify such a rule varies from one jurisdiction to another.

This general rule has been modified in some states by exceptions, such that one landowner may not needlessly, or carelessly, or maliciously do injury to other land by changing the movement of "his" diffused surface waters. That is, any such injury that substantially impairs the value of the land and its uses creates a legal liability. The injury might take the form of harmful erosion, sediment accumulation, waterlogging, or pollution (11).

In other words, the so-called absolute property right in land is not really absolute after all. It is qualified by exceptions in several states in the interest of the rights or needs of neighbors. These limited exceptions find fuller expression in the rule of reasonable use (discussed later) that has permeated our whole body of land and water law (i.e., the rule of reason). This rule could afford the opportunity for incorporating more science into the law because sound reasoning calls for consideration of relevant scientific fact, method and technique. Judges could, with confidence, open up the rules of evidence to let in more scientific facts and principles, especially when supported by local custom and practice.

9. Civil Law Rule.—In several other states and territories, the common enemy rule has been replaced by the civil law rule, or only the civil law rule was adopted there. This holds that the upper or higher landowner has an easement over the lower landowner for the natural flow of "his" diffused surface waters (12). Alabama, California, Colorado, Georgia, Illinois, Iowa, Kansas, Kentucky, Louisiana, Maryland, Michigan, Nevada, North Carolina, Ohio, Pennsylvania, South Dakota, Tennessee, Texas, Quebec, and Puerto Rico apparently have this rule, and possibly Australia, India, and Scotland. Presumably, the states of Mexico also have it. In a state of nature, this flow obviously would include some sediment. But when the land is improved, whether in rural or urban areas, the volume of runoff water and sediment may be greatly increased, especially under poor watershed management.

The problems arising out of improving land for various uses have led to the adoption of exceptions (really reasonable use rule) that the upper landowner may not unduly collect, concentrate, and discharge diffused surface waters on the lower land in increased velocity and volume, so as to do substantial injury to the lower lands. These exceptions have many ramifications.

Most state courts do not yet seem to recognize a right in the lower owner to have diffused surface waters (and sediment) come down to him, so he could use them beneficially, except for "dicta" in such states as Colorado and Utah, tending to give lower stream users some use rights in these waters. Presumably, the reasoning rests on the assumption that these waters do not have a defined and regular source or course.

But such is not really true for they do have a recurring source in the water cycle. And their courses often are relatively dependable, though not deeply chiseled out of the soil and bedrock. Though subject to overflow and spreading over the land in flat areas and under conditions of excessive runoff, the flow characteristics differ mainly in degree from that of streams that overflow their channels. In England it appears that the lower stream user has no tangible use claim to diffused waters in excess of what ordinarily does come down to him (13).

Thus, we see that from an old common law rule that permitted almost unlimited obstruction of these waters and sediment (favoring the individual interest of the lower owner to protect his land) or an old civil law rule that permitted little or no obstruction (favoring the individual interest of the upper owner to get rid of his excess waters), we find the law changing toward a rule of reason that tends to balance the relative interests of upper, lower, and adjacent landowners as to damage resulting from harmful runoff. Thus, their relative rights to be free from undue water damage caused by acts of man seems to have become more nearly correlative or reciprocal, as the land becomes more fully occupied and used. But the basic problem of the courts in defining these rights and the interests they are intended to serve, is the scientific one of how to measure, evaluate, and predict runoff and damage. And this is the task for engineers.

This change in the law merely reflects emphasis on the modern needs of society in relation to the needs of the individual; that is, the common need of the community to share the whole resource. However, the law has had little real development in regard to rights to use diffused surface waters. This has been left largely to local custom and practice of private landowners and local agencies, e.g., soil conservation districts, flood control districts, and drainage districts.

It has been brought out in various publications that the courts have made efforts to adjust the common enemy and civil law rules, working from the two extremes indicated previously toward the reasonable use rule. One aspect has dealt with the capture and use of these waters by artificial means (14).

A second aspect has dealt with the disposal of excess diffused surface waters by artificial means. These cases (in Arizona, Arkansas, Connecticut, District of Columbia, Indiana, Massachusetts, Mississippi, Missouri, Nebraska, New Jersey, New Mexico, New York, North Dakota, Oklahoma, Rhode Island, South Carolina, Virginia, Washington, West Virginia, and Wisconsin), discuss types of improvements that may be used, the consideration of good husbandry and improvement of the soil (15), and the volume of discharge into drainageways (16).

The third aspect has dealt with obstructions to the natural flow; a fourth concerns the so-called reasonable care rule. These seem merely to be steps toward adoption of the reasonable use rule but involve a wide variety of policy concepts that are somewhat conflicting. What is really needed is a new concept of using resources according to capability and treating them according to need.

10. Reasonable Use Rule.—The landowner, under modern conditions, is not unqualifiedly privileged to deal with diffused surface waters as he pleases. Nor is he absolutely prohibited from interfering with the natural flow to the injury of others. He is legally privileged to make a reasonable use of his land, even though the flow of water is altered thereby and causes some harm to others. He incurs legal liability only when his interference with the flow of water is substantially harmful and clearly unreasonable in the circumstances. The action shows a lack of respect for others and for the land.

This rule was first applied in New Hampshire. Apparently, only a few states have the rule as yet, e.g., New Hampshire, Minnesota, Alaska, and North Dakota (17).

The rule, in effect, is this: a landowner may use his own land as he pleased, provided he does not unreasonably interfere with like rights of others. Reasonableness and unreasonableness are questions of fact. It is the amount of harm caused in relation to the value of the improvements made; that is, the nature of improvements made or ought to be made, the nature and extent of the interference, the foreseeability of harm of one making the improvements and changes in flow, the purpose or motive with which he acts, and other relevant matters. Actually, these factors do not get at the real issue of capabilities and needs of the resource (18).

Every interference with the drainage of diffused surface waters, actually harmful to the land of another and not made in the exercise of reasonable use of one's own land, may be considered unreasonable. The rule does not lay down any specific rights or privileges but leaves the policy of the law to depend on the facts of each case and the principles of fairness.

This makes a lot of sense if scientific facts are used but it also provides limited guidance for landowners. Here is where local custom and practice become so important, for if they are scientifically established and then accepted, the courts have a sound basis for good decisions.

In some states it is recognized that the rule needs to be applied differently in agricultural areas, as compared with urban areas (19). The common law, then, has been modified in most states, even where the other two rules are applied, to this effect: a person may so use his own property as not unreasonably or unnecessarily to do substantial injury to his neighbor (20). The courts must reconcile the benefits with the harm caused.

Sometimes the terms "negligence," "trespass," and "nuisance" are applied to harmful changes in the natural flow of diffused surface waters, but without much consistency and precision. These remedies, in several different legal forms, were given a great deal of consideration to Roman Law (21). They could be used more fully today if the forms of action for redress of harm were built more around local custom and conservation experience, now receiving widespread consideration in local soil and water conservation districts. In other words, judges could lift up modern local experiences and incorporate these into local rules of law with real confidence that scientific facts and judgments had been used in their development. Such experiences are based on capability and need concepts of public resource policy.

11. Rules Governing Pollution Damage by Sediments to Lower Lands and Diffused Surface Waters.—Numerous court decisions deal with the liabilities of upper landowners for damage to land and water supplies below, resulting from sediment and other impurities introduced into diffused surface waters as a result of land-use practices. These cases may or may not speak of sediments, as such, but usually speak in terms of soil, rock, mud, sludge, harmful chemical or biological substances, including atomic or radioactive wastes.

It has been held that the upper landowner is not liable for damage to lower lands caused by diffused surface waters carrying soil and rock when they constitute part of the "natural formation of the land." But he is liable for resulting damage if he places other soil and rock where the natural drainage of such water will carry it to lower tracts or where it interferes with the normal drainage (22).

It has also been held that the riparian owner is not liable for impurities or pollution that find their way into a stream from the natural wash or drainage of his land within the upper watershed (23). But it is unreasonable use of one's riparian land negligently to dig up swamp mud and to sluice large quantities of loose mud into a stream and thereby cause (pollution) damage to a lower property owner.

Some states and counties in the United States are beginning to regulate erosion and sediment yield by disallowing certain poor land-use practices or requiring additional erosion and sediment control measures to reduce downstream damages.

12. Rules Governing Pollution Damage by Sediments to Navigable Waters and Adjacent Lands.—Many cases similar to these have resulted from mining and milling operations that sluiced sediment-type debris into streams, so as to pollute the water and land by acid and other harmful substances, cause overflow of the channel and flood damage, and reduce navigable channel capacity (24).

Hydraulic gold mining in California, where it leads to the reduction in navigable channel capacity due to sedimentation, came to full regulation after 1884, with only limited operations permitted (25). A situation something like this, involving iron and steel manufacturing, has arisen on the Calumet River in Illinois, that now empties out of Lake Michigan and connects with the Mississippi River system (26).

The 1975 high prices for gold and oil and the increased use of strip mining of coal, will see far greater pollution problems of the general type referred to in this section and also far greater controls under both state and Federal resource protection laws.

Here there are new legal questions to be answered in the light of changed needs of the public interest today, e.g.; What is a legal obstruction to navigation under the Rivers and Harbors Act? What is refuse matter? Where is the dividing line between sediment material that constitutes such an obstruction in fact and in law and that which does not?

One similarity in kind in these two situations lies in the fact that the debris was of grade size that could be carried in suspension by flowing water and later deposited with change in velocity in a navigable river channel. But beyond this the two problems differ in degree.

In the mining instance, the debris resulted from the washing of recent alluvial stream gravels and from sluicing the suspended silt and fine sand into the river after the gold particles had been removed. While in the manufacturing instance, the debris consisted of flue dust, iron oxide, slag, and calcine lime. This resulted from iron and steel making and from sluicing the very finely flocculated material, either directly into the river or into the sewers and thence into the river in which it deposited under relatively low-flow velocity (27).

The grade size of the latter was below 325 mesh screen. Expressed otherwise, 90% fell below 62-1/2μ and 80% fell below 31μ. A micron is 1/1,000 of a millimeter. In contrast, the largest molecule in solution is said to be about 0.001μ.

The companies used settling basins to remove most of the debris but enough passed on so as to create shoaling in the close-by channel. Some of the material may have been colloidal iron oxide (though the court record does not say so) because much of the suspended load could not be seen under the microscope, as indicated by expert testimony (28).

In the mining situation, there was specific legislation so stringent in regulation as virtually to force abandonment of the industry, except in rare instances, under the then prevailing prices of gold. In effect, this was also a drastic regulation of land use where the dredging takes place. In the manufacturing situation, there was only general and relatively old legislation that was subjected to rather strained legal interpretation by the Supreme Court largely in terms of certain definitions. This is shown by the 5–4 decision, detailed technical arguments of the dissenting judges, and sharply contrasting concepts of public policy as between the lower and higher courts. Here is an example of how a basic policy is changed merely by changing certain definitions (29).

The Republic Steel case reinterprets the broad general policy provision of Section 10 of the Rivers and Harbors Act of 1899, as amended, that says: "That the creation of any obstruction not affirmatively authorized by Congress, to the navigable capacity of any of the waters of the United States is hereby prohibited."

The words, "any obstruction," are, by this decision, thereby no longer limited to "works of improvement," as argued by the dissenting justices and the Circuit Court (referring to prior cases), but now also include industrial wastes of a very fine size when they reduce navigable capacity. The words, "navigable capacity," are already interpreted broadly enough in other court decisions to reach almost any stream capable of being made navigable by works of improvement.

It is of interest here that previous opinions of the Attorney General concerning the Act (Op. 21, 305, 1896, and 594, 595, 1897) had attempted many years ago to draw a line between suspended materials and materials in solution by defining an obstruction to include "other than works of improvement." The present decision seems to adopt the substance of those opinions (30). Thus, it is of special significance to engineers and lawyers that the interpretation of Section 10 of the Act now includes industrial solids in suspension but not in solution. And the interpretation also limits the exception clause, too, Section 13, to apply only to refuse flowing from sewers in a liquid state as sewage; i.e., organic wastes. The court implies that organic wastes react chemically on discharge into a stream, so as not to remain permanently as an obstruction in the form of a shoal deposit.

In subsequent situations, any industrial wastes sluiced into a navigable water of the United States or indirectly by way of tributaries to a navigable water body, whether direct or through municipal sewers, so as adversely to obstruct its navigable capacity, may be subject to the control of the Corps of Engineers, and to injunctive relief on failure to comply with Corps regulations. It could be a short step beyond this to apply these principles to other wastes and to silt from agricultural watersheds if the injury to navigable capacity could be traced with reasonable certainty to the source, "as a sequence of related physical events."

Thus, the Republic Steel case may have indirectly overruled the Brazoria District case in which silt resulting from faulty drainage ditch outlets and discharged into and adversely affecting a navigable channel were previously not held subject to injunction. In that case, the line defining an obstruction was drawn on the basis of a work of improvement, a structure, not on the basis of sediment in suspension that later became a deposit. At the same time, the Steel case tends to lend weight to the Sand and Gravel Case (31). It would seem worthwhile to relate these sediment problems, involving regulation, to those that follow, involving compensation under the 5th Amendment to the United States Constitution.

C. Rights in Land and Defined Surface Waters

13. Definitions of Supplies and Interests.—Waters accumulated in or flowing through channels with well-defined beds and banks are known as the waters of "watercourses." The courts tend to treat the channels and the waters in them as one and the same thing in the use of language but separates them, usually, as to ownership and jurisdiction. They should be treated in law as well as in science as separate and distinct, though closely related in terms of the external energy properties of water (32).

The law of watercourses, as it relates to sedimentation, involves riparian rights to land (involving deposited sediment and access to water), and to the flow and use of water (including storage, diversion, and disposal of waters containing sediments in suspension and solution). It involves rights to be free from undue damage caused by water and land use (involving injury to property done by overflow, erosion sediment deposited in channels and on the land, as well as other harmful consequences of interference with the flow of water), such as water-logging and salinization of lands and reduction in navigable capacity.

Few court decisions speak directly of these matters, as they relate to benefits from sediments carried in or deposited by these waters. They speak mainly of legal damages and compensation from interference with the flow or impairment of the quality of the water. Eventually, the law may recognize rights to sediment in transit.

Streams and rivers, as running waters, are not the absolute property of anyone, as previously noted. And one can acquire the rights referred to previously only by acquiring riparian land, or by making beneficial use of the waters, or by suffering injury or the threat of injury to property through the harmful acts of others, or by using public waters and channels as part of the public.

The conditions under which these rights are acquired and lost are important to engineers (and their clients) because engineers are called on to render services in measuring and appraising land and water resources and in evaluating property damage resulting from their control and use. Engineers need to have some familiarity with the legal rules of property ownership and boundaries and use values at the time surveys and appraisals are made (33).

14. Rights to Riparian Land as Deposited Sediment.—Rights to sediment deposited in stream channels or on flood plains may be gained or lost by changes in the position of the channel itself, especially its "thread," due to the action of water and sediment. The right of access to the channel and the right to the flow and use of water in it is the primary key to defining property boundaries and to the apportionment of the damage to land caused by erosion, overflow, and sediment. But there are also rights to be free from damage to property that has no frontage on the stream, e.g., railroads, bridges, and levees. Sediments affect the position of the channel in the flood plain by changing channel capacity and the topography of the surrounding flood plain (34).

Rights to deposited sediment depend on the law of the particular state in which they occur, unless the land is owned by or acquired from the United States. The rules vary somewhat from state to state, centering around the definition of navigable (or public) and non-navigable watercourses. The Federal rule defining navigability is controlling as to ownership of bed and banks of the channel where title stems from the United States (35).

15. Riparian Rights Gained or Lost by Accretion.—Accretion, as a process is defined in law as the slow, imperceptible addition of alluvial deposits (sediments) on the land margin or in the channel of a watercourse. When a stream or river recedes below the low watermark, the exposed deposits are also recognized as accretions. Apparently, the word "imperceptible" means to be unaware of sudden change (36).

The change in the land must be permanent, however. The rule does not apply to sediment alternately above and underwater, so long as the water substantially retains its old boundaries. This right is one of the bundle of riparian rights based on access to the stream that is the main guide to where the property lines are as the water recedes (37).

In the West, this right to accretions remains (despite the adoption of the prior appropriation law), so long as the stream has not been diverted by an appropriator. But the riparian owner has no vested right to have channel conditions remain the same, such that accretions will continue to be formed there in the future (38).

Under the common law of accretions, the water boundary and ownership shift with the stream (really the channel formed by it) and is not fixed. Thus, the lateral lines extending from the banks toward the thread of the stream change with the channel thread, so that they are always at right angles to the thread at any permanent position of the channel (to preserve the right of access). The riparian owner may lose or gain land in this process. But the common law may be changed by statute or words of an express grant, such as set forth in a deed. So it is important to anticipate these rules of the common law and of exceptions to it when surveys are made (39).

It is very important to keep in mind that where the United States granted title in fee of the beds of navigable streams to each new State as it was admitted to the Union, constituted a major statutory change in the common law of riparian rights in those stream valleys. Therefore, the natural change in the middle thread or "thalweg" of the stream did not thereafter control the private property lines coming out to the middle thread, as under the common law of riparian rights. This is brought out in Fig. 7.1(a)–7.1(d) where the lateral property lines are not necessarily at right angles to the middle thread of the Arkansas River. However, it appears that the courts have sought to assure each land owner in valleys of navigable streams a fair share of the area abutting on the stream channels.

When accretions are added to an island on one side of a channel, and to the mainland on the opposite side, and then by a change in the river they are brought together as one continuous body, the physical union of the two tracks does not make the island a legal accretion to the mainland (40).

An island "rising" in a river and unconnected with the bank belongs to the owner of the bed at that place. This owner may be a private person or the State. The general rule is that the State owns the bed of a navigable watercourse, unless that State permits the adjacent riparian owner to own the bed, subject to the navigation servitude, as in Illinois and Mississippi (41).

16. Riparian Rights Gained or Lost by Avulsion.—Avulsion is defined in law as the sudden, natural change in the stream channel, so that the water flows elsewhere than in its previous course. Usually, this takes place during a major flood when the flow breaks through the natural dikes formed by sediments along the previous channel and flows into lower alluvial areas. This may produce degradation in one reach of a stream and aggradation in another reach, such that a new equilibrium is established (42).

Generally speaking, the tract of land severed by a sudden change in the channel of the stream does not mean that the right to the tract has been lost. The basic riparian right may be lost by this method when the thread of the stream is no longer the natural boundary. But the original owner may ditch the stream back to its former channel if he does not delay beyond a reasonable time. This time is a question of fact for a jury to consider. The owner has a right to take precautions by strengthening the banks against sudden changes brought on by freshets and washouts, if he can do so without trespassing on the land of another and without doing undue harm to another's land. Here, again, the rule varies somewhat from state to state (43).

Where the channel is changed suddenly and the stream abandons its former channel, the respective riparian owners are only entitled to the possession and ownership of the soil formerly under the waters to the thread of the stream, as it previously existed (44). If the change is gradual instead of sudden, as previously indicated for accretion, the right is not lost because the accretion belongs to him with his won land and thus preserved his right of access. William K. Finefield, Chief of the Real Estate Division of the United States Army Corps of Engineers at Little Rock, Ark., has provided some excellent illustrations of both accretion and avulsion in the Arkansas River valley and their legal effects, if any. These are reproduced in Figs. 7.1(a)–7.1(d) from his publication entitled: "The Avulsion —Nature's Bad Boy" with due credit to him and to the United States Army Corps of Engineers:

> Let's look at the history of a typical avulsion which usually causes all the trouble. In Fig. 7.1 a typical situation has been sketched in which a build-up of land through an accretion followed by an avulsion can cause a great many problems. This is a map depicting four stages in the progressive change of the course of a river with the consequent changes in the property lines. Fig. 7.1(a) shows a piece of land lying on the right bank of a stream from which for all practical purposes, could be the Arkansas River, or any other rambunctious stream located in an alluvial valley, such as the Arkansas.
>
> Fig. 7.1(b) shows the area several years later and we notice that accretions have built up along the bank and the ownerships have slowly and imperceptibly increased. Fig. 7.1(c) shows that the accretion has now grown to a point where the river has started to create prominent bends which become quite susceptible to avulsions, either natural or man-made.
>
> Now, along comes a major flood and everything gets covered with water which is just tearing the heck out of things and, then behold, when the water goes down, we find the river bank in almost its old channel, as shown in Fig. 7.1(d).
>
> As a result of this action a large piece of Mr. Accretion is now on the other side of the river; or, more probably, it looks like an island in the middle of the

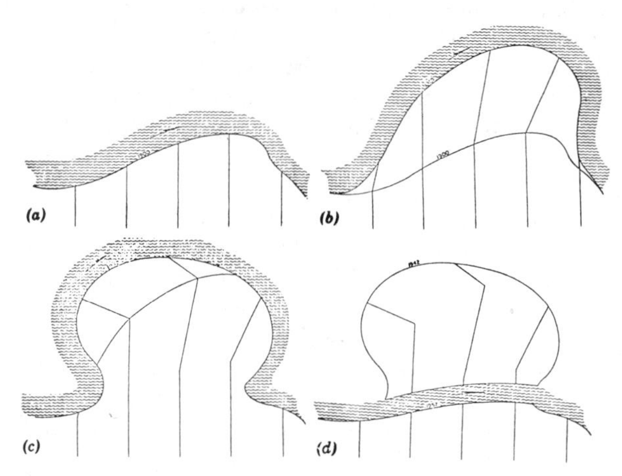

FIG. 7.1.—(a) Land Lying on Right Bank of Stream in Alluvial Valley; (b) Same Area Several Years Later; (c) Accretion Grown; River Has Started to Create Prominent Bends; (d) River Bank in Almost its Old Channel

river with the main stream at one point and a chute which used to be the river over at another point. Thus, we see that in the typical avulsion, the accretion or land lines do not change, neither does the ownership.

D. Rights to Be Free from Undue Damage Caused by Obstructions: Major Works of Improvement, Sediment Wedges, and Similar Causes

17. Definition of Influences Causing Undue Damages.—Environmental influences causing undue damage to lands, waters, and other resources are multiple in nature. This makes it difficult for the affected landowner to understand the cause of his injured property and to take appropriate action to protect his legal rights. The principal influences are briefly described under the following subject headings: (1) Construction of Major Works of Improvement; (2) Fluctuations in Reservoir and Other Surface Water Levels; (3) Severe Erosion and High Sediment Yields of Upstream Watershed Lands; (4) Unwise Use and Treatment of Upstream Watershed Lands; (5) Backwater Effects of Dams, Reservoirs, and Sediment Wedges; and (6) Combinations of Environmental Influences and Their Consequences. Each of these influences are discussed briefly hereafter.

Construction of Major Works of Improvement.—Two major types of improvement on, along, or across streams and rivers cause undue damage to lands because they intercept or accelerate the movement of coarse bed load sediments. The first and most prominent one is the dam and reservoir of which there are more than 1,000 large ones in the United States. Once constructed, these works permanently block the movement of coarse bed load to lower lying valleys, except in very unusual circumstances.

The latter process accelerates damage with time, particularly when the sediment in these wedges is alternately exposed to the air and then submerged. The body of American law on this subject is relatively new but it is increasing because of the large number of dams that have been built since the early 1930's.

The second type of improvement, i.e., levees and other bank protection works, has the opposite effect from the dam. This tends to accelerate flood flows and their water-borne sediments, so as to move the deltas or sediment wedges into the lower-lying reaches of the stream, river, lake, or reservoir. The legal problem here is very complex because it is difficult to prove the cause and effect relationships of the damages and the legal obligations of the parties, as required by the law of evidence. In preparing this chapter, the law on this particular subject was not searched.

There is also the effects of the release from reservoirs of clear water with high carrying capacity, e.g., to cause erosion of lower-lying stream channels and subsequent deposition of coarse sediments in reaches that have reduced gradients and in which the capacity to transport sediments is diminished. These effects may be combined with those of associated levees, so as to accelerate downstream damages. The Colorado River below Hoover Dam and the Sacramento River below Shasta Dam are good examples of the consequences of releasing clear water into erodible streams.

Fluctuations in Reservoir and Other Surface Water Levels.—It is recognized that a fluctuating water level exposes wedges to the air thus leading to more rapid consolidation of those materials. Thus they become hard and resistant to stream erosion and thereafter increase in height and extent to resemble low dams. The opposite effect occurs if the water body does not fluctuate appreciably in elevation and thus, the time required to consolidate those materials increases. The damaging consequences vary as to time, place, and intensity. This variability places a great burden upon the damaged landowner to show that the 6-yr statute of limitations has not run against him.

Reservoirs used primarily for irrigation purposes, tend to fluctuate more than reservoirs where storage and replenishment conditions are more stable from season to season and from year to year. Quite often fluctuating reservoirs exist in the higher sediment contributing watersheds of the hotter, dryer regions of the nation. However, many reservoirs built in recent years have multiple use purposes. The problem of reservoir level fluctuation cannot be considered in isolation from the other influences herein outlined, especially that of upstream control by dams or the lack of it.

Severe Erosion and High Sediment Yields of Watershed Lands.—Watersheds vary greatly in terms of their erosion and sediment yield characteristics. This is due to differences in geologic materials, in soils developed on those materials, in slope and drainage dissection of the land surfaces, in type, density, and condition of the vegetative cover, and finally in the management of the area by man.

Most volcanic rock materials weather into soils with moderate to high clay contents that support fairly good vegetative covers. Thus, for any given climatic zone, these factors combine to resist erosion and reduce sediment yield. However, the opposite is true of plutonic and some sedimentary sandstone rock materials that weather into soils with lower clay content, covered by sparse vegetation and yielding much more sand and gravel into stream channels. Of course, both of these situations are also dependent upon climatic influences and management practices. Volcanic rocks are composed of very fine-grained crystals while plutonic rocks are composed of large crystals. Each rock type weathers first along the cleavage lines of its respective crystal components.

Outstanding examples of high sediment-producing watersheds are the Payette River watershed in Idaho for coarse bed loads and the Rio Puerco River watershed in New Mexico for fine suspended loads. The former watershed is composed almost entirely of granite that contributes much sand and gravel into the river above Black Canyon Reservoir near Montour. The latter watershed is composed largely of fine-grained sandstones and clay shales that contribute much silt and clay into the Rio Grande River above Elephant Butte Reservoir.

Unwise Use and Treatment of Watershed Lands.—Probably the most variable and vital influences on erosion and sediment yields of watershed lands is that of use and management by man. This involves timber management in higher rainfall areas, management of cattle and sheep in areas where they are adapted, and management of vegetative cover on croplands. The type, density, and condition of the vegetative cover and the numbers, season of use, and distribution of livestock are crucial factors to be taken into consideration in the management of the range (see Chapter V, Section A). Since the early 1930's, this aspect has been improving steadily throughout the west. The management of dry-farm and irrigated farming areas is more complex than that of the range areas but this too has improved greatly in recent years.

The broad climatic belt of the southern half of the United States has the more erodible soils and the more critical climatic influences, so it deserves special attention in planning the development of watersheds with serious sediment problems. This also includes the northern plains of the midwest.

Backwater Effects of Dams, Reservoirs, and Sediment Wedges.—One of the most critical effects of dams and reservoirs or other works of improvement that cause damage to land is the hydraulic problem of backwater effects. This arises because the flow in the stream approaching the head of the reservoir is slowed down and consequently the flow depth is increased (see Chapter V, Section D.). This effect extends upstream to elevations that are higher than the lake surface. In the backwater zone, the water surface for a given river flow is higher than it was before the reservoir was filled, a situation that tends to cause flooding of valley lands and raising of the ground-water level.

Dam builders seek to acquire rights in land that may be adversely effected by the backwater. But all too often, they do not acquire enough land or do not take the full fee, and so in later years the land owners have to sue to recover damages. This is expensive to both parties. There are over 1,000 dams and reservoirs in the United States, each one intercepting the coarse bed load sediments. Many of these dams cause severe damages due to backwater.

However, this is only part of the story for there is a tendency for the coarse fractions of the sediment load to be deposited as the velocity is reduced in the backwater zone. This tends to further raise the water surface and to increase the potential for flooding. The severity and extent of such flooding depends upon fluctuations in level of the reservoir surface. Drawing down the reservoir allows the flows to move the sediment into the reservoir and minimizes the backwater effect. Continuous operation at maximum level will produce the greatest backwater effect. The accumulation of coarse bed loads in and above the upper portion of reservoirs leads to the formulation of sediment wedges that constitute low dams themselves. These low sediment dams may extend many miles upstream and then produce their own backwater effects at every major stream reach where the coarse sediments have accumulated.

Flooding due to backwater can be eliminated by levees, but this will not correct the problem of rising ground water. Levee systems must be designed so the sediment carrying capacity of the stream is uniform from section to section. If this capacity is not uniform, some sections may erode and others may aggrade and require expensive removal of sediment if exceeding the capacity of the channel and overtopping of the levee is to be avoided.

Combinations of Environmental Influences and Their Consequences.—It is a well-known principle in the earth sciences and ecology that certain combinations of environmental influences tend to accentuate damaging conditions or the reverse; depending upon whether the more dominant influences are destructive or constructive in their ultimate effects. These are commonly referred to as synergisms and as trigger, time lag, and threshold effects.

Out of all these influences, some of them are particularly important in this discussion because they relate to sedimentation. Some sediments arise out of erosion of unstable soils dominated by montmorillionite clays that tend to clog spawning gravel beds and adversely effect both fish and bottom life in streams, reservoirs, and other water bodies. Other situations involve mining debris, loaded with harmful chemicals that combine with clay particles to impair water quality and fish and bottom life.

Time lag effects are especially critical to this discussion for sediments accumulated in reservoirs, lakes, and stream channels may not reach critical levels for many years after the stream or river has been blocked. This makes the problem of the damaged landowner or water user who must sue to protect his rights, subject to the 6-yr statute of limitations, very difficult. An excellent example is the situation in the Payette River Valley of Idaho above Black Canyon Reservoir near the community of Montour. (The remarks made herein with respect to the Payette River Valley in Idaho, Walnut Creek Valley in California, and Verdigris River Valley in Kansas are based upon personal observation or detail study of the records by C.E. Busby as consultant to the parties at interest.)

How is the owner of land to know when damage due to the backwater effects of the dam and reservoir first come clearly to his attention? How can he distinguish these effects from other causes in the valley, e.g., sediments arising out of upstream road construction or channel maintenance work of railroads and highways? Of even greater difficulty is how to determine when the backwater effects of sediment wedges come clearly to his attention; in fact there is real doubt as to whether the 6-yr statute of limitations is really an equitable and constitutional policy of the law, for it does not comport with the time factors in the sedimentation process. It is time for the Congress to bring this old principle in line with the process character of reality and the rule of fairness already established by the courts.

Out of the sum total of these several effects, it is possible to erect a classification scheme for streams, reservoirs, and watershed areas that can be generalized as follows: (1) Fluctuating reservoir level, uncontrolled stream above; (2) fluctuating reservoir level, controlled stream above; (3) stable reservoir level, uncontrolled stream above; and (4) stable reservoir level, controlled stream above.

"Control" means that dams have been constructed in the valley above the lower dam, so as to effectively control flood runoff and coarse sediment movements. Of course, there are many degrees of fluctuations and stream control or the lack of them.

This type of classification scheme probably could be adapted to streams that are leveed or provided with concrete channels that tend to accelerate coarse sediment movements into areas of lower gradient, e.g., Walnut Creek in Contra Costa County, California. Furthermore, each of the four basic divisions previously mentioned, could be subdivided, so as to recognize differences in watershed sediment yield characteristics, i.e.: (1) Slight erosion and low sediment yield watersheds; (2) moderate erosion and moderate sediment yield watersheds; and (3) severe erosion and high sediment yield watersheds. "Sediment yield" here relates mainly to coarse bed load rather than to fine suspended debris. In many cases, the bed load source is stream bank erosion.

This type of subdivision scheme could also be applied to reflect the degree to which streams above lower reservoirs or other water bodies are actually controlled by dams. For example, the Verdigris River watershed in Kansas, above the Toronto reservoir, now contains 34 small dams on tributaries. The watershed is not a high sediment contributor, even without the small dams. However, the Payette River above Black Canyon reservoir in Idaho, has only two very small tributary streams controlled by dams. Yet the watershed is very erodible, producing large sediment yields of coarse gravel.

Key Decisions in Brief Form by Topic Headings.—Insofar as reasonably possible, based upon the facts, the statements made hereafter in this text on the succession of court decisions will seek to identify and relate key rulings of the courts to the influences or combinations of influences previously mentioned. Here are the key decisions in brief form:

Taking of Property by Overflow	*Cress* v. *United States* (45, 48) *Jacobs* v. *United States* (47) *King* v. *United States* (71)
Taking of Property by Overflow and Erosion	*Dickinson* v. *United States* (49,50)
Taking of Property by Overflow, Erosion, and Sediment Deposition	*Cotton Lands* v. *United States (55)*
Taking of Property by Overflow, Erosion, Sediment Deposition, and Rise in the Ground-Water Table	*Pashley* v. *United States* (66) *A.T. and Santa Fe Railway* v. *United States* (67) *Kenite Corp.* v. *United States* (69)
Taking of Property by Overflow, Erosion, Sediment Deposition, Rise in the Ground-Water Table, Obstruction of Land Drainage Systems, Loss of Water Rights, and Other Causes	*United States* v. *Gerlach Live Stock Co.* (65) *Richards* v. *United States* (68)

18. Rules Governing Recovery of Damages Caused by Obstructions and Sediment Wedges.—The law in regard to damage caused by sediments has had much of its recent development in the Federal courts and the United States Supreme Court. especially the United States Court of Claims (as to compensation under the 5th Amendment). But the rules in the early case law were developed mainly in connection with damage caused by water itself, e.g., overflow, and waterlogging of lands, resulting from dam construction. This has since been extended to certain other natural consequences of the building of structures and sluicing of debris into stream waters that change natural or existing water flow, sediment position and movement, and channel capacity. Eventually, it may include other related consequences, e.g., waste or loss of water supplies and rights.

Sedimentation and other forms of related damage are now recognized in law, as in science, as a result of the more recent court decisions. But the full extent of the legal aspects of sedimentation, salt, and water damage effects of artificial obstructions (including debris itself) have not yet been fully developed by the courts. Ultimately, this may require improvements in methods for determining the nature and extent of damage caused by the erosion-sedimentation-salinization process, especially where there are other causes involved, both related and unrelated to an obstruction placed in the stream channel. Damage consequences of change in the water table due to obstructions have only partially been settled in law.

Prior to the passage of the Tucker Act, there apparently was not an organized system whereby Congress could permit suits against the United States for damage resulting from dams and other improvements built by the government. With the passage of that Act, suits were permitted on the grounds of an implied contract to pay compensation for the taking of property under the 5th Amendment (due process clause). This principle was followed even though the property subject to damage was not specified and ascertained at the time of planning and construction (45).

But the argument was held for some time that the taking of property must be contemplated by government officials before an implied contract arose in law. If the legal complaint were based on tort (negligence, trespass, nuisance), there was no remedy at law under the Tucker Act because Congress had not authorized such remedy. Several cases were dismissed on these grounds. In fact, the proof of probable cause required by tort law could be very difficult when several miles of stream channel separate cause-and-effect areas, such as in the Southwest. And, also, where several industries are disposing of wastes into the same water body, as in the Chicago area.

So an implied contract was necessary to recover under the 5th Amendment, as then interpreted. But this method was complicated in the early cases by the lack of sound scientific methods, techniques, and information, particularly of an engineering nature, as to predictability of floods, rates of sedimentation, and related matters. In fact, this situation tended inadvertently at one time toward the legal conclusion that damage consequences were not contemplated, e.g., in the suit involving the Calaveras River, California (Sanguinette v. United States) (46).

Taking of Property by Overflow.—However, in 1930, in the suit on Jones Creek, Ala., a tributary of the Tennessee River, the Federal courts held that the Act of Congress authorizing the navigation dam and the findings made incident thereto were evidence enough that the taking was contemplated, even though the dam made only a slight increase in intermittent rises in streamflow and the overflows were only occasional, such that the full use of land for agricultural purposes was prevented for short periods. The dam caused increased legal burdens on the land in terms of more frequent overflows than normally occurred, and the resulting injury to use of the land called for compensation under the 5th Amendment (47).

The court organized its reasoning first on the basis of the riparian (property) right to the natural flow without burden or hindrance by artificial means. And, second, that no public easement for overflow can arise without government grant or compensation. The relationship here is between the private owner who is damaged, and the government. But note that this uses the natural flow rule, not the reasonable use rule (48).

The difference in the early cases is that in the Cress case the land had not been previously subject to overflow before the dam was built but was afterwards, whereas in the Jacobs case the land was subject to overflow before the dam was built, but the overflow was increased afterwards. The courts recognized the difference as one of degree, not in kind, but, nevertheless, fully compensable. Attempts to extend this "degree" theory of legal damage to sedimentation consequences have not been wholly acceptable, as yet, in some other cases, as shown by the close decisions and dismissal of some suits on grounds that the causes were legally too remote. King v. United States parallels the Jacobs case.

Taking of Property by Overflow and Erosion.—From overflow alone, as in the foregoing cases, the courts have moved to the next step of considering resulting erosion damage. This arose after construction of the Winfield Dam on the Kanawha River, leading to both intermittent and permanent flooding of property and to erosion of the banks of the pool. The government could have purchased the land and necessary easements (for intermittent flooding) but chose not to do so.

The suit was brought under the Tucker Act as on an implied contract. The dam caused a rise in the river level that led to both the overflow and the erosion damage. The court treated these as related aspects of a continuous series of events, not a single event. This very important case seems to have brought about a major shift in the basis of legal recovery away from implied contract and placed it squarely on a claim stemming from the 5th Amendment itself. Once established, this approach cannot help but be applied to other related aspects of the sedimentation process, so long as the damage can be fully ascertained and the cause of it proved with reasonable certainty. However, it is fraught with difficulty in proof where other watershed causes intervene, usually from upstream areas but not limited to such areas. For example, the effect of a dam and reservoir on ground water in the capillary fringe, and on sediments as they tend to spread surface water and build up ground water in this fringe, has not really been treated as yet.

The court established the new rule that property is taken in the Constitutional sense when inroads are made on an owner's use of his land to the extent that, as between private parties, a servitude would otherwise have been acquired by agreement or by the course of time. As a result, the 5th Amendment is now recognized in these cases as a basic rule of fairness (49).

The second significant part of the case was the court's answer to the government that a property owner is not obliged to sue when the flooding threatens his land, for to be required to do so would in effect force him to run many undue risks, e.g., several piecemeal law suits and also premature evaluation of what property is really taken. By the same token, it could impose undue obligations on the government. This, of necessity, prolongs the final decision in semi-arid valleys. Ice flows in northern latitudes do likewise.

As the overflow and erosion were merely parts of a continuous series of events in reality, there was no reason why the owner's suit should not be postponed until the river flow and channel situation had become reasonably well stabilized, such that a final legal account of the damage could be struck off in full and thereby discharged. Even though the owner may have subsequently sold the property, or have done protection work against future damage, the damage originally inflicted on the property was, nevertheless, fully compensable. The big problem, however, is how to measure and ascertain, within the requirements of the law of evidence, the actual damage due to the structures or other obstruction.

In other words, the 6-yr statute of limitations does not begin to run when the initial interference with the use of property starts (when reservoir filling started or when the pool first filled and the land was partly submerged); it runs concurrently with the sequence of subsequent related physical events. Thus time, quantity of flow, and level of flow are recognized, as well as their physical consequences. However, stabilization of a stream channel in high sediment-producing areas often requires many years (50).

This is important in engineering survey and appraisal work because the statute of limitations can prevent legal recovery, once it has run its full course in time. The problem here is how to determine when and to what extent the statute begins to work against the damaged party and when it has terminated or run its full course. The injured party must have legal notice of a harmful invasion of his property before the statute begins to "run" against him. What is legal notice is a question of fact that varies with the circumstances.

The third established point was that the secondary injury from flooding, namely, erosion of the banks, was also fully compensable. Mere level of land needed for reservoir capacity does not determine what property has been taken. But what is or has been taken includes land washed away by the water, even though incidental to the main part of the land taken. When any such land is left in a condition to be of less value than before, the owner is entitled to full compensation for that loss. The government must pay for all it takes but no more (51).

The fourth established point was that the measure of the erosion damage was the cost of preventative measures, if this could be accomplished by prudent action of the property owner. Here again, sound engineering practice comes into play. Even though the land was subsequently reclaimed by the owner, this would be no valid defense to his full recovery. It was the original taking that was compensable (52).

The fifth point was the rule that damage caused by intermittent flooding above the reservoir pool was also compensable (53). These points are reviewed here in some detail to help engineers and geologists understand the development of the law in the more complex cases, so they can better anticipate what the courts may do in the future.

Taking of Property by Overflow, Erosion, and Sediment Deposition.—The next stage in the development of the law took place in a high sediment-producing area, the Colorado River, above Parker Dam and Havasu Reservoir, where the causes and effects are separated by many miles of channel. This was a forerunner to possible fuller development of the law on the Rio Grande River above Elephant Butte Reservoir. In these cases, the very important principle of the continuous series of related physical events was extended to sedimentation and its direct consequences.

But the Colorado suit dealt with damage to land with a fairly stable reservoir level below and a major water control dam above (Hoover), while the Rio Grande suit dealt with damage to railway facilities with a fluctuating reservoir level below and lack of major control dam above. Thus, in the Rio Grande situation the intervening watershed causes and the instability of the stream channel, due to fluctuating reservoir levels and runoff from above, complicated the factual situation; so did the presence of salt cedar, one of the most important and little analyzed influences in these cases.

The Cotton Land Company sued to recover compensation for damage to (taking of) its property midway between Hoover Dam (86 miles above) and Parker Dam (53.4 miles below). The land occupied an 8-mile frontage on the river, though in alternate sections for 16 miles, and extending back for 1 mile to 5½ miles. It was 10 miles above Topoc and partially opposite the town of Needles. The greater space, time, and quantity relationships are significant here compared to the earlier cases, dealing with more localized problems mostly in the eastern states (See Fig. 7.2).

The river dropped sand and some coarse silt at the head of Havasu Lake, as a result of its scouring action below Hoover, beginning in 1939. This deposition "progressed" upstream, reaching Needles in 1939, and thereafter moved to a point 4 miles below the Company lands.

The sediment raised the river bed 9 ft, 3 miles above Topoc; 4 ft at Needles; and zero feet, 4 miles below the lands, as of the time of the suit. Thus, the water level in the river channel was also raised and the flood waters overflowed the Company lands. This action progressed up the river concurrently with the sediment deposition in the channel. The process, as shown by the survey records in this case, seems to reflect growing stability of the river channel.

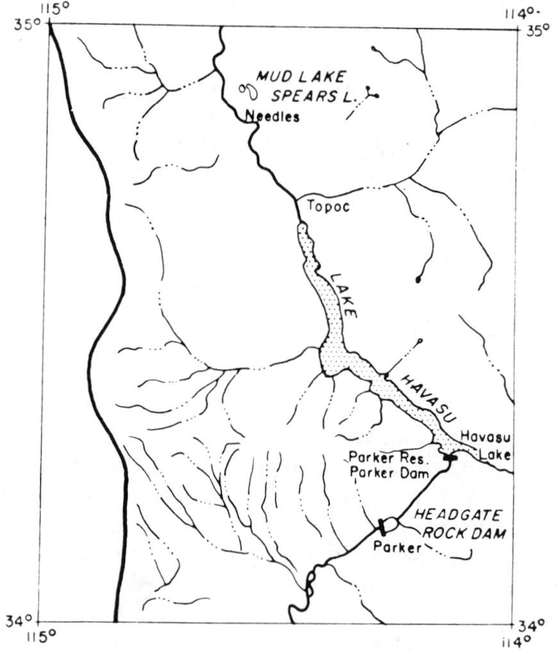

FIG. 7.2.—Sedimentation above Parker Dam and Havasu Lake

With heavy releases from Hoover Dam, the Company lands on the south were overflowed, and higher level of flows caused water to back into the sloughs on the north. There were 7,000 acres flooded, and 2,000 acres isolated (stipulation by the parties). Not much was mentioned by the court about waterlogging of lands but this might be anticipated in other cases of this type, e.g., in the old suit on the Savannah River in Georgia (United States v. Lynch, 188 U. S. 445, 1903), where established drainage ditches became no longer effective because of the dam. And, also, in the more recent case on Dardenne Creek, a non-navigable tributary of the Missouri River, in Missouri (United States v. Kansas City Life Insurance Co., 339 United States 799, 807, 1950). The latter was a 5–4 decision in which one judge argued that the common enemy rule in Missouri would prevent recovery in a private suit, so why should the government pay?

The series of physical events shown here in the Cotton Land Company case were: (1) Construction of dam; (2) filling of lake; (3) deposition of sediment in and at head of lake; (4) accumlation of sediment upstream therefrom, creating obstacles to flow and reduction in channel capacity; (5) rise in level of river flow in broad reaches; (6) overflow of banks; and (7) spreading of water over lands of the Company. The court does not seem to recognize, as a secondary cause, that the building of Hoover Dam and subsequent degradation of sediment below it by clear water affected deposition at Havasu. But these physical events all seem to stem from downstream causes. Upstream causes seem less significant here than in the Rio Grande situation. The events recognized by the court do not involve ground water and water-loving plants, as such.

The court held there was no taking under the 5th Amendment on initial filling of the lake. But subsequent physical events caused by the obstruction led to the taking and set in motion the chain of related and connected events, such that the injury could not be considered legally remote. Furthermore, this question should not be confused with the issue of the extent to which the United States has consented to be sued. The court held that the remedy in obtaining compensation was as broad as the requirements of the Constitution (54).

The court also held that it was not necessary to find that the agents of the United States were aware that their acts would result in a taking, adding up to a promise to pay. And that compensation was due, even so, whether or not United States agents were aware of the effect of their acts. This is significant to engineers when the question of available scientific information and predictability methods are taken into consideration. The loss is compensable, though it took considerable time to occur, resulting naturally from the improvements (55).

Taking of Property by Overflow, Erosion, Sediment Deposition, and Rise in Ground-Water Table.—The next step in development of the law has begun with the suit by the Santa Fe Railway against the government for damage alleged to have resulted to the Company right-of-way and facilities by the construction of Elephant Butte Dam and Reservoir. Although the physical situation is similar to that of the Colorado suit, there are important differences, e.g., fluctuating reservoir levels and streamflow and the intervening upstream watershed causes. It appears that the consideration of the causes by the Commissioner was restricted mainly to engineering aspects of surface phenomena (56).

The findings of fact and recommendations for conclusion of law by the Commissioner were not fully adopted, as such, by the court, though most of the findings were officially accepted. The petition of the Company was dismissed without prejudice to a future suit (on the condition of finding an ascertainable harmful effect that is not legally too remote).

Perhaps this is a wait-and-see determination directed to possible future effects of government programs, future action of the river, and more adequate and certain records as they relate to specific causes and issues. The relationship of salt cedar to sedimentation accumulation and flow damage and loss of water supply is an especially difficult and important factor in problems of this general type involving the consideration of soil moisture, capillary ground water, and sedimentation phenomena (57).

The determination of the court is that the suit of the Company was premature and that the plaintiff has failed to prove to the satisfaction of the court that the government dam and reservoir have increased or will increase the hazards to and expenses of maintaining the Company's properties. The court considered that the plaintiff had not proved with reasonable certainty, an ascertainable harmful effect in this regard.

It may be that the average annual sedimentation rates in this case tended to mask and obscure the cause and effect relationships shown by the cyclic fluctuations as carefully recorded at surveyed points. Variations of this type are sometimes more meaningful and certain than averages; both are abstractions, but averages are in a higher order of abstraction. Differences that make a difference are usually the more meaningful (58).

The court draws attention to the fact that the parties had agreed upon the rate of sediment deposition, had the dam and reservoir not been built, of 0.27 ft/yr, at or near Bridge 1006-A. And, of course, the findings show the actual recorded rates, year by year and cycle by cycle. The general method of finding the facts set out by the Commissioner and that adopted by the court are available in the official report for study by engineers.

The significance of the Santa Fe Railway case, had it not been dismissed, could be the recognition that damages by overflow, erosion, and sediment deposition are inextricably intermingled with other damaging upstream influences. These include past overgrazing on the watershed, increased irrigation that depletes the sediment carrying capacity of the stream, salt cedar infestation that slows flood flows, and increased sediment deposition on the flood plains. There is channel erosion after sediment loads are dropped, and artificial strictures in channels and valleys, produced by bridges, embankments, and dikes. This would have been fully compensable.

However, it does not appear that any real consideration was given to the effect of the dam and reservoir on soil moisture and capillary ground water in the valley, as affecting the growth and spread of salt cedar, in combination with their combined effect on sediment accumulation in broad areas of the valley of low gradient and of critical sediment grade sizes with high moisture holding capacity in specific reaches.

Finding No. 31 recognizes that the delta is an ideal area for the growth of vegetation (in discussing salt cedar), and also the areas above Bridge 1006-A but does not relate the two broad localities ecologically in reference to the dam and what took place sedimentation-wise between them. Had the suit been over lands and their water rights, this might have been a more vital factor.

These matters apparently led the Commissioner to the finding of an average "natural" rate of aggradation (parties agreed to this figure) and of an average artificial rate due to the dam. The interdependence of causes complicated the legal requirements of certainty and ascertainability.

The report of the case is presented, hereafter, in detail for discussion purposes because of its intensive treatment of the Rio Grande sedimentation problem. However, it should be pointed out that the suit did not really touch the massive problems of upstream watershed influences and lack of stream control above the Dam. Preliminary to that discussion is a very important addition to the subject of sedimentation, engineering, and law, prepared by J. W. Johnson (61) of the University of California in which he details the fact situation on the Rio Grande River and valley above Elephant Butte Reservoir. This stems from his unique experience as Consulting Engineer to and Expert Witness on behalf of the Santa Fe Railway Co., in the instant suit against the United States. It is therefore reproduced here with his permission, as his comments on the writer's paper appeared in the Journal of the Hydraulics Division of ASCE in 1961, but without the references. Immediately thereafter, the comments by C. E. Busby on the facts presented by Johnson are included herein as they appeared later in those proceedings.

J. W. Johnson.—The author has done the engineering profession a great service in presenting the important factors in an area of study involving many complex social and technical problems. One item of particular importance is whether certain types of sedimentation damage are "transitory" or "permanent" in character. One of the most common types of such damage is the delta deposit at the heads of reservoirs. This phenomenon has been studied extensively over the years by engineers and geologists. Two general types of delta deposits have been observed—one is the case in which a more or less single delta occurs as a result of a stable reservoir level made possible by upstream regulation, and the second case is one in which multiple deltas occur as a result of a fluctuating reservoir level due to no major upstream control. The author has described the conditions at Lake Havasu as an example of the first case and at Elephant Butte Reservoir as an example of the second case.

When a reservoir level does not fluctuate with time and the delta is more or less continuously wet, consolidation of the deposits probably is at a relatively slow rate compared with the multiple deltas that usually exist in a reservoir without appreciable upstream regulation. For example, H. M. Eakin and C. B. Brown, in examining the sedimentation in and above Elephant Butte Reservoir, state that "During low-water stages in the average year, most of the reservoir basin above the Narrows is uncovered. Sediment deposits dropped here during floods are exposed to air and hardened." Of particular importance in this connection is the fact that this effect of alternate wetting and drying of the sediments is to stabilize the delta. Flows during later years exert no appreciable scouring action or reassortment of the deposits. In fact, the delta (or deltas) act as barriers behind which deposition continues to occur over the years. Perhaps the best documented example of this condition is the deposition that has occurred in and above Elephant Butte Reservoir over the years following completion of the dam in 1916.

Delta deposits were formed at various locations in the reservoir as a result of major floods. The locations of the deltas, that later became stabilized and acted as debris barriers, were determined by the reservoir water-surface elevations at the time that the floods occurred. Fortunately, enough surveys of sediment deposition have been made over the years by the Corps of Engineers, Soil Conservation Service, and the Bureau of Reclamation to permit a reconstruction of the events and processes of delta formation. Basic data for the discussion to follow include chronological plots of monthly discharges at the San Marcial gaging station and the water surface elevation in the reservoir (Fig. 7.3), as well as sedimentation surveys of reservoir deposits in the years 1915, 1917–18, 1920, 1935, and 1947. An examination of sediment profiles from each of these surveys and an explanation of their characteristics in the light of the principles of the transportation and deposition of sediment are given.

Fig. 7.4 shows the thalweg profile, observed in 1915 prior to the construction of Elephant Butte Dam, and the bed profile observed in August, 1920. The location of Bridge 1006A near San Marcial, the Bosque del Apache, and the end of the leveed channel also are shown on this figure (also see Fig. 7.5). Fig. 7.4 shows the deposition occurring in Elephant Butte Reservoir during the period between completion of the dam in 1916 and 1920. The reasons for the location of the various humps or deltas occurring on the 1920 profile are readily apparent when the sequence of floods and the position of the water level in the reservoir at the time of such floods (as presented in Fig. 7.3) are considered. Thus, in Fig. 7.4 the water surface elevation at the beginning and end of the major flood flows occurring in the 1915–1920 period are shown, along with the deltas that were formed. The sequence of events are evident in this figure:

1. 1915 flows steadily raised the reservoir level to 4,360 and deposited sediment upstream from the dam.
2. During the 1916 flood, the reservoir level raised from about El 4,350 to 4,390, thereby creating a delta in the region of mile 14, 20 miles above the dam.
3. The 1917 high flows, occurring with a reservoir level between approximately 4,375 and 4,395, caused a delta to occur upstream from mile 20.
4. The 1919 flows filled the reservoir from elevation of approximately 4,340 to 4,400 and enlarged the whole delta deposit from about mile 12 to 25 above the dam.
5. A major flood in 1920 occurred with the reservoir level at the highest elevation since completion of the dam, with the result that sediment deposited upstream from the delta that was formed at, approximately, mile 25 during the 1917 and 1919 floods. These deposits apparently extended above the spillway elevation and beyond Bridge 1006A by backwater effect. The 1920 bottom profile, as appearing in Fig. 7.4 is, therefore, the net result of a series of floods that occurred with the reservoir water surface at various elevations, as well as some reassortment of the sediments during intervening periods of low flow in the rivers. The important floods and the resulting deposition that occurred during the other periods between sedimentation surveys will be discussed briefly:

Sediment Deposition During Period 1920–25.—The survey (Fig. 7.4)

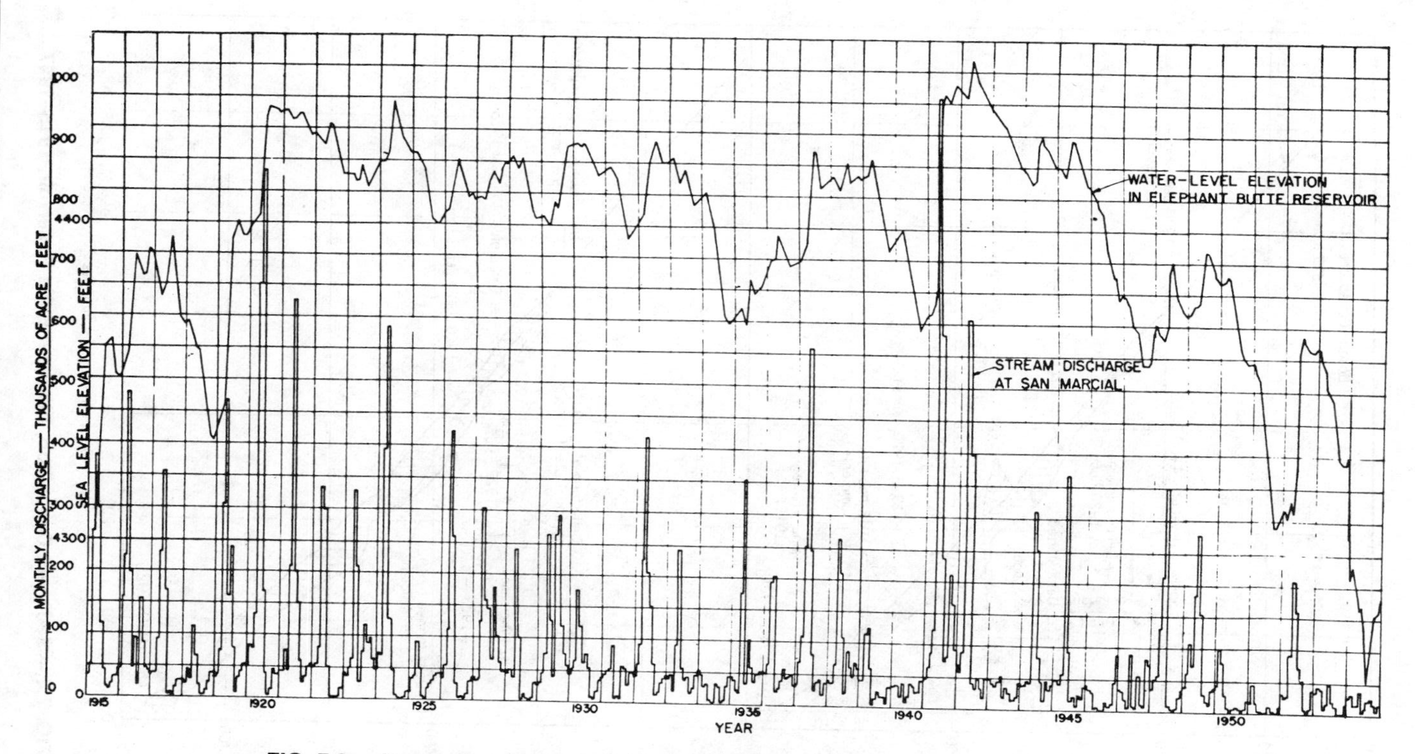

FIG. 7.3.—Streamflow in Rio Grande and Water-Level Elevation in Elephant Butte Reservoir from 1915–1954

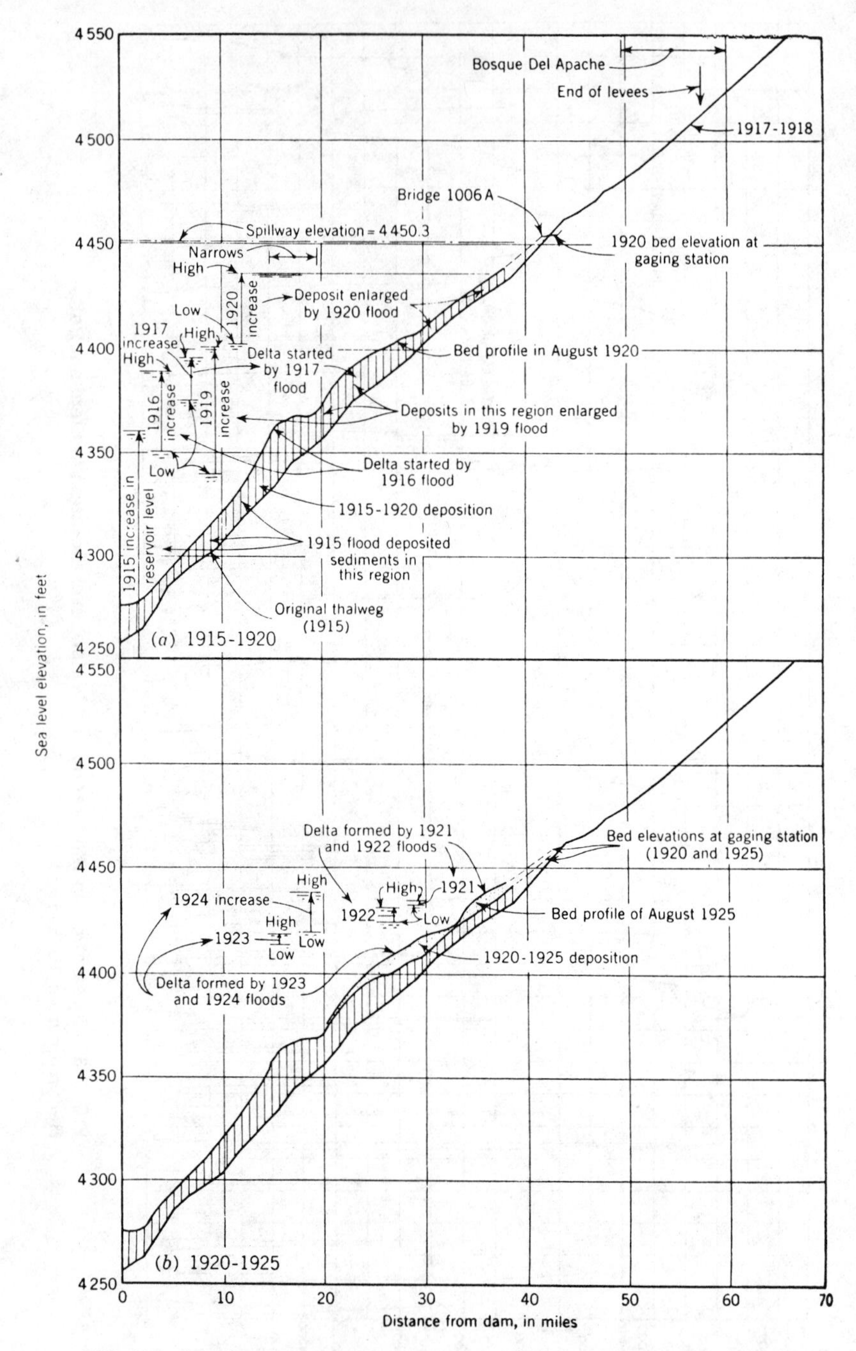

FIG. 7.4.—Sediment Deposition, Elephant Butte Reservoir, 1915–1947

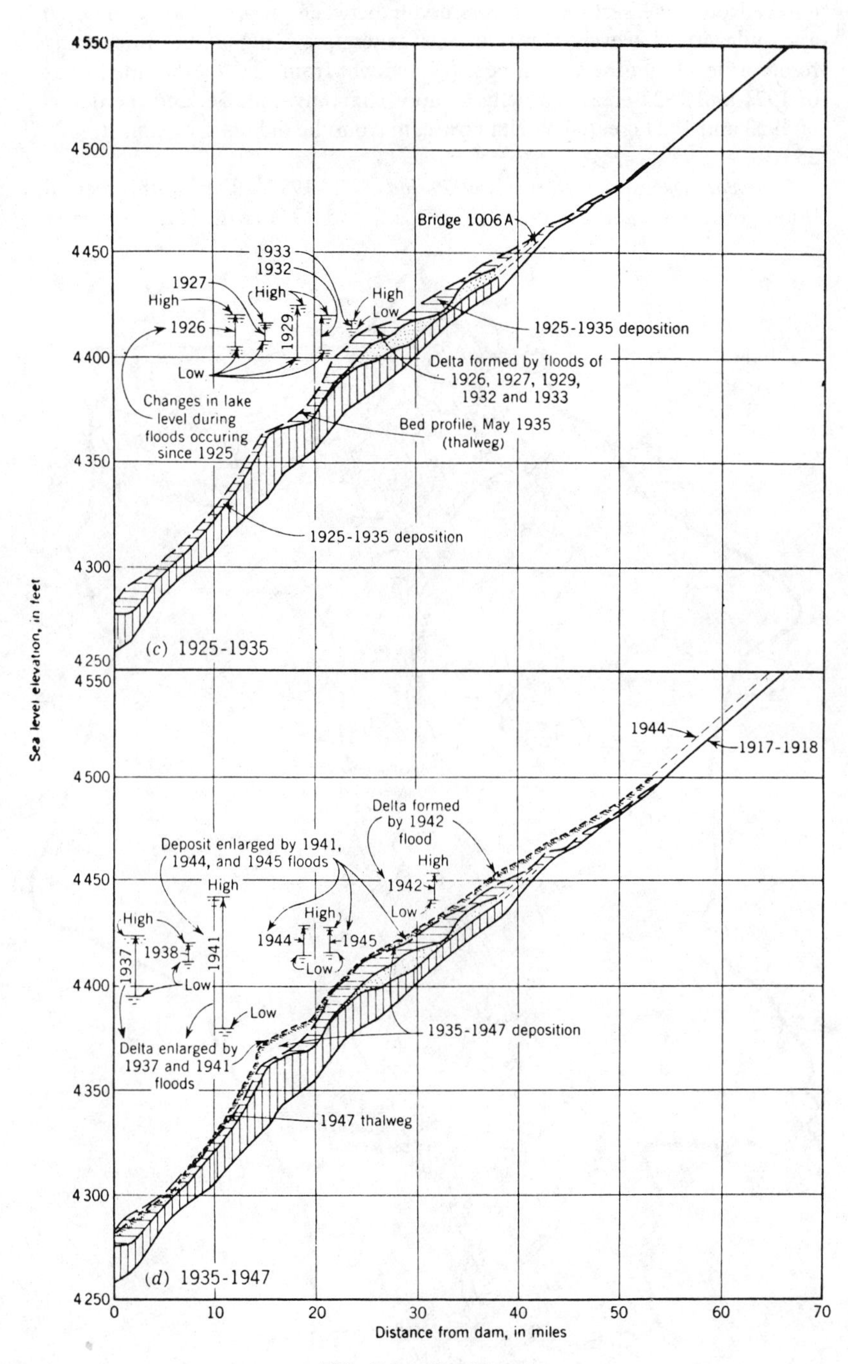

FIG. 7.4.—Continued

covered only the section of the reservoir between, approximately, mile 20 and mile 40. However, it was in this general region that the fluctuating reservoir level met the stream bed. It is evident from Fig. 7.4 that the floods of 1921 and 1922 created a delta at, approximately, mile 36, and the floods of 1923 and 1924 created a delta upstream from the old delta existing at mile 25.

Sediment Deposition During the Period 1925–1935.—During this period floods occurred in 1926, 1927, 1929, 1932, and 1933, with reservoir water

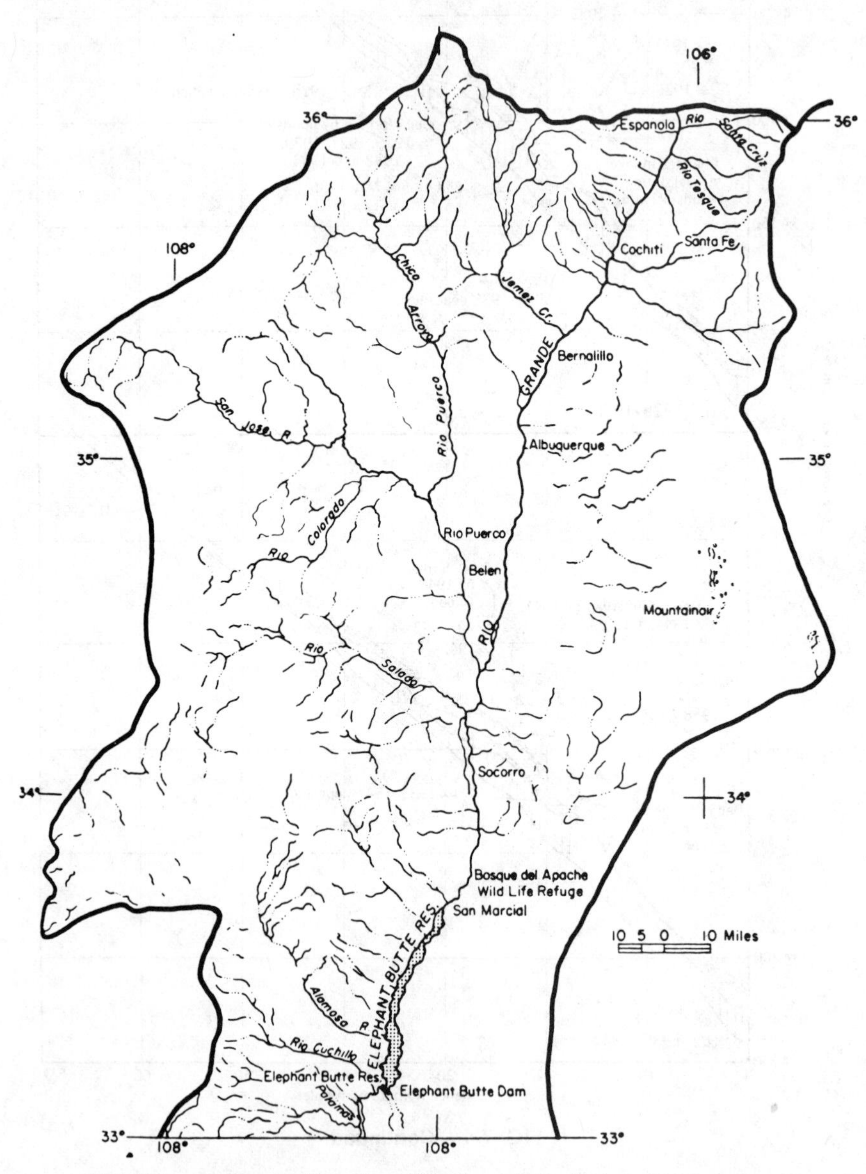

FIG. 7.5.—Sedimentation above Elephant Butte Dam and Reservoir

surface fluctuating only over approximately a 30 ft change in elevation. Sediments carried by these floods were effective in enlarging the delta in the general area of mile 25 (Fig. 7.4). It is evident from Fig. 7.4 that some sediments were deposited in the lower reaches of the reservoir during the period 1920-1935 (possibly by density currents).

Sedimentation Deposition During the Period 1935–1947.—Floods in 1937, 1938, and 1941 created a large deposit between, approximately, mile 14 and mile 21. The flood of 1941 was the largest on record and caused the reservoir to fill and flow over the spillway for the first time (Fig. 7.3). With the reservoir at this high level, the 1942 flood created a delta deposit in the upper reaches of the reservoir and undoubtedly well beyond Bridge 1006A. Smaller flows in 1944 and 1945 probably added to the wedge of sediment upstream from the old delta at mile 25.

Flows in the Rio Grande have been relatively small, and the water level in Elephant Butte Reservoir has steadily declined to the lowest elevations since completion of the dam. During these years of extended low flows, the river has trenched into the old deposits to some extent. However, this trenching only occurs over the width of the low water channel (generally less than 100 ft), whereas the sediment deposits extend over the entire width of the valley (several thousand feet). The sediment scoured by such trenching action and carried into the lower reaches of the reservoir is an extremely small portion of total sediment deposits. Also, the trenched low-water channel will quickly refill with stream-born sediments during a succeeding flood, due to the smaller stream gradient in the upper part of the delta, compared to that in the river further upstream.

It is evident from Figs. 7.4 as well as from the results of more recent sedimentation surveys in this area, that the multiple deltas in Elephant Butte Reservoir, particularly the deposits near mile 25, have become compacted over the years and are now stabilized barriers behind which a sediment wedge builds upstream at a greater than normal rate of aggradation. The volume of sediments removed from such deposits by low-flow trenching action in the future may be expected to be nil compared to the volume of stream-born material that will be deposited above these barriers.

C. E. Busby.—The writer appreciates Johnson's excellent statement on the multiple delta deposits above Elephant Butte reservoir, together with his related comments on single delta formation. This type of information is helpful to all interests in evaluating upstream damage problems stemming from sediments, especially as these damages relate to the statute of limitations. It would have been helpful to have had similar data on the Colorado above Havasu.

Time is important in application of the statute of limitations. In fact, its application in the field of damages is unique for the reason that the running of the statute and the natural processes are inextricably interwoven in time and space. Johnson points out that stable-reservoir deltas take more time to consolidate and presumably to perform their barrier funcitions. The reverse seems true of the unstable-reservoir deltas. Thus, the interval of time from initial filling of the reservoir until notice of damage comes fully and legally to the attention of the injured party varies accordingly.

In any event, the important question as to legal liability is proof of the sequence of related physical events, as one continuous causal process. Johnson brings this out in terms of precise records of flood flows, reservoir levels, and sediment deposits by years, periods of years, and reaches of the stream channel.

The writer sought to bring out from the official records the fact that the Commissioner, Court, and expert witnesses were not sure of the real cause of the injury to company property in the early years after completion of Elephant Butte dam. However, they were reasonably sure in the years after completion, when the damage effects became more evident many miles upstream. This points to the typically cautious official attitude, such as prevailed many years before in the Sanguinette case on the Calaveras River, Calif. It is submitted that to be sure in later years is also to be sure in the early years if the concept of related physical events makes any sense.

The problem of relating cause and effect in these matters is similar to that involved in the rule of "relation back" (principle of due diligence in putting water to beneficial use) in the law of appropriative water rights. However, in the damage situation, the time, distance, and quantity relationships are of greater magnitude. In these circumstances and because of the possible intervention of other causal factors, it would be helpful if the Court could adopt special criteria for formulation of a rule of "relation back" in water and sediment damage suits. Not only would this facilitate tracing the cause upstream to the actual points of damage but also back again to the point of origin. These criteria might also include the consideration of chemical and biological factors.

The question is how best to go about the development of such criteria? One way might be for the Court to appoint Boards of Commissioners. These Boards could be composed of both lawyers and engineers or of profession people trained and experienced in both fields. In this way there could be incorporated into the fact-finding deliberations and discretionary functions both legal and engineering concepts and precepts, thus fostering a new type of social-engineering conscience. Roscoe Pount, Dean Emeritus of Harvard Law School developed some ideas along this line many years ago. The California Water Resources Control Board is one modern example of the idea in effect.

Such a Commission could develop a body of experience, so as to come to grips with both artificial and natural conditions involved in the casual relationships. The Topoc gorge in the Colorado River above Havasu is one such example that influenced deposition in the Cotton Lands situation. Another is the San Marcial basin in the Rio Grande valley above Elephant Butte. These examples are primarily of a topographic nature and their influence can be predicted with some assurance. A more illusive condition is represented by the spread of salt cedar, a sort of ecological invader on new unweathered soils (sediments) with a good source of moisture at the root zone.

Salt cedar performs barrier functions also, especially on the flood plains. However, whether it is the spread of salt cedar or the removal of waterloving plants along stream channels, their place in the scheme of nature and their influence on water and sediment processes can best be understood from their sensitiveness to moisture and chemical conditions.

In all these matters, it would seem that the new rule of fairness under the 5th Amendment to the United States Constitution should be geared to water as the dynamic causal factor in the sequence of related physical events. This is precisely what Johnson has brought out. He implied that oxidation is involved in the consolidation of deltas.

The time may come when the new rule of fairness built into the due process clause will be the basic over-all resource policy in America. As of 1962 the rule is in skeleton form and needs more "meat on its bones." In addition to the criteria previously suggested, the rule should also recognize the wisdom of developing and using land and water according to capability and treating them according to need. To make this concept effective for water, it must first be applied to land. There is a long tradition of avoiding the regulation of land use except in urban areas.

However, such a basic policy is already in effect in the more than 3100 Soil and Water Conservation Districts across the nation, not as coercive rule of law, but as persuasive rule of local custom among land owners and occupiers. Because of its widespread acceptance among land users, it would be a simple matter to raise this broad policy principle and incorporate it progressively into the rules of use and damage law, as the issues arise and by way of legislation.

Such an expanded rule with its built-in fairness and conservation principles could be applied uniformly by local, state, and Federal jurisdictions, and without so much emphasis on the controversies over which jurisdiction owns what water and why.

The United States Court of Claims has rendered a real service to the American people by the adoption of the new rule of fairness. So has the Supreme Court of New Hampshire in adopting the principles of capability and need in the case of St. Regis Paper Co. v. N. H. Water Resources Board, 92 N. H. 164, 26 A. 2d 832 (1942). The engineering profession can take genuine pride in rendering valuable assistance to these two distinguished tribunals. (Busby, C. E., closure to "Some Legal Aspects of Sedimentation," *Journal of the Hydraulics Division*, ASCE, Vol. 88, No. HY3, Proc. Paper 3144, May, 1962, pp. 209–211.)

General Statements

The Commissioner, in his report to the Court, reviewed the Cotton Land, Cress, Jacobs, and Dickinson cases (referred to herein) to show that they differ on the facts only as to degree, not as to kind; and that the resolution of the conflicting interests depends on the available facts and their careful quantitative evaluation.

The Cotton Land case is more directly in point, though Havasu Lake had more permanent levels than Elephant Butte; Boulder Dam served the preceding regulation purposes, and the cycles of sedimentation seem less erratic in the lower Colorado River. As to future cases of this type, engineers might familiarize themselves with the old Lynch case on the Savanah River, Georgia, the Sanguinette case on the Calaveras River, California, and the more recent Life Insurance case on the Missouri River tributary in Missouri. These treat waterlogging damage as a physical event resulting from dam construction in both navigable and non-navigable situations.

Questions Raised

The Commissioner asked a series of questions preliminary to making his findings of fact. These are: (1) How large should flowage rights have been initially (in relation to periodic aggradation, fluctuating lake levels, and avulsion)? (2) When did the cause of action accrue (in relation to start of impoundment, intervening causes, and matters of legal notice to the complainant)? (3) What items should be compensated (in relation to maintenance and life of railway facilities, and other factors)? (4) Predictability of delta and reservoir influences? (5) Datum for surveys and evaluations? (The Court omitted these questions. Why? They were extremely pertinent.)

Pertinent Findings of Fact

(No. 29) Causes and Effects of Aggradation.—The Court recognizes artificial restrictions, salt cedar, tributary stream erosion, increased upstream irrigation, overgrazing in past years, as well as the dam and reservoir as damage causes. It then focused attention on Bridges 1002-A and 1006-A (with related facilities) in regard to injury suffered to the security and permanence of railway facilities. If the suit had been over damage to lands in the San Marcial area, rather than over damage to railway facilities, the problem of cause and effect might have been more like the Cotton Land case.

(No. 30) Artificial Restrictions in Flow.—The Court recognized that the river channel had lost considerable capacity, the overflows were confined by dikes directed toward bridge openings, and these together reduce discharge capacities. The valley development has been altered but this is a natural process unless it can be shown to adversely affect the railway facilities. Despite lack of statistics on total sedimentation, the sequence of restricted flow and increased velocity led to pressure on the dikes, aggradation above the embankments, and avulsion in flood seasons, thus adversely affecting railway property.

(No. 31) Salt Cedar Infestation.—The Court states that infestation has increased since 1930 in eight areas above Bridge 1006-A, and by 1947 became almost continuous in this stretch. It was also present below the bridge, but the Court does not define locations in respect to the delta.

This infestation causes sediment deposition on the flood plain more than cottonwoods and willows, and the clear water does channel erosion work thereafter. The Court seems to absolve the plant as a factor contributing to legal damage (59).

The Court emphasizes this problem, especially, but barely touches this cause ecologically in relation to other related causes. The relationship of salt cedar to dam construction and sedimentation would seem to call for the combined judgment of engineers and ecologists because the plant has a critical moisture need, satisfied mainly by soil moisture conditions and by ground-water conditions in the capillary fringe. Of critical importance to engineers concerned with the whole process is whether the dam and the sediments it caused led to accumulation of silt size particles as suitable seed beds in certain localities and to rises in the water table and capillary fringe favorable to the germination, growth, and spread of salt cedar (59). (See the United States Geological Survey *Professional Paper 655-H* for historical evidence of vegetation changes along the upper Gila River valley in Arizona where Thomas Maddock and C. E. Busby started their sedimentation work in 1934–1935).

(No. 32) Tributary Erosion.—The Court recognized the clear Rio Grande River above Rio Chama, but heavy sediment carrier after Chama, Jemez, Puerco, and Salado enter. This stems mainly from tributary channel erosion that started about 1895.

(No. 33) Irrigation and Overgrazing.—The tributary channel erosion synchronized with a period of increased aridity, grazing, and use of water in irrigation between 1880 and 1915. The first two reduced water control on the tributary flood plains and the third removed a third to half the river flow, especially in the summer months. It is presumed the government had legal notice of these events. What is legal notice in these circumstances to a private individual or corporation is not clear.

(Nos. 34, 35, 36) Influence of Elephant Butte Reservoir.—The reader should consult the series of tables included in the report for details on trends in aggradation and degradation at various reaches in the stream channel over the years, especially the periods from 1931–1937 and 1941–1949. On these facts, the Court recognized as a matter of scientific knowledge, that loaded streams tend to deposit sediments at or near the point where they flow into lakes, reservoirs, and other still waters because of reduced velocity; and that greatest aggradation takes place at times of heavy flows, also causing scouring and degradation in certain reaches. Of course, the stream deposits sediments or overflows in areas of lower gradient that may be suitable for the growth of water-loving plants.

The delta at the head of the reservoir takes time to form and is, itself, subject to erosion. But it compacts with age and then resists erosion more. With age it tends to cause a sediment wedge to build back upstream at greater rate than normal aggradation. As the reservoir level recedes by consumptive use (and low inflow), degradation sets in at the front of the delta and works back upstream seeking equilibrium. The influence of this factor differs significantly from the Colorado case. The Court recognized these scientific facts about the whole process and might have gone further in this regard if more reliable facts were available. This is the reason they are presented herein.

(Nos. 37, 38, 39, 40, 41) Sediment Deposition.—In the first 4 yr–5 yr after dam completion, more sediment was deposited than at any comparable period thereafter, especially in the 17-mile–27-mile reach below Bridge 1006-A. This was the area occupied by the reservoir head, but such did not take place at the bridge itself.

From 1920–1925 heavy sediment deposition took place 10 miles–18 miles and some 4 miles–8 miles below the bridge. The average annual rate was about 0.84 ft then, rather than 0.27 ft as "normal." (Expert witnesses seemed not to be sure this was due to the reservoir.)

From 1925–1935 the reservoir and valley received added sediments in the reach from the bridge down to the dam, especially 14 miles–20 miles below the bridge. This caused a wedge that later reached Bridge 1006-A. The court is more definite here by saying that the reservoir influence probably contributed to the increased aggradation from 1931–1937, together with other causes previously discussed. This was also the beginning of salt cedar infestation above the delta as shown by the record and other public reports to be referred to.

From 1937–1947 aggradation took place between 30 miles below and 10 miles above Bridge 1006-A. In 1941 and 1943 flood flows were high and the reservoir head extended closer to the bridge than at any time. Sediment filled the old erosion channels in the head delta. This caused aggradation to move upstream, involving also the area above the bridge. By 1937 salt cedar had spread into eight areas above Bridge 1006-A, and by 1947 into much of the valley below the bridge.

The Court takes up the quantitative aspects of the problem by comparing the average annual and the recorded rates, finding the latter higher for the period 1949–1954. Time and certain time periods seem to have been important in court recognition of the cause and the consequences, and of the application of the statute of limitations.

(No. 42) Effect of Reservoir.—The Court indicates that the reservoir may have accentuated the sediment process and interfered with the normal function of the river to dispose of its sediment load from the high contributing tributaries. This has been aggravated by natural and other artificial conditions.

The reservoir sediment rates predicted in 1916 were more than 50% in excess of what actually was deposited there. (This does not mean that there were not intervening influences that caused deposition above and prevented transport into the reservoir itself.) But the record leaves in conjecture the question of ultimate stabilization of sediments at or near the bridge.

(Nos. 43–50) Accrual of Cause of Action; Control and Correction of Aggradation.—The Court then disposes of the contentions of the Railway Company and arrives at the conclusion that the cause of action could not have arisen until about 1949 when the aggradation had moved far enough upstream to become reasonably well stabilized as a damaging factor at Bridge 1006-A, so as to bring the cause fully to the attention of the Company. Even then, the Company could not be sure it was permanent; for one thing, the silt survey was not available until then.

Legal notice, then, in the form of direct damage or the threat of damage, is necessary to give rise to a legal cause of action and start the statute of limitations to operate. But the completion of the legal process as cause, and its outlaw by reason of these limitations, is another matter. For here, again, there is a sequence of physical events running concurrently with the statute. Had there been a major control dam above and stable reservoir level below, together with information on the effect of the dam on ground water in the capillary fringe below the soil, the factual story might have been much different.

The control structures planned for tributaries may help solve the sedimentation problem as it presently exists. But much depends on future floods, effect of improvements, and sedimentation effects. (These findings, taken together, seem to have materially affected the judgment of the Court.)

In this case, it is not clear by any means as to the effect of the dam on the growth of and spread of salt cedar through the media of sedimentation of fine materials in strategic locations and the rise in the capillary ground-water fringe, if any. The tremendous expansion in salt cedar began about 1937 after establishment in the delta areas (Finding No. 31) (60), and this tends to coincide with the greatest increase in sedimentation (Finding No. 20). What was the concurrent position of the capillary fringe? Was this in any way due to the dam? What about deposition of fine sediments with high water-holding capacity? This presents a challenge to lawyers, engineers, and conservationists—a challenge of social engineering.

The writer's conclusion is that the law will eventually recognize all the harmful damage consequences from the building of dams, insofar as these can be proved as a sequence of related physical events and insofar as they involve the interests of the parties to the suit. Much depends on whether the damage is to lands and their water rights or to facilities and rights-of-way.

Time and reliable records may determine the issues. These should include vegetative as well as engineering considerations. Actually, the development of the law is only now unfolding, because the new rule of fairness was established and the new interpretation of what constitutes an obstruction to navigable waters. The same rules might well be considered by the respective states under their constitutions.

Taking of Property by Overflow, Erosion, Sediment Deposition, Rise in Ground-Water Table, Obstruction of Land Drainage Systems, Loss of Water Rights, and Other Causes.—The final step in the development of the law has barely begun for the physical processes in this step often extend over many years of time and many varying physical conditions in which water is the common denominator. This extended time period in nature tends to come into conflict with the short statute of limitations of 6 yr. Thus, the law is far behind the physical processes of nature and the people. The conflict raises some fundamental issues in constitutional law because local landowners who are severely damaged are at the mercy of the government. [See United States v. Dickinson, pp. 747–749, (49)].

One of the best examples of the complex physical processes that lead to all the damaging consequences defined under this subdivision of title is that of the Montour Valley in Idaho, above Black Canyon Dam and Reservoir on the Payette River (Figs. 7.6–7.8). This structure was built in 1924 by the Federal government for irrigation and power generating purposes. It has blocked the passage of all coarse sediments from 2,680 sq miles of watershed lands in which granite bedrock and the soils developed from granitic parent materials are almost the exclusive sources of those coarse sediments. Soils developed from granite are often very erosive, especially if they are immature in nature. And they often produce much sediment debris that finds its way into stream channels, particularly in the regions of low rainfall. There are only two very small tributaries above the reservoir area that are controlled by upstream dams, i.e., the Cascade Dam and the Deadwood Dam. The rest of the watershed is uncontrolled (62).

Erosion is very active in the tributary valleys and in road and railroad cuts. Sediment has accumulated in and above the Black Canyon Reservoir, so that the valley is choked, leading to more frequent overflows in the flood season, deposition of fine and coarse sediments in the river channel and upon the valley farm lands, rise in the water table, and obstruction of the land drainage system of the community. There is the strong possibility of loss of surface water rights for irrigation purposes by reason of nonuse due to flooding and other conditions that make irrigation impractical or impossible. This damaging consequence seems to be increasing with time.

There is a tendency for government officials to blame the damaging situation upon ice accumulation and flows in the winter and to ignore the erosion and sediment deposition processes (63). The area was examined in some detail by the writer. It was concluded that if the land owners were to take the matter to the Federal Courts it could produce a landmark case in the development of American Sedimentation Law (Figs. 7.6, 7.7, 7.8) for all of the causes previously listed.

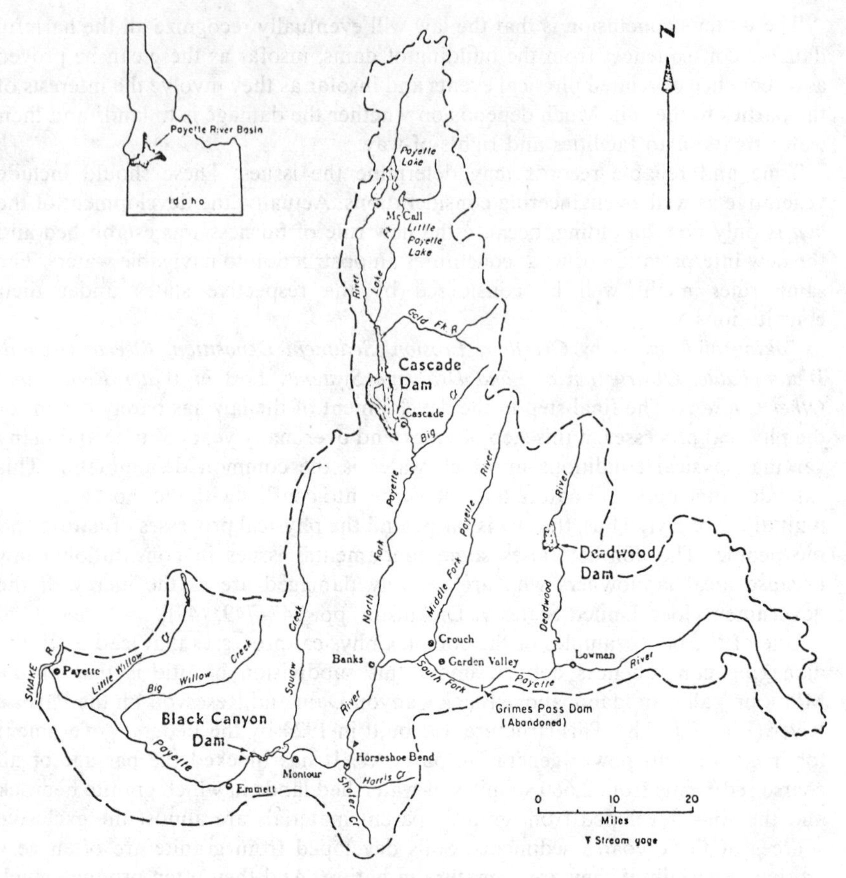

FIG. 7.6.—Payette River Basin (Map Courtesy of United States Bureau of Reclamation)

Comments on 55 Georgetown University Law Journal 631–646 (1967) in Relation to King v. United States (71).—The author Pittle brings out the fact that a taking under the Tucker Act presents a difficult problem as to whether the fee or lesser estate is acquired by permanent or intermittent flooding, citing the Dickinson case. He emphasizes that the intent of the government is evidenced by factors such as: (1) The nature of the project; (2) its permanency; and (3) the effect of the flooding upon the ability of the person in possession to use and enjoy the property. He says that the determinative factor is whether all practical uses to which the land might be put were destroyed. And, if they are destroyed, there is the full estate taken rather than a mere easement.

But this summation is weak for it fails to recognize the processes of stream overflow, erosion, sedimentation transport and deposition, and other synergistic as well as time lag effects involved when the dam and reservoir and sediment barriers combine to produce upstream damage. However, Pittle does cite the Dickinson, Cotton Lands Co., and A.T. and Santa Fe cases which, step by step,

FIG. 7.7.—Sediment Deposits have Partially Filled Upper End of Black Canyon Reservoir; Payette River Flow is to Left, Regan Bend is at Upper Left (Photograph Courtesy of United States Bureau of Reclamation)

bring out the time and process factors. Pittle is correct in stating that the intent of the government is shown by the authorization of the project for construction. The government recognizes the project's useful purposes, but fails to recognize its damaging effects (49, 55, 67).

Pittle also correctly brings out that the predictability requirement of the early cases on behalf of the government engineers, e.g., in Sanguinette v. United States, 264 U.S. 146, 44 S.CT. 264; 68 L.Ed. 608, (1924) no longer applies. He says it merely is sufficient for a claimant to establish causation, citing the recent Kenite Corp. and Pashley cases (69, 66).

However, it is not so easy to dispose of the cause and effect problem this way. It is not that simple. However, scientists and engineers can now do a much better job of predicting the probable effects of stream obstructions upon the natural processes, if they have a chance to develop the facts, especially with the use of their computer equipment. The need is for scientists to be permitted to put into evidence these complex mathematical calculations. Pittle cites the Horstmann case to attempt to show that damage to land caused by percolating waters or subsurface seepage from a government project does not constitute a taking, unless the course of the waters was predicted. But this was a 1921 case in the same

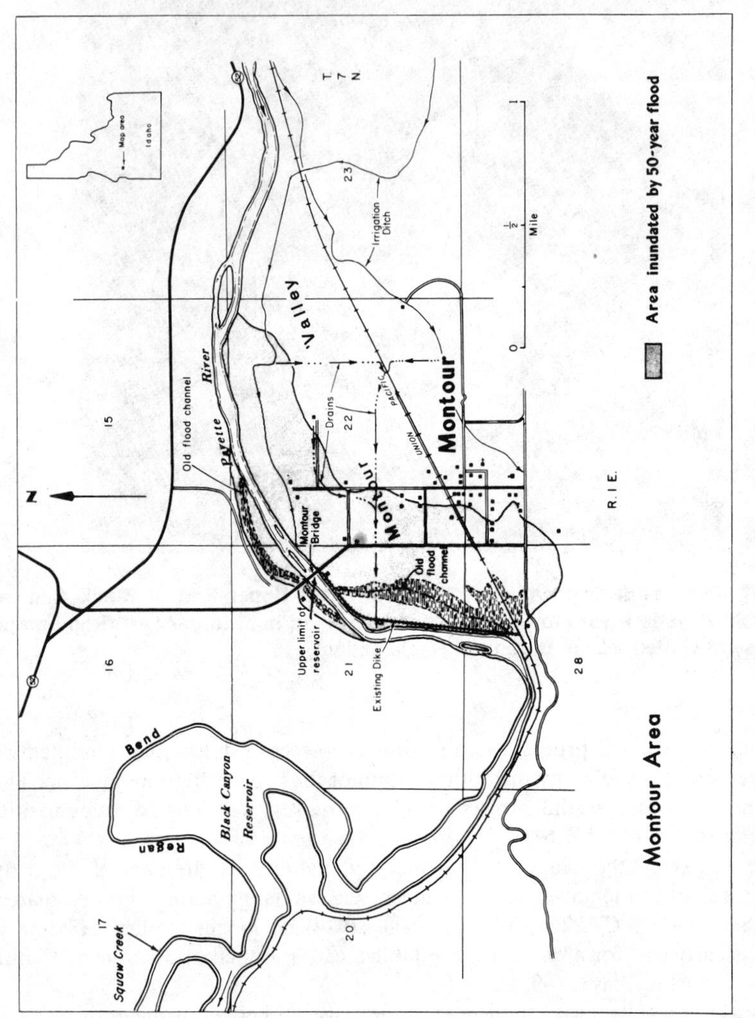

FIG. 7.8.—Montour Area (Map Courtesy of United States Bureau of Reclamation)

category of the older Sanguinette case, requiring predictability of the government agents; neither is any longer an authority on the point.

Pittle's citation to the North Counties Hydroelectric Co. cases supports the view that the damages must not be too speculative. Again, the point cannot be disposed of that easily. What is being too speculative when the damaging effects may take 10 yr–60 or more years to reach equilibrium? The questions of: (1) Extent of the taking; (2) cause and effect relationships; and (3) time of reaching equilibrium in a stream and reservoir system are not easy for judges and juries to understand and evaluate, let alone geologists, engineers, and scientists (70).

Pittle points out, by way of illustration, the difficulty of determining a full or partial taking in the case of airplane overflights, the reason being that the invasion by sound is not directly upon the property but upon the person. And furthermore, the application of the statute of limitations is different for the invasion varies in degree with the height of overflights. The original cause is the airplane but the immediate cause is the sound. This is similar to the dam as original cause and the sedimentation accumulation as immediate cause.

The cases cited have not been fully spelled out herein because they are set forth in 55 Georgetown Law Journal. The point made here is that Pittle sees mainly the law side—not the physical process side of the environmental problem. The same is true to some extent of the Briefs and the Opinion in King v. United States (64, 71).

E. Court Citations and Other References

1. Ehrick, E., "The Fundamental Principles of Sociology of the Law," Moss translation, (1936).
2. "Science and Sanity," by Albert Korzybski (1941); "People in Quandaries," by Wendell Johnson (1946); "The Language of Law," *Western Law Review* (1958).
3. "Water Resources and the Law," Legislative Research Center, University of Michigan, Ann Arbor, Mich., (1958), p. 84.
4. Baylor Law Review 892, 894, (1935) Note 15; and Ann, 40 A. L. R. 829 (1926). (For discussion of factual bases of these legal classes of water supply.)
5. The American Judge, 6, 8, (1924), Justice Andrew A. Bruce, North Dakota.
6. The theory back of this rule is the "cujus est solum" principle. See discussion on p. 85, together with references, "Water Resources and the Law," supra.
7. Preservation of Integrity of State Water Laws, 31–44, October, (1942), Natl. Reclamation Assn.
8. "Roman Contributions to the Law of Soil Conservation," by Karl F. Milde, 19 *Fordham Law Review,* (1950), p. 192.
9. 24 *Minnesota Law Review* 902, 903, 904, (1940).
10. "Water Resources and the Law," supra; Hutchins, Wells A., "Selected Problems in the Law of Water Rights," (1942), p. 3. See also, 28 *Harvard Law Review* 478; 2 *Dakota Law Review* 365; 11 *Virginia Law Review* 159, (1928); 15 *Virginia Law Review* 177, (1929); 5 *Wisconsin Law Review* 239; 51 Cent. Law Journal, 360.
11. 1 *Alabama Law Journal* 117; 8 Cal. Law Review 197; 13 *Illinois Law Review* 63, (1918); 15 *Illinois Law Review* 282, 462, (1920, 1921); 17 *Illinois Law Review* 454, (1923); 14 Ia. Law Review 547; "Law of Water Rights in the West," by Wells A. Hutchins, (1942), p. 115; C. J. S. 805–815; 24 *Minnesota Law Review* 899, (1940).
12. 24 *Minnesota Law Review* 399, (1940).
13. English court decisions prior to 1850 are not authority because the law there was not settled at that time. Shortly after that date, decisions held that the lower owner had no cause of action against upper owners for withholding the natural flow. (Greatex v. Hayward, 8 Exch. 291, 22 L. J. 137, (1853); Rawstrom v. Taylor, 11 Exch. 369, 25 L. J. Exch. 115, 4 W. R. 290, (1856).

14. For types of capture and intent to confer benefits by taking and putting water to use, see, Rawstrom v. Taylor, supra, domestic use; Benson v. Cook, 47 S. D. 611, 201 N. W. 525, (1924); Terry v. Heppner, 59 S. D. 317, 239 N. W. 759, (1931); Garns v. Rollins, 41 Utah 260, 125 Pac. 867, Ann. Case 1915C 1159, (1912); irrigation; Boynton & Moseley v. Gilman, 53 Vt. 17, (1880), commercial. State v. Hiber, 48 Wyo. 172, 44 P, 2d 1005, (1935), livestock.
15. Templeton v. Voshloe, 72 Ind. 134, 37 Am. Rep. 150, (1880). (Note the use of the terms "good husbandry" and "proper improvement of the soil." The courts might have said "sound land use and conservation practices prevalent in the community.")
16. Trigg v. Timmerman, 90 Wash. 768, 156 Pac. 846, L. R. A. (1916), F424, (1916), 226 Ala. 400, 147 So. 178; Baker v. Akron, 145 Iowa 485, 122 N. W. 926, 30 L. R. A., United States, 619; Boll v. Ostroot, 25 S. D. 513, 127 N. W. 577, (1910).
17. 59 Minn. 436; 191 Minn. 591, (1934); 2 Minn. L. Rev. 449, (1918); 6 *Michigan Law Review* 448, 460, (1908); 17 Cent. L. J. 62, 67, (1883); 43 N.H. 569, 577, (1862); 50 N.H. 439, 446, (1870); 71 N.H. 186, (1901); 384 Pac. (2d) 450, 452 (1963); 190 N.W. (2d) 1, 7 (1971).
18. Miles v. A. Arena & Co., 23 Cal. App. 2d 680, 683, 73 Pac. 2d 1260, 1262, (1937); Cook v. Haskins, 57 Cal. App. 2d 737, 740, 135 P. 2d 176, 177, (1943).
19. Vanderweile v. Taylor, 65 N.Y. 341.
20. Sheehan v. Flynn, 59 Minn. 496, 61 N.W. 462, 26 L. R. A. 632, (1894).
21. 19 *Fordham Law Review* 192–197, (1950).
22. Ashley v. Holbert, 265 N.Y.S. 138, 148 Misc. 45, (1933).
23. Griffin vs. National Light Co., 60 S.E. 702, 79 S.C. 351, (1908).
24. Samuel C. Wiel, 50 Harv. L. Rev. 254, (1936) and cases cited there; Marvin D. Weiss, Industrial Water Pollution, (1951).
25. S. C. Wiel, supra; Lindley, Mines, 3d Ed. (1914), S848; Woodruff v. North Bloomfield Gravel Mining Co., 18 Fed. 753, 756, 774, C. C. D. Cal. (1884); People v. Gold Run Ditch & Mining Co., 66 Cal. 138, 4 Pac. 1152, (1884); 27 Stat. 507, (1893); 34 Stat. 1001, ch. 2077, (1907); Cal. Resources Code, div. 2, Ch. 6.5, sec. 2551 to 2559, (1953); Calif. Water Code, Div. 7, Water Pollution; Wells A. Hutchins, "Water Law and Its Significance to the Mining Industry," February, (1959); 33 U.S.C.A. 661–687.
26. *Civil Engineering*, Vol. 29, No. 7, July, (1959), p. 80.
27. United States v. Republic Steel Corp., 155 F. Supp. 442, (1957). (The fact situation is to be found mainly in this report of the District Court.)
28. United States v. Republic Steel Corp., 264 F. 2d 289, (1959). (The report by the Circuit Court refers to expert testimony and in dismissing the suit, relies in part at least, upon State ex. rel. Dyer v. Sims, 341 United States 22, 24 in which the problem involved suspended solids.)
29. United States v. Republic Steel Corp., 362 U.S. 482, (1960). Note: The report by the Supreme Court draws attention to another recent case in which sand and gravel in waste water is treated in law as unlawful under S13, and as an obstruction under S10, of the Rivers and Harbors Act. Gulf Atlantic Transportation Co. vs. Becker County Sand and Gravel Co., 122 F. Supp. 13, (1954).
30. 59 *Colorado Law Review* 1065, (1959) for detailed analysis of the history of the Rivers and Harbors Act, as relating to these basic policy provisions, and to its consideration in the foregoing cases.
31. United States v. Brazoria County Drainage Dist. No. 31, 2 F. 2d 861.
32. Angell, Watercourses, 7th Ed., (1877); Hutchins, Wells A., Law of Water Rights in the West, 7–20, (1942); Water Resources and the Law, 177–182, (1958), 3 *Baylor Law Review* 473–476, (1951).
33. 24 *Minnesota Law Review* 305, (1940).
34. 10 *California Law Review* 86–89, (1922); 36 *Michigan Law Review* 346–348, (1937); 6 *Arkansas Law Review* 68–72, (1952); 36 *Texas Law Review* 299–321, (1958).
35. 24 *Minnesota Law Review* 305, (1940) and cases cited therein.
36. Calif. Civic Code, Sect. 830, 1014, 1015, and Code Civil Proc., Sect. 2077, McBride v. Steinweden, 72 Kansas 508, 83 Pac. 822; Fowler v. Wood, 73 Kansas 511, 117 Am. St. Rep. 534, 85 Pac. 763, 6 L. R. A. U. S. 162. Lammers v. Nissen, 4 Nebraska 245;

Hammond v. Shepard, 186 Illinois 235, 78 Am. St. Rep. 274, 57 N. E. 867; Wiel, Water Rights in the Western States, S862, 901, 902, 3d Ed., (1911).

37. 1 Wiel, Water Rights, S692–695, 3d Ed., (1911).
38. Dietrich v. Northwestern Ry. Co., 412 Wisconsin 262, 24 Am. Rep. 399; Sternberger v. Seaton Co., 45 Colorado 401, 102 Pac. 168, (1909); Hutchinson v. Watson & Co., 16 Idaho 484, 133 Am St. Rep. 125, 101 Pac. 1059, (1909); Western Pacific Co. v. Southern Pacific Co., 151 Fed. 376, 30 C.C.A. 606; Steers v. City of Brooklyn, 101 N.Y. 51, 4 N.E. 7; Wiel, supra, footnote 37.
39. New Orleans v. United States, 10 Pet. 717, 9 L. Ed. 595; Cook v. McClure, 58 N.Y. 437, 17 Am. Rep. 270; Wiel, Water Rights in the Western States, S902, 3d Ed., (1911); 24 *Minnesota Law Review* 305, (1940).
40. Haln v. Dawson, 137 No. 581, 590, 36 S. W. 233; Wiel, supra, footnote 37.
41. 26 *Texas Law Review* 223–226, (1947);14 *Louisiana Law Review* 267–273, (1953); 28 *Oregon Law Review* 385–390, (1949); 15 *Louisiana Law Review* 463–465, (1955); 10 *Rutgers Law Review* 738–741, (1956). See also United States v. 11.8 acres of land (1963) and R. A. Beaver et al. v. United States (1965) for the Lower Colorado River situation. See also County of St. Clair v. Lovingston, 90 United States (23 Wall) 46, 50–66 (1874) in regard to artificial structures placed in a stream that may cause accretion.
42. Missouri v. Nebraska, 196 U.S. 23, 25 Sup. Ct. Rep. 155, 49 L. Ed. 372; Fowler v. Wood, 73 Kansas 511, 117 Am. St. Rep. 534, 85 Pac. 763, 6 L.R.A., U.S. 162; Wiel, supra, footnote 37.
43. Paige v. Rocky Ford Co., 83 Cal. 84, 21 Pac. 1102, 23 Pac. 875; Wooley v. Caldwell, 108 Cal. 85, 49 Am. St. Rep. 64, 41 Pac. 31, 30 L.R.A. 820; York County v. Rollo, 27 Ont. App. 72; Morton v. Oregon Ry. Co. 48 Ont. 444, 120 Am St. Rep. 827, 87 Pac. 151, 7 L.R.A., U.S., 344; Cox v. Barnard, 39 Or. 53, 64 Pac. 860; Wiel, supra, footnote 37.
44. Kinkead v. Turgeon, 74 Nebr. 573, 104 N.W. 1061; Wiel, supra, footnote 37.
45. 28 USCA S41, 20; Art. V, United States Constitution; United States v. Lynch, 188 U.S. 445–466, 23 S. Ct. 349, 47 L. Ed. 539, (1903); United States v. Cress, 243 U.S. 316, 328, 37 S. Ct. 380, 61 L. Ed. 746, (1917).
46. Tempel v. United States, 248 United States 121, 39 S. Ct. 56, 63 L. Ed. 162, (1918); Sanguinette v. United States, 264 United States 146, 44 S. Ct. 264; 68 L. Ed. 608, (1924).
47. Jacobs v. United States, 45 F. 2d 34, C. Ct. 5th, (1930).
48. Cress v. United States, supra. (This involved the taking of riparian lands and water rights by backwater of a navigation dam on the Cumberland and Kentucky Rivers.) See also, United States v. Great Falls Mfg. Co., 112 U.S. 645, 656, 5 S. Ct. 306, 28 L. Ed. 846, 850, (1884); United States v. Lynch, supra. (Four dissenting judges advocated overruling the Cress decision in part, on the ground that "it would be incongruous to deny compensation to owners adjacent to navigable rivers and require it for others bordering its tributaries for like injuries caused by the single act of lifting the river's mean level to the high watermark.") United States v. Kansas City Life Insurance Co., 339 U.S. 799, 807, 812, 815, (1950).
49. United States v. Dickinson, 331 U.S. 745, 747–748, 67 S. Ct. 1382, 91 L. Ed. 1789, (1947). (District Court decision affirmed.) 4 Cir., 152 F. 2d 865, certiorari 328 U.S. 828, (1945); 28 U.S.C. S41, 40, 28 U.S.C. S4, 20.
50. United States v. Dickinson, supra, 748.
51. Bauman v. Ross, 167 U.S. 548, 574, 17 S. Ct. 966, 976, 42 L. Ed. 270, (1897); U.S. v. Dickinson, supra, 750, 751. See also, United States v. Welch, 217 United States 333; United States v. Grizzard, 219 United States (1801) compare Sharp v. United States, 191 United States 141; Campbell v. United States, 266 United States 368. (Congress has recognized damage to be assessed not only for part taken but also "for any injury to the part not taken.") S6 Act July 18, 1918, 40 Stat. 911, 32 U.S.C. S595, 33 USCA S595.
52. United States v. Dickinson, supra, 751.

53. United States v. O'Donnell, 303 United States 501, (1938).
54. Sec. 145, Judicial Code, 28 USCA S250. (The Court reviewed the question of consent based on implied contract or consent based on the 5th Amendment itself. The early cases (United States v. Lynch, supra; Tempel v. United States, supra; and United States v. North American Transport Co., supra) rested jurisdiction under the Tucker Act on a narrow basis of its construction. Recent cases rest this squarely on the 5th Amendment and do not require proof as in a tort claim (foreseeability and intervening cause aspects). The form of the remedy does not qualify the right. Yearsley, et al. v. W. A. Ross Const. Co., 309 U.S. 18, 22, 60 S. Ct. 413, 415, 84 L. Ed. 554, (1940).
55. Cotton Land Co. v. United States, 75 F. Supp. 232, 234–235, 109 C. Cts. 816, (1948).
56. Report, United States Court of Claims, Atchison, Topeka and Santa Fe Railway Co. v. United States No. 90–54, June 8, 1960; Report of Commissioner March 18, 1959.
57. "The Problem of Phreatophytes," by J. S. Horton, Symposium of Hannoversch-Münden, Internatl. Science Hydr. Assoc. Publication 48, September, (1959).
58. See footnote 2.
59. "Seed Germination and Seedling Establishment of Phreatophyte Species," by J. S. Horton, F. C. Mounts, and J. M. Kraft, Rocky Mt. Forest and Range Experiment Sta., April, (1960).
60. Report, Symposium on Phreatophytes, Pacific Southwest Interagency Committee, (1957), p. 20, (Fig. 2).
61. Johnson, J. W., discussion of "Some Legal Aspects of Sedimentation" by C. E. Busby, *Journal of the Hydraulics Division,* ASCE, Vol. 88, HY1, Proc. Paper 3044, Jan., 1962, pp. 149–154.
62. Summary Statement: Upstream Environmental Effects of Black Canyon Dam and Reservoir, Payette River Watershed, Idaho, by C. E. Busby, Consultant to Montour Flood Committee, Mar. 15, 1972.
63. Montour Flood Problem, Bureau of Reclamation, Feb. 1973. Note: A sedimentation survey of the reservoir has been made by the Bureau of Reclamation and Soil Conservation Service.
64. Memorandum Statement on King v. United States in Relation to the Upper Vergigris River Watershed, Kansas by C. E. Busby, Consultant to Kahrs, Nelson, Fanning, Hite & Kellogg, Attorneys at Law, Wichita, Kan., Dec. 10, (1970).
65. United States v. Gerlach Livestock Co., 339 U.S. 725, (1950).
66. Pashley v. United States 140 Ct. Cl. 535, 156 F. Supp. 737, (1957).
67. A. T. and Santa Fe Ry., v. United States 150 Ct. Cl. 339, 365, 278 F. 2d. 937, (1960).
68. Richards v. United States, 152 Ct. Cl. 225, 282 F. 2d. 90, (1960); rehearing denied 152 Ct. Cl. 266, 285 F. 2d 129, (1961).
69. Kenite Corp. v. United States, 157 Ct. Cl. 72. (1962).
70. North Counties Hydroelectric Co. v. United States 108 Ct. Cl. 470, (1947).
71. King v. United States, 192 Ct. Cl. 548-569 (1970), 504 F. 2d. 1138.

Note: References to the legal literature are made in the following order from left to right: (1) Court Decision Name or Law Review Name; (2) volume number of the set of reports; (3) name of the set of reports, in abbreviated form; (4) page number or numbers of the volume; and (5) date of the decision or law review. If there are citations to two or more sets of reports, then repeat these items for: (1) Each such report; (2) set of abbreviations; and (3) page numbers. And finally comes the date of the decision or report in parentheses. The same applies to the Law Reviews except that the volume number comes first, the full name of the Law Review, and then the page number and date, if any. Some Reviews are not quoted by date and page numbers because they are not always necessary. The whole Review may deal with the subject.

Appendix I.—Conversion Factors

To convert	To	Multiply by
1. Length (L)		
inches (in.)	centimeters (cm)	2.54
feet (ft)	meters (m)	0.3048
miles (miles)	kilometers (km)	1.609
meters (m)	inches (in.)	39.37
meters (m)	feet (ft)	3.281
kilometers (km)	miles (miles)	0.6214
2. Area (L^2)		
square inches (sq in.)	square centimeters (cm^2)	6.452
square feet (sq ft)	square meters (m^2)	0.09290
square miles (sq miles)	square kilometers (km^2)	2.590
acres (acre)	square meters (m^2)	4047
square centimeters (cm^2)	square inches (sq in.)	0.1550
square meters (m^2)	square feet (sq ft)	10.76
hectares (ha)	acres (acre)	2.471
square kilometers (km^2)	square miles (sq miles)	0.3861
3. Volume (L^3)		
cubic inches (cu in.)	cubic centimeters (cm^3)	16.39
cubic feet (cu ft)	cubic meters (m^3)	0.02832
cubic yards (cu yd)	cubic meters (m^3)	0.7646
gallons (gal)	liters (*l*)	3.785
cubic centimeters (cm^3)	cubic inches (cu in.)	0.06102
cubic meters (m^3)	cubic feet (cu ft)	35.31
liters (*l*)	cubic feet (cu ft)	0.03531
liters (*l*)	gallons (gal)	0.2642
4. Velocity (L/T)		
feet per second (fps)	meters per second (m/s)	0.3048
meters per second (m/s)	feet per second (fps)	3.281
5. Discharge (L^3/T)		
cubic feet per second (cfs)	cubic meters per second (m^3/s)	0.02832
cubic feet per second (cfs)	liters per second (l/s)	28.32
cubic meters per second (m^3/s)	cubic feet per second (cfs)	35.31
liters per second (l/s)	cubic feet per second (cfs)	0.03531

6. Mass (M)

pounds (lb)	kilograms (kg)	0.4536
kilograms (kg)	pounds (lb)	2.205

7. Density (M/L^3)

pounds per cubic foot (pcf)	kilograms per cubic meter (kg/m^3)	16.02
kilograms per cubic meter (kg/m^3)	pounds per cubic foot (pcf)	0.06243
kilograms per cubic meter (kg/m^3)	grams per cubic centimeter (g/cm^3)	0.001000

8. Force (ML/T^2)[a]

pounds (lb)	kilograms (kg)	0.4536
pounds (lb)	newtons (N)[b]	4.448
kilograms (kg)	pounds (lb)	2.205
kilograms (kg)	newtons (N)	9.807
newtons (N)	kilograms (kg)	0.1020
newtons (N)	pounds (lb)	0.2248
dynes (dynes)	newtons (N)	10^{-5}

9. Pressure (M/LT^2)[a]

pounds per square inch (psi)	kilograms per square meter (kg/m^2)	703.1
pounds per square inch (psi)	newtons per square meter (N/m^2)	6895
pounds per square foot (psf)	kilograms per square meter (kg/m^2)	4.882
pounds per square foot (psf)	newtons per square meter (N/m^2)	47.88
kilograms per square meter (kg/m^2)	pounds per square inch (psi)	0.001422
kilograms per square meter (kg/m^2)	pounds per square foot (psf)	0.2048
kilograms per square meter (kg/m^2)	newtons per square meter (N/m^2)	9.807

10. Specific Weights (M/L^2T^2)[a]

pounds per cubic foot (pcf)	kilograms per cubic meter (kg/m^3)	16.02
pounds per cubic foot (pcf)	newtons per cubic meter (N/m^3)	157.1
kilograms per cubic meter (kg/m^3)	pounds per cubic foot (pcf)	0.06243
kilograms per cubic meter (kg/m^3)	newtons per cubic meter (N/m^3)	9.807

11. Kinematic Viscosity (L^2/T)

square feet per second (sq ft/sec)	square centimeters per second (cm^2/s)	929.0
square feet per second (sq ft/sec)	square meters per second (m^2/s)	0.09290
square meters per second (m^2/s)	square feet per second (sq ft/sec)	10.76
square meters per second (m^2/s)	square centimeters per second (cm^2/s)	10^4

[a]The factors relating pounds of force, kilograms of force, and newtons are based on the standard value of the gravitational acceleration, $g = 32.174\ ft/sec^2 = 9.80665\ m/s^2$.
[b]$1\ N = 1\ kg \cdot m/s^2$.

Appendix II.—Density and Viscosity of Water 0°C–40°C

Temperature, in degrees Celsius	Temperature, in degrees Fahrenheit	Density, in grams per cubic centimeter	Density, in pounds-mass per cubic foot[a]	Kinematic viscosity, in square centimeters per second	Kinematic viscosity, in square feet per second X 10^5
0	32	0.9998	62.42	0.0179	1.92
0.6	33	0.9999	62.42	0.0175	1.89
1.	33.8	0.9999	62.42	0.0173	1.86
1.1	34	0.9999	62.42	0.0172	1.85
1.7	35	0.9999	62.42	0.0169	1.82
2	35.6	0.9999	62.42	0.0167	1.80
2.2	36	0.9999	62.42	0.0166	1.79
2.8	37	1.0000	62.43	0.0163	1.76
3	37.4	1.0000	62.43	0.0162	1.74
3.3	38	1.0000	62.43	0.0160	1.72
3.9	39	1.0000	62.43	0.0157	1.69
4	39.2	1.0000	62.43	0.0157	1.69
4.4	40	1.0000	62.43	0.0155	1.66
5	41	1.0000	62.43	0.0152	1.64
5.6	42	1.0000	62.43	0.0149	1.61
6	42.8	0.9999	62.42	0.0147	1.58
6.1	43	0.9999	62.42	0.0147	1.58
6.7	44	0.9999	62.42	0.0144	1.55
7	44.6	0.9999	62.42	0.0143	1.54
7.2	45	0.9999	62.42	0.0142	1.53
7.8	46	0.9999	62.42	0.0140	1.50
8	46.4	0.9998	62.42	0.0139	1.49
8.3	47	0.9998	62.42	0.0137	1.48
8.9	48	0.9998	62.41	0.0135	1.45
9	48.2	0.9998	62.41	0.0135	1.45
9.4	49	0.9997	62.41	0.0133	1.43
10	50	0.9997	62.41	0.0131	1.41
10.6	51	0.9996	62.41	0.0129	1.39

Temperature, in degrees Celsius	Temperature, in degrees Fahrenheit	Density, in grams per cubic centimeter	Density, in pounds-mass per cubic foot[a]	Kinematic viscosity, in square centimeters per second	Kinematic viscosity, in square feet per second X 10^5
11	51.8	0.9996	62.40	0.0127	1.37
11.1	52	0.9996	62.40	0.0127	1.37
11.7	53	0.9995	62.40	0.0125	1.34
12	53.6	0.9995	62.40	0.0124	1.33
12.2	54	0.9995	62.40	0.0123	1.32
12.8	55	0.9994	62.39	0.0121	1.30
13	55.4	0.9994	62.39	0.0120	1.30
13.3	56	0.9993	62.39	0.0119	1.28
13.9	57	0.9993	62.38	0.0117	1.26
14	57.2	0.9992	62.38	0.0117	1.26
14.4	58	0.9992	62.38	0.0116	1.25
15	59	0.9991	62.37	0.0114	1.23
15.6	60	0.9990	62.37	0.0112	1.21
16	60.8	0.9989	62.36	0.0111	1.20
16.1	61	0.9989	62.36	0.0111	1.19
16.7	62	0.9988	62.36	0.0109	1.17
17	62.6	0.9988	62.35	0.0108	1.17
17.2	63	0.9987	62.35	0.0108	1.16
17.8	64	0.9986	62.34	0.0106	1.14
18	64.4	0.9986	62.34	0.0105	1.14
18.3	65	0.9985	62.34	0.0105	1.13
18.9	66	0.9984	62.33	0.0103	1.11
19	66.2	0.9984	62.33	0.0103	1.11
19.4	67	0.9983	62.32	0.0102	1.10
20	68	0.9982	62.32	0.0100	1.08
20.6	69	0.9981	62.31	0.00991	1.07
21	69.8	0.9980	62.30	0.00980	1.06
21.1	70	0.9980	62.30	0.00977	1.05

Temperature, in degrees Celsius	Temperature, in degrees Fahrenheit	Density, in grams per cubic centimeter	Density, in pounds-mass per cubic foot[a]	Kinematic viscosity, in square centimeters per second	Kinematic viscosity, in square feet per second X 10^5
21.7	71	0.9978	62.29	0.00965	1.04
22	71.6	0.9978	62.29	0.00957	1.03
22.2	72	0.9977	62.29	0.00952	1.03
22.8	73	0.9976	62.28	0.00940	1.01
23	73.4	0.9975	62.27	0.00935	1.01
23.3	74	0.9975	62.27	0.00928	0.999
23.9	75	0.9973	62.26	0.00916	0.986
24	75.2	0.9973	62.26	0.00914	0.983
24.4	76	0.9972	62.25	0.00905	0.974
25	77	0.9970	62.24	0.00893	0.961
25.6	78	0.9969	62.23	0.00882	0.949
26	78.8	0.9968	62.23	0.00873	0.940
26.1	79	0.9968	62.23	0.00871	0.938
26.7	80	0.9966	62.22	0.00861	0.926
27	80.6	0.9965	62.21	0.00854	0.920
27.2	81	0.9965	62.21	0.00850	0.915
27.8	82	0.9963	62.20	0.00840	0.904
28	82.4	0.9962	62.19	0.00836	0.900
28.3	83	0.9961	62.19	0.00830	0.893
28.9	84	0.9960	62.18	0.00820	0.883
29	84.2	0.9959	62.17	0.00818	0.881
29.4	85	0.9958	62.17	0.00811	0.873
30	86	0.9956	62.16	0.00801	0.862
30.6	87	0.9955	62.15	0.00792	0.852
31	87.8	0.9954	62.14	0.00785	0.844
31.1	88	0.9953	62.14	0.00783	0.843
31.7	89	0.9951	62.12	0.00774	0.833

Temperature, in degrees Celsius	Temperature, in degrees Fahrenheit	Density, in grams per cubic centimeter	Density, in pounds-mass per cubic foot[a]	Kinematic viscosity, in square centimeters per second	Kinematic viscosity, in square feet per second X 10^5
32	89.6	0.9950	62.12	0.00769	0.827
32.2	90	0.9950	62.11	0.00765	0.824
32.8	91	0.9948	62.10	0.00756	0.814
33	91.4	0.9947	62.10	0.00753	0.811
33.3	92	0.9946	62.09	0.00748	0.805
33.9	93	0.9944	62.08	0.00740	0.796
34	93.2	0.9944	62.08	0.00738	0.795
34.4	94	0.9942	62.07	0.00732	0.788
35	95	0.9940	62.06	0.00724	0.779
35.6	96	0.9938	62.04	0.00716	0.771
36	96.8	0.9937	62.03	0.00710	0.764
36.1	97	0.9937	62.03	0.00708	0.762
36.7	98	0.9935	62.02	0.00701	0.754
37	98.6	0.9933	62.01	0.00696	0.749
37.2	99	0.9933	62.01	0.00693	0.746
37.8	100	0.9931	61.99	0.00686	0.738
38	100.4	0.9930	61.99	0.00683	0.735
38.3	101	0.9929	61.98	0.00679	0.731
38.9	102	0.9926	61.97	0.00672	0.723
39	102.2	0.9926	61.97	0.00670	0.722
39.4	103	0.9924	61.96	0.00665	0.716
40	104	0.9922	61.94	0.00658	0.708

[a]Mass in pounds is numerically equal to weight in pounds for $g = 32.17$ ft/sec^2.

Appendix III.—Symbols

The following symbols are used in this manual:

A = cross-sectional area of channel;
= land surface area;
= coefficient or factor;
A' = portion of A deriving its resistance from bed skin friction;
A'' = portion of A deriving its resistance from bed form drag;
A_b = area of water prism deriving its resistance from bed (determined by sidewall correction procedure);
A_w = portion of water prism deriving its resistance from walls or banks;
a = characteristic length;
= length of longest axis of particle;
= distance from stream bed at which characteristic concentration, C_a occurs;
= factor of proportionality;
a_r = dimensionless factor in logarithmic velocity equation;
B = constant in von Karman velocity equation used by Hunt;
B_s = constant in Hunt equation for distribution of suspended sediment;
b = channel width or water surface width;
= length of intermediate axis of particle;
= characteristic length;
= coefficient;
C, C' = Chezy coefficient;
$\overline{C}$ = mean value of suspended-sediment concentration in vertical;
C_a = suspended-sediment concentration at distance a above bed;
C_D = drag coefficient;
C_{hm} = Chezy coefficient for water flows with suspended sediment;
C_{hw} = Chezy coefficient for clear water flows;
C_m = coefficient in Mostafa-McDermid velocity relation;
C_m, C_w = G_s/Q = sediment discharge concentration;
C_o = maximum sediment concentration in vertical in fraction by volume;
C_v, c_v = sediment concentration, fraction by volume;
c = instantaneous suspended-sediment concentration at a point;
= length of shortest axis of particle;
$\bar{c}$, C = mean suspended-sediment concentration at a point;
c' = fluctuating portion of suspended-sediment concentration at a point;
c_1,c_2 = dimensionless constants;
D = diameter of cylinder or pipe;
D_1, D_2 = depths of sediment deposit;
d = diameter of sphere;
= flow depth;

d' = depth of stream in Engelund's velocity equation (analogous to r')
d_g = geometric mean size of sediment;
d_n = nominal diameter of particle;
d_s = size of particle;
d_s^* = equivalent or characteristic size of sediment;
d_{si} = mean size of ith size fraction of sediment;
d_{50} = median size of sediment (letter d with numerical subscript denotes particle size in sediment for which percentage by weight corresponding to subscript is finer, e.g., $d_{84.1}$ is size for which 84.1% by weight of sediment is finer);
F = force, submerged weight of particle or sphere;
F = $V/\sqrt{gd}$ = Froude number;
= densimetric Froude number;
F_1 = factor in Rubey equation for fall velocity of particle;
f = Darcy-Weisbach friction factor;
= constant factor;
f' = bed friction factor for sand grain roughness or skin friction;
f'' = bed friction factor for form drag;
f_b = friction factor for bed determined by sidewall correction procedure;
f_e = friction factor used by Engelund;
f_f = friction factor for flat beds (Lovera-Kennedy);
f_m = friction factor for pipes carrying sediment;
f_w = friction factor for walls or banks;
G_s = discharge of bed sediment;
G_{sb} = bed load discharge of bed sediment;
G_{ss} = suspended load discharge of bed sediment;
g = acceleration of gravity;
g_s = sediment discharge per unit width of channel;
g_{sb} = bed load discharge of bed sediment per unit width of channel;
g_{sbi} = portion of g_{sb} in ith size fraction of bed sediment;
g_{ss} = suspended load discharge of bed sediment per unit width of channel;
g_{ssi} = portion of g_{ss} in the ith size fraction of bed sediment;
$\overline{H}$ = mean height of ripples or dunes;
h = height of ripple or dune;
h_L = head loss;
hp = horsepower;
I_p = plasticity index for cohesive sediment;
I_w = soil erodibility index;
i = energy gradient for pipe flow;
i_m = energy gradient for pipe flow with suspended particles;
J = dimensionless parameter;
K = resistance factor;
K' = factor;
K_L = bend loss coefficient for flow in curved channels;
k = von Karman universal constant;
= $2\pi/L$ = wave number;
k_s = roughness length;
= constant in Hunt's equation for distribution suspended sediment;

$k_{1\text{-}5}$ = exponents;
L = wave length of ripple or dune;
= length of pipe;
l, l_1, l_2 = mixing lengths;
N_I = dimensionless number;
n = Manning friction factor;
= exponent;
n_s = exponent in power relation between Q and G_s;
n_v = exponent in power law relation for velocity profile;
p = pressure;
= wetted perimeter of channel;
p_b = wetted perimeter of bed;
p_i = fraction by weight of the ith size fraction of sediment;
p_w = wetted perimeter of walls or banks;
Q = water discharge;
Q_s = sediment discharge by volume;
q = flow rate or discharge of fluid per unit width of channel;
q_c = critical discharge per unit width;
q_{ci} = critical discharge per unit width for ith size fraction of bed sediment;
q_s = sediment discharge in volume per unit width;
q_{s*} = $q_s/U_* d_s$ = dimensionless sediment discharge;
R = Reynolds number of flow;
R = dimensionless soil-cover factor;
R_* = $U_* d_s/\nu$ = bed Reynolds number;
r = hydraulic radius;
r' = portion of r_b due to grain roughness or skin friction;
r'' = portion of r_b due to form drag;
r_b = hydraulic radius of bed (determined by sidewall correction procedure);
r_w = hydraulic radius of walls or banks;
S = channel slope;
S' = portion of channel slope due to grain roughness;
S'' = portion of channel stop due to form drag;
SF = shape factor;
S_f = slope of energy gradient;
S_v = vane shear-strength of soil;
s = $\rho_s/\rho = \gamma_s/\gamma$ = specific gravity sediment;
T = temperature;
t, T = time;
$\overline{U}$ = mean velocity at vertical;
U_* = $\sqrt{\tau_o/\rho}$ = shear velocity;
U'_* = $\sqrt{\tau'_o/\rho}$ = grain-roughness shear velocity;
U''_* = $\sqrt{\tau''_o/\rho}$ = form drag shear velocity;
U_{*c}, U_{*t} = critical shear velocity;
U_{max} = maximum velocity at vertical;
U_{oc} = critical mean velocity near bed;
u = $\bar{u} + u'$ = instantaneous fluid velocity at a point;
$\bar{u}, U$ = mean fluid velocity at a point;

u' = instantaneous turbulence fluctuation of u at a point;
u_o = mean velocity near bed;
$V = Q/A$ = mean velocity;
= volume of sediment;
V_B = mean velocity of sediment laden pipe flow separating flow with moving bed from heterogeneous flow;
V_c = critical velocity for entraining sediment;
V_H = critical velocity of sediment laden pipe flow dividing homogeneous and heterogeneous flow;
V_L = limiting deposit-velocity in pipe;
V_o = characteristic velocity;
$v = \bar{v} + v'$ = instantaneous component of velocity in y direction;
$\bar{v}$ = mean value of y component of velocity;
v' = instantaneous turbulence fluctuation of v at a point;
W = specific weight of sediment deposit;
w = fall velocity of particle;
$= \bar{w} + w'$
= instantaneous component of velocity in z direction;
$\bar{w}$ = mean value of velocity in z direction;
w' = instantaneous turbulence fluctuation of w at a point;
X = variable length or distance;
x = dimensionless factor in Einstein velocity relation; coordinate, usually in flow direction;
y = coordinate, usually normal to water surface (vertical);
z = coordinate, usually transverse to flow (horizontal);
= dimensionless exponent in Rouse equation for distribution of suspended sediment;
z_i = dimensionless exponent in Rouse equation based on sediment size d_{si};
z_1 = dimensionless exponent in Ippen equation for distribution of suspended sediment;
α = angle or dimensionless factor;
β = ratio, ϵ_s/ϵ_m;
$\beta_1\beta_2$ = correlation coefficient;
γ = specific weight of fluid;
γ_m = specific weight of water-sediment mixture;
γ_s = specific weight of sediment grains;
δ = Kennedy lag distance;
= lower limit of integration;
= factor in Wilson equation for velocity;
δ' = thickness of laminar sublayer;
ϵ_m = diffusion coefficient for momentum (eddy viscosity);
ϵ_s = diffusion coefficient for sediment;
ϵ_w = diffusion coefficient for water;
ϵ_x = coefficient for diffusing sediment in x direction;
ϵ_y = coefficient for diffusing sediment in y direction;
η = elevation of bed;
= efficiency of pump;
θ = angle of repose of sediment;
= dimensionless number or exponent;

λ = wave length of dune;
μ = dynamic viscosity of fluid;
μ_m = dynamic viscosity of particle-fluid mixture;
ν = kinematic viscosity of fluid;
ν_m = kinematic viscosity of particle-fluid mixture;
ρ = density of fluid;
ρ_m = density of particle-fluid mixture;
ρ_s = density of sediment grain;
σ = standard deviation of particle sizes;
σ_g = geometric standard deviation of particles sizes;
τ = shear stress;
$\tau_* = [\tau_o/(\gamma_s - \gamma)d_s]$ = dimensionless shear stress or Shields parameter;
$\tau'_* = [\tau'_o/(\gamma_s - \gamma)d_s]$ = portion of τ_* due to grain roughness;
$\tau''_* = [\tau''_o/(\gamma_s - \gamma)d_s]$ = portion of τ_* due to form drag;
τ_b = shear stress on bed of channel;
τ_c = critical value of τ_o;
τ_{ci} = critical value of τ_o for sediment of size d_{si};
τ_i = shear stress at interface between two stratified fluids;
τ_o = bed shear stress;
τ'_o = portion of τ_b due to grain roughness;
τ''_o = portion of τ_b due to form drag;
τ_w = shear stress on walls or banks of channel;
ϕ = slope angle;
= indicates, "function of";
= Einstein's dimensionless sediment discharge;
$\phi_D = (i_m - i)/iC_v$;
ϕ_e = Engelund's dimensionless sediment discharge;
$\psi = 1/\tau_*$ dimensionless variable in Einstein $\psi - \phi$ relation;
$\psi' = 1/\tau'_*$;
= particle form coefficient;
$\psi'' = 1/\tau''_*$; and
ψ_d = factor in DuBoys' sediment discharge relation.

Index

A

B

C

D

F

G

I

J

K

L

M

N

O

P

Q

R

S

T

U

V

W

X, Y, Z